Cosmic Noise
A History of Early Radio Astronomy

Providing a definitive history of the formative years of radio astronomy, this book is invaluable for historians of science, scientists and engineers. The whole of worldwide radio and radar astronomy is covered, beginning with the discoveries by Jansky and Reber of cosmic noise before World War II, through the wartime detections of solar noise, the discovery of radio stars, lunar and meteor radar experiments, the detection of the hydrogen spectral line, to the discoveries of Hey, Ryle, Lovell, Pawsey and others in the decade following the war, revealing an entirely different sky from that of visual astronomy.

Using contemporary literature, correspondence and photographs, the book tells the story of the people who shaped the intellectual, technical, and social aspects of the field now known as radio astronomy. The book features quotes from over 100 interviews with pioneering radio astronomers, giving fascinating insights into the development of radio astronomy.

WOODRUFF T. SULLIVAN III is Professor of Astronomy and Adjunct Professor of History at the University of Washington, Seattle. Trained as a radio astronomer, his research has included studies of the interstellar medium in our own and other galaxies, the search for extraterrestrial intelligence, and astrobiology.

Cosmic Noise
A History of Early Radio Astronomy

Woodruff T. Sullivan, III
University of Washington

CAMBRIDGE UNIVERSITY PRESS
Cambridge, New York, Melbourne, Madrid, Cape Town, Singapore, São Paulo, Delhi

Cambridge University Press
The Edinburgh Building, Cambridge CB2 8RU, UK

Published in the United States of America by Cambridge University Press, New York

www.cambridge.org
Information on this title: www.cambridge.org/9780521765244

© W. T. Sullivan 2009

This publication is in copyright. Subject to statutory exception
and to the provisions of relevant collective licensing agreements,
no reproduction of any part may take place without
the written permission of Cambridge University Press.

First published 2009

Printed in the United Kingdom at the University Press, Cambridge

A catalogue record for this publication is available from the British Library

Library of Congress Cataloguing in Publication data
Sullivan, Woodruff Turner.
 Cosmic noise : a history of early radio astronomy / Woodruff T. Sullivan.
   p. cm.
 Includes bibliographical references and index.
 1. Radio astronomy–History. I. Title.
 QB475.A25S85 2009
 522′.68209–dc22   2009027558

ISBN 978-0-521-76524-4 Hardback

Cambridge University Press has no responsibility for the persistence or
accuracy of URLs for external or third-party internet websites referred to
in this publication, and does not guarantee that any content on such
websites is, or will remain, accurate or appropriate.

For Barbara,

who has been married to this book project
almost as long as to me

I had the opportunity only yesterday of watching Sagittarius rise in broad daylight on the needle of a millivoltmeter ... It is certainly gratifying to see gunlaying radar apparatus put to such uses!

Alan Hunter[1]

---

[1] A. Hunter (Royal Greenwich Observatory): J. L. Greenstein, 8 October 1946, box 39, GRE.

Contents

Annotated table of contents	*page* xi
Foreword	xxvii
Preface	xxix
Acknowledgments for figures	xxxii

1 Prologue — 1
 1.1 A new sky — 1
 1.2 Organizing the story — 3
 1.3 Analysis — 10

2 Searching for solar Hertzian waves — 18
 2.1 Hertz — 18
 2.2 Edison and Kennelly — 19
 2.3 Lodge — 20
 2.4 Wilsing and Scheiner — 21
 2.5 Nordmann — 23
 2.6 Why did no one discover solar radio waves until four decades after Nordmann? — 24

3 Jansky and his star static — 29
 3.1 Jansky's early years — 29
 3.2 The setting for Jansky's work — 30
 3.3 Jansky's investigations — 31
 3.4 Jansky's later years — 43
 3.5 Reaction of the scientific community to Jansky's work — 44
 3.6 Was Jansky "stopped" by Friis? — 49
 3.7 Why did Jansky succeed? — 51

4 Grote Reber: science in your backyard — 54
 4.1 The man and his dish — 54
 4.2 Searching for Milky Way signals — 57
 4.3 First publications — 60
 4.4 The ups and downs of 1941 — 63
 4.5 All-sky surveys — 65
 4.6 Reber beyond Wheaton — 69
 4.7 The Reber phenomenon — 74

5 Wartime discovery of the radio sun — 79
 5.1 Development of radar — 79
 5.2 Hey's discovery — 80
 5.3 Other wartime incidents with the sun — 83
 5.4 Prewar observations — 85

	5.5	Controversy with Appleton	90
	5.6	Southworth and the quiet sun	91
6	**Hey's Army group after the war**		**100**
	6.1	Wartime problems lead to postwar success	100
	6.2	Galactic noise and Cygnus intensity variations	101
	6.3	Meteor radar	105
	6.4	Solar observations	111
	6.5	End of an era at AORG	112
7	**Radiophysics Laboratory, Sydney**		**118**
	7.1	Radio research in Australia before 1945	118
	7.2	Radiophysics Laboratory, 1945–1952	121
	7.3	Early solar studies	126
	7.4	Radio stars	138
	7.5	RP's early years	143
8	**Ryle's group at the Cavendish**		**155**
	8.1	The setting at Cambridge and TRE	155
	8.2	Transition to peacetime	156
	8.3	Solar observations	159
	8.4	Radio stars	163
	8.5	Overview	169
9	**Lovell at Jodrell Bank**		**178**
	9.1	Cosmic ray showers	178
	9.2	Meteor radar	181
	9.3	The 218 foot dish and the Andromeda nebula	186
	9.4	Other projects	191
	9.5	Plans for a huge steerable dish	192
	9.6	Jodrell Bank after five years	193
10	**Other radio astronomy groups before 1952**		**200**
	10.1	United States	200
	10.2	Canada	211
	10.3	Soviet Union	214
	10.4	France	221
	10.5	Japan	225
	10.6	Other small early groups	226
11	**Meteor radar**		**231**
	11.1	Pre-1945 intimations of radio and meteors	232
	11.2	The 1946 Giacobinids	236
	11.3	Stanford and Ottawa	239
	11.4	Scientific results before 1952	242
	11.5	The meteoric rise and rapid decline of a field	253
12	**Reaching for the moon**		**260**
	12.1	Prewar thinking	260
	12.2	Wartime calculations and observations	261
	12.3	Project Diana	264

	12.4	Bay in Hungary	271
	12.5	Australia	274
	12.6	The 1950s	280
13	**The radio sun**		**284**
	13.1	The quiet sun	285
	13.2	The active sun	297
	13.3	Overview	311
14	**Radio stars**		**315**
	14.1	First steps by Bolton and Ryle: 1948–49	317
	14.2	Scintillations	324
	14.3	New discrete sources	327
	14.4	Optical identifications	335
	14.5	Angular sizes	351
	14.6	Radio stars or radio nebulae?	360
	14.7	Status of radio sources in 1953	363
15	**Theories of galactic noise**		**366**
	15.1	Early surveys	366
	15.2	First theories	367
	15.3	What *are* radio stars and how do they emit?	374
	15.4	Synchrotron radiation and cosmic rays	378
	15.5	The beginnings of radio cosmology	389
	15.6	The radio sky and cosmic rays	389
16	**The 21 cm hydrogen line**		**394**
	16.1	Prediction	394
	16.2	Postwar developments	396
	16.3	Search and discovery at Harvard	398
	16.4	The Dutch quest	404
	16.5	Confirmation from Australia	409
	16.6	Initial astronomical results	410
	16.7	No race, no serendipity, but international cooperation	414
17	**New astronomers**		**418**
	17.1	Development of early radio astronomy	418
	17.2	Radio astronomy and (optical) astronomy	423
	17.3	National influences in the US, Britain, and Australia	438
	17.4	Radio astronomy as technoscience	449
	17.5	The new astronomers	453
18	**A new astronomy**		**457**
	18.1	New science	457
	18.2	First of many new spectral windows	462
	18.3	A New Astronomy	467
	18.4	Four major historical themes in early radio astronomy	470
	18.5	Closing	471
Appendix A	**A primer on the techniques and astrophysics of early radio astronomy**		**472**
	A.1	Electromagnetic radiation	472

A.2	The earth's atmosphere	472
A.3	Thermal radiation	475
A.4	Radiation transfer	475
A.5	Radiation mechanisms	476
A.6	Astronomical coordinates	477
A.7	Basic astronomy of the early 1950s	478
A.8	Radiometry	480
A.9	Receivers of early radio astronomy	482
A.10	Antennas (filled apertures) of early radio astronomy	485
A.11	Interferometers of early radio astronomy	488
	Index of terms	490

Appendix B   The Interviews   492

B.1	Doing oral history	492
B.2	The collection	493
B.3	How the interviews have been used	495

Appendix C   Bibliographic notes and archival sources   503

C.1	Bibliographies of early radio and radar astronomy	503
C.2	Archival collections used in this study	503
C.3	Collections of biographies	505
C.4	Literature on radar development through 1945	505

References (also an index)   506
Index   527

Annotated table of contents

(an asterisk indicates key sections of the book)

Foreword (Francis Graham Smith)   page xxvii
Preface   xxix
Acknowledgements for figures   xxxii

1 Prologue*   1
   1.1 A new sky*   1
      The radio sky is very different from the familiar sky seen with our eyes
   1.2 Organizing the story   3
      1.2.1 Defining radio astronomy   3
          Considering what to include before the term *radio astronomy* existed in the late 1940s
      1.2.2 Structure of the book*   4
      1.2.3 Narrative summary*   5
          Summary of the field's development through ~1953; why 1953 for an endpoint
   1.3 Analysis   10
      1.3.1 Earlier studies   10
          Relationship to earlier studies of the history of radio astronomy, in particular Edge & Mulkay's *Astronomy Transformed* (1976)
      1.3.2 My historiographic style*   11
          How the author thinks about history of science
      1.3.3 Historical issues*   13
          Summary of the book's main historical issues and themes: World War II and Cold War effects; material culture and technoscience; "visual culture"; the twentieth century's "New Astronomy"
   Tangent 1.1 Conventions used in this book*   16

2 Searching for solar Hertzian waves   18
   2.1 Hertz   18
      Radio waves are discovered in the laboratory in Germany in 1886–88
   2.2 Edison and Kennelly   19
      A proposed US experiment in 1890 to search for solar electric disturbances
   2.3 Lodge   20
      The first attempt to detect solar Hertzian waves, in 1894 in Liverpool
   2.4 Wilsing and Scheiner   21
      An elaborate 1896 experiment at an astrophysical observatory in Potsdam
   2.5 Nordmann   23
      Another astronomer searches, from a glacier on Mt. Chamonix in 1901

2.6	Why did no one discover solar radio waves until four decades after Nordmann?*	24
	A long delay because of a lack of sensitive, directional radio equipment, discipline specialization, and a reliance on Planck's blackbody theory	
Tangent 2.1	Nordmann's sensitivity to solar bursts	27
Tangent 2.2	Signal levels for the quiet and disturbed sun	27

3 Jansky and his star static — 29

- 3.1 Jansky's early years — 29
  Jansky graduates in 1927 and goes to work for Bell Labs as a research physicist
- 3.2 The setting for Jansky's work — 30
  - 3.2.1 Radio communications research in 1928 — 30
    Shortwaves (λ < 200 m) and radio telephony are the latest technology
  - 3.2.2 Bell Telephone Laboratories in 1928 — 30
    Shortwave telecommunications at the nation's premier industrial research lab
- 3.3 Jansky's investigations* — 31
  - 3.3.1 Phase One (1928–30): orientation and building — 31
    Large rotating 20 MHz antenna and sensitive, stable receiver are built to study sources of static
  - 3.3.2 Phase Two (1930–31): diversions and first shortwave observations — 32
    A weak, steady static is first picked up in August 1931
  - 3.3.3 Phase Three (1932–33): the astronomical discovery — 34
    Continued observations show the static not to be the sun, but coinciding with sidereal time (Dec. 1932); first ascribed to galactic center alone, later realized to be extended along the entire galactic plane; major public announcement
  - 3.3.4 Phase Four (1934–37): practical work, with occasional star static — 42
    Other projects occupy Jansky's time and he does little follow-up on the "star static"
- 3.4 Jansky's later years — 43
  Wartime work, failing health, and death at age 44
- 3.5 Reaction of the scientific community to Jansky's work — 44
  - 3.5.1 Other contemporary investigations — 44
    Potapenko & Folland make a few (unpublished) follow-up observations at Caltech; Whipple & Greenstein at Harvard attempt to explain Jansky's static as hot dust
  - 3.5.2 Reactions from astronomers — 47
    The astronomy world learns of this radio static, but doesn't know what to do about it
- 3.6 Was Jansky "stopped" by Friis?* — 49
  Friis should not be faulted for not encouraging further work on the new effect; Jansky was always the loyal team player at Bell Labs
- 3.7 Why did Jansky succeed?* — 51
  A combination of the world's most sensitive receiver, a large directional antenna, dogged detective work, and a minimum in the 11-year solar activity cycle
- Tangent 3.1 An all-sky contour map based on Jansky's data — 53

4 Grote Reber: science in your backyard — 54

- 4.1 The man and his dish* — 54
  Radio engineer Reber reads Jansky's papers and in 1937 with his own funds builds a 31 ft dish in his backyard in Wheaton, Illinois
- 4.2 Searching for Milky Way signals* — 57
  Unsuccessful attempts at 3300 and 910 MHz, but at last Reber picks up cosmic static from the Milky Way at 160 MHz in late 1938 and extensively observes in 1939

	4.3	First publications*	60
		Reber explains cosmic static as free–free radiation in 1940 papers in *Proc. IRE* and, after consultation with Struve and others at Yerkes Observatory, in *Ap. J.*; Henyey & Keenan also work on free–free theory	
	4.4	The ups and downs of 1941	63
		Many more observations, another paper in *Proc. IRE*; Struve balks at second *Ap. J.* paper, yet assists Reber in attempts for funding	
	4.5	All-sky surveys*	65
		4.5.1 160 MHz survey (1943–44)	65
		Two hundred all-night traces assembled into a contour plot of the entire northern Milky Way and published in *Ap. J.*; also, first published detection of the radio sun	
		4.5.2 480 MHz Survey (1946–47)	68
		A new receiver and contour map; first review of the field written with Greenstein	
	4.6	Reber beyond Wheaton	69
		4.6.1 Postwar attempts to find funding	69
		Frustrations with failure to obtain support from Yerkes, industry, or the Navy to move his dish or to build a new 200 ft dish	
		4.6.2 Reber at the National Bureau of Standards	71
		Reber moves his dish to Virginia in 1947 for solar monitoring, but never fits in to the government agency	
		4.6.3 Hawaii	73
		Reber spends 1951–54 running a sea-cliff interferometer from the summit of Mt. Haleakala	
		4.6.4 Tasmania	73
		Moves to the location of an ionospheric "hole" in order to observe at 0.5 to 2.1 MHz	
	4.7	The Reber phenomenon*	74
		Reber's contrarian philosophy, engineering skill, scientific intuition, and initiative before 1947 created pioneering science which was ironically of little influence, especially outside the US, in the postwar decade	
	Tangent 4.1 Reber's quoted antenna properties and derived intensities		77
	Tangent 4.2 The Würzburg antenna		78
	Tangent 4.3 The fate of Reber's dish after 1952		78
5	**Wartime discovery of the radio sun***		**79**
	5.1	Development of radar	79
		First operational radar system is deployed in England in late 1930s; during the war all major combatants stage massive development efforts vital to postwar radio astronomy	
	5.2	Hey's discovery	80
		British coastal radars accidentally detect 55–85 MHz radio waves in Feb. 1942; Hey convincingly demonstrates them to be of solar origin and writes secret report	
	5.3	Other wartime incidents with the sun	83
		5.3.1 Schott in Germany	83
		Likely solar detection in 1943 by a German coastal radar in Denmark	
		5.3.2 Alexander in New Zealand	84
		New Zealand radars pick up the sun in March 1945; Alexander studies the effect for six months	
	5.4	Prewar observations	85
		5.4.1 (Brief) purposeful tries	86
		Adel and Kraus at Michigan in 1933; Piddington and Martyn in Sydney in 1939	

xiv   Annotated table of contents

    5.4.2 Heightman and the radio amateurs   86
            In 1936–39 ham radio operators (mainly in England) study a shortwave hiss
            phenomenon associated with activity at solar maximum, but remain puzzled
            as to its origin
    5.4.3 Professional radio physicists and ionospheric activity   89
            Noise associated with sudden shortwave communications fade-outs is studied
            in the late 1930s, but again no conclusions as to its origin
  5.5 Controversy with Appleton   90
        Appleton (characteristically) tries to gain priority over Hey and the amateurs for
        discovery of solar radio waves
  5.6 Southworth and the quiet sun   91
        Radio physicist at Bell Labs studies microwave (1 to 10 cm) solar emission with
        small dish during 1942–43 and publishes results on the sun's temperature in 1945
  Tangent 5.1   Southworth's miscalculation of solar brightness temperature   98

6 **Hey's Army group after the war**   **100**
  6.1 Wartime problems lead to postwar success*   100
        V-1 and V-2 rocket attacks on London in 1944 lead to detection and later study by Hey,
        Parsons & Phillips of meteor echoes and galactic noise
  6.2 Galactic noise and Cygnus intensity variations*   101
        The Milky Way is mapped and a small region in Cygnus is found in 1946 to vary in
        intensity – the first discrete radio source (later called Cyg A)
  6.3 Meteor radar*   105
    6.3.1 Observations   105
            After the war Hey shows that anomalous ionospheric echoes encountered
            in 1944 are caused by meteors; Hey and Stewart detect daytime meteor showers
            in 1945 with 3-station observations
    6.3.2 Publications   109
            Appleton again tries to claim credit for Hey's group's work, this time on meteor echoes
  6.4 Solar observations   111
        Huge radio bursts in 1946 are shown to be polarized and correlated with solar
        flares and sunspots
  6.5 End of an era at AORG*   112
        In 1948 the Cold War brings an end to radio and radar astronomy in Hey's Army group
  Tangent 6.1   DeWitt (1940) and Fränz (1942) measure galactic noise   113
  Tangent 6.2   British work on galactic noise (1944–46)   114
  Tangent 6.3   Hey's antenna beam and intensities   116
  Tangent 6.4   Why Hey's group and others did not find Cas A   116
  Tangent 6.5   Radio source scintillation   116
  Tangent 6.6   Phillips's method for improving angular resolution   116

7 **Radiophysics Laboratory, Sydney**   **118**
  7.1 Radio research in Australia before 1945   118
    7.1.1 Prewar: the Radio Research Board   118
            Australia develops a strong community of ionospheric and radio
            communications researchers
    7.1.2 Wartime: the Radiophysics Laboratory   120
            One of the world's premier radar labs develops during World War II
  7.2 Radiophysics Laboratory, 1945–1952   121

			Annotated table of contents	xv

	7.2.1	Transition to peacetime	121
		Radiophysics Lab stays intact after the war; Bowen and Pawsey chart out nonmilitary radio research directions	
	7.2.2	Research program	123
		Radio astronomy and cloud physics grow to dominate two-thirds of the lab's efforts; other fields die out	
	7.2.3	Growth of research on extraterrestrial radio noise	124
		Pawsey and Bowen foster extraterrestrial research by a dozen young physicists and engineers	
7.3	Early solar studies		126
	7.3.1	Wartime efforts*	126
		Payne-Scott and Pawsey observe the microwave sky in 1944, but do not try the sun; 1945 reports by Alexander of "Norfolk Island effect" (solar bursts detected on New Zealand military radars) and reports of other solar observations reach RP	
	7.3.2	Solar bursts and the sea-cliff interferometer*	129
		McCready, Pawsey and Payne-Scott develop the sea-cliff interferometer and in 1945–46 establish that solar bursts come from <10′ sunspot regions with T_b (200 MHz) > 3 × 10^9 K	
	7.3.3	The million-degree corona*	135
		In 1946 Pawsey measures a base value of $T_b = 1 \times 10^6$ K and Martyn shows this is due to opacity of a hot solar corona at radio wavelengths; priority dispute ensues	
	7.3.4	Mt. Stromlo	137
		Allen, Martyn and Woolley at Mt. Stromlo Observatory support radio noise studies and even make their own observations, but relationship with RP is sometimes strained	
7.4	Radio stars*		138
	Bolton and Stanley in 1947 use a sea-cliff interferometer to measure the size of Hey's source Cygnus A as <8′; survey reveals several more sources; New Zealand data in 1948 allow three possible optical identifications, including the Crab nebula		
7.5	RP's early years		143
	7.5.1	The isolation factor*	143
		Australian radiophysicists suffer from isolation Down Under; long trips abroad critical; 1952 URSI meeting in Sydney	
	7.5.2	The field stations	146
		RP style is for groups of 2–3 researchers at each of seven field stations near Sydney; Pawsey holds it all together	
	7.5.3	Management of radio noise research	148
		Bowen and Pawsey's styles and abilities make an ideal team to nurture RP research	
	7.5.4	RP evolves during the 1950s*	151
		Radio noise studies become radio astronomy; solar work declines relative to non-solar; antennas become fewer and far larger; researchers mature	
Tangent 7.1	Vacuum and high-energy physics in the Radiophysics Division, 1945–48		154

8 Ryle's group at the Cavendish — 155

8.1	The setting at Cambridge and TRE	155
	The Cavendish Lab is the world's best in nuclear physics; TRE is the premier British wartime radar lab	
8.2	Transition to peacetime	156
	Ratcliffe recruits Ryle, who, after initial uncertainties, establishes a lab and settles on extraterrestrial noise research	

xvi   Annotated table of contents

    8.3   Solar observations*   159
Ryle and Vonberg develop the Michelson variable-spacing interferometer and a new type of receiver; radio emission comes from solar active regions of size <10′ with T_b (175 MHz) > 2 × 10^9 K

    8.4   Radio stars*   163
Ryle invents the phase-switched interferometer; Cas A discovered and Cyg A's position seems variable; Ryle, Smith & Elsmore's preliminary survey of the northern sky yields 23 new sources by mid-1949

    8.5   Overview*   169
        8.5.1   Ryle, Ratcliffe, Bragg   169
Leadership, research, and personality qualities of Ryle and his superiors Ratcliffe and Bragg

        8.5.2   Group style and development   172
A close-knit, informal group under a charismatic leader; strained relationships with other groups, theorists, and optical astronomers

        8.5.3   A turning point   175
1950–54: Ryle's attention shifts from the sun to radio stars, from small antennas to major structures, from the Galaxy to cosmology, and toward testing the concept of aperture synthesis

    Tangent 8.1   An example of camaraderie in Ryle's group   177

9 Lovell at Jodrell Bank   **178**
    9.1   Cosmic ray showers   178
Blackett brings Lovell to Manchester; at Jodrell Bank field station no success in 1946–47 in detecting radar echoes from cosmic ray showers

    9.2   Meteor radar*   181
Radar echoes from 1946 Perseid and Giacobinid meteor showers observed by Lovell, Clegg, Banwell & Prentice; summer daylight showers discovered in 1947; Ellyett and Davies develop method to measure meteor velocities

    9.3   The 218 foot dish and the Andromeda nebula*   186
Fixed dish is built in 1947, by far the largest in world; Hanbury Brown (arrives 1949) & Hazard map emission from Andromeda nebula (M31) in 1950

    9.4   Other projects   191
Briefly: auroral echoes, lunar radar, intensity scintillations of radio stars, Hanbury Brown's intensity interferometer

    9.5   Plans for a huge steerable dish*   192
From 1949 on, Lovell and Blackett seek support for a 250 ft steerable dish

    9.6   Jodrell Bank after five years*   193
A large group with excellent leadership becomes oriented to astronomy and moves from radar to radio astronomy

    Tangent 9.1   Attempts to detect audio-frequency solar radio waves   198

10 Other radio astronomy groups before 1952   **200**
    10.1   United States   200
        10.1.1   Dicke and the Radiation Laboratory   200
At the premier US wartime radar lab, Dicke develops microwave radiometry and in 1945–46 observes the moon and a solar eclipse

		10.1.2	US Naval Research Laboratory	206
			Hagen's group at well-financed military lab pioneers solar observations at microwavelengths (1946 on), travels to three total eclipses, and builds a 50 ft dish that first detects sources in 1953	
		10.1.3	Cornell University	210
			Seeger leads Navy-financed university group in solar and galactic observations over 1947–50	
	10.2	Canada		211
		Covington begins long-term microwave solar observations at government lab (NRC) in 1946		
	10.3	Soviet Union		214
		10.3.1	Ginzburg and Shklovsky	214
			Ginzburg (Lebedev Physics Institute) and Shklovsky (Moscow State University) start longtime leadership of Soviet theory in 1946 with studies of radio sun and interstellar medium	
		10.3.2	Lebedev Institute observations	217
			Khaykin leads 1947 eclipse expedition to Brazil where 200 MHz radio sun shown to be much larger than optical sun; Crimean field stations established in 1948–49 for both applied and basic radio research	
		10.3.3	Gorky	220
			Troitsky develops microwave techniques from 1948 onwards	
		10.3.4	Overview	220
			Soviet radio astronomy strong in theory, but lacking in observational results during postwar decade; strong isolation from the West	
	10.4	France		221
		Rocard of Ecole Normale Supérieure (Paris) starts group in 1946 using German and American radar equipment with Denisse (solar theory), Steinberg and Blum; solar eclipses observed from France (1949) and Africa (1950–51). Laffineur in 1947 starts solar observations at Meudon Observatory with Würzburg dish		
	10.5	Japan		225
		Solar radio groups begin in 1949 at Tokyo Observatory (Hatanaka), Osaka University (Oda) and Nagoya University (Tanaka); all isolated from West		
	10.6	Other small early groups		226
		10.6.1	Germany	226
			Radio experiments forbidden until 1950, but Unsöld (Kiel) and Kiepenheuer (Freiburg) do early radio theory and later encourage observations	
		10.6.2	Sweden	228
			Rydbeck and Hvatum adapt Würzburg dishes at Onsala Observatory in early 1950s	
		10.6.3	Norway	228
			Eriksen (Solar Observatory, Oslo University) leads effort on sun in early 1950s	
		10.6.4	Soviet Union	229
			Sanamyan (Byurakan Observatory) uses meter-wavelength interferometers in early 1950s; Molchanov (Leningrad University) does microwave solar observations	
		10.6.5	United States	229
			Tuve (Dept. of Terrestrial Magnetism) initiates 21 cm hydrogen line work in 1952; Kraus (Ohio State University) invents a helix-array radio telescope in 1951.	
		Tangent 10.1	Shklovsky's free–free radiation calculations	229

xviii   Annotated table of contents

11 Meteor radar — 231
    11.1 Pre-1945 intimations of radio and meteors — 232
        11.1.1 Nagaoka* — 232
        Japanese radio physicist suggests in 1929 that meteors could affect the ionosphere and radio propagation
        11.1.2 Skellett* — 232
        Over 1931–33 Bell Labs researcher finds a tentative connection between visual meteors and inceases in ionization of the ionosphere
        11.1.3 The situation before 1945 — 234
        Various studies give evidence of possible meteor effects, but role of meteors is still uncertain before Hey & Stewart make seminal discoveries in 1944–45 (Chapter 6)
    11.2 The 1946 Giacobinids* — 236
    Spectacular meteor shower observed around the world with radar; launches many groups including Lovell's at Jodrell Bank, largest by far
    11.3 Stanford and Ottawa — 239
    Group in Stanford Electrical Engineering Dept. develops techniques and theory and by late 1940s develops meteors for radio communications. In Canada Millman and McKinley combine optical and radar meteor studies.
    11.4 Scientific results before 1952 — 242
        11.4.1 Daytime showers* — 242
        Previously unknown strong daytime meteor showers are discovered and mapped at Jodrell Bank by Lovell, Clegg, Davies, Hawkins, Almond *et al.*
        11.4.2 Meteors from interstellar space?* — 245
        Radio data enter the astronomers' debate over the origin of sporadic meteors; 1948–52 velocity measurements at Jodrell Bank and in Canada find no hyperbolic orbits, indicating solar system origin
        11.4.3 Ionosphere physics — 251
        Ionospheric winds and trail formation mechanisms studied
    11.5 The meteoric rise and rapid decline of a field* — 253
    Meteor radar astronomy distinct from most radio astronomy, closely tied to (optical) astronomy, dominated by Jodrell Bank, short-lived (not fruitful beyond mid-1950s)
    Tangent 11.1 Pre-1945 ionosphere/meteor studies in Japan, India, and the United States — 255
    Tangent 11.2 Pre-1945 ionosphere/meteor studies in Britain — 257
    Tangent 11.3 The Doppler method and Fresnel theory of meteor echoes — 258
    Tangent 11.4 The velocity cutoff for interstellar meteors — 259

12 Reaching for the moon — 260
    12.1 Prewar thinking — 260
    Claims and calculations of radio contact with Mars and Martians from earliest days of radio; lunar radar also considered
    12.2 Wartime calculations and observations — 261
    Many radar operators and scientists do feasibility studies or try to make lunar contact; best case for success is that of Stepp in Germany in 1943–44
    12.3 Project Diana* — 264
    US Army Signal Corps team led by DeWitt detects lunar echoes in January 1946
    12.4 Bay in Hungary* — 271
    In war-torn Hungary Bay mounts experiment to do lunar radar; success in 1946

Annotated table of contents   xix

12.5 Australia   274
   12.5.1 Lunar radar   274
          Kerr & Shain (RP) use 20 MHz Radio Australia as transmitter in 1947–48;
          aim is to study ionosphere's effects on echoes
   12.5.2 Passive lunar observations*   277
          In 1948 Piddington & Minnett (RP) find monthly phase-lag in 1.25 cm lunar
          emission and explain it as due to dust layer
12.6 The 1950s   280
          Trexler (NRL) does secret moon-link communications experiments
          from 1951 onwards with huge fixed dish; Jodrell Bank starts major lunar radar program
Tangent 12.1   Lunar ranges implied by Bay's published data   282

13 The radio sun   284
13.1 The quiet sun   285
   13.1.1 Observed spectrum   285
          Pawsey & Yabsley (1949) establish the quiet solar spectrum
   13.1.2 Theory   286
          Martyn (1948), Smerd (1950) and Denisse (1949) work out expected
          radio intensities based on various solar atmosphere models; limb-brightening
          predicted at shorter wavelengths
   13.1.3 Eclipses and interferometers*   290
          Observations of eight eclipses 1945–52 allow detailed checks of
          emission theories; Ryle and his Cambridge students (Stanier, Machin, O'Brien)
          develop Fourier techniques to map the sun using Michelson interferometry,
          culminating in 2-D solar map in 1953; in Australia Christiansen simultaneously
          develops grating arrays for 2-D maps
13.2 The active sun   297
   13.2.1 Meter wavelength bursts   297
          13.2.1.1 Payne-Scott's work   298
                   Timing of bursts at different frequencies and a swept-lobe
                   interferometer reveal fast motions in the corona (1946–50)
          13.2.1.2 Wild's work*   302
                   Wild's group at Penrith (1949) and Dapto (1952) develops
                   antennas and receivers for wideband dynamic spectra that reveal basic
                   types of radio bursts: Types I, II and III
   13.2.2 Burst radiation theory   307
          Inconclusive arguments for and against synchrotron radiation,
          bremsstrahlung, and plasma oscillations, as well as debate over the source of
          high-energy radiating particles
   13.2.3 Slowly varying component   310
          Covington's 10.7 cm monitoring data over years, starting in 1947, shows excellent
          correlation with total area of sunspots; Denisse, Waldmeier, and Piddington &
          Minnett interpret this new component to originate in dense regions above sunspots
13.3 Overview   311
   13.3.1 Acceptance of a hot corona   311
          Following the prewar identification of coronal lines as coming from
          high-energy ions, radio evidence for a million-degree corona is critical for
          some solar researchers, not so for others, in clinching the acceptance of a hot corona

xx   Annotated table of contents

      13.3.2 Imaging* 312
                Radio mapping techniques at Cambridge and Sydney eventually lead to
                two Nobel Prizes and applications in medicine
      13.3.3 The radio and optical suns* 313
                Radio sun is more dynamic and indicative of the corona than is the optical sun,
                but overall solar radio astronomy does not revolutionize its sector of astronomy
                as do meteor radar, radio sources, and radio galactic structure
  Tangent 13.1 Proposals to observe lunar occultations of radio sources 314
  Tangent 13.2 Solar radio bursts and plant growth 314

14 Radio stars **315**
14.1 First steps by Bolton and Ryle: 1948–49* 317
      Bolton's and Ryle's groups work on first radio stars, mainly Cyg A.
      Scintillations are bothersome. First suggested optical identifications for
      Tau A (Crab nebula), Vir A (M87) and Cen A (NGC 5128); contact with astronomers.
14.2 Scintillations 324
      Groups studying radio star intensity scintillations argue, but reach
      consensus by 1950–51 that scintillations are caused by the ionosphere, not intrinsic
      to the radio stars.
14.3 New discrete sources 327
      14.3.1 The 1C Survey at Cambridge 328
                50 radio stars measured by Ryle, Smith & Elsmore in 1949–50 with
                the 81 MHz "Long Michelson" interferometer; positions do not correlate
                with bright stars or galaxies; no detectable proper motions or parallax;
                first log N–log S plots; radio stars are very nearby, dark stars
      14.3.2 The Mills Survey 330
                Two-baseline 101 MHz interferometer in 1950–52 finds 77 "discrete
                sources"; establishes Class I sources of high intensity and near the galactic
                plane (and within the Galaxy) and Class II sources uniformly distributed,
                either very close or very far (extragalactic)
      14.3.3 The Jodrell Bank survey 331
                Hanbury Brown & Hazard (1953) list 23 158 MHz sources visible to
                their large, fixed dish and agree with Mills's two classes, but find no
                agreement in overlap areas of the three surveys
      14.3.4 Bolton's group's surveys 332
                Stanley & Slee (1950) publish 18 100 MHz discrete sources based on
                sea-cliff interferometry (1947–49); Bolton *et al.* (1954) publish 86 more
                gathered in 1951–53; statistical interpretation mostly agrees with Mills,
                except Class II sources' log N–log S plot slope is much steeper
      14.3.5 A galactic center source 334
                Building on work by Piddington & Minnett (1951), McGee & Bolton (1954) use
                "hole-in-ground" dish at 400 MHz to establish existence of source at galactic center
14.4 Optical identifications* 335
      14.4.1 Cygnus A 335
                Precision interferometry (1′ accuracy) by Smith (1951), with similar work by Mills,
                allows Baade & Minkowski to identify Cyg A with a distant, peculiar extragalactic
                object visible on a Palomar 200 inch photo; object taken to be two galaxies in collision;
                implied radio luminosity is enormous

| | | 14.4.2 | Cassiopeia A | 341 |

Again, Smith's precise position allows identification with a network of peculiar filaments – a supernova remnant?

| | | 14.4.3 | Baade and Minkowski | 344 |

Two eminent California astronomers work closely with all the radio groups; by 1953 100 and 200 inch telescope photos and spectra make fundamental contributions to the optical identifications and astrophysics of over a dozen radio sources

| | | 14.4.4 | The nature of optical identifications | 348 |
| | | | 14.4.4.1 Techniques | 348 |

The basic problem is that there exist too many stars and galaxies within the positional error box of a typical radio source; how does one choose?

| | | | 14.4.4.2 Identity and reality | 348 |

The visual wins out over radio chart recordings; finding an optical correlate gives a radio source a stamp of approval and a reality that it otherwise does not have

14.5	Angular sizes			351
	14.5.1	The intensity interferometer *		351
		14.5.1.1	Theory	351

Hanbury Brown and Twiss invent the intensity interferometer in 1950, a radical departure from the Michelson interferometer

| | | 14.5.1.2 | Equipment | 353 |

Students Jennison & Das Gupta at Jodrell Bank build the electronics and antennas needed in 1951 and test the concept on the sun

| | | 14.5.1.3 | Observations | 355 |

Jennison & Das Gupta spend a year measuring the visibility curve of Cyg A and find it to have two components with separation of 1.5′

| | 14.5.2 | Results from Cambridge and Sydney | | 359 |

The major groups all measure angular sizes of the strong sources over 1951–53, finding that most are resolvable and not at all starlike

| 14.6 | Radio stars or radio nebulae?* | | | 360 |

Arguments over the nature of radio stars/discrete sources brought on by differing observing and analysis techniques, group styles, and samples of the sky

| 14.7 | Status of radio sources in 1953* | | | 363 |

Over 200 sources catalogued, but only ~10–15 with good optical identifications; different surveys profoundly disagree, but nevertheless an exciting field because of the unprecedented nature of many of the identified sources

Tangent 14.1	The sea-cliff interferometer	364
Tangent 14.2	Interstellar dispersion and bursts in 1950	364
Tangent 14.3	Early log N–log S analysis	365
Tangent 14.4	The intensity interferometer	365

| **15** | **Theories of galactic noise** | | 366 |
| | 15.1 | Early surveys | 366 |

Galactic noise distribution in Bolton & Westfold's (1950) 100 MHz map becomes the standard for theorists to explain

| | 15.2 | First theories* | 367 |

xxii Annotated table of contents

       15.2.1 Hot interstellar gas   367
                  Free–free radiation is popular, but fails badly to explain intensities at the
                  lowest frequencies
       15.2.2 Combined effect of radio stars   369
                  Galactic noise explained as a huge population of dark radio stars by Unsöld (1949),
                  Ryle (1949), Bolton & Westfold (1951), Westerhout & Oort (1951), and
                  Hanbury Brown & Hazard (1953); isotropic component (extragalactic?) also needed
  15.3 What *are* radio stars and how do they emit?*   374
        Radio stars must have strong magnetic fields and be part of Baade's Population II, but no details
        are agreed; Ryle argues for common, nearby stars, while Gold and Hoyle argue that
        they could all be extragalactic
  15.4 Synchrotron radiation and cosmic rays   378
        Cosmic rays had been long studied by geophysicists and nuclear physicists
       15.4.1 In the West*   378
                  Alfvén & Herlofson (1950) and Kiepenheuer (1950) suggest synchrotron
                  radiation from cosmic ray electrons to explain radio emission from radio
                  stars and from the general Galaxy
       15.4.2 In the Soviet Union*   380
                  Ginzburg develops synchrotron theory from 1951 on and exploits it for
                  explaining the origin of cosmic rays as much as for galactic noise;
                  in 1952 Shklovsky switches from Ryle-like model (sum effect of dark
                  radio stars) to synchrotron emission from cosmic rays in a galactic halo;
                  argues supernovae important both as radio sources and as originators of
                  cosmic rays; in 1953 suggests that *optical* radiation from Crab nebula is also
                  synchrotron, soon detected
       15.4.3 Why was the synchrotron mechanism unpopular in the West?   385
                  Russian literature mostly unknown in the West; theoretical arguments (Fermi)
                  that electrons must be totally absent from cosmic rays; not accepted in the
                  West until optical polarization of the Crab confirmed in 1956
  15.5 The beginnings of radio cosmology   389
        Residual galactic noise, evidence for radio galaxies, and log N–log S plots allow radio
        data to test cosmological models
  15.6 The radio sky and cosmic rays*   389
        Radio stars and galactic noise, though still poorly understood, provide new avenues
        to the origin of cosmic rays; high-energy processes gain importance in astronomy
  Tangent 15.1 The integrated brightness of a distribution of radio stars   391
  Tangent 15.2 The multi-frequency Galaxy model of Piddington (1951)   391
  Tangent 15.3 The galactic plane mapping of Scheuer & Ryle (1953)   391
  Tangent 15.4 Minor synchrotron studies in the West (1951–54)   392

16 **The 21 cm hydrogen line**   **394**
  16.1 Prediction*   394
        In wartime occupied Holland van de Hulst studies the possibilities of spectral
        lines in the radio and in 1944 finds a candidate for atomic hydrogen at 21 cm
  16.2 Postwar developments   396
        Shklovsky (1949) independently does a more thorough study of radio lines;
        interest of many persons piqued, but no one searches for the 21 cm line
  16.3 Search and discovery at Harvard*   398

		16.3.1 Background	398
		Purcell & Ewen at Harvard team up for an all-out attempt	
		16.3.2 Equipment	399
		Ewen, backed by excellent radio electronics and many experts, designs a fixed 21 cm horn and complex sensitive receiver using frequency-switch technique	
		16.3.3 Observations	401
		Finally successful in March 1951 with broad line in the Ophiucus part of the Milky Way	
	16.4	The Dutch quest*	404
		16.4.1 Background	404
		Oort struggles to get radio astronomy going at Leiden from 1945 on	
		16.4.2 Trials with Hoo	405
		Engineer Hoo (1948–50) makes little progress; a fire burns all equipment	
		16.4.3 Success with Muller	406
		Engineer Muller hired and detects the line in 1951 shortly after Ewen, using Würzburg dish at Kootwijk	
	16.5	Confirmation from Australia	409
		Radiophysics Lab notified; Christiansen & Hindman quickly build a receiver and detect the line at Potts Hill field station	
	16.6	Initial astronomical results	410
		16.6.1 First interpretations*	410
		In *Nature* Purcell & Ewen discuss physics of the line, while Oort & Muller describe signs of galactic structure and rotation visible in their first month of data; Christiansen & Hindman quickly survey entire southern sky, find double lines in some places	
		16.6.2 Astrophysical theory	412
		Discussions about the transition probability and excitation mechanisms of the line	
		16.6.3 Follow-up observations	412
		None at Harvard until Bok gets 24 ft dish in 1953; Muller rebuilds Dutch receiver and Oort and van de Hulst begin systematic survey of northern galactic plane; Kerr starts large effort in Sydney with 36 ft dish	
	16.7	No race, no serendipity, but international cooperation*	414
		"International Dutch School" of galactic structure, headed by Oort, creates very different climate than in rest of radio astronomy – more cooperation among radio groups and integration with optical astronomy; one of few non-serendipitous discoveries in radio astronomy	
	Tangent 16.1	Other pre-discovery considerations of the 21 cm line	417
17	New astronomers*		418
	17.1	Development of early radio astronomy	418
		17.1.1 Origin in war	418
		The crucible of World War II shaped the technology, the skills, styles of work, and personalities of those who became radio astronomers	
		17.1.2 Early growth and the 1952–53 watershed	420
		Statistics for growth of the field; 1952–53 was watershed in terms of size of projects, type of research questions, and amalgamation with optical astronomers; by 1953 radio astronomy size still only 3–6% of astronomy overall	
	17.2	Radio astronomy and (optical) astronomy	423
		17.2.1 Terminology	423
		Insight gained from studying the introduction and usage of terms such as *radio astronomy*, *radio telescope*, *radio observatory*, *astronomical*, and *optical astronomy*	

            17.2.1.1   Is a radio telescope a telescope?   426
                      Attitudes towards radio interferometers – do they *see*?
      17.2.2   Styles of radio and optical astronomers   427
               Radio astronomers' styles of doing research were very different from traditional
               astronomy: much more engineering and electronics, faster pace, in teams
      17.2.3   Interactions between radio and optical astronomers   429
               Some traditional astronomers supportive, others patronizing or dismissive,
               most neutral; radio astronomers' knowledge of basic astronomy often lacking
      17.2.4   IAU versus URSI   432
               Both international organizations set up Commissions on Radio Astronomy
               early (1946–8); radio astronomers debated which was the best forum for them and
               gradually shifted to IAU
      17.2.5   A new discipline of radio astronomy?   435
               Radio astronomy never became its own discipline – instead, a subfield or
               specialty of traditional astronomy; radio researchers looked to astronomy for
               their intellectual framework, valued optical images and identifications, and
               developed their own "visual culture"
17.3 National influences in the US, Britain, and Australia   438
      17.3.1   Influence of Jansky and Reber   439
               Despite first observations of extraterrestrial radio waves being made by
               two Americans, after the war their influence on the field was minimal
      17.3.2   Astronomers vis-à-vis radio researchers   439
               The strength of American (optical) astronomy acted as a deterrent to
               development of US radio astronomy; in England much more favorable reception
               by astronomers to radio work; Australian (optical) astronomy very weak and
               therefore a minimal factor
      17.3.3   Discipline structures   441
               Ionosphere radio research in England and Australia associated with physics, but
               in US with electrical engineering; jump to a radio cosmos was easier for physicists
      17.3.4   Group styles   442
               Intensity of wartime research in England led after the war to tight-knit, can-do teams;
               Radiophysics Lab in Sydney remained intact with its wartime camaraderie; US postwar
               groups not as affected by wartime styles of work
      17.3.5   Military patronage   442
               Postwar situation in US was of large military funding for any research related to
               national needs in Cold War; radio groups at Stanford, NRL and Cornell well funded,
               but nevertheless lagged overseas groups in radio astronomy efforts; military needs
               distorted their research towards microwave technology and communications systems;
               despite far smaller budgets, British and Australian groups, working primarily at longer
               wavelengths, accomplished far more
17.4 Radio astronomy as technoscience   449
      Often impossible to answer whether (a) instrumentation and techniques, or
      (b) scientific goals were more important; early radio astronomy best described
      as *technoscience*; gradual shift in emphasis from "blind" empiricism with available
      tools to astronomically-defined research programs
17.5 The new astronomers   453
Tangent 17.1   Prewar uses of the term *radio telescope*   453

Tangent 17.2	Comparison with the introduction of the electron microscope into biology	454
Tangent 17.3	Testing Forman's "distortionist" ideas	454
Tangent 17.4	Photons and apertures in the radio and optical regimes	455

18 A new astronomy — 457

    18.1 New science* — 457
        Summary of the main scientific results and the new type of universe revealed by radio astronomy as of ~1953

    18.2 First of many new spectral windows — 462
        Radio astronomy was the first of many other spectral windows to be opened. These required access to space via rockets, balloons and satellites
        18.2.1 Beginnings of infrared, ultraviolet, γ-ray and X-ray astronomies — 463
            Brief accounts of how these astronomies began (all in the US) over 1945–75, often with many surprises
        18.2.2 Comparing early X-ray and radio astronomies — 465
            Hirsh (1983) study of early X-ray astronomy reveals many scientific and historical similarities with early radio astronomy

    18.3 A New Astronomy* — 467
        18.3.1 Was radio astronomy a revolution? — 467
            Early radio astronomy was not a revolution as defined by Kuhn or by Hacking
        18.3.2 The twentieth century's "New Astronomy" — 467
            Radio astronomy was comparable to other major introductions of technology into astronomy, including those by Galileo (telescope), William Herschel (large reflectors), and nineteenth century astrophysicists (spectroscope and photography). Radio astronomy, or better the entire opening of the electromagnetic spectrum (of which radio was the harbinger), was the twentieth century's "New Astronomy," a major event in the long history of astronomy

    18.4 Four major historical themes in early radio astronomy* — 470
        (1) The twentieth century's "New Astronomy"; (2) World War II and Cold War effects; (3) Material culture and technoscience; (4) "Visual culture"

    18.5 Closing* — 471

Appendix A A primer on the techniques and astrophysics of early radio astronomy — 472

    A.1 Electromagnetic radiation — 472
    A.2 The earth's atmosphere — 472
    A.3 Thermal radiation — 475
    A.4 Radiation transfer — 475
    A.5 Radiation mechanisms — 476
        A.5.1 Free–free radiation — 476
        A.5.2 Gyromagnetic and synchrotron radiation — 476
        A.5.3 Spectral line radiation — 477
        A.5.4 Plasma oscillations — 477
    A.6 Astronomical coordinates — 477
    A.7 Basic astronomy of the early 1950s — 478
    A.8 Radiometry — 480
    A.9 Receivers of early radio astronomy — 482
    A.10 Antennas (filled apertures) of early radio astronomy — 485
    A.11 Interferometers of early radio astronomy — 488

Index of terms — 490

Appendix B	**The interviews**	**492**
B.1	Doing oral history	492
B.2	The collection	493
B.3	How the interviews have been used	495
Appendix C	**Bibliographic notes and archival sources**	**503**
C.1	Bibliographies of early radio and radar astronomy	503
C.2	Archival collections used in this study	503
C.3	Collections of biographies	505
C.4	Literature on radar development through 1945	505

References (also an index)   506
Index   527

Foreword

Not long ago, before the birth of radio astronomy, the starry sky was observed by astronomers looking through telescopes, using their eyes and photography. Now we call them optical astronomers, using optical telescopes, admitting their new colleagues who detect radio, X-rays, and gamma-rays. The radio astronomers were the first of these, and their radio telescopes have developed over half a century into complex and sophisticated instruments that reveal a new universe. This has been no less than a revolution in both astronomy and its instruments: as in all revolutions, it has a history whose early beginnings are at least as interesting as the explosive growth in which we are now immersed.

Woody Sullivan's history takes the subject up to 1953. This is perhaps the latest date for which a comprehensive history can be contained in a single volume, but it is a good date to mark the emergence of radio astronomy as an integral part of modern astronomy. There was by this time a basic understanding of the origin of cosmic radio waves, and the techniques of radio telescopes, spectrometers, and interferometers. Funding for large projects was becoming available, and research groups were consolidating. The following half century saw the extension of the visible spectrum into the ultraviolet and infrared, and the exploitation of the new windows of radio, X-rays, and gamma-rays, with many discoveries which changed our view of every aspect of the universe. In more recent years we have seen the scale of radio telescopes expand to international proportions, both physically and in cooperation between many observing groups. Nevertheless, the elements were all there in 1953; furthermore there was extensive documentation of the early steps (notably in Australia), and even where the written record was patchy most of the original players were available for interview when this history was undertaken.

As an experienced radio astronomer Sullivan is well placed to relate the history of the technical advances of the early years. He is also gifted with an understanding and interest in people that enables him to give a balanced account of some difficult relationships between the ambitious, enthusiastic, and sometimes competitive research groups of the time. His study and his interpretations will be of interest not only to the participants, many of whom are still alive, but to historians of science and sociologists, who will doubtless argue whether or not this was, in their terms, a revolution. In my terms it certainly was, and Sullivan has done us a service in writing this excellent historical account of it.

F. Graham Smith FRS

Sir Francis Graham Smith is Emeritus Professor at Jodrell Bank Observatory, where he was Director (1981–88). He also served as Director of the Royal Greenwich Observatory (1976–81), and as the 13th Astronomer Royal (1982–91).

Preface

Freshly minted as a Ph.D. in astronomy, I began this project in 1971 with the observation that almost all of the pioneers of radio astronomy, including my advisor (Frank Kerr[†1]), were still available as sources for a book on the worldwide history of radio astronomy. World War II, during which radio astronomy and I were both born, had ended only a quarter-century before and memories were relatively fresh. Armed with a cassette tape-recorder, I naively began interviewing "old-timers." But when I learned more about doing history and about interviewing, I eventually repeated those early interviews, added many more (see Appendix B), and became serious about archival research (Appendix C). Guided by Urania and Clio, I gathered data from around the world as I could, mostly during 1972–88. The bulk of the initial writing followed in 1984–89, but then the mostly-finished book stalled as other projects intervened. Scattered efforts were sometimes possible, but in the end it took a sabbatical year in 2006 to resurrect the book and finally bring it to completion. The 24 year span of writing triples the 1945–53 period that the book mainly covers, and also far surpasses the stewing period of nine years for writing that the Roman poet Horace famously advised. As another measure of the time that has passed, 60% of the interviewees whose materials have been used for the present volume have now passed away.

Along the way I did produce two other books (still in print) that I consider handmaidens to the present volume: *Classics in Radio Astronomy* (Sullivan 1982) and *The Early Years of Radio Astronomy* (Sullivan 1984). The former is a collection of reprints, with extensive commentary, of 37 seminal papers in radio astronomy and the latter a collection of 21 articles discussing the pre-1960 period by early radio astronomers and historians. Furthermore, over the years portions of the present book appeared in the form of articles, abstracts, and talks; the principal contributions among these can be found in the list of references.

In 1984 I signed a contract with Cambridge University Press to produce this book in two years. Although that contract expired in the last millennium, I am delighted that the Press nevertheless has been willing to publish this opus. Simon Mitton has been encouraging all along and I also thank Richard Ziemacki, Helen Wheeler, and Vince Higgs for their support and advice.

The book is a monograph designed to appeal to astronomers and historians of science, as well as to others with some background in the physical sciences who have a serious interest in the development of twentieth-century science. I cover the entirety of worldwide radio and radar astronomy through the year 1953. By the word "cover" I mean the best story I can assemble about the intellectual, technical, and social aspects that shaped early radio astronomy. This story has been based on (1) the published literature of the time (including lab reports), (2) correspondence and other items found during archival research, (3) over 115 interviews with the early radio astronomers themselves, and (4) photographs of the time (about one-half of the book's 180 figures). Quotations from the interviews are an important feature of the book – they create liveliness and provide insights, although I am well aware of the pitfalls of memory. I thank interviewees, publishers, and photographers for permissions to use their materials.

The first chapter sets out the structure and organization of the book and the conventions that I have used. Unusually for a history book, I have strived to make indexes and cross-references such that the volume acts as an efficient reference book. The first chapter also sets this volume in the context of other studies and discusses my approach to doing history. Section 1.2.3

---

[1] The superscript † after a name in the Preface indicates that the person is known to have died.

gives a précis of the book's narrative and Section 1.3.3 summarizes the main historical themes.

This enterprise has had a large supporting cast. I thank sincerely my history of science colleagues at the University of Washington (UW) and the University of Puget Sound whose friendship, mentoring, and criticisms over the decades have been fundamental to my education in history of science. Chief among these have been Keith Benson, Jim Evans, Mott Greene, Tom Hankins, Bruce Hevly, Karl Hufbauer, and Jody Yoder. Further afield in the history of science community, I have profited tremendously in general or in terms of specific reviews of chapters by David DeVorkin, Steve Dick, David Edge†, Paul Forman, Peter Galison, Stewart Gillmor, Owen Gingerich, Rod Home, Michael Hoskin, Wayne Orchiston, Simon Shaffer, Robert Smith, and Spencer Weart. I am truly sorry that David Edge, who died in 2003, will not see this work. David, a radio astronomer turned sociologist of science, was a good friend and fellow lover of cricket and baseball. From the start, he was supportive of my efforts and generous with advice even though I was horning in on his own research that eventually resulted in *Astronomy Transformed* (Edge and Mulkay 1976) (see Section 1.3.1). Through the years we had a marvellous correspondence that greatly enriched the present study.

The cooperation and advice of the community of radio astronomers (and related researchers) has been indispensable to this project. I am thankful for the willingness of many to be interviewed (see Appendix B for the full list), to review draft chapters, to answer myriad follow-up questions, and to supply copies of archival materials, photographs, and reprints. It is perhaps odious to pick out those who have been the most generous with their time, but the following indeed went the extra mile: John Baldwin, John Bolton†, Taffy Bowen†, Ron Bracewell†, Arthur Covington†, Chris Christiansen†, John DeWitt†, Bruce Elsmore, Harold "Doc" Ewen, Vitaly Ginzburg, Jesse Greenstein†, Robert Hanbury Brown†, Gerald Hawkins†, Denis Heightman†, Stanley Hey†, Roger Jennison†, Ken Kellermann, Frank Kerr†, Bernard Lovell, Ken Machin†, Bernie Mills, Harry Minnett†, Lex Muller†, Jan Oort†, James Phillips, Grote Reber†, Alexander Salomonovich†, Peter Scheuer†, John Shakeshaft, Bruce Slee, Graham Smith, Gordon Stanley†, Henk Van de Hulst†, Gart Westerhout, and Paul Wild†.

In addition I similarly thank the following people most heartily for their cooperation and information: Mary Almond, Zoltán Bay†, Emile-Jacques Blum, Henry Booker, M. K. Das Gupta†, John Dickey, John Findlay†, Kurt Fränz†, Frank Gardner†, Tommy Gold†, Cyril Hazard, Tony Hewish, Jim Hindman†, Vic Hughes†, George Hutchinson, Nik Kardashev, John Kraus†, Laurence Manning, Connie Mayer†, Ed McClain†, Kenichi Miya, Fumio Moriyama, George Mueller, Vivian Phillips, Jack Piddington†, John Pierce, Wolfgang Priester, Ed Purcell†, J. J. Riihimaa, Jim Roberts, Peter Robertson, Olof Rydbeck†, Boris Schedvin, Jean-Louis Steinberg, Gordon Stewart, King Stodola, Tatsuo Takakura, Charlie Townes, James Trexler†, Derek Vonberg, Kevin Westfold†, Fred Whipple†, and Don Yabsley†.

One contemporary of mine who has played a huge role in improving this book and bringing it to fruition is Miller Goss, who carefully reviewed the entire manuscript and raised many issues both of detail and broader impact. I thank him for his labors and wish him well on his own forays into the history of radio astronomy.

I have enjoyed significant institutional support over the years, starting with the Kapteyn Laboratory of the University of Groningen for a postdoc, then the UW Department of Astronomy since 1973, supplemented by sabbatical stays at the Institute of Astronomy, Cambridge University and the Observatoire de Meudon near Paris. I am grateful to the directors of these institutions for backing my historical pursuits. I also especially thank Arthur Whiteley and his eponymous Center for establishing a marvellous scholarly retreat in the San Juan Islands where for the past five years major portions of this book have been written and rewritten.

Financial support, primarily in the form of partial summer salary, has come from the Dudley Observatory (thrice), the UW Graduate Student Research Fund (once), and the National Science Foundation (eight times). NSF's Program in History and Philosophy of Science, headed by Ron Overmann, also awarded a major grant in 1976–79, which allowed, for instance, a three-week visit to the Radiophysics Laboratory in Sydney for archival and oral history research. In 1980 I also was privileged to visit the main Soviet radio astronomy sites and groups on an exchange sponsored by the US National Academy of Sciences. And I would be remiss if I did not mention my first-ever "grant" to do history: a 100-guilder gift in 1973 from my colleagues at the Kapteyn Lab – I have never forgotten this endorsement of my fledgeling efforts.

Processing the interviews has been a laborious task. I thank NSF and the Center for History of Physics (American Instutute of Physics.) for funding transcriptions. In particular, the skill and attention to detail of transcribers Bonnie Jacobs and Pamela Jernigan is acknowledged. Furthermore, during the early 1990s Karen Fisher provided excellent secretarial services.

Librarians have often worked wonders for me, whether via Interlibrary Loan, locating obscure reports and journals, or allowing special access after hours. I also thank those who have made the NASA Astrophysics Data System a powerful and vital bibliographic tool for the historian of astronomy.

This history would be impoverished and stale without access to well-organized archives, or sometimes to a person's papers before placement in an official archive. For the latter privilege I thank in particular Lady Rowena Ryle (widow of Martin Ryle), as well as Alice Jansky[†] and David B. Jansky (widow and son of Karl Jansky). A special, huge thanks goes to Sally Atkinson, longtime chief administrative assistant of the Radiophysics Division in Sydney, and unofficial archivist after her retirement. Sally was tireless in fulfilling my requests and providing access to everything from scrapbooks to photographs to official correspondence in the rich Division records. Other archivists too, from around the world (Appendix C), have rendered superior service.

I have been aided in translations over the decades by Helga Byhre, Karl-Heinz Böhm, Tom Hankins, Larry Sandler[†], Jim Naiden[†], Vlad Chaloupka, Bob Schommer[†], Julian Barbour, Dave Jenner, and Joke Huizinga[†].

My family has sustained this project in many ways, for instance helping with bibliographic and archival tasks, tolerating warped holiday trips, and encouraging me to stick with it. Daughters Rachel and Sarah grew up with "The Book" ever present and my wife Barbara's support has been continuous for decades. To her I dedicate this volume.

Acknowledgments for figures

The author thanks the following persons, publications and institutions for the use in this book of the indicated figures from their publications or collections. I apologize if I have unwittingly, despite my efforts, omitted any required permissions.

American Astronomical Society (*Astrophysical J.*): 4.5, 4.6, 4.7, 10.3, 10.6, 11.3, 11.10, 14.4, 14.13, 14.14

American Institute of Physics (*Reviews of Scientific Instruments*): 10.1

Astronomical Society of the Pacific (*Publications of the Astronomical Society of the Pacific*): 7.8

Bulletin of the Astronomical Institutes of the Netherlands: 14.3, 15.4

CSIRO Publishing (*Australian Journal of Scientific Research; Australian Journal of Physics*): 7.12, 12.12, 13.4, 13.5, 13.13, 13.14, 14.8, 14.11, 14.16, 15.1, 16.8

Elsevier (*J. Franklin Inst.*): 5.8

IEEE (*Proc. IRE*): 3.2, 3.4, 3.5, 3.6, 3.7, 3.8, 4.7, 11.1

Institute of Physics Publishing (*Proc. Physical Soc.; Reports on Progress in Physics*): 6.5, 6.6, 6.7, 6.8, 6.9, 14.8

Nature Publishing Group (*Nature*): 6.3, 7.11, 8.4, 13.8, 13.12, 14.9

Wiley–Blackwell (*Monthly Notices of the Royal Astronomical Society*): 9.6, 11.6, 11.8, 11.9, 13.9, 14.6

Royal Society (*Proc. Royal Soc.*): 6.1, 6.4, 7.9, 7.10

ScienceCartoonsPlus.com (Sidney Harris): 14.17

Taylor & Francis (*Philosophical Magazine*): 11.4

University of Chicago Press (from *The Sun*, ed. G. Kuiper [1953]): 13.16

Cavendish Laboratory, Cambridge University (Radio Astronomy Group): 8.2, 8.5, 8.6, 8.7, 8.8, 8.9, 10.9, 13.7, 14.12

Combined Arms Research Library, Ft. Leavenworth, Kansas – from p. 323 of *Army Radar* by A. P. Sayer (1950), a volume in the series *The Second World War, 1939–1945, Army* (London: The War Office): 5.2

Historic Photographic Archive, Australia Telescope National Facility: 7.1, 7.3, 7.4, 7.6, 7.7, 7.13, 12.9, 12.10, 12.11, 13.10, 13.15, 13.17, 13.18, 14.7, 14.9, 16.9

Public Record Office, Kew, UK (Crown copyright): 5.3, 5.4

Reber papers, Archives, National Radio Astronomy Observatory: 4.1, 4.2, 4.3, 4.4, 4.8

Ryle papers, Churchill College Archives: 15.3

Southworth collection, Niels Bohr Library, American Institute of Physics: 3.1

Southworth papers, AT&T Archives: 5.6, 5.7

Z. Bay: 12.6, 12.7, 12.8; E.-J. Blum: 10.11, 10.12; R. N. Bracewell: A.3; A. E. Covington: 10.7; J. H. DeWitt & US Army: 12.2, 12.3, 12.5; R. H. Dicke: 10.2; H. I. Ewen: 16.2, 16.3, 16.4; V. L. Ginzburg: 10.8; J. L. Greenstein: 14.15; J. P. Hagen: 10.2; D. W. Heightman: 5.5; J. S. Hey: 5.1; R. C. Jennison: 14.18, 14.19, 14.20; J. Katgert-Merkelijn: 16.1; J. D. Kraus: 1.1; A. C. B. Lovell: 9.1, 9.2, 9.3, 9.5, 9.7, 9.8, 9.10, 11.2, 11.5, 11.7; A. C. Muller: 16.6; J. H. Oort: 16.5, 16.7; J. W. Phillips: 6.2; G. W. Potapenko: 3.9, 3.10, 3.11; A. E. Salomonovich: 10.10; I. S. Shklovsky: 10.8; H. Tanaka: 10.13; F. Trenkle: 12.1; J. H. Trexler & US Navy: 12.13; US Navy: 10.4, 10.5

Unable to locate
Electronics: 12.4
Hochfrequenztechnik und Electroakustik: 6.10
Radio News: 4.9
Science Progress: 9.4

1 • Prologue

1.1 A NEW SKY

> My own suspicion is that the universe is not only queerer than we suppose, but queerer than we *can* suppose.[1]

British biologist J. B. S. Haldane penned this aphorism in the 1920s. Two decades later, in a 1948 popular article, he cited new results on radio noise from the sun as an example of what no one had imagined, with undoubtedly "much queerer things" to follow.[2] Casting over millennia, he talked of how people had first looked to the heavens as a source of heat and light, as well as of religious beliefs and astrological guidance. Later, they made other connections with the sky, such as the ocean tides and meteorites. And now, he argued that radio waves were of similarly great "philosophical importance." Haldane overstated the case, but there is no doubt that the coming of the radio telescope in the mid-twentieth century fundamentally shifted how extraterrestrial realms were viewed. It was not unlike the situation in the seventeenth century when the natural philosopher Robert Hooke remarked that the (optical) telescope had led to

> a new visible World discovered to the understanding. By this means the Heavens are open'd, and a vast number of new Stars, and new Motions, and new Productions appear in them, to which all the ancient Astronomers were utterly Strangers.[3]

Hooke also emphasized that this new world could only be attained by the "adding of artificial Organs to the natural." In this spirit, and jumping ahead three centuries, let us imagine a being with a radio antenna instead of eyes – call him *Homo radio*. What new aspects of the sky did he come to know after a decade of radio astronomy following World War II? How striking were they?

Figure 1.1 illustrates several startling differences from the familiar optical (or visual) sky, which of course was all that had been known for millennia. The signals received with *Homo radio*'s antenna, at wavelengths roughly one million times longer than those in the optical band, would be noise-like radiation first called *cosmic noise* – "noise" because when detected and impressed on a loudspeaker, it would sound like the *shhhh* heard on an analog radio or television channel when the station is *not* broadcasting.[4] The only information to be gained as he scanned the sky at a given frequency would be the intensity of the radio noise at different locations. Hooke's contemporary Blaise Pascal had been frightened by the eternal silence of infinite space,[5] but now the heavens were full of "mysterious sounds, a haunted Universe forever emitting ghost-like wails."[6]

Homo radio's sky would be seen at any time of day or night, through clouds or rain – very different from optical astronomy.[7] Radio astronomy was sometimes even called "blind astronomy":

---

[1] J. B. S. Haldane (1927), *Possible Worlds and Other Essays* (London: Chatto & Windus), p. 286 in "Possible Worlds."

[2] Haldane, "Radio from the sun," *Daily Worker*, p. 2, 16 February 1948. Haldane was inspired to write this article after hearing a paper at the Royal Society, later published by Ryle and Vonberg (1948).

[3] R. Hooke (1665, London), *Micrographia*, quotes from the Preface.

[4] *Noise* is a technical term in physics and engineering, confusing because it is used for much more than sound waves, in fact for any signal characterized by random fluctuations. In early radio astronomy it could refer to the desired radio signal (as in *cosmic noise* or *solar noise*), or just as well to competing noises from the sky or from receiver electronics (see Sections A.8 and A.9 for further explanation).

[5] "Le silence éternel de ces espaces infinis m'effraie." [Pascal (1623–1662, *Pensées*)]

[6] W. L. Laurence, "Radar yields new world of sound; brings 'music of spheres' to Earth," 6 October 1948, *New York Times*, p. 1.

[7] If *Homo radio* had happened to live on a permanently beclouded planet, then this view would have been the *only* available one of

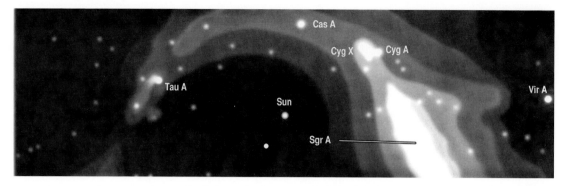

Figure 1.1 The northern sky as seen with "radio eyes." The map extends from −40° to +70° in declination; north is up and east to the left. This artistic rendering is based on a map, observed with a beam of 1° × 8° at a frequency of 250 MHz (wavelength of 1.2 m), taken with a helix array at Ohio State University in the mid-to-late 1950s. The broad curving band, defining the radio Milky Way, corresponds in position to the optical Milky Way (the galactic plane). The position of the radio sun varies through the year and is shown for the spring equinox; its intensity also varies greatly on various time scales, but normally is far fainter than the radio Milky Way. The strongest radio stars are labelled (see Table 14.1 for details of sources); Cyg X is a large, diffuse region and Sgr A, the galactic center, is lost in the surrounding region of strong emission.

> The biggest telescope in the world now stands on a site near (of all places) Manchester [England]. Here its 200 ft reflector gazes blindly upwards, heedless of mist and cloud and even of daylight.[8]

The siting of radio telescopes was thus not as critical as for optical telescopes (although manmade radio interference could be a problem), a fact that brought British and Dutch astronomy into renewed prominence in defiance of their rainy homelands.

Homo radio's sun, although a strong emitter, would be vastly outshone (especially at the lowest frequencies) by a broad band of radio Milky Way stretching across an otherwise dark sky (also see Fig. A.5).[9] His main sources of illumination would be neither sun nor moon, but this Milky Way swath, with a strong concentration toward the constellation Sagittarius. And if our creature tuned his radio receiver to the specific frequency of 1420.4 MHz, the entire sky would take on an additional radio glow, girdled by a thinner Milky Way strip.

"Radio daytime" would correspond to when the brightest and largest part of the Milky Way was above the horizon, and have nothing to do with the sun – but even in this "daytime" most of the sky would be dark. Moreover, the contrast between day and night would be far less than we optical beings experience. At shorter wavelengths, a very faint radio moon, complete with phases, would shine. Strikingly, the radio sun, rather than shining steadily, would sometimes suddenly become thousands of times brighter than usual, and on rare occasions millions of times brighter. These outbursts would usually last seconds or minutes, but sometimes persist for days in groups called "noise storms." Rather than a simple *shhhh*, these sounded like "a combination of gravel falling on the roof and the howling of wolves"[10] – this was no music of the spheres!

Scattered around the sky (not unlike at optical wavelengths) would be hundreds of discrete *radio stars*,

---

the universe, at least until rockets and satellites could breach the clouds and gain access to optical and other wavelengths (see Section 18.2).

[8] "An unsuspected universe revealed," *The Times* (London), 4 October 1952, p. 7. (The mentioned reflector was the Jodrell Bank fixed dish of 218 ft diameter [Section 9.3].) The term "blind astronomy" was first used by P. M. S. Blackett, during a Royal Astronomical Society meeting in Manchester on 1 July 1949 [*Observatory* **69**, 122 (1949)]. Also see note 17.31.

[9] In 1959 optical astronomer Otto Struve called this fact "one of the most sensational discoveries in astronomy during the present century" [MIT Compton Lectures, as published in *The Universe* (1962) (Cambridge: MIT Press), p. 98]. At a frequency of 20 MHz the total radio radiation from the Milky Way is ~10^4 times more than that from the (quiet) sun; the comparable ratio at optical wavelengths is 10^{-8}.

[10] As in note 6; quotation from D. H. Menzel.

the brightest handful (Fig. A.2) comparable to the quiet sun, but without the huge bursts. Each radio star would appear as an ill-defined, fuzzy blob on the sky, not a pinpoint of radiation. *Homo radio* would realize that this was due to an inherent defect in his antenna apparatus, but for technical reasons would not yet be able to create sharper radio images except for the few strongest radio stars. Some of the radio stars would twinkle (or scintillate) in their intensity; at first it would not be clear whether these changes were intrinsic to the radio stars or caused by the earth's atmosphere.

After ten years of applying powerful physics and mathematics, what would *Homo radio*'s theoretically minded colleagues make of all this, even if we should grant them full knowledge also of the optical sky? The biggest quandary would be what to do with the radio stars, for, despite their name, not a single radio star's position on the sky agreed with positions of the bright optical stars, although a handful agreed with the positions of faint, peculiar optical nebulosities. The only consensus as to mechanisms of emission would be (a) the bright glow at 1420.4 MHz arose from ubiquitous cold hydrogen atoms, and (b) the quiet sun's radio emission originated in a hot coronal gas surrounding the sun. All other explanations would be tentative. As one early radio astronomer later recalled this situation:

> There appeared to be an optical universe and a radio universe which were utterly different, which coexisted. So there was obviously a need to tie them together somehow.[11]

And one early textbook of radio astronomy flatly stated:

> The primary source of radio waves in the universe is not known; it is probable that [optical] stars supply only a small fraction of the total.
> (Pawsey and Bracewell 1955:210)

This remarkable sky posed numerous puzzles even as it led to expanded views of many old astronomical questions. It not only created a new sky, but necessarily meant that the previously known sky now became an *optical* sky. There were now *two* ways to view the heavens. Indeed, another important effect of the establishment of radio astronomy was to create *optical* astronomy, *optical* astronomers, and *optical* telescopes.

In the following section, I lay out the overall structure of the book and summarize my basic narrative for early radio astronomy, as well as explaining the rationale for an ending point of 1953. In Section 1.3 I discuss my approach to history, compare this study with other treatments of the field, and lay out the four major historical themes of the book. These deal with the profound influence of World War II and the Cold War on the development of the field; the field's intimate entanglement of technology and science; early radio astronomers' ironic drive toward a "visual culture" in terms of their quest for tie-ins with astronomers and their (optical) sky; and finally, radio astronomy as the twentieth century's New Astronomy, one of the key developments in the millennia-long history of astronomy.

1.2 ORGANIZING THE STORY

1.2.1 Defining radio astronomy

Before relating a narrative, I must first define "radio astronomy." The question is vexed because there is the danger of engaging in "tunnel history," in which the standards of a later time, in this case how science organizes itself into research areas and disciplines, are imposed on an era that did not know them. To avoid such an ahistorical exercise, I tell the story in terms that would be understood by the participants at any given time, to allow the reader to experience the world-picture (*Weltbild*) of that day. To do otherwise might cause us to miss fascinating and insightful aspects of the history, in particular the decade-long transition from (1) disparate physicists and radio engineers (with a few astronomers) making discoveries about the earth's ionosphere, the sun and moon, the Milky Way, and beyond; to (2) a research community of "radio astronomers" who, although still distinct from traditional astronomy, were slowly becoming integrated. Note, however, that the term *radio astronomy* did not have currency until 1948–49 (and *radio astronomer* even later[17.2.1],[12]), so

---
[11] J. G. Davies (1971:40T), also cited by Edge and Mulkay (1976:268). The notation "Davies (1971:40T)" is explained at the end of Section 1.2.2.

[12] In the remainder of this chapter, superscripts of the form [X] indicate that Chapter or Section X contains the relevant topic. The book's organization, key episodes, and main arguments can also be telegraphically followed using the Annotated Table of Contents.

how can one legitimately talk about a history of radio astronomy beforehand?

My approach to this conundrum is akin to that of Good (2000), who dealt with the same issue when analyzing the development of geophysics before 1900, at which time the term *geophysics* and a discipline began to emerge. He discusses an "assembly" of geophysics at that time from components contributed by many fields and recombined in various ways. Thus I also will follow the story of the motley components before 1948 that were eventually to emerge as something called radio astronomy (although I will argue that this emergence was not as a separate discipline, but as a research specialty).[17.2.5] These components included trans-Atlantic radio telephony, studies of the ionosphere and its effects on communications, military radar development during and after World War II, radio waves from the quiet and active sun, radar reflections from meteor trails and from the moon,[13] sky surveys with various antennas and receivers, special studies of regions of discrete emission (radio stars), and (optical) astronomical studies of Milky Way rotation.

1.2.2 Structure of the book

The book relates activity before World War II in Chapters 2–4, the wartime discovery of the radio sun in Chapter 5, and in Chapters 6–10 the beginnings, first major results, and distinguishing characteristics of each major group. Chapters 11 and 12 examine the subfields that studied meteors and the moon via radar reflections. Chapters 13 through 16 then follow in detail, through ~1953, the development of each major radio astronomy topic: the sun, radio stars, theories of galactic noise, and 21 cm hydrogen-line studies of the interstellar medium and Galaxy. Chapter 17, "New astronomers," is a detailed historical analysis of various aspects of early radio astronomy. Topics include its origin in war, the watershed in its character circa 1952–53, interactions between radio and (optical) astronomers, whether or not it was a new discipline, national and Cold War influences on how the field developed, and the role of technology. Chapter 18, "A New Astronomy," summarizes the new science that emerged from radio astronomy and appraises where radio astronomy fits in the larger history of astronomy. In particular, Section 18.2 briefly relates for comparison the openings after World War II of each of the *other* spectral windows: infrared, ultraviolet, X-ray and γ-ray (summarized in Table 18.1). Radio astronomy preceded and paved the way for these. Finally, I argue that the appearance of radio astronomy (or more broadly, the opening of the entire electromagnetic spectrum) was not a "revolution" as historians usually define it, but more a transformative event, a *New Astronomy*, similar in nature and importance to earlier major developments in the long history of astronomy. The issues of Chapters 17 and 18 are further introduced in Section 1.3.3.

In the back matter, Appendix A is a primer on the techniques and astrophysics of early radio astronomy (with its own index of terms) for readers who would like to learn more of the basic science and engineering involved in early radio astronomy. Appendix B discusses many aspects of interviewing and how it was done for this history, as well as listing the interviews conducted during this project. Appendix C gives details of the archival sources that I consulted and acts as an index for the three-letter codes (e.g., GRE, RPS) seen at the end of archival citations in footnotes. Tangent 1.1 (at the end of this chapter) discusses conventions in this book (such as using the unit MHz when describing Jansky's work, although he used Mc/s). "Tangents," located at the end of individual chapters, include peripheral topics and technical details. Finally, besides the Index, note that the References list, which is arranged chronologically, also acts as an index in that each listed item includes the primary page(s) where the item is discussed. The References list gives full information for all items cited in the text with "Jones (1950)" or Jones (1950:472) formats (the latter referring specifically to p. 472). Quotations from interviews are cited with the formats "Smith (1976:14T)", "Smith (1976:107B:420)" or "Smith (1976:2N)," each referring to a 1976 interview with Smith (see Appendix B for details).[14]

---

[13] At a later stage the term *radar astronomy* came to be applied to astronomical studies involving radar reflections. However, during the period of this book (before 1953), meteor and lunar radar studies were considered aspects of radio astronomy. For example, the term *radar astronomy* does not appear in two early books on radio astronomy despite their extensive chapters on radar techniques and results (Lovell and Clegg 1952; Pawsey and Bracewell 1955).

[14] Additional information relating to the topics and specifies of this book is available at www.astro.washington.edu/woody.

1.2.3 Narrative summary

Here I present a précis of my narrative, which extends through ~1953, a year that was a watershed for radio astronomy for several reasons.[17.1.2] By that time the bulk of the radio astronomers were no longer astronomical novices, at a mid-career stage in their 30s, and some even recipients of significant professional honors. Furthermore, integration with astronomy as a whole was gaining strength; for instance, to aid communications, the radio astronomers at this time officially voted to discourage use of terms like *solar noise* and *cosmic noise* (as in this book's title) in favor of *solar radio emission* and *galactic radio emission*. In addition, the field's most important sectors had migrated from solar to galactic and extragalactic questions (although solar research still quantitatively dominated). The first two textbooks of radio astronomy also appeared about this time. Finally, the field was moving from empirical exploration to a more programmatic approach, entailing enormous antennas whose correspondingly larger budgets meant that they could no longer be built in-house by the individual groups. Radio astronomy was moving into a Big Science phase.

Figures 1.2 and 1.3 illustrate the broad flow of the history by following, respectively, the major groups and the primary research subjects through ~1953. These charts should be consulted while reading the following narrative. In addition, Tables 10.1 and 11.1 present basic data for each of the postwar groups in radio and radar astronomy. Almost all of this research was a direct outgrowth of the development of radar and other radio techniques for myriad military purposes during World War II. Radar research and development during the war was huge, second (in the US) only to the atomic bomb project. Its importance to the war effort, caricatured in Fig. 1.4, was sometimes expressed as "The atom bomb only ended the war; radar won it" (Kevles 1977:308).[5.1, 17.1.1]

Although the first detection of extraterrestrial radio waves was not until 1932, in the period 1894–1901 there were several attempts in Europe to detect Hertzian (radio) waves from the sun.[2] Yet in the ensuing four decades, as radio technology steadily improved, no one made serious attempts to detect the sun. This was because of a lack of sensitive, directional radio equipment, discipline specialization, and a reliance on Planck's blackbody theory (1900) to calculate expected signal levels. So it was that Karl Jansky, a radio physicist working at the Bell Telephone Laboratories in New Jersey, only serendipitously discovered extraterrestrial radio emission in 1932 – and it came from the Milky Way, not the sun.[3] With his beam size of ~30° at a frequency of 20 MHz, he established that the emission (he called it "star static") was strongest toward the center of our Galaxy, yet appeared all along the galactic plane.

The hard times of the economic Depression and the bizarre nature of Jansky's discovery meant that, although his results were generally known, little serious follow-up occurred throughout the 1930s. The lone exceptions were investigations carried out over a decade by radio engineer Grote Reber.[4] Living in a Chicago suburb, in his spare time and with his own funds, Reber took it upon himself to build a 31 ft (9.4 m) diameter dish in his backyard in order to check Jansky's findings at much higher frequencies. He ended up making two maps over the northern sky of what he called "cosmic static" (Fig. 4.7). Along the way, Reber also detected solar radio emission (but did not know of earlier unpublished work by others) and established working relations with the astronomers at Yerkes Observatory. Yet despite this pioneering and ingenious work through 1947, Reber had little influence on the postwar development of radio astronomy, especially outside of the US.[4.7]

The man who was seminal for the beginnings of radio astronomy in England was Stanley Hey. As a radar-operations researcher during the war, in 1942 he deduced that apparent jamming of British coastal radar installations was in fact radio radiation picked up from the sun.[5.2] This was the earliest conclusive detection of the radio sun (and kept secret until war's end), although before the war amateur radio operators and ionospheric physicists had garnered some evidence of the same.[5.4] Later during the war, the sun was independently studied by George Southworth of Bell Telephone Labs at centimeter wavelengths,[5.6] as well as detected during wartime operations in Germany and in New Zealand.[5.3]

Before the end of the war, Hey also encountered and investigated both the Jansky/Reber galactic noise and anomalous echoes that he determined originated from the ionized trails of meteors.[6.1] Until the Cold War heated up in 1948, Hey was able to hold together his small Army group and conduct significant research

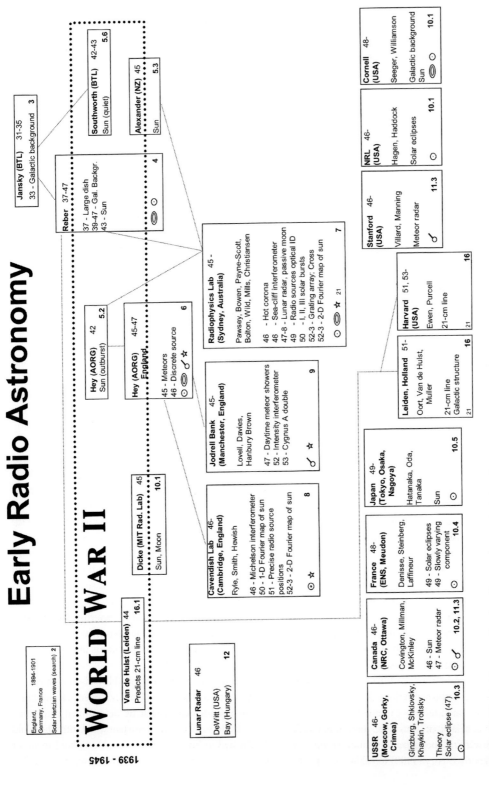

Figure 1.2 The development of early radio astronomy tracked via major groups and persons. Research highlights in each box are incomplete and emphasize the earlier years. Dates are indicated by, for example, 45 = 1945. The boxes are arranged to show a rough chronology; in particular note their placements and extents relative to the large World War II box. Connecting lines indicate direct influences on the origin of a group. Symbols at the bottom of each postwar box indicate to which of five main subfields a group contributed: for example, at the bottom of the "Hey (AORG)" box, the symbols represent (l. to r.) the sun, galactic background radiation, meteors, and radio stars. The fifth subfield symbol is "21" for 21 cm hydrogen line study. Bold numbers in the lower-right corner of each box refer to chapters or sections where each group is discussed.

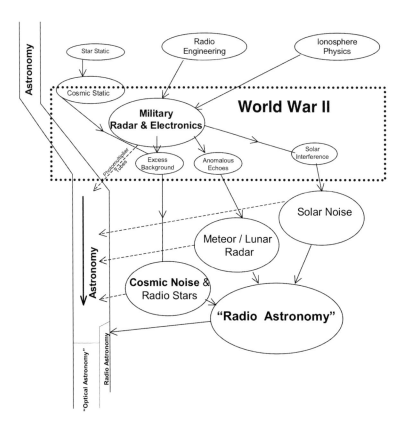

Figure 1.3 The development of early radio astronomy tracked via subject areas. Note the primacy of military radio and radar electronics and engineering, as well as the influence of ionosphere physics. Note also the secondary influence of Jansky's star static and Reber's cosmic static, along with the influences of wartime anomalous radio phenomena, two of which J. S. Hey discovered, that led to the three primary postwar areas of inquiry. These in turn became "radio astronomy" in 1948–49, which eventually joined the ongoing astronomy "stream," although keeping its own identity. Later came the newly defined "optical astronomy" (see Section 17.2). The photomultiplier tube's effect on postwar optical astronomy is discussed in Section 17.2.2.

in all aspects of radio astronomy, most importantly establishing the existence of the first discrete "radio star," later known as Cygnus A. [6.2]

Two other major postwar groups also began in England at this time as an outgrowth of radar research at the wartime Telecommunications Research Establishment. Martin Ryle started at the Cavendish Laboratory of Cambridge University in an ionospheric research group, but soon was investigating radio waves from the sun and radio stars, of which by 1949 his group had found 23 (including Cassiopeia A).[8] In the process he invented many techniques fundamental to radio astronomy, such as the Michelson interferometer and the phase switch, and assembled the core of a powerful research group featuring Graham Smith and Antony Hewish. The second British group, led by Bernard Lovell, studied radar echoes from meteor trails at Jodrell Bank, a field station of the Physics Department of Manchester University.[9] Over the postwar decade Lovell's group developed techniques to measure the trajectories and velocities of meteors and dominated the world in this field. For instance, new techniques enabled study of *daytime* meteor showers. In 1949 Lovell was joined by Robert Hanbury Brown, another leading wartime radar veteran. Hanbury Brown focused on using a huge fixed dish of diameter 218 ft (66 m), and with his student Cyril Hazard detected for the first time radio waves from an external normal galaxy, the Andromeda nebula.[9.3]

The largest of all the postwar groups was led by Edward "Taffy" Bowen and Joseph Pawsey of the Radiophysics Laboratory (RP), Sydney.[7] This

8   Introduction

Figure 1.4 World War II cartoon showing the effects of radar on the Axis powers represented by cowering Hitler, Mussolini, and Hirohito. (Philco Co. advertisement, *Life* magazine, 9 August 1943, p. 3)

Australian laboratory was one of the world's premier wartime radar labs and remained intact after the war, switching its electronic know-how to investigation of cosmic noise. With Bowen's management skills and nose for good projects, and Pawsey's inspired scientific leadership of teams of radio engineers and physicists at field stations scattered around Sydney, RP developed techniques such as the sea-cliff interferometer and Fourier-synthesized maps. Basic discoveries in the early years included a hot radio solar corona at a temperature of 1×10^6 K (Pawsey and David Martyn)[7.3.3]; angular sizes of <10′ [15] for solar radio burst regions (Pawsey, Lindsay McCready, and Ruby Payne-Scott)[7.3.2]; motions and frequency-drift structure of solar bursts (Payne-Scott; Paul Wild)[13.2.1.2]; suggested identifications of three radio sources with optical objects: Taurus A with the Crab nebula (remnant of a supernova explosion), Centaurus A with a weird nebula catalogued as NGC 5128, and Virgo A with the distant elliptical galaxy Messier 87 (John Bolton, Gordon Stanley and Bruce Slee)[7.4, 14.1]; and pioneering measurements of radio source properties and distributions across the sky (Bernard Mills)[14.3.2, 14.4.1]. The success of the RP group was all the more remarkable given that Australia at this time was geographically and scientifically isolated, and not known for its science. The esteem of RP's colleagues can be gauged by the decision to hold the 1952 meeting of the International Union of Radio Science (URSI) in Sydney, the first time *any* major international scholarly society had met in Australia.

The story of early Dutch radio astronomy is anomalous in that it was not driven by radar veterans and in fact was directed by (optical) astronomers.[16] Upon learning of Reber's results, astronomers Jan Oort and Hendrik Van de Hulst, at Leiden University in occupied Holland, as early as 1944 considered the possibilities of using radio spectral lines to study the structure of the

---

[15] 60′ = 60 arcmin = 1°; the angular size of the sun is 32′. See Section A.6 for more information.

Milky Way, a specialty of Oort's since the 1920s. Once the war ended, resources were scarce and radar veterans were nonexistent, but Oort persisted and eventually succeeded when he found Alexander Muller, a talented radio engineer fresh from undergraduate study. Muller built a receiver that in 1951 detected a spectral line of atomic hydrogen at a wavelength of 21 cm (closely following its discovery in the US [see below]). This technique made accessible an entirely new component of the interstellar medium. There quickly followed the first of many Galaxy surveys exploiting the line's Doppler shifts in order to detect spiral arms composed of cold hydrogen. An "international Dutch school" of 21 cm Milky Way studies was soon established.

Besides the above five major groups in the postwar decade, many other efforts sprung up around the world.[10] In the United States[10.1] Robert Dicke, a physicist at the Radiation Laboratory run by the Massachusetts Institute of Technology, the largest of the American wartime radar labs, developed his eponymous microwave receiver and with it in 1945 was the first to observe the moon and a solar eclipse. At the US Naval Research Laboratory (Washington, DC), another leading wartime radar lab, John Hagen's group, starting in 1946, undertook many solar projects at microwavelengths, including eclipse expeditions, and also built a 50 ft (15 m) diameter dish.[10.1.2] Smaller groups were at Cornell University[10.1.3] and Stanford University[11.3]. Finally, a one-time effort (a Ph.D. thesis project) represents the most significant US contribution to radio astronomy in the postwar decade. This was the discovery at Harvard University in 1951 of the 21 cm hydrogen line by physicists Harold Ewen and Edward Purcell (both former wartime radar experts).[16.3]

Radio astronomy in the USSR was strongest in theoretical studies and dominated from the start by physicist Vitaly Ginzburg (Lebedev Institute [FIAN], Moscow) and astronomer Iosef Shklovsky (Moscow State University).[10.3.1] In 1946 both made seminal contributions to solar coronal radio theory. From 1951 onwards, Ginzburg argued in detail for a connection between cosmic rays and the still-unexplained galactic background radiation, namely that relativistic electrons in the Galaxy's magnetic field copiously produced radio waves via the synchrotron mechanism.[15.4] Shklovsky worked out a theory for interstellar radio lines of hydrogen, hydroxyl (OH) and CH in a 1949 paper, and in 1953 suggested that *both* the enigmatic radio and optical radiation from the Crab nebula arose from the synchrotron mechanism.

In France, groups at the Ecole Normale Supérieure in Paris and Meudon Observatory undertook solar observations both at home and on eclipse expeditions to exotic locales.[10.4] Jean-François Denisse was a leader in solar radio theory and recipient in 1949 of the first doctorate awarded for a radio astronomical topic. Canada produced important observations of the sun[10.2] and of meteor echoes[11.3] and in Japan several isolated solar groups made contributions[10.5].

Study of the moon using radio techniques occupied a small but fascinating niche for a few years.[12] In early 1946 John DeWitt of the US Army Signal Corps in New Jersey led Project Diana, which marshaled a powerful transmitter to first detect radar echoes from the moon.

Meteor radar was a fruitful subfield during the postwar decade, although unique in early radio astronomy in that by the mid-1950s it had peaked and gone into decline.[11] There had been several studies before the war suggesting that the ionized trails of meteors could effectively reflect radio waves, but the effect was not confirmed until 1945 when Hey's group undertook a three-station triangulation experiment on anomalous echoes that had often been seen during wartime operations.[6.3] The October 1946 Giacobinid meteor shower, spectacular (as predicted) for visual observers, also represented a milestone for the many radar teams who were inundated with echoes.[11.2] The major scientific question to which meteor radar contributed was the issue of whether there existed a component of meteors arriving from interstellar regions, completely outside the solar system. Astronomers had been debating this issue for decades, and in the end statistics from huge numbers of echoes gathered at Jodrell Bank and in Canada convinced most that very few, if any, meteors had hyperbolic orbits indicative of interstellar origin.[11.4.2]

Studies of solar noise constituted fully ~70% of the efforts of early radio astronomers, and yet this was the subfield where the fewest fundamental astronomical discoveries were made.[13] Radio observers did, however, contribute important evidence for the million-degree corona that astronomers had been suspecting since the 1930s.[13.3.1] Tremendous efforts went into monitoring radio burst activity and correlating it with various optical properties – the main outcome was the recognition of a "slowly varying component," a long-term oscillation in total microwave intensity that followed the 11-yr sunspot cycle.[13.2.3] In other studies, expeditions were mounted to eight solar eclipses over the period 1945–52 (Fig. 13.6) in order to gain higher angular resolution by using the moon as a moving knife edge.[13.1.3] Perhaps the greatest contribution of the sun, so to speak, to radio astronomy was as a handy test bed for new high-resolution mapping techniques developed by Ryle at Cambridge and Wilbur "Chris" Christiansen in Sydney. By 1953 both groups had made two-dimensional maps of the sun using Fourier analysis of data gathered with interferometric portable antennas and grating arrays.[13.1.3, 13.3.2]

Radio stars, or radio sources as they eventually became known, were the central puzzle in early radio astronomy.[14] Four major surveys were carried out by the groups in Cambridge, Manchester, and Sydney over 1949–53,[14.3] but only about 5% of the ~250 known radio stars correlated with any optical object, i.e., had what became known as an "optical identification."[14.4] Furthermore, different surveys in areas of overlap often seriously disagreed (Fig. 14.5). Two astronomers with access to the world's largest telescopes, Walter Baade and Rudolph Minkowski, were central to taking the deep photographs required to make identifications, which then bolstered the degree of reality of any associated source. On the radio side, positions accurate to a few arcminutes or better were required for any chance at an identification; one particularly pivotal effort was the precision interferometry of Smith at Cambridge. Positional accuracy of ~1' allowed the two strongest sources in the sky, which had stubbornly resisted previous attempts, finally to be identified with (a) a strange network of fast-moving filaments that was perhaps a supernova remnant (Cas A, Fig. 14.14), and (b) a faint, distant peanut-shaped object, taken to be a pair of galaxies in collision (Cyg A, Fig. 14.13). The latter especially was significant, for it implied that if Cyg A's intrinsic properties were at all typical, then the bulk of the *un*identified radio sources were at distances beyond what (optical) telescopes could explore. Thus did radio astronomy become relevant to cosmological inquiries at an early stage.[15.5] Together with the earlier identifications of Tau A, Vir A and Cas A, it was also clear that radio astronomers were revealing violent, high-energy regions hitherto unsuspected by (optical) astronomers.

Another desired measurement for a radio source was its angular size.[14.5] But the intrinsically poor resolution of most early antennas and interferometers made such observations a difficult task except for the strongest few sources. These turned out to have sizes of a few arcminutes, which argued that they were not radio *stars* at all, but *radio nebulae*. One novel technique for measuring sizes, the intensity interferometer, was developed at Jodrell Bank by Hanbury Brown and Richard Twiss. With it Roger Jennison and M. K. Das Gupta not only measured the angular extent of Cyg A, but found that it mystifyingly consisted of two blobs separated by 1.5′, one on either side of the optical object.

The source of the galactic background radiation that Jansky had originally discovered, and its connection, if any, with the enigmatic radio stars, were vigorously debated in the postwar decade.[15] By 1953 there was still no consensus as to how radio power originated in radio stars and in the general background. One approach was to consider the background to be the integrated effect of a huge number of faint radio stars, but such a hypothesis implied that this new type of (radio) star was not confined to the galactic plane. Others considered radio stars to be very close-by and of low intrinsic power, whereas the background radiation originated either in hot gas or in cosmic ray particles. Soviet theorists in particular pushed for the synchrotron idea, but in the West it was little accepted until after the mid-1950s.[15.4]

By 1953, when the present history ends, we thus have a burgeoning and maturing field generating more questions than answers regarding the origins of the various types of celestial radio emission visible in Fig. 1.1. It was a revolutionary new sky, albeit with most radio sources (and the galactic background) still a mystery despite collaborations with (optical) astronomers that had yielded a handful of key optical identifications.

1.3 ANALYSIS

1.3.1 Earlier studies

The major prior study of the history and sociology of radio astronomy is *Astronomy Transformed: the Emergence of Radio Astronomy in Britain* (1976) by David Edge (1932–2003) and Michael Mulkay, the former a student of Martin Ryle's at Cambridge (Ph.D. in 1959) who then became a founder of the "Edinburgh school" of the sociology of scientific knowledge, and the latter also a leading sociologist of science.[16, 17] Their book is roughly half history and half sociology. Its historical aspects, which have informed the present study, cover to some extent the entire world, although with much more depth for Ryle's Cambridge team and Lovell's Jodrell Bank group. Furthermore, *Astronomy Transformed* (*AT*) focuses not only on Britain, but also particularly on radio source distribution and cosmology controversies of the 1955–65 era that are beyond the up-to-1953 period of the present study. Much attention is paid to the leadership styles and internal social structures of the Cambridge and Jodrell Bank teams, examining cooperation and competition between and within groups.

I cite *AT* in this book whenever appropriate regarding its authority or priority on historical facts or conclusions, or whenever I feel the reader may wish a different approach. However, the worldwide scope and historical detail of the present study (Section 1.2.3) is far greater than that of *AT*, and my conclusions regarding the development of early radio astronomy (Section 1.3.3) are mostly of a very different, less sociological, nature.

As part of their research Edge and Mulkay conducted 20 interviews in the early 1970s, almost all with members of the Cambridge and Jodrell Bank groups. Transcripts of these interviews, kindly made available to me, have also been useful for the present study, but in fact I re-interviewed all of these persons because I found that my historiographic approach markedly contrasted with their sociological aims. Two examples illustrate our very different overall approaches: (1) in *AT* there is not a single photograph, and (2) quotations from interviews are liberally sprinkled throughout the text of *AT*, but are all unattributed (e.g., only to "an early Jodrell researcher").[18]

I have also profited from other book-length treatments on aspects of the pre-1953 history of radio and radar astronomy. These include *The Evolution of Radio Astronomy* (1973), a short, but thorough book by Stanley Hey; several books by Bernard Lovell on the contributions of Jodrell Bank, of which *The Story of Jodrell Bank* (1968) and his autobiography *Astronomer by Chance* (1990) are most relevant; autobiographies by Robert Hanbury Brown (*Boffin*, 1991) and by Hey (*The Secret Man*, 1992); Peter Robertson's history of Australian radio astronomy up until ~1970, focusing on the Parkes dish (*Beyond Southern Skies*, 1992); *To See the Unseen* (Butrica 1996), a history of solar system radar through the 1980s; and *Science and Spectacle* (Agar 1998), a sociological examination of the Jodrell Bank Observatory. Numerous other articles and books by historians, sociologists and scientists, many not dealing with astronomy at all, have also been invaluable and are cited throughout the book.

An important and delightful meeting on the history of radio astronomy was held at the National Radio Astronomy Observatory on the occasion of the fiftieth anniversary of Karl Jansky's discovery. The resulting proceedings, *Serendipitous Discoveries in Radio Astronomy* (1983, eds. K. I. Kellermann and B. Sheets), include transcribed comments (even *during* the talks!) chronicling the thoughts and recollections of many pioneering radio astronomers.

Two earlier books of mine (both still in print) act as handbooks to the present volume. *The Early Years of Radio Astronomy* (1984a) is a collection of 21 articles discussing the pre-1960 period by pioneers of radio astronomy and historians. *Classics in Radio Astronomy* (1982) is a collection of reprints, with extensive commentary, of 37 seminal papers in radio astronomy over the period 1896 to 1954.[19] Along the same lines is Lang and Gingerich's *A Source Book in Astronomy and Astrophysics, 1900–1975* (1979), which comprises 150 seminal articles (although many are only partially reprinted), of which 30 directly relate to radio astronomy.

1.3.2 My historiographic style

In this section I present the guiding principles in my historiographic approach, while the following section summarizes the particular historical questions addressed in this study.

Firstly, this study is colored by the fact that I was trained as a radio astronomer, beginning graduate

---

[16] I recommend Edge (1977), which is based on a talk about his findings in *Astronomy Transformed* and which reveals Edge's memorable personality and style.
[17] Mulkay (1974) discusses many of the methodological issues that he and Edge encountered during their project.
[18] See Appendix B for details on my interviews.

[19] These 37 papers are indicated in the References list with an asterisk.

study in 1966, 13 years and a scientific generation after this book's cutoff date of 1953. I thus came to know most of the key persons in this history – they were professional colleagues, and in many cases personal friends. I interviewed them and they contributed in many other ways such as answering follow-up questions, commenting on drafts, and supplying materials.[20] My frequent usage in this book of quotations from interviews should not be construed as indicating that I feel that oral histories are superior *per se* to other types of historical evidence, but rather that quotations enliven the story and present a perspective different from the historian's usual sole reliance on written documents. See Appendix B for a full commentary on the interviews and their uses.

"History will judge us kindly," Churchill is said to have quipped to Roosevelt and Stalin at the Tehran Conference in 1943, "because *I* shall write the history." So has my own career undoubtedly produced biases relative to a history written by a non-radio astronomer, but on the other hand intimate knowledge of the underlying science and technology has also immeasurably aided in probing and analyzing the history. Mulkay (1974:110–11) and in particular Brush (1995) have pointed out the advantages of history of science studies carried out by those with scientific backgrounds. I also note an apt analogy between history and astronomy set forth by art historian George Kubler:

> Knowing the past is as astonishing a performance as knowing the stars. Astronomers look only at old light ... Hence astronomers and historians have this in common: both are concerned with appearances noted in the present but occurring in the past ... Like the astronomer, the historian is engaged upon the portrayal of time. The scales are different: historic time is very short, but the historian and the astronomer both transpose, reduce, compose, and color a facsimile which describes the shape of time ... Both astronomers and historians collect ancient signals into compelling theories about distance and composition.[21]

This study is paradoxically both incomplete and complete. It is *incomplete* in that any historian works with limited evidence and must select and evaluate sources in an attempt to present a coherent story. On the other hand, I have attempted a history that is *complete* in the sense that it covers *all* aspects of *worldwide* radio and radar astronomy through ~1953.

My understanding of the scientific enterprise and how we learn about the natural world has been significantly informed by the scholarship of sociologists of science, even by the constructivists, but only occasionally do I explicitly refer to their writings. Although I myself am uncomfortable using most of their language, I have little quarrel with many of their concepts as expressed in the cant of *tacit knowledge, inscriptions, blackboxing, closure, privileging knowledge claims, knowledge negotiation, vested interests,* etc. Perhaps this reluctance stems from my basic belief that explanations for empirical data or phenomena are as severely constrained by an existing natural world as they are by a given cultural setting and historical epoch. Yes, the practice of science is intrinsically social, but it is hardly arbitrary. Yes, the reported phenomena have been shaped by many technologies and social influences, but they have no less been shaped by the interaction of instruments with the natural world. Yes, the possible explanations for phenomena are always underdetermined, but nevertheless science as a whole does seem to progress toward a more powerful and reliable understanding of the natural world.[22]

Martin Rudwick's study of early nineteenth-century British geology, *The Great Devonian Controversy* (1985:5–16, 450–6), has many similarities with the present work. Because we both believe that the only sound way to build up a "big picture" of how science works is to examine *in extenso* the actions and choices of persons and institutions, we both employ a detailed narrative format. *History* is at heart a good *story*, as the etymologies of the two words show. I also resonate with Rudwick's compromise between the extremes of whether scientific claims are artifactual or reflections

---

[20] The 37-year duration of this project, however, has meant that ~60% of those interviewees whose materials I have used are now deceased (see Appendix B).

[21] G. Kubler (1962), *The Shape of Time* (New Haven: Yale University Press), pp. 19–20.

[22] A useful overview of some of the arguments between the rationalists and the constructivists can be found in rationalist Allan Franklin's *Selectivity and Discord* (2002:chap. 1, Pittsburgh: University of Pittsburgh Press). For a constructivist's overview, see Jan Golinski's *Making Natural Knowledge* (1998:chap. 1, Cambridge: Cambridge University Press).

of an independent reality. His apt metaphor is that of a map, which is never the same as the territory. Although a map is full of conventions and negotiated decisions guided by empirical data, when constructed by a team of skilled cartographers it gains verisimilitude and becomes a reliable piece of "shaped knowledge" about the territory. In the end, I believe that one need not make an epistemic choice between "reality" and "convention" when analyzing the scientific enterprise – both aspects continually and intensely interact.

My narrative style has been heavily influenced by historian Raymond Aron's precept: "The investigation of cause by the historian is directed at ... restoring to the past *the uncertainty of the future*" (italics mine).[23] In other words: What did each historical actor *see himself* as doing *at that time*? What were his options? what were the paths *not* taken? What were the ambiguities of the time? Why were certain problems, theories, bits of evidence, and techniques deemed important, and others not? Why were some theories and observations accepted at first and then abandoned? What were the norms of that period in language, in customs, and in beliefs? Only by understanding the "then-norms" can we recognize and analyze the "then-strange." This is why I do not retrospectively "update" 1950-era science throughout the narrative – to do so would both distract and unacceptably distort the story; heated debates over "wrong" (as viewed today) theories or data often better reveal the scientific process than results that everyone harmoniously accepted.

There are many approaches to writing history of science. This book is predominantly "internalist" history, but not ignoring the larger setting – both text and context have been examined. My emphasis is on science as knowledge, what used to be called intellectual history. But I also care about material culture, especially how the equipment and technology interacted with the science. Furthermore, social and institutional factors, including the politics and economics of World War II and the Cold War, were often vital in determining which science got done and how well. These factors play prominent roles in my story. A more unusual aspect of this particular history is my concern with *visual* culture, especially the drive of radio astronomers to make images, and their concern with optical identifications of discrete radio sources. In a related vein, the abundant images and plots in the following pages are integral to the book, designed to bring the reader better understanding of the times, the science, the techniques, and the people, not to speak of relief from endless pages of text.

1.3.3 Historical issues

In the course of following the origin and adolescence of radio astronomy, many interesting historical issues and questions have arisen. Some are specific to radio astronomy, such as: why was there such a long delay in the discovery of the radio sun, despite rapidly improving radio technology over the four decades before World War II?[2.6] Others have broader interest to historians of science, such as: how did traditional astronomers react to their disciplinary domain being invaded by radio interlopers unschooled in astronomy?[17.2] In this section, I first mention two particular historical points, and then summarize the four major themes of the book.

Jansky, Reber, and Hey. It is surprising that Jansky and Reber, despite their prewar discoveries, little influenced the postwar decade of radio astronomy, especially outside the US. Jansky's personality and health problems and Reber's idiosyncrasy meant that neither played a significant role after the war.[4.7, 17.3.1] If one were to name a single person who most influenced the course of early postwar radio astronomy, it would be the less-well-known Stanley Hey, who not only made three distinct seminal discoveries in the years 1942–46,[5.2, 6] but also directly influenced the initial activities of the other two British groups.[8, 9] Figure 1.3 is designed to illustrate the muted influence of Jansky's star static and Reber's cosmic static, as well as the more direct influences of wartime anomalous effects, two of which Hey discovered, that led to the major postwar areas of inquiry.

Contingencies. Historians are paid to locate rational trends, but it is sobering to realize that the past has been profoundly suffused with historical contingencies. The histories of radio astronomy groups around the world have all heavily depended on the specifics of institutional settings, sources of advice and patronage, and the caprices of individuals making particular

---
[23] R. Aron (1961), *Introduction to the Philosophy of History* (Boston: Beacon Press), p. 179.

choices. Attempting to craft a unified historical analysis thus becomes all the more challenging. How might radio astronomy have developed *if*:

- researchers had detected the radio sun in the 1920s?[2.6]
- Lovell had not (futilely) sought radar echoes from cosmic rays, which led him to meteor echoes?[9.1]
- Ryle had quit science at war's end?[8.2]
- Dicke or Purcell had chosen to start an American research group in cosmic noise right after the war?[10.1.1, 16.3, 17.3.5]
- Australian radio astronomers had been located much closer to their British rivals?[7.5.1]
- Sydney's sea cliffs had not existed, forcing Australian interferometry from the beginning to be of the more tractable Michelson type?[7.3.2, 14.3.4, 17.4]
- Any leading theorist in the West had at an early stage enthusiastically taken up the Soviet idea of synchrotron radiation (to explain galactic radiation)?[15.4]
- Computer capabilities had been sufficient much earlier to allow image-making (high-resolution mapping) of celestial sources?[13.1.3, 13.3.2]
- Hydrogen, the most ubiquitous element in the Universe, had not had a handy spectral line at 21 cm wavelength?[16]
- Optical emission from the strongest radio sources had been even fainter, making identifications impossible until many years later?[14.4]
- Radio astronomers had decided to found their own journals and societies and become a separate discipline from (optical) astronomy?[17.2.5]

Despite the problems posed by such historical contingencies, this study identifies and characterizes four major themes in the history of early radio astronomy: (1) World War II and the Cold War; (2) material culture and technoscience; (3) a "visual culture"; and (4) a twentieth-century "New Astronomy."

World War II and the Cold War. The first theme is how World War II and the ensuing Cold War spawned and shaped the field of radio astronomy, albeit with important national differences. In the first place, every major postwar group (except Holland) was filled with war veterans of radar research or operations, often skilled in ionosphere research or other aspects of radio communications (Fig. 1.3).[17.1.1] Especially in England and Australia, these researchers not only had knowhow and ample surplus equipment, but also a can-do attitude and camaraderie that continued to permeate their postwar groups. Also forged during the war were management skills vital for the success of the new field's leaders, men such as Hey, Ryle, Lovell, Hanbury Brown, Bowen, Pawsey, Bolton, Wild, Mills, and Christiansen. Furthermore, programs after the war to support research and train new scientists were funded at much higher levels than before.

The strong military postures of the Cold War, especially in the United States and the Soviet Union,[10.3.4] also influenced the type and amount of radio astronomy that was done. The highly successful activities in radio astronomy of one military group, Hey's Army Operational Research Group in England, were in fact terminated in 1948 due to Cold War pressure.[6.5] In the US money flowed from military agencies to university physicists and engineers, but the direction of their research in radio astronomy (toward shorter wavelengths) was inappropriate for the most fruitful results at that time. US groups thus long lagged behind Australia and the two university groups in England (at Cambridge and Manchester), who, despite much smaller budgets, achieved pioneering discoveries at longer wavelengths. Postwar US radio astronomy provides another example of historian Paul Forman's thesis that military funding during the Cold War substantially warped the contents of seemingly "pure" university research.[17.3.5] A topic such as radio emission from sources thousands of light-years distant may at first seem of no military consequence, but in fact the sensitive radio technology necessary to find and track such sources was of direct use to the military. In contrast, the Australian and English groups' relative isolation from "Cold War science" contributed to the success of their postwar radio astronomy.

Material culture and technoscience. The second principal historical theme deals with the role of instrumentation and technology in early radio astronomy. As will be seen throughout this volume, this *material culture* was intimately intertwined with radio astronomy's development and scientific claims.[17.4] The practitioners themselves had a mixture of engineering and scientific skills, continually switching between building radio instrumentation and making calculations about the sun or Milky Way. The field itself grew out of military radio and radar engineering (Fig. 1.3), and new technical capabilities (especially those that improved angular resolution or sensitivity to weak signals) often led to serendipitous discoveries as investigators took a raw empirical approach and tried out the latest equipment at hand. However, for other episodes (especially later), astronomical goals dictated experimental design. And in many other cases it is hopeless to ask whether the technology led or followed the science, or sometimes even whether the two can be separated. A more useful formulation may be that radio astronomy was *technoscience*, an amalgam activity in which technology and science were intimately related. For example, based on cost, experimental design, and happenstance, different antenna systems and receivers were favored by the various groups conducting radio star surveys. Largely because of these choices, serious discrepancies arose between the surveys and serious arguments arose between the surveyors, casting doubt on the legitimacy of their putative sources. This in turn led to revisions of the equipment and follow-up surveys.[14.3, 14.6]

A "visual culture." The third theme is that radio astronomers ironically drove themselves to more and more of a "visual culture" even as they continued to record their data as very "un-visual" jiggly ink lines on rolls of strip chart paper (for example, Fig. 4.6). By *visual culture* I mean concerns with (a) finding optical galaxies and nebulae that were also radio emitters, (b) creating radio images that one could begin to compare with photographic images, and (c) seeking the expertise, company and advice of (optical) astronomers. Regarding (a), a continual aim of radio astronomy was for better angular resolution in order to pinpoint more finely the locations and sizes of discrete sources.[17.4] And the main reason for wanting better angular resolution was in hope that an unambiguous optical correlate at the same sky location could be found, a so-called *optical identification*.[14.4] Such an optical identification, carrying the imprimatur of astronomy and appealing to epistemically superior visual information, conferred on a radio source a degree of legitimacy and reality that it could not otherwise have.[14.4.4] As Bolton wrote to an optical astronomer in 1952: "Photographs are much more satisfying than the evidence we get out of our machinery."[24] His Sydney colleague Mills agreed and was prompted to produce "radio pictures," pseudo-photographs representing radio emission (Fig. 14.16). Regarding (b), by 1953 two groups had worked out the details of how in principle to create a radio image, i.e., a map of high resolution (by sampling its Fourier components and then mathematically calculating the image), although it would be many more years before enhanced computer power would make the technique tractable.

The final aspect of the visual culture of radio astronomers was their desire *not* to set up a new discipline, but rather to become part of the very old discipline of astronomy.[17.2] Although their backgrounds immensely differed from those of astronomers and their research style was much more hurried and involved more teamwork, by 1953 radio astronomers had learned enough astronomy and gained enough legitimacy that they were welcomed at astronomy meetings, often collaborated with astronomers, and were viewed as making significant contributions to astronomy.[25] In the process they invented the research specialty of *radio astronomy*, and made astronomers comfortable with the notion that, for example, what looked like a field full of clothes lines (e.g., Fig. 8.7) was in fact worthy of being called a telescope, albeit by the new term *radio telescope*.[17.2.1.1] It was also radio astronomy that created the obverse category of *optical* astronomy, which until then had not been necessary.[26] This study therefore traces specialty formation and amalgamation into an existing discipline (astronomy), not discipline formation.[17.2.5] Figure 1.3 indicates this evolution from anomalous radio phenomena before and during the war, to the three main subfields of what became radio astronomy in 1948–49, to radio astronomy's eventual assumption into the separate "stream" of traditional astronomy. Nevertheless, as the chart shows, radio astronomy and optical astronomy remained distinct well beyond the period of this book.

---

[24] J. Bolton:R. Minkowski, 11 January 1952, cartons 4–5, MIN.

[25] The degree to which the various subfields of radio astronomy connected to traditional astronomy varied markedly. The connection was high in studies of the sun, meteors, and 21 cm hydrogen, and much lower in studies of the galactic background radiation and the radio stars, both of which exhibited minimal correlation with the optical sky.

[26] I have found it revealing to trace the introduction and usage of terms like these.[17.2.1]

A twentieth-century "New Astronomy." The fourth main historical theme concerns the place of radio astronomy in the long history of astronomy. Was it a Kuhnian or other type of scientific revolution?[18.3.1] I argue that it does not fit this category – there were no new paradigms, no desire to overthrow the *ancien régime*, no sustained resistance from the astronomers, no overhaul of astronomy overall. Rather it was the start of a transformation, what I call the twentieth-century *New Astronomy*.[18.3.2] The radio sky was astoundingly at variance with the familiar optical sky, including new aspects of previously known objects, wholly new phenomena, and novel ways to attack old problems.[1.1, 17.1] This transformation of the sky by a New Astronomy was not unlike earlier ones also brought on by the introduction of novel technology. These included Galileo with his telescope, William Herschel with his large reflectors, and the late-nineteenth-century astrophysicists with their spectroscopes and photography.

In fact, the definition of the twentieth century's New Astronomy should be expanded byond just radio astronomy. Radio astronomy's role should be viewed as the harbinger of a broader twentieth-century phenomenon: the openings to astronomy, mostly made possible by access to outer space, of the infrared, ultraviolet, X-ray and γ-ray windows.[18.2] In particular, the first years of X-ray astronomy in the 1960s had many close analogies with early radio astronomy.[18.2.1] The emergence of radio astronomy thus heralded the era of this broader New Astronomy, namely the opening of the *entire* electromagnetic spectrum over the period 1930–80.

Each of these New Astronomies – Galileo, Herschel, astrophysics, and radio astronomy (or more broadly, the opening of the electromagnetic spectrum) – radically changed their contemporary astronomical worldview. Each was enabled by innovative technologies and came to fruition through a combination of material culture, faith in empiricism, and key intellectual players. Moreover, each encountered and overcame difficulties with the establishment of the day, eventually insinuating itself into and redefining mainstream astronomy. Such a broader context provides a gauge of the significance of this New Astronomy of the twentieth century: I conclude that, even within the compass of the millennia-long history of astronomy, the emergence of radio astronomy must be ranked as a major event.

TANGENT 1.1 CONVENTIONS USED IN THIS BOOK

UNITS

In general I use units and terms contemporary to the postwar period, but for convenience and brevity have violated this policy by using the units *hertz* (Hz), *kelvin* (K), and *jansky* (Jy).

Frequency. MHz, kHz, Hz, etc. are uniformly used despite the fact that only the Germans and Soviets were using this convention before and during the postwar decade; everyone else used the more cumbersome Mc/s, kc/s, c/s, etc. (cycles per second).

Absolute temperature. The convention before the 1970s was °K (degrees Kelvin), but this book uses K (kelvins).

Flux density. At various places the book notes a great variety of units used to express the intensity of signals, but eventually radio astronomers standardized on W m^{-2} Hz^{-1}, sometimes for convenience defining 1 flux unit = 1 f.u. as 10^{-25} or 10^{-26} W m^{-2} Hz^{-1}. In the 1970s the latter was adopted by the International Astronomical Union as the standard and named after Karl Jansky: 1 jansky = 1 Jy = 10^{-26} W m^{-2} Hz^{-1}.

Lengths and distances. For the English-speaking world I use the units in use at the time for dimensions of antennas and telescopes ("25 ft diameter dish," "200 inch telescope") and distances ("15 miles west of Sydney"). Otherwise I use metric units.

1 foot = 12 inch = 0.30 meter; 1 mile = 1.61 km.

TERMS

Aerial versus *antenna*. In some specific cases I use the British term *aerial*, although in general I use *antenna*.

[Electronic] *valve* versus *tube*. I use *tube*.

Galaxy, galactic. Capitalized *Galaxy* refers to our own Milky Way , and *galactic* is the adjectival form ("the galactic plane"). Uncapitalized *galaxy* is the general term, e.g., "our Galaxy is like many other spiral galaxies."

Microwavelength. I employ this useful word, obviously meaning the wavelength of a microwave (shorter than, say, 30 cm), though it appears in no dictionaries.

NAMES OF PERSONS

At a person's first mention, a complete first name (or middle name if used as his/her familiar name) is given along with other initials, even though this format may never have been used by the person, e.g., "Peter A. G. Scheuer" or "A. C. Bernard Lovell." Likewise, I use the form "Van de Hulst" and alphabetize it under *V* despite the Dutch practice of often spelling it "van de Hulst" and putting it under *H*. Birth and death dates are given for main characters who are known to have died. Finally, I use the original forms of two particular men's names, as they themselves used at the time they appear in my story: "Robert Watson Watt" (double last name later hyphenated to "Watson-Watt") and "F. Graham Smith" (single last name later changed to "Graham-Smith").

2 • Searching for solar Hertzian waves

The existence of radio waves was established by Heinrich Hertz in 1887. Intense interest in these Hertzian waves ensued, and this chapter recounts simple attempts to detect them from the sun in the United States and England, as well as more complex experiments in Germany and France at two leading centers of the new science of astrophysics. These attempts over a fifteen-year period surprisingly contrast with the succeeding four decades, during which no one seriously tried despite ever-advancing radio technology. The reasons for this inattention are discussed in the chapter's final section. In the end the sun's radio emission came to be discovered accidentally, in 1942 (Chapter 5). Meanwhile, discovered by Karl Jansky a decade earlier, extraterrestrial radio waves were picked up not from the most obvious source of radiation in the sky, but from the more subtle Milky Way emission (Chapter 3).

2.1 HERTZ

A key issue in late nineteenth-century physics centered on the correctness of James Clerk Maxwell's (1831–1879) theory of a dynamical electromagnetic field. By introducing the concept of a "displacement" current, a new type of current proportional to the rate of change of an electric field, Clerk Maxwell had been able in 1865 to produce an elegant mathematical theory. But the theory was vigorously debated, as it was not at all clear what the displacement current was or in what kind of a medium it flowed.[1] Sound experimental checks of the theory's chief predictions were needed, and this was what Hertz (1857–1894) accomplished. Working in a large lecture hall at the Karlsruhe Technische Hochschule, he conducted a series of experiments in 1886–88 that were immediately accepted by most of his contemporaries as definitive. Hertz validated a central aspect of the theory: changing electrical currents generate electromagnetic waves that propagate through space at a speed equal to that of light.

Previous experimenters had been unable to obtain convincing results largely because of their inability to produce electromagnetic waves shorter than 10 meters, far easier to work with in the confines of a building. But Hertz overcame several obstacles in generating, launching, and detecting waves as short as 60 cm. His equipment was simple: the transmitting antenna was two brass rods separated by a small gap across which sparks jumped when energized by an induction coil, the receiving antenna was a loop with an adjustable gap, and the "detector" was the human eye observing a spark across this gap. With this apparatus he was able to show convincingly that electricity and light were different manifestations of the same phenomenon, and that the ethereal "electric" waves completely corresponded to the familiar light waves in terms of reflection, refraction, interference, and polarization. The clincher was his determination of their speed of propagation, derived from a measured wavelength and a calculated frequency.

Hertz won immediate fame for this discovery, but died only six years later. Although he appears never to have concerned himself with any possible geophysical, astronomical, or commercial applications of the electric waves, these followed soon after his death.[2]

---

[1] On the development of Maxwell's theory see D. M. Siegel (1991), *Innovation in Maxwell's Electromagnetic Theory: Molecular Vortices, Displacement Current, and Light* (Cambridge: Cambridge University Press). For an insightful account of British work on electrodynamics in the late nineteenth century, see B. Hunt (1991), *The Maxwellians* (Ithaca: Cornell University Press).

[2] The best entry into studies of Hertz, including a complete bibliography of the historical literature, is *Heinrich Hertz: Classical Physicist, Modern Philosopher*, eds. D. Baird, R. Hughes and A. Nordmann (1998) (Dordrecht: Kluwer). His important publications are available in translation in *Electric Waves* (Hertz 1893) and in *Heinrich Rudolf Hertz (1857–1894): a Collection*

2.2 EDISON AND KENNELLY

Within three years of Hertz's work on radio waves, Thomas A. Edison (1847–1931) and Arthur E. Kennelly (1861–1939) considered the possibility of detecting "violent disturbances" of long wavelength from the sun. Edison's eclectic electric interests had long included astronomy – in 1878 he had joined an eclipse expedition to Wyoming and tried to detect infrared radiation from the solar corona with a recent invention, the "tasimeter" (Eddy 1972). Kennelly, at the start of a notable career in electrical engineering and ionospheric physics, was principal assistant to Edison in his New Jersey laboratory.

In 1890 Kennelly began corresponding with Edward S. Holden, Director of the new Lick Observatory on Mt. Hamilton in northern California. Holden had decided that their principal telescope, a 36 inch refractor, would be more efficient if the setting circles were illuminated with electric lamps. In the end he persuaded Edison, as wealthy as he was inventive, to donate an entire plant, including a small steam engine and boiler, dynamo, and storage batteries. Kennelly was handling details from New Jersey and in his letters frequently discussed scientific matters with Holden. On 2 November 1890 these included an experiment to test the effects of temperature on gravity, as well as the following.

> I may mention that Mr. Edison, who does not confine himself to any single line of thought or action, has lately decided on turning a mass of iron ore in New Jersey that is mined commercially to account in the direction of research in solar physics. Our time is of course occupied at the Laboratory in practical work, but on this instance the experiment will be a purely scientific one. The ore is magnetite, and is magnetic not so much on its own account like a separate steel magnet but rather by induction under the earth's polarity. It is only isolated blocks of the ore that acquire permanent magnetism in any degree. Along with the electromagnetic disturbances we receive from the sun, which of course you know we recognize as light and heat (I must apologize for stating facts you are so conversant with), it is not unreasonable to suppose that there will be disturbances of much longer wave length. If so, we might translate them into sound. Mr. Edison's plan is to erect on poles round the bulk of the ore a cable of seven carefully insulated wires, whose final terminals will be brought to a telephone or other apparatus. It is then possible that violent disturbances in the sun's atmosphere might so disturb either the normal electromagnetic flow of energy we receive, or the normal distribution of magnetic force on this planet, as to bring about an appreciably great change in the flow of magnetic induction embraced by the cable loop, enhanced and magnified as this should be by the magnetic condensation and conductivity of the ore body, which must comprise millions of tons.
>
> Of course it is impossible to say whether his anticipation will be realised until the plan is tried as we hope it will be in a few weeks. It occurred to me that supposing any results were obtained indicating solar influence, we should not be able to establish the fact unless we had positive evidence of coincident disturbances in the corona. Perhaps if you would, you could tell us at what moments such disturbances took place. I confess I do not know whether sun spot changes enable such disturbances to be precisely recorded, or whether you keep any apparatus at work that can record changes in the corona independently of the general illumination.[3]

---

of Articles and Addresses, ed. J.F. Mulligan (1994, New York: Garland Publishing). The latter also contains an extensive biographical article. Details of his life and science can be found in *The Creation of Scientific Effects: Heinrich Hertz and Electric Waves* by J.Z. Buchwald (1994, Chicago: University of Chicago Press), *Heinrich Hertz: Memoirs, Letters, and Diaries* by J. Hertz (1977, San Francisco: San Francisco Press), *Hertz and the Maxwellians* by J.G. O'Hara and W. Pricha (1987, London: Peter Peregrinus), and Aitken (1976).

[3] Kennelly:Holden, 2 November 1890, Mary Lea Shane Archives of Lick Observatory, University of California, Santa Cruz. This letter was originally found by C.D. Shane (1958), although an earlier reference to Edison's idea was in a popular article, "What is the music of the spheres?," by G.P. Serviss (pp. 18–20 in *The Mentor* for December 1927). Serviss discussed the plausibility of sound waves generated either in the solar atmosphere or by planets moving through the ether. These motions would modulate electric waves that in turn could be picked up by a receiver on earth. Hearkening back to the Pythagorean harmony of the spheres, Serviss asked: "If a kind of celestial uproar is possible close around such bodies as the sun, how about smooth, delightful harmonies issuing from the soft gliding of the planets

Edison's audacious idea was thus to use the mountain of magnetite ore as the core of a huge induction coil, and thereby detect changes in the local magnetic field. Holden replied that unfortunately neither Lick nor any other American observatory could supply him with sunspot data and that he would have to apply to the European solar observatories. On 21 November Kennelly commented that "there may be some little difficulty in getting the poles set up this winter, although they have just arrived at the spot." But this is the end of the archival record: we do not know whether or not the experiment was ever carried out.[4]

2.3 LODGE

One of Hertz's principal exponents was Oliver J. Lodge (1851–1940), Professor of Physics and Mathematics at the University of Liverpool. At the time of Hertz's experiments Lodge was in fact doing very similar work, although his studies were limited to electric waves confined to wires. His chief contributions to the radio art of the 1890s were development of a sensitive detector (see below) and work on resonant, or "syntonic," circuits for communication. Lodge's lifelong interest in astronomy may have arisen from his studies of the ether through which electric waves travelled. And not only did he pursue the ether's physical properties, but (with several other notable Victorian physicists) also its psychical aspects, including mental telepathy and direct communication with the departed.[5]

When Hertz died prematurely of cancer in 1894, Lodge gave a lecture-demonstration in his honor at the Royal Institution in London, during which he stated:

> I hope to try for long-wave radiation from the sun, filtering out the ordinary well-known waves by a black-board or other sufficiently opaque substance.
>
> (Lodge 1894a:29, 1894b:138, 1900:33)

Over the next decade his original lecture was expanded and brought out in several editions. The third edition of 1900, titled *Signalling Across Space Without Wires*, was the first to follow the above remark with a new inclusion:

> I did not succeed in this, for a sensitive coherer in an outside shed unprotected by the thick walls of a substantial building cannot be kept quiet for long. I found the [associated galvanometer's] spot of light liable to frequent weak and occasionally violent excursions, and I could not trace any of these to the influence of the sun. There were evidently too many terrestrial sources of disturbance in a city like Liverpool to make the experiment feasible. I don't know that it might not possibly be successful in some isolated country place; but clearly the arrangement must be highly sensitive in order to succeed.
>
> (Lodge 1900:33)

Lodge also gave an illustration (Fig. 2.1) of one of his portable detectors of Hertzian waves, presumably similar to the one he actually used. His setup consisted of a copper cylinder, 3 × 2 inches (8 × 5 cm), in which he plugged a few-inch-long wire acting as an antenna. The tube contained a battery, galvanometer, and coherer. The *coherer*, named and developed by Lodge after its invention in France by Edouard Branley, was at that time the workhorse detector of radio waves until supplanted by the cat's whisker and Fleming valve in the next decade. It took many forms, but most commonly

---

through the ether?" I thank Paul Luther for bringing this article to my attention.

[4] In any case the scheme certainly would not have worked given that 1890 was at minimum phase in the 11 year solar activity cycle (Fig. 13.11) and that the arrangement was sensitive only to extremely low frequencies, which we now know are not in general admitted by the earth's ionosphere. Finally, it would have been extremely difficult to disentangle any direct solar electromagnetic waves from the frequent fluctuations in the earth's magnetic field.

[5] Lodge's many pursuits in science, engineering, parapsychology, and academic administration are discussed at length by Aitken (1976) and by W. P. Jolly (1975) in *Sir Oliver Lodge* (Rutherford, NJ: Farleigh Dickinson University Press). Also see J. Oppenheim (1986), "Physics and psychic research in Victorian and Edwardian England," *Physics Today* **39**, No. 5, 62–70. B. Hunt (1986) describes a series of experiments in the early 1890s, contemporaneous with the solar episode, that

nicely set Lodge's work into his time: "Experimenting on the ether: Oliver J. Lodge and the great whirling machine," *Hist. Stud. Phys. and Biol. Sci.* **16**, 111–34; also see Hunt's *The Maxwellians*, as in note 1. Lodge's work is further detailed by P. Rowlands (1990) in *Oliver Lodge and the Liverpool Physical Society* (Liverpool: Liverpool University Press); and in Chapter 2 of S. Hong (2002), *Wireless: From Marconi's Black-box to the Audion* (Cambridge: MIT Press).

Figure 2.1 An early radio detector of Oliver Lodge. The coherer is contained within the cylinder A, the short wire B acts as an antenna, and the galvanometer lamp and scale are seen on the right. This is probably similar to the equipment Lodge used in his attempt to detect radio waves from the sun in 1894. (Lodge 1900:33)

consisted of a tube of metal filings or steel balls in loose contact. In its normal state it presented a high electrical resistance, but when a Hertzian wave (or a DC voltage) was applied, the resistance dropped dramatically and the loose particles were said to cohere. It was quite sensitive for its day, but had the disadvantages of responding strongly to thermal and mechanical stresses and requiring mechanical tapping to reset it after detection (Phillips 1980).[6]

Although the second edition of Lodge's book in 1897 did not mention the above experiment, it seems certain that it was in fact carried out within two weeks of his original lecture. In one of Lodge's research notebooks lies the following entry for 12 June 1894.

> Tried sunshine today on coherer in hole in dark room. Reflected in through ? window by tin plate outside. Spot of light did not keep still, but couldn't swear to proper sun effect.

> Tried next, filings tube. Spot steadier. On admitting sunlight the spot crawled slowly but distinctly up and did not return. Effect repeated every time copper screen was removed. Probable heat effect. Interposed extra screen of paper. Now effect was less.

> Evidently no strong constant sun effect but it may be intermittent or it may be weak.[7] [underlining in original]

The first known attempt to detect extraterrestrial radio waves was thus in June 1894.

2.4 WILSING AND SCHEINER

Two countrymen of Hertz carried out the first detailed, properly published effort to detect electric waves from the sun. Johannes Wilsing (1856–1943) and Julius

---

[6] The physics behind the coherer's action is discussed at length in Phillips (1980) and is still being investigated, e.g., F. Falcon and B. Castaing (2005), "Electrical conductivity in granular media and Branley's coherer: a simple experiment," *Amer. J. Physics* 73, 302–7.

[7] 12 June 1894, Notebook No. 3–17, p. 230, Lodge papers, Sydney Jones Library, University of Liverpool. Another short entry, on 24 June, appears to describe an inconclusive second attempt. I am indebted to M.R. Perkin, Curator, and M.A. Houlden, Dept. of Physics, for their indispensible assistance in locating this record.

Scheiner (1858–1913) were senior staff members under H. C. Vogel at the Royal Observatory for Astrophysics, Meteorology, and Geodesy in Potsdam. Founded in 1874, the observatory was the first in the world dedicated to astrophysics, specifically a physical understanding of the sun and more distant stars and nebulae, largely through the new technique of photographic spectroscopy. In the late nineteenth century, when mathematical and positional astronomy still held sway, in the upstart field of astrophysics Potsdam was the acknowledged leader and a mecca for advanced training (as was much of German science). Wilsing, Scheiner, Vogel, and their colleagues obtained unprecedentedly accurate spectral classifications, radial velocities, and temperatures for hundreds of stars, as well as the brighter nebulae.[8]

The sun, an ideal testing place for both new stellar theories and new techniques, rated special attention.[9] This combination of a strong interest in the physics of the sun and familiarity with the latest technology led to Wilsing and Scheiner's attempt in 1896 to detect Hertzian waves. They were familiar with Samuel P. Langley's work in America showing that the solar spectrum extended to wavelengths as long as 6 μm, and were confident that the solar photosphere emitted electric oscillations of all wavelengths, depending on the size of the oscillating region.[10, 11] But Langley had also shown

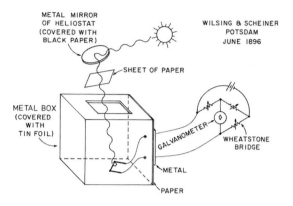

Figure 2.2 A schematic representation of the setup employed by Wilsing and Scheiner to search for solar radio waves in 1896. The detecting element consists simply of one wire resting on top of two others. (Drawing based on the description given by Wilsing and Sceiner [1896])

that the earth's atmosphere heavily absorbed these longer "caloric" waves. It was thus worrisome that the electric waves too might not penetrate to the earth's surface. It had been demonstrated in the laboratory that electric waves could be absorbed by a rarefied gas and cause it to glow. Wilsing and Scheiner reasoned that the same process could occur in the thin air of the upper terrestrial atmosphere (perhaps even causing aurorae and magnetic disturbances, as regularly measured at Potsdam), or perhaps even in an upper layer of the *solar* atmosphere.

Despite these misgivings, Wilsing and Scheiner felt an experiment with high sensitivity worthwhile. Their setup employed an extremely simple form of the coherer as the detector – one wire resting on top of two others. They argued, based on extensive laboratory testing and calibration, that this was a sensitive and reliable method of discerning the presence of electric waves from distances as large as 25 meters. During tests the resistance at the contact points of the wires would drop from thousands of ohms to a few ohms upon the sparking of a distant induction coil. The detector was isolated from manmade interference and reflections off room walls by placing it inside a cubical metal box with a hole of size ~10 × 10 cm on the top, through which the signal entered (Fig. 2.2). But a serious problem arose when directing the solar radiation into the box using a metal mirror as a heliostat: the intense solar heat caused the wires to deform, sending the galvanometer off-scale. This was finally solved by a combination of black matte paper covering the mirror, a second sheet

---

[8] On Potsdam, see D. B. Herrmann (1984), pp. 130–2 in *Astrophysics and Twentieth-century Astronomy to 1950* (ed. O. Gingerich, Cambridge: Cambridge University Press). For a comparison of the rise of astrophysics with that of radio astronomy much later, see Section 18.3.2.

[9] An excellent overview of the entire history of solar research is *Exploring the Sun* by K. Hufbauer (1991).

[10] Scheiner (1899:83). Scheiner even pointed out that the maximum permissible wavelength stems from considering the entire sun as a single, giant resonator! He calculated 2×10^6 km for this wavelength, corresponding to a frequency of 0.2 Hz. German physicist Hermann Ebert (1893:806) also pointed out that the entire sun should oscillate at 0.15 Hz and emit electromagnetic waves of this frequency, which could then affect the earth's magnetic field.

[11] Confidence in the sun emitting radiation of all kinds also led at this time to at least one search for X-rays. This was by George Ellery Hale, who in 1897 used a pinhole (covered with cardboard) in a sheet of lead in front of photographic film as an X-ray camera, but found no effects from the sun. [D. E. Osterbrock (1993), p. 45 in *Pauper and Prince: Ritchey, Hale and Big American Telescopes* (Tucson: University of Arizona Press)]

of paper intercepting any further infrared rays which had managed not to be absorbed, and modifying the coherer by bending the top wire and welding it down on one end.

They hauled this arrangement outdoors for eight days spread over a three week period in the summer of 1896. Experiments were tried on both clear and cloudy days, with and without the mirror covered – all were negative. A full article in *Annalen der Physik und Chemie* (Wilsing and Scheiner 1896), as well as a shorter report in *Astronomische Nachrichten* (Wilsing 1897), carefully emphasized that their results did not necessarily imply a lack of electrodynamic waves emitted by the sun, only that any such waves did not fall on the surface of the earth with an intensity sufficient to be measured.[12]

2.5 NORDMANN

Five years after the Potsdam attempt, a French graduate student named Charles Nordmann (1881–1940) decided to make his own search.[13] Accepting that the main reason for the German failure was the atmosphere, he ran his observations on the slopes of Mont Blanc in the French Alps. Although it would seem that he had far more sensitivity, he too was unsuccessful.

In 1901 Nordmann was a student in Paris Observatory's astrophysical group, which P. C. Jules Janssen (1824–1907) had led at nearby Meudon since 1876. Under the influence of Janssen and Henri Deslandres (1853–1948), a senior staff member who himself became Director in 1908, Nordmann became enchanted with the Hertzian waves. He not only thought them important on the sun, but wrote many papers in the 1902–04 period arguing that Hertzian waves were a panacea "solving" aurorae, terrestrial magnetic storms, tails of comets, the glowing of spiral and planetary nebulae, and even the spectra of the recent 1901 nova in Perseus.[14]

Nordmann tested for a Hertzian sun by setting up his apparatus at a station of Janssen's designed for geomagnetic and atmospheric observations. This was near Chamonix on the Mont Blanc massif at 3100 meters altitude – a railroad to the town had just opened to meet the demand for tourists who wanted to climb, take in the beauty, and ski. In his report of the experiment Nordmann (1902a) remarked that bad weather prevented him from ascending to the very summit. He felt that the Glacier des Bossons was an ideal site because of its high altitude and the fact that its surface held the antenna at a large distance from the ground (~25 m). The radio transparency of the ice meant that it was as if his antenna were suspended in space, unbothered by reflections of the solar waves from the ground. He had much more sensitivity than Wilsing and Scheiner because of a wire fully 175 m long, attached to a conventional coherer with either nickel filings or thirty ball bearings in a tube (Fig. 2.3). The wire was strung out on an east–west line and data were taken throughout the perfectly clear day of 19 September 1901. The galvanometer never gave an indication of Hertzian waves and Nordmann apparently never again tried the experiment.

In his first report Nordmann (1902a) came to the same conclusion as had Wilsing and Scheiner: electric waves must be emitted by the sun, but were absorbed by either the upper atmosphere of the sun or that of the earth. He did not comment on whether he was expecting a steady radiation from the sun, or something connected with solar activity. But only one month after Nordmann read this paper to the French Academy, his colleague Deslandres and

---

[12] A pioneering contemporaneous science fiction novel, *Auf Zwei Planeten* by Kurd Lasswitz (1897), may well have made use of this Potsdam experiment. Referring to a giant Martian space station hovering above the Earth's pole, Lasswitz wrote:

> On the surface of the earth terrestrials employ mainly only the heat and light of solar energy. But here in empty space it became evident that the sun emits incomparably greater quantities of energy, especially rays of very large wave-length, such as electric, as well as those of still much smaller wave-length than those of light. [Fischer, W. B. (1984:165), *The Empire Strikes Out: Kurd Lasswitz, Hans Dominik, and the Development of German Science Fiction* (Bowling Green, Ohio: Bowling Green State University Popular Press)]

[13] Débarbat *et al.* (2007) give details of Nordmann's career.

[14] Reviews of these ideas were published in 1902 and in his 1903 thesis (Nordmann 1902b, 1903). This was only the beginning of a gadfly career of unusual ideas and popular articles and books. Van de Hulst (1984:386–9) has discussed another of these ideas, namely Nordmann's claimed discovery in 1908 of a dispersive effect on light by the interstellar medium.

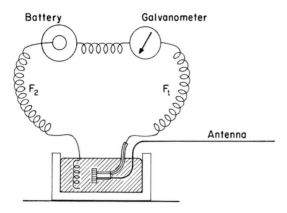

Figure 2.3 The apparatus used by Charles Nordmann in an attempt to detect solar radio waves in 1901. The hatched area indicates a container of mercury, used as a control to short-circuit the 175 m long antenna. During operation the mercury was drawn off to below the level of the coherer (the bolt-shaped object in the center). (Drawing based on one given by Nordmann [1901])

physicist Louis P. Décombe (1902) wrote that what Nordmann would have detected, if anything, were the giant eruptive prominences ejected from the solar surface. They reasoned that the quiet sun was a poor source of Hertzian waves since incandescent bodies in the laboratory emitted only light and heat. But Deslandres had long held that solar activity, as well as the corona and chromosphere themselves, had an electric origin, and further that the sun acted as a giant Crookes tube copiously emitting cathode rays.[15] He felt that the solar prominences were sources of electric waves that *did* reach the earth's surface, but nevertheless went unrecognized by commercial wireless telegraphers of the day because of their inability to separate them from interference caused by thunderstorms. Deslandres and Décombe therefore proposed the establishment of a worldwide network of radio monitoring stations with the purpose of sorting out local effects from global ones. They even suggested that the Eiffel Tower would be excellent as an antenna for one of these stations!

Immediately following this paper, Nordmann (1902c:531) published another in which he emphatically stated:

> The surface of the sun must emit Hertzian waves, and this emission must be particularly intense from those regions which produce violent surface eruptions and at those times when the intensity of these eruptions is maximum, i.e., in the vicinity of spots and faculae and at the time of maximum in [the cycle of] solar activity.

In his 1903 thesis for the University of Paris, "On the role of Hertzian waves in physical astronomy," he developed various Hertzian ideas at length. In Chapter 1 Nordmann noted that although he was at a high altitude, two other factors may well have led to his negative result. The first was that the September sun was not that high in the sky, thereby leading to more absorption by the earth's atmosphere. The second was that photographs revealed not even the smallest sunspot on that day. As Nordmann pointed out, 1901 was in fact the exact minimum in the eleven year cycle of solar activity (Fig. 13.11). He concluded that the experiment should be repeated in the early summer of 1904, when the sun would be high and solar activity should next be maximum. But there is no evidence that this was ever done, despite Nordmann's continued research into solar effects on the earth's atmosphere.[16]

2.6 WHY DID NO ONE DISCOVER SOLAR RADIO WAVES UNTIL FOUR DECADES AFTER NORDMANN?

It is remarkable that after the above attempts at the turn of the century, and despite a continued belief that the sun emitted electric (radio) waves,[17] radio waves from

---

[15] Although Deslandres here discussed Hertzian waves from the sun, he was in fact far more enthusiastic about the general importance of *cathode rays* (energetic electrons) in astrophysics. During this period he and Nordmann feuded over whether Hertzian waves or cathode rays were the end-all explaining aurorae, comet tails, nebulae, etc. Deslandres's career was a curious mixture of speculative ideas, basic astrophysics, and first-rate instrumental developments; for instance, he developed the spectroheliograph in 1894, independently of G. E. Hale, who did it the year before.

[16] See Tangent 2.1 for an analysis of the sensitivity of Nordmann's setup to solar radio waves.

[17] For example, an article in *Scientific American* in 1909 refers to a possible receiver on a high-altitude balloon "responding to electrical waves sent out from the sun." ["More about signalling to Mars," *Sci. Amer.* **100**, No. 20, 371 (15 May 1909)]

the sun were neither purposely sought[18] nor accidentally discovered *for over four decades*. This was a long time – for instance, aviation moved from the Wright Brothers to World War II bombers. At the risk of upsetting those who consider it unfruitful and even ahistorical to ask why something did *not* happen, here are some thoughts on this lacuna.

First, why did no one *accidentally* detect the sun's radio emission as radio technology and sensitivity steadily improved over the decades? Despite the extensive work before 1900 in the microwave region (frequencies greater than 500 MHz), the focus over the following 25 years changed to the other end of the spectrum, to frequencies less than 5 MHz. The widely-used Austin–Cohen formula (1910–13) predicted that the longest waves were best for long-distance circuits, and in any case high power transmitters could much better be made for long waves. Guglielmo Marconi led the radio industry in developing the technology needed at these lower frequencies, and little research was done outside this range.[19] Furthermore, the same ionosphere that aided communications through reflections off its bottom side also caused any putative incoming extraterrestrial signal to bounce off the *top* side and never reach the ground. Not until the late 1920s, when shortwaves (5 to 30 MHz) came into use for communications, was there much equipment operating at frequencies that (at higher elevation angles) could indeed penetrate the ionosphere (Chapter 3).

The normal, quiet sun, however, emits far less radiation than that detectable by even the most sensitive shortwave antennas (such as Jansky's) in the pre-World War II era. If the sun had been found, it would have been through the agency of one of its great outbursts of radiation. During the most active period in the 11 year solar cycle, such radio outbursts are common and can reach levels typically a thousand times the normal level (Section 13.2). Such a burst would have produced a noticeable hissing sound to a radio operator,[20, 21] but then how would he have discerned that this sound originated with the sun? He heard a tremendous variety of strange noises all the time – from his own electronics, from the earth's atmosphere, and from manmade interference. Most operators would have dismissed it, but what if a curious fellow *did* follow it up? He would have then wanted to pinpoint its origin, but soon would have grown frustrated because of the low directivity and lack of steerability of his antenna.[22] In fact this kind of thing was precisely what happened in 1936–38 in England to radio amateur D. W. Heightman, who was not quite able to make the solar connection despite extensive investigation (Section 5.4.2). To establish that the sun emitted powerful radio bursts required not only reasonable sensitivity in the shortwave range, but also a relatively large and steerable antenna. Add to this the advantages of several widely spaced antennas making observations and one has in every respect the situation of J. S. Hey when he ended the long hiatus in February 1942 (Chapter 5).

If not an accidental discovery, why did no one after Nordmann make a *purposeful* effort to search for radio waves from the sun? There are three primary reasons. The first is that long-distance communications

---

[18] One exception is the (unpublished) attempt by J. D. Kraus and A. B. Meinel in 1933 (Section 5.4.1).

[19] Aitken (1976:267–73) discusses reasons for Marconi's steady push to ever lower frequencies. Aitken (1976, 1985) also gives a nice overall treatment of radio technology and radio business up until 1932, especially in the US.

[20] J. J. Riihimaa (*CQ* (May 1968), 52–5) has carefully considered the sensitivity of amateur radio sets in the late 1920s and concludes that shortwave listeners most likely *did* hear "swish" sounds from solar bursts, and perhaps even from Jovian bursts, but were unaware of their origin. He estimates that solar bursts strong enough to produce a tenfold increase in receiver noise level probably occurred as often as every week or two at that time of maximum activity in the solar cycle.

[21] Donald H. Menzel (1976:56B) recalled that in the late 1920s he was unable to publish a manuscript in which he speculated that rapidly changing magnetic fields associated with sunspots might generate radio waves detectable on earth. According to Menzel, both *Science and Invention* and a professional journal turned his paper down, but unfortunately the manuscript has not been located. (Also see Tangent 9.1 for an instance in 1948 when Menzel resurrected this idea.)

[22] A case in 1909 illustrates both this point and the perceived feasibility of solar radio waves. The editors of *Scientific American*, criticizing a scheme by David P. Todd to search from a balloon for intelligent radio signals from Mars, stated that Todd had no way to "tell whether his signals come from Mars, or whether the receivers have not simply responded to electrical signals sent out from the sun." [*Scientific American*, 15 May 1909, p. 371 – quoted by Dick (1996:406)]

reinforced the notion that the earth's atmosphere was opaque to most (all?) impinging radio waves from outside. It was not until the use of higher frequencies, starting in the late 1920s, that some indications were found of a transparent atmosphere at these frequencies. As late as 1931, just as Jansky began his measurements at Bell Labs (Chapter 3), a meteorologist wrote of a radio-opaque atmosphere in *Scientific American*. He proposed the use of rockets to study the upper atmosphere, as well as non-visual portions of the spectra of the sun and stars:

> We know nothing of static from the sun or other celestial body, and could not, or at best, know but little, however full of such strays [radio interference] interplanetary space might happen to be, because the very high atmosphere is so good a conductor as to be opaque, or nearly so, to radio of every wavelength. Possibly we some day may be able to send a rocket, with a radio-recording attachment, beyond the conducting limits of our own envelope, and thus know definitely whether static of appreciable intensity is or is not knocking at our outer door.[23]

The second reason discouraging a purposeful search has to do with the rapidly increasing specialization of the physical sciences after the turn of the century. The era of Wilsing, Scheiner, and Nordmann was the last where it was feasible for a physicist to stay reasonably informed over the field's gamut. Sweeping changes (e.g., X-rays and radioactivity at century's turn, Planck's quantum in 1900 [see below], Einstein's relativity starting in 1905) led to such complexity that it was a rare person with sufficient command of radio technology, radiation laws, and solar astrophysics to put them all together. But in fact such a rare person's knowledge of these three areas would probably have weighed *against* him trying, unless he were also of a speculative and imaginative bent. This follows from the third main reason why no one tried.

The radiation theory introduced by Max Planck in 1900 introduced the concept of the quantum in order to explain many long-puzzling aspects of the spectrum of a heated body. It allowed one precisely to calculate the intensity emitted at any given wavelength by a body of a certain temperature (Appendix A). In fact Nordmann did mention Planck's theory in his 1903 thesis as support for his contention that the sun emitted very long waves, but then in characteristic fashion he made no *quantitative* calculation of expected solar radiation or of his equipment's sensitivity. Over the ensuing decades the sun was found to obey Planck's theory very well (for a temperature of 6000 K) over a range of wavelengths from 0.3 to 3 μm. Consequently, it was natural to accept 6000 K as *the* temperature of the sun and Planck theory as a reliable description of solar radiation.

At radio wavelengths then in use, Planck theory, however, yielded an answer for the expected solar intensity far below the sensitivity of any antenna–receiver combination until well into World War II (Chapter 5).[24] Those few persons who knew both how to make the calculation and how to interpret its result in practical terms were undoubtedly discouraged – in the entire scientific literature such a calculation does not emerge until 1944 (Section 5.6). The only early (unpublished) solar calculations that have come to light were by Grote Reber in 1937 (Chapter 4) and by Frederick B. Llewellyn in 1934.[25] Llewellyn, a radio physicist at Bell Telephone Laboratories in New Jersey, perhaps prompted by Jansky's recent discovery, actually worked out the expected signal not directly from the Planck theory, but from the equivalent ideas of J. B. Johnson and H. Nyquist on thermal noise in radio circuits (Appendix A). He considered the sun and a receiving antenna as two coupled circuits. The sun, approximated as an "agglomeration of doublets" (an array of transmitting half-wave dipoles), was taken to generate a certain noise power by virtue of its temperature, and these random solar voltages then induced currents in any antenna on earth. But the calculation indicated that the measured noise power from the sun would be hopelessly minuscule, ~10^{-4} as much as the noise inevitably present in even the best available receivers.[26]

---

[23] "Mining the sky for scientific knowledge," W. J. Humphreys (1931), *Sci. Amer.* **144**, No. 1, 22–25 (quote from p. 23). I thank Peter Abrahams for pointing this article out to me.

[24] See Tangent 2.2.

[25] Llewellyn, 6 December 1934, Lab Notebook 14174, p. 1, BTL; a copy of this, dated 26 February 1944, is in file 51–01–02–02, SOU.

[26] A modern calculation indicates that Llewellyn's estimate of the solar radiation, even accepting his temperature of 6000 K, was about ten times too low. But his conclusion would not have been affected by a more accurate calculation, such as he himself

In the face of all this discouragement, the only reason that someone might nevertheless have mounted a prolonged experiment would probably have involved a speculation that (a) the effective temperature of the sun might be much higher during times of great optical activity, (b) the atmosphere would allow radio waves to reach the ground, and (c) monitoring during such times might catch a radio "eruption." We should not be amazed that in fact no one both reasoned this way and put thought into action.[27]

TANGENT 2.1 NORDMANN'S SENSITIVITY TO SOLAR BURSTS

If Nordmann had used his apparatus for an extended series of observations, say several months, at solar maximum, there is a small possibility that he might have detected a solar radio burst. I have tried to make a quantitative estimate of the sensitivity of his setup despite many unknowns. If we assume that ~0.5 volt was needed for the coherer action (Phillips 1980), then Nordmann could perhaps detect a flux of $\sim 10^{-7}$ W m^{-2}, depending on losses in the antenna, mismatching of the antenna and receiver, and the relative position of the sun in the lobes of the long wire's response pattern. (For most wavelengths less than ~1/4 of the wire's length of 175 m, the primary lobes were more aligned with the wire than normal to it; Nordmann's east–west alignment was therefore not by any means ideal.) His antenna's frequency response is also problematic, perhaps ranging from 5 or 10 MHz (the ionospheric cutoff) to 100 MHz, where solar bursts rapidly become weaker, but only portions of this band were resonant with the antenna. Adopting a bandwidth of 100 MHz leads to a detectable flux density of

$$\sim 10^{-15} \text{ W m}^{-2} \text{ Hz}^{-1} = 10^{11} \text{ Jy}.$$

How strong can solar bursts be? In this frequency range the record appears to be one which occurred four solar cycles later, on 8 March 1947, with a 60 MHz flux density of $> 10^{11}$ Jy (Payne-Scott, Yabsley and Bolton 1947; Section 13.2.1.1). But such a burst is truly exceptional – a typical one at solar maximum is 10^6–10^7 Jy and even then not at this level over such a large bandwidth as 100 MHz (for reference, the quiet sun at 50 MHz is ~4000 Jy). We can thus conclude that even if Nordmann had followed his own advice to observe at solar maximum, and even if he had been very patient and very diligent, he would also have had to have been very lucky to catch a solar radio burst that could have set his galvanometer swinging.

Jorma J. Riihimaa has also analyzed Nordmann's setup.[28] His detailed calculations consider possible effects on the received signal due to a sagging antenna wire (assuming it to have been supported by poles), antenna contact with the ice, a bad mismatch between antenna (impedance ~300 ohm) and coherer (10^5–10^6 ohm), and the fundamental problem as to whether or not a coherer with one end disconnected could work at all (perhaps so, if one considers the self-capacitance of the equipment or its stray capacitance to ground). In the end Riihimaa finds Nordmann's sensitivity to be fully 1000 times less than my estimate, most of the difference resulting from the antenna–coherer mismatch and an assumed bandwidth of 20 MHz. If this estimate is correct, Nordmann would have had no chance for success, even at solar maximum.

The sensitivities of Lodge's and Wilsing and Scheiner's setups were far inferior to that of Nordmann, not only in an absolute sense, but also because their response peaked at shorter wavelengths, where quiet solar radiation and bursts are much weaker.

TANGENT 2.2 SIGNAL LEVELS FOR THE QUIET AND DISTURBED SUN

Planck's blackbody radiation law predicts a flux density of 1000 Jy/(wavelength in meters)^2 from a body with a temperature of 6000 K and the sun's angular size. As an

---

did in 1943 (using Planck's formula), probably motivated by Southworth's observations (Section 5.6) [Southworth papers, as in note 25].

[27] At the 1922 General Assembly of the International Astronomical Union (IAU) in Rome, Deslandres proposed that the Solar Atmosphere Commission "study the best means to…reveal the other radiations emitted by the sun, such as cathode rays, anode rays, X-rays, ultra X-rays, and even Hertzian radiation" (*IAU Transactions* [1922], p. 14). Nothing more is known about this; I thank Henk Van de Hulst for pointing this proposal out to me.

[28] J. J. Riihimaa (unpublished ms., 14 pp., 1986). I thank Dr. Riihimaa for his communication of these results.

example, Jansky's setup (Section 3.3.1), one of the most sensitive of its day, could not detect any source weaker than $\sim 2 \times 10^5$ Jy at its wavelength of 15 m, 50,000 times above the expected solar level of 4 Jy. We know today, however, that the effective temperature of the sun at these wavelengths is the $\sim 10^6$ K of the corona, not the 6000 K of the photosphere. Thus the quiet sun in fact yields a flux density of ~ 1000 Jy at 15 m wavelength, and on top of that bursts that can be as much as 1000 times stronger (Tangent 2.1). A solar burst of 10^6 Jy, five times higher than Jansky's threshhold, was thus detectable.

3 • Jansky and his star static

On 27 April 1933 a small audience in Washington, DC heard a talk on "Electrical disturbances of extraterrestrial origin." Today we view Karl Jansky's paper as the beginning of what became radio astronomy, but at the time of its delivery it was hardly recognized as the birth of a new vista on the universe. A week after his talk, Jansky wrote to his father:

> I presented my paper in Washington before the U.R.S.I., or International Scientific Radio Union, an almost defunct organization, … attended by a mere handful of old college professors and a few Bureau of Standards engineers … Not a word was said about my paper except for a few congratulations that I received afterwards. Besides this, Friis [Jansky's supervisor] would not let me give the paper a title that would attract attention, but made me give it one that meant nothing to anybody but a few who were familiar with my work. So apparently my paper attracted very little attention in Washington.[1]

In the context of contemporary research into radio communications and astronomy, Jansky's fundamental discovery was a misfit. Neither fish nor fowl, it was unable to be appreciated by either the scientists or engineers, and therefore lay untouched as an isolated curiosity.

This chapter first lays out Jansky's background and the setting of radio communications and Bell Telephone Laboratories (Bell Labs) in the late 1920s when he began working. I then relate the story of his discovery of extraterrestrial radio waves and his follow-up observations and interpretations. Finally, I discuss various issues such as the reception by astronomers to this discovery, why Jansky did not do more follow-up, and the factors leading to his success.[2]

3.1 JANSKY'S EARLY YEARS

Karl Guthe Jansky[3] was born in 1905 in the Territory of Oklahoma. His father's parents had emigrated from Czechoslovakia in the 1860s, while his mother was of French and English background. His father taught electrical engineering at several Midwestern schools, finally settling at the University of Wisconsin for three decades. Karl grew up in a competitive academic environment, attended the local university,[4] and in 1927 obtained his B.A. in Physics (Phi Beta Kappa) with an undergraduate thesis involving experimental work with vacuum tubes.

Following a year of graduate study in physics, Jansky applied for a job with Bell Labs in New York City. A physical exam, however, revealed a kidney ailment that would eventually lead to his early death. The Labs were reluctant to assume such a risk in a new employee, and only through the intervention of his older brother C. Moreau were they finally persuaded that his professional promise outweighed any medical problem.[5] Partly for reasons of health, Jansky was not assigned to the main laboratories in New York City, but to a radio field station in rural Cliffwood, New Jersey.

---

[1] Karl Jansky:Cyril M. Jansky, his father (hereafter KJ:CJ), 5 May 1933, JA1.

[2] Earlier versions of parts of this chapter have appeared as Sullivan (1983) and Sullivan (1984b).

[3] The name honors Karl E. Guthe, a German-American physicist with whom Karl's father trained at the University of Michigan at the turn of the century.

[4] Despite his small size (5 ft 7 inches tall and 140 pounds weight), he played varsity ice hockey at Wisconsin. Throughout his life he was a fierce competitor in a wide variety of sports and games.

[5] C.M. Jansky (1973:4T). Karl was ten years younger than C. Moreau (1895–1975) and his early career closely paralleled his brother's: both studied physics at Wisconsin, took jobs at Bell Labs, and pursued careers in radio engineering. His brother, however, only worked at the Labs for a few years and then joined the electrical engineering faculty at the University of Minnesota. From 1930 onwards he headed the prominent radio consulting firm of Jansky & Bailey in Washington, DC.

And so in August 1928 he found himself working on the problem of shortwave static for $33 per week.

3.2 THE SETTING FOR JANSKY'S WORK

3.2.1 Radio communications research in 1928

When Jansky began at Bell Labs, techniques of radio communications were rapidly expanding. G. Marconi had first sent radio signals across the Atlantic in 1901. This startling feat prompted O. Heaviside and A. E. Kennelly in the following year to propose the existence in the upper atmosphere of an electrified layer which guided radio waves around the curve of the earth. But not until two decades later was strong experimental evidence put forth for this "Kennelly–Heaviside layer" (later known as the ionosphere) at a height of ~100 km. This came through propagation experiments by Edward V. Appleton and Miles A. Barnett (1925) in England and reflected-pulse studies by Gregory Breit and Merle A. Tuve (1925) in the US.

Intercontinental radio circuits were typically at the lowest frequencies (below 100 kHz) before 1920, but thereafter pressure for more message channels and better reliability steadily pushed operations to higher frequencies, which led to exploitation of the high frequency, or long wave, region up to 1.5 MHz. After 1920 amateur operators found that even the frequencies above 1.5 MHz (wavelengths of "200 m down" or "shortwaves"), previously thought useless, worked surprisingly well for long-distance contacts. Although at first only a curiosity, these shortwaves soon were intensively developed, and their technology became dominant after the introduction of the high-vacuum, oxide-coated triode, valuable both as a power tube in transmitters and as a sensitive amplifying tube for reception.[6]

Besides the push to higher frequencies, another major migration during the 1920s was from radiotelegraphy, which employed Morse code, to radiotelephony,

the transmission of voice. The latter was a much more difficult technical proposition, requiring increased bandwidth, simultaneous two-way service, greater reliability and fidelity, and 24-hour service. By 1927 American Telephone & Telegraph (AT&T) could offer its first radiotelephone service between New York and London for $75 per three-minute call. But this was at long waves (60 kHz) and it soon became clear that shortwaves were superior. Shortwaves required smaller antennas and transmitters to achieve the same signal levels, allowed many more voice channels, and did not suffer as much interference from the atmospheric noise that plagued long waves, especially from summertime tropical thunderstorms.

In 1929 a shortwave, trans-Atlantic radiotelephone service opened to the public and quickly superseded the long-wave circuits. It operated at frequencies between 9 and 21 MHz and employed a 15 kW transmitter with air-cooled power tubes and a quartz crystal oscillator. Within a few years this circuit was handling an average of 50 calls per day. But there were now new sources of interference such as automobiles, intrinsic noise in the receiving electronics, and magnetic storms that needed investigation. Moreover, other bothersome phenomena, such as rapid fading due to changing signal paths, called for new designs in receivers and antennas.

These then were the kinds of problems facing the radio engineers whom Jansky joined at Bell Labs in the late 1920s. Their constant goal was to maximize, within economic limits, the ratio of wanted signal to unwanted noise in the telephone customer's ear.

3.2.2 Bell Telephone Laboratories in 1928

In the first decade of the last century several major industrial research laboratories were established in the United States, in particular at General Electric, DuPont, Eastman-Kodak, and AT&T. The last of these evolved into the Bell Telephone Laboratories in 1925. Its first director, Frank B. Jewett, placed strong emphasis on precision of measurement and step-by-step attacks on problems in communications engineering. Bell Labs quickly established an international reputation in a wide variety of scientific fields – witness the work in the 1920s of C. J. Davisson and L. H. Germer on electron diffraction by crystals and that of Harry Nyquist and John B. Johnson on the noise generated by electronic components. The strength of the

---

[6] Many aspects of the development of ionospheric research before 1935 are covered by C. S. Gillmor (1982). Swords (1986:146–161) describes the critical 1925 ionospheric experiments, as does Gillmor in "The big story: Tuve, Breit, and ionospheric sounding, 1923–8," pp. 133–41 in *The Earth, the Heavens and the Carnegie Institution of Washington*, (1994, ed. G.A. Good) (Washington, DC: American Geophysical Union).

Labs in physics can be gauged by the fact that over the period 1925–28 it ranked among the top ten institutions in terms of number of articles published in the prestigious *Physical Review* (Hoddeson 1980). In fact the Labs were very much like a large technical university, but with no students, much better equipment, and a greater concern for achieving practical results on a schedule.[7]

In 1914 Carl R. Englund began AT&T's research in radio telephony with work on long wave signal and static levels. Five years later Ralph Bown and Harald T. Friis joined him in investigating problems of radio propagation, measurement methodology, and receiver and antenna design. As radio communications steadily moved to higher frequencies, research at Bell Labs became more quantitative with detailed, synoptic studies of the level and character of both signals and noise. The Radio Research Division, one of eight in the Labs, was run by William Wilson in the early 1930s and had five branches. The branch of 15 to 20 men joined by Jansky was concerned primarily with problems of reception and was jointly headed by Friis and Englund.

3.3 JANSKY'S INVESTIGATIONS

Jansky's work relevant to the discovery of extraterrestrial radio waves can be divided into four phases over nine years. From 1928 to 1930 he was "learning the ropes," recording long wave static, and beginning to build a rotatable antenna and shortwave receiver. In the midst of several diversions during the following two-year phase, he finished testing and obtained his first shortwave observations with hints of a new source of static. In the climactic years of 1932 and 1933 the "hiss type static" was recorded and studied in detail and a full astronomical explanation developed. In the last phase, over the next three years, Jansky only occasionally made measurements of his "star static" while he primarily worked on more practical aspects of radio noise.

3.3.1 Phase One (1928–30): orientation and building

After only six weeks on the job at Cliffwood, Jansky was deeply involved in Friis's assignment to build a receiver to record the electric field intensity of shortwave atmospheric interference ("atmospherics"). But it is also clear that the young graduate was spending much time simply adjusting:

> I have been building apparatus for the last few weeks for my new shortwave recorder. It will be several months yet before I get any actual results. ... When I first came here the language they spoke was almost foreign to me, but I am beginning to get used to it now. At Madison I had never heard of such things as attenuators, T.U.'s, gain controls, double detection, etc., but that is what I get for not taking engineering.[8]

Jansky found that many components needed for his study already were available, but others required considerable modification and some simply did not exist. For instance the superheterodyne shortwave receiver (Fig. 3.1) was basically a Friis design, and had a sensitivity and stability unsurpassed for its day.[9] A novel requirement for the receiver was a long integration time, i.e., averaging the output over a considerable time (30 seconds) in order to detect weaker signal levels. This recording circuit was developed largely by W. W. Mutch, another newcomer to the Labs. Jansky's shortwave antenna was also a mixture of old and new ideas. Friis had already built a long, manually rotatable

---

[7] Details of the early development of Bell Labs are in Fagen (1975) and Hoddeson (1980, 1981). A dry uncritical account of radio research at Bell Labs in the 1925–45 period can be found on pp. 193–207 of *A History of Engineering and Science in the Bell System: Communications Sciences (1925–1980)* (1984, ed. S. Millman) (Short Hills, NJ: Bell Labs).

[8] KJ:CJ, 23 September 1928, JA1. *T.U.* refers to a transmission unit, later known as a decibel. Given the state of education in electrical engineering at most universities in the 1920s, it is not clear at all that Jansky would have been better served by such a degree. Most electrical engineering training concentrated on power engineering, in essence not worrying about any frequencies higher than 60 Hz. In fact a good argument can be made that an important component in Jansky's discovery was his training as a physicist rather than as an electrical engineer (Section 3.7).

[9] Information given by Jansky (1937) enables an estimate of 5 db to be made for the noise figure of the receiver, corresponding to a system temperature of ~600 K. However, since the sky itself always contributed at least 50,000 K (see Fig. 3.12 and Appendix A), the receiver contribution to the effective noise figure was in fact negligible. Receiver *stability* over long periods was much more critical than high sensitivity.

3.3.2 Phase Two (1930–31): diversions and first shortwave observations

Jansky lost several months on the shortwave static project when in February 1930 the entire group at Cliffwood moved to a new station a few miles away in Holmdel. In addition to the usual disruptions accompanying such a move, the concrete foundation and track for his antenna also had to be re-done. But now the growing group had proper room to carry out a wide variety of experiments. The Holmdel site afforded 440 acres of rolling farmland and woods, complete with a trout stream and skating pond. Activity centered on a large laboratory nicknamed the "turkey farm," surrounded by an array of shacks for individual experimenters.

In the spring Jansky searched for a quiet band on which to observe, eventually settling on a wavelength of 14.5 m. With this datum in hand, the final dimensions of his Bruce array were determined and the carpentry shop built it in the summer from 400 ft of 7/8 inch brass piping supported by glass telephone-wire insulators on 2×4 inch fir lumber. The array consisted of two almost identical crenellated curtains of quarter-wavelength sections, spanning in total two wavelengths, or 29 m (Figs. 3.2 and 3.3). The curtains were separated by 1/4 wavelength and one of them, the reflector or passive element, was about 15% taller. The receiver was connected to the center of the smaller element and the net effect was that the antenna responded only to radiation arriving from a single direction perpendicular to the array. The whole thing was mounted on four Ford Model T front wheels and automatically rotated completely around every twenty minutes through the agency of a 1/4 horsepower motor, speed reducer, and 10 ft sprocket wheel and chain.[11]

By the autumn of 1930 Jansky had carefully tuned the antenna, measured its azimuthal antenna pattern (using an oscillator at a distance of 1000 ft), removed the bugs from the motor and chain arrangement, and

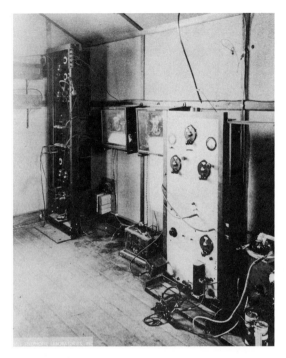

Figure 3.1 The long-wave (left) and shortwave receivers with which Jansky recorded the intensities and times of occurrence of atmospherics on 43 kHz and 20.5 MHz. Two strip chart recorders can also be seen between the receivers. (Jansky 1932)

platform with a loop antenna at each end which operated at 43 kHz and allowed the rough azimuthal direction of any long wave static to be determined. Although Jansky's focus was to be on shortwave static, one purpose of his study was to find correlations with the better-known long wave static. Thus he began recording long wave static early on, and adapted the idea of a rotating array to shortwaves. The shortwave antenna was a new type known as a Bruce array, recently developed for trans-Atlantic commercial circuits by Edwin H. Bruce, also a member of Friis's group. In August 1929 Jansky recorded the start of construction in his notebook: "Mr. Sykes [a carpenter] will start work on the merry-go-round next Monday."[10]

---

[10] 24 August 1929, p. 60 of a photocopy of portions of Laboratory Notebook 10136 of K. Jansky, BTL. Most of Jansky's laboratory notebooks and monthly work reports are available at the Bell Labs Archives in Warren, NJ; extensive quotations can also be found in an article about Jansky by Southworth (1956). During the critical 1932–33 period Jansky did not keep a detailed notebook concerning data analysis or scientific reasoning, but rather used it for more mundane purposes such as calibrations and circuit details.

[11] A working replica of Jansky's antenna was built in the 1960s at the National Radio Astronomy Observatory in Green Bank, West Virginia, and can still be viewed today (Beck 1983). In 1998 a sculpture based on the antenna design was placed at the original site of Jansky's antenna in Holmdel.

Figure 3.2 The 29 m (two wavelengths) long rotating Bruce array at the Holmdel, New Jersey field station of Bell Labs. With this antenna Jansky discovered extraterrestrial radio noise in 1932–33. The active element is the farther from the camera; it was connected through the small white box and an underground copper pipe to the receiver shack about 100 m away. (Jansky 1932)

occasionally recorded shortwave static. But as the winter approached he switched to other projects because he took it for granted that "there is practically no short-wave static in winter" (although he did occasionally check). In fact he spent most of that winter using the array to study the direction of arrival of signals from a transmitter in South America. In the spring of 1931 he was diverted yet again, this time to build an *ultra-shortwave* (4 m wavelength) receiver for static studies at yet higher frequencies. Observations of thunderstorms for this project continued until autumn,[12] but in the summer he also commenced regular recording of shortwave static at 14.6 m wavelength (20.5 MHz)[13] over a bandwidth of 26 kHz. For this he had overhauled his recording system so as to obtain a permanent quantitative record of the very faintest static, which had not actuated his first recorder although he could often hear it in headphones. This permitted recording even of the receiver's own noise and turned out to be the final critical increment in sensitivity needed for his discovery. In his August 1931 work report we find what in retrospect appears to be the first recognition of a new, weak component of static. Faintly audible with headphones above first circuit noise (intrinsic receiver noise), it followed a daily east-to-west pattern:

> Static was strongest during the month just before, during, or just after an electrical storm; however, nearly every night that the receiver was run, static was received from a source that apparently always followed the same path. Early in the evening, about 6 PM, this static (it has always been quite weak) comes from the southeast; by about 8 PM it has slowly moved to the south; by midnight it comes from the southwest; and by 3 AM it comes from the west. The reason for this phenomenon is not yet known, but it is believed that a study ... of

---

[12] Jansky, 29 December 1931, "Notes on ultra-shortwave static," Memo for File MM-327-273, Vol. P, Case 16916, BTL.

[13] The antenna was designed for 14.5 m wavelength, but Jansky found that in fact 14.6 m gave a slightly higher antenna gain.

the known thunderstorm areas of the world will reveal the cause.[14]

In August it was a night-time phenomenon, but when it persisted through the autumn and began to shift to different times of the day, Jansky became intrigued.

3.3.3 Phase Three (1932–33): the astronomical discovery

As Jansky continued irregular monitoring of the shortwave atmospherics through the autumn and winter of 1931–32, he began to isolate a component which he called "hiss type static." At one point it appeared to be interference transmitted from some unmodulated carrier, but he later became convinced that it was of natural origin. His January 1932 work report talks about a "very steady continuous interference … that changes direction continuously throughout the day, going completely around the compass in 24 hours." In a letter home the same month he reported not only on this noisy static ("Sounds interesting, doesn't it?"), but also on the wails of his newborn first child.[15] At first Jansky also used the term "sun static," for the direction of arrival seemed to coincide quite closely with the sun's position. But by February he had studied his records closely enough to see that the mysterious static was no longer aligned with the sun, but preceded it "by as much as an hour." Practical problems intervened: in February two weeks were lost while the motor and chain assembly were repaired and in April another three weeks as a consequence of windstorm damage. But enough data were obtained in the spring to establish that the hiss type static was exhibiting a continual shift in time "in accordance with the approaching summer season and the lengthening day." Although Jansky now could see that the radio waves certainly were not coming directly from the sun, he still felt it likely that they were somehow being controlled by the vernal northerly swing of the sun, perhaps through the changing angle of incidence of sunlight on the atmosphere

Figure 3.3 Karl Jansky, here posing at age 28 by his "merry-go-round" for a publicity shot in 1933.

or through the shifting position of the sub-solar point. He thus fully expected the shift in the daily signal to reverse itself when the sun retraced its path back south after the summer solstice in June. But only more data would settle the point.

In February Friis told Jansky to write up his results on both the long wave and shortwave static for publication in the *Proceedings of the Institute of Radio Engineers* (*Proc. IRE*) and for presentation at an April meeting in Washington, DC of the US section of URSI. Jansky spent much of the spring working on this, his first publication, entitled "Directional Studies of Atmospherics at High Frequencies." The paper was largely concerned with a description of his equipment and techniques, as well as with data on thunderstorm static, but he did spend three pages discussing the steady hiss type static. As he summarized:

> From the data obtained it is found that three distinct groups of static are recorded. The first group is composed of the static received from local thunderstorms and storm centers. Static in this group is nearly always of the crash type …

---

[14] Jansky, August 1931 Work Report, contained in R. Kestenbaum (1965), "Karl Jansky and radio astronomy," unpublished report, BTL. All other work reports cited in this chapter are also included in this report – see note 44.

[15] KJ:CJ, 18 January 1932, JA1.

The second group is composed of very steady weak static coming probably by Heaviside layer [ionosphere] refractions from thunderstorms some distance away. The third group is composed of a very steady hiss type static the origin of which is not yet known. [Jansky 1932:1925] .

At this point, just when the scientific problem was proving fascinating, the reality of the Great Depression fell hard on Bell Labs. In June 1932 the work week was cut back from 5.5 to 4 days with a corresponding reduction in pay which was not to be restored until 1936. The overall budget was cut to 60% of its peak in 1930 and 20% of all employees were fired. Morale amongst the engineers at Holmdel sagged, and in one letter to his father Jansky even worried about the entire field station being shut down. He asked his father about possible teaching positions at colleges or high schools in Wisconsin, but also allowed that "I can't think of a better company to work for."

Radio work at Holmdel nevertheless continued. In the second half of 1932 Jansky was again distracted from studying the origin of the shortwave static as he began investigating the general methodology of measuring noiselike signals. What were the effects of changing the bandwidth? Should one record the effective, peak, or average voltage? Was a linear or square-law detector better? Through this he managed to take shortwave data several times each month, but the daily 12 ft lengths of strip chart recordings were only cursorily examined. Even so, by August he had noticed that the pattern on the daily traces was *not* shifting back to its springtime position, as he had anticipated, but rather kept steadily arriving earlier. Also then, a fortuitous partial (94%) solar eclipse on 31 August allowed him to test any effect on the hiss type static attributable to the temporary covering of the sun: records on three consecutive days centered on the eclipse proved to be indistinguishable. The enigmatic static seemed even less likely to be associated, directly or indirectly, with the sun.[16]

The idea of the static arriving on a *sidereal* schedule, and therefore being fixed in celestial coordinates, has an uncertain genesis, but December 1932 was the date. Jansky's correspondence and work reports indicate that George C. Southworth (then working for AT&T in New York City; also see Chapter 5) asked him to plot up his entire year's data in order to see if it correlated with "diurnal changes in the directions of earth currents" that Southworth was then studying.[17] This request apparently caused Jansky for the first time to see the precision in the shift of the overall pattern and that after one year the pattern had slipped exactly one day, that is, the peak signal was now in the south at the same time of day as it had been the previous December. To an astronomer this kind of shift is a fact of life – a star or other source fixed in celestial coordinates each day rises four minutes earlier (with respect to the sun) as a result of the earth orbiting the sun; after a year, this slippage amounts to one day (see Appendix A). But to a communications engineer, the connection was not at all obvious. A. Melvin Skellett (1901–1991), a good friend and bridge partner of Jansky's, much later recalled that it was he who provided the key suggestion.[18] Skellett was leading a highly unusual life as simultaneously a radio engineer for Bell Labs (at the Deal Beach field station) and a graduate student in astronomy at nearby Princeton University (working on meteors; see Chapter 11). In any case, by the end of December Jansky had consulted Skellett and had learned about astronomical coordinates. His December work report even cites a preliminary direction of arrival: right ascension 18^h, declination $-4°$. In a Christmas letter home he could hardly contain his excitement:

> Since I was home [early November] I have taken more data which indicates definitely that the stuff, whatever it is, comes from something not only extraterrestrial, but from outside the solar system. It comes from a direction that is fixed in space and the surprising thing is that…[it] is the direction towards which the solar system is moving in space. According to Skellett (our friends in Deal) there are clouds of "cosmic dust" in that direction through which the earth travels.

---

[16] Jansky, 13 September 1932, "Results of the studies of atmospherics made at Holmdel during the eclipse of the sun," Memo for File MM-327–427, Vol. Q, Case 16916, BTL.

[17] *Earth currents* are variable currents measured between two grounded terminals separated by distances upwards from 100 m. Their variations are controlled largely by changes in the earth's magnetic field.

[18] Skellett (1977:3T); Skellett 1975 questionnaire, p. 2, Sources for the History of Modern Astronomy, CHP. In his publication, Jansky (1933b:1398) acknowledged Skellett's help "in making some of the astronomical interpretations of the data."

There is plenty to speculate about, isn't there? I've got to get busy and write another paper right away before somebody else interprets the results in my other paper in the same way and steals the thunder from my own data.[19] [underlining in original]

In the first two months of 1933 Jansky continued his attack along two fronts. First he tried to analyze past records more carefully in order to pin down the direction of arrival of the extraterrestrial radio waves. Further study, however, proved as confusing as it was helpful, for the daily time of arrival was not behaving as regularly (on the assumption of a *single* source) as it should have. Secondly, he made several attempts with another antenna on the site to determine the *vertical* angle of arrival of the hiss type static. In particular he varied the height above the ground of a "horizontal antenna setup" available on site, thus shifting its response in elevation angle, but found no detectable change in intensity of the source.[20]

That spring Jansky also got busy writing a second paper, although apparently without much enthusiasm from his boss:

My records show that the "hiss type static" … comes … from a direction fixed in space. The evidence I now have is very conclusive and, I think, very startling. When I first suggested the idea of publishing something about it to Friis, he was somewhat skeptical and wanted more data. Frankly, I think he was scared. The results were so very important that he was timid about publishing them. However, he mentioned them to W. Wilson, the department boss, and Wilson discussed it with Arnold, who is in charge of the whole Research Department of the Bell Labs (he reports directly to Jewett), and Arnold wanted the data published immediately.[21]

Jansky presented this paper in Washington on 27 April (as described at the opening of this chapter) and on 27 June at the national convention of the IRE in Chicago. A short note was sent to *Nature* in May and the full paper was submitted in June for publication in the *Proc. IRE*. Until the end Jansky was wrestling with Friis over the tone of the scientific claims:

I haven't the slightest doubt that the original source of these waves, whatever it is or wherever it is, is fixed in space. My data prove that, conclusively as far as I am concerned. Yet Friis will not let me make a definite statement to that effect, but says I must use the expressions "apparently fixed in space" and "seem to come from a fixed direction," etc., etc., so that in case somebody should find an explanation based upon a terrestrial source, I would not have to go back on my statement. I am not worried in that respect, but I suppose it is safer to do what he says.[22] [underlining in original]

Thus it is that Jansky's 1933 paper in *Proc. IRE*, unlike his two talks (or the *Nature* article, entitled "Radio waves from outside the solar system"), has the cautious title "Electrical disturbances apparently of extraterrestrial origin."

This paper, one of the most important in twentieth-century astronomy, ironically devotes fully half of its twelve pages to explanations for its engineer audience of elementary astronomical concepts such as right ascension, declination, and sidereal time (e.g., Fig. 3.4); the only citations in the entire article are to introductory astronomy texts of the day and to Jansky's first paper. Once past these fundamentals and a presentation of the observations (Figs. 3.5 and 3.6), Jansky derived that the radio waves arrived from a position of right ascension $18^h00^m \pm 30^m$, declination $-10° \pm 30°$. The right ascension was determined in a rather straightforward manner from the sidereal time when the maximum signal was in the south, on Jansky's meridian (Fig. 3.7). The determination of the declination, however, was more complex. It was deduced from a comparison of (a) the observed curves showing azimuth of arrival of peak signal versus time with (b) the curves expected for various declinations on the supposition that the signal was received *even*

---

[19] KJ:CJ, 21 December 1932, JA1.
[20] Given the expertise in antennas at Holmdel, I do not understand why Jansky apparently never appreciated that the modest vertical directivity of his Bruce array could have been profitably used to locate the source of emission (see Tangent 6.3). Perhaps it was the communications engineer's overriding concern with the *azimuthal* direction of arrival of long distance transmissions. See Tangent 3.1 for details of the all-sky contour map (Fig. 3.12) that can be constructed when advantage is taken of the vertical directivity (Sullivan 1978).
[21] KJ:CJ, 15 February 1933, JA1.
[22] KJ:CJ, 10 June 1933, JA1.

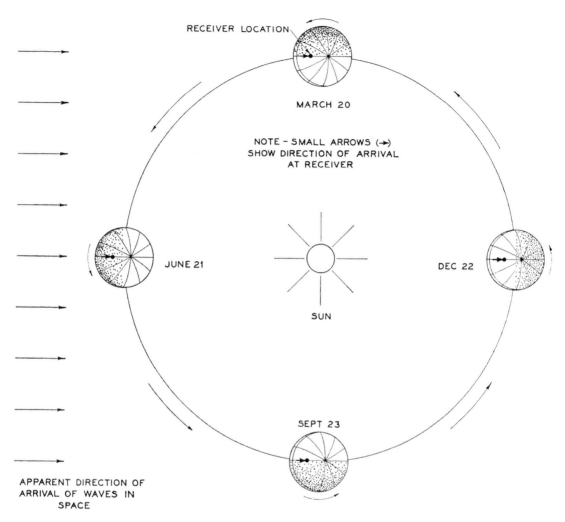

Figure 3.4 An example of Jansky's efforts to explain astronomical fundamentals to his engineering audience. (Jansky 1933b)

when the source was below the horizon. Jansky had somehow to reconcile the idea of a single source of emission with his receiving the hiss type static for typically twenty hours every day (Fig. 3.5). Thus his notion was that when the source was below the horizon, the radio waves struck the limb of the earth (as seen from the source) and travelled to New Jersey along the shortest path on the surface, i.e., in such a way that the azimuth was preserved. The last figure in his paper (Fig. 3.8) presented a comparison between actual and calculated curves and illustrated that no *single* value of declination seemed to work – Jansky could only suggest that this might be due to effects of severe refraction and attenuation near the horizon.

Jansky also had a few thoughts about the origin of the hiss type static. In the first place he was not certain that he was observing the actual *primary* phenomenon: "It may very well be that the waves…are secondary radiations caused by some primary rays of unknown character … striking the earth's atmosphere" (Jansky 1933b:1397). These "primary rays" would be high-speed particles, i.e., cosmic rays. But whatever the origin, he noted without further discussion the approximate agreement (within 4°) of his derived position with two significant directions in the sky: (1) the direction in Hercules toward which the sun moves with respect to nearby stars, and (2) the direction in Sagittarius toward the center of our Milky Way galaxy.

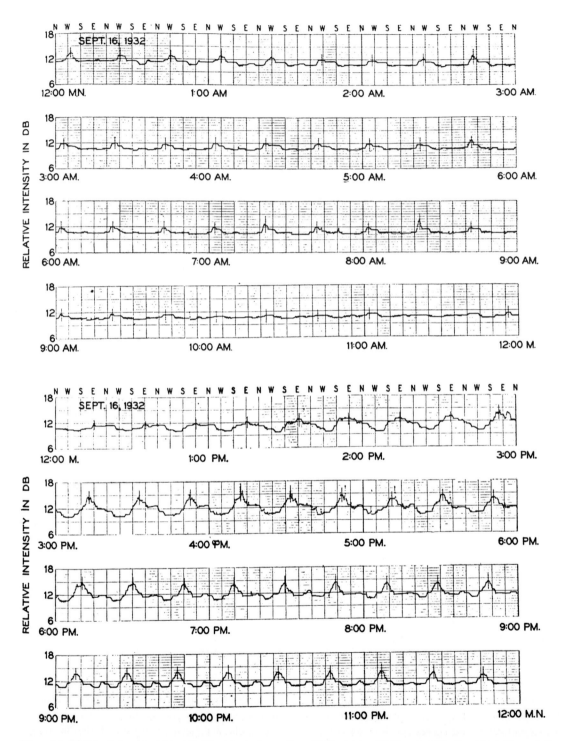

Figure 3.5 Of all Jansky's original data, only this strip chart recording of 16 September 1932 is extant. The logarithmic "intensity" scale refers to measured electric-field strength; 3 db corresponds to a factor of two. The cardinal direction of the rotating antenna is indicated at the top of each half day. Depending on the orientation of the Milky Way with respect to the local horizon, each 20 min rotation yielded either one or two prominent humps. On this day the galactic center crossed the meridian 21° above the southern horizon at 5:56 PM Eastern Standard Time. (Jansky 1935)

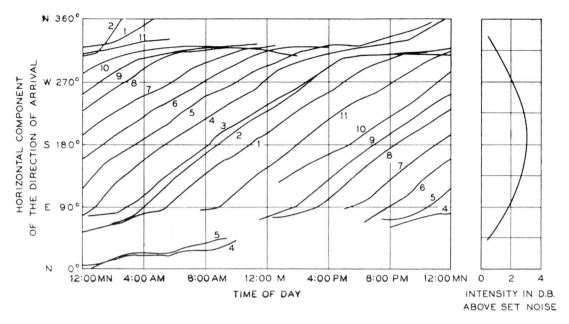

Figure 3.6 Plot illustrating the monthly slippage in time of day when the maximum signal from the "hiss type static" was detected from a given azimuth. The numbers labelling each curve refer to data gathered on (1) 21 January 1932, (2) 24 February, (3) 4 March, (4) 9 April, (5) 8 May, (6) 11 June, (7) 15 July, (8) 21 August, (9) 17 September, (10) 8 October, and (11) 4 December 1932. (Jansky 1933b)

Although Jansky's talk in Washington garnered little reaction, a press release issued a week later by Bell Labs made him an instant celebrity. *The New York Times* for 5 May 1933 heralded the discovery with an entire front-page column: "New radio waves traced to centre of the Milky Way – mysterious static, reported by K. G. Jansky, held to differ from cosmic ray – direction is unchanging – recorded and tested for more than year to identify it as from Earth's Galaxy – its intensity is low – only delicate receiver is able to register – no evidence of interstellar signalling."[23] The remainder of that front page reported on other world events: Japan invading China, F. D. Roosevelt inaugurating the policies of his New Deal in the famed "One Hundred Days" (Jansky was a rabid anti-New Dealer), Nazi demonstrations in Germany, and financial troubles for the League of Nations. On the lighter side, that spring the film *King Kong* was released, while in baseball Babe Ruth hit the winning home run in the first All-Star game in Chicago. Articles about Jansky's discovery appeared all over the world. On the evening of 15 May he appeared on a weekly science program on the (NBC) Blue radio network, following Lowell Thomas with the news and Groucho and Chico Marx. A direct connection had been rigged over the fifty miles from New York City to Holmdel and listeners heard for themselves the Galaxy's hiss. One reporter described it as "sounding like steam escaping from a radiator." On the utilitarian side, Hugo Gernsback, a radio amateur promoter and science fiction publisher, was intrigued that the galactic center was producing "40,000 billion, billion, billions of horsepower" and concluded that "there is no question that some use – an important one – will be found one of these days, perhaps not very far in the future, for these celestial waves."[24] The *Times* editorialized that poets would link this hissing from the constellation of Hercules with the serpent that Hercules strangled in his cradle, or with the hydra that he slew as one of his twelve labors.[25] All this brouhaha caused the staid *New Yorker* magazine for 17 June to harrumph: "It has been demonstrated that a receiving set of great delicacy in New Jersey will get a new kind of static from the Milky

---

[23] See Section 12.1 for a brief account of the previous twenty years of speculation and experiments on radio communications with Martians.

[24] H. Gernsback, "Celestial short waves," *Short Wave Craft* **4**, No. 3, p. 135 (July 1933).

[25] "Galactic radio waves," *New York Times*, 6 May 1933, p. 12.

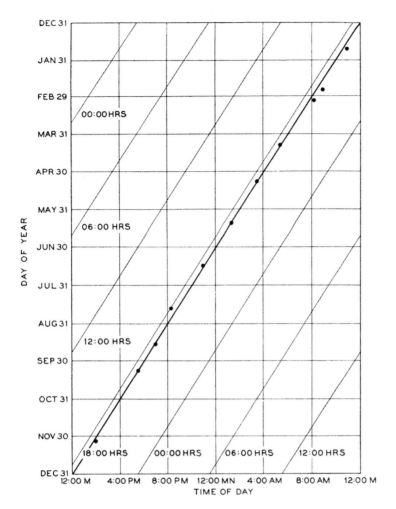

Figure 3.7 Jansky's determination of the right ascension of the maximum signal. The plotted points represent the (solar) time of transit of the signal through the year. The diagonal lines correspond to different right ascensions – from these data Jansky derived a value of $18^h \pm 30^m$. (Jansky 1933b)

Way. This is believed to be the longest distance anybody ever went to look for trouble."

Jansky was naturally eager to follow up on his discovery. To him the next question was how the intensity of the radiation varied with frequency, and so in the second half of 1933 he tried to measure something of the spectrum. His own antenna could not be tuned very much, but he found no detectable change in intensity when the frequency was changed by 5–10% from its usual 20.5 MHz. Next he checked higher frequencies with a 75 MHz ultra-shortwave receiver connected to a "fishbone" antenna (an array of 1 m rods) mounted on top of the merry-go-round. But he and colleague Alfred C. Beck never were able to secure any reliable data. Other antennas on the site were used at frequencies of 11 and 19 MHz, in particular a large rhombic which had been developed by Bruce; but although the star static could be detected, it proved virtually impossible to make the desired quantitative comparisons between different frequencies.

New thinking, however, not new observations, led to the next major step in 1933. In his August work report, only two months after submitting his paper to *Proc. IRE*, Jansky for the first time mentioned that the

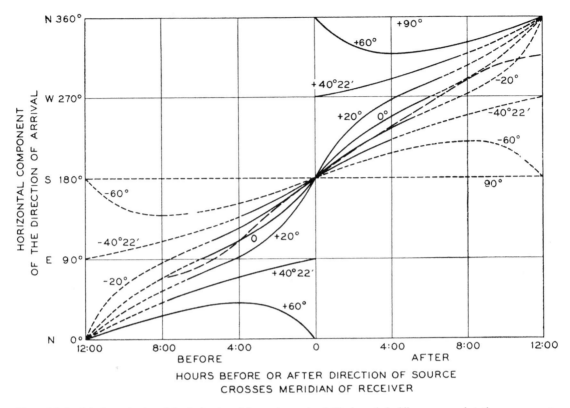

Figure 3.8 Jansky's determination of the declination of the maximum signal. The long-dashed line corresponds to the measurements and the other curves to expected behavior for various declinations; short dashes indicate that the source of emission was below the horizon. From these data at first Jansky derived a declination of $-10° \pm 30°$, later revised (Jansky 1933c) to $-20°$. (Jansky 1933b)

data in hand possibly indicated that the static came from *the entire Milky Way* rather than just the direction of its center. By October he had this theory fully worked out and was convinced of its superiority. One no longer had to suppose that the galactic center was being detected even when well below the horizon since the pattern of bumps obtained over any day (Fig. 3.5) nicely fit the concept of an entire band of emission. When the antenna's sweep along the horizon was also along the Milky Way, a broad bump resulted; when perpendicular, a narrow peak. Sometimes a single sweep even intersected the Milky Way twice, in which case one could directly see how much stronger the emission was from the general direction of the galactic center than from the anti-center.

In September Jansky (1933c) attempted to communicate with the astronomical community through an article in *Popular Astronomy*, the leading popular astronomy magazine in America. The article was very similar to his *Proc. IRE* paper except that it briefly mentioned the above new "fascinating explanation of the data."

Jansky effectively ceased working on the star static at this point and not until two years later were these new ideas fully published. In a short 1935 *Proc. IRE* article, "A note on the source of interstellar interference," he set forth the interpretation outlined above and wrote that the signal strength seemed to be proportional to the number of stars in the beam at any given time. But then he reasoned: if the noise somehow is emitted by stars, why is the sun not an overpowering source? Is it "some other [unknown] class of heavenly body" which emits radio waves? Or could the emission be due to "thermal agitation of electric charge"? This latter, prescient suggestion was made because of the remarkable similarity between the audible characteristics of the star static, as monitored on a receiver headset, and the sound of static produced by thermal effects in electronic circuits. Jansky further noted that such hot charged particles were also found in the "very considerable amount of interstellar matter that is distributed throughout the Milky Way" (Jansky 1935:1162). He obviously had been learning considerable astronomy in an attempt to explain his results, although he still only cited a popular book of Arthur Eddington's rather than primary literature.

3.3.4 Phase Four (1934–37): practical work, with occasional star static

After the magnificent results of 1933, Jansky did little more on the star static from early 1934 onwards:

> I am not working on the interstellar waves anymore. Friis has seen fit to make me work on the problems of methods of measuring noise in general – a fundamental and interesting work, but not near as interesting as the interstellar waves, nor will it bring me near as much publicity. I am going to do a little theoretical research of my own at home on the interstellar waves, however.[26]

Here begins a period of several years in which Jansky's correspondence indicates that he very much wanted to continue work on the star static, but was allowed only sporadically to make any measurements. His work focused on measuring shortwave and ultra-shortwave static of all kinds, without any special attention paid to the hiss from the Milky Way. The little star noise work he was able to do was concentrated in two periods. In the first half of 1935 interest in his work was rekindled and led to the publication of the 1935 paper discussed above, and at the end of 1936 a brief period of data collection occurred.

Jansky's letters to his father speak for themselves:

> Nothing new has developed in the field of interstellar noise.... Friis came out with a surprising remark the other day. He said, "You don't want to forget about that 'hiss static,' Jansky..." and yet he keeps me so busy doing other things that I don't have time to think much about it. [10 December 1934]
>
> I have finally succeeded in stirring up considerable interest among the men in the New York Laboratories on my work. It all started late last fall when Mr. [Oliver] Buckley, the present Director of Research of the Bell Labs, called me in to give him some pointers on static and noise in general for a speech he was giving in Toronto in January. Later, at a meeting of the Colloquium (an organization of the physicists of the Bell Labs and a few outsiders) he discussed some of the points of his Toronto speech. At the end he attached considerable importance to my work. In fact his whole talk was pointed towards a discussion of the importance and implications of my data. He concluded his speech with the statement that he thought it was the most interesting discovery made in recent years! Well, anyhow, a short time afterwards Friis came around and suggested (he had heard about Buckley's talk) that I write another paper for publication setting down my ideas on the subject, as well as giving certain other deductions I had made from my data. This I did, and with the consent and approval of Friis and Bown (Friis's boss). [19 February 1935]
>
> I am going to attempt [to receive the radiations on a wavelength of one meter] if the powers that be will ever give me time enough from my other jobs. [9 July 1935]
>
> During the last hour of work this last week I got my ultra-shortwave apparatus for measuring star static working and immediately detected the static on 10 meters. I will now make a study of it in the range of 3.5 to 12 meters.[27] Also they have discovered that they get it on their new big antenna system with which they are studying the direction of arrival of signals.[28] In fact it appears that this star static, as I have always contended, puts a definite limit upon the minimum strength signal that can be received from a given direction at a given time,

---

[26] KJ:CJ, 22 January 1934, JA2.

[27] No record of such a study has been found.

[28] The "big antenna system" was the Multiple Unit Steerable Antenna (MUSA), a 3/4 mile long array of six rhombics operating over a range of 5 to 25 MHz. MUSA was able to quickly change its elevation angle of maximum response (through automatically adjustable relative phasing of its elements) and thus follow signals varying in arrival angle as a result of ionospheric fluctuations. Friis and C. B. Feldman (1937) published a detailed description of this system, including also a few measurements of 10 and 19 MHz star static in the autumn of 1935.

and when a receiver is good enough to receive that minimum signal, it is a waste of money to spend any more on improving the receiver. Friis is really beginning to show a little interest! [20 September 1935]

I have just recently finished a few experiments on the effect of various receiving bandwidths on the peak voltages of static crashes and Friis thinks I am getting data enough to make a real good paper for publication soon. Well, anyhow, it is the end of the work that has kept me away from the star noise, so maybe after I get the paper written I'll be able to go back to it. [29 March 1936]

I am considering the possibility of taking a job at Ames [Iowa State University] … I am at present making $2825 and in spite of the advantages don't think I should accept much less than $3000 … What do you know of Ames? Of course I would ask for the time and facilities to carry on my research, which would be more than I have had for the last two years here. [18 May 1936]

At present I have one of my measuring sets set up on Friis's front yard at the edge of the river where I am making a study of the interference caused by motor boat ignition systems … I was given the job of obtaining this information along with all my other jobs. I don't mind, though; it is a very pleasant place to work and the work is interesting, but I would like to put in some time on the star noise problem. [4 August 1936]

Finally, after about two years, I am working on star static again. The reasons why and the manner in which I got back on the job are interesting … [For all] antenna system…sites along the sea shore … I pointed out that [for] the weakest signal that can be picked up, the star noise and not the ignition noise will be the limiting factor. This, of course, led to a request for more information on star noise … It begins to appear that some very definite practical importance can be attached to my discovery after all. [13 December 1936][29]

The study mentioned in the last quotation led to a few weeks of measurements of star static. Jansky (1937) incorporated these into a paper whose abstract gave the theme: "On the shorter wavelengths and in the absence of man-made interference, the usable signal strength is generally limited by interstellar noise." He emphasized that in all his measurements he had never found a noise level less than 6 db above the intrinsic receiver noise, and that most of this excess was interstellar in origin. Also reported were some absolute measurements (in micromicrowatts) of received power at frequencies ranging from 9 to 21 MHz, obtained both with simple half-wave dipoles and a large rhombic antenna. But there was little directional information and most of his data were deemed inconclusive because of the large and variable absorption of the ionosphere's D region at this time of maximum in the 11-year solar sunspot cycle.

This then was the muted end of Jansky's involvement with star static. But almost a thousand miles away and within months of the submission of this 1937 paper, Grote Reber began construction of a 31 ft dish in his back yard – radio studies of the heavens were to stay alive, and in a most unlikely setting (Chapter 4).

3.4 JANSKY'S LATER YEARS

Until his death in 1950 Jansky's career at Holmdel continued its theme of investigating and minimizing the sources of noise in radio communications, whether they were internal to the electronics or external to the antenna, whether manmade or natural. He published only three more papers: in the *Proc. IRE* (Jansky 1939) on intermittent investigations over the years of the best methods for measuring noise and in 1941 with C. F. Edwards on shortwave echos arising from ionospheric scattering, and in a Bell Labs publication (Jansky 1948) on a microwave repeater system for television transmission. Long after the 1932–33 era, his merry-go-round was used as a mount for a wide variety of antennas, including even microwave radar reflectors during World War II. Jansky's own wartime work involved developing systems for locating and identifying German U-boats through interception of their shortwave transmissions.

After the war Jansky of course followed with great interest the myriad discoveries in the field which came to be called radio astronomy. But his lifelong kidney ailment (then called Bright's disease, now nephritis) led over the years to very high blood pressure, in the end necessitating a diet of little more than rice and fruit.

---

[29] All quotations in this series except the last are from JA2. The last is quoted in a 7 March 1954 letter of Helen Jansky, Karl's sister, to John Pfeiffer, PFE.

He died of a stroke in 1950 at age 44, never having received any scientific award for being the first man to detect and study radio waves from outside the earth.[30]

3.5 REACTION OF THE SCIENTIFIC COMMUNITY TO JANSKY'S WORK

3.5.1 Other contemporary investigations

Few radio engineers or physicists looked upon Jansky's work as more than an interesting curiosity. Jansky did try to stir up activity by proposing to the 1934 URSI General Assembly in London that the star noise be studied at a variety of stations scattered around the globe and at differing frequencies. But although Commission III on "Atmospherics" formally adopted such a resolution, it came to nought.[31]

The only other observations of any depth[32] in Jansky's era were carried out in 1936 by Gennady W. Potapenko (1895–1979) and Donald F. Folland, a professor of physics and his student at the California Institute of Technology. Potapenko was a colorful Russian émigré who worked on a variety of problems in physics, their only common denominator being that they involved radio technology, the shorter the wavelength the better. He read Jansky's articles,[33] decided he should try to confirm these results, and asked Rudolph M. Langer, a theorist in the same department, to think about possible mechanisms of emission. Langer (1936) worked up a theory which he presented at the end of 1935 to a meeting of the American Physical Society. He argued that dust grains in the interstellar medium become highly ionized by starlight, and could be considered to have quantum energy levels. When an electron recombined with such a grain, discrete spectral lines would be emitted; for substantial emission at Jansky's wavelength of ~15 m, the theory required grains about 1 μm in size.

Potapenko assigned Folland the task of thinking about the extraterrestrial radiation and the best way to detect it. In February 1936 Folland presented a seminar in which he summarized Jansky's results and roughly estimated the signal levels implied by Jansky's published data. Comparison with that expected for blackbody emission at a temperature of 10,000 K then revealed that the star noise was of much higher intensity and therefore undoubtedly not arising from thermal processes. Folland built a sensitive receiver operating at precisely Jansky's wavelength of 14.6 m, since, if Langer were right, the radiation was narrowband in nature. Potapenko and Folland's first antenna (Fig. 3.9) consisted of two one-meter diameter loops at a spacing of ~2.5 m on an equatorial mount. They were plagued by manmade interference on the Caltech campus in Pasadena and on a nearby farm in Arcadia, but finally were successful in the spring of 1936 in a remote valley of the Mojave Desert. As they rotated the two loops about the sky, a weak peak in signal was found from the general direction of Sagittarius, a peak which moved in the expected manner as the night wore on. But after only one night of observations, a strong wind blew down the antenna and mount and they decided to switch to a less cumbersome and more portable setup. This was simply a 10 m long wire attached to the top

---

[30] In 1947 Jansky was made a Fellow of the IRE, a modest recognition. If he had survived another decade or more, he might well have received the Nobel Prize in Physics. Shortly after his death radio astronomers began discussions of naming a unit of flux density after him [J. L. Pawsey, presidential report (1955) to IAU Commission 40, in P. H. Oosterhoff (ed.), *Trans. IAU* 9, 563–586 (1957, Cambridge: Cambridge University Press), p. 573.]. Finally, in the 1970s, the International Astronomical Union honored him by officially adopting a unit of flux density defined as the jansky = Jy = 10^{-26} W m^{-2} Hz^{-1}, now used throughout astronomy.

Contrast Jansky's lack of formal recognition with the case of Southworth, who in 1942 first detected microwaves from the sun (Chapter 5), also at Holmdel, and received for this work the 1946 Louis Levy Medal of the Franklin Institute. Also see Section 5.6 on Southworth's attempt after the war to have Jansky join him in radio astronomy research.

[31] Resolution No. 7 of Sub-commission 1 of Commission III, *Proc. URSI General Assembly* 4, 116 (1935). Also see Edge and Mulkay (1976:72).

[32] An early, unsuccessful experiment was carried out in May 1935 by John H. DeWitt, a radio engineer working in Nashville, Tennessee. Using a hand-held yagi with reflector on a few nights, DeWitt was not able to detect any 300 MHz signals from the general direction of the Milky Way. He finally succeeded, however, in 1940 at 111 MHz (Tangent 6.1) and later won fame for heading the US Army team that in 1946 first bounced radar off the moon (Section 12.2).

[33] In fact on 27 October 1933, soon after Jansky's (1933b) discovery paper, Potapenko gave a talk at Caltech entitled "The work of the Bell Laboratories on the reception of shortwave signals from inter-stellar space." [*Caltech Weekly Calendar*, 20 October 1933; I thank Karl Hufbauer for referring this to me.]

Figure 3.9 The double-loop antenna with which Potapenko and Folland first detected galactic radiation in the spring of 1936. The pier, about 2 m high, provided an equatorial mount for the rotatable loops, each ~1 m in diameter. Photograph taken on the roof of the Norman Bridge Physics Laboratory at the California Institute of Technology.

of a 8 m mast and carried about in azimuth, May-pole fashion, by one man while the other took readings on the receiver (Fig. 3.10). With this rig, which also had the advantage of higher directivity, they obtained more definite results over several nights in the desert. During the summer Folland also repeated these experiments with the same setup in the mountains east of his native Salt Lake City, Utah.[34]

These results were too rough to publish, however, and it was clear that a large antenna was needed to do the job properly. Potapenko therefore enlisted the expert help of Russell W. Porter, one of the key men then working on the design of the Palomar 200 inch telescope. Porter produced the concept shown in Fig. 3.11, a 180 by 90 ft rhombic antenna mounted on a rotating assembly very similar to Jansky's. Potapenko was eager to launch into a full-scale project, but was not able to convince Robert A. Millikan, head of the Physics Department and President of Caltech, that the required $1000 would be well spent.[35] Stymied in his desire to look upwards, Potapenko switched his energies to a more lucrative venture – using radio waves to probe for subterranean oil fields! Radio astronomy would not return to Caltech until twenty years had passed.[36]

---

[34] Since Potapenko and Folland were observing at a time of maximum solar activity, it is not clear how strongly their measurements were influenced by variations in ionospheric attenuation.

[35] Caltech astronomer Fritz Zwicky later wrote that he also was part of this group seeking funds for an antenna; he remembered the requested amount as $200, while Potapenko (1975:13T) recalled $1000. [F. Zwicky (1969:90–1), *Discovery, Invention, Research* (New York: Macmillan)]

[36] Sources for this episode are primarily interviews with Potapenko (1974–5) and with Folland (1974, 1983). Both men also supplied useful documentation such as notes, photographs, and newspaper clippings. Minkowski (1974:3T) also recalled trying to revive Potapenko's interest after the appearance of Reber's (1940b) paper, but to no avail.

Figure 3.10 The single-wire antenna (wire not visible), pictured here at Big Bear Lake in the San Bernardino Mountains, with which Potapenko and Folland carried out further measurements near Los Angeles. The wire was carried about the mast, Maypole fashion, while readings of the received intensity were taken inside the 1924 Chevrolet.

Only one other effort paid attention to Jansky's strange hiss, and this was on the theoretical side. Fred L. Whipple (1906–2004) and Jesse L. Greenstein (1909–2002), respectively instructor and graduate student at the Harvard College Observatory, were both experienced with amateur radio. They learned of Jansky's work in the mid 1930s when Greenstein was in the last stages of his thesis work (under Bart J. Bok) involving calculations of the scattering of starlight by interstellar dust grains. Only in the previous decade had the enormous importance of the interstellar medium, both dust and gas, been brought home to astronomers and much research at Harvard was being directed towards an understanding of interstellar astrophysics. Greenstein (1979:7–8T)[37]

much later recalled why he became fascinated with Jansky's results:

> I guess it was probably because cosmic static was a rather romantic idea. [But even] for a sensible young astronomer – I was 27 – the idea of learning anything about the center of the Galaxy directly with cosmic static was exciting. My thesis work, for example, indicated that we would never see the center of our Galaxy … that nothing would come through. Interstellar absorption blinded the astronomer at optical wavelengths towards the center, and here was this challenge that the *only* thing Jansky saw *was* the center.

Whipple and Greenstein (1937) set out to explain Jansky's signal in terms of heated dust concentrated at the galactic center. Current ideas indicated that normal interstellar dust had a temperature of only 3 K (Eddington [1926]; based on heating by the ambient stellar radiation field), so they supposed that fully one-tenth of the mass of the galactic center consisted of dust grains a good bit larger (~100 μm in diameter) than those in the solar neighborhood. In this way virtually all of the starlight in the galactic center was absorbed by grains that were large enough to radiate with some efficiency at the required wavelength of 14.6 m. The emitted intensity of the dust was then calculated using S. Chandrasekhar's new theory of radiative transfer.

It was one thing for astrophysicists to do the astrophysics, but quite another to interpret the signal levels published by Jansky. Their first attempt at converting Jansky's (1932) value of "0.39 microvolts per meter for 1 kHz bandwidth" went far astray, essentially through neglecting the factor of 120π for the impedance of free space. Then, making very optimistic assumptions about conditions in the galactic center, they derived a temperature of ~800 K, which produced a radio intensity at the earth agreeing with their mistaken value for Jansky's intensity. Excited about this agreement, they even arranged for a press release to the local newspapers, but soon discovered their error after consultation with the well-known electrical engineer George W. Pierce.

---

[37] The notations "Smith (1976:14T)", "Smith (1976:107B:420)" or "Smith (1976:2N)" refer to a 1976 interview with Smith: either page 14 of a transcription, or Tape 107, Side B, position 420 of an untranscribed tape, or page 2 of my notes of an unrecorded interview.

Figure 3.11 1936 sketch by Russel W. Porter of Potapenko's proposed rotatable, rhombic antenna (never built).

In the end, their paper published in 1937 in the *Proc. Natl. Acad. Sci.* stated that the dust could not be heated to more than 30 K, yielding radio waves a factor ten thousand shy of that needed.

Although this first detailed attempt to understand the origin of the Milky Way radiation ended in frustration, Whipple (1979:2T) nevertheless entertained thoughts about checking Jansky's results at other wavelengths:

> I did talk Harlow Shapley, who was the Director, into giving me $50 to buy an acorn [tube] set. My thought was to put a rhombic antenna on top of the 61 inch telescope dome [at Harvard's Oak Ridge station]. Then one would have a rotating antenna that could look for sources. That was the idea … but it was just too hard to get going on it, and I didn't really make a serious attempt to bring in the electronics people at Harvard.

3.5.2 Reactions from astronomers

If Jansky's work on star noise did not excite the physicists or his fellow engineers, then perhaps it might have fallen to the astronomers to investigate this new phenomenon that stretched the wavelengths of the observed electromagnetic spectrum by a factor of a million. But although at first there was some enthusiasm, it quickly waned. Jansky tried to reach the astronomical community through his 1933 *Popular Astronomy* article (Section 3.3.3). There was also correspondence with leading astronomers such as Henry Norris Russell at Princeton University and Joel Stebbins of Wisconsin; Stebbins even enthused in one popular talk about the discovery of the "Jansky center of the Milky Way" as the greatest achievement since Lindbergh's solo flight across the Atlantic six years earlier.[38]

Jansky's discovery did receive prominent treatment in the 1934 popular book *Earth, Radio, and the Stars* by Harlan T. Stetson (1885–1964), a maverick astronomer who was then Director of the Perkins Observatory in Ohio. Furthermore, in 1933 Stetson suggested that his Observatory's new 69 inch reflector be fitted with a microwave receiver to do a systematic search of the heavens![39] The following year, when unsuccessfully seeking to establish an interdisciplinary geophysical

---

[38] The positive reaction from Stebbins is consonant with his pioneering work on photoelectric photometry at this time; his observatory equipment then was far more electrical in nature than usual.

[39] O. H. Caldwell (Editor of *Electronics* magazine):F. Conrad (Westinghouse Corp.), 20 November 1933, JA2. In the same letter Caldwell refers to existing "data on similar [to Jansky's] apparent concentrations of 15 cm waves from the stars by using parabolic reflectors." I have no idea to which data this refers.

institute at Harvard, he proposed experiments to follow up Jansky's results.[40]

In October 1933 Jansky shared an evening program at the American Museum of Natural History in New York City with the eminent Harvard spectroscopist Annie Jump Cannon – he spoke on "Hearing radio from the stars" (once again with a direct hook-up to Holmdel for the hiss) and she on "Unravelling stellar secrets." A few months later Jansky went to New York to hear a similar popular talk by Harlow Shapley, Director of the Harvard College Observatory, and he took the occasion to introduce himself to Shapley after the talk. Only a few weeks before, Shapley had heard about Jansky's work in the form of a report, by one Mr. King in the Harvard physics department, which had "incited vigorous comments, pro and con."[41] Shapley asked Jansky about the cost of doing such radio work and was somewhat disheartened by the answer, which included several years of salaries and development effort. Jansky later had second thoughts and in a follow-up letter emphasized that he had overestimated the cost, outlined the minimum equipment needed to study the star static, and closed by saying:

> I would be very much interested in seeing my experiments tried by someone else and would be willing to help in any way possible.

Shapley's reply was noncommittal:

> Possibly I can arouse interest with some of the local radio people. I may write you again about details if the local programs are not already too overcrowded.[42]

But things went no further.

One other sign of interest was a letter that Jansky received circa 1935 from a radio engineer in Chicago asking if Bell Labs were planning to do any more work on the star noise. Jansky had to reply in the negative, but his correspondent, Grote Reber, did not let that dampen his enthusiasm for the subject (Chapter 4).

---

[40] Stetson:H. Shapley, 23 February 1934, box 10, Stetson papers, Massachusetts Institute of Technology Archives.
[41] Shapley:Jansky, 16 January 1934, JA2.
[42] Jansky:Shapley, 13 March 1934; Shapley:Jansky, 23 March 1934, both from the Shapley papers, Pusey Library, Harvard University Archives.

Regarding the astronomers, Reber (1975:2T) gave his opinion many years later:

> I wouldn't say the astronomers were shortsighted. You have to remember, in that day even the photoelectric tube was a mysterious black box; when it came to vacuum tubes and amplifiers, tube circuits and all the rest of it, they just didn't have any comprehension of these matters. And they didn't build radio sets – they weren't even radio amateurs. If they needed a radio, they went out to a store and bought one. And consequently, well, from their point of view it would be foolish to embark on anything like this. The chances of them going wrong would be about a hundred to one … These kinds of electromagnetic waves just weren't part of their repertoire.

Similarly, Skellett (1977:3–8T) recalled the reactions of the faculty at Princeton:

> The astronomers said, "Gee, that's interesting – you mean there's radio stuff coming from the stars? I said, "Well, that's what it looks like." "Very interesting." And that's all they had to say about it. Anything from the Bell Labs they had to believe, but they didn't see any use for it or any reason to investigate it any further … It was so far from the way they thought of astronomy that there was no real interest.

Astronomy in the 1930s was only just exploiting the full power of the reflector over the refractor telescope, of photographic techniques over traditional visual observations. Electronics were not part of the observatory and no observatory director would think of hiring a radio engineer in place of a conventional astronomer. Only at a few observatories were instruments such as the thermopile and photoelectric cell being adapted from physics and attached to telescopes. Albert E. Whitford, a pioneer in using photoelectric cells, recalled that most astronomers understood only the rudiments of electricity and that he was viewed at Mt. Wilson Observatory as a "wild man who marched around the mountain carrying a soldering iron" (DeVorkin 1985:1212). Even distinguished astronomers had misconceptions about electronic techniques. In 1931 George Ellery Hale, doyen of American astrophysics, said of measurements of stars made by radiometers: "The radiometer … provided the means of detecting the last wave-lengths

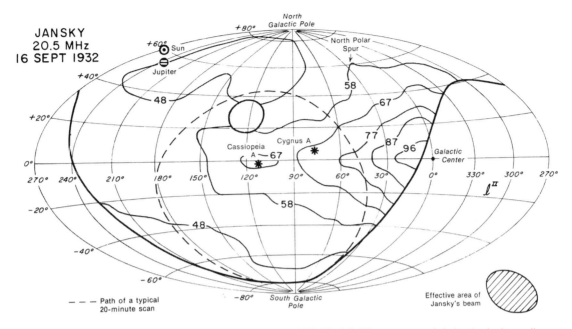

Figure 3.12 A modern reduction of Jansky's traces of 16 September 1932 (Fig. 3.5). The contour map is in (new) galactic coordinates in which 0° latitude corresponds to the plane of the Milky Way and 0° longitude is toward the galactic center. Contours are normalized to a peak value of 100, corresponding to ~100,000 K in brightness temperature at 20.5 MHz. The positions of the sun and Jupiter on that day, as well as several other radio sources discovered later, are indicated. See Tangent 3.1 for further details. (Sullivan 1978)

missing in the long range from the gamma rays to radio waves 20,000 meters in length."[43] It appears that Hale incorrectly thought that any radio waves from a star were already being measured by radiometers.

Perhaps if Jansky had drawn the astronomers more of a *visual* representation of his data, such as a contour map of the sky similar to Fig. 3.12, he might have elicited more interest. But even then, if he could specify the source of his star static only barely better than either "Milky Way" or "non-Milky Way," what could an astronomer do with such gross accuracy?

The world of decibels and superheterodyne receivers was simply too far removed from that of binary star orbits and Hertzsprung–Russell diagrams. The supreme night-time quiet of the observatory dome was the antithesis of the rumbles, clicks, crashes, fluttering, grinds, grunts, grumbling, and hiss of radio communications.

3.6 WAS JANSKY "STOPPED" BY FRIIS?

A controversy has existed over whether Friis hindered Jansky from continuing his investigations on the star static. The evidence given above would seem to indicate that the answer is "Yes," but the situation deserves more discussion.[44] The controversy began with the publication of the first popular book on radio astronomy by a science writer. In *The Changing Universe* John Pfeiffer

---

[43] G. E. Hale (1931:11), *Signals from the Stars* (New York: Charles Scribner's Sons). Usage of the word *signals* in the title of this popular book on astronomy indicates the influence of radio on American culture of that period.

[44] About 1965 the Bell Labs Publication Department conducted an investigation into Jansky's work and the reasons why he discontinued his work on star noise; this resulted in the report by R. Kestenbaum cited in note 14. For this purpose many of Jansky's former colleagues were interviewed. I have also conducted interviews with Jansky's former colleagues and relatives. The present analysis leans heavily on these two sets of interviews.

(1956:16) argued that "Rarely in the history of science has a pioneer [such as Jansky] stopped his work completely, at the very point where it was beginning to get exciting." He then painted a picture of Jansky wanting very much to continue, but being thwarted by his practically-minded superiors.[45] Jansky's brother Moreau, in a eulogistic article in 1958, also stated that his brother was "transferred to other activities" and would have preferred to continue his work in "radio astronomy." Friis (1965:842) replied to all this in an article in *Science*:

> In 1938, Karl dropped the study of star noise and, some 17 years later, I was criticized by people who thought that I had stopped him. This was not true. Karl was free to continue work on star noise if he had wanted to, but more than five years had passed since he had made his epochal discovery, and not a word of encouragement to continue his work had appeared from scientists or astronomers. They evidently did not understand its significance. Also, Karl would have needed a large steerable antenna to continue his work, and such antennas were unknown to us at that time. Radio astronomy, as such, did not then exist, and neither Karl nor I had the foresight to see it coming ten years later.

Harald T. Friis (1893–1976) was a Danish immigrant with a highly successful career in research and development in a wide variety of radio fields.[46] Over forty years at Bell Labs he acquired thirty patents, several medals from engineering societies, and even a Danish knighthood. Besides his technical expertise, he was known for techniques of management that accounted in large measure for the productivity of the Holmdel group. He had an aggressive personality and ruled Holmdel with an iron hand (Friis 1971:36).

On the other hand Jansky, although fiercely competitive in sports and games, did not have a forceful character when it came to dealing with conflicts in professional matters. This attitude probably was instilled by his father, who often gave his son advice such as the following:

> I do not know what to make of Friis' attitude ... Do not antagonize him. Keep on consulting him as formerly. He is your boss, and loyalty ultimately pays no matter whether it is deserved.[47]

Jansky was not one to rock the boat. This 1958 recollection of his widow gives a likely picture of a typical interaction regarding the question of resuming work on the star static:

> Periodically, over the years that Karl worked under Friis, he would come home and say, "Well, Friis and I had a conference today to discuss what my next project should be, and, as usual, Friis asked what I'd like to do, and, as usual, I said, 'You know, I'd like to work on my star noise,' and, as usual, Friis said, 'Yes, I know, and we must do that some day, but right now I think such-and-such is more important, don't you agree?'"[48]

A poll of Jansky's colleagues reveals a large majority agreeing with Friis.[49] They testify to the practical aspects of the group's research in the 1930s and furthermore state that they never saw Jansky disgruntled or frustrated, which they reason certainly would have happened if he had been making repeated, unsuccessful requests to continue on his star noise. But the lack of overt discontent is undoubtedly explained by Jansky's attitude of being a "team player." One final piece of evidence which has come to light is a 1965 letter from Lloyd Espenschied, a senior Bell Labs engineer in the

---

[45] Pfeiffer asserted, among other things, that Jansky at one stage proposed to build a 100 ft diameter dish, but was turned down on grounds of cost. His source for this was Grote Reber, who recalled a coversation with Jansky to this effect [Reber:author, 18 July 1983; Reber:Pfeiffer, 31 January 1955, PFE; Reber (1983:74)]. There is no evidence of such a proposal in the Bell Labs archives for Case 16916 ("Shortwave Radio") over the years 1928–37.

[46] Friis (1971) tells his story in an autobiography.

[47] CJ:KJ, 9 May 1933, JA2.

[48] Alice Jansky:C. Moreau Jansky, 11 February 1958, JA2. Jansky's widow Alice was here reacting to the following statement of Friis's made in a 23 January 1958 letter to C. Moreau: "Karl never expressed to me or any of his associates a desire to continue his work on star noise. He felt that astronomers should follow up his discovery and it disappointed him that they did not." [Container 3, Friis papers, Library of Congress]

[49] These colleagues included A. C. Beck, R. Bown, A. B. Crawford, D. H. Ring, J. C. Schelleng, and A. M. Skellett. Viewpoints on the other side were expressed by C. B. Feldman and G. C. Southworth.

1930s, objecting to Friis's assertion that Jansky was free to continue his star noise research:

> I have a distinct remembrance of Ralph Bown [Friis's boss] having discussed with several, including myself, the question of the continuance of Karl's star-noise observations. Ralph could not see that the Bell System could ask its subscribers to pay for an investigation that had proven to be so definitely in the realm of Science. I was inclined to agree with him, but thought we were obligated to see to it that such work was continued by an appropriate [outside] agency ... I had the impression that Karl's star work was called off so far as Bell was concerned; probably he was led to agree to it; too bad Karl isn't here to speak for himself![50]

It thus seems that yes, Friis did not allow Jansky to continue his work in any major way, but also that Jansky did not press the point on those occasions when he was turned down. Given Friis's style, he may well have felt that something did not even qualify as a request unless detailed plans were forcibly advanced. Furthermore, from Friis's point of view they were all working for the Telephone Company in the middle of a Depression, and Bell Labs was oriented towards a specific mission – applied research and development bearing on electrical communications. This probably influenced Jansky's stance in later years, exemplified in his 1937 article, that the star static was of *practical* import. One of Friis's management doctrines was that the worthiness of any research job could be evaluated by asking: "If I owned the entire Bell System, would I pay for it?" (Friis [1971:49]; also recall the Espenschied letter above). On such a criterion, investigation of the Milky Way did not stand much of a chance, especially with no urgent petitions from the astronomical community. The Depression was strongly molding Jansky's attitude, too. He had chronic health problems and a young family to support and was not going to jeopardize his job – he realized that there were few other places, if anywhere, where he could find such a first-rate, well-paying position.

And so for about four years Friis and Jansky went back and forth like partners in a dance, but there never was any doubt as to who was leading.

---

[50] Espenschied:Friis, 24 August 1965, container 3, Friis papers.

3.7 WHY DID JANSKY SUCCEED?

While there can be no doubt that Jansky's discovery was serendipitous, it was more than just lucky – it also possessed another aspect of any true instance of serendipity. Horace Walpole, inspired by the ancient tale *The Three Princes of Serendip*, coined the word *serendipity* in the eighteenth century and his original definition involved a happening that was not only accidentally felicitous, but also sagacious. In Jansky's case we have seen that such sagacity was expressed in a tenacious inquisitiveness that led him to delve into an incidental phenomenon of no importance to the practical problem at hand. His training as a physicist, rather than as an electrical engineer, undoubtedly strengthened this trait. Jansky himself commented along these lines in a letter written only five months before his death:

> If there is any credit due me, it is probably for a stubborn curiosity that demanded an explanation for the unknown interference and led to the long series of recordings necessary for the determination of the actual direction of arrival.[51]

There was of course also the direct contributions to the discovery from the advanced antennas and receivers available at Bell Labs. And this equipment in turn resulted from the strong staff of research engineers assembled by AT&T to attack the technical problems of telephony.[52] The quality of the components in Jansky's setup was therefore no matter of luck, although his particular combination fortuitously was precisely that needed to detect and recognize extraterrestrial radio waves: a steerable antenna with reasonable directivity, a stable and sensitive receiver, a recording apparatus and methodology ideally suited for the eventual discovery,

---

[51] Jansky:E.V. Appleton, 28 September 1949, JA2. This was a letter of appreciation for remarks Appleton made in his Presidential address to the 1948 URSI General Assembly in Stockholm. Appleton devoted most of his talk to Jansky's discovery and the post-war rise of radio astronomy, and stated that "Jansky's work has all the characteristics of a fundamental discovery." [*Proc. 8th URSI General Assembly* (1948), Vol. 7, 18–20]

[52] Note also the remarkable circumstance that A.M. Skellett, who conducted experiments at Bell Labs involving detection of radio echoes off ionized meteor trails in 1931–32 (Chapter 11), did in effect the first *radar* astronomy in precisely the same milieu where Jansky began what became *radio* astronomy.

and an operating frequency that (we recognize today) yielded about as strong a signal as possible from the Milky Way.[53] Note that it seems certain in retrospect that others in the same era also detected "star static," but none had the combination of excellent equipment, supportive institutional circumstances, and methodical curiosity necessary to consummate the discovery. Three examples:

(1) A study of atmospherics on 5 to 20 MHz published by Ralph K. Potter in 1931 flirted with having these essential ingredients. The difference was that Potter employed antennas of little directivity, as well as a method whereby he only recorded the *peak* values of the "crash" atmospherics, thus largely ignoring any low-level signals. Nevertheless, he did make reference in his paper to "exceptional cases of atmospheric noise of a hissing character" and to occurrences of a "brief hiss" at times; whether these were extraterrestrial in origin is problematic.

(2) Jansky's colleague Alfred C. Beck (1983:35) recalled that it was common knowledge at Holmdel that an excess was often measured when comparing the noise level from an antenna with that from a matched resistive load. In retrospect it seems certain that much of this was galactic radiation, but this could only be established with the aid of a steerable antenna and long-term study.

(3) In the period 1930–31 Gordon H. Stagner, an engineer at an R.C.A. trans-Pacific shortwave communications station near Manila, Philippines, had another near-miss. As part of his job Stagner measured total circuit noises on a variety of antennas (of little directionality) at frequencies of 5 to 12 MHz. At one stage he noticed that the background hiss seemed to be varying with time of day and consulted with Charles Deppermann, a Jesuit friend and Chief Astronomer at Manila Observatory. The priest, who himself happened to be then studying long wave static, became convinced that the excess shortwave noise followed sidereal time, but had no idea of the cause. He persuaded Stagner to seek funds from his superiors for a rotatable antenna that would allow a proper investigation, but Stagner was told to confine himself to his assigned duties.[54]

The final ingredient of Jansky's serendipity was that the 1931–33 period of his investigations occurred at the minimum phase in the 11-year cycle of solar activity (see Fig. 13.9). This meant two things. First, the sun itself was not emitting powerful radio bursts, bursts that at times of solar maximum are often much stronger than the Galaxy's radiation. Thus it was the Milky Way, and not the sun, which was first detected.[55] Second, and more importantly, ionospheric effects such as refraction and absorption were relatively small and stable. This allowed the extraterrestrial hiss component to be much more easily sorted out on the 1932 strip chart recordings; this point is particularly brought home by an inspection of Jansky's solar-maximum data published in 1937, where the situation is far more confused.

Jansky therefore succeeded through a mix of fortune and sagacity and the right environment. But although he was at one of the few institutions in the world where in 1932 the equipment was at hand for such a discovery, the irony is that that there was no home for significant follow-up. Where further research could be done, the problem was not deemed of sufficient interest; and where a desire for new results might have logically lain, that is, in the astronomical observatories, there was not the technical expertise nor much interest. The discovery of extraterrestrial radio waves lay neglected as a misfit in the scientific order of the day.

There can be no doubt that Jansky is the father of what eventually became radio astronomy – but (shifting metaphors) only in the sense of finding and sowing the seed, not in raising the crop. His discovery fell on stony ground and he played no role in developing the burgeoning field. As it happened, the technical demands of a war soon were to create a new generation of men and equipment who quite independently

---

[53] Many of these ingredients were also important in the discovery thirty years later of the 2.7 K cosmic background radiation by Arno Penzias and Robert Wilson, also at the Holmdel site. In this case, Penzias and Wilson *did* receive the Nobel Prize in Physics. Another interesting parallel is the role that Princeton student Skellett played in interpreting Jansky's data, just as Princeton physicist Robert Dicke first interpreted Penzias and Wilson's weak signal in a cosmological context.

[54] Stagner (1984) interview. I thank the late Francis J. Heyden, S.J., for information on Deppermann.

[55] The quiet sun produced for Jansky's antenna a signal ~1% of that from the galactic center.

reaped their own harvests after the war and did not work in a tradition of "following Jansky." The one exception, to whom we now turn, was Grote Reber, who, still in the prewar era, was indeed immediately inspired by Jansky's work.

TANGENT 3.1 AN ALL-SKY CONTOUR MAP BASED ON JANSKY'S DATA

An analysis (Sullivan 1978) of the characteristics of Jansky's antenna indicates that its effective half-power beamwidth was 24° in azimuth (as he measured) and 36° in elevation (a vertical directivity that he never commented upon), with maximum gain at an elevation of 24° (also see Tangent 6.3). When this information is applied to the one day's worth of extant Jansky data (Fig. 3.5), an all-sky contour map emerges. The caption for this map (Fig. 3.12) gives further details. Note that this map is offset by about 5° (much less than the beam size) because an incorrect value was used for the time constant of Jansky's receiver: 13 sec rather than the correct value of 30 sec.

4 • Grote Reber: science in your backyard

If the Bell Telephone Laboratories was an unlikely venue for a discovery that would fundamentally alter astronomy, what can one say for the backyard of a home in the suburbs of Chicago? Yet it was here that the only significant follow-up to Jansky's work took place. In the late 1930s a radio electronics engineer designed and built in his own time and with his own funds a radio "dish" that for over a decade remained the largest in the world. With painstaking development of equipment and methodical observation, he mapped the northern sky at two frequencies, detected and monitored the sun, and made an important suggestion concerning the origin of the galactic radiation. Over the period 1937–45 he was the only person in the world spending considerable time studying extraterrestrial radio waves. He also established close links with astronomers at nearby Yerkes Observatory and in particular influenced two of them who much later helped lead the development of American radio astronomy. It is hard to decide which is the more remarkable story, Karl Jansky's or Grote Reber's.

4.1 THE MAN AND HIS DISH

Grote Reber (1911–2002; Fig. 4.1) lived for his first 36 years in Wheaton, Illinois, a neat, middle-class town 25 miles west of Chicago. His father was senior partner in the Reber Preserving Co. (jams and jellies) and a prominent local lawyer. The family lived in a large frame house on a half acre of land with many trees, ideal for teenager Grote to use for stringing ever larger antennas for his hobby of amateur radio; station W9GFZ eventually established contact with over sixty countries. After high school[1] Reber studied electrical engineering at nearby Armour (now Illinois) Institute of Technology, there also learning many non-electronic skills that would later serve him well: foundry, forging, machine design, and pattern making. Upon graduation in 1933 his first job paid $25 for a six-day week of constructing test equipment for General Household Utilities, a manufacturer of radio sets for home and auto. Over the ensuing years Reber migrated among several radio firms in the Chicago area such as Belmont Radio and Stewart Warner, plus later stints at a laboratory associated with his *alma mater* and for part of the war at the Naval Ordnance Laboratory near Washington, DC. He specialized in designing test apparatus for the manufacturing of electronic equipment, and during the war worked on production of IFF ("Identification Friend or Foe") aviation transponders.

As an avid reader of the professional literature, Reber was fascinated from the start with Jansky's reports. Realizing that this unorthodox discovery would likely lie dormant, he took it upon himself to pursue what he came to call "cosmic static":

> Here was a scientific opportunity – a first-class discovery had been made and nobody was going to do anything about it. No matter what I did in this new area, it seemed that I just couldn't go wrong.[2]

With steady employment and few family responsibilities – he lived at home with his mother through almost this entire period (his father died in 1933) – Reber had the financial wherewithal, the time, and the space for a major initiative. Furthermore, his access to well-equipped electronics and machine shops, as well as to salesmen of all the major electronics manufacturers, provided an indispensible support system. These circumstances, coupled with his fiercely individualistic

---

[1] Twenty-three years before Reber's graduation was that of none other than Edwin Hubble from the same high school; in fact Reber's mother had been Hubble's 7th–8th grade teacher.

[2] Reber, 5 April 1976, lecture at University of Washington, Seattle, Tape 44A.

Figure 4.1 Grote Reber, ca. 1940.

personality and scientific, electronic and building skills, proved an ideal recipe for success.³

Reber pored over the radio and astronomy literature in search of the best design for an antenna and receiver. Gradually he conceived the idea of a giant paraboloidal reflector. The rationale for the big "dish" relied on Jansky's radiation being thermal in nature; if so, Planck's blackbody relation predicted that signal level would rise as frequency squared, favoring the highest frequency possible (Appendix A). Jansky's beam size and signal strength would be enhanced by going to shorter wavelengths and a large collecting area. A giant reflector also had the distinct advantage of allowing operation over a wide range of frequencies. Reber calculated that Jansky's signals corresponded to (in modern terms) brightness temperatures of as high as 500 K, and that with his planned high angular resolution at 3000 MHz, he should be able to detect an expected 6000 K from the sun, 600 K from the moon, and even 6 K from the dark nebulae of the Milky Way.⁴

Although one could conclude that a large dish for microwavelengths was the antenna of choice, it was then sobering to notice that no one had ever built one. In the early 1930s the French had demonstrated 18 cm wavelength communication across the English Channel using 3 meter diameter dishes, in Italy Marconi had done experiments at 50 cm wavelength with parabolic cylinder reflectors, and RCA had done experimental tests with small dishes operating at 9 cm wavelength,⁵ but most work with microwaves was concerned with waveguide propagation or experimental physics (see Section 5.6). An even greater trouble at these very short wavelengths was the electronics: there simply were no tubes available of sufficient power, reliability, sensitivity, and stability. To generate characteristic oscillations at such high frequencies, the dimensions of tube innards necessarily shrunk, leading to a host of problems such as high inductance of the leads, difficulty in dissipating heat, and large travel time (relative to the period of a wave) of electrons from cathode to anode. Despite these frailties, Reber concluded that the highest frequencies were worthwhile, especially since he could often wrangle the choicest individual tubes out of RCA salesmen.

At first Reber thought of hiring an outside contractor for the antenna, but inquiries yielded such high estimates that he soon took matters into his own hands. The design of the dish was constrained by a marvellous blend of the science that needed to be done and the practicalities of life in Wheaton. *What size?* Well, the longest available 2-by-4 inch boards were 20 feet. Since the backup structure of the dish was based on a square, that square became 20 ft on a side, or 28 ft on the diagonal; allow for a little overhang and the final diameter

---

³ For details of Reber's life see the obituary by K. I. Kellermann in *Pub. Astron. Soc. Pacific* **116**, 703–11 (2004), as well as Kellermann (2005).

⁴ Reber:Otto Struve, 15 May 1937, box 169, Struve papers, Yerkes Observatory Archives, Williams Bay, Wisconsin [YKS]. This is the earliest record of any contact between Reber and the astronomical community. Reber presented Struve, Director of Yerkes Observatory, with Jansky's basic results and his own calculations and asked for any comments or information Struve might have. Struve's reply simply referred to the 1937 study of Whipple and Greenstein (Section 3.5.1) as ruling out that dark nebulae caused the radiation.

⁵ There were, however, much larger reflectors used by the military for detecting *sound* waves from aircraft. In addition, in 1924 Gregory Breit of the Dept. of Terrestrial Magnetism (Washington, DC) had laid out detailed plans for a large fixed dish (~20 m diameter) to operate at 3 m wavelength, but abandoned the project in favor of longer wavelengths and another type of antenna for his pioneering work on the height of the ionosphere (Breit and Tuve 1925). ["Tuve, Breit, and ionospheric sounding," C. S. Gillmor (1994), pp. 137–8 in *The Earth, The Heavens, and the Carnegie Institution of Washington*, ed. G. A. Good, (Washington, DC: American Geophysical Union)]

of the dish was 31.4 ft (9.6 m). *What focal length?* It was clear that one too long would be mechanically unwieldy, while one too short would make it difficult to illuminate the dish with the feed. Reber settled on 20 ft. *What kind of mounting?* A traditional astronomical equatorial mounting would have been ideal, but the cost of such a civil engineering headache was prohibitive. Considerations of cost and terrain finally led to a transit design, where the dish faced due south and could only move up and down along the meridian. Such a restrictive mounting was primarily used in astronomy for precise measurements of position, but here its use was forced by the dish's huge size. Observations would have to be made by moving the dish north or south to the appropriate declination and then just waiting as the sky drifted by. *How accurate a reflector surface?* Optical principles indicated that it should follow a paraboloidal form to within a fraction of a wavelength, or an accuracy of ~1–2 cm for his planned operating wavelength of 9 cm.[6] This required extremely tight tolerances in the backup structure and in the surface itself, but Reber was able to achieve this through careful design and workmanship. *What kind of structure at the focus?* In many ways this was the trickiest question of all and over the years Reber ended up experimenting with many solutions (see below).

In the summer of 1937 Reber was between jobs and for four months of twelve-hour days he built the world's first radio telescope[7] in his backyard with the parttime assistance of two laborers (Fig. 4.2). It had a trellised, semi-circular wooden framework for its support structure, mounted on four 6 ft concrete piers. The dish rolled north or south on crane wheels in response to a hand crank driving the rear axle of a Model T truck. The reflector surface comprised 45 0.02 inch galvanized iron sheets, fastened with screws to 72 radial, parabolic ribs. Distortions in the parabolic shape were minimized by a large depth in the support framework. The dish was held together with 1320 bolts and 1665 screws. Cost of the foundations, steel, wood, and paint (two coats before assembly and two afterwards) was no more nor less than $676.96, about one-third of Reber's annual salary.[8,9]

One can imagine the reaction of townsfolk as this machine rose some 50 feet into the air behind the house at 212 West Seminary Avenue – perhaps akin to those of Noah's neighbors when he started on the Ark. Eventually it became a local landmark, along with the water tower and courthouse, but at first rumors flew. Although Reber assured town officials that nothing sensational would happen once it was working, he was reluctant to say more about its purpose than that it was for experiments in the propagation of radio waves. A Chicago *Daily News* article about it at the end of the summer of construction quoted Reber as saying:

> There it is, and you can make your own guesses. I'm against talking about inventions before they work out. All I'll say is that it has to do with radio, and I don't expect to get rich from it. …I hope to have it finished by the end of the month – that is if I can dodge some of the people who ask me questions about it.

But if it was supposedly for radio, why did it not have any familiar aerials associated with it? And what were those loud snapping and banging sounds ringing out at sunrise and sunset?[10] Speculations included that it was an atom smasher, a device for launching rockets to the moon, or a source of death rays for plane engines (Army-funded defense work). Local housewives thought it would make a dandy rug dryer, and in fact Reber's mother often did use it to dry the wash. A rhubarb patch grew well under it. Kids climbed all over this semi-Ferris wheel, and curious strangers passing by in autos frequently stopped and rang the

---

[6] Reber's estimate of the required surface accuracy was based on the traditional Rayleigh criterion, used for optical reflectors, of 1/4 wavelength as the maximum permissible deviation of any errors in optical paths.

[7] The term *radio telescope* was not used, however, until a decade later (Section 17.1.1).

[8] In response to a letter of inquiry from the Dutch astronomer Jan Oort as to how to get started in radio astronomy (see Section 16.4.1), Reber sent a detailed list of materials and exact costs. This was reminiscent of another American individualist a century earlier: in *Walden* (1854) H. D. Thoreau toted up the cost of his self-built cabin as 28.12\frac{1}{2}$. [Reber:J. H. Oort, 17 September 1945, file 2.1.1945 or 5.1.1, OOR].

[9] The greatest detail concerning the dish is found in a delightful retrospective by Reber (1958a) (also reprinted in Sullivan [1984a]). This article also is my source for many details concerning the various receivers used with the reflector.

[10] These sounds were caused by the reflector panels adjusting to solar heating gradients.

Figure 4.2 The 31 ft paraboloidal reflector with which Reber conducted all his observations in Wheaton, Illinois. This photograph (ca. 1939) shows the 160 MHz resonating-cavity drum at the focus.

doorbell. During heavy rains water gushed through a two foot hole at the base of the dish (as provided for a [never-used] Gregorian focus) and this led to ideas that it was a rainwater collector or a weather controller. Reber's reticence to discuss his machine was well-founded, for its actual purpose – to establish "contact" with the Milky Way – would indeed have appeared preposterous.[11]

---

[11] Samuel G. Lutz, an expert on high-frequency tubes at Purdue University in the late 1930s, recalled that he was contacted by Reber for assistance, but refused to do so when Reber would not divulge the nature of his project. [Lutz, 14 April 1975 telephone interview]

4.2 SEARCHING FOR MILKY WAY SIGNALS

The highest frequency feasible for operations with the technology of the mid-1930s was of the order of 3000 MHz, and even there available receiver components were highly experimental. Reber acquired an RCA end-plate magnetron, able to generate all of 0.5 watt, and used it for bench tests in his basement. For detection he found that a crystal of sphalerite (ZnS) worked better than other possibilities such as a tungsten-pyrex diode. For a feed structure he was impressed with the recent work of W. L. Barrow and G. C. Southworth (Section 5.6) on circular waveguides, concluding that their properties were ideal for capturing the radio waves that bounced off the dish and delivering them to a dipole pickup placed within the waveguide. In the end he placed this waveguide inside a drum with an aperture at one end located at the paraboloid's focus. Many experiments, the results of which Reber published in engineering magazines, established how the efficiency of this setup varied with the dimensions of the drum and its aperture and with the positioning of the dipole. Reber also experimented with conical feed horns and for his first observations, at 3300 MHz in the spring and summer of 1938, used a short section of horn attached to a small drum. Any signal picked up by the dipole in the drum was fed to a crystal detector and thence to a four-stage audio amplifier from which it was conducted to a microammeter and headphones.

There in his basement sat Reber, intently listening and writing down a new meter reading every minute as the sky continually passed through the antenna beam at a rate of 15° per hour. His main objective was of course the Milky Way, in particular the portion nearest the center where Jansky's signal had peaked, but he also tried for the sun, the moon, Jupiter, Venus, Mars, and various bright stars – all to no avail. Sometimes there were tantalizing suggestions of signals at about the right times, but they would not repeat on subsequent nights. This was indeed discouraging, for his operating frequency was fully 160 times Jansky's, and he therefore expected a signal $(160)^2$ times as strong. On the other hand this failure meant that the galactic radiation had a very unusual spectrum, making it all the more alluring. Deciding he needed a stronger physics and astronomy background, in 1937–38 he took ten courses at the University of Chicago, including a graduate course entitled "Interstellar Matter" taught by the Dane Bengt Strömgren.

With Jansky's results in front of him and a dish in his backyard, Reber had no thoughts of quitting. He determinedly built an entirely new receiver at a lower frequency, although still well beyond the highest frequencies then used in communications. The exact frequency of 910 MHz (33 cm wavelength) was determined by the drum acquired for the focus – a barrel for 100 pounds of white lead. Dispensing with the horn, he used the drum alone with a halfwave dipole placed 1/4 wavelength from the back end. The longer wavelength dictated the use of Lecher wires (parallel wires with sliding connections for tuning circuits) to conduct

Figure 4.3 Control box for one of Reber's early receivers (ca. 1938), with 900 MHz oscillator on top. Hanging on the wall is the power supply.

the signal from the dipole to a preamplifier (a pair of RCA acorn triode tubes) and thence to a crystal detector and audio amplifier as before. Despite this receiver (Fig. 4.3) having much greater sensitivity than the first one, all results remained negative. Reber even rotated the dipole pickup with the idea that perhaps Jansky's radiation was strongly polarized. As before, the lack of signal was both frustrating and intriguing, for now he estimated that the galactic radiation must in fact decline in intensity at frequencies higher than Jansky's.

Undeterred, Reber pushed yet one further step lower in frequency. He ended up at 160 MHz (1.9 m wavelength), the exact frequency again being determined by drum availability. Since the drum size scaled roughly with wavelength, the weight and bulk supported at the focus was now becoming a problem. The drum therefore was made of 1/16 inch aluminum, available from Alcoa in sheets of 6 by 12 ft; hence the length of the drum, one wave, was 6 ft with a diameter of ~12 ft/π (Fig. 4.2). At 160 MHz it was now possible to think about a superheterodyne design for the receiver, but Reber decided against this; in fact, all of Reber's receivers in Wheaton were of the tuned radio frequency variety, not superheterodynes.[12] This new receiver featured a five-stage amplifier employing RCA acorn pentode tubes with coaxial line resonators (Fig. 4.4). It was tunable over 150–170 MHz and the detector, an acorn tube diode, was superior to earlier ones. From its first tests it was clear that this third receiver was by far more sensitive, for now one could detect the system's own internal noise, never before measured. When it was mounted on the back of the drum, however, another problem immediately arose: external interference so common during the day that decent observations could only be obtained between midnight and 6 a.m. It soon became apparent that the sparking of passing automobile engines interfered worst, but Reber was also plagued at times by lightning, police and taxi transmitters, home oil burners, and small home appliances such as cake mixers (see Fig. 4.6 for later examples of interference). Daytime observations, except under thick clouds, were also bothered by gain variations caused by the sun's direct heating of the front-end amplifier. But these same new capabilities that uncovered interference also disclosed, for the first time, after a year of frustration, cosmic static from the Milky Way.

By late November 1938 Reber had enough data that he excitedly wrote to his brother Schuyler, then a student at Harvard Business School, asking him to communicate his results to Fred Whipple (whom he had learned about through Jesse Greenstein at Yerkes). Schuyler did in fact spend an hour with Whipple on 16 December and sent back a detailed report of the meeting. Reber had found about twenty "point sources" in the galactic center region. Whipple first compared their positions

Figure 4.4 End view of Reber's first 160 MHz receiver (diameter is ~20 cm), constructed of copper water piping and plates. The five successive stages start at the upper right and move counterclockwise, ending with the detector compartment on the right. In each stage note the acorn tube, the large capacitor (tunable by a screw leading to the exterior), and the end of its associated coaxial line (white circle). (Reber 1942)

with photographic plates in his office, and said they seemed to correspond more to dark regions (dust) than bright clouds. Next Whipple read another Reber letter to his brother, wherein Reber explained that he thought that free–free radiation (see below) was a good bet for the source of the radio radiation. Schulyer reported:

> I noticed a change of attitude whereby he [Whipple] mentioned that he had done some work in this field and had decided he didn't have time enough to go on with it. However, he had made some original hypotheses and was about to publish a paper on it, but decided that he would lay himself open to criticism and attack ... Later on, he again tried to rationalize or console himself.

---

[12] Reber did not require the easier tunability that a superheterodyne receiver afforded, and felt that high sensitivity was better served by the larger bandwidth of the tuned-radio-frequency type, as well as by its lack of a superheterodyne's troublesome local oscillator and mixer insertion loss.

Apparently Whipple had done some follow-up to his paper with Greenstein (1937; Section 3.5.1), wherein they tried unsuccessfully to account for Jansky's signal levels using hot dust. He must have been toying with free–free radiation, but then dropped the idea, and now saw that he'd been "scooped" by this unknown guy with a huge dish in his backyard. He soon wrote to Reber, congratulating him for his "galactic static," and said he had not been able to get the free–free mechanism to produce an effective temperature of more than 10,000 K (the presumed temperature of the ionized regions of the Galaxy), whereas Jansky's levels required an intensity fully thirty times higher. Whipple invited Reber to meet with him and Harlow Shapley, Director of Harvard College Observatory, at the following week's meeting of the American Astronomical Society. Reber could not get to New York City on such short notice, but instead on Christmas day sent a four-page letter to Whipple, going through details of the free–free theory and how it applied to his and Jansky's data. He also lamented that absolute sky positions determined with his "machine" were not very good – he wanted someday to use an airplane to properly calibrate its pointing.

After this episode, Reber decided that his 160 MHz receiver needed improvement. In the early spring of 1939, just before dawn when Sagittarius crossed the beam, he began to get signals "more within reason": these new signals did not indicate point sources, but rather a broad band of radiation corresponding to the optical galactic plane. In June he wrote to Whipple explaining that the previous point sources were incorrect, and gave details about his new data (similar to what he published in 1940).[13] Launching into a survey program, Reber over the summer traced the 160 MHz radiation along the plane from its peak toward the galactic center to as far as 90° away in Cygnus. He also made a special effort to detect the noontime sun, but found neither it nor bright stars such as Vega, Sirius, Antares, and Deneb. A typical day of observations involved nine hours at work in Chicago plus a 45 minute commute, supper, sleep from 7 to midnight, data taking until dawn, and a short nap before heading off to work again. Once the dish was set to its appointed declination, receiver warmup required about an hour. Reber recorded microammeter readings by hand every minute while also monitoring with headphones for interference, whose staccato or clicking nature was quite different from the steady shhhhhh of cosmic static.

4.3 FIRST PUBLICATIONS

By the end of the summer of 1939 Reber figured that he had sufficient data in hand to merit publication and it was natural that he sent a paper to the journal that housed all of Jansky's articles, the *Proceedings of the Institute of Radio Engineers*. This paper, submitted in September and published with no revision in February 1940, contained only a brief description of equipment (despite its likely audience), presented many plots of signal level recorded as the Milky Way drifted through his beam (as in Fig. 4.5), and concentrated on an astronomical interpretation of the radiation (see below). Simultaneously, Reber felt that he should be getting the word to the astronomical community and so began to build on earlier contacts with staff members of Yerkes Observatory made while he took courses at the University of Chicago. One of the staff enthusiastic toward Reber was Philip C. Keenan (1908–2000), a young instructor who had come to know Reber in an undergraduate astronomy class. Keenan was in fact the first member of the astronomical community to visit Wheaton, in May 1939, at which time he took notes on Reber's preliminary results and interpretations.[14]

Keenan arranged a colloquium at Yerkes for Reber during the last week of October 1939, and, oh, how one yearns in vain for a tape recording of the proceedings. Reber recalled a mixed reception to these unorthodox results presented by a person with little training in astronomy:

> [Gerard P.] Kuiper and some others thought my work was at best a mistake and at worst a hoax.

---

[13] G. Reber:S. Reber, 21 November 1938; GR:SR, 29 November 1938; S. Reber, "Talk with Dr. Whipple at Harvard Observatory 12/16/38" (2 pp.); F. Whipple:GR, 2 December 1938; GR:FW, 25 December 1938; GR:FW, 16 June 1939 – all Reber archives, National Radio Astronomy Observatory, Charlottesville, Va. I thank Ken Kellermann for informing me of these letters.

[14] P. C. Keenan, "Data from Reber," 20 May 1939, one page of notes, photocopy from Keenan. Reber later recalled that Keenan was "interested but bewildered by the equipment" [Reber:N. Ben-Yehuda, 8 January 1984, photocopy from Ben-Yehuda].

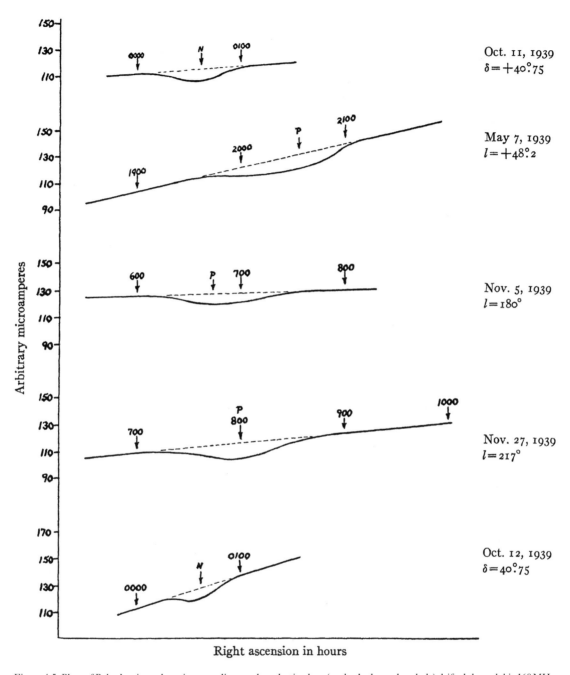

Figure 4.5 Plots of Reber's minute-by-minute readings as the galactic plane (or the Andromeda nebula) drifted through his 160 MHz beam. A negative deflection corresponded to a positive signal. P indicates the expected passage of the galactic plane and N that of the Andromeda nebula (see note 23). On the right are the galactic longitudes (old system) of the passage or the declination of the dish. (Reber 1940b)

Otto Struve, the Director, took a more enlightened view, but still could not support me [in any financial way]. You have to remember that in that day anything as complicated as a photocell or a DC amplifier was considered by the astronomers to be a magic black box. Futzing around with my kind of equipment was not considered a sensible thing to undertake.[15]

Kuiper's initial skepticism apparently was mainly based on the subjectivity inherent in Reber's editing of the frequent interference spikes (see Fig. 4.6) from his series of readings.[16] On the other hand Struve (1897–1963), a distinguished astrophysicist at the peak of his career (Section 4.8), was impressed – in the discussion following the talk he stressed that the radio work was so important that Reber should place high priority on moving his setup to a more isolated location where interference would be minimal. A decade later Struve (1949:28) recalled:

> Mr. Reber brought with him a large stack of tracings on which his instrument had recorded the intensity of the radiation ... There was clearly indicated a general bulge of the red line ... at the exact time the Milky Way passed over his antenna. An annoying factor was the presence of numerous violent and sharp disturbances which Mr. Reber explained to us were the result of various electrical appliances, such as a dentist's drill a block or two away.

Struve even offered to cooperate with Reber if he should wish to use their new site at McDonald Observatory in Texas. Reber replied that his current job responsibilities ruled out extensive efforts at a remote site, but that he was indeed interested in the prospect. Finally, he indicated that he would be pleased to receive "a delegation from the Observatory" when he was finished with improvements then being made to his receiver.[17] This visit took place, probably during the second half of 1940 and probably including Struve, Kuiper, Keenan, and Greenstein. Over the next seven years calls to Wheaton became a regular side trip for many astronomers visiting Yerkes, for example, B. J. Bok, M. G. Minnaert, and H. C. Van de Hulst.[18]

With Struve's favorable reaction to his work, Reber was encouraged to seek communication with a wide astronomical audience by submitting a paper in December 1939 to the *Astrophysical Journal*, for which Struve was editor. This manuscript was substantially similar to the one earlier sent to the *Proc. IRE*, but Struve, although willing to publish Reber's basic results, imposed major changes. In the final version published in June 1940 virtually nothing is left of either a description of the equipment or of Reber's extensive interpretation of his observations as free–free emission from the disk of our Galaxy. Struve apparently felt that (a) astronomers did not want to learn about electronics, and (b) others more competent should carry out the astronomical interpretation.[19] For this latter purpose he inserted, immediately following Reber's curtailed paper, a study by Louis G. Henyey (1910–1970) and Keenan in which the free–free interpretation was worked up in "proper" style.

In his 1940 *Proc. IRE* article Reber pointed out that his largest signals corresponded to only $4 \times 10^{-21}\,\mathrm{W\,cm^{-2}}$, equivalent to a bolometric magnitude of 22.1 "which indicates some of the technical difficulties involved."[20] He presented the run of measured intensity (units of $\mathrm{W\,cm^{-2}}$ (circular degree)$^{-1}$ kHz^{-1})[21] over 110° of galactic longitude, and argued that it agreed roughly with a model in which the sun was 8.4 kpc from the center of a Galaxy of radius 12.6 kpc. The cosmic static was taken to be caused by free–free emission, or thermal *bremsstrahlung*, in which hot interstellar electrons radiate while closely passing ions (Appendix A). Reber borrowed this concept from optical astronomy and was the

---

[15] Reber, as in note 2.

[16] Within a year, however, after a visit to Wheaton to see things for himself, Kuiper came to believe Reber's data. [Reber 1975:6T, 8T]

[17] Struve:Reber, 2 November 1939 and Reber:Struve, 8 November 1939 – both box 181, YKS.

[18] The question of who visited Reber in Wheaton, and when, is greatly confused. Reber did not keep a log and various written accounts and interviews of mine give conflicting evidence. What does seem clear is that the first was Keenan in May 1939 and that the dish was a popular attraction over the period 1940–47. My statement (Sullivan 1982:43–4) that Struve specifically sent out a delegation to "check out" Reber *before* accepting his 1940 paper for publication in the *Astrophysical Journal* is probably wrong.

[19] Reber (1975:5,10T).

[20] Astronomers would have been impressed: 22nd magnitude stars were then entirely too faint to study.

[21] See Tangent 4.1 for a discussion of Reber's units and calibrations.

first to propose that it was also important at radio wavelengths. He particularly liked the idea because of its similarity to noise caused by thermally agitated electrons in receiver circuits. His model of the interstellar medium, based on the classical work of Eddington (1926), supposed an electron density of 0.045 cm^{-3} and temperature of 10,000 K. He calculated the expected intensity of free–free radiation at 160 MHz, but his correction for stimulated emission was off by a factor of 50. Furthermore, all effects of reabsorption of the radiation were ignored, that is, the medium was implicitly taken to be optically thin (see Appendix A). Despite these errors, over the next decade thermal radiation became one of the leading explanations, albeit a troublesome one, for the galactic background (Chapter 15). Reber (1942:367) during these years had the following marvellous conception of the connection between Wheaton and the Milky Way, despite a separation of thousands of light years:

> It is suggested that cosmic static is the equivalent of thermal agitation in which all space is the conductor and the input terminals of the detecting equipment are projected by means of an antenna system to some far-distant part of space.

In contrast to the *Proc. IRE* paper,[22] the four-page 1940 *Astrophysical Journal* version only showed plots of radio intensity (Fig. 4.5) and had no mention whatsoever of the idea of free–free radiation, nor any reference to Henyey and Keenan's (1940) following paper (although that paper referred to both of Reber's 1940 papers). Other results were identical except for some further measurements toward the galactic anticenter and a claimed detection of the Andromeda nebula.[23]

It was natural that Struve would assign Henyey and Keenan (Reber called them the "quantum mechanics boys") the task of doing a proper job on applying the theory of free–free radiation to the radio regime, since Henyey had long worked on various theoretical aspects of the interstellar medium and Keenan took great interest in Reber's observations. Using an electron density of 1 cm^{-3} and temperature of 10,000 K, they extended the theory a full seven orders of magnitude from the familiar visual band to a wavelength of ~2 meters. For a line of sight extending 16 kpc, they calculated a brightness agreeing reasonably well both with Reber's maximum value and with recent optical observations at Yerkes of the diffuse Milky Way light. But they found that Jansky's datum at 20 MHz, as converted into usable units for them by Reber, could not be reconciled with the others – as Whipple had found, it required a much higher electron temperature of ~150,000 K. This discrepancy at the low frequencies was to remain outstanding, despite the efforts of many, until the development of synchrotron theory in the early 1950s (Chapter 15).

4.4 THE UPS AND DOWNS OF 1941

With the solid but incomplete results of 1939 in hand, Reber felt justified in pouring more money into his apparatus. He built a tunable signal generator for receiver calibration, testing, and alignment. Tired of monitoring a meter all night long, he spent $292.81 on an Esterline–Angus strip chart recorder – now he had the luxury of being able to set things up at the start of the night and in the morning, after a good night's sleep, finding a four-ft long ink tracing of the night's drift curve. Receiver stability was greatly improved by a first-class General Radio DC amplifier and new AC power supplies, replacing a bank of 25 Edison storage batteries. This new system was working by October 1940 and over the next eight months recorded over fifty all-night drift curves. As each declination's drift scan was obtained, Reber built up a sky map by transferring his radio intensities to a celestial globe. In June 1941 he visited Yerkes again with more stacks of strip chart recordings full of red squiggles and discussed his new results with the staff. By late August he had written

---

[22] Robert Hanbury Brown, at the time in the thick of radar development in wartime England, recalled "Reber's article was sandwiched between conventional papers on radio engineering and was so intriguingly odd that I left it open on the bench for several days until, inevitably, someone put a hot soldering iron down on it and it was burnt." Nine years later this nascent interest in Reber's cosmic static flowered when he joined the Jodrell Bank group and became one of its leaders (Chapter 9). [Hanbury Brown 1991:96]

[23] This detection would seem to be incorrect, based on later measurements of the 160 MHz intensity of the Andromeda nebula as well as the too-narrow width of the drift scan illustrated by Reber (Fig. 4.5). This claim, however, led H. C. Van de Hulst

in Holland to do some preliminary radio cosmology in 1944 (Section 16.1).

up an extensive paper, mostly concerning his apparatus, for *Proc. IRE*,[24] as well as a short note that he submitted to Struve for the *Astrophysical Journal*. Struve gave it to two referees (one of whom was Greenstein) and a month later wrote Reber that "after considerable effort…we have finally come to the conclusion that it would probably be best to postpone the publication of your material."[25] Until additional data could be secured at a site entirely free of interference, Struve felt that there was not enough new material to warrant publication. But at the same time he went out of his way to offer Reber assistance in working up a proposal for a grant, suggesting the National Academy of Sciences as a possibility. He also outlined what he thought was needed for Reber to gain credibility – letters of reference from "distinguished radio engineers who have seen your equipment and who can from personal experience and knowledge support the application." Finally, Struve offered to send letters "to explain the situation" to whomever Reber requested.

Over the last quarter of 1941 Struve was continually supportive of Reber, but nevertheless never dared wholly to commit his prestige to any proposals because he was never able to satisfy himself that he had secured a solid independent opinion of Reber's technical ability. In October he wrote to Mervin J. Kelly, Director of Research at Bell Labs, explaining Reber's situation and asking if someone from Bell could visit Wheaton and in essence "certify" Reber:

> Because Mr. Reber is somewhat outside the profession of astronomy, I feel a little reluctant to give him my unqualified support, and I know of no astronomers who would be qualified to pass upon the soundness of the technique. On the other hand, I feel a certain amount of responsibility in this case because it is entirely possible that the neglect of this phase of astronomical inquiry would retard progress in other fields.

Struve also wrote to Shapley, asking for advice as to how to proceed:

> It is my own opinion that Mr. Reber's work is sufficiently important to be carefully scrutinized by competent astronomers and especially by competent radio technicians. … Because of Reber's lack of professional astronomical training we are perhaps overlooking a development in observational technique, the importance of which should be very great provided the results are correct.

Kelly replied that Bell Labs was too busy with defense work to look into Reber's case, but Shapley said he was willing, if Reber proved technically sound, to support an application to either the American Philosophical Society or the National Science Fund, for which he sat on key committees. With these replies in hand Struve sent Reber an encouraging letter in early November. He suggested that Reber secure the best engineering references he could, work up a full proposal, check it out with Keenan and himself, and send it within a month to Shapley, who could then present it for consideration. Struve suggested that an appropriate request would be for $2000 plus salary over a period of two years.

But in late November, just when things seemed to be rolling, Reber suddenly put a damper on matters by announcing that his work was closely related to the national defense effort and he didn't see how he could devote full time to the study of cosmic static for at least the next year. Struve nevertheless urged him to produce a report, and so Reber sent one out on 21 December 1941.[26] In four pages he outlined his results and future

---

[24] This paper (Reber 1942) contained many photographs of the apparatus and sample strip chart recordings, as well as extensive analysis of the receiver circuits and how signal-to-noise ratio depended on various parameters such as the time constant of the crystal diode, biasing of tubes, etc. Reber also discussed the concepts of antenna radiation resistance and effective temperatures of various receiver elements and of the receiver as a whole. Although he tabulated many values for the intensity of the cosmic noise, there was little astronomical interpretation.

[25] Struve:Reber, 30 September 1941, box 197, YKS. I have not been able to locate a copy of the manuscript that Reber submitted, nor of the referees' remarks.

[26] Struve:M. J. Kelly, 9 October 1941; Struve:Shapley, 9 October 1941; Kelly:Struve, 20 October 1941; Shapley:Struve, 23 October 1941; Struve:Reber, 6 November 1941; Reber:Struve, 20 November 1941; Reber:Struve, 21 December 1941, including "Cosmic static" report (4 pp.) – all box 197, YKS. In his letter of 23 October Shapley mentioned that "a graduate student at the Cruft Laboratory began to work in the same field [of cosmic static], but got completely interrupted and sidetracked by national defense obligations." It is uncertain who this student was, but it may have been Albert W. Friend. Olof E. H. Rydbeck, then also a student at Harvard, recalled that in 1939 or 1940 Friend discussed with him a project designed to check

needs, in particular the requirements of a suitable location in terms of isolation from interference ("at least two miles from any major automobile highway"), weather, roads, power, machine shop, etc.[27] But the Japanese had attacked Pearl Harbor two weeks before and the United States was at war – not until 1945 would the question of funding for Reber again be addressed.

4.5 ALL-SKY SURVEYS

4.5.1 160 MHz survey (1943–44)

In the second half of 1941, Reber again improved his receiver. First, the system was made far more stable, both electrically and thermally. Secondly, for greater sensitivity it was now designed for a bandwidth of 8 MHz, over 40 times the previous bandwidth. To accommodate this broader range of frequencies the entire front plate of the focal drum was removed and the conventional dipole was replaced with a double-cone dipole. This new 160 MHz receiver turned out to be so superior that Reber abandoned the fifty nights of data he had acquired up through June 1941 with his automatic recorder and the old receiver. The traces (Fig. 4.6) were now so stable and free of noise, except for the cursed interference spikes, that a signal as weak as ~0.1 percent of the total system noise (~11,000 K) could be detected as it drifted through the beam.

Reber set out methodically to map the entire northern sky at 160 MHz and from early 1943 spent a year recording two hundred all-night traces of cosmic static. With these he assembled a contour map that beautifully limned the entire Milky Way and showed nothing over the rest of the sky (Fig. 4.7). This was the key result in a manuscript submitted to the *Astrophysical Journal* in May 1944 and quickly accepted by Struve. Although his astronomical audience might not have been able to understand the radio technology, they could certainly make sense of his new representation of the data, an intensity contour map in equatorial coordinates. In this paper Reber explained that "whenever the mirror passes over a cosmic static disturbance, energy is collected." He then pointed out several concentrations of emission in Cygnus, Cassiopeia, and Canis Major, which he interpreted as "projections" from the Milky Way, perhaps similar to the arms that had been photographed in external spiral nebulae.[28] Inspection of Fig. 4.7 also reveals contours extending as far south as -48° declination, corresponding to the dish looking precisely along the southern horizon. The normal minimum elevation angle was 15° (declination of −33°), but in March 1943 such a heavy load of wet snow built up on the reflector that the brakes failed and the dish went crashing down, completely off its track and looking at the horizon. Unable to fix it for a while, Reber resourcefully turned the situation to his advantage and took tracings (dotted on the plot to indicate some degree of uncertainty) at this usually inaccessible position.[29]

Modestly tucked away at the end of the paper (and not even mentioned in the abstract!) is the first *published* radio detection of the sun (although see Chapter 5 regarding the earlier (1942) detections by Southworth and J. S. Hey). He neatly illustrated his solar detection with a series of eight days of drift curves over a 10-week period when the sun's position was passing through the galactic plane – one saw the solar bump at first clearly, then melded with the galactic radiation, then emerging on the other side. Reber had long tried for solar radiation, but daytime interference and receiver problems had hampered his efforts.[30] But on 2 October 1943 he succeeded, although monitoring

---

(at ~15 MHz) Jansky's results. The plan was to build a coarse-mesh paraboloid supported by telephone poles. [Rydbeck 1978:1–5T; Rydbeck:author, 19 January 1988]

[27] In preparation for a possible new location for his antenna Reber at this time designed and had built at a local machine shop a turntable for his dish. This would allow movement in azimuth, faster surveying, and cleaner, more reliable records by scanning at a much faster rate than afforded by the earth's rotation. But the war intervened and the turntable went into storage, not to be used until after 1947 when Reber and his dish moved to the National Bureau of Standards (Section 4.6.2). [Reber (1975:64T)]

[28] See Tangent 4.1 for a modern interpretation of these concentrations.

[29] If it had been a different month, then Reber's nighttime trace with the antenna looking far south would have been over different right ascensions and he would have had a good chance to pick up the strong discrete source Centaurus A (Table 14.1). Since Cen A is located at a galactic latitude of ~20°, the course of early postwar radio astronomy might have been significantly affected by a much earlier discovery of a non-galactic plane source.

[30] The solar signal shown in Reber (1944:286) is very similar in strength to the galactic center region signal. It is difficult to understand why Reber did not readily detect the radio sun after

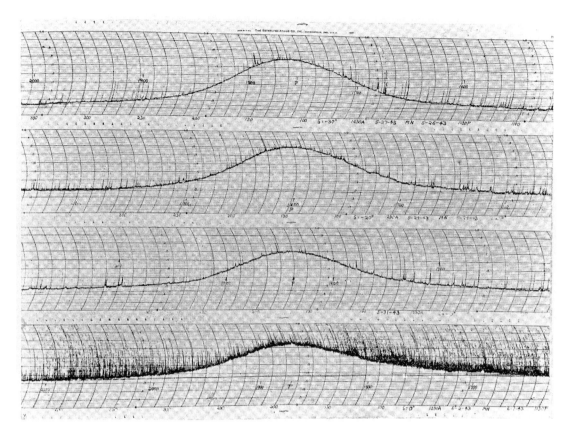

Figure 4.6 Sample strip chart recordings taken by Reber in 1943 as the Milky Way drifted through the telescope beam at declinations of (top to bottom) −30°, −20°, −10°, and 0°. Chart deflections are proportional to received electric field at 160 MHz. The bottom recording was particularly afflicted with interference from an electrical storm; most of the remaining interference arose from automobiles. (Reber 1944)

for five months thereafter yielded no changes in intensity. In his article he remarked on the "rather surprising intensity" of the sun, and further noted that similar stars causing the Milky Way radiation would not work since then the central part of the Milky Way would be optically as bright as the sun. This reasoning was incorrect, but the relative radio and optical intensities of the sun and of the Galaxy nevertheless remained a dilemma until long after the war (Chapter 15).

If Reber had expressed his observed solar intensity as an equivalent brightness temperature for the optical solar disk, he would have calculated a value of $\sim 3 \times 10^6$ K, which would have been an amazing, even incredible figure for a sun known to be at 6000 K.

Whether Reber or any astronomer would then have tied such a value in with the recently proposed million-degree corona (Section 13.3.1) is an open question.[31]

he had detected the Milky Way at 160 MHz in 1939, unless solar heating of his front-end electronics vitiated his attempts.

[31] Reber (1944:287, also 1942:377) did say that "it has been suggested that this long-wave radiation could be set up in the corona of the sun" and, in a letter to George Southworth: "It is my hunch that considerable of the observed radio emission is not generated in the photosphere but rather in the corona" [Reber:Southworth, 9 May 1945, folder 51–01-02–05, BPS-1, SOU]. Reber's idea was mainly based on interpreting his data to indicate a radio sun much broader than the 30′ optical sun, but it apparently did not germinate further before the discovery of the radio-bright corona in Australia in 1946 (Section 7.3.3). In fact at this time Reber felt that his 160 and 480 MHz solar intensities were in agreement with Southworth's microwave data (Section 5.6) as showing a blackbody spectrum consistent with a temperature of 6000 K. [Reber:Southworth, 17 February

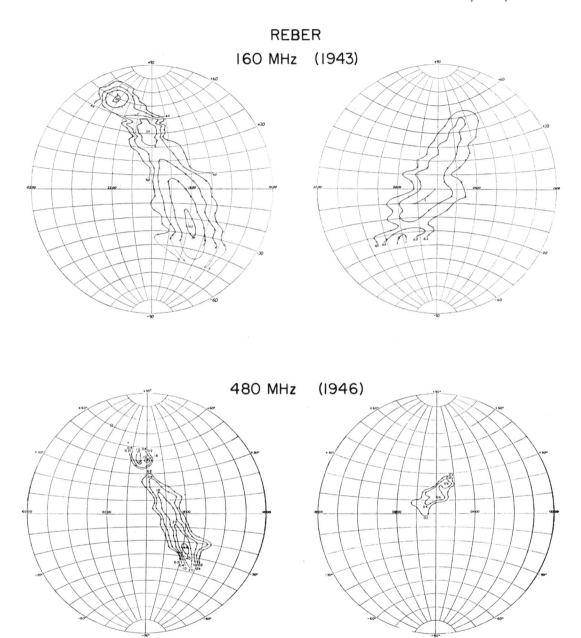

Figure 4.7 (Top) Contour map of 160 MHz intensities over the entire northern sky in equatorial coordinates. One contour unit of brightness corrresponds to 10^{-22} W cm^{-2} (circular degree)$^{-1}$ MHz^{-1}. The beamwidth is approximately 12°. For later identifications of the intense regions and for calibration details see Tangent 4.1. (Reber 1944) (Bottom) Similar contour map of 480 MHz intensities. Contour units are the same as above, but the beamwidth is now about 4° and the map was completed only up to +44° declination. Comparison of the two maps indicated the steep spectrum of the galactic radiation. (Reber 1948b)

Although fully aware of the concepts of antenna and brightness temperature, Reber found them uncomfortable and not in accord with his intuitive style. As he wrote in 1945:

> To me the concept of temperature in these discussions is a bit artificial and rather difficult for me to think in terms of. ... Nowhere in the apparatus can this [radiation] resistance be visualized, much less inspected or measured for temperature. ... Personally, I would rather work with watts and let the temperature fall where it may.[32]

4.5.2 480 MHz Survey (1946–47)

By the end of 1943 Reber had done just about all that he could at 160 MHz. He decided to move back up in frequency in hopes that a smaller beam would reveal more detail in the Milky Way – this despite his expectation of weaker signals and greater receiver problems. Over 1944 he fashioned a 480 MHz receiver and modified his focal apparatus, replacing the drum with a hemispherical reflector. But tests and observations in the spring and summer of 1945 went badly – the main tubes in his four-stage amplifier were so noisy that they picked up the sun faintly at best and the Milky Way not at all. In fact they were so poor that they only rarely detected interference from automobiles! Not only were the tubes insensitive, but their lifetimes turned out to be only a few nights, making the cost of observing a rather extravagant several dollars per hour. Indeed, by this point Reber had spent about $2000 on development of 480 MHz equipment and had little more to show for it than the statement that the radio spectrum of the sun seemed different from that of the Milky Way.[33]

As the war wound down Reber abandoned this receiver and instead acquired, through his contacts at General Electric, superior tubes that he knew were then being manufactured for military use. These GE 446B lighthouse triodes were sensitive, long-lived, and stable – just what was needed. They formed the heart of

a six-stage, wide-band (8 MHz) amplifier with a system temperature of ~1300 K. By mid-1946 Reber began a year-long program of over 200 days of solar traces at noon and galactic traces through the night. He soon discovered that the spectra of the solar noise and of the galactic noise markedly differed. Whereas the sun and the Sagittarius peak had given about equal signal levels at 160 MHz, he now found that at 480 MHz the sun was ~100 times stronger than the Galaxy. He also established that on some days the sun's intensity suddenly rose to ten to several thousand times its "normal" value (see below) and that this background level itself was variable over weeks and months (Reber 1948a). The radio sun was clearly different at the solar maximum of 1947 than it had been during solar minimum (Fig. 13.9) when Reber found no activity at 160 MHz in 1943–44.

Reber's map of the Milky Way emission at 480 MHz (Fig. 4.7), published in the *Proc. IRE* in 1948, nicely complemented his earlier map at 160 MHz. At the higher frequency his beamwidth was only ~4° and the radiation was feebler; consequently, new detail appeared and the detectable width of the Milky Way considerably diminished. For instance, the previous concentration in Cygnus now broke up into two peaks. Reber was able to deduce, with the aid of interferometric measurements in Australia by John Bolton and Gordon Stanley (Section 7.4), that they were quite different in either spectrum or angular size (these peaks later became known as Cyg A and Cyg X). Furthermore, his position for the maximum in Sagittarius coincided nicely with that of a recent infrared measurement by J. C. Stebbins and A. E. Whitford.

Two other major projects occupied Reber during 1946. The first was the building of 21 cm equipment with the idea of searching for the spectral line of hydrogen recently predicted by Van de Hulst (Section 16.1), an endeavor (described in Section 16.2) that never came to fruition. The second was writing with Greenstein a review of radio astronomy published in *Observatory* for February 1947, the first of its kind and in fact the first collaboration between an astronomer and a radio observer. By that time 48 citations were necessary for an exhaustive coverage of the worldwide literature; in addition, Reber and Greenstein reported many ongoing unpublished studies of cosmic noise. Greenstein was particularly concerned to put all of the measurements on a common intensity scale, as well as to discuss their

---

1945, folder 51-01-02-04; Reber:C. H. Townes, 18 July 1946, folder 51-01-02-07 – both SOU; also see the article cited in note 40]

[32] Reber:Southworth, 23 February 1945, folder 51–01–02–4, SOU.

[33] Reber:Struve, 23 September 1945, box 221, YKS.

reliability, but the disappointing conclusion after many calculations and much correspondence was that "we can expect absolute calibration of the work of different investigators to show only order-of-magnitude agreement" (Reber and Greenstein 1947:15; Greenstein 1980:132B). The review dealt not only with the radio data, but also with various attempts to explain its origin in the Galaxy and the sun. At the time, it provided a welcome introduction and overview for many astronomers; today, it provides a valuable benchmark in the history of a rapidly changing, fledgling field.

4.6 REBER BEYOND WHEATON

4.6.1 Postwar attempts to find funding

As the years went by Reber became less and less satisfied with his situation in Wheaton. The manmade interference grew worse, his observational successes made him yearn for a fulltime position as a researcher, and he grew to appreciate the advantages of a remote, southern site. In August 1945 matters came to a head when his mother died and it eventually became apparent that the family home would have to be sold. Reber again, as he had in 1941, began a concerted effort to secure independent funding. As before, he leaned heavily on the advice and support of the staff at Yerkes, in particular Greenstein and Struve. In early 1945 he wrote to both Struve and Shapley, asking them about possible sources of support after the war, perhaps at a site in Texas and perhaps with some funding from Bell Labs, which he had just learned was doing solar radio work in the person of Southworth (Section 5.6).[34] Struve's reply was positive but non-committal – before he was willing to seek support, he again, as in 1941, sought expert radio opinion about Reber's technical abilities. This time he wrote for advice to a fellow editor, Alfred N. Goldsmith of the *Proc. IRE*:

> Since the technical aspects of his work are completely outside the competence of an astronomer, I am wondering whether you can direct me to a person who could give me his opinion as to the soundness of Mr. Reber's experimental procedure. If his work is experimentally sound, the results are probably very important. On the other hand, there are certain weak points in his astronomical interpretation which are very obvious to me and which tend slightly to undermine my confidence in the soundness of his work. This is one of those cases where an editor does not want to take the responsibility for rejecting what might conceivably be of utmost importance, but where he does not have of his own knowledge sufficient evidence to feel convinced that the work is good.[35]

Goldsmith replied that he had "been in touch with one of the major experts in the field of cosmic static" and that this expert saw nothing wrong with Reber's data, but was not willing to comment on Reber's techniques without inspecting his equipment firsthand. Such a visit, however, was said not to be possible under the pressure of war work. This was hardly the definitive reply for which Struve must have hoped – all he could muster in his note of thanks was that he was "glad to know that there is at least nothing obviously wrong with Mr. Reber's procedure."[36]

At this point there is a gap of over a year in the correspondence extant in the Struve papers. But it is clear that by July 1946 Reber had had considerable discussions with Greenstein and Struve about how to proceed, culminating in a "Program for the Investigation of Cosmic Static," an 11-page detailed proposal.[37] In this document Reber gave a detailed inventory of his equipment on hand (everything from "one mirror 31.4 ft in diameter" to "over fifty Weston meters"), outlined his record over the previous decade ("a cash outlay of approximately $11,473 and untold hours of labor"), and argued the need for a remote, southern site. He split the program into two parts: the first involved setting up the new site, continuing his present studies, and doing development work for the second part, the construction of a 200 ft diameter, fully steerable dish within, say, five years. The overall astronomical goals were for more sensitive and detailed studies at 480 MHz, sky maps at a suite of frequencies ranging from 20 to 3000 MHz, and improved interpretation through cooperation with

---

[34] Reber:Struve, 21 February 1945; Reber:Shapley, 15 March 1945; Struve:Reber, 26 February 1945, all box 221, YKS. Also Reber:G. C. Southworth, 23 February 1945, folder 51–01–02–04, SOU.

[35] Struve:A. N. Goldsmith, 3 April 1945, box 221, YKS.

[36] Goldsmith:Struve, 23 April 1945; Struve:Goldsmith, 26 April 1945 – both box 221, YKS.

[37] Attached to Reber:Struve, 16 July 1946, box 227, YKS.

"any observatory or university with skilled scientists who can do theoretical work." Capital expenses for the new observatory, six truckloads to move his antenna and equipment, and two years of operating expenses, mostly for his own salary and that of a helper, totaled $26,450.

In this 1946 proposal he envisioned the huge dish operating at wavelengths as short as 10 cm and argued that the required surface accuracy, which he estimated as ~3 cm, was achievable.[38] Because the ability to track a source was seen as vital if surveys at a variety of wavelengths were to be carried out in a reasonable time, Reber wanted a fully steerable dish. But an equatorial mounting for such a behemoth was out of the question, so the problem of how to track a sky position with an altitude-azimuth mount arose. His solution: an equatorially mounted shaft, rotating at the sidereal rate, whose motion would be transferred to the altitude and azimuth drives through a suitable arrangement of cams and levers. The price tag for this precocious idea was pegged at ~$100,000.[39]

Reber then sent Shapley a modified version of this proposal, emphasizing the initial two-year program. Shapley replied that he would keep Reber's needs in mind and mentioned that he had consulted Donald H. Menzel, an astrophysicist on the Harvard staff who had many contacts in the military-industrial network and who advised trying for money from the Army, Navy, or RCA. Reber had already made overtures, with no success, to electronic firms such as Sperry Gyroscope, Bendix, and those where he had worked, but the military was a new possibility. Soon thereafter Shapley came up with a contact in the Navy and Reber sent his proposal to the Pentagon on 19 August. In September he was able to discuss these matters in person with Struve and Shapley at a meeting of the American Astronomical Society in Wisconsin.[40] At last, in October the Office of Naval Research replied, saying they were indeed interested. Reber was elated and wrote to Greenstein on the day of the news: "Maybe by spring something concrete will come of it. I HOPE!" The evidence indicates that the Navy only moved after Struve convinced Frank B. Jewett, former head of Bell Labs, Vice-President of AT&T, and at that time President of the National Academy of Sciences, to send a letter of support.[41]

Reber now began negotiating with the Navy for a contract. Knowing nothing about overhead rates and the other arcana of accounting, he consulted with Greenstein. The question immediately arose as to whether or not the contract should be sought through the University of Chicago. Struve was on a visit to Copenhagen at the time and Greenstein wrote to him for advice. Struve replied that it was fine with him to do it through the University, by appointing Reber as a visiting assistant professor, but that they should be careful: "Reber must not think that we work to swipe his results or glorify ourselves by them." His other concern was that it be made clear that Yerkes itself had no money whatever to apply to this project, whether or not the Navy contract came through. This thinking may have been foremost in Struve's mind when he also did not act on a suggestion from Kuiper at this same time that Reber be appointed resident astronomer at McDonald Observatory in Texas. Kuiper's letter to Struve says a lot about how Reber was perceived:

> We visited Reber and saw his large installation. It is certainly impressive what he has done and accomplished, all with his own means, single-handedly. It occurred to me that he might be a suitable man to run the McDonald Observatory for you. He would know all about electricity, electronics and construction jobs; is interested in work that could be done in daytime; is jolly; is not married and would be generally popular. He would be a great help to such astronomers as myself who would want to use electronics equipment but are dependent on technical assistance in that field. Such assistance would not take much of his time, but be invaluable when the need arises. All in all I

---

[38] See note 6.

[39] Reber's 200 ft dish was never built, but some measure of his precocity in 1946 can be gleaned by noting that such a dish for US radio astronomy only came into existence in 2000, and that only three others exist in the entire world, the first (the 250 ft reflector at Jodrell Bank) not completed until over a decade after Reber's proposal.

[40] For a detailed report of Reber's talk at this meeting, see *Science News Letter* **50**, 166 (1946).

[41] Reber:Shapley, 27 July 1946; Shapley:Reber, 11 August 1946; Reber:Navy Department, 19 August 1946; T. J. Killian (Office of Naval Research):Reber, 7 October 1946; Reber:Greenstein, 8 October 1946; Adm. H. G. Bowen (Chief of Naval Research): F.B. Jewett, 23 October 1946 – all in box 227, YKS except 8 October 1946 (box 39, GRE). More evidence of Jewett's support of Reber is given by Needell (1991; 2000:263–4).

would think he might be just the man. At Madison [at the AAS meeting mentioned above] he was pleased to be regarded as a competent scientist who had achieved something quite remarkable. Apparently his electronics colleagues do not give him the satisfaction astronomers do.[42]

Greenstein visited Wheaton with the news that Struve was willing to give Reber a nominal visiting position but was surprised to find him unenthusiastic about any connection with the University. Instead he had apparently sold the bosses at his current radio firm on doing the cosmic static work as a company contract. Greenstein warned Reber that the project's much greater cost under such auspices might upset the Navy. Indeed, for a while the project expanded to a five-man operation at about three times the cost, but the Navy rejected this and it quickly shrank back to two men, with Reber again proposing to operate on his own. In early 1947 the Navy put Reber off, saying it would not be able to take further action until the new fiscal year beginning in July.[43] This delay, however, meant that the Navy never got its chance, for by then Reber had found other fish to fry.

In the autumn of 1946 Reber had heard a talk in Chicago by Edward U. Condon, Director of the National Bureau of Standards (NBS). Condon described the kind of "pure" research done at NBS and inspired Reber to inquire whether NBS might be interested in sponsoring work on cosmic static. In response Condon sent a trio of NBS radio engineers to Wheaton on 21 November, a cold rainy day, to see the operation. Reber set up for a standard 480 MHz noontime drift curve, but 30 minutes before the expected arrival of the sun in his beam, the pen on his strip chart recorder surprisingly rose and fell in a slow, regular manner. (This, he later figured out,

was the first sidelobe of the antenna pattern, something he had never before seen.) Then as noon approached and the sun came into the main beam, the expected orderly curve never materialized and instead the pen began dancing around, finally pinning at the edge of the chart. As the four men watched in amazement, Reber switched to a less sensitive meter, but it too soon went off scale after jumping around. Had the receiver gone haywire? Was it incredibly strong interference? Or was it really the sun? Reber ran from the basement to the backyard and cranked the dish far north of the sun, whereupon the signal returned to a normal level – indeed, it was the sun.

Reber and his guests monitored the nature of this baffling solar emission with headphones and found it to be a loud hiss, varying rapidly from second to second with puffs and whistles and grinding, sometimes sounding like wind whistling through the trees. As to its intensity, all Reber could say was that on that day the sun radiated over 400 times its normal amount. He wrote to Greenstein: "Needless to say, my visitors were quite impressed and I was rather astonished."[44] Three days later Reber (1946) reported this remarkable phenomenon, "a great radio storm," in a letter to *Nature*. The NBS men went back to their lab with a dazzling tale and very soon negotiations for NBS to take on Reber's operations began. After precisely a decade the Wheaton era ended in June 1947 when Reber disassembled his antenna, sold it to the government for $18,570, packed his bags, and became a civil servant in Washington.

4.6.2 Reber at the National Bureau of Standards

The Central Radio Propagation Laboratory of NBS operated a field station in Sterling, Virginia, 40 miles

---

[42] Kuiper (from Texas):Struve, 1 October 1946, folder 3, box 226, YKS. I thank Derral Mulholland and David Evans for leading me to this letter, whose information was used for p. 112 of their book *Big and Bright: a History of the McDonald Observatory* (Austin: University of Texas Press, 1986). Their statement that Reber was on the University of Wisconsin staff, however, is incorrect.

[43] Reber:Greenstein, 10 November 1946 (box 6, GRE); Greenstein:Reber, 13 November 1946 (box 6, GRE); Struve (from Copenhagen):Greenstein, 19 November 1946 (box 39, GRE); Greenstein:Struve, 27 November 1946 (box 6, GRE); Reber:Greenstein, 1 February 1947 (box 233, YKS); Reber:Greenstein, 15 March 1947 (box 6, GRE).

[44] Reber:Greenstein, 21 November 1946 (box 39, GRE); Reber (1975:9–10T, 1978:112–4T); *Popular Science Monthly* **152**, 153–4 (January 1948). Reber promptly examined his previous records and found a similar, though less spectacular burst on 17 October. He had passed this off as automobile interference, although why it should occur just at noontime had been somewhat mystifying. When Arthur Covington read Reber's article in *Nature*, he realized that he too had encountered the great outburst of 21 November during his routine 11 cm solar monitoring at Ottawa (Section 10.2). At the time he had ascribed it to his apparatus going berserk [Reber:Greenstein, 16 May 1947, box 6, GRE].

Figure 4.8 Reber standing next to one of the three 7.5 m diameter Würzburg reflectors that he mounted equatorially and brought into operation for NBS at Sterling, Virginia in 1947–48.

west of Washington, DC (now the site of Dulles Airport). At the time NBS had many programs studying the ionosphere and radio propagation, and its interest in cosmic noise (and Reber) was twofold. First, solar observations might well allow improved predictions of ionospheric conditions vital to the performance of communications. Second, solar and galactic noise had a direct effect on the sensitivity of equipment used in communications and broadcasting (e.g., FM and television frequency bands were then being chosen).

Reber's first task was to re-assemble his dish and give it an all-sky pointing ability by at last putting it on the azimuthal turntable which he had constructed back in 1941.[45] Next, he equatorially mounted three 7.5 m German Würzburg antennas[46] acquired from the US Army Signal Corps (Fig. 4.8), Americanized them with paint jobs of red (160 MHz), white (480 MHz), and blue (51 MHz), and supplied them with receivers for monitoring the sun. These antennas collected vast amounts of data over the years, but little was ever done with it, either in the form of internal or external reports (Reber 1975:75T).

Reber soon grew weary of the routine of solar monitoring and pressed for other researches. He made detailed designs for his 200 ft dish, now estimated to cost ~$300,000, but got nowhere in trying to sell the idea to NBS or others.[47] In 1948 he also proposed to survey discrete radio sources with a 10 MHz Michelson interferometer, each element consisting of a ~150 m × 150 m array of dipoles, but this too went unsupported.[48] What *was* supported at NBS tended to be of a more prosaic nature. For example, Herman V. Cottony and Joseph R. Johler used single dipoles at four frequencies from 25 to 110 MHz over a two-year period to measure a daily, all-sky average of the absolute level of galactic noise.[49] In short, Reber was not being listened to and he was uninspired by the prevailing brand of science. As he later recalled:

> NBS was a peculiarly reactionary and backward agency. Only sure (not pure) science was possible. Anything speculative was stamped on because a failure would bring discredit to the Bureau![50]

When it was announced that everyone had to change offices and that eventually the entire outfit would be moving to Colorado, Reber rebelled against the bureaucracy. In the spring of 1951 he signed out for a vacation in the Territory of Hawaii and never came back.[51,52]

---

[45] See note 27.

[46] See Tangent 4.2.

[47] Reber:Oort, late 1949 (undated copy attached to Oort:Kuiper, 6 January 1950), G. P. Kuiper papers (photocopy supplied by Ron Doel).

[48] "Memo on Galactic Radiation," 12 November 1948, photocopy supplied by Reber.

[49] Details of radio astronomy at NBS are given by W. F. Snyder and C. L. Bragaw in *Achievement in Radio* (1986, Washington, DC: Government Printing Office), pp. 595–606.

[50] Reber:author, 3 January 1984.

[51] A visitor to NBS in June 1951 reported: "Reber is away and I can't find anyone who can really say what he is doing. It is reported that he has said he does not intend to move to Colorado … it is therefore suspected that he is away because he is trying to find a new home for his work." [J.A. Ratcliffe (in US):M. Ryle, 6 June 1951, file 4 (uncat.), RYL]

[52] See Tangent 4.3 regarding the fate of Reber's dish.

4.6.3 Hawaii

Reber had learned in 1950 of a private foundation in New York City called Research Corporation. The man and the foundation were an ideal match, for the philosophy of Research Corporation was to support research by the "little guy" who had a promising idea. He had made contact with them before he left NBS, but not until early 1952 did he receive the first of a long string of their grants, for $15,000 over several years.[53] Reber's plan centered on a sea-cliff interferometer (Appendix A), as successfully developed in Australia first by Joseph Pawsey on the sun and then by John Bolton for measuring the sizes and postions of discrete sources (Sections 7.3.2 and 7.4) and, after extensive consultation through the mails with the Sydney group,[54] he decided to seek the ultimate "cliff" in order to achieve the highest possible angular resolution. He sought resolution of ~1' over most of the sky, which implied an effective baseline of ~3000 wavelengths and a tropical location.

He came up with Mt. Haleakala, an extinct volcano towering 3100 m above the island of Maui. From its summit, which was accessible since other scientific stations were already there, the Pacific Ocean could be seen to a distance of 230 km over a sweep of 340° of azimuth. Any source was thus observable at both rising and setting, allowing its dimensions to be measured at two different crosscut angles. A check of past weather records had indicated that minimal trouble should be expected from variable atmospheric refraction, one of the chief bugaboos of the sea-cliff interferometer, which always observed very near the horizon where refraction was extreme. Finally, minimum phase in the 11-year solar cycle was approaching, which meant that fewer problems from the ionosphere were expected and that lower frequencies could be profitably used.

But this was one case where practice proved far different from theory. Reber spent three years working on Haleakala and ended up with little to show for it. He fabricated a 25 m rotating turntable topped by a redwood framework for support and steering of various Yagi and rhombic antennas operating on 20 to 100 MHz. Over 1953–54 he made observations of the eight strongest discrete sources, but found that the only useful data came at the higher frequencies and for Cas A, Cyg A, and Hydra A, whose angular size and structure could roughly be measured.

In a frank evaluation of his results and his woes published much later, Reber (1959) laid out the whole story and concluded that "radio astronomy observatories should not be located atop high mountain peaks." This remark was specifically directed at the problem of manmade interference at such an exposed site – he was continually hampered by radiation from spot-welding machines at a pineapple cannery 20 miles away, as well as from the Honolulu area, 120 miles away and on a direct line of sight. Even worse, variable conditions in both the troposphere and ionosphere (especially "spread-F" irregularities, which surprisingly turned out to be *worse* at solar minimum) meant that the interaction between the direct ray from the source and the ray reflected from the sea was highly unpredictable. Minima and maxima of the fringes on the chart records became so lost in general variability from a host of causes that Reber complained that their identification was often more an exercise in psychology than in science.

4.6.4 Tasmania

With things not going well in Hawaii, Reber cast around for a new scientific adventure and soon hooked the idea of extremely low frequency observations. The lowest frequency anyone had used for radio astronomy was 9 MHz by C. S. Higgins and C. A. Shain in Australia and Reber decided to try to push to frequencies ten times lower, in the opposite direction but with a similar philosophy to that in the 1930s when he had gone to unprecedented frequencies at the *high* end. He planned to take advantage of the current period of solar minimum activity and scoured ionospheric monitoring records for the very best observing site in the world. He found two "holes" in the ionosphere, places where the electron density was consistently much lower than average and which therefore would allow lower frequency radio waves to reach the ground. One was near Hudson's Bay in Canada, the other was Tasmania, just off the southeast coast of mainland Australia. Tasmania sounded more agreeable as a place to live and so Reber

---

[53] Over the next thirty years, Research Corporation supported Reber with over $200,000 in grants. Reber balanced his personal budget mainly through extreme frugality and income from sound investments extending back to the 1930s.

[54] Reber:Bolton, 17 January 1950; Pawsey:Reber, 20 March 1950; Reber:Pawsey, 22 April 1950 and 4 July 1950 – all file A1/3/1a, RPS.

continued his wanderings south and west (Reber 1975:86–93T).

Reber teamed up with a young ionospheric physicist in Hobart, George R. Ellis, and in 1955–57 made observations at frequencies over the range 0.5 to 2.1 MHz (Reber and Ellis 1956; Reber 1958b). On those winter nights when the "hole" opened up particularly widely, they found that not only did a strong signal from the Galaxy gush in, but also that the usual prohibitively intense manmade interference ceased – the same hole which let the Galaxy *in* also let the interference and any atmospherics *out*. Thus were they able to measure the intensity of the extraterrestrial sky at frequencies as low as ~1 MHz. During the succeeding solar minima of ~1964 and ~1975 Reber continued with low-frequency observations in Tasmania. In addition he pursued a variety of other topics having nothing to do with radio astronomy, such as the dating of Aboriginal sites in Tasmania. The most notorious was botanical research into whether or not the sense of twining of a bean plant, clockwise or counterclockwise, affected the yield of beans (Reber 1960). His conclusion, after extensive experiments with bean plots in Maui, Tasmania, and West Virginia, was that yield (defined as the ratio of beans to shucks) could indeed be increased – one only needed every few days to manually "unwrap" the bean plants and force them to spiral oppositely to their natural predilection!

4.7 THE REBER PHENOMENON

Grote Reber is an anomaly in the history of the physical sciences in the twentieth century. There must be very few, if any, other cases where someone from outside the scientific establishment has made such important experimental contributions: the technique of the large parabolic reflector, a detailed mapping of the radio intensity and spectrum of the Milky Way, the suggestion of the free–free mechanism for this radiation, and the independent discovery of radio waves from the sun. Reber was not an anomaly because of his expertise in electronics or in designing radio antennas. Many others have also had those skills. Nor was he an anomaly because of various failed experiments and misinterpretations of how his equipment worked or what his measurements meant. Others have also made such mistakes. What made him different, first of all, was his extremely individualistic style, which meant in practice that he could only work well alone, and which led to his boldness in pushing into new avenues of inquiry. Received opinion often branded this unwise:

> I got a lot of advice from the pundits that I was wasting my time. ... But when you started to question these people about the basis of their pessimism, they really didn't have anything. ... They're interested in the things they're interested in, whatever they are. ... When you come and suggest something else, they haven't given it any thought and off-hand it's very different from anything they have ever thought of. ... There's a philosophical point here: if these people thought it was any good, they'd already be doing it!
>
> (Reber 1975:89T)

Wall Street investors would call Reber a contrarian, one who believes that the best action is to go against what the crowd is doing, since by the time the crowd learns of something, profits lie elsewhere. Such a philosophy tends to be speculative, with higher risks of failure and bigger payoffs, and Reber's career demonstrates this aspect: his many forays into uncharted waters yielded only one major success, but it alone was far more than most scientists achieve in a lifetime. Reber won big when he gambled several years' worth of salary and labor on nothing more than Jansky's results and a hunch.

Reber's second outstanding characteristic was that he was the true, unfettered *amateur*, or lover, of scientific experimentation. He was not a professional, in the sense of being paid for his cosmic static work; this work was an avocation, not a vocation, until he went to NBS in 1947. Note that although not at first a professional *astronomer*, he was certainly a professional *radio engineer* in his Wheaton days; it is entirely wrong to refer to Reber, as is often done, as "a radio amateur," implying little more than a weekend tinkerer or at best a "ham" radio operator. His amateur-astronomer status did not make his scientific contributions any less valuable,[55]

---

[55] For example, Lankford (1981:277) has studied the vital role amateurs played in the emergence of astrophysics in the late nineteenth century. He notes that H. Draper, A. A. Common, and J. Franklin-Adams were "true entrepreneurs of science who, given intense intellectual curiosity, talent and wealth, were free to develop instrumentation and pursue research projects on the frontiers of astronomy." Certainly this is also an

although it did lead to problems with acceptance from many professional astronomers (see below) who rather focused on the pejorative sense of "amateur" and ignored or dismissed these strange, un-astronomical results. Williams (2000:chapter 2) has studied the roles of amateurs in twentieth century science and emphasizes that there are two types: the recreational amateur who may know a lot, but does not contribute to science, and the few who, although unpaid, nevertheless fully contribute. Reber started off as the latter type of amateur astronomer, but after he left Wheaton in 1947 he joined the ambiguous status of other investigators of extraterrestrial radio waves of that era: they were not astronomers, nor even "radio astronomers" (the term had not yet been invented – see Section 17.1.1), although they clearly were contributing to astronomy. They were in a nebulous state that would persist for many years (Chapter 17).

These two characteristics, his idiosyncratic boldness and his amateur spirit, coupled with Reber's propitious circumstances in Illinois (Section 4.2), created his success. On the other hand, World War II was ill-timed for Reber's research. It hobbled his investigations when at their zenith, and by the end of the war he no longer had the field to himself. He was faced with the irony that the same war that had slowed him down had in effect trained all of his new overseas competitors.[56]

In considering Reber's interaction with the astronomical community, we encounter many of the same issues discussed previously for Jansky, but with important differences. Because of his assertiveness and greater degree of independence, he was of a mind to, and could afford to, persist in his approaches to the staff at Yerkes Observatory, eventually winning several of them over. He took courses, read the astronomical literature, and attended astronomical meetings in order to learn the appropriate language. And he found in Otto Struve, as well as in his lieutenants Greenstein and Keenan, a basically open-minded person in an influential position in American astronomy:

> In some ways Reber is the kind of person who does not fit into a regulated scientific institution. ... In astrophysics he is almost entirely "self-made", and his ideas are occasionally naive. But considering that he has only had one or two war-time courses in astrophysics..., he has an excellent grasp of many advanced phases of the science.[57]

Struve (1949:27) wrote that "to Grote Reber, more than to anyone else, belongs the credit for having convinced astronomers that here was a new and fruitful field of investigation." While this overstates the case for the astronomical community as a whole (many other examples will be discussed in later chapters), it nevertheless points out Reber's importance in providing a "radio-wave education" for many US astronomers of the time. Reber also planted two seeds at Yerkes that bore fruit for American radio astronomy much later. In the mid-1950s Greenstein would establish radio astronomy at the California Institute of Technology when he brought in Bolton and Gordon Stanley from Australia. And in 1959 Struve became the first director of the National Radio Astronomy Observatory in Green Bank, West Virginia.

Struve's role with regard to support of Reber is remarkable. He had arrived at Yerkes in 1921, a refugee from the White Russian Army and the intellectual heir to three previous generations of Struves who had been directors of Pulkovo and Kharkov Observatories.[58] Indeed, in 1944 he became the fourth in his family to receive the Gold Medal of the Royal Astronomical Society, one of astronomy's premier prizes. Besides his extensive work on stars, he had also long been interested in the distribution of interstellar gas as revealed through optical spectral lines. It is therefore perhaps not surprising that he would pay attention to radio evidence for a hot gas permeating the Milky Way. But it took great courage to publish Reber in the *Astrophysical*

---

excellent description of Reber, and the assimilation of Reber's results by astronomers, as for the nineteenth-century amateurs, suffered from his wholly different background. But in contrast to Lankford's study on (optical) astronomy, amateurs in radio astronomy never were otherwise important and Reber's case remained unique.

[56] Note that although the war moderately hindered Reber's astronomy research, those who would lead radio astronomy after the war were in overseas situations where such a huge effort on cosmic noise during the war was altogether out of the question.

[57] Struve:Shapley, 12 August 1946, Box 227, YKS. I thank David DeVorkin for supplying this letter.

[58] For Struve's career, see the memorial article by K. Krisciunas (1992) in *Biog. Mem. Natl. Acad. Sci.* 61, 351–87; also "Otto Struve: scientist and humanist" by V. Kourganoff (1988), *Sky & Tel.* 75, 379–81.

Journal – one need only peruse other articles in Volume 100 of 1944 to appreciate how utterly foreign was the title "Cosmic static" – topics include observation of stars and theory of stellar interiors and atmospheres by the likes of Chandrasekhar, Swings, the Gaposchkins, Russell, Van Maanen, Nicholson, and Wildt.[59]

Although Struve published Reber and championed him in a search for funding, he himself was not willing to go so far as to supply substantive support. This was despite Struve's recognition that during the postwar era electronics was becoming a vital tool for astronomy (see quotation in Section 17.2.2), and his attempt at one point to obtain funding for a faculty member expert in electronic astronomical devices (DeVorkin 1985:1216). In August 1946 he wrote to Shapley:

> My feeling is that we have here a new process for studying cosmic phenomena which is technically so different from anything that is normally done at an observatory that there is considerable danger that no one will [be] responsible for its development. … It is clear that [Reber] should not be expected to spend his own money on this project. At the same time there are certainly no funds available in our organization to finance the undertaking.[60]

The idea of Reber, say, in a regular position at Yerkes or McDonald was out of the question to Struve. The closest possibility would have been some sort of temporary title if Reber had been able to secure full outside funding. Struve referred to Reber as a "radio engineer" or "radio technician," in contrast with the "astronomers" on his staff. Reber also made this distinction; for example, in a 1945 letter he stated that "at the present state of the art, this work is primarily a radio and not an astronomical investigation." He felt the reverse would be true only when the radio sky could be studied in "minute detail."[61]

Reber thus found himself explaining his work to two entirely different audiences.[62] For example, for

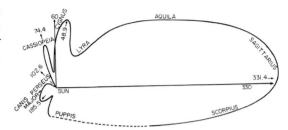

Figure 4.9 A representation of the 160 MHz survey presented by Reber to an engineering audience. The strength of radio emission as a function of longitude in the galactic plane is indicated as a "galactic antenna pattern." Galactic longitudes and constellations are also given. (Reber 1948c)

astronomers he presented his two major surveys in the form of contour maps in astronomical coordinates (Fig. 4.7), but in an engineering magazine he displayed the results in a polar-coordinates format that made the Milky Way look like nothing so much as an antenna pattern with small sidelobes (Fig. 4.9). Again, note the shifts of emphasis in the following quotations from abstracts in the early 1940s, the first for radio engineers and the second for astronomers:

> Cosmic static is defined as electromagnetic radiation which may be detected by radio receiving equipment and which has extraterrestrial origin.
> (Reber 1942:367)

> Cosmic static is a disturbance in nature which manifests itself as electromagnetic energy in the radio spectrum arriving from the sky.
> (Reber 1944:279)

These examples are symptomatic of the difficulties Reber encountered for a decade because there was no niche for someone studying radio waves from the heavens. By the time such a niche began to develop in the late 1940s, his work could be accepted, but now he found incompatability between his own style of science and the organizations increasingly needed to support the burgeoning field. Reber on the worth of organizations:

> I've worked at a variety of places. Usually the higher up one goes, the less anybody knows what is

---

[59] Two other prominent results in this noteworthy volume were Kuiper's detection of methane on Saturn's moon Titan (the first satellite known to have an atmosphere) and Walter Baade's recognition of two basic stellar populations in nearby galaxies.

[60] As in note 57.

[61] Reber:Struve, 23 September 1945, box 221, YKS. See Section 17.1 for more on these semantic distinctions.

[62] Reber (1983:76) later discussed his being caught between the engineers and the astronomers, as well as his contrarian

philosophy. Also see Reber (1958a) for more on his style and situation in the 1940s.

going on. The fellow at the very top sits in an air-conditioned office with a flashy secretary. He reads vacuous reports and lives in an unreal pink haze. If you want to know what is going on, see the janitor. *He* is down where the work is.[63]

One place possibly suitable was the Radiophysics Lab in Sydney (Chapter 7), for although it was a large government laboratory, its style afforded a degree of independence and it supported men from a wide range of academic backgrounds. In fact, while touring the US in 1948 Pawsey, the Lab's chief radio astronomer, observed:

> Reber impressed me very favourably. I forgive him the imperfections in his papers when I consider how he worked, alone with no encouragement, working in his spare time and buying equipment with his own money. He lacks the research background which many of us have, but I believe he "has what it takes" to make a success of things. He is a young bachelor and has a delightfully direct personality. My feeling is that if there is anything we can do to help him along, let us do it. He will give back as much as he gets.[64]

Reber's story has interesting parallels and differences with that of another amateur of astronomy who astonished the establishment of his day, 150 years earlier. This was William Herschel, who metamorphosed from a musician to court astronomer for George III after his discovery of Uranus in 1781. Schaffer (1981) has nicely analyzed Herschel's situation. Like Reber, Herschel for many years had two fulltime careers and little time to sleep. Like Reber, Herschel was not trained in the astronomical techniques of his day and invented a new astronomical practice based on technology (although Herschel's superior telescopes were closer to those of his contemporaries than Reber's radical new "telescope"). Like Reber, Herschel encountered much skepticism and received many delegations wishing to inspect the instruments for which he claimed such powers.[65] Like Reber, Herschel did not at first know how to present his results in an acceptable manner, but soon learned. Unlike Reber, however, Herschel quickly joined the establishment and became one of its leaders in the following decades.[66]

The final irony of Reber's story, like Jansky's (Section 3.7), is that despite his pioneering science and remarkable achievements, he and his work turned out to have limited influence on the flowering of radio astronomy in the postwar decade, especially outside the United States. Before 1946 Reber essentially had the field to himself, but after the war his unique personality and the world's changed circumstances meant that others became the leaders. As we will see in the next seven chapters, the origins and development of the key postwar groups in England and Australia were little connected with Reber or his studies. Although aware of and accepting Reber's (and Jansky's) discoveries, these groups grew instead out of the development of wartime radar and had entirely different social, technical, and intellectual roots (Figs. 1.2 and 1.3). In only a very limited sense were they "following in the footsteps" of Jansky and Reber (Section 17.3.1).

TANGENT 4.1 REBER'S QUOTED ANTENNA PROPERTIES AND DERIVED INTENSITIES

In retrospect we can recognize that the concentrations in Reber's (1944) 160 MHz map (Fig. 4.7) are due to the (later recognized) discrete sources and regions Cas A (see Tangent 6.4), Cyg A plus Cygnus X, Sagittarius A plus the galactic center region, and Vela X (see Table 14.1). In his 480 MHz map are signs of all of the 160 MHz sources except Vela X; in addition a suggestion of Tau A (the Crab nebula) can be seen at exactly the right location (5.5^h, $+22°$).

The best measures now available of Reber's beam are his map of Cas A and the solar drift curves in Reber (1944:286). These yield a half-power beamwidth of 12.5°. Reber, however, quoted a beam size of 8° by 6°

---

[63] Reber:author, 19 July 1983.
[64] Pawsey:E. G. Bowen, 15 April 1948, file A1/3/1b, RPS.
[65] Although many of Reber's visitors went home amazed, none was affected so marvellously as the savant Jérôme Lalande, who wrote to Herschel after a visit in 1788: "Je n'ai jamais passé de nuit plus agréable, sans en excepter celles de l'amour". [quoted by Schaffer (1981:21)]
[66] Another interesting parallel to Reber's case is that of George Ellery Hale, who in his early twenties, also in the Chicago area, operated his own state-of-the-art backyard observatory in the early 1890s, in this case for solar spectroscopy, part of the "new astronomy" of astrophysics. The relation of this new astronomy (as well as Herschel's work) to the rise of radio astronomy is discussed in Section 18.3.2.

in his 1944 paper. This type of error, plus others dealing with incorrect antenna efficiencies, means that one must proceed with caution in interpreting the intensities and angular sizes given in Reber's various papers; for instance, Reber incorrectly thought that he could measure angular sizes (for solar bursts) of ~0.1°, far smaller than his beam (Reber 1948a:88). Sullivan (1982:42–3) discusses Reber's errors in detail, but itself contains errors in the "Bandwidth" line of its Table 1, which should read "1.0 (<0.2), 8.0, 8.0." My best estimates are: (1) Reber's measured 160 MHz intensity of Cas A is ~12,000 Jy, (2) the correction factors to be applied to the 160 and 480 MHz maps of Fig. 4.7 are 0.25 and 0.41, and (3) a quoted value of 1.0×10^{-22} W cm^{-2} (circular degree)$^{-1}$ MHz^{-1} corresponds to 130 K and 24 K of brightness temperature on the 160 and 480 MHz maps, respectively.

TANGENT 4.2 THE WÜRZBURG ANTENNA

The Würzburg precision reflectors developed in Germany during World War II were used extensively by a number of radio astronomy groups for a decade after the war. They were one of the most well constructed and reliable radars of the war. They played a key role in radar defenses, serving (with a three-man crew) primarily for anti-aircraft gun-laying (flak) or as guidance for night fighters. The accurate, punched-hole aluminum surface (made by the Zeppelin firm) allowed operation to wavelengths as short as 10 cm. The standard wartime operating wavelength was 54 cm (560 MHz) and the antenna's feed rotated conically so as to quickly and accurately acquire a target. Sizes of 1.5, 3.0, and 7.5 m were built, the largest (called the *Riese*, or "giant") being highly prized after the war. (The Allies had developed few such large dishes, since their microwave radar was mainly airborne.) The Würzburg Riese was altitude-azimuth mounted, had its own rotating control cabin, and was equipped with slew motors for fast motion and hand cranks for fine adjustment. Groups in England, France, Holland, the Soviet Union, Sweden, and the United States used the Würzburgs for astronomical observations (for photos see Figs. 4.8, 10.12, 14.12 and 16.4). For example, see Van Woerden and Strom (2006:12–16) for detailed information on the suite of Dutch Würzburgs used for scientific research.

TANGENT 4.3 THE FATE OF REBER'S DISH AFTER 1952

Shortly after Reber's departure from NBS, his dish was used as one end of a pioneering moon-relay communications experiment. "What hath God wrought?" was sent at 418 MHz from a 22 ft wide horn antenna in Iowa via the moon to the dish in Virginia.[67] The dish continued to be used for routine observations until the NBS Radio Division moved to Colorado in 1954. It was then dismantled and carted out west, but never used there. Instead, Reber eventually retrieved it and assembled it for the third time and in the third state in 22 years. This time the state was West Virginia and the year was 1960 when it was installed for interference monitoring and historical display at the Green Bank site of the new National Radio Astronomy Observatory, where it can be seen to this day. A stone's throw away is a precise reproduction of Jansky's original antenna, constructed in 1966 under the supervision of his former colleague Al Beck (1983).

---

[67] P. G. Sulzer *et al.* (1952), *Proc. IRE* **40**, 361. Also see note 12.54.

5 · Wartime discovery of the radio sun

A hiatus spanning four decades followed the various attempts at the turn of the century to detect Hertzian waves from the sun (Chapter 2), and in the end giant solar radio bursts forced a serendipitous discovery. Despite tantalizing evidence gathered by both amateurs and professional radio physicists in the late 1930s, not until February 1942 were solar radio waves recognized, by J. S. Hey as he tracked down coastal radar anomalies in embattled England. This was shortly followed by purposeful attempts at higher frequencies in America by G. C. Southworth and Grote Reber,[1] both of whom detected a much weaker signal from the "quiet" sun. Other wartime detections occurred in Germany and New Zealand, the latter significant in spurring initial solar observations at the Radiophysics Lab in Sydney.

All of these discoveries, and indeed postwar radio astronomy as a whole, came as a direct outgrowth of the development of radar under the pressure of World War II (Figs. 1.2 and 1.3). This chapter therefore begins with a brief look at the overall development of radar.[2] Later chapters discuss specific aspects of radar's wartime history as it relates to the various postwar groups, and Section 17.1.1 discusses more generally radio astronomy's origin in wartime radar.

5.1 DEVELOPMENT OF RADAR

Although the basic notion to bounce radio waves off metallic objects at large distances was set forth even in the early 1900s, not until the 1930s were practical systems developed for showing through darkness or clouds the positions, ranges, and speeds of ships and aircraft. These systems required many of the same techniques then being developed for television, for instance, generation of μs pulses and improved cathode ray tubes (Brown 1999:458–9). As with television, no one is *the* inventor of radar: simultaneous and independent programs were carried out in the mid to late 1930s in at least eight countries. In any case, the definition of *radar*[3] is ambiguous. But there *is* general agreement that the effort led in England by Robert Watson Watt from 1935 to 1938 developed the first truly *operational* radar system, by which is meant not just a laboratory testing of an idea or a prototype demonstration, but the production and deployment of an extensive network integrated into military strategy. Watson Watt's work was commissioned and strongly supported by the Air Ministry, which had come to the conclusion that Britain was helpless against an air attack from the growing Luftwaffe. By the start of the war in 1939 there thus existed twenty "Chain Home" stations covering the southern and eastern coast of England. These constituted a floodlight-type system sending out $\sim 10^{-5}$ s pulses with a peak power of a few hundred kilowatts at wavelengths of ~ 12 m. Mounted on 200 ft tall towers were a stacked array of transmitting dipoles and for reception two pairs of crossed horizontal dipoles (at different heights). The elevation angle of any reflections was determined by the ratio of signal strengths from the different receiving heights, and the azimuthal angle by comparing signals from each member of a crossed dipole set as it was rotated. In this way approaching bombers could be picked up out to 50–100 miles with

---

[1] Reber's detection of the radio sun at 160 MHz is covered in Section 4.5.1.
[2] A brief bibliography on the history of early radar is given in Appendix C.
[3] *Radar* is a particularly felicitous acronym coined in the US in 1940 for "radio detection and ranging" – its palindromic nature neatly reflects the nature of the process. The British code name was RDF, for radio direction finding, which was a purposely misleading term since finding the direction of a transmitter was a well-established technique in communications and military work. The term *radiolocation* was also used for a while in England. [Swords 1986:2–3]

a position good enough to deploy Spitfire fighters, as in the crucial Battle of Britain in 1940.[4] With its early start, its strong physics community, and the urgency of its nearby enemy, Britain remained in the forefront of radar development throughout the war. The emergence in England after the war of three of the four leading groups in radio and radar astronomy was no accident.[5]

In America, the Army and Navy also undertook radar development in the late 1930s. There was extremely close cooperation among the Allies, as initiated and exemplified by the British "gift" in 1940 of the newly invented cavity magnetron, a revolutionary source of microwave power for transmitters, fully one thousand times stronger than anything previous. By the end of the war the US had created a huge radar effort ($3 billion spent on about one million equipments), but this did not lead to a matching amount of important radio astronomy; this seeming puzzle is discussed in Section 17.3.[6]

The Germans too had a strong radar capability at the start of the war, and in fact were well ahead of the Allies in use of higher frequencies (up to ~600 MHz), although behind in integrated operational systems (Beyerchen 1998). During the early part of the war they chose not to commit their scientific and engineering talent to further major development, since radar, viewed primarily as a defensive weapon, was deemed incompatible with their offensive posture. Despite major efforts once this policy was reversed in 1943–44, their systems remained in general far less effective than those of the Allies. After the war radio astronomy, or indeed any kind of research with radio, was impossible for any German researcher, *verboten* by the occupying authorities until ~1950. Ironically, however, captured German equipment, in particular components requiring precise mechanical tolerances, was extremely helpful to many of the early radio astronomy groups in England, France, the Netherlands, and America – the best examples were the excellent large dishes (7.5 m diameter) known as Würzburgs (Tangent 4.2).

Once the secret of radar was revealed to the British Dominions of Australia, New Zealand, South Africa, and Canada (Section 10.2), they too set up radar laboratories. The Radiophysics Laboratory in Sydney remained intact after the war and became the largest postwar group in radio astronomy, led by the same E. G. Bowen who had been one of the initial members of Watson Watt's team; Section 7.1.2 describes the Australian wartime radar effort. In other countries such as the Soviet Union, Japan, and France (until 1940), smaller radar efforts existed and the resultant radio astronomy is discussed in Chapter 10. Finally, there is the anomalous example of the Netherlands: nothing significant in radar happened before the war and nothing *could* happen during the war, and yet there were strong efforts, albeit unsuccessful until 1951, to study extraterrestrial radio noise. Here the impetus came from an astronomer, J. H. Oort, and this story is told in Chapter 16.

5.2 HEY'S DISCOVERY

After early years in Lancashire, James Stanley Hey (1909–2000) (Fig. 5.1) studied physics at Manchester University, obtained a Master's degree in 1931 in X-ray crystallography, and went into teaching. After the outbreak of war he learned the principles of radar through a six-week course in 1940 led by J. A. Ratcliffe, the leading radio physicist at Cambridge University (Chapter 8). He then joined the Army Operational Research Group (AORG), one of several elite British military units based on the relatively new concept of "operational research."[7] Hey led a group of ten scientists in Petersham (near London) that advised on the performance and field problems of anti-aircraft radar systems. In early 1942 they were focusing on ways to counteract the increasing ability of German stations on the northern French coast to "jam" the protective chain of radars on the home side of the English Channel. The Army particularly feared that airborne

---

[4] For an overview of the Chain Home system, see B. T. Neale (1985), "CH – the first operational radar," *GEC J. of Research* **3**, 73–83.

[5] Further information on British radar development during the war is given at the starts of Chapters 6, 8 and 9.

[6] More information on US radar development during the war is given in Sections 10.1 and 11.4

[7] The concept of operational research, largely born with British radar, referred to scientists conducting field investigations of military operations, with the goal of improving the conduct and efficiency of those operations. For more on AORG, see J. D. Cockcroft in *IEE Proc.* **A132**, 338–9 (1985). Also lecturing in Hey's introductory course were two future Nobelists in Physics, N. F. Mott and P. M. S. Blackett (Chapter 9).

Hey's discovery

Figure 5.1 J. Stanley Hey (photo at Jodrell Bank, 1953).

jamming during bombing attacks might effectively neutralize anti-aircraft radars. Matters came to a head when on 12 February 1942 the German battle-cruisers *Scharnhorst* and *Gneisenau*, aided by radar jamming, successfully slipped through the English Channel into the North Sea. *The Times* of London called it "the most mortifying episode in our naval history since the Dutch got inside the Thames in the seventeenth century."

The vexed British redoubled their efforts to counter jamming, AORG was charged to investigate, and Hey set up a system for operators to report details of all suspected jamming incidents. Almost immediately a remarkable event was reported. On 27 and 28 February,[8] about ten anti-aircraft gun-laying radars (GL Mark II; Fig. 5.2), widely scattered around the coast (Fig. 5.3), suffered excessive noise-like interference that could not be "tuned out," that is, it occurred over the entire 55–85 MHz range of operating frequencies. Although there were no enemy raids on those days, this apparent new jamming capability was alarming. Perplexed radar operators could do little but watch their cathode-ray-tube displays overload from the excessive noise.

Checking the reports, Hey found that the interference on both days came almost continuously and exclusively between dawn and sunset. By swinging their antennas (small arrays of half-wave dipoles) around in azimuth, stations were also able to determine the azimuth from which the peak interference originated. When Hey compared all reported azimuths with those of the sun over the day, the agreement was striking (Fig. 5.3). Although most of the units only measured azimuths, those at Dover and Hull also succeeded in getting good elevation angles for the peak interference: indeed, their sighting telescopes then revealed the sun to be dead "on target." The clincher came when a check with the Royal Greenwich Observatory disclosed that a huge sunspot, one of the largest ever recorded, had crossed the solar meridian on 28 February and had been the seat of an exceptionally intense chromospheric eruption (flare) lasting for several hours in the early afternoon.[9] This in turn had led to a long-lived sudden ionospheric fade-out (Section 5.4.3) and to a magnetic storm on the following morning.[10]

Hey (1942) wrote a secret report[11] (Fig. 5.4) within days:

> The interference appeared on the C.R.T. [cathode-ray tube] in the form of a very high noise level in all cases except Site 13 (35th Brigade) where

---

[8] Coincidentally it was also on 27–28 February 1942 that the British staged a celebrated commando raid against a radar station at Bruneval on the French coast, successfully bringing back large parts of a Würzburg radar set (Brown [1999:229–31]; also see Tangent 4.2).

[9] H. W. Newton, *Observatory* **64**, 260–4 (1942); F. J. M. Stratton, *Nature* **157**, 48 (1946).

[10] This solar event was also important for cosmic ray research. Scott E. Forbush recorded an unprecedented sudden increase in cosmic rays on 28 February 1942 and suggested that these might have been "actually emitted by the sun." [*Phys. Rev.* **70**, 771–2 (1946)]

[11] Hey, March 1942, "Notes on G.L. interference on 27th and 28th February" (Secret, 5 pp.), file AVIA 7/3544 (6486), Public Record Office, Kew, England; cited after the war as "ADRDE (ORG) Memo J4." For ten years I searched for this report, and I am indebted to T. R. Padfield, Assistant Keeper, who in the end located it for me. Copies of this, as well as Hey (1945), have now been placed in the Radiophysics Library, Sydney, and the Jodrell Bank Library. Hey's handwritten version of this report is available in the Science Museum, London, Item No. 2001–749.

interference of pulse form was observed ... The whole G.L. [Gun Laying] frequency band appears to have been covered by the interference and a considerable portion of the sets were rendered inoperative ... There was no enemy activity during this time.

The intensity of the interference varied at different times. It was only present during the day on February 27th and 28th, and there has been no subsequent recurrence.

The bearing [azimuth] of the interference provides what appears to be at present the only clue to its origin ... [Reported bearings from four sites] gradually changed and followed the course of the sun.

Although no explanation of how the interference was caused can be given, it does not appear possible to arrive at any other conclusion ... than that the interference was associated with the sun and the recent occurrence of sun spots.

Although the report does not explicitly mention electromagnetic radiation directly from the sun, Hey (1973:16) later recalled that it seemed eminently

Figure 5.2 A gun-laying Mark II radar receiving set similar to the ones that detected the sun in February 1942 along the English coast; a full system consisted of this receiver, a transmitter unit, and a generator. A rotating cabin holds two dipoles (not visible) to the left and right for azimuth determination, as well as a dipole above (at the location of the highest man) that was adjustable in height in order to vary the position of the vertical lobes for better altitude angle determination. On the ground to the left one can see short supporting pickets for a large ground screen that was also often deployed.

5.3 Other wartime incidents with the sun

of the war no additional incidents came to light in England.[13]

Three years later as the war was winding down, Hey mentioned at a radio committee meeting that he intended to publish his 1942 findings in *Nature*. The head of this committee was Edward V. Appleton (1892–1965), the renowned radio physicist who was to win the 1947 Nobel Prize in Physics for his theoretical and experimental work on the structure of the ionosphere (Clark 1971). Hey later recalled that Appleton was astonished at the news of Hey's solar radio waves and requested that he work up his results for presentation at a later meeting.[14] This led to a widely circulated, more formal report in which Hey (1945) concluded that "the noise radiations were emitted by the sun." And finally, once wartime secrecy subsided, he submitted this fundamental result, "Solar radiations in the 4–6 metre radio wave-length band," to *Nature* in late 1945 (Hey 1946). In this he further pointed out that the estimated intensity of the solar signal was an astounding 10^5 times that expected from a blackbody at a temperature of 6000 K.[15,16]

5.3 OTHER WARTIME INCIDENTS WITH THE SUN

5.3.1 Schott in Germany

After reading the many reports on solar radio work in *Nature*, a German named E. Schott (1947) wrote a short note to *Physikalische Blätter* describing his experiences as a radar operator. The following account is

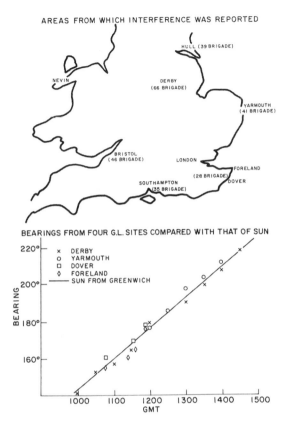

Figure 5.3 The first definite evidence for radio emission from the sun, as illustrated in a restricted report by Hey (1945); the map is very similar to one appearing in the original report (Hey 1942), while the data of the graph were originally only in tabular form. The azimuths (bearings) of the 55 to 85 MHz interference on 27–28 February 1942, as reported by coastal gun-laying radars, agreed remarkably well with those of the sun.

reasonable to him at the time that charged particles moving in the intense magnetic fields of sunspots could generate radio waves not unlike the manner in which a manmade magnetron copiously produced microwaves. This 1942 report had an extremely limited circulation, initially only six copies, although word of its contents spread widely in radar circles as the war wore on. Hey and his AORG colleagues[12] later recalled, however, that several radio scientists gave it a skeptical reception, for they wondered how such a striking effect could have so long lain unnoticed by radio researchers. Furthermore, over the remainder

---

[12] S. J. Parsons (1978:4T) and G. S. Stewart:author, 4 September 1985.

[13] The period 1942–44 was a minimum in the 11-year solar activity cycle (Fig. 13.11). If World War II had occurred during a sunspot maximum period, there would have undoubtedly been far more accidental radar discoveries of the sun.

[14] Hey (1973:18) and Hey:author, 21 January 1985; also see Clark (1971:154).

[15] Sources for the 1942 solar discovery are various documents and accounts of Hey (1942, 1945, 1946, 1973:14–19, Hey:author, 21 January 1985), Appleton and Hey (1946:75–6); and G. S. Stewart:author, 4 September and 2 October 1985. Also see E. Eastwood:Appleton, 27 November and 4 December 1945, file C.335, APP. Hey (1992) also published a short, informative autobiography, *The Secret Man*, which contains many details of his wartime radar investigations.

[16] Priority disputes regarding the radio sun between Hey, Appleton and others are discussed in Section 5.5.

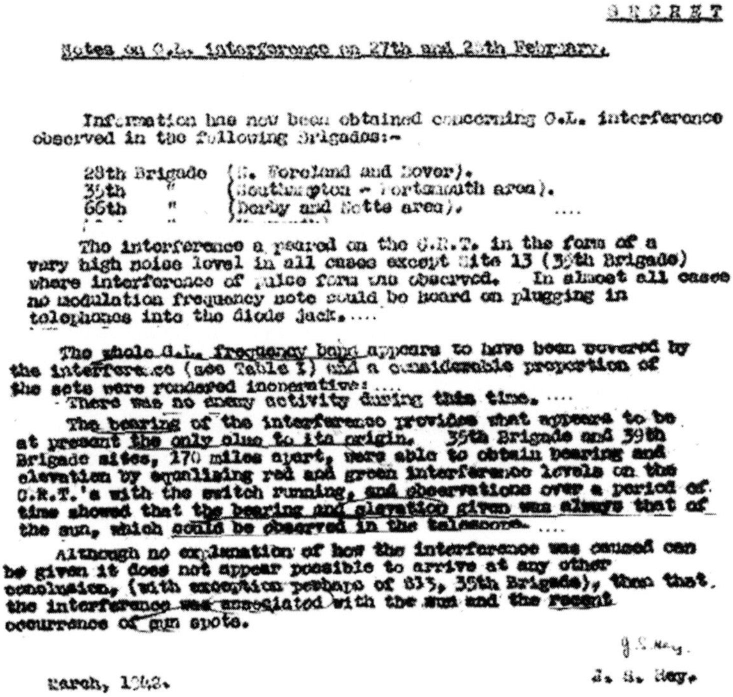

SECRET

Notes on G.L. interference on 27th and 28th February.

Information has now been obtained concerning G.L. interference observed in the following Brigades:-

28th Brigade (N. Foreland and Dover).
35th " (Southampton - Portsmouth area).
66th " (Derby and Notts area). ....

The interference appeared on the C.R.T. in the form of a very high noise level in all cases except Site 13 (35th Brigade) where interference of pulse form was observed. In almost all cases no modulation frequency note could be heard on plugging in telephones into the diode jack. ....

The whole G.L. frequency band appears to have been covered by the interference (see Table I) and a considerable proportion of the sets were rendered inoperative: ....

There was no enemy activity during this time. ....

The bearing of the interference provides what appears to be at present the only clue to its origin. 35th Brigade and 39th Brigade sites, 170 miles apart, were able to obtain bearing and elevation by equalising red and green interference levels on the C.R.T.'s with the switch running, and observations over a period of time showed that the bearing and elevation given was always that of the sun, which could be observed in the telescope. ....

Although no explanation of how the interference was caused can be given it does not appear possible to arrive at any other conclusion, (with exception perhaps of 833, 35th Brigade), than that the interference was associated with the sun and the recent occurrence of sun spots.

J. S. Hey.

J. S. Hey.

March, 1942.

Figure 5.4 A mosaic image of portions of the original report in which Hey (1942) described the interference of 27–28 February 1942 and his conclusion that it originated with the sun. Much of this text is quoted in Section 5.2. Ellipsis dots indicate omitted text in this image; hand underlining seen here is not on the original. The original document contains only Hey's printed name; his signature has been added as taken from a document written two weeks later.

based solely on this note. In the winter of 1939 Schott apparently first noticed solar radiation at a frequency of 175 MHz, but it was during May and June 1940 that he made his main observations, on a frequency of 125 MHz at a coastal radar site in northern Denmark at Skagen. His antenna, a standard "Freya" air surveillance radar, consisted of six vertical dipoles steerable in azimuth and had a beam size of about 20° by 6°. The solar disturbances were as much as "six to eight times normal receiver noise" and completely disabled the radar set. They occurred most strongly when the sun passed through the lobes of the antenna pattern while rising or setting, and were variable from day to day. In addition, the excess noise was observed in all types of weather, and never at night. Schott further stated that over the following years "we also made observations at the other wavelengths used by German equipment (1.50 m, 80 cm, 50 cm, 9 cm, and 3 cm)." Although these later observations apparently were very brief, he did say that in the spring of 1945 the solar effect was observed several times at 9 cm wavelength.[17]

5.3.2 Alexander in New Zealand

Another wartime observation of the sun by radar installations occurred in the South Pacific, and in late 1945

[17] Also in Germany about 1942–43, Kurt Fränz, a radar development engineer who at this time briefly published galactic noise observations at 30 MHz (Tangent 6.1), used a 3 m diameter Würzburg dish near Berlin in an attempt to detect the sun at 50 cm wavelength, but was unsuccessful. He also recalled hearing a report from a German nightfighter pilot concerning radio interference from the sun [Fränz (1973:1T) and Fränz:author, 28 March 1985]. In addition, Horst Wille (Section 10.6.1) recalled seeing a German report, sometime during the 1940–42 period, describing an accidental detection by a Würzburg dish of solar radio waves [Wille:author, 25 April 1979]. Today's story of the radio sun's discovery would undoubtedly have a distinctly different slant if the outcome of World War II had been different.

initiated Australian scientists' involvement in solar radio research. The first report came from a tiny island about 1000 miles northeast of Sydney and the phenomenon thereafter was eponymously dubbed the "Norfolk Island effect." The commander of a New Zealand Air Force radar station reported that over the period 27–31 March 1945 his operators observed a large increase in 200 MHz noise on their "range tubes" for about 30 minutes as the sun either rose or set. The noise level was maximum at the azimuth of the sun and varied with time as expected for the antenna's beam. The astonished New Zealanders seemed to be observing what several ancient cultures said of the sun, namely that it made a mighty hissing noise when it nightly descended into the sea.

This report was referred to Director Ernest Marsden of the Radio Development Laboratory (RDL) of the Department of Scientific and Industrial Research in Wellington, New Zealand.[18] He assigned F. E. S. (Elizabeth) Alexander (1908–1958), leader of an Operations Research Section, to investigate the effect. Alexander had a Ph.D. in geology from Cambridge, and had been working on radar for the entire war. She coordinated observations over a ten-day period in April at the Norfolk Island site and four similar stations on the northern coast of New Zealand. Each station managed to observe the effect at least a couple of times, although sometimes only slightly, and in general they located the azimuth of peak noise very close to the sun's position. But there were nagging exceptions where the noise peak seemed to be well away from the sun, or where the weather or local terrain seemed to influence its strength. And why was the effect never observed once the sun was more than 8° above the horizon? And why had the radar stations never previously seen the effect?

The measurements in April were crude. Only one station used as much as a meter; the rest simply eyeball-estimated the amount of increase in the "animated grass" on a cathode-ray-tube display (increases of anywhere from 0.5 to 2 inches were reported!). But in her report Alexander (1945) was reasonably confident that the effect was real, that it represented radio radiation emitted by the sun at a level far above that expected from blackbody theory, and that it deserved further study.

Alexander next arranged for a more quantitative watch, and all stations received calibrated signal generators and voltmeters.[19] This monitoring, which began in September 1945, revealed a further period of excessive noise around 5 October. By this time, however, Joseph. Pawsey and his group at the Radiophysics Division in Sydney had gotten underway with their own solar observations using similar equipment. Indeed, their initial impetus had come in July from Marsden's and Alexander's news of the Norfolk Island effect (Section 7.3.1). By December both Alexander and Pawsey were convinced of a correlation between sunspots and the radio noise. But unlike Pawsey and his colleagues, Alexander never published any results in the scientific literature and the New Zealand effort ended with the shutdown of all radar sites and the closing of RDL at the end of 1945.[20, 21]

5.4 PREWAR OBSERVATIONS

Although Hey is fairly credited with first establishing that solar radio waves impinge on earth, already during the previous solar activity maximum around 1937 (Fig. 13.11) we can recognize in retrospect that radio

---

[18] RDL was the New Zealand equivalent of Australia's Radiophysics Division (Section 7.2) for radar research and development, although much smaller. [R. Galbreath (2000), "New Zealand scientists in action: the Radio Development Laboratory and the Pacific war," pp. 211–27 in *Science and the Pacific War* (ed. R. M. MacLeod) (Dordrecht: Kluwer)]

[19] Alexander was encouraged to study the solar hiss by Appleton, who learned of the effect when the New Zealanders replied to his general request of June 1945 asking about any observations of solar hiss or related phenomena (Section 5.5). The New Zealand work, however, was *not* cited in Appleton (1945a).

[20] In a separate one-page report, Alexander also mentioned that two American-operated radar stations had picked up excess noise on 26 March 1945 (the same period of activity as for Norfolk Island, but curiously one day earlier). The noise corresponded to the setting sun's position. These stations were on the coast of Lingayen Gulf (northern Phillipines) and near Darwin, Australia. [F. E. S. Alexander (17 December 1945), "Long wave solar radiation," DSIR/RDL report, 1 page, file C.344, APP]

[21] My primary source for the story of the Norfolk Island effect is the restricted 1945 report by Alexander, found in file C.344, APP. Relevant correspondence of Bowen and Pawsey is in file A1/3/1a, RPS. A few facts in this account are from Alexander (1946) and from Orchiston (2005a), which contains many more details of Alexander's life and this episode.

amateurs and professionals around the world were frequently encountering solar noise. No one convincingly interpreted this noise as direct solar radio waves, but at least one case came within a hairbreadth.

5.4.1 (Brief) purposeful tries

In the forty years between Charles Nordmann's adventure on Mont Blanc and Stanley Hey's discovery, there are only three known purposeful attempts to detect radio waves from the sun. Besides Reber in 1939 (Section 4.2), the others took place in America (1933) and in Australia (1939).

In the Physics Department at the University of Michigan Arthur Adel and John D. Kraus were both just obtaining their Ph.D. degrees, respectively on the infrared spectrum of carbon dioxide and on the propagation of 60 MHz waves. Another student finishing up his thesis work was Claude E. Cleeton, who had painstakingly built several vacuum-tube oscillators (magnetrons) that radiated at wavelengths as short as 1 cm, shorter than ever before. He and his advisor Neil H. Williams then used these to measure the wavelength of a predicted absorption line of the ammonia molecule at 1.25 cm (Cleeton and Williams 1934) – an elegant experiment that was the first in microwave spectroscopy. Adel had the idea to use this equipment for an attempt at detecting radio waves that he thought might be emitted by sunspots, with their strong magnetic fields and probable energetic charged particles. With Kraus providing the radio expertise, they used a 3 ft brass paraboloid (a World War I searchlight) as an antenna and no more than an iron-pyrite phosphor-bronze crystal and galvanometer as a receiver. But 1933 was at the minimum in the solar activity cycle and, even if they had been lucky and caught an outburst, we now know that the sun at 1 or 2 cm emits only a trace of what it does at meter wavelengths. In a word (or four): they had no chance. Nothing was ever published, but this was the first astronomical experiment for Adel and Kraus, who would both later conduct many more.[22]

A second attempt was made in 1939 by Jack H. Piddington and David F. Martyn of the Radiophysics Laboratory, Sydney. Both men were involved in ionospheric research using various antennas at Liverpool, a field site near Sydney. They had worked out that a 6000 K blackbody such as the sun was probably not detectable, but they decided to speculate anyway with a brief experiment. Planck's relation indicated that the highest possible frequency should be used, and that meant for them a 60 MHz array of half-wave dipoles. The array had some directionality and its main beam was centered on the zenith. The summer sun over Sydney comes within $\sim 10°$ of the zenith, and so they looked for any increase in receiver noise at noontime on a few summer days. They might have been able to detect the 10^6 K quiet corona with an extensive observational effort, and of course a solar burst would have been startlingly obvious (1939 was ~ 2 years after sunspot maximum), but they found nothing in their quick look, nor did they publish any results.[23] Notwithstanding this modest beginning, both Piddington and Martyn later made important contributions to postwar radio astronomy (Chapters 7 and 13).

5.4.2 Heightman and the radio amateurs

Denis W. Heightman (1911–1984) graduated from technical high school in 1927 and over the years worked his way up as an engineer in the British radio and television industry. As a teenager he obtained the radio amateur call letters G6DH and for the following decades operated out of his home in Clacton-on-Sea, Essex, 50 miles northeast of London. He soon became fascinated with operating on the shortest possible wavelength, which was the 10 m band in 1930 and the 5 m band starting in the mid-1930s. At that time no more than twenty British amateurs worked on such high frequencies, where equipment was necessarily all home-built. Heightman's receiver used acorn tubes and his antenna (Fig. 5.5) was a rotatable Yagi, which allowed a modicum of directionality, perhaps one-third of the sky, as opposed to the usual omnidirectional long-wire antenna. His interests ranged far beyond chats with fellow "hams," and over the years he undertook several methodical investigations of shortwave and ultra-shortwave propagation phenomena. In particular Heightman

---

[22] Adel later did some of the earliest infrared astronomy observations (Sinton 1986) and Kraus led a radio astronomy group at Ohio State University in the 1950s (Section 10.6.5). Sources for the Adel–Kraus story are Adel:author, 8 October 1981 and Kraus (1975:1–3T, 1976:21–2).

[23] Piddington (1978:1–4T) is the sole source for this episode.

Figure 5.5 Denis Heightman in his radio room in 1938, and his 10 m wavelength rotary "beam" antenna in Clacton-on-Sea, Essex. Heightman studied the properties of the hissing sound often accompanying shortwave fade-outs during the late-1930s solar maximum.

contributed to the work of the Radio Society of Great Britain and its Research and Experimental Section. In early 1937 he reported on his 28 MHz monitoring in the Society's magazine *T & R Bulletin*:

> A strange phenomenon, first observed by the writer in late 1935, was the appearance, at irregular times, of a radiation which took the form of a smooth hissing sound, when listened to on a receiver. It was pointed out by G2YL [Nell Corry] that on the days when hiss was heard that there had frequently been fade-outs or poor conditions on the high frequencies … On one day the hiss may last for only a few seconds and not be heard again that day. On other occasions it will last as long as five minutes, then disappear and reappear again within a few minutes, repeating the process several times. It invariably starts at a weak strength and gradually builds up to a maximum, then gradually fades away again. The phenomenon apparently originates on the sun, since it has only been heard during daylight, and it has been suggested that it is caused by a stream of particles shot off from the sun during abnormal activity.
>
> (Heightman 1937:497)

Heightman (1936, 1938) also mentioned the "hiss phenomenon" to a wider audience in both a letter and an article in the leading British radio journal *Wireless World*.

Besides these public notices Heightman briefly corresponded on the topic with Appleton. In a June 1936 letter Heightman described the phenomenon and asked if Appleton had any ideas as to its cause. Heightman stated that he first thought it was "some strange fault in my receiver," yet was still not convinced of an atmospheric effect. He noted that the hiss sound was similar to the "fizzlies" experienced when charged raindrops fell on an antenna, although not so staccato, and he wondered if the hiss might be similarly caused by a shower of charged particles. Appleton replied with interest that he had tried to check for any other peculiarities on the specific days when Heightman had heard the hiss, but so far had turned up nothing. This correspondence ended with two further letters where Heightman gave more details, in particular his inference that the "source" was the sun (from the fact that the hiss only occurred in daytime) and a correlation in one case with a solar eruption. Furthermore, he now saw a good correlation between the hiss and shortwave fade-outs (see below), in which signals suddenly disappeared and gradually returned after up to an hour. In the spring of 1937 Heightman sent a letter to *Wireless World* regarding his idea that both the hiss and fade-outs were possibly caused by solar particles entering the earth's atmosphere.

Another English radio amateur of this era, Nelly Corry (G2YL), coordinated reports and kept detailed logs for the 28 MHz amateurs. At the same time as Heightman, she also commented on the hiss:

> This takes the form of a sudden increase of background noise over the whole band, and usually lasts from ten seconds to two or three minutes … It has been heard on … 56 Mc., as well as 7, 14, and 28 Mc., but it is usually strongest around 28 Mc. It is definitely due to some form of intense solar activity, and on at least seven occasions has been heard on the same day as a Dellinger Fade-out.
>
> (Corry 1937:513).

Ham (1975) has analyzed Corry's compilations and over the 1936–39 period finds that fully 24 amateurs in Britain, America, India, and South Africa reported the hiss phenomenon on 107 different days, on frequencies ranging from 7 to 56 MHz. As it became clear to the amateurs that there existed a good correlation with radio fade-outs and solar activity, Eric J. Williams (G2XC) conducted a detailed study (1939) of other possible solar-terrestrial relations and concluded that the hiss was closely correlated with chromospheric eruptions. But it was not a one-to-one relation, for although the hiss would commonly precede a radio fade-out and happen within an hour of a solar event, there were also many cases where this was not true: a radio fade-out or a solar flare with no hiss, hiss without any fade-out, or hiss not coinciding with the commencement of a fade-out, but only occurring on the same day. Heightman also looked into the solar data, and at various times wrote for information to the Astronomer Royal at Greenwich and to Robert S. Richardson at Mt. Wilson Observatory. In his reply of 19 May 1938 Richardson expressed interest and a limited knowledge of the phenomenon, but tellingly and ironically (to us today) concluded that "I am not familiar with the radio end of this investigation myself, since our work has only to do with the sun."

In the first half of 1938 Heightman's correspondence indicates that he began to lean towards the hiss as resulting from "a radiation of some form." First, in December 1937 he had obtained solid evidence that at least one hiss event was heard within minutes on both sides of the Atlantic – how could a shower of particles arrive nearly simultaneously at widely separated points? Second, observations indicated that the effect was strongest at frequencies of ~25–30 MHz and much weaker (usually absent) below 10 MHz – should not a particle shower be equally audible at all frequencies (as were the fizzlies? These ideas, however, were never further developed or published, and Heightman never was able to establish the hiss as direct electromagnetic radiation from the sun.[24]

In the end, many characteristics of the hiss were established, but interpretation of the evidence at hand was ambivalent. Reception of the hiss was often localized to only a portion of England, and this argued that it was not a global phenomenon. Heightman and his amateur colleagues were never quite sure whether the hiss was a secondary or a primary effect of the sun. Seeking correlations in *time* was suggestive but not convincing. What was critically needed was a highly directional antenna that could pinpoint the *direction* of arrival of the hiss – this is what Karl Jansky had had five years earlier and what Hey would have five years later. Such a large antenna was beyond the resources of amateurs. Moreover, their amateur status

---

[24] Sources for the Heightman story are a 1981 interview and photocopies of his correspondence that he kindly provided me – Heightman (DH):Appleton, 22 June 1936; Appleton:DH, 11 July 1936; DH:Appleton, 11 November 1936; DH:editor of *Wireless World*; 28 April 1937 (published in Vol. 40, p. 458 [7 May 1937]); DH:J. H. Dellinger, 21 February 1938; DH:Richardson, 21 April 1938; Richardson:DH, 19 May 1938. Most of these, plus DH:Appleton, 22 July 1936, are also in file C.337, APP.

undoubtedly contributed to the fact that no professional adjudged their results of sufficient reliability and interest to follow up. The radio amateurs also did not have the background in physics needed to make rough calculations of an expected solar intensity at shortwaves or of an inferred temperature on the supposition that the hiss was direct electromagnetic radiation. On the other hand, as we will now see, many radio professionals who *did* know the requisite physics also knew this phenomenon full well and yet likewise missed the opportunity.

5.4.3 Professional radio physicists and ionospheric activity

In 1937 J. Howard Dellinger of the US National Bureau of Standards published a major study in which he presented voluminous evidence for the existence of "sudden ionospheric disturbances," also known as shortwave (or radio) fade-outs. He had compiled material on more than a hundred instances where shortwave communications, always on the day-side of the earth, had been suddenly and absolutely cut off, gradually returning after ten minutes to an hour. The fade-outs were found to be well correlated with variations in the earth's magnetic field, earth currents, radio atmospherics at ~30 kHz frequency, and solar eruptions. And furthermore "the radio fade-out is on some occasions preceded by a short period of unusually violent fading, echoes, and noise (of a type different from atmospherics). ..." (Dellinger 1937:1271). This noise clearly seems today the same phenomenon as Heightman's "hiss." Dellinger said no more about it, but instead discussed an explanation for the fade-out itself, namely that the sun emitted an intense, high-energy radiation that ionized an atmospheric layer lower than the usual ionosphere at ~100 km altitude. This short-lived "D layer" was then taken to completely absorb all shortwaves.

Another near-miss occurred as a result of a study in Tokyo by Minoru Nakagami and Kenichi Miya (1939), members of a radio propagation research group in the International Telecommunications Company. They had become interested in the effects of fade-outs as reported by Dellinger and their countryman Daitaro Arakawa (1936), who also reported excess noise often accompanying fade-outs. Working at ~15 MHz, they measured the elevation angle of arrival of various distant signals during a fade-out by comparing received power between two horizontal half-wave dipoles, one 1/2 and the other 5/4 wavelength above the ground. Altogether they caught five fade-outs over a year, including one case where intense noise also occurred. Relative levels of the noise on the two dipoles indicated an elevation angle of ~70–80°. Since the time was about 10 a.m. on 1 August 1938, the sun was high in the sky and Miya, who was observing, thought there was a definite possibility that he was detecting direct radiation from the sun. His senior Nakagami, however, was more cautious in his interpretation and the resultant article said no more than that they believed the noise was not local, but probably originated in the ionosphere's E layer.[25]

Dellinger's study also stirred up interest among solar astronomers and several such as Richardson (mentioned above) examined correlations between events on the sun and those in the earth's ionosphere. This continued a long tradition of studying solar activity and terrestrial effects such as aurorae and deviations of the compass needle; as early as 1872, for instance, C. A. Young had demonstrated the simultaneity of two solar flares and strong disturbances in the earth's magnetic field (Young 1895:166–70).

Prior to the late 1930s, radio communications had experienced three solar maxima, but solar radiation simply did not penetrate the ionosphere at the low frequencies previously used (see Section 2.6). The solar maximum of 1937 was the first with any significant amount of traffic in shortwaves and therefore a decent chance for detection and identification of solar radio waves.[26] Yet this did not happen. Even though radio specialists often had the sun on their minds, they too did not have the directional antennas needed to pinpoint the sun, and they tended to think in terms of indirect or particle effects. And although solar astronomers in turn often also had radio on *their* minds, they knew insufficient radio physics to pay attention to the hiss as anything more than an ancillary phenomenon.

---

[25] Source for the Nakagami–Miya story is Miya as told to Tanaka (1984:335–7) and Miya:author, 25 January 1985.

[26] It has been said that other radio amateurs, even as early as the solar maximum of 1927–9, sometimes heard a strange hiss on their shortwave sets, but I have found no specific citations other than those mentioned in this chapter. I would be grateful for full information on any such occurrences with which readers are familiar.

5.5 CONTROVERSY WITH APPLETON

In September 1945, three weeks before Hey's letter with the announcement of his 1942 radio detection of the sun was submitted to *Nature*, Appleton (1945a) sent his own short note to the same journal entitled "Departure of long-wave solar radiation from blackbody intensity." In this he first referred to the data of Reber and Southworth (see below) that apparently confirmed the expected blackbody radio radiation, and then pointed out that there was also good evidence for the sun occasionally emitting greatly enhanced radiation. He credited Heightman and other amateurs for sending him excellent reports on this hiss during the previous solar maximum, and said that from these "I concluded that the noise was due to the emission of *electromagnetic* radiation from active areas on the sun" (italics in original). He then went on to calculate that in order for the hiss to have been noticeable, an active area on the sun must have radiated $\sim 10^4$ times as much as normal blackbody emission at a temperature of 6000 K. No where in the article was there a mention of Stanley Hey.

It was an important realization that the meter-wavelength radio noise implied that the sun was not at all behaving in the expected manner, but Appleton's attempt to establish priority for himself in the discovery of solar radio waves belies the facts. As recounted above, Heightman (and undoubtedly others, for Appleton acted as a clearinghouse for ionospheric reports from both amateurs and professionals) had indeed told Appleton all about the hiss. But Appleton was a master at playing the scientific game – if he had deduced the true meaning of the hiss with any certainty in 1936, one can be sure that *Nature*'s readers would have been promptly informed of such a fundamental discovery.

The evidence of Heightman's and Appleton's papers indicates the following. Appleton apparently did have the hiss on his mind during the war, for in July 1942 a BBC engineer sent him information on a hiss event at 15 MHz, in response to a recent expression by Appleton of interest in the hiss. Perhaps his curiosity at this time, six years after contact with Heightman, had been rekindled by reading Hey's 1942 report. But, as previously mentioned, Hey recalls that Appleton was "astonished" when informed at a 1945 meeting that Hey was planning to publish his 1942 results in the open literature. Somehow, despite his prominent position in British science and its application to the war effort, as well as his central place in radio research, Appleton's actions imply that he had never heard of the 1942 incident. Be that as it may, in June 1945 Appleton asked Heightman (and others around the world)[27] for any available information on the hiss, whereupon Heightman sent him details of the many events he had observed over 1936–39, as well as his correspondence on the subject.

Appleton digested all this and worked on the problem more quantitatively, but the archival evidence indicates that what finally appeared in *Nature* was in fact largely not his own work, but that of Ronald E. Burgess (1917–1977), an expert on radio noise in antenna systems. In the Appleton papers is a two-page document dated 19 September 1945 (five days before his submittal to *Nature*) and titled "Comments by R. E. Burgess on note entitled 'Calculation of solar noise developed in radio system' by Sir Edward Appleton." It begins:

> The method of calculating the radiation flux at the earth must be carried out somewhat differently, for the volume of the sun is not the relevant quantity in determining its radiation.

It thus appears that Appleton's original note (unfortunately not extant), which Burgess must have been asked to criticize, followed Southworth's fundamental error (see below) of calculating the solar radiation as proportional to the *volume* of the sun, rather than to its projected area. After pointing this out, Burgess proceeded with a correct analysis of the expected solar blackbody radiation and worked out its detectability with shortwave receivers. This analysis was remarkably similar in many particulars to what ultimately appeared in Appleton's published article, yet Burgess was never credited. In fact, Burgess (1946) had already in 1944 worked out the expected solar signal in a confidential report that could not be published until after the war. This article dealt with the fundamentals of signals and noise in antennas, and concluded that the directivity of an antenna must be at least 10^4 for any chance to detect the sun (taken as a 6000 K blackbody).

---

[27] Besides many private communications, Appleton issued a general appeal for information concerning any past occurrences of the the hiss phenomenon. [September 1945 issue of the *Bulletin of the Radio Society of Great Britain*]

We do not have the reactions of Burgess to Appleton's article, but Heightman was upset enough to send a letter to *Wireless World* claiming that his work had been wronged, a letter that Appleton tried to persuade Heightman to withdraw. Hey too was upset, for Appleton had requested him to channel his own short paper to *Nature* through Appleton so that they could appear together. But Appleton sent Hey's off to F. J. M. Stratton at Cambridge Observatories for comment and the final result was that the Hey letter arrived at *Nature*'s offices three weeks after Appleton's and was published fully *ten* weeks later.[28,29,30]

Appleton, then a government science administrator, had had a long and distinguished scientific career, but not without earlier involvement in battles over priority (Clark 1971). These had been with Robert Watson Watt (two cases, including the very invention of radar) and Balth Van der Pol. Furthermore, Gillmor (1982, 1991) has documented three other such incidents, including one in the late 1920s where Appleton never acknowledged a fundamental contribution by a junior collaborator, Wilhelm Altar, to the development of Appleton's magneto-ionic theory of radio wave propagation in a plasma.[31] Appleton's behavior with regard to his 1945 letter to *Nature* was thus no isolated incident (also see Sections 6.3.2 and 6.4 for more Hey/Appleton disputes). As E. G. Bowen remarked in a letter in April 1946, prompted by Appleton's concern over the citations in the first Australian paper in *Nature* on solar noise (Section 7.3.2):

> I am sorry that Appleton is making a song and dance about our letter to *Nature*, but I suppose he is just expressing his well-known "ownership" of all radio and ionospheric work.[32]

5.6 SOUTHWORTH AND THE QUIET SUN

Only four months after Hey's discovery in England, George C. Southworth (1890–1972) became the first to succeed in a purposeful search for solar radio waves. Southworth obtained his Ph.D. in Physics at Yale University in 1923 and spent the rest of his career in the Bell Telephone System. His research variously involved dielectric properties of materials, radio propagation, earth currents, and antenna arrays. His greatest technical contribution was the development of waveguides (pipes for transmitting radio waves) for use at frequencies above 1000 MHz. Southworth began this work in the early 1930s and from 1935 on used a converted farmhouse at Holmdel as a laboratory, only 1 km from where Jansky worked. Because microwaves promised to become the new medium of choice for a variety of tasks in communications, Bell Labs through the 1930s increasingly supported research on these techniques. When World War II began, microwaves came to the fore in a wide variety of military radars, in particular in airborne systems that demanded small antennas and sharp beams. Bell Labs was thus in position to become a wartime center for radar research in the United States, along with the Naval Research Laboratory and the MIT Radiation Laboratory (Section 10.1). By 1942 Southworth

---

[28] Sources for this controversy include Hey:author, 21 January 1985, and three handwritten letters from Appleton to Heightman (photocopies from Heightman) of 25 June 1945, 11 February 1946, and 18 February 1946. Also, L. W. Hayes (BBC):Appleton, 9 July 1942, file C.331; Heightman:Appleton, 6 July 1945, file C.337; "Comments by R. E. Burgess..." (see text), 19 September 1945, file C.323; Heightman:Editor of *Wireless World*, 6 February 1946 (published on p. 99 of the March 1946 issue), file C.337; and Heightman:Appleton, 16 February 1946 – all APP. For information about Appleton's overall career, the biography by Clark (1971) is useful; Appleton also had a daughter Rosalind ("Wanda Alpar") who starred in the Anglo-Polish ballet company and the famous "Windmill Girls" after the war [Scottish Theatre Archive Catalogue, University of Glasgow].

[29] Things went even worse for a competing Australian paper submitted to *Nature* by Pawsey, Payne-Scott, and McCready (1946) (Section 7.3.2). Air-mailed to England in late October, it did not appear until early February 1946. We do not know the reasons for the inordinate delay, but the Australians were suspicious of shenanigans by Appleton and the British establishment. Also see Orchiston (2005a:83).

[30] Appleton's role as a collector and synthesizer of data from colleagues in distant locales was not unlike that played a century before by Louis Agassiz in Switzerland in the 1830s–40s. Agassiz also became embroiled in priority disputes as he pieced together geological evidence for the existence of a great, prehistoric ice age. [Hallam, A. (1989), pp. 92–4 in *Great Geological Controversies* (New York: Oxford University Press)]

[31] For another view on the Appleton/Altar case, see M.V. Wilkes (1997), "Sir Edward Appleton and early ionosphere research," *Notes & Records Roy. Soc. London* **51**, 281–90.

[32] Bowen:F. W. G. White, 26 April 1946, file A1/3/1a, RPS.

Figure 5.6 George Southworth and a Navy officer standing beside a 5 ft diameter dish with a ~1 cm wavelength feed in early 1944 at Bell Telephone Labs.

(Fig. 5.6) headed a sizable group developing radar at wavelengths as short as 1 cm. When the press of wartime demands allowed, this was the setting for Southworth's solar researches.[33]

In Southworth's laboratory notebook for 30 June 1942 we find the following entry:

> Some time ago Mr. Friis suggested that star noise might be detected by pointing a sharply directive receiver at the sun. It so happened that [Archie P.] King was starting some directional experiments at a wavelength of 3.2 cm and was set up with a sensitive double detection set coupled by waveguide to the focus of a 60 inch searchlight mirror. It was therefore very easy to conduct this experiment. We did so several weeks ago and found a small but positive effect. Mr. King repeated the experiment yesterday under better conditions and verified the result.[34]

Although Southworth had been interested in the sun as far back as 1925,[35] note that again Harald Friis, Jansky's boss (Section 3.6), played a role in early radio astronomy. Over the first days of measurements King and Southworth became convinced that they were measuring *direct* solar radiation because the ~2° beamwidth of their 5 ft dish allowed them to pinpoint the origin of the slight noise excess. Their mean value for the measured 3.2 cm solar power was expressed as 127 db below one watt ($10^{-12.7}$ W). Southworth then immediately calculated the expected intensity from a blackbody radiating at 6000 K and obtained 126.4 db below one watt. This he considered simply too good

---

[33] General background on the development of microwaves and waveguide technique, as seen from Southworth's perspective, is contained in his autobiography *Forty Years of Radio Research* (1962). A more even-handed account, also describing the contributions of Wilmer L. Barrow, has been given by Packard (1984). Information about Southworth's solar work is available in his retrospective article in *Scientific Monthly* (1956); this article, however, has been found to be at variance in many details with archival evidence.

[34] Southworth, 30 June 1942, lab notebook no. 18123, pp. 109–10, file 51–08–02–01, SOU.

[35] While working on ionospheric influences on radio propagation and on earth currents, Southworth designed an experiment at the time of a solar eclipse in 1925. Furthermore, over the period 1926–28 he photographically monitored sunspots, and in 1926 he even made a few calculations on solar radiation, but never worked out the expected radio intensity. [Southworth, "Radiation – Vol. VII" (personal notebook), pp. 1–3, file 51–04–01, SOU]

Figure 5.7 James Ward operating the 3.2 cm receiver and 5 ft dish with which the bulk of Southworth's group's solar observations were made at Bell Labs over 1942–46 (photo: autumn, 1946). Ward is aligning the antenna with the sun using a telephoto lens connected to a Graphlex camera. The signal was led from the focus through a waveguide to the left where it was mixed with a local oscillator to produce a 60 MHz intermediate frequency. The strip-chart recorder on the right was available only for the 1946 observations; previously, visual readings were taken from a microammeter.

an agreement: "I am inclined to say that much of this is a matter of chance."[36] But data over the next year continued to support such an intensity and he concluded that the microwave sun indeed closely obeyed Planck's radiation law. The irony is that this agreement *was* a matter of chance, obtained despite a serious conceptual error that persisted until 1945 (see below).

Over the period July to October 1942 Southworth and King made about 800 measures of the 3.2 cm (9370 MHz) intensity of the sun on fifteen different days. Sometimes they simply peaked up manually on the sun and read a microammeter, and other times after peaking up they let the sun drift out of the antenna beam. Their dish (Fig. 5.7) was altitude-azimuth mounted and had a surface coated with dark colloidal carbon since they found that the original highly polished reflector led to a red-hot waveguide at the focus! Through use of a telephoto lens they studied whether differences existed between the optical and radio directions to the sun. Several times they monitored the sun all day long and established that the radio signal was not affected by clouds and was recorded only when the optical sun was above the horizon. These all-day sessions also seemed to show that the noontime intensity of the sun was several decibels lower than when the sun was near the horizon – the opposite of that expected from any atmospheric absorption. But in all of this they were dealing with very low signal-to-noise ratios – the sun produced at most only a 6% deflection on the output meter.[37] Southworth filled his notebook with drift curves and all-day plots that ranged between ragged and well behaved (Fig. 5.8). This inconsistency led him to propose that the direction of arrival of the solar microwaves might be variable with time

---

[36] Note 34, pp. 111–13.

[37] With today's knowledge of what the solar signal likely was for Southworth, namely a mean 3.2 cm brightness temperature on the solar disk of ~12,000 K, this would have produced an antenna temperature of ~600 K. We can therefore estimate his system temperature as ~10,000 K.

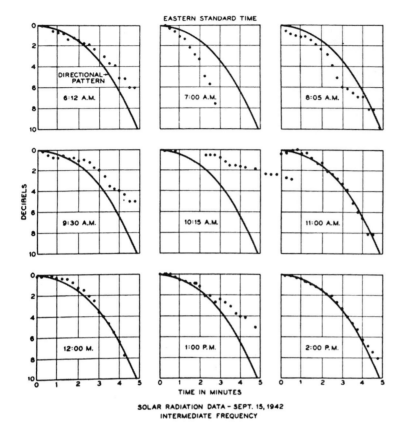

Figure 5.8 Drift curves of the 3.2 cm sun on 15 September 1942, taken with a 1.4° beam by Southworth's group. The dish was first pointed to the position of maximum signal and the sun then allowed to drift out of the beam. The solid curves represent the calibrated antenna pattern and the dots are solar data. (Southworth 1945)

and thereby cause erratic results dependent on conditions in the earth's atmosphere. The apparently lower noontime intensities might then be due to increased ionization and atmospheric turbulence during midday, leading to more extreme or more rapid changes in direction of arrival. But Southworth was cautious in forming any explanations – as an old hand at ionospheric radio propagation work, he knew that months and years of data were generally required to sort out causes and effects in solar-terrestrial relations.

In the summer and again in November 1942 Southworth (1942) wrote up the above findings in two memoranda-for-file.[38] These memos, which were circulated to his radio colleagues at Bell Labs, including Jansky and Friis, were classified "Confidential" for it was thought that such scientific results could help the enemy deduce "the state of the radio art in this country." Southworth further stated that Jansky had informed him that no solar signals "of any considerable magnitude" were noted at 20.6 MHz.[39] Likewise, he cited Reber (1942) regarding the absence of solar radio waves at 160 MHz (recall that Reber did not succeed in detecting the sun until late 1943 [Section 4.5.1]). He reasoned that his use of much higher frequencies

---

[38] Southworth, 14 July 1942 (although containing data of August), "Evidence for microwave radiation in the sun's spectrum," memo-for-file MM 42–160–73; 20 November 1942, "Evidence for microwave components in the sun's spectrum," MM-42–160–140 – both in file 51–05–03–01, SOU.

[39] This is correct in the sense that Jansky never ascribed any effect to the sun. I have never, however, found any evidence of Jansky doing a specific analysis of signal levels received when the sun was in his antenna's beam.

had led to success simply because the Planck relation predicted that solar radiation should increase as frequency squared. Southworth stressed the preliminary nature of his results and stated a desire for extensive follow-up work on atmospheric absorption, studies at other frequencies, checks on polarization, general monitoring, and recording with a shorter time constant. But he also conceded that such studies were difficult to justify in a time of war.

Southworth's first memo so impressed Lloyd Espenschied, a friend and radio engineer with AT&T, that Espenschied fired off an equally long letter (four pages) bubbling with ideas:

> This discovery impresses me as nothing less than thrilling ... and of very real significance to natural scientists. ... [It] should therefore be published upon as soon as reasonably possible ... Due to the war stimulation ... it seems only a question of time before the same observation will be made by others. Hence the need for early publication if you and Friis in particular, and the Laboratories in general, are to be credited with the discovery.[40]

Espenschied then went on to outline many possible applications of the solar microwaves, including illumination through clouds, signalling à la traditional naval heliograph, and probing the properties of the atmosphere and ionosphere.

Over the winter Southworth took a break from solar observing, but resumed from May 1943 through the fall, obtaining data now also at two additional frequencies. His main receiver engineer and principal observer at this point became George E. Mueller, a young electrical engineer trained at Purdue University. Mueller had been designing and building microwave equipment needed for studies of propagation through rainfall and now the sun afforded him an excellent "test source" outside the earth's atmosphere. He substantially improved the stability and sensitivity of the 3.2 cm apparatus and adapted existing receivers for two new wavelengths.[41] With these Southworth and Mueller first detected the sun in June at 1.25 cm (24,000 MHz) and in July at 9.8 cm (3060 MHz). As expected, the 9.8 cm intensity was much lower than that at 3.2 cm, in fact almost exactly as predicted by the Planck relation. But the 1.25 cm sun was only 1/9 as strong as they expected; instead of *gaining* a factor of 7 over the 3.2 cm signal by going to this higher frequency, they were surprised actually to *lose* some signal.[42] Southworth could only speculate that maybe the diameter of the 1.25 cm sun was much greater than that of the optical sun or of their 0.5° antenna beam, thus reducing the received signal. Or perhaps rapid variations in the direction of arrival of the 1.25 cm waves lowered the measured mean intensity. Furthermore, the measured size at 1.25 cm was indeed more than expected (Southworth 1945:294). Data on the shortest wavelength also differed in that clouds *did* noticeably affect the received signal, and atmospheric absorption seemed evident, especially near sunrise and sunset. In contrast, the new, better data at 3.2 cm showed no such diurnal effects.[43]

On one June night Southworth and Mueller slowly scanned the Milky Way "between Cygnus and Scorpio," but found nothing detectable at 3.2 cm wavelength.[44] Nor did the moon, planets, and a few bright

---

[40] Espenschied:Southworth, 27 August 1942, file 51–01–02–04, SOU. The reference to Friis came about because Southworth mentioned in his first memo (as in his notebook) that Friis made the first suggestion to try for the sun. In all subsequent publications, however, Southworth never again mentioned Friis's role in the discovery.

[41] The receivers used in 1943 were of standard superheterodyne ("double detection") design with an intermediate frequency of 60 MHz and bandwidths of 3 to 4 MHz. The noise figures for the crystal detectors ranged from 13 to 18 db. Stability was a major problem (note the unkempt nature of some of the data in Fig. 5.8) since no sort of differencing technique such as Dicke switching was used. The aperture efficiency of the 5 ft dish, as measured using a distant test oscillator, was 0.60.

[42] Southworth's extremely low 1.25 cm value is difficult to understand in light of our present knowledge that the mean brightness temperature of the sun at 1.25 cm wavelength is usually only ~50% less than that at 3.2 cm.

[43] Southworth and Mueller's 1.25 cm observations of absorption by the atmosphere, and in particular by clouds, were understood by the end of the war to result primarily from an absorption band of water vapor. Also see note 10.3.

[44] Southworth, 5–6 June 1943, notebook 18123, p. 165, file 51–08–02–01, SOU. In the summer of 1946 Mueller again attempted to detect 3.2 cm radiation from the Milky Way, this time with more sensitive equipment, but still found nothing ("less than 40 K") [Southworth:Reber, 6 August 1946, file 51–01–02–07, SOU].

stars yield anything. Other general scans of the sky, however, revealed the curious effect of considerably more received power when the dish was pointed at or just below the horizon. After puzzling over this "earthshine" for several weeks, their preferred explanation was that the radio receiver was acting as a pyrometer, a heat measuring device. As Southworth (1945:285) put it:

> When the receiver is pointed at the sun, equilibrium conditions are such that more energy is absorbed than is radiated, and when it is pointed at interstellar space ["blank sky"], conditions are reversed and more energy is radiated than is absorbed. In the latter case, the result is almost as though a sensitive pyrometer calibrated to work at high temperatures had been turned suddenly on a block of ice.

His notion involved direct thermal equilibrium between the receiver and whichever target the antenna was pointing toward. By analogy to the classic experiment in which two mirrors were set to face each other and heat transfer occurred between objects placed at their foci, Southworth conceived that radiation would actually be emitted into space if the receiver were warmer than the dish's target.[45]

As an employee of an industrial lab, Southworth from the beginning considered possible uses of solar microwaves. In late 1943 he applied for a patent on a "System for and method of utilizing microwave radiation from the sun." But he soon regretted having done so, for the Patent Office promptly issued a Secrecy Order, which effectively forbade any kind of open publication. This frustrated his plans, for he was indeed working on a paper designed to establish priority for his discovery, albeit shorn of sensitive technical details. The secrecy order ended up delaying the paper for a full year, and in the meantime all he could do was to issue yet another technical memorandum.[46] His patent application meanwhile languished in bureaucracy's depths and US Patent No. 2,458,654 did not emerge until 11 January 1949. Moreover, the basic ideas of the patent never did pan out: to use reflected solar microwaves as a means of (a) "artificial vision" in daytime fog or smoke, or (b) creation of a microwave beam to guide landing aircraft.

On and off throughout 1944 Southworth battled the censors – he actually obtained the permission of the Army and Navy without much trouble, but the Patent Office remained stubborn. He was also unhappy with lackluster support from the Bell Labs administration, but at last a memo of displeasure in January 1945 appears to have done the trick: "Unless you can find a way to rescue the paper on solar radiation, it looks as though it will die of neglect."[47] Wheels spun faster, a month later the Labs acquired a security release, and Southworth at once submitted his paper to the rather obscure *Journal of the Franklin Institute* in Philadelphia, where he had recently given a major public lecture. They immediately accepted the paper and rushed it into their April issue. And so, just as the war was coming to a close, and now five months after Reber had published a small mention of his 160 MHz solar detection (Section 4.5.1), the three-year-old discovery was announced. But Southworth was still constrained not to mention such basic quantities as size of dish, frequencies of operation, or details of receiver. He nevertheless craftily sneaked much of this information past the censors through his illustrations, which in essence conveyed many of the needed parameters. The article made no mention of the considerable contributions of King and Mueller.

During this same period Southworth also made many contacts with the astronomical community. As early as April 1943 he had made enquiries to Harvard

---

[45] Southworth's idea of direct thermal equilibrium between a receiver and its antenna's target is incorrect (see Appendix A). His concept implies that if a targeted region had the same temperature as the receiver, one would measure only the noise power arising from the receiver itself (not the receiver plus the target's noise); and if the region were cooler, one would measure *less* than the receiver noise [Southworth:E. V. Griggs, 12 November 1943, file 51–01–02–02, SOU]. After Grote Reber questioned this concept in a letter, Southworth inserted a footnote (to the statement in his 1945 paper here quoted) to the effect that his idea of reversible energy flows might need revision [Reber:Southworth, 17 February 1945, file 51–01–02–04, SOU].

[46] Southworth, "Microwave radiation from the sun," 1 June 1944, memo-for-file MM-44–160–30, WTS3-16D, SOU. Despite its date, near-final drafts of this memo (which is very similar to the 1945 paper of the same name) were in circulation as early as February.

[47] Southworth:R. Bown, 11 January 1945, file 51–01–02–04, SOU.

College Observatory regarding his need to switch to an equatorial mount for equipment that security compelled him to describe only as an electrical apparatus of 5 by 2 ft cylindrical shape and weighing 250 pounds. In the autumn of 1944 both Donald Menzel and Harlow Shapley supported his efforts to secure a suitable mounting. Southworth visited Cambridge in October and they even considered removing the mirror of Harvard's 61 inch reflector (coincidentally the same diameter as Southworth's dish), but did not because the requirement to observe right at the horizon could not be satisfied. Shapley obviously was excited by these new results, for in March 1945 he had his son Alan, an ionospheric physicist at the Department of Terrestrial Magnetism in Washington, DC, review the findings of Southworth's preprint at an informal astronomy meeting in Washington. Besides the Shapleys, others in attendance included Philip Keenan, Robert d'E. Atkinson and Menzel. As the elder Shapley described the affair to Southworth:

> Alan presented your paper in pretty good form and it awakened both objections on the part of one or two and much enthusiasm on the part of several who had never heard of the business before. In fact, the consensus is that the microwave telescope merits much further experiment and refinement.[48]

As the war wound down, Southworth had plans for more concerted solar research (although he never did obtain an equatorial mount) and at one point was preparing for detailed observations of an upcoming solar eclipse in July 1945. But in early 1945 he was persuaded by his superiors to commence work on a massive textbook on waveguide techniques and this occupied most of the next year.[49] When his duties shifted, Southworth proposed that Jansky be transferred to his group to continue the solar studies and to resume his earlier work on galactic radiation, but this was turned down.[50] In the spring of 1946 Friis, who had become his supervisor, allowed him one technician for solar work and so James E. Ward (Fig. 5.7) took many 3.2 cm observations. Mueller and Ward had developed an entirely new receiver using a Dicke "chopper" comparison method (Section 10.1) for superior sensitivity. But none of these data were ever published and solar work at Bell Labs came to an end in late 1946.[51]

Southworth's 1945 paper, his only one in radio astronomy, presented the basic results already discussed, but, as mentioned above, contained a fundamental error. The agreement obtained between observations at 3.2 and 9.8 cm and predictions from the blackbody relation turned out to be fortuitous, the result of a cancelling of mathematical and conceptual errors. The error, which is present in Southworth's notes right from the summer of 1942, was only caught in late 1945 by fellow Bell Labs researcher Charles H. Townes (Section 15.2.1), and in Australia by Ruby Payne-Scott (Section 7.3.2).[52]

The clue to the error's existence can be found where Southworth (1945:288) stated that "assuming that the sun's … volume is 1.41×10^{33} cm^3, we may calculate the power radiated.…" But the *volume* of the sun should not at all enter the problem, rather only its radiating surface *area* (Appendix A). Southworth's method involved calculating the total energy of 6000 K blackbody radiation occupying a volume equal to the sun's, and then deriving from this the energy flux falling on the earth. It is first of all amazing that such a

---

[48] H. Shapley:Southworth, 14 March 1945; A.H. Shapley: Southworth, 15 March 1945 – both file 51–01–02–04, SOU.

[49] Southworth, 3 February 1945, p. 11, notebook no. 19594, file 51–08–02–01, SOU: "Messrs. [M. J.] Kelly and [R.] Bown at Holmdel Feb. 3. Discussed matter of book. Strong urge to drop all other work." The next entry in this notebook for the solar work is not until 21 January 1946. Southworth (1956:65) later stated simply that at this time he was "assigned to another task." The book was published in 1950 as *Principles and Applications of Waveguide Transmission* (New York: van Nostrand).

[50] Southworth:J. Schelleng, 20 November 1956, file 51–02–01–04, SOU. In this same letter Southworth stated that much later, shortly before Jansky's death in 1950, he and Jansky *were* given the opportunity to work together on radio astronomy, but that at that late date neither of them were enthusiastic to compete with well-established groups.

[51] In 1946 Mueller left Bell Labs to pursue a Ph.D. in physics at Ohio State University. In the 1960s he was the Associate Administrator of NASA for Manned Spaceflight, including the Apollo 11 moon landing in 1969. [Mueller:author, 30 July 1985; also used for other parts of this section]

[52] J.L. Pawsey:Southworth, 7 December 1945 (but not received until 29 January 1946), file 51–01–02–05, SOU. Townes is acknowledged in the erratum that Southworth sent to the *Journal of the Franklin Institute*.

basic error remained undetected for so long, despite internal reviews of Southworth's results.[53] And it is equally amazing that in the face of the error anything like the "expected" answer was obtained.

First, how did this error persist for three years? The primary answer must be in the extreme plausibility of Southworth's results – if an expert has carried out a calculation yielding an answer agreeing well with one's expectations, then why check it, especially under the demands of wartime research? Moreover, in none of his publications did Southworth actually show his calculation – the only clues that something was amiss were in his mention of the *volume* of the sun and in two equations which have incorrect units. Second, how did this error give such a plausible result? Here my analysis in Tangent 5.1 indicates that it was happenstance that the numbers came out as they did. Overall, this seems an excellent example of how a scientist can too easily obtain just the answer he or she expects.

Despite its shortcomings, the 1945 paper nevertheless deserves attention for its report of a basic astronomical discovery and for several passages that reveal Southworth's general thinking. First of all, he summarized the previous half-century of solar *infrared* observations on ever-longer wavelengths and clearly considered himself, in this same tradition, simply to be working at the longest wavelengths yet – as it were, in the far-far-far-infrared. In fact, Southworth had extended observations of the thermal sun by a factor of 4000 in wavelength – the previous longest wavelength had been 24 μm, by the same Arthur Adel who had himself tried for the microwave sun in 1933. Southworth also noted that his microwave frequencies were far beyond those for which the Planck theory was originally conceived, and thus his data could even be considered as more confirmation of that theory. He was in fact so fixed on the Planckian idea that he never considered that a variable sun might influence his results; in this sense his work was diametrically opposite to all the other early solar work discussed in this chapter. He also pointed out that it might be only a matter of personal preference whether one regarded the solar microwaves as blackbody radiation or as a form of resistance noise, although "superficially at least" there seemed to be a difference between energy originating in a local circuit and that from an incandescent mass ninety-three million miles distant. He closed by calling for the future cooperation of radio engineers and designers of astronomical equipment, suggesting that there may lie ahead a new field of inquiry every bit as rich in scope as the previous decades of ionospheric research. Southworth could not have been more correct.

TANGENT 5.1 SOUTHWORTH'S MISCALCULATION OF SOLAR BRIGHTNESS TEMPERATURE

In April 1943, Southworth's Bell Labs colleague F. B. Llewellyn (also see Section 2.6) carried out calculations on the expected solar radio emission, at first making the same "volume" mistake as did Southworth, but then immediately correcting this with an "area" approach. Llewellyn's motivation to do these calculations is not known, but in February 1944 he sent the corrected calculations to Southworth, where they evidently had no influence.[54]

Both John R. Pierce and Friis questioned aspects of Southworth's concepts of antenna noise in response to his 1944 memo, but neither touched on the calculation of expected radiation. Friis further commented: "Jansky is very interested in the subject of noise from space and I hope you soon can release a copy of your paper for his study."[55]

A comparison of Southworth's calculations (note 37) with a correct approach reveals that his final result for the solar flux was too high by a factor of $1.33R/c$, where R is the solar radius and c the speed of light. It turns out that the ratio of R to c is 2.3 sec, a number which in the usual scientific unit of time (seconds) happens to be close to unity. Any other unit of time would have greatly magnified the numerical error and undoubtedly led to the exposure of the conceptual error. But Southworth also had to make a mistake in the units of his equations in order for his final result to appear dimensionally correct. This he indeed did, through a misinterpretation of the Planck equation. For these reasons, his final expected flux was a factor 3.1 higher than he should have calculated. But does

---

[53] See Tangent 5.1.

[54] Llewellyn, lab. notebook 17786, pp. 185–6, BTL; Llewellyn:Southworth, 26 February 1944, file 51-01-02-02, SOU.

[55] Friis:Southworth, 2 March 1944; Pierce:Southworth, 27 March 1944 – both file 51-01-02-03, SOU.

this then imply that Southworth's measurements were also off by a factor of ~3, thus making them jibe with his incorrect theoretical calculation? No, and again it is a coincidence that created this circumstance. We know today that a typical value during solar minimum for the sun's brightness temperature at a wavelength of 3.2 cm is ~12,000 K, twice what Southworth assumed. If this was indeed the value in 1942–43, then his measurement, which is equivalent to a value of ~15,000–20,000 K, was ~25–70% too high.[56]

---

[56] Results based on my own analysis, as well as an explanation given by Southworth in a letter to J. H. Dellinger of 3 May 1946, file 51–01–02–07, SOU.

6 • Hey's Army group after the war

Because the war's last years were at a minimum in the 11-year solar activity cycle (Fig. 13.11), radio bursts occurred rarely and never became of consequence for wartime operations. Stanley Hey's 1942 discovery of the radio sun (Section 5.2) thus remained no more than an isolated incident, known by few.[1] But Hey and his radar investigators in the Army Operational Research Group (AORG) became involved with astronomy twice more before war's end. Trying to counter German rocket attacks, in 1944 their operations led to detection of echoes later identified as caused by meteors, and in 1945 excess noise in radar receivers was identified with the Milky Way. After the war Hey kept his small group together and skillfully led them in astronomical investigations. For a while they made major discoveries and competed on a par with much larger groups, but in 1948, with the Cold War heating up, the Army brought such nonrelevant science to an end.

The next section gives an overview of the group's efforts over 1944–48, followed by sections with details of individual projects.

6.1 WARTIME PROBLEMS LEAD TO POSTWAR SUCCESS

In the summer of 1944 the Germans attacked London with their new low-flying pilotless aircraft, the "buzz-bomb" or V-1. Hey's group hurriedly responded by improving existing radars to gain a few minutes of warning. These V-1 attacks were eventually neutralized through a combination of an American microwave radar system, proximity fuses on shells (Guerlac 1987:857–9), and capture of their launch sites, but worse followed in September 1944 with the devastating V-2 missile. To counter its unprecedented range, speed, and height, AORG and others quickly worked on major modifications to 60 MHz Army anti-aircraft radar. By searching at higher elevation angles (~55°) and longer ranges (more than 100 miles), these radars could still give only a one-minute warning to the London area. It was this field experience that revealed two operational problems whose solution unexpectedly led Hey into astronomical research.

The first problem was that the radars reported a tremendous number of false alarms for objects at heights of ~100 km. This, it soon became clear, was a phenomenon called "short scatter" echo that had been studied before the war. Many ionosphere researchers considered these echoes to be caused by meteors entering the earth's atmosphere, but the evidence was disputable (Section 11.1). By the end of the war Hey had convincingly demonstrated that meteors were indeed the culprits, and meteor radar astronomy began in earnest.

The second problem arose when more sensitive receivers were installed on the V-2 warning radars in hopes of improving their range, only to find no improvement in actual performance. Hey was puzzled until J. M. C. Scott, a physicist working on radar at another laboratory (see Tangent 6.2), suggested that system performance might not be limited by electronics, but by the cosmic noise that Jansky and Reber had studied. And this in fact *was* the answer – technology had reached the point predicted in Jansky's (1937) admonition that "star noise" was becoming the limiting factor in receiver performance (Section 3.3.4). When Hey began looking into the directionality of this new external factor, he soon found himself not only working on a technical problem, but also substantially contributing to research on our Galaxy.[2]

---

[1] For instance, the Radiophysics Lab in Sydney did not learn of the discovery until the end of the war when Hey's later 1945 report was widely circulated (Section 5.2).

[2] Hey's wartime activities are described by Hey (1973:19–20) and in more detail by Hey (1992:19–34).

There followed a detailed survey of Milky Way emission during which the sun, then nearing its activity maximum of 1947–49, intruded to such an extent in February 1946 that the survey had to be temporarily abandoned. Instead great solar bursts were scrutinized – after four years, Hey was back in the solar business. Finally, an even more remarkable bit of serendipity struck when Hey's group found that one portion of the galactic emission, a secondary peak in Cygnus, rapidly varied in intensity. This peculiar behavior led to the recognition of a *discrete* source of radio emission, the first example of a phenomenon that long remained the most mysterious and hotly debated in radio astronomy.

Thus within a year of war's end Hey and his colleagues, working in a military environment and without the least bit of astronomical training, had made seminal contributions to both radar and radio astronomy. Figures 1.2 and 1.3 illustrate the importance of Hey's research to the nascent field. How had this come about? It basically arose from a combination of good equipment, a well-honed team with a "can-do" attitude stemming from the war, abundant support personnel,[3] and a military hierarchy that, given AORG's excellent track record and the sudden lack of wartime exigencies, was willing to allow them considerable freedom. Moreover, Hey (Fig. 5.1; Section 5.2) recognized and seized the opportunity to conduct "pure" research by retaining the best of his wartime team. He tended to the group's welfare in the parent Ministry of Supply, participated in observations, led the various research projects in a loose, but always watchful fashion, and sparked the discussions and publication of results. Hey was the only physicist on the team, but the skills of others nicely complemented his own. S. John Parsons, an Army Major who had studied electrical engineering at Birmingham University and had worked with Hey since 1940, was the hardware man. Major James W. Phillips had been a teacher of mathematics and had known nothing about radio techniques before the war; but by 1945 he also was expert and primarily worked on data taking and analysis of the galactic noise. And Gordon S. Stewart, an electrical engineer, concentrated mainly on the meteor radar equipment and observations.

Their antenna site was in Richmond Park, a 2500 acre preserve whose majestic oaks were first enclosed by Charles I three hundred years earlier. Although isolated enough that herds of deer grazed nearby, its location only eight miles from central London led to troublesome manmade interference. The park had been used throughout the war by Anti-Aircraft Command for research and training and much of the needed equipment was already on hand. And, as Parsons (1978:20T) later recalled, what wasn't already there was easily acquired:

> A lot of materials started coming back from Europe and there were huge dumps of radar and radio equipment of various kinds. We used to take a lorry down to these and had *carte blanche* to collect whatever we thought was useful. By these means we were able to make the conversions we needed. We didn't call for new materials – [instead] there was cannibalization of equipment.

6.2 GALACTIC NOISE AND CYGNUS INTENSITY VARIATIONS

As the war in Europe wound down in the spring of 1945, Hey initiated detailed investigations of the galactic noise.[4] By V-E Day when Phillips returned to Richmond Park, he found Hey and Parsons observing and plotting signal levels of 64 MHz galactic noise on a large globe. With his mathematical background Phillips soon became enmeshed in the recording, reduction, and analysis of these data.

The AORG antenna was a twin-Yagi array aimed horizontally and mounted on a cabin[5] that rotated completely around the horizon in a few minutes (Fig. 6.1). A reflective "ground screen," a wire mesh suspended over the grass out to a distance of ~60 m

---

[3] Laborious calculations in Hey's group were carried out with the services of several computers, frequently young women originally trained as predictor operators for tracking sights on antiaircraft gunnery [Parsons 1978:14–17T]. Many military and civilian technicians also assisted with observations.

[4] Tangent 6.1 describes other brief measurements of galactic noise, by DeWitt in the US in 1940 and by Fränz in Germany in 1942. Tangent 6.2 describes other British measurements of galactic and solar noise during 1944–46, mostly unpublished and all carried out as military research.

[5] The cabin (Fig. 6.1) was that of a GL Mk II anti-aircraft radar (Fig. 5.2), the type that serendipitously led to the discovery of the sun's radio emission by Hey in February 1942 (Section 5.2).

Figure 6.1 The 64 MHz antenna of Hey's group at Richmond Park, near London. Notice the screen supported just off the ground; the vertical structure is a vestige of the wartime radar configuration (compare Fig. 5.2). This setup (with only two Yagis) was used in 1945 for a first survey of galactic noise and then again in 1946 (with four Yagis as pictured) for a second survey, during which fluctuations from Cygnus led to the deduction of a discrete source. (Hey, Parsons, and Phillips 1948a)

from the cabin, controlled ground reflections of the incoming waves. The main beam was measured, with the aid of an oscillator suspended from a balloon, as 12° in elevation by 30° in azimuth, peaking at an elevation of 12°. Parsons made two required modifications to a standard radar receiver: replacement of the cathode-ray-tube display with a galvanometer showing detector current, and provision for calibration of the received noise intensity.[6] Calibration allowed much more confidence in measuring both relative and absolute intensities ($\pm$ 50% errors were quoted for the latter), and had been notably missing in the work of Jansky and Reber. Observations were carried out by sweeping the antenna around the horizon, recording the receiver output every 10° in azimuth, and then waiting thirty minutes for the sky to move a bit before repeating. In this way they obtained 48 great circle scans of the northern sky every 24 hours; note that continuing this routine on a second day led to no additional sky coverage, only an exact repeat. This schedule wreaked havoc with any attempts to catch sleep on night shifts. Since there was no room in the cabin to recline, each half hour the observer was obliged not only to record data, but also to twice cross a long catwalk suspended above the ground screen, barely allowing fifteen-minute chunks of sleep.

The late spring and early summer of 1945 were spent in sorting out any possible solar influences (in the end none were found[7]) and in developing observational

---

[6] See Tangent 6.3 for further details.

[7] Hey, Parsons, and Phillips also wrote a greatly extended version of their 1946 *Nature* article on the galactic radiation ["An investigation of cosmic noise at 64 Mc/s," undated (ca. late 1945), AORG Memo No. 641 (Restricted)]. In this memo they

techniques. A week's worth of good data then led to the production of a map that was the highlight of a short note submitted the following January to *Nature* (Hey, Phillips, and Parsons 1946). With a frequency and beam size intermediate between those of Jansky and Reber, this third-ever map of the Milky Way confirmed the earlier results of a strong concentration toward the galactic center (although from England's high latitude they could not properly map the galactic center itself), a secondary peak in Cygnus, and enhanced radiation all along the galactic plane.[8]

For better angular resolution, Hey's group next added two more Yagis to their array and in 1946 repeated the survey of galactic noise with a 13° beam. This new survey, however, was twice delayed: first, when intense "interference" from the sun was encountered (see below), and second, when Phillips noticed a curious phenomenon in the spring of 1946. It was common during observations to experience manmade interference, manifest as a dancing needle on the output meter; if this happened, all one could do was to log its occurrence and later repeat that patch of the sky. But during one period Phillips began to notice that a string of interference events was occurring not only at about the same time of day and same azimuth, but, more precisely, four minutes earlier each day. This sidereal signature indicated that the cause of the jiggling needle was not manmade, but originated in a fixed part of the heavens, which turned out to be in the constellation of Cygnus. Many found this at first hard to believe, including the Dutch astronomer Jan Oort, who wrote at this time: "In England Appleton told me a curious, nearly unbelievable thing about the Milky Way radiation in Cygnus. He says that it varies with an amplitude of about 20% in periods of 1 sec to 1 minute!"[9]

Attention immediately shifted to this discovery and for weeks Cygnus was carefully monitored at the times of its daily rising in the northeast and setting in the northwest. Assuming that a fluctuation at a given time occurred at the sky position of the electrical axis of the beam, Phillips worked up a map (Fig. 6.2) of the amplitude of the intensity fluctuations. This revealed that the source of the "disturbance" seemed to subtend an angle of no more than 2°. Furthermore, its position of 20^h00^m, +43° coincided with the second strongest peak in the overall galactic emission. The fluctuations averaged about 15% of that peak's intensity, and were not to be found elsewhere in the sky.

But what caused these unprecedented fluctuations? The AORG group soon came to the conclusion that their fast time scale (a few seconds – see Fig. 6.3) necessarily implied that the emission mechanism involved small, individual sources rather than an extended medium somehow varying. As they stated in a second note to *Nature* in the summer of 1946: "It appears probable that such marked variations could only originate from a small number of discrete sources. This suggests at once the analogy with the radio-frequency sunspot radiation" (Hey, Parsons, and Phillips 1946). It was natural to make this connection for they had often seen the sun cause similar variations in their receiver's output. Perhaps a small number of variable "super suns" produced the Cygnus phenomenon while the rest of the galactic emission originated in the combined effects of millions of lesser radio stars. This idea was attractive in that a single type of object (radio-emitting stars) then explained three separate discoveries, but even in 1946 Hey and his colleagues knew that this picture of special radio stars had grave difficulties, since the required ratio of radio to optical power for the purported stars was orders of magnitude greater than the same ratio for the sun (see Chapter 15). On the other hand how could one account for the Cygnus fluctuations as thermal radiation from a dispersed interstellar plasma (the generally accepted picture)? Neither way satisfied.

The AORG group continued to follow Cygnus on and off through the end of 1946. They discussed further findings in a later paper (Hey, Parsons, and Phillips 1948a), but the picture did not clear up – if anything, it grew murkier. There now was a suggestion in the data that the disturbed region had become more extended in angle and that its autumn position had shifted ~10° from that in the spring. This seasonal effect led them to note that the fluctuations might be imposed on the radiation "as it passes through some near attenuating medium." All bets were now off on how to interpret what was happening out there in the Galaxy until one could confidently understand any effects caused by the

---

mentioned that observations of the galactic noise were made during a partial solar eclipse in July 1945, as well as on days before and afterwards, but that no effects were seen. The sun itself was apparently not detected during the eclipse.

[8] See Tangent 6.4 on why this survey did not reveal the presence of a later-discovered and stronger source, Cassiopeia A.

[9] Oort:B. van der Pol, 27 July 1946, OOR.

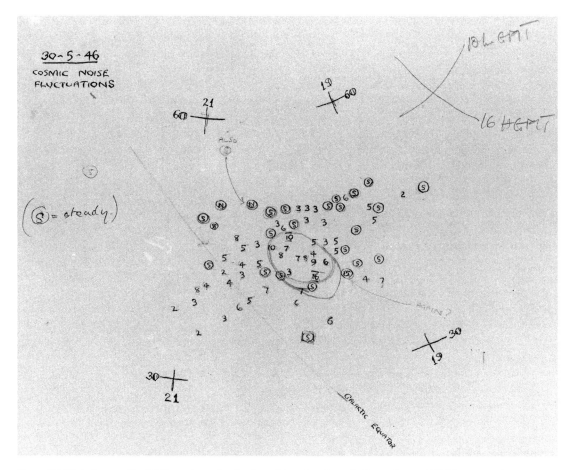

Figure 6.2 Work sheet used by Phillips to determine the sky position where intensity fluctuations on 64 MHz were maximum. This led to the discovery of the first discrete source, later known as Cygnus A. Numbers represent the amplitude of the observed fluctuations on different days in May 1946, and occur along arcs representing the constant-elevation scans along the horizon as the source either rose (upper left to lower right) or set (lower left to upper right). The crosses indicate the region between 30° and 60° declination and 19^h and 21^h right ascension.

nearby earth's ionosphere. As it turned out, not until about 1950 did a consensus develop that the ionosphere indeed caused "radio stars" like that in Cygnus to fluctuate (Section 14.2). Hey's group had indeed established the existence of the first discrete radio source by showing that the fluctuations came from a region of less than 2° in size, but their argument that its intensity variations implied a far smaller source had an incorrect premise.[10] They did no more along these lines, but subsequent investigations were soon carried out by John Bolton in Sydney (Section 7.4) and Martin Ryle in Cambridge (Section 8.4). Numerous further examples of radio stars were found, often less than 15′ in size, but their fundamental nature remained unknown for many more years (Chapters 14 and 15).

Eventually, after all the excitement of the solar bursts and the Cygnus source had subsided, the second 64 MHz survey of much of the northern sky was completed in the summer of 1946 and a year later a full-scale paper, notable for its astronomical orientation, was submitted to the *Proceedings of the Royal Society* (Hey, Parsons, and Phillips 1948a). Hey and Phillips were by then reading astronomy journals, attending meetings of the Royal Astronomical

---

[10] See Tangent 6.5 on ionospheric scintillations.

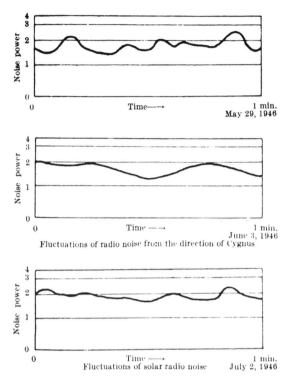

Figure 6.3 Two examples of intensity variations over one minute observed on 64 MHz from the discrete source in Cygnus. Solar intensity fluctuations are given in the bottom panel for comparison. (Hey, Parsons, and Phillips 1946)

Society, and talking to astronomers at the nearby Royal Greenwich Observatory. Their improved map of the Milky Way[11] was now plotted in galactic coordinates and detailed comparisons were made with a multitude of optical objects, including novae and O stars (because the high optical luminosities of these stars allowed their detection to large distances in the Galaxy), H_α regions (indicating high-temperature gas), and the nearest stars (Fig. 6.4). But what frustratingly emerged from all this radio-optical intercomparison was that there simply were no good detailed correlations, only the general one that the prominent regions of sky in both wavelength regimes were in Sagittarius and Cygnus. Moreover, as before, Hey *et al.* were unable to settle on a cause for the galactic radiation: the conflicting evidence of the Cygnus fluctuations, the general emission, and the

relative amounts of radio and optical intensity proved intractable (see Chapter 15).[12]

6.3 METEOR RADAR

6.3.1 Observations

Ionospheric researchers had for a decade studied irregular ionization effects in the E layer at a height of ~100 km. The ion density of the normal E layer was known to be regularly controlled by ultraviolet radiation from the sun, but there were also sudden increases in ionization lasting anywhere from less than a second to hours. When studied using radars of wavelength greater than 10 m, several hundred anomalous echoes per hour could often be picked up in this "abnormal E" or "sporadic E." Shorter wavelength radars sometimes picked up even stronger echoes, persisting for less than a few seconds each. Various arguments indicated that these phenomena might well be related to meteors, but no consensus had emerged by the end of the war (Section 11.1.3) before Hey's group persuasively established in 1945 that meteor trails indeed were the "targets" being detected. They then developed the radar technique into a potent new probe of the meteors themselves through a combination of high transmitter power, good angular resolution, a relatively short wavelength of operation, and clever experimental designs.

AORG's first study of these bothersome "short scatter" echoes was in October and November of 1944. Analysis of echoes recorded at twelve of the V-2-watching radar stations of Anti-Aircraft Command revealed that they came from a height of ~90 km and, despite overlapping coverage by the beams of adjacent stations, were seldom seen by more than one station. This was Hey's first contact with the phenomenon and he determined to investigate more fully as soon as possible. Immediately upon the European ceasefire, he took advantage of fully-staffed radar stations with nothing to do, and orchestrated, with Gordon Stewart as chief assistant, dedicated observations. The antennas at the three stations shown in Fig. 6.5 were aimed so that the beams intersected at a common point in the E region and the basic idea was to triangulate and find the exact locations and movements of any bursts of ionization

---

[11] See Tangent 6.6 for details of how this map was calculated.

[12] This section has profited especially from correspondence and an interview (1978) with Phillips.

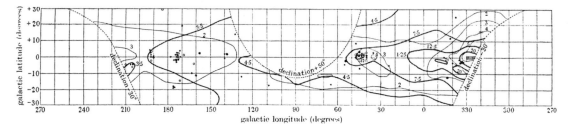

Figure 6.4 Map of galactic noise at 64 MHz made in 1946 using the antenna shown in Fig. 6.1 with a beam of 14°, although the map has been processed in order to gain better angular resolution (see Tangent 6.6). The horizon-sweeping antenna could not map outside the indicated declination limits; this caused the source Cas A (at a longitude of 79°) to be missed (see Tangent 6.4). The light contours represent the appearance of the optical Milky Way, filled circles are O-type stars, and open circles are regions of hydrogen emission. The shaded box (longitude 327°) is the position estimated by astronomers for the galactic center. Radio contour unit is 1×10^{-21} W m^{-2} Hz^{-1} ster^{-1}, or a brightness temperature of ~800 K. The original caption read "Comparison of cosmic electro-magnetic radiation at 64 Mcyc./sec. with astronomical data." (Hey, Parsons and Phillips 1948a)

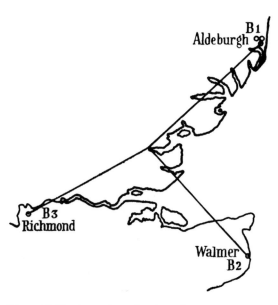

Figure 6.5 The three radar stations in southeastern England used by Hey and Stewart in June and July of 1945 for triangulation on the anomalous E-layer echoes. The beam axes intersected at an altitude of ~100 km at the position indicated. Separation of the stations was ~125 km. (Hey and Stewart 1947)

observed in common. Conversely, if events were not observed in common by all radars, it would provide important information about the scattering properties of the regions of enhanced ion density.

The antennas were arrays of dipoles or Yagis with beamwidths at 73 MHz of 30° to 50°. In addition to looking obliquely at an elevation of ~55°, Hey and Stewart also used several antennas aimed straight up at the zenith. The transmitters put out 500 3-μsec pulses per second with peak powers of 150 kW. Observations were made by either visually or photographically monitoring the radar's standard cathode-ray-tube display (Fig. 6.6).

After six weeks of observations in June and July of 1945 on the three-station network it became clear that only rarely did even two, let alone three, stations observe the same anomalous echo. Clearly there were strong aspect effects to the phenomenon, that is, the reflectivity of the burst regions was different in different directions. Direct triangulation was thus not possible. What did show up, however, was a strong diurnal dependence of the hourly echo rate, which furthermore was not at all identical at the three stations (Fig. 6.7). To understand this behavior Hey made a key assumption, namely that the non-coincident daily peaks at each station represented the very same disturbance. Further, he drew on the work of John A. Pierce (1938) (Section 11.1.3), who had emphasized that one should expect a strong reflection only when looking *perpendicular* to the long, thin column of ionization created by passage of a meteor. If this were true, the relative orientation of radar beam and direction of a meteor trail would be critical in determining whether or not a healthy echo was picked up.

Hey hit upon the idea that each diurnal peak might represent a meteor shower[13] whose time of detection

---

[13] A century of visual observations of meteors had shown, in addition to a random ("sporadic") component, the existence of streams or showers occurring at specific times of the year, that

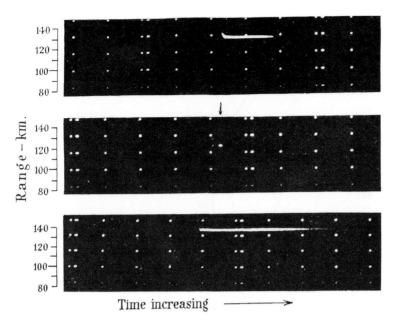

Figure 6.6 Typical records of 73 MHz "transient ionospheric echoes," shown to be reflections off meteor trails by Hey and Stewart. The film continuously moved horizontally; the regular marks indicate 6 second intervals. The ordinate displays delay time between pulse and echo, calibrated as range. (Hey and Stewart 1947)

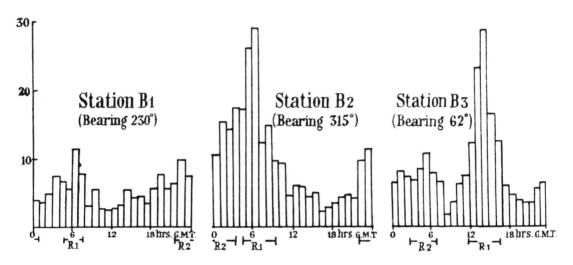

Figure 6.7 Diurnal variation of hourly echoes obtained at the three stations shown in Fig. 6.5, averaged over the week of 6–13 June 1945. "Bearing" refers to the azimuth of each station's beam. Also shown are the times when two particular meteor stream radiants, called R1 and R2, were favorably situated for echoes to be received. (Hey and Stewart 1947)

is, at specific points on the earth's orbital path. During a *meteor shower*, swarms of tiny (mm to cm sized) meteoric particles collide with the earth's atmosphere on parallel paths, and the visible streaks produced as they burn up appear to come from a single direction in the sky, called the *radiant*. Over the course of a day a radiant appears to move across the sky just as does a constellation; in fact, the astronomical convention is to name a shower by the constellation or star in the direction of its radiant.

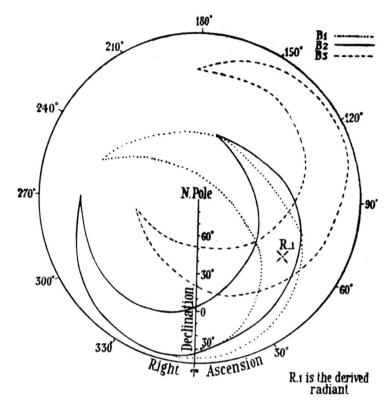

Figure 6.8 Determination of the position of radiant R1, whose echo peaks are visible in Fig. 6.7. Each radar station's peak determined the locus on the sky of a long arc taken to include the radiant. The derived radiant, at the intersection of the arcs, is shown with an X and represents the first observation of a daytime meteor shower. (Hey and Stewart 1947)

depended on the exact azimuth and elevation angle of each radar beam. The radars would tend to pick up echoes when looking by chance perpendicular to the majority of meteor trails, that is, perpendicular to the direction of the radiant of any shower. He could now use a radar beam's position at the time of an echo peak to plot the locus of all directions on the sky 90° distant – the radiant would have to be somewhere on this circle. By following the same procedure for all stations, the radiant could be found, to within about 10°, as the common intersection of the circles (Fig. 6.8). Of course this method might not produce consistent results at all if the assumption of a meteor shower were in error, but Hey and Stewart were able to locate three distinct radiants, one of which coincided with the previously known δ Aquarid shower of late July. This splendid result clinched the meteoric hypothesis for the echoes as well as Pierce's idea of perpendicular reflections from the trails. But it was one of the other radiants whose implications were most exciting, for it was what astronomers had known all along must exist, but had been powerless to observe – a *daytime* meteor shower. It would now be possible with radar to catalog and study the *other* half of the meteoric debris continually raining on the earth.[14]

Bolstered by this success, Hey and Stewart set up a meteor program at Richmond Park. They found a marked increase in long-duration (greater than 5 s) echoes during the Perseid shower of August 1945. From December 1945 through June 1946, for five hours on each of ~150 days, Stewart and his assistants visually kept watch on a cathode-ray-tube display connected to a zenith-looking radar. They recorded the times, ranges, durations, signal strengths, and any

---

[14] Hey and Stewart's (1947) R1 radiant turned out to be part of the massive complex of summer daytime showers studied from 1947 onwards at Jodrell Bank (Sections 9.2 and 11.4.1).

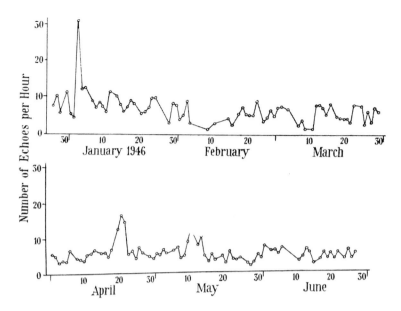

Figure 6.9 Plot of approximately 5000 echoes logged over a six-month period by Gordon Stewart and his assistants with a zenith-looking 73 MHz radar at Richmond Park. The two prominent peaks corresponded well to the visually known Quadrantid (2–3 January) and Lyrid (20–22 April) meteor showers. (Hey and Stewart 1947)

unusual characteristics of about 5000 echoes: the basic result of this persistence is shown in Fig. 6.9. Besides the usual fluctuating rate of 5–10 echoes per hour, two peaks were prominent and these neatly coincided with the well-known Quadrantid and Lyrid meteor showers. Moreover, for the Lyrids they watched overhead on three nights (with an overhead window cut into a celluloid screen) for simultaneous visual and radio events, using buzzers for timing. Eight of thirteen visual meteors were accompanied by echoes, but another fifty radar echoes not seen visually presumably demonstrated that the radar technique was sensitive to much smaller meteors than detectable by traditional means; in fact, overall they found an exponentially increasing number of echoes as signal levels approached their limit of sensitivity. Hey and Stewart also analyzed the height distribution of the meteoric echoes and found 90% to originate in a "meteoric layer" between 90 and 110 km, ~20 km below the E layer but in good agreement with astronomers' estimates of the heights of visible meteor trains.[15]

---

[15] Sources for these early results of meteor radar are Hey and Stewart (1947), Hey (1973:20–23), and correspondence with Hey and with Stewart.

The final AORG work on meteors was to observe the magnificent display of the Giacobinid shower of October 1946, also observed by many other groups (Section 11.2). A new high-speed camera developed by Parsons enabled measures for the first time of the speeds of individual meteors (Hey, Parsons, and Stewart 1947; Section 11.2).

6.3.2 Publications

Despite the important new results on meteors from the summer of 1945, Hey recalls that he was in no rush to publish:

> We did [our post-war astronomical research] in just the same spirit that we tackled the problems of wartime operational research. We were completely unaware then of the competition for scientific priorities and prestige that we subsequently encountered.[16]

Only when he learned that Bernard Lovell had just gone on the air with his own observations of the August Perseid shower at Jodrell Bank (Section 9.2) did he

---

[16] Hey:author, 21 January 1985.

hurriedly put together a short paper for *Nature* (Hey and Stewart 1946). Hey said in a letter at the time:

> The main reason which has stirred me to preparation of the *Nature* letter…was through hearing that Manchester University were pressing ahead with their work on the same wavelength and similar equipments. I have visited them in the last few days to find out what is happening.[17]

But again Hey ran into an attempt by Appleton to claim priority for a scientific feat (see Section 5.5), in this case establishing the meteoric nature of the sporadic E phenomenon. To be sure, Appleton had long considered the meteor idea, but had remained unconvinced, as had most others (Section 11.1.3). In early 1944 he had started a long-term monitoring program of the anomalous E layer echoes with Robert A. Naismith at the National Physical Laboratory, Slough. After collecting data for a year, he wrote to the Astronomer Royal, Harold Spencer Jones, asking about the basic properties and times of occurrences of meteors. Spencer Jones replied in detail about the expected meteor showers in 1945, and, noting that the Geminid shower for December 1944 had given a very good display, inquired whether Appleton's records gave any evidence of correlated ionization bursts.[18] There must have been no convincing correlations, for in a lecture in April 1945 before the Institution of Electrical Engineers, Appleton (1945b:349) stated that the meteor idea had a "good deal of evidence in its support," but the only evidence that he in fact cited were prewar results showing that the echoes centered on a height of 110 km. Appleton and Naismith (1947) were not able to make the necessary correlations with meteor showers until the October 1946 Giacobinids (Section 11.2), primarily because their 27 MHz frequency led to a harvest of *too many* echoes, whereas frequencies two or three times higher detected only the largest meteor events and therefore produced data easier to sort out and correlate with visual data. This was a case where Appleton's bird-dog instincts had him on the right trail, but with too many confusing scents.

Hey and Stewart sent a letter to *Nature* on 20 August 1946. In it they concluded that most, if not all, of the echoes were of meteoric origin, and thus supported A. M. Skellett's suggestion of same a decade earlier (Section 11.1.2). Hey also sent this to Appleton for his comments and, although the reply is not extant, it is clear Appleton strongly suggested that his 1930s work should receive prominence in the introduction. Hey, however, instead of adding Appleton's work, chose to eliminate all specific citations in a revised version of 27 August and only referred to a recent news note in *Nature*. He explained to Appleton that he had taken this course because he felt that the past work could not be fairly treated in a brief communication. But in the end the version published in *Nature did* have references to both Appleton and Skellett – Hey had been persuaded to change it yet again.[19]

A full account of the meteor results was ready in October, but again Appleton intervened and delayed release of this manuscript until his own could be presented alongside at a special session on meteor radar held four months later by the Physical Society.[20] Hey and Stewart's (1947) paper was seminal and extremely well written as it methodically presented overwhelming evidence that the bulk of the short scatter echoes were the direct result of ionization by meteors. The abundant references to previous astronomical investigations of meteors also indicate that the authors were fast acquainting themselves with the astronomical literature. Hey had done no less than take what was at first a military problem, turn it into the solution to a

---

[17] Hey:Appleton, 16 September 1946, file C.340, APP. At some time in 1946 it also appears that Hey obtained Lovell's agreement that Hey should be allowed to publish his meteor results first [Hey:Lovell, 26 August 1946, file R9 (uncat.), LOV].

[18] Spencer Jones:Appleton, 8 February 1945, file C.231, APP.

[19] Hey:Appleton, 27 August and 16 September 1946, plus two (undated) drafts of Hey and Stewart (1946), files C.229 and C.340, APP. Also see Hey (1973:32–33) for comments on Appleton's relation to AORG meteor work.

[20] Three papers on meteor radar were all read at this meeting of the Physical Society on 31 January 1947 and submitted for publication in its *Proceedings* within five weeks. They were by Appleton and Naismith (1947), Hey and Stewart (1947; Section 6.3), and Eric Eastwood and K. A. Mercer (1948); the last was instigated by Appleton and based on data gathered at various Royal Air Force stations in 1945–46. These papers provide a telling example of how the handling of manuscripts for publication can depend on who you are. Hey and Stewart's paper (which in my judgment is far the superior) appeared four months later than Appleton and Naismith's, and Eastwood and Mercer's fifteen months later (including a nine month delay for revision).

longstanding ionospheric dilemma, and then further transform it into a powerful technique for plumbing a traditional area of astronomy.

6.4 SOLAR OBSERVATIONS

As mentioned above, Hey's group was compelled to return to the radio sun, four years after the 1942 discovery, when radio bursts occurred of such strength that their survey of galactic noise became impossible. Whereas the Milky Way would typically double or triple a receiver's inherent noise, the largest of the solar bursts caused the measured noise to rise by a factor of 10^4 to 10^5! Such a high (and variable) solar signal inevitably sneaked into sidelobes and masked the galactic noise, even if one's antenna were pointing far from the sun. Over the following year the upshot was that galactic observations were often in recess waiting for the sun to calm down.

The initial event of February 1946 was no ordinary incident, for the sunspot group that caused it turned out to be the largest *ever* recorded in a century of accurate observations. It was twice as large as Hey's 1942 sunspot group and covered almost one per cent of the solar disk (see Fig. 7.8 and Section 7.3.2, where Australian work on this same sunspot group is discussed). The AORG team monitored the sun for two weeks as the sunspots moved across the solar disk, and Appleton and Hey (1946a) wrote a paper a few months later for *Philosophical Magazine*. They found the greatest radio noise enhancements to occur when the sunspot group was passing across the sun's central meridian. Correlations of radio, optical, and geomagnetic activity were often good, but not complete. The main new result was a rough estimate of a burst's radio spectrum, garnered from reports fed to Appleton after he alerted military organizations all over the country – the most thorough observations were those of A. C. Copisarow aboard *HMS Collingwood* while docked on the Hampshire coast.[21] Data over a wavelength range from 3 cm to 15 m indicated that the solar radiation was enhanced by factors of 10^4 to 10^7 above that expected from a 6000 K blackbody radiator. The peak of the spectrum seemed to be at about 5 m, although it was realized that the ionosphere probably absorbed longer wavelengths. Furthermore, the burst was not at all detected on 50 cm or shorter, for example using a small dish at 3 cm.

In July 1946 another giant sunspot group, only a bit smaller than that of February, appeared and created new opportunities. Hey's group this time searched for any polarization in the radio burst. By analogy with ionospheric propagation, it was argued that radio waves passing through strong magnetic fields in the outer solar atmosphere should become circularly polarized. And indeed their setup, involving two dipoles connected in and out of phase and variously oriented and separated, conclusively demonstrated that the enhanced solar radiation was ~100% circularly polarized. These results were passed on to Appleton, who then immediately submitted a joint paper to *Nature* (Appleton and Hey 1946b).[22] As it turned out, other measurements of circular polarization on the same giant radio bursts had just been submitted to *Nature* by David Martyn working in Australia (Section 7.3.4) and by Martin Ryle and Derek Vonberg at Cambridge (Section 8.3). Martyn (1946b) noted that the sense of the circular polarization changed from right-handed to left-handed as the spot group passed the central meridian, as expected from Appleton's magneto-ionic theory of radio propagation. Ryle and Vonberg's (1946) results differed primarily in that they were using a spaced interferometer, which allowed the additional deduction that the enhanced radiation emanated from a region of the sun less than 10′ in size.[23]

By this time Hey and his group were growing weary of Appleton's persistent attempts to receive credit for work with which he was minimally involved: discovery of the radio sun (Section 5.5), meteor radar echoes, solar radio bursts. Although at first they had

---

[21] A. C. Copisarow, "Radar observations during sunspot activity," 1 March 1946, file C.334, APP.

[22] Hey:Appleton, 18 August 1946, "Measurement of the polarisation of solar radio noise" (a much longer report than Appleton and Hey [1946b] published in *Nature*), file C.340, APP.

[23] Ryle and Hey appear to have been working entirely independently, despite their separation of only 60 miles. F. J. M. Stratton, Professor of Astronomy at Cambridge and solar expert, having been told Hey's polarization results by Appleton, responded on 31 July: "I mentioned Hey's results to our radio experts today and found that they had already got it ... from the great flare of Thursday of last week. ... It is all very exciting. ... It will be to the advantage of all concerned if Hey and Ryle get together on this." [Stratton:Appleton, 31 July 1946, file C.351, APP]

been honored by Appleton's overtures for collaboration, the partnership had grown sour and so the connection was at last severed. Appleton was then Secretary of the government's Department of Scientific and Industrial Research, the major agency for support of civilian research, and, despite a continued interest, had no time to carry out research. The one time he visited the AORG site in Richmond Park was on the occasion of an organized "show" for the press at the time of the February 1946 sunspot activity. As Hey later diplomatically stated:

> With reluctance I became aware of the pitfalls of an undefined association. Appleton's incursion into our research findings began to arouse discontent within my own team and the liaison could not survive.[24]

Once galactic mapping was completed in the summer and meteor work ended in the autumn of 1946, Hey's group commenced daily solar monitoring that lasted for over a year. The results centered on a detailed statistical analysis of the coincidences of (optical) solar flares, radio bursts, communications fade-outs, and geomagnetic storms (Hey, Parsons, and Phillips 1948b). They concluded that these events often occurred together and that flares usually preceded bursts by 5–15 minutes. This was taken as strong evidence that a common solar disturbance moved outwards in the solar envelope, from the photosphere where a flare originated to the corona where a subsequent radio burst started. Another solar initiative during this period was to join a British expedition heading to Brazil for a May 1947 solar eclipse. Parsons built a special Dicke-switched 60 MHz receiver, but the expedition was cancelled after a plane crash at Dakar in West Africa killed two (optical) astronomers in an advance party and destroyed critical instruments.[25] Thus the British were not there when American and Soviet groups became the first radio observers of a total solar eclipse (Sections 10.1.2 and 10.3.2).

6.5 END OF AN ERA AT AORG

Despite great success, by the middle of 1947 radar and radio techniques were no longer being applied to astronomy at AORG. With the Cold War intensifying (for instance, the Berlin blockade began in April 1948), AORG was transferred from the Ministry of Supply to the War Office – such a line of pure research could no longer be justified. The Army also felt that such investigations were clearly in good civilian hands at Jodrell Bank and at Cambridge, as well as overseas. Furthermore, the AORG radar unit, despite the "OR" in its name, had always worked more on issues of technical development than on operational research, and it was decided that a separate radar group was no longer needed. Finally, there was no one left to carry on the astronomy. Parsons joined the television industry and Phillips resumed teaching mathematics – this was easy to do since each thought of himself more as a practically trained person following up on wartime activities than as a professional scientist. When in 1949 Hey himself became head of all of AORG, he was able to continue with some solar monitoring and burst observations with a swept-frequency receiver, but this effort was never competitive. As John Bolton reported home to Australia after a visit to AORG in April 1950:

> I was very disappointed in the lack of funds and time for research work at AORG and I think they may soon fade out of the picture entirely. I found Hey very pleasant, energetic and enthusiastic. … As superintendent, he has very little time to spare from his administrative duties and [Stewart and V. A. Hughes] are mainly concerned with Army problems and regard the radio-astronomy as a spare time job … Hey told me the story of AORG's foundation, their great hopes when the war ended, and the subsequent collapse of same. Parsons is now at Mullards and Phillips is fruit farming![26]

It was the end of a short, fruitful period in which a small group took full advantage of propitious conditions for discovery, albeit under unlikely auspices. The

---

[24] Hey (1973:18). The press visit was on 5 February 1946 [*The Times* (London), 6 February, p. 7d]. The *Times* reporter clearly was impressed, but not exactly for the right reason: "The [solar] hiss was loud enough to be heard some yards away from the cabin." Other information on Appleton's relations with the AORG group comes from Parsons (1978:11–12T); Phillips (1978:30T); Phillips:author, 23 August and 16 October 1978; and Hey:author, 21 January 1985.

[25] Parsons (1978:26–30T); details of the mishap are given in *Nature* **159**, 666–7 (1947).

[26] Bolton:Bowen, 14 April 1950, file F1/4/BOL/1, RPS.

AORG group may have died early,[27] but its influence on the course of early radio astronomy was lasting. The next decade was to be dominated by researchers building on the fundamental discoveries of Hey, Parsons, Phillips, and Stewart: solar bursts, meteor radar reflections, and discrete sources.

TANGENT 6.1 DEWITT (1940) AND FRÄNZ (1942) MEASURE GALACTIC NOISE

The only investigations of galactic noise after Jansky's and before those described in Tangent 6.2 were by Potapenko and Folland at Caltech in 1936 (Section 3.5.1), Reber (Chapter 4), DeWitt briefly in 1940, and Fränz in 1942.

John H. DeWitt, Jr. (1906–1999) had made a brief, unsuccessful attempt at 300 MHz in 1935 (note 3.32), but five years later he succeeded at 111 MHz, although never publishing his results. DeWitt was head engineer at a broadcast station in Tennessee and an ardent amateur astronomer. He had high quality radio equipment and a large antenna site at his disposal, and with these he built a 111 MHz acorn tube receiver and a double rhombic antenna supported on a 30 ft tower. The main purpose of this setup was to try to bounce a signal off the moon (which in 1946 DeWitt was the first to do – see Section 12.3), but DeWitt also used it briefly to study Jansky's reported noise from the Milky Way. On the evenings of 20–23 September 1940 he detected radiation from the region of the galactic center and found a contrast of about a factor of two between maximum and minimum signal. His antenna's beamwidth was ~10°, and by pointing it over a grid of directions and reading a volume meter on an audio amplifier, he found the peak to be in Sagittarius at about 18^h, $-22°$.

With the same antenna, but now with some improvements in the receiver, DeWitt on 25 September also made a single attempt to detect solar radio waves. Despite no signal, he was nevertheless convinced that the sun in fact emitted radio waves and so recorded in his notebook:

> The sun may cause such intense ionization of the upper atmosphere that waves of 111 Mc cannot come through. This can be tested in December when the sun is in Sagittarius. At this time the strength of the cosmic noise should be measured in order to determine whether the intense ionization blots out cosmic noise from this direction.[28]

But the press of mundane matters meant that he never looked again at the sun or Milky Way.

In wartime Germany Kurt Fränz was a radar engineer specializing in minimum noise levels in receivers and antennas at the laboratories of Telefunken[29] outside Berlin. He therefore knew about Jansky's work and was not surprised when his colleagues reported excess noise in their 30 MHz receivers. He undertook a short study of this noise and in 1941 recorded the signal level as the sky passed overhead. His so-called Knickebein system was used as a navigational aid for directing the paths of bombers to England, and consisted of a rotatable array of two lines of vertical half-wave dipoles separated by ~40 m. As with Jansky, an observed daily shift of four minutes in the time of maximum signal soon convinced him that it was of cosmic origin and he published a short supplement to a recent paper of his in *Hochfrequenztechnik und Electroakustik* (1942). In this note he mainly emphasized the deleterious effects of the "disturbing radiation" for shortwave sensitivities. Using noise diodes to calibrate the system, he reported what in retrospect are the first reliable intensities of galactic radiation, although unfortunately the direction of his antenna beam was not specified. Fränz's plot (Fig. 6.10) showed that for three-quarters of the day the signal level was constant at ~12,000 K of antenna temperature (termed "radiation resistance temperature") but also exhibited a strong (~120,000 K), ~25°-wide peak (his antenna's resolution in azimuth was ~15°). This measurement was felt to be consistent with Jansky's at 20 MHz, but of significantly higher intensity than

---

[27] Sources for the reasons for the end of astronomy in the AORG group are Hey (1973:48–9), Parsons (1978:13–4, 32–4T), and Phillips (1978:17, 28–30T). In the mid-1950s Hey did re-enter radio astronomy with a group at the Royal Radar Establishment, Malvern (successor to the Telecommunications Research Establishment).

[28] DeWitt, 25 September 1940, Lab. Notebook no. 3, p. 205 (from a photocopy courtesy of DeWitt). Information also from an interview (DeWitt 1978:107B).

[29] The Telefunken firm at this time was the largest center of research, development, and manufacture of German radar systems. [Swords 1986:241–57]

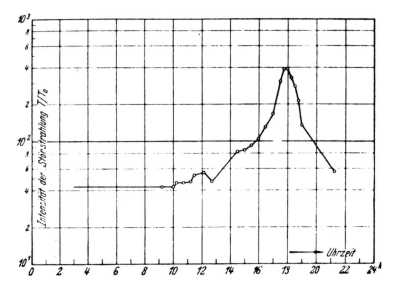

Figure 6.10 Galactic noise at 30 MHz as measured with a dipole array by Fränz at Telefunken Laboratories near Berlin on one day in 1941. On the ordinate, T is an antenna temperature and T_0 refers to an ambient temperature of 290 K. The abscissa is local German time. (Fränz 1942)

Reber's (1940a) at 160 MHz. Fränz also tried to investigate the spectrum of this radiation with further measurements at ~150 MHz, but over a day (with a ~30° beam) he found no signal "greater than room temperature" (Fränz 1943:692).[30]

TANGENT 6.2 BRITISH WORK ON GALACTIC NOISE (1944–46)

Toward the end of the war and immediately afterwards several British investigators other than Hey's group, mostly in military research laboratories, became interested in the "Jansky noise," as it was often called. Although none of these efforts produced important scientific results, the resulting reports provide insight into the milieu in which the AORG group operated and from which emerged the major postwar groups at Cambridge and at Jodrell Bank.

The earliest study was a review and critical analysis of available data on radio noise made in 1944 at the National Physical Laboratory by H.A. Thomas and Ronald E. Burgess (1917–1977).[31] The goal of this report was to determine noise effects on communications links and the emphasis was on atmospherics, but noise of all types – including manmade and cosmic – was treated. Thomas and Burgess reviewed the results of Jansky, Reber, and Fränz in great detail, concluding that on 15 to 30 MHz cosmic noise was indeed the main contributor to received noise. They also stated that "although cosmic noise has been well established as a phenomenon, its source is still a matter of theoretical speculation."

Next was Kenneth F. Sander of the Army's Radar Research and Development Establishment (RRDE) at Malvern in Worcestershire. In the early spring of 1945 he investigated galactic noise at 60 MHz and in May issued a confidential report[32] that was subsequently turned into an open article (Sander 1946). Sander's motivation, as for Hey, was that new radar

---

[30] For this episode I have relied on Fränz (1973:1–7T) and Fränz:author, 28 March 1985, as well as the publications of Fränz (1942, 1943).

[31] H.A. Thomas and R.E. Burgess, "Survey of existing information and data on radio noise over the frequency range 1–30 Mc/s," 1944, Radio Research Board Report RRB/C.90; later published as Radio Research Special Report No. 15 of the Dept. of Scientific and Industrial Research (1947, London: HMSO). For an earlier discussion in the open literature of the basic thermodynamics of an antenna's signal, in which "Jansky noise" and Reber's first observations are also mentioned, see Burgess (1941), "Noise in receiving aerial systems," *Proc. Phys. Soc.* **53**, 293–304.

[32] K.F. Sander, "Measurement of cosmic noise at 60 Mc/s," 31 May 1945, RRDE Research Report No. 285 (Confidential).

receiver pre-amplifiers (~7 db noise figure) when in the field had not delivered the improvement promised by bench tests. The Milky Way was thought the culprit, but Sander needed to find out more, so he set up four half-wave dipoles mounted on a cabin that rotated around the horizon every twenty minutes. This gave a beam of 30° by 20° and allowed him to confirm the earlier results, although he never made an all-sky map. His conclusion was that the *effective* noise figure of a radar system with even an excellent 4 db receiver would be in the range 9 to 13 db (~2000 to 5500 K system temperature) because of the galactic noise. Sander also took a 3.2 cm look with a 3 ft dish at a July 1945 partial solar eclipse, although he did not write up his results until later (Sander 1947); these radio observations were, along with those of Dicke and Beringer (1946) on the same day (Section 10.1.1; Figs. 10.3 and 13.6), the first ever during an eclipse. Despite problems with rain and clouds, he found the radio intensity to fall and increase in a fashion roughly agreeing with theoretical expectations that the sun had a bright microwave limb (Section 13.1.2). It was probably Sander's data that inspired his RRDE colleague, mathematical physicist John M. C. Scott (1911–1974) in the summer of 1945 to study in detail the galactic noise's spectrum.[33] It was extremely tricky to place the rough data available on a common scale, but Scott nevertheless tried, finding the mean brightness temperature over the whole northern sky to vary inversely with wavelength cubed, although much steeper if Reber's 1940 results at 160 MHz were taken at face value (which he was not inclined to do). In either case Scott emphasized that the spectral shape of the Milky Way signal did not at all look Planckian.

The last studies of this type were carried out under the aegis of the Admiralty Signal Establishment Extension at Witley in Surrey. L. A. Moxon wanted to learn how the effective range of shipboard radars at various frequencies was affected by galactic noise. He carried out his own 90 MHz observations beginning in August 1945,[34] and later organized more observations on land and at sea. The most interesting of these were done by Sub-lieutenant D. H. Cummings while cruising to Australia on *HMS Formidable* from March to May 1946. Since the Radiophysics Group in Sydney (Chapter 7) worked exclusively on the sun in the early years, it turns out that Cummings's measurements of the *southern* Milky Way, although never formally published, stood by themselves for a full three years![35] He made 900 horizon scans with 40 and 90 MHz radar equipments as the ship sailed through a variety of latitudes. In his report[36] he presented the run of noise completely round galactic longitude and estimated, from apparent beam broadening, that the true width of the galactic plane emission was about 25°. To address the detrimental effects of this noise on naval radars, Cummings supplied a host of charts for estimating the expected reduction in a radar set's range depending on one's latitude, antenna azimuth and local sidereal time. The worst case was when a radar set happened to be pointing toward the galactic center in Sagittarius, at which time it was estimated that sensitivity was about halved, implying that maximum range was reduced by ~20%.[37,38] Working with Cummings's surveys, Hey's, and his own, Moxon published an article in *Nature* (1946) in which he estimated the 40–200 MHz spectral index of the galactic noise, not averaged over the entire sky as Sander had done, but separately at its peak and at its minimum (toward the galactic pole). He estimated that brightness temperature varied as the -2.7 (-2.1) power of frequency at the peak (pole). Given the rough calibrations and large beams of these early surveys, it is amazing that this value required minimal modification in later years.

---

[33] J. M. C. Scott, 13 July 1945, "The intensity of cosmic noise. A survey of the data available," RRDE Research Report No. 286 (Restricted).

[34] L. A. Moxon, "Galactic noise measurements on 90 Mc/s," 26 November 1945, ASEE Report XRC3/45/9 (Confidential).

[35] The one exception was the 200 MHz galactic background map quickly obtained in late 1945 by a Radiophysics Lab team in Sydney and included in an internal report by Payne-Scott (1945) (Section 7.3.2).

[36] D. H. Cummings, "The effect of galactic noise on radar performance," 16 July 1946, ASEE Report XRC3/46/3 (Restricted).

[37] Because radar signals suffer an inverse-squared loss both going out and returning, signal is proportional to (range)$^{-4}$ and maximum range is proportional to (weakest signal detectable by the receiver and antenna)$^{-1/4}$ (see Section 12.3).

[38] A question of military tactics: just as it is standard practice for an airplane to approach its target with the sun at its back, did the military ever train pilots to approach with the constellation Sagittarius at their backs?

TANGENT 6.3 HEY'S ANTENNA BEAM AND INTENSITIES

The effective raising of Hey's antenna beam above the horizon (as in the case of Jansky's antenna) was a result of ground reflection of the incoming wave interfering with that portion which fell directly upon the antenna. When ground reflection is important, an antenna acts as an interferometer with an antenna "image" located below the ground (as for the sea-cliff interferometer in Fig. A.6). While this leads to better vertical directivity, it also causes the beam to be lifted from the horizon by an amount that depends on any phase shifts introduced by imperfectly reflecting soil or other surface material.

Calibration of intensities for Hey's survey was effected through matching receiver input not to the antenna, but to a resistor in circuit with a tungsten-filament diode running in a saturated condition. This technique allowed a good calibration because one could calculate (using Schottky's equation) the diode's expected noise power once one measured the diode current, the load resistance, and the bandwidth of operation.

TANGENT 6.4 WHY HEY'S GROUP AND OTHERS DID NOT FIND CAS A

As a further check on the nature and constancy of the galactic noise, Hey's group employed a Yagi setup that could be oriented to various elevation angles. These data underwent preliminary reductions, but were taken no further because of a broad beam and poor sky coverage. But in a later paper (Hey, Parsons, and Phillips 1948a:439) it was stated that these observations revealed "some evidence of peaks at galactic longitudes 85° and 110°." From present knowledge it would seem that Cassiopeia A, at a longitude of 79° and the strongest discrete source in the sky, was probably contributing to these peaks. But because circumpolar Cas A never dips near the horizon in England (its lowest elevation is 20°) and Hey's group had an antenna fixed at 12°, they missed discovering a second discrete source. Note the gap in longitude coverage in Fig. 6.4, precisely where Cas A lies!

Cas A is clearly visible as a marked concentration on Reber's 1943 160 MHz map (Fig. 4.7 and Tangent 4.1), but was not recognized as a discrete source because its substantial apparent size (relative to what Reber thought his beam size was) jibed with Reber's notion of interstellar radiation from spiral arms. Also note that John Bolton of course had *no* chance in Australia – Cas A never rose above his horizon. Thus it was not until three years later that Cas A was found, by Ryle and Smith at Cambridge in 1949 (Section 8.4).

TANGENT 6.5 RADIO SOURCE SCINTILLATION

Later studies showed that fluctuations in a discrete radio source come about not because the source is relatively small in *linear* extent; rather, they arise because the source is so small in *angular* extent that fast-changing ionospheric irregularities (in the F-region at a height of ~400 km) act as a multitude of variable refracting lenses causing the image of the radio source at the earth's surface both to move about and to scintillate in intensity. Sources of larger angular size must of course also pass through the same gauntlet, but individual ray paths from different portions of the source largely cancel out each other's effects. This ionospheric phenomenon is directly analogous to the familiar "twinkling" of visible stars, except that twinkling arises from density fluctuations in the *lower* atmosphere. Analogously, twinkling is absent for the planets and moon, which have larger angular extents than stars.

TANGENT 6.6 PHILLIPS'S METHOD FOR IMPROVING ANGULAR RESOLUTION

The maps published by Hey's group did not represent contours of intensity as observed, but underwent considerable further analysis. Phillips developed a mathematical method of successive approximations designed to alleviate their coarse angular resolution. By the time of the second survey the method had grown complex, but in essence it worked in the following manner (Hey, Parsons, and Phillips 1948a:430–4). Phillips first put the observed intensities on a grid drawn on a large sheet of plastic. He then moved around a transparent overlay having the contours of the measured antenna pattern and estimated, at each sky position, the contribution arising from the side of the primary lobe and from sidelobes. In this way the measured intensity value could be iteratively corrected for the responses of the outer portions of the beam in a trial-and-error fashion (Phillips 1978:9T). Results at each stage were checked by convolving the derived model distribution with the antenna pattern and seeing whether the result was converging to the original observation. As an example, this methodology meant that a measured width of 31° for the galactic plane emission was transformed into a final map value of 15°. See Bracewell (1984:171–5) for extensive comments on this and other methods of restoration that were later used.

7 • Radiophysics Laboratory, Sydney

The undisputed leaders in postwar radio astronomy were England and Australia. While it is not surprising to find England at the forefront of a scientific field in the middle of the twentieth century, Australia's presence cries for explanation. How was it that an isolated country succeeded so impressively in such an arcane field? The answers revolve around Australia's relationship with the mother country, the course of World War II, and the Australian government's policies toward its scientific laboratories. First, a strong community of radio physicists developed in Australia in the 1930s, based on intimate ties with the ionospheric community in England. Second, Britain shared the secret of radar with its Dominions as the war began, nurturing intense radar research, development, and manufacture in Australia. Third, the team of scientists and engineers that grew out of that effort at the Radiophysics Laboratory in Sydney remained intact at war's end, and soon put their new skills to use in developing peacetime research ventures. And finally, dynamic and skillful leadership was provided by E. G. Bowen and J. L. Pawsey – two men whose styles of science and complementary personalities produced a potent mix for exploring and exploiting the radio sky.

This chapter follows the origins and early years of radio astronomy in the Radiophysics Laboratory.[1] Soon after war's end a multi-faceted program, by far the largest of its kind in the world (Figs. 1.2, 17.2 and 17.3), was well established and continually producing pioneering results. Here I tell the story of the Laboratory as a whole up until 1952–53, and also relate several salient studies during the first three years. Pawsey, R. Payne-Scott and L. L. McCready developed fundamental methods of interferometry even as they found relationships between sunspots and solar radio bursts and established the existence of a 10^6 K solar corona.

Astronomers at Mt. Stromlo Observatory, including C. W. Allen and R. v.d.R. Woolley, often collaborated with the radio researchers. J. G. Bolton, G. J. Stanley and O. B. Slee began by studying the enigmatic Cygnus A, but also searched the sky and discovered many other radio sources. They proposed the first optical identifications in 1948: for Virgo A, Centaurus A and Taurus A, the latter as corresponding to the Crab nebula. The chapter concludes with an analysis of the factors that led to such stunning scientific success in the unlikely location and time of Australia in mid-twentieth-century. Further scientific results in the postwar decade are covered in later, topical chapters.

7.1 RADIO RESEARCH IN AUSTRALIA BEFORE 1945

7.1.1 Prewar: the Radio Research Board

In 1926 the Council for Scientific and Industrial Research (CSIR) was established as a semi-autonomous body of the Australian government. Its guiding lights for two decades were chemist A. C. David Rivett (1885–1961) and George A. Julius (1873–1946), a consulting engineer. To produce the scientific research needed by Australia, they set up independent divisions headed by the best chiefs they could find, with each chief allowed a large measure of freedom in how he ran his program. Research divisions such as Animal Nutrition, Economic Entomology, and Plant Industry reflected the agricultural economy of prewar Australia. Despite severe budget problems during the depression years, CSIR steadily developed a reputation for excellence in applied research. Successes included developing a pleuropneumonia vaccine for cattle, controlling blue mold on tobacco, and finding ways to maintain the quality of beef shipped to England.[2]

---

[1] An earlier version of portions of this chapter has appeared in Sullivan (1988a).

[2] Information on the early years of CSIR and on Rivett comes from Schedvin (1982, 1987) and from *David Rivett: Fighter*

Figure 7.1 "Taffy" Bowen (center) and Joe Pawsey (right) greet Edward Appleton on board ship after his arrival in Sydney in 1952 for the URSI General Assembly.

Amidst all this, then, CSIR's Radio Research Board (RRB) was an anomaly. The RRB was founded and chaired for more than twenty years by John P.V. Madsen (1878–1969), first professor of electrical engineering in Australia. Madsen and Rivett considered that radio research was vital for communications within a large, sparsely populated land, as well as for shortwave links overseas. Furthermore, research in Australia had the inherent advantage of access to regions of the ionosphere different from the mid-latitude, northern hemisphere locations of Europe and North America. Problems defying solution or ignored in the north often could be profitably attacked in Australia by filling in the missing tropical and southern parts of the picture. RRB had minimal facilities of its own, but funnelled both CSIR and Post Office funds to a Melbourne group for research on atmospherics and Madsen's group at Sydney University for ionospheric work. A third strong group arose at the Sydney research laboratory of Amalgamated Wireless (Australasia), Ltd. (AWA), the country's dominant electronics and communications firm. These three groups of radio physicists, however, were only partly trained in Australia. Without exception they either took their doctorates or gained experience in England, primarily with Edward Appleton (Fig. 7.1) at King's College, London or with John A. "Jack" Ratcliffe at the Cavendish Laboratory, Cambridge (Chapter 8). Australian universities and Australian science still operated in a colonial mode[3] and did not then offer advanced training – the first Ph.D. in Australia was not awarded until 1948. Nevertheless, the skills learned in England transplanted well and by the end of the 1930s the Australian ionospheric community had developed an international reputation, no mean feat given Australia's isolation (Home 1983). Members included Victor A. Bailey, Geoffrey Builder, Alfred L. Green, George H. Munro, John H. "Jack" Piddington, Owen O. Pulley, and Frank W. Wood.[4]

The intellectual leader of this community was David F. Martyn (1906–1970), a Scot who obtained his Ph.D. in 1928 at the Royal College of Science in London and then emigrated to Australia. Over the next decade he made many major contributions, including the theory (with Bailey) of the "Luxemburg effect" (a powerful broadcast signal modulating other ionospheric radio signals) and a detailed model (with Pulley) of the composition and conditions of the upper atmosphere.

---

for Australian Science by Rohan Rivett (1972, Melbourne: Rivett).

[3] For overviews of the development of Australian science in the context of the British Empire, see the introduction in Home (1988) and R. MacLeod (1982) in *Hist. Recs. of Austral. Science* 5, No. 3.

[4] Gillmor (1991) provides an authoritative overview of the ionosphere research community in Australia for the period 1925–70.

His influence fomented many other investigations, too, and led Rivett in 1935 to write:

> Martyn is a first-class worker, deserving of the designation "brilliant." I am inclined to think that the radio work done by the C.S.I.R., though it does not receive much local recognition, is of a higher scientific standard than that attained in any of our other lines, and much of its excellence is due to Martyn.
>
> (Evans 1973:128)[5]

Gillmor (1991:188) found in a 1972 survey that Martyn was deemed the fourth most important founder of worldwide ionospheric physics (after Appleton, Sydney Chapman, and Ratcliffe).

7.1.2 Wartime: the Radiophysics Laboratory

In February 1939, as war in Europe seemed increasingly certain and Britain felt ever more threatened, the Australian Prime Minister received the following cable from his High Commissioner in London:

> Conversations have been carried on from time to time with the Air Ministry on the subject of secret research. They culminated today in the disclosure…of a new development in defence applicable particularly to air. … [We and the other Dominions] have been informed that if [we] send our best qualified physicist to England all information will be placed at his disposal. … Utmost secrecy is essential and the choice of a man of the greatest discretion important. I am satisfied that the new development, which is the product of the best scientific brains here, is of great significance and that the Commonwealth of Australia should be fully advised in relation to it.
>
> (Mellor 1958:423)

After consulting Rivett at CSIR, the Prime Minister soon picked Martyn as the man who would carry the secret of radar to the Antipodes.

After a short stay in England Martyn returned laden with the documentation that spelled out what radar was and what it could do; complete radar equipments followed on. In August the Radiophysics Laboratory (RP) was created as a secret branch of CSIR with Martyn as its head and a startup capital budget of £80,000 ($300,000), as compared with the ~£5000 annual budget of the entire RRB during the 1930s. The agreement was that RP would do radar research and development in close liaison with British laboratories and with the Australian military, and also that Australia would manufacture resulting radar sets. RP found a home in the new building of the National Standards Laboratory on the grounds of Sydney University, and by mid-1940 had a staff of twenty. RRB's groundwork over the previous decade in helping establish a radio physics community was paying off, and Madsen now pumped up his training of physicists and electrical engineers – by war's end the staff numbered three hundred,[6] of whom sixty were professionals and fourteen bore names that would become notable in radio astronomy after the war. This new self-reliance in education was only one example of how World War II jolted Australia. Indeed, Australia's agrarian economy rapidly expanded and diversified to include a panoply of industries never before seen. The hectic changeover led to problems for RP and the military, for example, in the fabrication of vacuum tubes and other radar components, but eventually a sophisticated electronics industry was centered on AWA.

RP both designed wholly new radar systems and adapted British radars (Section 5.1) to Australian needs; as the Lab's expertise and confidence grew, it also moved into fields such as operational research and tube design. Radar problems in the Southwest Pacific theatre were considerably different from those in Europe. First, electronics had to operate reliably in jungle heat and humidity; this led, for instance, to hermetic seals for components and spare parts. Second, the enemy was the Japanese and their tactics and electronic capabilities differed from those of the Germans. Third, distances were far greater, stressing navigation and communications

---

[5] An obituary for Martyn is given by H.S.W. Massey in *Biog. Mem. Fell. Roy. Soc.* **17**, 497–510 (1971), and an entry by R.W. Home is in *Dictionary of Scientific Biography*, Vol. 18, pp. 599–601 (1990), ed. F.L. Holmes (New York: Scribners). Also see Gillmor (1991:187–9).

[6] The subsequently most famous member of the wartime staff was certainly eighteen-year-old Joan Sutherland, who worked as a typist for £2 per week before resigning to pursue her career as one of the premiere sopranos of the twentieth century. [April 1944 to January 1945, file G23/11, CSIRO Archives, Canberra]

systems. Perhaps the outstanding radar design by the Lab was the LW/AW Mk. IA(Aust.), a lightweight air-warning radar designed for installation in the tropics within hours of an amphibious assault.[7] Consisting of a 10 kW transmitter and a 32-element, 200 MHz rotating array protruding from a tent, it weighed only three tons (including diesel generator) and could be packed into a C-47 (DC-3) aircraft.[8]

7.2 RADIOPHYSICS LABORATORY, 1945–1952

7.2.1 Transition to peacetime

As the war closed, RP was able to concentrate more on radar research and less on development and manufacturing problems. Various memoranda began to circulate on potential peacetime roles for the laboratory, culminating in an agenda paper put together by Welshman Edward G. "Taffy" Bowen (1911–1991) (Fig. 7.1) for a meeting of the CSIR Council in July 1945. Bowen, who would soon take over as Division Chief at age 34, had been working on radar for over a decade. Trained in physics at the University of Wales, he then studied atmospheric physics for his Ph.D. under Appleton at the University of London in 1933. Two years later Robert Watson Watt coopted him into the initial team of four that developed the first operational military radar systems, systems that became vital in the defense of Britain against the Luftwaffe (Section 5.1). He led the development of 200 MHz airborne radars (Swords 1986:539–57), for which he flew thousands of hours (often with Robert Hanbury Brown, who would also later become a central figure in radio astronomy [Section 9.3]). In 1940 Bowen was a member of the famous Tizard Mission that delivered radar secrets to the United States, including the cavity magnetron, the first source of power sufficient to make microwave radar a feasible proposition (Section 10.1.1). He remained in the US for three years, eventually developing airborne radar systems at the MIT Radiation Laboratory. In early 1944 he went to RP as its Deputy Chief, and, although still officially on loan from the British, soon took a liking to RP and to Sydney and spent the remainder of his career (until 1971) there as Chief.[9]

Bowen's proposals for RP's peacetime role were warmly received and quickly endorsed[10] by his CSIR bosses Rivett and Frederick G. W. White, a New Zealander who had taken over from Martyn as RP Chief for three years and who knew Bowen from common years at Kings College, London. Rivett felt strongly that each CSIR division should achieve a roughly even balance between free-running basic research and applied research – a vital element in the culture of CSIRO (Schedvin 1982). RP's proposed program accordingly emphasized new scientific possibilities as well as areas where Australian commerce and industry would more immediately benefit. Bowen and his staff were as excited about the potential of radar techniques in peacetime as they were weary of applying them to warfare. It mattered not that RP's original *raison d'être* had disappeared – fresh

---

[7] A photograph of this radar set is in Sullivan (1984a:84), and H. Minnett (1999) describes its development. ["Light-weight air warning radar," *Historical Recs. Austral. Sci.* **12**, 457–68]

[8] In this section I have drawn from the following sources: the development of Australian physics and science in general (Home 1983); studies of the Australian ionospheric community and its ties with Britain (Gillmor 1991; White 1975); and histories of RRB (Evans 1970) and wartime RP (Evans 1973; Mellor 1958; Schedvin 1987:232–80). The Evans manuscripts are available from the Radiophysics Division Library, Epping [RPL]. Further background comes from several interviews: Bowen (1973:1–15T and 1978:63–4T), Bracewell (1980:130B), Kerr (1971:1–4T), Minnett (1978:1–8T), and Piddington (1978:4–8T). Additional information on RP's wartime work and non-astronomical postwar work is given by Robertson (1992:chapter 2). One wartime researcher's account is in chapter 5 of Joan Freeman (1991), *A Passion for Physics: The Story of a Woman Physicist* (Bristol: Adam Hilger). R. MacLeod (1999) gives an overview of RP's wartime history and historiography in "Revisiting Australia's wartime radar programme," *Historical Recs. Austral. Sci.* **12**, 411–8; also see Minnett, as in note 33.

[9] Bowen's career particularly well illustrates the close cooperation among the Allies on the wartime development of radar (although see Guerlac [1987:1106–11] for a counterexample). Also see C. Eldridge (2000), "Electronic eyes for the allies: Anglo-American cooperation on radar development during World War II," *History and Technology* **17**, 1–20.

Bowen's (1987) memoirs of the period before 1944 are in *Radar Days* (Bristol: Adam Hilger). For his obituary by R. Hanbury Brown, see *Biog. Mem. Fellows Royal Soc.* **38**, 41–65 (1992).

[10] E. G. Bowen, 2 July 1945, "Future programme of the Division of Radiophysics," 30 pp.; White:Bowen, 15 June 1945; Rivett:Bowen, 3 July 1945; G. A. Cook:Bowen, 26 July 1945 – all in file D1/1, Radiophysics Division Archives [RPS].

directions could now be charted. The new radar techniques were "perhaps as far-reaching in themselves as the development of aircraft [during World War I] or the introduction of gunpowder in a previous era." They laid out a long shopping list of possible projects in radio propagation, vacuum research (directed toward generating power at millimeter wavelengths), radar aids to navigation and surveying, and radar study of weather. These topics, along with producing *Textbook of Radar*, which incorporated RP's knowledge and was edited by Bowen (1947), were to form the initial postwar program.[11]

Radiophysics was CSIR's glamour division, arguably containing within its walls the densest concentration of technical talent on the continent, and CSIR was eager to keep this winner intact. As one of RP's early staff members later recalled:

> [Basic radio research] was thought of as a good subject for the Lab to get into, partly in order to keep the Lab in being because it was a collection of good people, well trained in the arts of radio. Especially at that time there was a feeling that it had been a great national value to have had the Lab, and so it was possible to sell the idea to the authorities that the group should be kept in existence as a "national asset."
>
> (Kerr 1971:7T)

Keeping the best of the research staff at RP was also immeasurably aided by the fact that research in physics and engineering at Australian universities after the war was minimal (Home 1983); government funds for university scientific research were thirty times less than for CSIR (Mellor 1958:678). In contrast to the situations in England and the United States, the young cadre of RP researchers saw their wartime laboratory as the best place to continue their peacetime careers. Despite a considerable reduction of staff at war's end, most who remained were oriented toward research. Few researchers had come from the universities (except as recent graduates) and fewer returned. Academia's role in Australian physics research would not strengthen until the 1950s when students were first able to do their postgraduate work at home.

RP's assets included not only its scientists and engineers, but also its significant support staff of technicians, its camaraderie molded during the war, ample laboratory space and workshops, and bulging stores of the latest radio electronics. This last was considerably augmented shortly after the war's end by an extraordinary bonanza. A large amount of American and British equipment (including whole aircraft!) was being discarded by loading it on the decks of aircraft carriers, taking it a few miles out to sea, and bulldozing it into the Tasman Sea. Bowen got wind of this, and for two or three weeks was allowed to take RP trucks down to the Sydney docks and load them up with radar and communications gear, often in unopened original crates. For several years thereafter RP researchers drew on this surfeit (Bowen 1984:86).

RP's direction changed from developing top-secret military hardware to a completely unclassified research environment. Rivett felt strongly that military research should be removed from the postwar CSIR. In retrospect Bowen (1984:85) named this policy as a key ingredient in RP's postwar success; in a 1945 document Bowen stated that peacetime military work by CSIR "stifles research and seldom produces effective assistance to the Armed Forces."[12] Rivett's policy, however, was not without political repercussions (Schedvin 1987:332–50). In the atmosphere of a deepening Cold War and its attendant espionage episodes, the loyalties of many prominent figures around the world were questioned. The most famous was J. Robert Oppenheimer in the United States, but Rivett too was attacked by right-wing politicians, in particular over the loyalty of CSIR workers[13] and over his policies on secret research. This

---

[11] Munns (1997:302–6) has more details on Bowen's early career and his guidance of RP from war to peacetime.

[12] Bowen, as in note 10, p. 23. Bowen did not, however, propose a complete break from military ties. He suggested that RP personnel should serve as scientific consultants to the various services, and that research work on *defensive* radar against ballistic missiles ("likely to form a great bulk of the offensive weapons in the next war") would be proper and important.

[13] For example, there was considerable trouble in gaining permission to hire W. N. "Chris" Christiansen in 1948 (Section 7.2.2) because of his Communist Party affiliations [Haynes *et al.* 1996:207–8]. A sample of Christiansen's Marxist history of radio astronomy: "Radio astronomy was not born with a silver spoon in its mouth. Its parents were workers. One parent was the radio-telephone, the other was radar" [W. N. Christiansen, "Two Sydney scientists saw first 'radio star,'" (Sydney) *Daily Telegraph*, 25 August 1952].

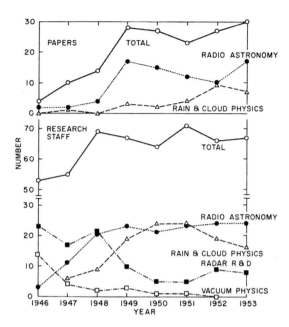

Figure 7.2 The growth and decline of different research areas at RP over the period 1946–53, as gauged by the number of published papers per year and the number of research staff.

and other troubles eventually led to his resignation. In 1949 CSIR became CSIRO, the Commonwealth Scientific and Industrial Research Organization, and White became its number-two man – over the next two decades he was a major force in fostering the growth of his old division (Bowen 1984:110).

7.2.2 Research program

Major programs at RP waxed and waned over the years 1946–53. The plots of Fig. 7.2 show the bare trends,[14] but one of the leading researchers more elegantly likened these years to the Biblical parable:

> A sower went out to sow his seed, and as he sowed some fell by the wayside and it was trodden down and the fowls of the air devoured it. And some fell upon a rock, and as soon as it sprung up it withered away … And other fell on good ground, and sprang up and bare fruit an hundredfold.[15]
>
> (Luke 8:5–8)

Vacuum physics fell on a rock and died away within two years, while work on radar applications steadily lessened over the first five years. The two research programs that prospered were radio astronomy (although note that this term was not used until the 1949 report) and rain and cloud physics. Between 1946 and 1949 their share of the professional staff increased from 6% to 63%. Because the total staff grew by only one-quarter over this same period, there were clearly many reassignments of personnel. In terms of papers published in the scientific literature, radio astronomy and rain and cloud physics also dominated, accounting for 71% of the papers by 1949 and 65% over the eight-year span. The radio astronomy staff, however, produced more than double the number of papers per person.

Rain and cloud physics, in which Bowen came to specialize and which he personally oversaw, relied on microwave radar measurement of clouds and rain, often from aircraft. Buoyed by one of the first successes in seeding clouds,[16] the RP group hoped that rainmaking for the dry Australian climate would ultimately become a reliable and economic proposition. Although this never happened, Bowen's group became one of the international leaders in the field.[17] This effort, as well as the development of radar systems for commercial aviation, such as a distance-measuring equipment allowing airliners to locate themselves relative to beacons, were important as the practical areas balancing off fundamental research in astronomy, fast becoming

---

[14] The data of Fig. 7.2 come primarily from annual reports and lists of publications issued by RP [file D2, RPS]. Papers are plotted for the year of publication and staff for the year of issuance of the annual report (published mid-year). Staff levels not plotted include those for ionosphere (which fell in a similar manner to vacuum physics) and mathematical physics and electronic computing (which rose to a level of about 10 by 1951–53 and then died away).

[15] J. P. Wild, 15 October 1965 lecture delivered to CSIRO Division of Plant Industry, "Origin and growth of radio astronomy in CSIRO," file D12/1/5, RPS.

[16] Kraus, E. B., and Squires, P. (1947), "Experiments on the stimulation of clouds to produce rain," *Nature* **159**, 489–91.

[17] For an excellent article on the history of RP's cloud physics and rainmaking work, see R. W. Home (2005), "Rainmaking in CSIRO: the science and politics of climate modification," pp. 66–79 in *A Change in the Weather: Climate and Culture in Australia*, eds. T. Sherratt *et al.* (Canberra: National Museum of Australia Press).

RP's most visible sector (Bowen 1984:105–9, Minnett 1978:66–8T). But even radio astronomy was sometimes shoehorned into the role of practicality, as in the 1949 Annual Report:

> Radio astronomy has already made important contributions to our knowledge and, like any fundamental branch of science, is likely to lead to practical applications which could not otherwise have been foreseen. For example, attempts to explain how certain types of radio waves arise in the Sun are already leading to new techniques for the generation and amplification of radio waves.

Other projects included a mathematical physics section and (after the late 1940s) a group developing an early electronic computer (CSIRAC).[18] And there were always a few ionospheric radio projects going on. The one example of a major effort that failed was vacuum physics, when costs of a desired linear accelerator became too great.[19]

7.2.3 Growth of research on extraterrestrial radio noise

Buried in the twenty-four pages describing RP's postwar plan is a fraction of a page under "Radio propagation" called "Study of extra-thunderstorm sources of noise (thermal and cosmic)":

> Little is known of this noise and a comparatively simple series of observations on radar and short wavelengths might lead to the discovery of new phenomena or to the introduction of new techniques. For example, it is practicable to measure the sensitivity of a radar receiver by the change in output observed when the aerial is pointed in turn at the sky and at a body at ambient temperature. The aerial receives correspondingly different amounts of radiant energy (very far infrared) in the two cases. Similarly, the absorption of transmitted energy in a cloud can be estimated in terms of the energy radiated to the receiver by the cloud. None of these techniques is at present in use.[20]

It was this enigmatic paragraph, with its heading designed primarily to indicate that it was *not* talking about thunderstorm noise (atmospherics), that would develop into RP's radio astronomy program! It surprisingly did not explicitly mention *solar* noise, but instead proposed an exploratory program of "very far infrared" radiometry wherein antennas would be pointed to different parts of the sky.

But when reports of anomalies arrived from radar stations (next section), Pawsey and his colleagues jumped on solar observations in October 1945 and never turned back. Joseph L. Pawsey (1908–1962; Fig. 7.1) by this time had become the linchpin and recognized leader of RP's fundamental investigations through his Propagation Research Group. He had studied physics at Melbourne and obtained his Ph.D. in 1934 under Ratcliffe at Cambridge, with a dissertation on radio waves reflected off the abnormal E layer of the ionosphere (Chapter 11). For five years he then developed equipment needed to make television a viable reality at the prewar BBC station at Alexandra Palace. Pawsey's main contributions, which involved no fewer than 29 patents, were in designing transmission lines and antennas necessary for television's broad bandwidth. After the outbreak of war he hastened home and joined the RP staff early in 1940. He became the local wizard on antennas and transmission lines, but by war's end had also gained new skills working on receivers, operational aspects of radar systems, and atmospheric propagation. Just as importantly, the intense wartime environment had cultivated and honed his abilities to lead scientific research teams.[21]

With Bowen as Chief and Pawsey as his right-hand man (in charge of most research activities), investigations of extraterrestrial noise steadily grew as

---

[18] CSIRAC, which was used at RP from 1951 on, employed 2000 tubes, used 30 kW of power, had a 2 msec cycle time, and stored 1K 20-bit words in mercury acoustic-delay lines. See Beard, M., and Pearcey, T. (1984), "The genesis of an early stored-program computer: CSIRAC," *Annals of the History of Computing* **6**, 106–15; and Pearcey (1988), *A History of Australian Computing* (Caulfield East, Victoria: Chisholm Institute of Technology).

[19] See Tangent 7.1 for details.

[20] Bowen, as in note 10, p. 6.

[21] Biographical information on Pawsey comes from obituaries by A. C. B. Lovell (*Biog. Mem. Fellows Roy. Soc.* **10**, 229–43 [1964], including a section on wartime work by H. C. Minnett) and by W. N. Christiansen and B. Y. Mills (*Australian Physicist* **1**, 137–41 [December 1964]).

resources were shifted toward those persons showing superior results or great promise. This flexibility was the CSIR style (Schedvin 1982), largely molded by Rivett, who believed that research programs should be based on people, not topics – getting the right people and then letting them loose. As Bowen (1978:42T) later recalled:

> We tried many things, but the criterion for going on with any program was, of course, success. And the things that Pawsey was trying on the sun and Bolton on point sources were so outstandingly successful that that's the way we went. ... With our first-rate staff as a handout from the war, we had the freedom and the encouragement to find new projects.

Or as Pawsey (1961:182) put it:

> [Scientific directors must] very quickly make decisions and supply facilities for the really promising developments. In all too many cases elsewhere the energies of scientists are taken up in advertising the potentialities of their prospective investigations in order to obtain any support at all.

As the years passed, work on solar and cosmic noise grew in importance at RP and a circle of group leaders emerged (most are pictured in Fig. 7.3). Besides his overall supervision, Pawsey led a large group studying numerous aspects of the radio sun (next section and Chapter 13). In 1947 John G. Bolton began his pioneering work on discrete radio sources (Section 7.4 and Chapter 14) and soon had an active group around him. J. Paul Wild arrived in 1947 and, after a year languishing in the instrument test room, moved into research on solar radio bursts with a swept-frequency receiver (Section 13.2.1.2). Bernard Y. Mills, an electrical engineer who went straight from Sydney University to RP during the war and became expert on receivers and displays, worked on a variety of projects before permanently switching in 1948 to radio astronomy, briefly on the sun and then into his own program on discrete sources (Chapter 14). W. N. "Chris" Christiansen arrived at RP in 1948 from AWA and immediately plunged into a solar research program (Section 13.1.3). Trained at Melbourne University in the mid-1930s, he was unique among this group in that, despite his longish career as a top antenna engineer, he had long wanted to be an astronomer (Christiansen 1976:1,6T).

Jack Piddington and Harry C. Minnett began a program of microwave research in 1948 (Sections 12.5.2 and 14.3.5), and Frank J. Kerr and C. Alexander Shain started on lunar radar in 1947 (Section 12.5.1). Finally, Stefan F. Smerd and Kevin C. Westfold complemented all of the observational work by working on the theory of solar radio emission (Section 13.1.2).

Among all these successes the RP archives also give evidence of one important (in retrospect) missed opportunity,[22] that of discovering the 21 cm spectral line arising from interstellar hydrogen. The line had been predicted in 1944 in Holland, and its 1951 discovery at Harvard and Leiden Universities was to be one of the major turning points in early radio astronomy (Chapter 16). Pawsey first got wind of the idea in early 1948 while on a tour of the United States. His report home triggered two years of intermittent activity at RP. Wild published a theoretical analysis, Bolton and Westfold translated a Russian article and were eager to search for the line, and Mills also gave the hunt serious consideration. But depite all this activity, in the end the decision of Bowen, Pawsey, and their staff was to not pursue the line. Details are discussed in Section 16.2, but it is interesting to discuss the case of Mills, who was looking for an independent line of research on cosmic noise in 1949. He and Pawsey discussed two main avenues:

> One was a search for the hydrogen line. Pawsey was very interested in it at the time. And the other was trying to locate very precisely the positions of radio sources. And it was a difficult decision to make. I eventually chose the precise positioning because I was more familiar with some of the techniques, and it looked as if it was something that would lead to an immediate result, whereas the other was extremely speculative.
>
> (Mills 1976:6–7T)

---

[22] Another "miss," of a different sort, only came to light after Americans Bernard F. Burke and Kenneth L. Franklin (1955) discovered very strong, low-frequency (22 MHz) radio bursts originating in Jupiter. Upon learning of this startling phenomenon, Alex Shain (1956) realized that, during 1950–51 when he had conducted an 18 MHz sky survey (Shain 1951), he may well have fortuitously also caught such Jovian bursts. And indeed, when he checked his old strip-chart recordings, evidence of Jupiter was not hard to find – Shain had dismissed the clearly visible signals as manmade interference (see Orchiston [2005a:131–3] for further information).

Figure 7.3 Radio astronomers at Sydney University for the 1952 URSI General Assembly. Left to right, ground level: W. N. Christiansen, F. G. Smith (England), J. P. Wild, B. Y. Mills, J.-L. Steinberg (France), S. F. Smerd, C. A. Shain, R. Hanbury Brown (England), R. Payne-Scott, A. G. Little, M. Laffineur (France), O. B. Slee, J. G. Bolton. First step: C. S. Higgins, J. P. Hagen (USA), J. V. Hindman, H. I. Ewen (USA), F. J. Kerr, C. A. Muller (Netherlands). Second step: J. H. Piddington, E. R. Hill, L. W. Davies.

If I had been a trained astronomer and therefore aware of the possible great importance of the H line, no doubt this would have been my choice. But I looked on it as merely a technical challenge, whereas I was intrigued by the mystery of the discrete sources and had no hesitation in choosing this option.

(Mills 2006:3)

Mills's decision not to pursue the hydrogen line can hardly be called a managerial mistake, for he went on to do leading research on discrete sources. Nevertheless, given its resources and technical expertise, the fact remains that RP surely would have soon succeeded in detecting the interstellar 21 cm line if it had ever made a serious effort. As it turned out, RP *did* make first-rate contributions to 21 cm hydrogen observations in the early 1950s (Section 16.5), but only after others had taken the initiative.

7.3 EARLY SOLAR STUDIES

7.3.1 Wartime efforts

As radar receivers during the war became more sensitive and moved to higher frequencies, concepts of receiver noise, background noise, and antenna temperature gained currency:

Receivers were getting more and more sensitive and we were concerned with the whole thermodynamic theory of their noise level and its relationship, through the antenna, to space – if the antenna were in an enclosure at three hundred degrees, what would be the noise level? This was different from the purely circuit approach that had been worked up by Nyquist and others. … And it obviously occurred to Ruby Payne-Scott and Joe Pawsey that radiation from objects might possibly be seen. I remember that Ruby had a small paraboloid poking out a window at certain objects in the sky to see how the noise level varied.

(Minnett 1978:10T)

Ruby V. Payne-Scott (1912–1981; Fig. 7.4) was the only woman to make a substantial contribution to radio astronomy during the postwar years. She trained before the war as a physicist at Sydney University, worked on cancer radiology, spent two years at AWA, and from 1941 on at RP mainly worked on display systems and calibration of receivers. She soon became known around RP for her considerable intellectual and technical prowess, forthright personality, "bushwalking" avocation, and left-wing politics. A pioneer for women's rights within RP and CSIRO, Payne-Scott fought regulations, often successfully, that specified things such as no smoking or wearing of shorts for women, as well as less pay for women than for men in the same position. When she married in 1944, she had to keep it secret because of CSIRO's policy forbidding married women on permanent staff. Then in 1951, despite being one of the leading handful of RP radio astronomers, her research career summarily ended when she chose to become a mother and had no choice but to resign in order to raise a family.[23,24]

It was in March and April 1944 that Pawsey and Payne-Scott (1944) looked at the microwave sky. In their subsequent RP report they discussed various contributions to the noise power measured by a receiver–antenna combination and cited Jansky's and Reber's (1940a) work on cosmic static. But their operating wavelength of 10 cm was far shorter than that of earlier reported work on noise from either terrestrial or extraterrestrial sources. They used a 20 × 30 cm horn connected to a receiver with a system temperature of ~3500 K, one person pointing the horn around the room or out the window in various weathers, the other taking readings from a meter. Changes of 20 to 300 K (in antenna temperature) were noted, and they

Figure 7.4 Ruby Payne-Scott as a young woman.

were particularly struck by the apparently low absolute temperature of the sky, less than 140 K. Moreover, they noted a "most unusual" consequence of this: inserting attenuation between the horn and receiver actually *increased* the measured signal!

They also tried to detect microwaves from the Milky Way with the same receiver and a 4 ft diameter dish pointing first in the vicinity of the constellation of Centaurus and then away. There was no detectable difference, that is, less than ¼ % (<10 K), "very much less than that observed by Jansky and Reber." Appealing to Eddington's (1926:430) work (about which they undoubtedly learned from Reber's citation), they ascribed the low signal to a very low temperature for the material in space.[25]

---

[23] I thank W. Miller Goss for information about Payne-Scott's life. His biography of her (with R. X. McGee) is *Under the Radar: Ruby Payne-Scott, First Woman Radio Astronomer* (2009).

[24] Another female physicist on the wartime RP staff who considered a postwar career in radio astronomy was Joan Freeman, but in 1946 she was advised by Pawsey that conditions at the field stations were not suitable for a woman. She therefore decided on nuclear physics, obtained a Ph.D. at the Cavendish Laboratory, and pursued a career as a physicist in England. [J. Freeman, *A Passion for Physics*, as in note 8]

[25] Pawsey and Payne-Scott misinterpreted Eddington's statement. Eddington had said that the energy density of starlight at a typical point in interstellar space was about the same as that of a blackbody radiation field of temperature 3.2 K. This does not at all imply that, as they stated, "the radiation from the matter in space is equivalent to that of a blackbody at 3.2 K." The spectrum of interstellar material could have been quite non-blackbody, depending on the details of its interaction with the starlight. (Eddington's

These Milky Way results were accompanied by a single sentence stating that they did not try for any solar radiation. At that time they were apparently unaware of Hey's and Southworth's secret reports (Chapter 5) on the sun (and Reber's 1944 publication [Section 4.5.1] was still six months away), but given that they mentioned the sun at all, why did they not try for it? If they had, they probably would have easily detected a change in power output.[26]

These kinds of idea were thus in the air around RP and therefore, as already discussed, merited a short paragraph in Bowen's postwar program proposed in July 1945. But over the next two months overseas reports all of a sudden were to make looking at the sun a high priority. How did this come about? There is some ambiguity in the archival evidence as to exactly in which order RP learned about the solar observations of Hey, Southworth, Reber, and Alexander, and the subsequent Australian reactions. Nevertheless, my reading of the evidence is that what galvanized Bowen and Pawsey into jumping onto solar noise was the "Norfolk Island effect" – solar radio bursts observed by New Zealand military radar stations from as early as March 1945 and analyzed by Elizabeth Alexander, an English radar researcher at an RP-type lab in New Zealand (and a personal friend of Pawsey); details of this episode are in Section 5.3.2. When Bowen learned of these observations in July 1945, he was entranced:

> We will attempt to repeat [these very interesting] radar observations here in Sydney. We are quite mystified by the results because it appears that while thermal noise from the sun is expected at radio frequencies and is actually received on 10 and 3 cm equipment, one would not expect to be able to detect it on C.O.L. equipment at 200 Mc/s.[27]
>
> I have heard rumours of the same thing happening in England, but as far as I am aware, the subject has never been followed up.[28]

These letters testify that in August 1945 Bowen and Pawsey knew about thermal microwave radiation from the sun, from Southworth's two restricted reports (*not* from his early 1945 article since this was not cited at RP even as late as 23 October 1945 by Pawsey *et al.* [1946]), but were unaware (except by rumor) of Hey's low-frequency solar bursts (either from his 1942 report or later June 1945 version; see Chapter 5).

Full information about the "Norfolk Island effect" arrived at RP with a 1 August 1945 letter to Pawsey from Alexander (1945), containing her formal report.[29] This letter also intimates that Pawsey knew something of the New Zealand radar data as early as the previous June. We also know that Payne-Scott (1945, see below) referred in a December report to having been "inspired by the almost simultaneous arrival of three reports in the laboratory": Hey, dated 13 June 1945; Alexander, dated August 1945; and Reber, journal issue of November 1944). Finally, in the original version of the group's major paper written six months later (see below) we find:

> Shortly after Reber's publication two restricted reports reached us. ... The first report ... [was] ... Royal New Zealand Air Force radar stations. ... The second ... [was] ... English radar stations. ... On receiving the New Zealand report we began similar observations.[30]

1926 value of 3.2 K is only by coincidence close to the cosmic microwave background temperature of 2.7 K discovered in the early 1960s.)

[26] If we assume a brightness temperature for the sun (at solar minimum at 10 cm wavelength) of 35,000 K, then Pawsey and Payne-Scott would have detected an antenna temperature with their 4 ft dish of ~150–200 K, well above their sensitivity to changes of ~20–30 K. This type of dish and microwave receiver was in fact very similar to that employed by Southworth in 1942–43 (Section 5.6). Minnett has speculated, from his memory of the room used, that the March sun was not easily observable from any window (Minnett:author, 22 October 1986).

Although there is no written record of anyone at RP trying for the sun before the end of the war, Kerr later recalled a brief attempt he made at a wavelength of 1.5 m (1971:7T, 1976:53T) with a small antenna. Also note the brief try in 1939 by Piddington and Martyn (Section 5.4.1).

[27] Bowen:E. Marsden (DSIR, New Zealand), 27 July 1945, file B51/1, RPS. "C.O.L." was "Chain Overseas Low-flying," in contradistinction to the original Chain Home system in England (Section 5.1; also see note 38).

[28] Bowen:F. G. W. White, 1 August 1945, file A1/3/1a, RPS.

[29] E. Alexander:Pawsey, 1 August 1945, A1/3/1, RPS. Other useful letters are Marsden:White, 9 July 1945, B51/1, RPS; Bowen:White, 17 October 1945, B51/14, RPS.

[30] J. L. Pawsey, R. Payne-Scott, and L. L. McCready, "Solar radiation at radio frequencies and its relation to sunspots," 16 June

Early solar studies 129

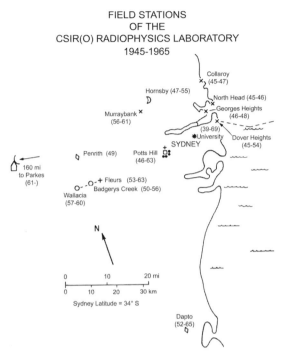

Figure 7.5 Sites of chief RP field stations. Each station has the years of operation indicated. "University" refers to the main RP site on the campus of the University of Sydney.

Figure 7.6 A 200 MHz wartime radar antenna for shore defense, consisting of a dipole array rotatable only in azimuth, at Dover Heights, Sydney (1941). This and a similar antenna at Collaroy were used by Pawsey's group in 1945–46 as sea-cliff interferometers for observations of the rising sun.

On balance, it seems that the first RP solar investigations in October 1945 were mainly triggered by the New Zealand observations, but certainly also influenced by the other solar reports.

7.3.2 Solar bursts and the sea-cliff interferometer

Pawsey swung into action and mounted an observing program on a frequency of 200 MHz using existing Air Force radar installations along the coastline near Sydney; Pawsey himself had played a major role in the development of these type radars. Working with him were Payne-Scott and Lindsay L. McCready (1910–1976), a receiver expert, prewar AWA engineer, and RP veteran who at the time was Pawsey's deputy and eventually became head of engineering services at RP. The first observations were on 3 October from Collaroy, fifteen miles north of Sydney on a hilltop one-half mile inland (Fig. 7.5). The antenna was an array of 40 half-wave dipoles (Fig. 7.6) and the tedious observations were carried out by Air Force as well as RP personnel. Because the antenna was restricted to looking

---

1946, Report No. RPR 24, typescript submitted for publication in *Proc. Roy. Soc.* I thank Ron Bracewell for supplying a copy.

The introductions to the original and final versions of this paper (McCready, Pawsey, and Payne-Scott 1947) provide an especially good example of how historical information is usually lost in the formalism of a scientific paper. In this case it appears to have happened because of referee comments rather than in the initial writing. The submitted version (for some reason also with a different author order than the published paper) contained historically interesting material about what the Sydney group knew from overseas reports and when they knew it. The finally published version, however, was modified in several places to merely recite who published what and what they said. In particular, the phrase "In a prior letter, not available here until our initial work was completed, Appleton (1945) …" was changed to simply "Appleton (1945) …"

only at the horizon, observations were only possible by sweeping back and forth along the horizon for an hour at dawn or dusk, writing down microammeter readings every 2° of azimuth. After only a week or two of data, Pawsey noticed that the daily maximum of "this noise effect" was highly variable from day to day and seemed to correlate with the number of visible sunspots.[31] For the latter information he had made contact with Clabon W. "Cla" Allen (1904–1987), longtime solar astronomer at the Mt. Stromlo Commonwealth Observatory near Canberra. After only three weeks of monitoring, Pawsey, Payne-Scott, and McCready (1946) rushed a letter to *Nature* showing a close correspondence between the total area of the sun covered by sunspots and the daily noise power from the sun. Despite various large corrections that had to be made for ground and sea reflections, they were confident that the daily values of solar noise varied by as much as a factor of 30 over the three weeks. They also pointed out that, for a thermally emitting disk the size of the optical sun, their detected levels implied "equivalent temperatures" ranging from 0.5 to 15×10^6 K, much higher than the sun's "actual" temperature of 6000 K. Such incredible signals, they reasoned, could not come from atomic or molecular processes, but more likely from "gross electrical disturbances analogous to our thunderstorms."[32]

With such a promising start, Bowen and Pawsey decided to increase their efforts on solar noise, and continued monitoring for another ten months. Gradually Air Force personnel and equipment were phased out as RP took over. Pawsey's group observed at a variety of frequencies, mostly at meter wavelengths and with antennas (such as Fig. 7.7) at a variety of coastal radar sites around Sydney including Collaroy, North Head, Georges Heights, and Dover Heights (Fig. 7.5).[33]

Figure 7.7 Pairs of 100 MHz (left) and 60 MHz Yagi antennas used for monitoring the sun and measuring its circular polarization at Dover Heights (1947). The same pairs, with the Yagi elements parallel and pointing toward the horizon, were used for the first studies and surveys of discrete sources by John Bolton (pictured) and his group.

While these data rolled in, the group also educated themselves about the solar atmosphere and began thinking about how it might emit radio waves. At the end of the year Payne-Scott (1945) wrote an insightful lab report summarizing all "knowledge available and measurements taken at RP."[34] After deriving various basic relationships about blackbody radiation, antenna temperature, and noise, she concluded that with available antennas and receivers the Milky Way should be more easily detectable at long wavelengths and the sun at short wavelengths, in agreement with known data. She also discovered Southworth's incorrect calculations of solar blackbody radiation (discussed in Section 5.6). Over the previous two months Payne-Scott and her colleagues had conducted many

---

[31] Pawsey:C. W. Allen, 15 October 1945, file A1/3/1a, RPS.

[32] Orchiston, Slee, and Burman (2006) also discuss these Collaroy observations and related events, as well as including many valuable photographs of equipment and people.

[33] Information on the wartime history of the Dover Heights site, which would become one of the most important in RP's postwar efforts and where during the war many radar sets had been tested, is in H. Minnett (1999), "The Radiophysics Laboratory at the University of Sydney," *Historical Recs. Austral. Sci.* **12**, 419–28.

[34] I thank Wayne Orchiston for a copy of this report.

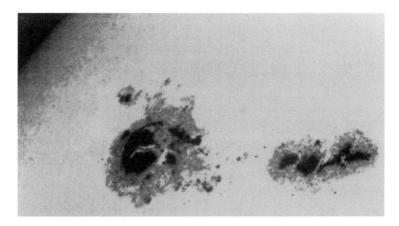

Figure 7.8 The sunspot group of February 1946, the largest ever recorded up to that time. Its associated radio bursts were studied by Pawsey's group with a sea-cliff interferometer and by Hey's group at AORG (Section 6.4).

tests to convince themselves that the radiation indeed originated on the sun and not in some secondary ionospheric effect that might be especially strong when observing at the horizon. One clincher was to use four 200 MHz Yagi antennas mounted on a steerable searchlight to track the sun high in the sky over an entire afternoon – and indeed the signal persisted at high elevation angles. However, they also frequently observed "kicks" in their meter readings – strong fluctuations over only a few seconds (superimposed on the day-to-day variations) – and Payne-Scott could not tell whether or not these were intrinsic to the sun or an atmospheric effect. Finally, she plotted a rough map of the entire visible sky that they made one day with the 200 MHz searchlight system. The contours of intensity showed a "fairly diffuse area" agreeing with the direction to the galactic center, just as in Reber's published map.

Payne-Scott ended the report with some ideas as to the cause of the observed radiation. Because of similarities in the Milky Way and solar radiation, she felt it best to assume that they both had the same cause, i.e., that the Galaxy should be considered as a "collection of stars."[35] But the origin of the sun's radiation, especially at their low frequencies, was very puzzling, whether one took it to originate beneath the (optical) surface of the sun or perhaps in the hot corona, as she notes that Martyn had recently suggested (see Section 7.3.3). Another idea was "disturbances in the sun's atmosphere analogous to our atmospherics and linked with sunspot activity." Because they had observed that the 200 MHz intensity not only correlated with total sunspot area, but also was stronger when large sunspots were located centrally on the solar disk (meridian transit), she speculated that perhaps "the radiation may emerge from, say, a crater on the sun's surface, and so be highly directional."

Payne-Scott's report set the thinking for the RP group. For instance, Bowen wrote to Appleton in January 1946 in order to comment on the latter's letter in *Nature* on radio noise and sunspots (in the 3 November issue, but only just arrived in Sydney; see Section 5.5). Bowen pointed out that RP had now obtained the "first direct experimental verification of this effect," and that, unlike in the still unpublished *Nature* letter sent from RP two months earlier,[36] the Australian team now felt that the solar noise was not "electromagnetic," but of thermal origin, either "from the depths of the sun or in some way from the corona."[37]

The climax of this initial breathtaking period came in early February 1946, when by good fortune the largest sunspot group ever seen (until then) chose to make its appearance (Fig. 7.8; also see Section 6.4). When

---

[35] Payne-Scott did not, however, point out one key difficulty with this idea, mentioned already by Reber and Jansky: if sun and Galaxy radiation are of the same origin, why is the sun/Galaxy ratio of optical intensity so much greater than that for radio?

[36] The *Nature* paper (Pawsey, Payne-Scott, and McCready 1946) was sent off on 23 October (and contains data through that date!), but did not appear in print until the 9 February issue.

[37] Bowen:Appleton, 23 January 1946, file A1/3/1a, RPS.

Allen phoned with news of the giant group (covering about 1% of the sun's visible disk), the RP group intensified their monitoring and realized that they now had the opportunity to take advantage of a property of their antenna system that had previously been more bother than help. A single antenna situated on the edge of a cliff or a hilltop, looking near the horizon over a relatively smooth terrain or over the sea, in fact acts as an interferometer and can achieve far better angular resolution than would otherwise be possible. The interference in this case is between that portion of a wavefront directly impinging on the antenna and that portion reflected from the sea, which must travel an additional length equal to twice the cliff height times the sine of the source's elevation angle. In classical optics this arrangement is known as "Lloyd's mirror" and the fringes obtained are equivalent to those with a conventional interferometer consisting of the antenna and an imaginary mirror image located under the base of the cliff (see Fig A.6b and Section A.11). With the antennas at Dover Heights and Collaroy located 85 and 120 m above the sea, the respective fringe lobes at 1.5 m wavelength were spaced by 30′ and 21′. In principle, then, one could locate objects with an accuracy of ~10′, far better than the ~6° beam of the antenna considered by itself.

This phenomenon was nothing new to those who had been developing radar systems, for during the war radar beams often pointed near the horizon, as with search radars on a ship or a coastline. The reflected signal from a distant aircraft was well known to oscillate as it passed through the fringes or lobes of such a radar. This effect was both a blessing and a curse to the radar systems designer, for it could be used to gather precise information on a target's height, but on the other hand it meant that low-flying aircraft could sneak in "under" a radar, since the first lobe was *not* at the horizon, but above it by a considerable amount, especially for longer wavelengths and antennas that were not high above their surroundings.[38]

So Pawsey and his colleagues used this sea-cliff interferometer[39] to advantage as the bespotted sun rose over the Pacific. The general level of solar emission was far above normal for several days and often interspersed with bursts. As before, they found that the solar signal appeared at sunrise and gradually faded as the sun rose above the antenna beam, but now superimposed were striking oscillations, the interferometric fringes (Fig. 7.9). And the exciting thing was that the very presence of these oscillations implied that the source of the solar signal was a good bit smaller than the spacing of the fringes (20–30′) and therefore smaller than the 30′ size of the optical sun. Exactly how much smaller the emitting region was, as well as its location, could be worked out through details of the oscillations' amplitudes and phases. This led to Fig. 7.10 where they inferred that the emitting region on any given day had a width of 8–13′ and coincided with the giant sunspot group being carried along by the sun's rotation. Even though the fringes of the sea-cliff interferometer were oriented parallel to the horizon and thus could give no information about the azimuth of the emitting region, it seemed eminently reasonable that the sunspot group itself originated the enhanced radiation. They had thus directly confirmed what Hey (1946) and Appleton (1945a) had earlier only surmised.

In a paper submitted to the *Proceedings of the Royal Society* in July 1946, McCready, Pawsey and Payne-Scott (1947) reported the above results and much more. They expanded on their first results in *Nature* and now characterized the solar radiation as consisting of two components: (1) a slowly changing type that could vary by a factor of 200 in intensity over many days, and (2) intense bursts, lasting from less than a second up to a minute, that could be tens of times more powerful than the general level on a given day. These results were so unexpected that they worried at length that the ionosphere might somehow induce the burst phenomenon, but various arguments, principally the fact that separate sites observed the bursts at the same time (to within a second), convinced them that this indeed was an extraterrestrial "and presumably solar" phenomenon. As in

---

[38] The "Chain Home" radars defending the British coast during World War II (Section 5.1) for the same reason had lowest lobes considerably above the horizon and were therefore much less effective against aircraft flying at low altitudes. But the Germans apparently were not aware of this basic property, for if they had been, their bombers during the Battle of Britain in 1940 could have avoided early detection simply by flying at lower altitudes

(Bowen 1947:516). This defect was later corrected by using shorter wavelengths in the "Chain Home Low-flying" (CHL) system.

[39] In this book I use the term *sea-cliff interferometer*, although at the time the arrangement was called either a *sea interferometer* or a *cliff interferometer*.

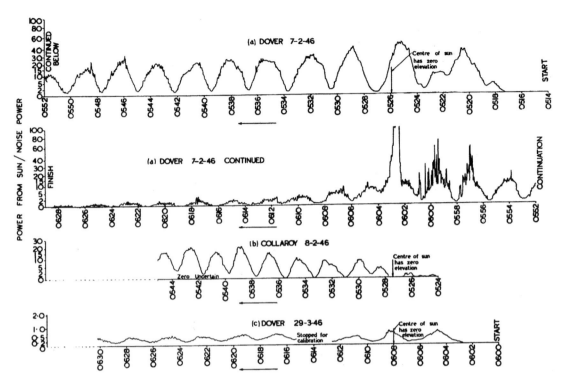

Figure 7.9 Sea-cliff interferometer fringes obtained at 200 MHz in February and March 1946. When the sun rose, the fringes suddenly appeared (note that "radio sunrise" came earlier than optical sunrise) and then gradually faded away an hour later as the sun moved above the antenna beam. Note the greater ratio of fringe maximum to minimum for 7 and 8 February compared to 29 March, indicating that the radiation originated from a smaller region of the sun in February. The very fast variations and intense signals recorded at 0600 on 7 February indicate a solar outburst. Note the closer spacing of the fringes for the Collaroy observations, taken from a higher elevation above the sea. (McCready, Pawsey, and Payne-Scott 1947)

their previous letter to *Nature*, they pointed out that the equivalent brightness temperatures for these bursts were extraordinarily high, as much as 3×10^9 K.

This seminal paper also explained many basics of the sea-cliff interferometer, considering effects such as refraction (the worst uncertainty),[40] the earth's curvature, tides, and imperfect reflection from a choppy sea. As mentioned above, many of these effects had already been worked out during the war, for instance in a 1943 RP report by J. C. Jaeger.[41] But Pawsey's group also introduced a vital *new* principle, namely that their interferometer was sensitive to a single Fourier component (in spatial frequency) of the brightness distribution across the sun, and that a complete Fourier synthesis could be achieved if one had enough observations with interferometers of different effective baselines (Appendix A):

---

[40] Bolton (1982:351) much later claimed that Pawsey's group used incorrect values for refraction in this study, and that they *assumed*, rather than *established*, that radio positions corresponded to optical sunspot positions. Unfortunately, the McCready et al. (1947:204–6) paper is ambiguous regarding their procedure for determining refraction, but an analysis by Miller Goss, Barry Clark, and the author concludes that Bolton's claim is incorrect. We find that their basic method must have been to determine mean radio refraction by combining all days of observation, and then apply this correction to each individual day, as modified by a model that accounted for day-to-day variability in refraction due to changes in temperature, pressure, and humidity. For details see Appendix A13 in Goss and McGee (2009).

[41] Jaeger, J. C., 17 March 1943, "Theory of the vertical field patterns for R.D.F. stations," Report No. RP.174, 15 pp. and 11 figs., RPL. See note 15.20 for another wartime study of this effect (in England by Fred Hoyle).

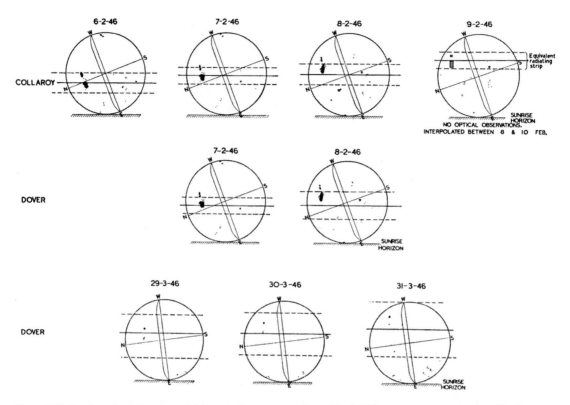

Figure 7.10 Sketches of optical sunspots visible on the days corresponding to Fig. 7.9. The top two rows are dominated by the great sunspot group (Fig. 7.8), while the March observations show much less activity. "N-S" indicates the rotation axis of the sun. The three horizontal lines on each sketch indicate the center and estimated width of the "equivalent radiating strip" causing the radio fringes. (McCready, Pawsey, and Payne-Scott 1947)

Since an indefinite number of distributions have identical Fourier components at one [spatial] frequency, measurement of the phase and amplitude of the variation of intensity at one place at dawn cannot in general be used to determine the distribution over the sun without further information. It is possible in principle to determine the actual form of the distribution in a complex case by Fourier synthesis using information derived from a large number of components. In the interference method suggested here … different Fourier components may be obtained by varying the cliff height h or the wave-length λ. Variation of λ is inadvisable, as over the necessary wide range the distribution of radiation may be a function of λ. Variation of h would be feasible but clumsy. A different interference method may be more practicable.

(McCready, Pawsey, and Payne Scott 1947:367–8).

Much of the subsequent technical development of radio astronomy was to be concerned with this method of making high-resolution cuts across sources, and eventually complete maps. Through the 1950s and later their suggested type of Fourier synthesis became central to radio astronomy. But the last two sentences of the above quotation were prophetic, for it was not sea-cliff interferometry, but the more tractable and flexible conventional interferometry with separate, movable antennas (Chapter 8), that made such mapping a reality.[42]

---

[42] See notes 8.61 and 8.62 for Martin Ryle's first (unpublished) mention of the synthesis principle. In the same August 1946 document, Ryle also considered the feasibility of having in essence a sea-cliff interferometer with variable fringe spacing – simply observe a low-elevation source from an aircraft flying at different heights above the sea!

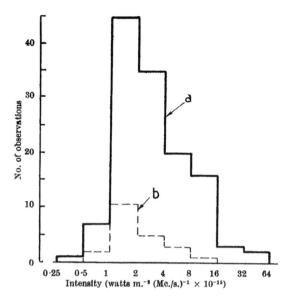

Figure 7.11 Histograms of daily 200 MHz solar intensity: (a) October 1945 to March 1946, observed by Air Force personnel, and (b) March to May 1946, RP staff. Note the marked lower limit, corresponding to an effective (brightness) temperature for the solar disk of ~1×10^6 K. (Pawsey 1946)

7.3.3 The million-degree corona

Sometime in mid-1946, Pawsey extracted another jewel from his wealth of data. He noticed that his large set of daily values of the 200 MHz solar flux density had a peculiar distribution (Fig. 7.11) with a sharp lower limit corresponding to an equivalent brightness temperature for the solar disk of about 1×10^6 K. This was drawn from the same data presented earlier, but looked at in a new way: first, with a histogram of values (~150 values over seven months) rather than a plot against time, and second, using single-day values rather than three-day averages. Pawsey had earlier argued that three-day averages were necessary because the solar bursts frequently vitiated daily observations, but now he saw that this averaging had also tended to mask the marked lower limit of intensity, since about two-thirds of all days exhibited enhanced levels.

At this same time David Martyn, who was now at Mt. Stromlo and very interested in this hot new field, introduced a theory that could explain a million-degree base level for the solar radiation. He had learned that recent studies of ionization states and spectral-line widths strongly suggested that the solar corona had a temperature of about 1×10^6 K (Section 13.3.1). *Why* the corona was so hot was not at all understood, but the evidence was there. Martyn realized he could apply standard techniques in ionospheric theory to calculate the expected radio emission from the sun. Once he had adopted likely values for the electron densities in the corona, he found that the corona was opaque at Pawsey's kind of frequencies. The observed radio waves were therefore emanating not at all from the 6000 K optical surface (photosphere) of the sun, but from well above, out in the million-degree corona. When the sun was quiet, this coronal thermal emission constituted the entire solar signal; when active, the coronal emission was dwarfed. Furthermore, at shorter wavelengths, the observed emission would come from deeper in the corona, and eventually even from the chromosphere. This powerful idea thus explained why the measured brightness temperature of the quiet sun always seemed greater than 6000 K and sharply increased at longer wavelengths. It also meant that one could now study the corona without the inconvenience of having to chase down a total eclipse. As it turned out, in the Soviet Union a few months earlier Vitaly L. Ginzburg (1946) had independently made similar calculations while considering the possibility of reflecting radar off the sun (Sections 10.3.1 and 12.2). And the basic ideas were again independently presented, yet a third time, in the Russian literature in a late 1946 paper by Iosif S. Shklovsky (Section 10.3.1). But Martyn had access to better confirming data and was positioned more in the mainstream of postwar radio astronomy. His paper, in *Nature* for 2 November 1946, had far more influence,[43] and is discussed in detail in Section 13.1.2.[44]

---

[43] It appears that Ginzburg's (1946) paper, although written in English, was not "discovered" at RP until mid-1948, only after a summary appeared in *Physics Abstracts*. [Bowen:Pawsey (in London), 26 July 1948, file A1/3/1a, RPS]

[44] At Bell Labs in 1945–46, Charles Townes also recognized that thermal radiation from a 10^6 K corona could account for the excess radio emission measured by Reber (1944), Southworth (1945), and Dicke and Beringer (1946). The final version of his paper (Townes 1947) was not submitted to the *Astrophysical Journal*, however, until September 1946 and the section on solar radiation was required to be eliminated by the editor and referees. (The paper dealt primarily with free–free radiation as the explanation for galactic noise and is discussed in

The above description of Pawsey's and Martyn's work seems fairly well established, but there is controversy over whether Martyn first predicted the million-degree corona and then suggested to Pawsey to seek it in his data, or whether Pawsey first found it empirically and so instigated Martyn's working on the problem. The archival evidence does not speak with certainty. It does show that Pawsey and Martyn were planning a joint publication on this subject in July and August of 1946, but that Martyn then backed out since he and Woolley had decided to do their own theoretical study. Pawsey then persuaded Martyn to change his mind, but in the end Martyn (1946a) sent off his own note to *Nature* early in September, apparently without Pawsey's knowledge. Pawsey got wind of this, however, and within a week convinced Martyn to agree with Pawsey (1946) sending in his own short note and suggesting to *Nature*'s editor that it follow Martyn's.[45] The collaboration had clearly gone sour, resulting in two adjacent notes: Martyn's did not mention Pawsey's base-level data at all (citing only Reber's and Southworth's measures of the solar intensity), while Pawsey's acknowledged his indebtedness to Martyn for "pointing out to me the probable existence of high-level thermal radiation."

Within a few months, however, arguments developed over who had priority over "the million-degree corona." In January Bowen wrote to Martyn because of "your insistence on the importance of the written as against the spoken word." Bowen cited a year-old press release, which referred to the RP work as indicating that the usual "apparent temperature" of the sun was a million degrees, as clearly antedating Martyn's note in *Nature*, and said that RP knew about million-degree radiation from the sun long before Martyn came along.[46] Further direct evidence lies in 1948 letters commenting on a draft of a radio astronomy review then being written by Appleton for the International Union of Radio Science (URSI). Martyn claimed to be holding the epistemological high ground:

There is a natural tendency now to look on my theory as one designed to explain the observed facts, which followed rapidly upon its heels. In point of fact it was developed (see the internal evidence in Pawsey's *Nature* letter) before the facts were known. It is a theory of prediction rather than explanation, and perhaps has correspondingly greater weight because of that.

And Pawsey separately wrote:

The actual sequence of events … was as follows: (a) observation of considerably high and very variable effective temperatures, 10^6–10^8 degrees on 200 Mc/s – J.L.P. and colleagues. (b) Suggestion of high-temperature coronal thermal emission – D.F.M. and colleagues. (c) Successful search for 10^6 degree base level on 200 Mc/s – J.L.P. (d) Detailed theory – D.F.M.[47]

Pawsey also wrote from overseas to Bowen about this time:

I think Martyn might get a mention in [Appleton's] section on the discovery of thermal radiation. I am all for a quiet life and the theory was a vital part of the discovery.[48]

From this evidence it would appear most likely that Pawsey's own recounting of events best tells the story, although it should be noted that he was at times self-effacing. It is clear that Martyn's withdrawal from collaboration and lack of any mention of Pawsey's base-level work upset the RP staff. Nonetheless, it seems that Martyn was indeed the one who brought in the previous astronomical evidence of a million-degree corona and who pointed out that the million-degree "effective" or "apparent" temperatures cited by the RP group could actually represent *thermal* emission from the solar atmosphere.[49] Pawsey and his colleagues had calculated these temperatures,

---

Section 15.2.1.) [Townes 1981:156B and archival material from Townes]

[45] Correspondence between Pawsey, Martyn, Woolley, and Bowen, July to September 1946, all in file B51/14, RPS.

[46] Bowen:Martyn, 28 January 1947, file A1/3/1a, RPS.

[47] Martyn:Appleton, 27 October 1948, APP; Pawsey:Appleton, 8 September 1948, file F1/4/PAW/1, RPS. Pawsey's step (b) regarding Martyn's suggestion would appear to have happened at some time before December 1945, as Payne-Scott's report that month (Section 7.3.2) mentioned a suggestion by Martyn that the 200 MHz radiation might originate in the hot corona.

[48] Pawsey:Bowen, 15 August 1948, file F1/4/PAW/1, RPS.

[49] The relationship between the radio discovery and the general acceptance by astronomers of a million-degree corona is discussed in Section 13.3.1.

but thought of them only in a formal sense. In fact, to them these incredibly high values were at first prima facie evidence of *non*thermal phenomena.

7.3.4 Mt. Stromlo

It is remarkable that the Commonwealth Observatory at Mt. Stromlo worked so closely with RP right from the start – such active collaboration between astronomers and radio investigators occurred no where else in the world in the first few years after the war (Section 17.1). Since 1930, however, Mt. Stromlo had been doing a small amount of RRB-funded ionospheric research (in particular by Arthur J. Higgs, who after the war became RP's Technical Secretary). Moreover, during the war Cla Allen had worked on the effects of sudden solar disturbances on ionospheric conditions and optimum communications frequencies (Mellor 1958:502–9). As we have seen, Allen from as early as October 1945 was feeding optical solar data to RP and indeed over the years his ties with RP remained strong (Smerd 1978:92A:720). Since Mt. Stromlo was already in the solar monitoring business at optical wavelengths and RP did not want to maintain a daily radio patrol, the idea soon developed of RP installing a radio system at Mt. Stromlo. This happened in early 1946 and from April onward Allen oversaw regular 200 MHz solar monitoring with a steerable array of four Yagi antennas.[50] This led to a paper in which he presented a year's worth of radio data and correlated it with optical and ionospheric activity (Allen 1947; Section 13.2.1 and Fig. 13.1). He used the same array in 1949 to make a map of galactic noise (Allen and Gum 1950).

In addition to Allen, Martyn worked on extraterrestrial noise as a sideline to his ionospheric research. Besides his important work on the million-degree corona, he also pointed out in his 1946 *Nature* note that at wavelengths of ≤ 60 cm the quiet sun should appear brighter at its edges than in the center. This prediction of "limb brightening" turned out to be qualitatively correct, although it took more than five years before observations of sufficient detail seemed to settle the question (Section 13.1). Further, when a large sunspot group appeared in July 1946, Martyn had Allen's four-Yagi array modified so that it could search for polarization in radio bursts (Martyn 1946b). This led to detection of a high percentage of circular polarization and the discovery that the handedness of the polarization neatly flipped when the sunspot group crossed the solar meridian. Similar and simultaneous results were also obtained in England and were valuable for elucidating the influence of the sunspot group's magnetic field on the emitted radio waves (Section 13.2).

This radio activity could not have flourished without the encouragement of the observatory director and Commonwealth Astronomer, Richard v.d.R. Woolley (1906–1986). Woolley was a stellar and dynamical theorist, an Englishman who had come to Australia to take over Mt. Stromlo in 1939 and who would return to Britain in 1955 to become Astronomer Royal.[51] He had a personal interest in the radio work; for instance, he authored an early paper on the theory of galactic noise and others on solar models incorporating radio data. In late 1946 Woolley suggested that Bolton should check for radio emission from the nebulosity near the bright star Fomalhaut, and in 1947, after Martyn had speculated that the Cygnus source (next section) might be a distant comet, Woolley searched for such an object.[52] Moreover, relations between Woolley and RP were cordial enough that Bowen first checked with Woolley before sending off the first RP paper on the Cygnus source:

> Much as I agree with you about letters to *Nature*, it seems to have become the recognised channel for publication of new results on solar and cosmic noise. We are therefore perpetrating another one giving a brief outline of Bolton's results on Cygnus. … I should be glad if you

---

[50] Orchiston et al. (2006:51–3) give further details of Allen's monitoring program.

[51] See Woolley's obituaries by D. Lynden-Bell (1987) in *Quarterly J. RAS* **28**, 546–51 and by W. McCrea (1988) in *Biog. Mem. Fellows Royal Soc.* **34**, 922–82. Further information on Woolley and non-radio work at Mt. Stromlo during his era may be found in S. Davies (1984), *Historical Records of Austral. Science* **6**, 59–69; S.C.B. Gascoigne (1984), *Proc. Astron. Soc. of Austral.* **5**, 597–605; Gascoigne, "Australian astronomy since the Second World War," pp. 345–373 in Home (1988); and Haynes et al. (1996).

[52] Martyn:Bowen, 16 April 1947 and 25 June 1947, file A1/3/1a, RPS.

would read it and let me know if you have any objection ...[53]

In 1948 Woolley also was elected the first chairman of the International Astronomical Union's new Commission 40 on Radio Astronomy and shortly thereafter became vice-chairman of Australia's national URSI organization.

Yet although there were fruitful exchanges of ideas, data, and know-how between the astronomers and the radio physicists, tension also existed between Woolley and Martyn on one side and Bowen and Pawsey on the other. Much of this stemmed from Martyn and his status as an "exiled" RP staff member, seconded to Mt. Stromlo from Sydney. Martyn had been removed as RP Chief late in 1941 after two years of continual problems – despite his scientific excellence, he did not have the managerial skills or temperament needed to run a large organization developing new technology under the threat of Japanese attack and invasion. By 1941 his relations with the military, with industry, and with his own staff were abysmal. On top of this, in early 1941 he was viewed as a security risk because of his liaison with a German woman recently emigrated to Australia (Schedvin 1987:253–9). With this background, one can understand that his postwar relations with RP often went less than smoothly, if not bitterly.

Woolley too appears to have developed an ambiguous relationship with RP in particular and with radio astronomy in general. For instance, in a major address on the solar corona, he mentioned Allen's and Martyn's work, but none of RP's results (Woolley 1947a). In another talk the same year on "Opportunities for astrophysical work in Australia," radio was not mentioned once, although this may have resulted from his definition of *astrophysical* (Section 17.1.1).[54] Several interviewees from RP and from Mt. Stromlo have testified to Woolley's lack of support for radio astronomy. Even as late as 1954, when asked about radio astronomy after a popular talk, Woolley apparently replied that in a gathering of "real" astronomers it was not considered decent to mention radio astronomy![55,56] On the other hand Woolley was part of a proposal for an independent department of radio astronomy at Mt. Stromlo. The matter culminated in 1951–52, after the departure of Allen to take up a professorship in England. Woolley and Mark L. Oliphant, head of physical sciences at the new Australian National University in Canberra, pushed to acquire a large radio telescope, but were beaten down by Frederick White at CSIRO headquarters and by Bowen and Pawsey.[57]

7.4 RADIO STARS

In August 1946 Bowen, then visiting England, excitedly sent Pawsey a reprint of the recent letter in *Nature* by Stanley Hey, John Parsons, and James Phillips of the Army Operational Research Group (Section 6.1).[58] While mapping the general distribution of galactic noise, they had accidentally discovered that the noise from one particular spot in the constellation of Cygnus fluctuated in intensity on a time scale of minutes. Although they could measure with their beam only that the fluctuating region was less than 2° in size, they argued that such rapid changes must originate in a small number of discrete sources, perhaps only one. These sources were taken to be stars, by analogy with the sun and its radio bursts. Pawsey jumped on this. As he wrote:

[Within a few days of receiving Bowen's letter] we immediately made some confirmatory measurements

---

[53] Bowen:Woolley, 11 November 1947, file D5/4/35, RPS.

[54] Woolley's talk was the Section A Presidential address at a meeting of the Australia and New Zealand Association for the Advancement of Science (ANZAAS). [pp. 64–80, *Report of the 26th Meeting of ANZAAS, Perth 1947*, ed. A. G. Ross, Canberra: Australian Government Printer]

[55] Bok (1971:11–12T), Bowen (1984:92–3, 1973:23–4T), deVaucouleurs (1976:1N), Kerr (1971:15T), Kerr:author, 6 April 1987), Stanley (1974:24T), and Wild (1987:100). The 1954 remark is mentioned in Mills:Pawsey, 27 August 1954, F1/4/PAW/3, RPS. See note 17.59 regarding W. McCrea's more sympathetic view of Woolley's conservatism.

[56] One traditional local astronomer who looked more favorably on the upstart radio observers, and who offered advice when asked, was Harley Wood (1911–1984), an astrometrist in charge of the small Sydney Observatory. [See Wild (1987:95) and the obituary for Wood in *Proc. Astron. Soc. Australia* **6**, 111–12 (1985)]

[57] Woolley:Bowen, 12 April 1947; Martyn:Bowen, 16 April 1947, Woolley:White, 12 February 1951; White:Woolley, 23 February 1951; Pawsey:Bowen (in London), 30 March 1951; Pawsey:Oliphant, 10 September 1951 – all file A1/3/1n, RPS. Further details of Woolley's stormy relationship with RP are in Robertson (1992:107–113) and Munns (2002:73–80).

[58] Bowen (in London):Pawsey, 28 August 1946, file A1/3/1, RPS.

on 60 and 75 Mc/s, obtaining similar fluctuations, of the same form as the "bursts" observed in solar noise. We have no hint of the source of this surprising phenomenon.[59]

This early success, however, was apparently followed by a period of conflicting observations, during which the reality of the Cygnus fluctuations came into question. In the end Pawsey's group gave up, no longer knowing what to make of Hey's claim (Bolton 1976:3–4T, 1982; Stanley 1974:4–5T).

Cygnus investigations thus lay dormant for several months until resumed by John G. Bolton (1922–1993), who had joined the RP staff as its second postwar recruit in September 1946.[60] Bolton (Fig. 7.4) was a Yorkshireman who had studied undergraduate physics at Cambridge before joining the Royal Navy, where he first developed radar at the Telecommunications Research Establishment for several years and then served as a radar officer before demobilization in Sydney Harbour. Assigned to the solar noise problem at Dover Heights, Bolton built two 60 MHz Yagi aerials to follow up on Martyn's earlier detection of circular polarization, and was soon assisted by technician O. Bruce Slee, a former Air Force radar mechanic who also had just joined RP.[61] But the sun was not cooperating with much activity, and so Bolton decided to check for radio emission at the positions of various well-known astronomical objects, as listed for instance in the venerable *Norton's Star Atlas*. His inattention to solar monitoring, however, got him in trouble:

> After a week or two our efforts were cut short by an unheralded visit from Pawsey, who noted that the aerials were not looking at the sun. Suffice it to say that he was not amused and we were both ordered back to the Lab for reassignment.
> (Bolton 1982:349–50)

Notwithstanding this setback, a few months later Bolton managed to resume at Dover Heights, where he was joined by electrical engineer Gordon J. Stanley (1921–2001), a New Zealander who had come to RP in 1944 after eight years in the electrical industry starting at age 14.[62, 63] This time the goal was to follow up recent studies by Payne-Scott and Donald E. Yabsley (a graduate of Sydney University and an RP radar researcher since 1944) on simultaneous observations of solar bursts at widely spaced frequencies. On 8 March 1947 the sun obliged with a remarkable burst exhibiting delays of a few minutes between signals arriving first at 200 MHz, then 100 MHz, and finally 60 MHz. This and earlier bursts were taken to arise from emission at the varying critical frequencies as successively higher coronal layers were excited; with a model of electron densities in the corona, it was even possible to infer a speed of ~600 km/s for the ejected material (Payne-Scott, Yabsley, and Bolton 1947; further details in Section 13.2.1.1). Here indeed was a dramatic confirmation of Martyn's model of different coronal levels effectively emitting different radio frequencies.

Bolton again grew tired of solar monitoring, however, and together with Stanley returned to the Cygnus phenomenon in June 1947. The antenna was nothing more than a pair of 100 MHz Yagis (similar to Fig. 7.7) connected to a converted radar receiver and operated as a sea-cliff interferometer. This allowed a three-week reconnaissance of the southern sky, during which they at last reliably found the Cygnus source, as well as

---

[59] Pawsey:Woolley, 11 September 1946, file B51/14, RPS.

[60] For obituaries, see Wild, J.P. and Radhakrishnan, V. (1995), *Biog. Mem. Fellows Roy. Soc.* **41**, 71–86; and Kellermann, K.I. (1996), *Pub. Astron. Soc. Pac.* **108**, 729–37.

[61] Although Slee did not join RP until November 1946, he had made an independent discovery of the radio sun while operating a radar set near Darwin in early 1946, a discovery which he duly reported in detail to RP (Slee:Radiophysics Division, 4 March 1946; J.N. Briton:Slee, 18 March 1946, file A1/3/1a, RPS). Orchiston, Slee, and Burman (2006:49–50) give further details of the Darwin observations. Slee (1994) has published his memories of the years 1946–54 at Dover Heights, and Orchiston (2004) provides a detailed look at Slee's career.

[62] Eventually Stanley complemented all his practical experience with a high school diploma and undergraduate degree earned at night. Further details about Stanley's career can be found in Kellermann, K.I., Orchiston, W., and Slee, B. (2005), "Gordon James Stanley and the early development of radio astronomy in Australia and California," *Pub. Astron. Soc. Austral.* **22**, 1–11.

[63] The first order of business after Stanley's arrival was to install a "normal" radio at Dover Heights in order to listen to broadcasts of ongoing England/Australia cricket test matches, the first since before the war. Besides cricket, after-work body surfing at nearby, world-famous Bondi Beach was another favorite activity for Bolton, Stanley, and Slee.

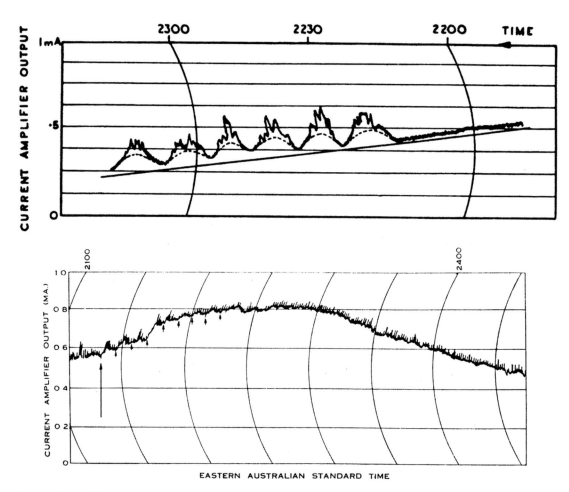

Figure 7.12 Sea-cliff interferometer patterns obtained at 100 MHz at Dover Heights for Cygnus A in June 1947 (top) and for Taurus A in November 1947 (bottom, discovery recording). Note the scintillations superimposed on the Cygnus fringes and the sloping baselines from changing galactic noise; the dashed line shows the adopted base level of the variable component. For the Taurus A record, the rising point and probable minima of the weak fringes are shown by arrows. The numerous short vertical lines on this record are due to interference from a timing mechanism. (Bolton and Stanley 1948b, 1949)

hints of two weaker ones.[64] They spent several months checking out the Cygnus source, and by the end of the year submitted papers to *Nature* and the very first issue of the *Australian Journal of Scientific Research* (Bolton and Stanley 1948a,b). Cygnus usually gave a workably strong set of fringes as it rose (Fig. 7.12), and this directly implied that the radiation came from a very small, single region of the sky. But the source, which crosses close to the zenith in England, never rose more than 15° above the northern Sydney horizon and observations were continually harassed by the strong intensity fluctuations that had led to Hey's discovery in the first place. By analogy to the sun, Bolton and Stanley's analysis partitioned the signal into a constant component (which they estimated as 6000 Jy) and a variable component that added (never subtracted) amounts that fluctuated over times of 5 to 60 sec. Through auxiliary observations made with other Yagis they found a

---

[64] Bolton, "Summary of Cygnus Results – 4th–25th June, 1947," 1 page, undated, file A1/3/1a, RPS. Details in Bolton (1982) have also been used in this section.

maximum in the spectrum of the constant component at ~100 MHz, whereas the variable component's intensity increased sharply at lower frequencies.

The heart of their study was concerned with the size and position of the source. Size came from the solar technique worked out before, namely from measuring the ratio of fringe maximum to minimum. As the "equivalent radiating strip" became broader, the fringes would wash out in a predictable manner (Appendix A). But Cygnus gave difficulties with (1) subtracting off a considerable baseline slope caused by strong galactic noise in its vicinity, (2) isolating the constant component from the variable, and (3) determining a proper upper limit for the fringe minimum, for it appeared that the best records in fact showed minima that were not distinguishable from zero (Fig. 7.12). They estimated that the maximum-to-minimum ratio was at least 50, implying that the source size was less than 8′ (about one-eighth of the lobe separation). Hey's group had inferred that the Cygnus fluctuations must arise from a discrete source or collection of sources, perhaps scattered over two degrees of sky, but here was strong evidence for a single, small source. In fact they thought the source even smaller than their published limit:

> Careful examination of the records suggests a much smaller source size than stated above [8′]. Further experiments using improved receiver stability and greater aerial height will probably substantiate the authors' belief that the source is effectively a "point."
>
> (Bolton and Stanley 1948b:64).

With a source size in hand, they moved on to the even trickier task of a position. This involved analyzing the timing and spacing of fringes in terms of sky geometry and radio wave propagation. One had to find the sidereal time when the source was highest in the sky (culmination) and the length of the arc travelled by the source between rising and culmination. But Bolton and Stanley were forced to tie together observations at three different cliffs around Sydney[65] in order to secure reliable data; furthermore, the necessary corrections for refraction were large and, as it turned out, uncertain (Section 14.4.1). In the end, the derived position was $19^h 58^m 47^s \pm 10^s$, $+41°47′ \pm 7′$ (see Fig. 14.1). With this first well-defined position for the enigmatic Cygnus source (recall that Hey's group had been able to give its position only to within 5°), their next step was of course to consult the optical catalogs and photographs. But this was disappointing:

> Reference to star catalogues, in particular the Henry Draper Catalogue, shows that the source is in a region of the galaxy distinguished by the absence of bright stars and objects such as nebulae, double and variable stars, i.e., the radio noise received from this region is out of all proportion to the optical radiation. … The determined position lies in a less crowded area of the Milky Way and the only obvious stellar objects close to the stated limits of accuracy are two seventh magnitude stars.
>
> (Bolton and Stanley 1948b:68)

They did, however, request Woolley to take a special photograph of that portion of the sky, and this appeared as a plate in their paper, along with a tracing-paper overlay indicating their source position and error box. It certainly appeared a nondescript patch of sky.[66]

Given that there was no optical counterpart, could one nevertheless put any constraints on the distance to the object? Since they had been observing the source for three months, the changing position of the orbiting earth might have caused an apparent shift in position if the object were nearby. But they had detected no shift greater than 2.5′ (corresponding to their 10 sec accuracy in timing the sudden appearance of the source at rising), and this meant the source was at least ten times the 50 light-hour distance to Pluto, that is, well outside the solar system. But *how* far outside? Bolton and Stanley could only suggest that the farthest imaginable would be if somehow the radio object were a star with total power output similar to that of the sun, but all channeled into the radio spectrum. That distance

---

[65] Because Dover Heights was not suitable to follow Cyg A's entire track low across the northern sky, Bolton and Stanley used two other sites besides Collaroy: Long Reef and West Head (Fig. 7.5 shows Collaroy).

[66] As it turned out, Bolton and Stanley's first Cyg A position was a full degree north of the correct position (Fig. 14.1) and so there was no chance of finding the eventual optical counterpart, although note that even positions measured years later to accuracies of a few arcminutes were still unable at first to disclose an optical identification (Section 14.4.1).

worked out to 3000 light years, so the source had been pinned down to be somewhere between 0.06 and 3000 lt-yr from Sydney! But no matter what the distance, the cause of the radio radiation was not at all understood. They could only say it had to be a nonthermal mechanism, for the measured effective (brightness) temperature was $> 4 \times 10^6$ K.

Just as Bolton and Stanley were writing up these results, they received an interesting communication from Pawsey, who was then on the first leg of an around-the-world tour. He had visited Mt. Wilson Observatory in Pasadena and there found Rudolph Minkowski and Seth B. Nicholson[67] "intensely interested" in the Cygnus results and willing to undertake observations directed toward finding an optical counterpart. Pawsey then described optical objects that Minkowski had showed him near the Cygnus position, mentioning that in the process they had had to convert Bolton's derived position to account for "the change of axes due to 'precession of the equinoxes'."[68] Pawsey's letter ended with a raft of suggestions from Minkowski for possible places to look for radio noise:

> The Magellanic Clouds [are] the nearest external galaxies, abnormal with much dust and blue stars. ... If we are interested in interstellar dust, etc. the "Crab Nebula", NGC 1952, is a good sample. If white dwarfs are of interest, the companion of Sirius is a convenient sample. The Orion region is a region of emission nebulae. [But] I do not think these ideas get us very far. I should recommend the method of empirical searching; our tools are not too fine to prevent this.[69]

With the Cygnus case temporarily closed, Bolton and Stanley, assisted by Slee, indeed set out in November 1947 to search the sky in Pawsey's "empirical" fashion. Stanley had made significant improvements to their receiver's short-term stability, in particular through constructing power supplies able to provide voltages stable to a part in a few thousand. Even weak fringes could now be reliably detected. They methodically took records at different points along the eastern horizon, and were delighted when fringes for several new sources appeared over the next few months. As it became clear that the sky had a lot more to offer than just the Cygnus source, Bolton introduced a nomenclature still used today: in the tradition of John Flamsteed's early eighteenth century α, β, γ, ... notation for stars, the strongest source in a constellation would be called A, the next B, etc. And so their second source became Taurus A, one-sixth as strong as Cygnus A, followed by Coma Berenices A at a similar level. The uncertainties of this work can be appreciated by noting that Taurus A appeared nicely on one November night (Fig. 7.12), but it took another three months for confirmation of its existence and measurement of a position good enough to assign a constellation.

By February 1948 Bolton had surveyed about half of the southern sky (manmade interference made daytime observations nearly worthless) and had good cases for six new discrete sources. He wrote up a short note to *Nature* (Bolton 1948) announcing that a new class of astronomical object existed: Cygnus was not unique, either in its existence or in its lack of association with "outstanding stellar objects." Upper limits on the new sources' sizes were no better than 15–60′, but Bolton was becoming convinced that all these discrete sources were truly stellar, "distinct 'radio-types' for which a place might have to be found in the sequence of stellar evolution" (Bolton 1948:141). Since even the most powerful solar-style bursts would not do the trick, he appealed to either pre-main-sequence, collapsing, cool objects or, at the other end of a star's life, to old, hot objects related to planetary nebulae.[70] He felt, too, that a large portion of the general galactic noise probably originated from the aggregate effect of solar-burst type

---

[67] A year earlier Bowen had also visited Mt. Wilson and found considerable interest in solar noise work. This led to his sending Nicholson a write-up summarizing RP solar research, as well as suggesting additional projects. Bowen urged that Mt. Wilson get involved in radio observations and assured Nicholson that there was room for all. [Bowen:Nicholson, 21 January 1947, file A1/3/1a, RPS]

[68] The phenomenon of precession of coordinates, covered at the start of any basic astronomy text (and in Appendix A), had apparently not been known to Pawsey before. It also almost slipped the attention of Bolton and Stanley (1948b) in constructing their photographic overlay (Stanley 1974:6–7T). Also see Section 17.2.3.

[69] Pawsey:Bowen and Bolton, 11 November 1947 (from Lincoln, Nebraska), file A1/3/1a, RPS.

[70] It was only fitting that an "ale-ing" Australian star should radiate its Swan song in the form of Cygnus-type radio noise.

emissions. In retrospect this was a pioneering paper, but Bolton later recalled:

> I guess when we found the first half dozen of these things [radio sources], we were very thrilled about it. Stanley and I published our observations and sort of sat back waiting for the acclaim that would follow. Well, it did not, because nobody believed us ... nobody understood it.[71]

To improve chances for acclaim, Bolton chose to improve his source positions, in particular to eliminate systematic errors, by observing source *setting* as well as rising. High westward- and northward-facing cliffs were needed and so Bolton and Stanley headed off to New Zealand in the southern winter of 1948. As Bolton (1976:8T) recalled:

> [Just before the New Zealand trip] I remember Taffy Bowen asking me what I really thought of the positions of my sources, and I said, "Well, they're the best I can do at the moment, but I'd like to be the first to correct them." And indeed the corrections were absolutely massive when they came in.

The superior observations in New Zealand put an even tighter limit on Cygnus A's size (<1.5′), and, together with simultaneous observations by Slee in Sydney, provided strong evidence that most of the intensity fluctuations originated in the earth's atmosphere, not in the source itself. Many new sources also turned up and and it became apparent that incorrect refraction corrections and other problems had thrown most previous positions 5° to 10° off. Some even changed names as when Coma Berenices A migrated into Virgo. But Cygnus A was still vexing, as neither its new position (shifted about 1° south from earlier) nor its old one agreed with that measured by Martin Ryle at Cambridge (privately communicated in June 1948; Section 8.4 and Fig. 14.1). For a while it seemed that the source might actually be moving, but after six months of sorting out, both hemispheres admitted earlier errors and came to agree on a common position. Further details of the New Zealand expedition and the subsequent development of RP's radio source work are in Chapter 14.

The beautiful outcome of the new positions of ~10′ accuracy was that for the first time optical counterparts could be tentatively suggested (Bolton, Stanley, and Slee 1949). And these turned out to be no ordinary objects. Taurus A was associated with the Crab nebula, the expanding shell of a supernova known to have exploded 900 years before (Bolton and Stanley 1949); Centaurus A was found to coincide well with one of the brightest and strangest nebulosities in the sky, so peculiar that astronomers were not even sure whether or not it was part of our Galaxy; and Virgo A's position correlated with that of a bright elliptical galaxy six *million* light years away. Although it took several years for these identifications to be ratified, they quickened interest in the study of discrete sources, and several optical astronomers, among them Minkowski, took serious note (Section 14.4.3).[72]

7.5 RP'S EARLY YEARS

7.5.1 The isolation factor

Almost any analysis of things Australian must consider the geographical isolation of Australia from the other centers of Western culture. Geoffrey Blainey (1968) speaks of "the tyranny of distance," that is, the overwhelming importance of distance, isolation, and transport in molding the general history of Australia. In the sciences also, isolation has played a major role. Although the vestiges of a subservient colonial relationship between British and Australian radio science created an asymmetry in status and power that was independent of the distance from London, many problems of Australian science during the postwar years were notably exacerbated by the antipodal separation.[73]

The early RP years are rife with examples of things that would have gone differently if RP had not been located 10,000 miles from its sister institutions, but instead 100, or even 1000. The best airline connections to Europe required a gruelling three days (or a civilized week) and more common passage by ship took about four weeks; moreover, the cost of a ship's

---

[71] Bolton interview in Bhathal (1996:108–9).

[72] The Dover Heights cliff-edge site (now called Rodney Reserve) now has an antenna-shaped memorial structure and plaque, dedicated in 2003 with Bruce Slee present.

[73] Illuminating discussions of these issues, especially during the nineteenth century but lingering into the twentieth, are found in two papers by D. W. Chambers and D. Knight, pp. 19–53 in *International Science and National Scientific Identity*, eds. R. W. Home and S. G. Kohlstedt (1991, Dordrecht: Kluwer).

berth amounted to one or two months' pay for an RP staff member. The RP staff (and Australian science in general) were constantly bedeviled by their inability to have frequent contact with colleagues from other institutions, the long interval before learning about research conducted elsewhere, the delays (sometimes inordinate) in publishing Australian results in the prestigious British journals, and the lack of foreign readership of Australian journals. One counterforce was the maintenance of Australian Scientific Research Liaison Offices in London, Washington, and Ottawa. These had been originally set up during the war to coordinate radar research, and served as scientific embassies to increase the flow of information to and fro. But a far better solution was to send an RP researcher on an extended jaunt through North America and Europe. In the six years after the war the primary overseas stays or trips of importance for the development of radio astronomy were taken by Bowen (in 1946), Ronald N. Bracewell (1946–49), Pawsey (1947–48), Westfold (1949–51), Bolton (1950) (see Bolton 1982:353), and Kerr (1950–51). The RP correspondence files resulting from these trips are particularly good sources for understanding the influence of the isolation factor on RP's work. What emerges is that such trips served four primary purposes: (1) intelligence (in the military and political sense of the term), (2) education, (3) publicity, and (4) establishment of personal contacts.[74]

The first purpose of the overseas trips was simply to find out what was going on. The RP visitor to an overseas laboratory typically sent back a detailed report of recent and ongoing research, and this report (as evidenced by multitudinous initials on the original documents) was widely circulated back home. Pawsey and Bowen in particular were masters at picking up what was being done elsewhere and analyzing its effects on RP's current research program and future plans. To give but two examples, in August 1946 Bowen cabled back that British work on the sun was ahead of RP's at observing frequencies less than 200 MHz and that therefore RP should concentrate on higher frequencies. And in April 1950 Bolton sent word home that the solar work by Hey's group lagged Wild's by at least eighteen months.[75]

Lack of knowledge of overseas research led of course to frequent duplication of experiments. In the first few years after the war, Pawsey frequently felt that poor communications were leading to needless repetition by British and Australian researchers. Somewhat idealistically, he therefore tried to coordinate radio noise work on opposite sides of the earth, but this worked little in practice. As an example, in September 1946 Pawsey wrote to his mentor Ratcliffe at Cambridge:

> I got rather a shock when I received Ryle's note enclosing a copy of the letter to *Nature* contributed by himself and Vonberg[76] ... The Cavendish and Radiophysics Laboratories have unfortunately succeeded in duplicating a very considerable part of the work. ... I do not know what we can do about this duplication or how we can avoid it in the future. My only suggestion is that you have a talk with E. G. Bowen, Chief of this Division, who is at present in England.
>
> I sent you a food parcel in August so that with any sort of luck you should get it about Christmas time. According to accounts here your menu is pretty dull.

Ratcliffe replied with a rundown of Ryle's plans for solar observations and noted that not all duplication should be avoided. He also stated that "now that the Air Mail works so quickly [1–2 weeks], we will make a special attempt to keep you fully in touch."[77] Even as late as 1951, other attempts by Pawsey at coordination took place, including one case where he persuaded Ryle not to publish a source position that disagreed with Bolton's until things could be sorted out (Section 8.4). But eventually it became clear that Australian science, like so many other aspects of postwar Australian society, was becoming an independent entity, not just an extension of the mother country.

---

[74] Groundwork was also being laid for a fifth purpose of overseas trips, namely fundraising for large antennas, but here the payoffs began only in the mid-1950s.

[75] Bowen (from London):Pawsey, 23 August 1946, file A1/3/1a, RPS; Bolton (from London):Bowen, 14 April 1950, file F1/4/BOL/1, RPS.

[76] Ryle and Vonberg (1946; Section 8.3).

[77] Pawsey:Ratcliffe, 10 September 1946; Ratcliffe:Pawsey, 17 September 1946 – both file A1/3/1a, RPS.

A second purpose of the overseas trips was education. Sometimes this was in the formal sense, as when Bracewell, an RP researcher since the war years, took a Ph.D. at Cambridge and Westfold one at Oxford, and Kerr a Master's degree at Harvard; but more often it was simply the knowledge to be garnered from overseas contacts. The background of the RP staff was of course far weaker in astronomy than in radio physics, and thus it was the visits to observatories that were particularly valuable. As Kerr (1971:19T) recollected:

> Bolton and also Pawsey did some touring at that time and learned something of what generally were the interesting problems in astronomy, acquiring some of the attitude of astronomers toward astronomy, instead of just the electrical engineers' and physicists' attitudes.

On the other hand Mills (1976:20T) later pointed out that the paucity of astronomers in Australia may have helped more than it hindered:

> Our isolation did help us develop with an independent outlook. We had no famous [astronomer] names to tell us what we should believe, and to some extent we just went ahead following our noses.

Overseas voyagers also served to spread the word about RP research – Pawsey called them "ambassadors for Australian science." The RP archives are full of instances where Bowen and Pawsey sent reprints, complained about Australian work being neglected in reviews overseas, and urged people to subscribe to the *Australian Journal of Scientific Research*, started by CSIR in 1948 and a further sign of the growing independence of Australian science from British hegemony. RP sent 30 full articles to the journal in its first four years, but only eight to British journals (plus eight letters to *Nature*).[78] Although this corpus probably lent more stature to the journal than did any other single field, it took years to foster a world readership even among radio astronomers. For example, Jodrell Bank did not subscribe until 1950; before then, the only copies they could locate were in London.[79]

Direct word-of-mouth, when possible, was of course also important. After attending a 1948 URSI meeting in Stockholm, Pawsey wrote back, "Martyn and I, to put the matter rather bluntly, attempted to put Australia on the map, and I think were fairly successful."[80] And Bracewell (1921–2007) recalled that while he was at the Cavendish as a postgraduate student, Ryle's group thought Sydney work way behind, but when he returned to RP he found that they thought the same of the Cambridge work – each side was simply acting on dated information (Bracewell 1980:131A). Preprints were not common in those days, and in any case were sent by sea mail (as were journals, even *Nature*), taking two to three months for the passage.[81] Bracewell also remembered wanting to act as a link between the two groups: "Being young and idealistic, I felt that I should try to close the gap, that a freer flow of information was a blow struck against entropy, as well as my duty." As he wrote to Bowen in early 1948:

> Publication [of Australian work] is slow and the diffusion of advance news by word of mouth does not occur. It results that ideas of priority are fixed before Australian work filters through. This is the case with solar noise. The attitude in the Cavendish Lab. is that nothing much of value is done elsewhere. … [Since] I am in an effective position for informal dissemination of news from Radiophysics, I recommend for your consideration the transmission of this news.[82]

Despite his request, it appears that Bracewell himself remained little better informed than others in Cambridge. Upon his return to Australia in late 1949, he wrote back to Ryle:

> There is a lot of good work going on, and the people are very keen. Very little pre-publication

---

[78] Kerr has remarked that one of the factors in the founding of the Australian journal was the suspicion that letters and articles sent to British journals were not being treated fairly, either through premature dissemination of their contents to rivals, or through delays in publication. Also see note 36. [Kerr:author, 6 April 1987]

[79] Lovell:Bowen, 18 May 1950, file D5/7, RPS.
[80] Pawsey:Bowen, 12 August 1948, file F1/4/PAW/1, RPS.
[81] A check of accession dates for *Nature* in the Sydney University library (which was used by RP) revealed that each weekly issue was received 5 to 11 weeks after its date of issue. This situation continued until 1954, when the delay became only one week, presumably because of airmail delivery. I thank J. Threlfall for this information.
[82] Bracewell:Bowen, 19 January 1948, file F1/4/BRA, RPS.

news seemed to filter through to me in the Cavendish from Australia … Do not hesitate to let me know if … I can make enquiries which you may think can be better done informally through me. There is a lot of interest in your work – I have had to ward off quite a barrage of minor queries about your set-up since arriving.[83]

This induced Bracewell over succeeding years to send three short papers to the informal British journal *Observatory* to advertise RP's research.

It is important to note that although RP as a laboratory was isolated, its individual researchers were not. There were enough first-rate radio physicists in one place that they did not suffer from lack of intellectual intercourse. RP thus avoided the phenomenon that Home (1986) has coined "the isolation of the elite," referring to scientists (such as W. H. Bragg while in Australia) whose careers were debilitated by lack of peers within hundreds of miles.

One of the grandest opportunities for interchange and to advertise RP's work was the URSI General Assembly that met in Sydney in August 1952. This was a feather in the hat not only for Australian radio research, but for all of Australian science, for it marked the first time that a major international scientific union had met outside Europe or North America.[84] In 1948 URSI had created a new Commission V on Extraterrestrial Radio Noise with Martyn as its first president and Pawsey as secretary. Martyn in particular engineered the General Assembly coming to Sydney and masterminded the organization and funding (Bracewell 1984:169–71). Sir Edward Appleton (Fig. 7.1) was the patriarch among the fifty foreigners in attendance, of whom about a third were active in radio astronomy (Fig. 7.4). At last the RP staff could associate faces with names like Jean-Louis Steinberg from France, Robert Hanbury Brown from Jodrell Bank, F. Graham Smith from Cambridge, C. Alexander Muller from Holland, and H.I. "Doc" Ewen from the United States. RP of course put on its best show for the guests with a detailed, glossy "Research Activities" booklet and full tours of all field stations. The home team were greatly stimulated and the visitors went away impressed.[85]

7.5.2 The field stations

The RP radio astronomy work took place at individual field stations, some as far as 30–50 miles from home base on the grounds of Sydney University (Fig. 7.5).[86] These sites provided sufficient land and isolation for observations free from manmade electrical interference. But why not just one or two sites well removed from Sydney? Because many small sites also provided freedom from a second type of "manmade interference": other people. The independently-minded RP scientists and engineers simply preferred to spend most of their time alone at the field stations, not in a central laboratory, and management too found this a productive style of operation. By the late 1940s RP's research in radio astronomy was divided into many teams of two or three: leaders and sites about 1948–50 were Piddington and Minnett (University grounds), Kerr and Shain (Hornsby), Bolton (Dover Heights), Wild (Penrith), Mills (Badgery's Creek), Payne-Scott and Christiansen (Potts Hill; Fig. 7.13), and Yabsley (Georges Heights). Christiansen (1984:113–4) has evoked the atmosphere of these stations:

> Each morning people set off in open trucks to the field stations where their equipment, mainly salvaged and modified from radar installations, had been installed in ex-army and navy huts. … The atmosphere was completely informal and egalitarian, with dirty jobs shared by all. Thermionic valves were in frequent need of replacement and old and well-used coaxial connectors were a constant source of trouble. … During this period there was no place for observers who were incapable of repairing and maintaining the equipment. One constantly expected trouble.

Although groups had little day-to-day contact, Pawsey's skill as roving monitor and coordinator gave cohesion to the laboratory's radio noise work. This was achieved first through meetings every 2–4 weeks

---

[83] Bracewell:Ryle, 6 January 1950, box 1/1, RYL.

[84] The only possible earlier contender was in 1923 when the new Pan-Pacific Science Congress met in Australia for its second conference. In 1914 the British Association for the Advancement of Science conducted a huge "meeting" in Australia, but, as for an earlier visit to South Africa, it was actually more a case of ~300 British visitors touring Australia by special trains for five weeks and wowing the colonials with public lectures, etc.

[85] Wild (1978:20T), Mills (1976:19T), Smith (1976:24–6T); Ryle:Pawsey, 27 September 1952, A1/3/1a; Pawsey:Ryle, 12 September 1952, A1/3/1a; Pawsey:Lovell, 10 September 1952, A1/3/1a.

[86] Orchiston and Slee (2005) describe in detail the various field stations and their programs.

Figure 7.13 One of the more active RP field stations, Potts Hill in October 1953, adjacent to a large Sydney water reservoir (Fig. 7.5). Antennas visible include a 97 MHz Yagi element of Little and Payne-Scott's solar interferometer (upper left; see Section 13.2.1), a 97 MHz "model" (120 × 120 ft) Mills cross (lower left) (Mills and Little 1953), a 10 ft dish (right center) used by Christiansen for 50 cm observations of a 1948 solar eclipse (Section 13.1.3), a 16 × 18 ft rectangular paraboloidal section (top right) used for solar work and in 1951 (Section 16.5) by Christiansen and J.V. Hindman for 21 cm hydrogen line observations, and a 36 ft dish (top center) used by Kerr and Hindman for later hydrogen line work, with a small dish near it used for testing and as the reference signal for a switched receiver.

of his "Propagation" Committee (changed to "Radio Astronomy" in 1949). These meetings provided a forum for progress reports, discussion of astronomical results and technical problems, floating of new ideas, coordination of experiments, and arguments about priorities. They also served to counter the danger that isolated groups would develop too narrow scientific or organizational perspectives. Several interviewees commented on the value of these sessions, for instance Christiansen (1976:19–20T):

> Despite the fact that we were independent groups, we used to have these sessions, sort of what Americans call "bull sessions," thinking of every conceivable sort of aerial … A really good one would last all day. Joe Pawsey was one to stimulate that.

Pawsey's second device for holding the radio noise research together was to frequently visit the field stations to see for himself what was happening and to give advice. As Wild (1972:5) recollected:

> On some days he would arrive unexpectedly at one's field station, usually at lunch time (accompanied by a type of sticky cake known as the Lamington,

which he found irresistible), or else infuriatingly near knock-off time. During all such visits one had to watch him like a hawk because he was a compulsive knob-twiddler. Some experimenters even claimed to have built into their equipment prominent functionless knobs as decoys, especially for Pawsey's benefit[87] ... [But] when one ran into problems, half-an-hour's discussion with Joe tended to be both soothing and rewarding.

These visits, however, sometimes led to Pawsey seeing things he didn't like:

> Pawsey was in direct linkage with the little isolated groups. He'd try and make sure they didn't clash. And he stopped us working at times when Jack [Piddington] had had some idea and we'd started in a new direction. ... For instance, one day he found me [working on a radio analogue to a Fabry–Perot interferometer] and I was stopped. He said there are other people already there, and they've got a prior claim.
>
> (Minnett 1978:29–30T)

But although Pawsey usually assigned exclusive turf to each small group, he sometimes encouraged two groups to plow ahead on the same problem if he felt their approaches differed enough. For example, Mills and Bolton for many years both observed discrete sources, albeit with different types of interferometers:

> Bolton's group and mine each felt rather strongly that our own technique was the best. Although we saw each other sometimes, Bolton lived out at Dover Heights and didn't come into the Lab very often and I spent most of my time out at Badgery's Creek. So we didn't actually have very much contact, and there were quite a few arguments about interpretation of the results.
>
> (Mills 1976:26–7T)

---

[87] Pawsey was in good company with his compulsive knob-twiddling. Philadelphia Orchestra conductor Leopold Stokowski was so notoriously unable to resist turning knobs of audio equipment during recording sessions in the 1930s that the frustrated engineers finally set up an entire control booth that was nothing more than an electronic placebo. [McGinn, R. E. (1983), *Technology and Culture* **24**, 38–77 (p. 45)]

7.5.3 Management of radio noise research

Bowen turned over scientific leadership for radio noise investigations to Pawsey, who was the Division's number-two man from the start (although the office of Assistant Chief was not created until 1951). Pawsey thus had a free hand in running the radio noise side of things while Bowen took on the general administrative burden and concentrated on the rest of RP's program, taking a particular interest in the rain and cloud physics research to which he himself made several contributions. Bowen, however, minimized the number of his collaborations and so through 1951 published only seven papers. His career had seen more than its share of scientific directors (such as Appleton, Watson Watt, and Martyn) who claimed credit for too much of what happened in their laboratory:

> When I became Chief, I was going to be quite certain of one thing. ... I was not going to jump in and claim credit when somebody else did the work. ... My previous experience of some pretty hard cases was that the best way to get first class work out of people was to give them the credit.
>
> (Bowen 1973:27–8T)

Along this line, Bolton (1978:118T) recalled:

> Bowen was on our side in terms of letting people have their head – giving you a pat on the back when you did well and commiserating with you when something failed.

Bowen's philosophy also was expressed in a 1948 letter to Pawsey after the latter had been overseas for eight months:

> It is true that those of us who have had a fair amount of experience can give a lot of help in choosing problems for the younger people, keeping their sights on the target and helping them snatch the odd pearl out of the tangled mass, but I am quite sure that what we are suffering from in the Lab. is not that there is too little of this help but too much. With few exceptions our youngsters have not learnt to stand on their own feet and go for a line of their own ... The boys in the Radio Astronomy Group are feeling your absence quite keenly, but I am taking

the view that their present gropings are part of their education.[88]

Bowen and Pawsey's leadership styles very much fit in with Rivett's philosophy discussed earlier: get the best people possible, give them the needed resources, and then let them run free (Bowen 1984:110). But there were bounds to this freedom, as we have seen, leading to a creative tension between tight control of the laboratory's work, as it had necessarily operated during the war, and the kind of individual freedom one might find in a university department. This delicate balance is well illustrated by the juxtaposition of allowing workers to be scattered all over the countryside, while still keeping close tabs on what they did.[89] Other strong limits existed. For example, most scientific correspondence was routed through either Pawsey or Bowen. More significantly, RP maintained a system of rigid internal reviews of all proposed publications, involving one or more of Bowen, Pawsey, and Arthur Higgs (Technical Secretary). The RP archives are replete with internal memoranda shuttling drafts back and forth between authors and management (and sometimes anonymous third-party RP referees), often to the frustration of the authors.[90] As Bowen wrote in 1948:

> As I keep on telling the chaps, there is no doubt about the excellence of the work being done in the Laboratory, but the writing up is awful.[91]

But once a paper surmounted this first hurdle, a journal's referees usually seemed easy. The extent of Pawsey's influence on the radio noise papers can be gauged by the fact that fully half of them from the 1946–51 era explicitly acknowledge his assistance with either preparation of the paper itself or the project in general. Yet he himself, like Bowen, published only seven papers through 1951.[92]

Bowen and Pawsey agreed on the basic policies needed to run RP, but their differing temperaments led to differing contributions to RP's success in radio astronomy:

> [Pawsey's scientific style] set the tone completely. … but he was a very, very unworldly fellow. … Bowen was the man who got the money, the tough businessman, while Joe was the rather academic scientist. And it was an excellent combination.
> (Christiansen 1976:22T)

Bowen knew how to deal with the CSIRO hierarchy, how to pull off the necessary balance of applied and fundamental research, and how to manage RP as a whole. On the other hand, interview testimony of numerous RP staff members indicates that Pawsey by nature was not suited for such things. For instance, he abhorred (and avoided) making managerial decisions that he knew would cause upset.

Pawsey played a vital role, however, as scientific father figure and mentor. He was ten to fifteen years older than most of the radio noise researchers, who in 1950 had a median age of 30 years (Section 17.1.2), and he quickly gained their respect and confidence. The words of his protégés speak for themselves:

> He had the ability to develop the latent powers in other people. All of the people who came out of that group – Christiansen, Mills, Wild, and so on – I also count myself in it – were made independent and skillful in their subject, experienced and self-reliant, quite largely because of Pawsey's way of drawing people out. He was not the kind of research leader who'd insist on claiming everything himself. But he fed in the ideas that other people developed – he was a teacher as much as anything. In the written record you don't find his name on

---

[88] Bowen:Pawsey (in England), 17 June 1948 (#4), file F1/4/PAW/1, RPS.

[89] Stanley (1994:507–13) describes this tension well (in the case of Bolton).

[90] A particularly illustrative example of RP's internal refereeing system is provided by the torturous career of a paper eventually published by Piddington (1951). Pawsey, Westfold, Bolton, and Piddington traded sharp criticisms, including tidbits such as "The author…is, like Procrustes, apparently prepared to mutilate the data to conform to his preconceived hypothesis." Also Piddington (1978:21–2T). [September-November 1950, file D5/4/110, RPS]

[91] Bowen:Pawsey (in England), 23 July 1948, file D5/7, RPS.

[92] Donald Yabsley has pointed out a typical example of Pawsey's keeping his name off publications. With regard to the paper by Payne-Scott, Yabsley, and Bolton (1947), Pawsey contributed much to the project and to the paper, and originally he was intending to be a coauthor. But in the end he withdrew his name because he felt that three authors were quite enough for a *Nature* letter. [Yabsley:author, 10 October 1986]

many papers, but he was the inspiration behind an awful lot.

(Kerr 1971:41–2T)

There were, and are, few scientific groups of comparable size where the head of the group had such a detailed knowledge of the work of each member and where every paper was criticised in detail by him. Yet this ... did not lead to any authoritarian regime. Pawsey's criticisms were usually accepted not only because they were sound but because they were so clearly and intelligibly expressed that acceptance was inevitable.

(Christiansen and Mills)[93]

It was in his research direction that Pawsey made his greatest contribution: patient, kindly, selfless direction; always probing for the simple vital question and significant experiment; and subtly transforming inexperienced research workers into leading contributors to their respective fields.

(Wild 1968:118)

Pawsey's style of science grew out of his training in the Cavendish Laboratory of Ernest Rutherford. He inherently loved the simple, inexpensive experiment and distrusted anything coming from complex setups (see Section 8.1). He also had an innate distrust of theory and mathematics (Westfold 1978:97B), complemented by a faith in experimentation. As he himself wrote in 1948 (regarding the possibility of solar bursts at frequencies less than a few hundred hertz):

> My present guess is that the theory is wrong in general, and consequently I do not advise any time-consuming observations which are based on the theory. On the contrary, the observation of low-frequency noise is a fundamental scientific observation which is of value independent of the theory. Positive or negative results are of use. Hence this investigation is in order, and it is up to the experimenters to decide how far they go.[94]

The experimental style that Pawsey inculcated was particularly striking to Doc Ewen (1979:42T), accustomed to much larger American budgets, when he visited Sydney for the 1952 URSI meeting:

> Their equipment was shoestring stuff, but there were a lot of cute tricks. ... They didn't waste much time with hardware where it wasn't all that important, [or with] trying to make it look pretty. But wherever a part was critical to the operation of a device, they spent a lot of time thinking about it.

And from the other perspective Christiansen (1976:31–2T) remembered how Ewen reacted upon seeing his 21 cm hydrogen line receiver (Section 16.5):

> Ewen came out and said he had to see how these damn Australians did in three weeks what took other people eighteen months to do. And when he saw our gear, lying all over the room and on the floor, he just about passed out.

The words of his colleagues best capture Pawsey's scientific style:

> He had an enormous enthusiasm. It was always a delightful experience to bring to Pawsey some new idea or some interesting new observation. His immediate reaction would be one of intense interest, followed by suspicion as he looked for some mistake or misinterpretation, or what he called "the inherent cussedness of nature." Finally, if convinced that all was well, his face would shine with boyish pleasure. ... He never forced his opinions on a younger colleague; if the matter was open to doubt he was willing to leave it to experiment. He was, in fact, the arch-empiricist. "Suck it and see" was one of his favourite expressions.

(Christiansen and Mills)[95]

Pawsey had a childlike simplicity about him, a childlike curiosity. He was not a sophisticated man in the least. I find this is a talent that a lot of people who are truly great have in common – retaining a feeling that science is not a business, that it's a

---

[93] As in note 21.
[94] Pawsey (from Ottawa):Bowen, 13 February 1948, file A1/3/1a, RPS. See Tangent 9.1 for details on the claimed low-frequency solar bursts that instigated these remarks. RP responded with a limited number of observations (never published) by McCready.
[95] As in note 21.

game. … If Joe had been a businessman, you would have called him a sucker, but [for science] I think that's actually an important characteristic.

(Stanley 1974:25–6T)

I think you could say Pawsey was a very simple soul. … But he could floor a speaker: there'd be a fellow turning up a great piece of astrophysics and Joe would get up at the end and say quite innocently, and it *was* innocent, "I can't reconcile this with Ohm's Law." It would absolutely torpedo the speaker.

(Christiansen 1976:21T)

Pawsey did not have much of a mathematical background – he once asked me what [statistical] "variance" meant – but he thought in physical terms. … He once proposed what he called the Sausage Theorem: "If the error bars on a set of visibility measurements fit inside a certain sausage, then the calculated source distribution [from the Fourier transform of the visibilities] runs down the middle of another sausage." Pawsey very reasonably wanted to know how fat this other sausage was and my job was to find out. It is a very good question.

(Bracewell 1984:171)

Pawsey was a wonderful fellow to work with and obviously knew his subject, but he was strictly a *non*mathematical physicist. He knew simple equations, but he didn't trust mathematics. For instance, when Kevin Westfold wrote something down, any mathematical formula was simply part of the sentence and you had to read it that way. But Pawsey always read it without appreciating the mathematical part of it, and had to say, "Look, I can't make heads nor tails of it the way it is. Can you explain the physics? What happens there? What is it? Is the electric field oscillating? How does it transform into electromagnetic waves?" And Kevin would say, "It's all there, it's all there!" And they didn't get anywhere.

(Smerd 1978:92A:525)

Finally, a dissenting view from Francis F. Gardner, an RP ionospheric colleague of Pawsey's during these years:

The impression of Joe as a naive, unworldly type is misleading. To some extent this was a pose, which contributed to his ability to "draw people out". … Nor was he opposed to theory. … In discussions he was able to grasp immediately what was said to him, even if poorly expressed, and he also was able to concentrate one's attention on the problem under consideration. Occasionally he would suggest solutions to some degree with tongue-in-cheek. His suggestions might not be appropriate, but enabled others to see the solutions.[96]

7.5.4 RP evolves during the 1950s

When an institution is created for one specific mission and then, because of changed circumstances, tries to adapt to a different role, the results are often less than satisfactory. RP's shift from war to peace, however, constitutes a striking counter to this. Through skillful leadership, scientific expertise, and good fortune (for instance, how might the fledgling solar noise efforts have gone if the "sunspot group of the century" had not shown up in February 1946?), RP put Australia at the forefront of radio astronomy over the postwar decade. By the early 1950s, RP was also clearly the scientific leader of CSIRO (Schedvin 1987:360). In fact, in no other natural science did such international stature come to Australia during these years – perhaps the closest was the immunology research led by F. MacFarlane Burnet at Melbourne's Walter and Eliza Hall Institute of Medical Research, or the neurophysiology led by J. C. Eccles.[97]

Many factors important in RP's success have already been discussed, but there were others. One was the sheer size of the radio noise group, far larger than other institutions in the field – with so many projects going on simultaneously, RP was much more likely to have at least one winner at any given time.[98] The radio physicists were also supported by invaluable assistance from the large staff of technicians for electrical and mechanical work. A mild climate also conferred distinct

---

[96] Gardner:author, 12 November 1986.
[97] Courtice, F. C., "Research in the medical sciences: the road to national independence," pp. 277–307 in Home (1988).
[98] The emphasis on radio astronomy might not have happened if RP had operated in a US-style, Cold War atmosphere lavishly funding nuclear physics and short-wavelength radio technology (Section 17.3.5); for example, the high-energy physics projects that in fact died out (Tangent 7.1) might instead have been well funded and come to dominate.

advantages for research involving outdoor construction and experimentation (Bolton 1978:107T). We can dismiss, however, one possible factor for the Australians' success, namely that they had the southern sky to themselves and therefore had no competition and only needed to mimic northern observers. Although for over a century Australia had been a fertile outpost for research precisely because of its unique flora and fauna and non-European skies, the evidence of this chapter shows that for radio astronomy this notion is patently untenable. After all, the same sun is shared between north and south. In fact, Bolton took the view that any new setup should work the reachable northern sky first so as to beat the northerners – the southern regions would always be there later.[99] Witness the trouble Bolton made for himself by observing the notably northern source Cygnus A as it barely scraped his horizon, even though the competition saw it overhead in England.

As work on radio noise developed in the Radiophysics Laboratory over the years, there was a gradual integration of the research into astronomy proper and radio physicists and engineers slowly morphed into radio astronomers. Even from the beginning Pawsey recognized that this new radio technique was fundamentally altering *astronomical* knowledge. As he stated during a talk to an August 1946 meeting in Adelaide:

> This [solar noise] work is a new branch of astronomy. ... New observational tools [in astronomy and astrophysics] have an unusual importance. The last outstanding development in solar instruments was probably the spectroheliograph (developed at the turn of the century). Consequently it is reasonable to expect that the discovery of this radiation will come to be recognised as one of the fundamental advances in astrophysics.[100]

Yet although the RP staff realized that they were essentially doing astronomy, albeit of a wholly different type and not well understood by astronomers, their astronomical education proceeded in a checkered manner. Whereas Bolton (1978:30, 36–7T) chose to methodically read volume upon volume of the *Astrophysical Journal* during long observing nights, most just picked up what they deemed necessary as they went along. Books such as George Gamow's *The Birth and Death of the Sun* were read and Bolton undertook a partial translation of Max Waldmeier's 1941 treatise on the sun. The exposure to Mt. Stromlo, including occasional joint colloquia, was also important. But RP had nary an astronomer on its staff and its orientation during the first postwar years was as often toward technique as astronomy:

> We were simply radio people trying to provide another tool for detecting what these astronomers said was likely to be there. ... We didn't consider ourselves to be astronomers – our primary interest was in the equipment. In fact we'd just left a wartime situation and we knew that our success in radar stemmed from having people who were very well trained in the techniques.
>
> (J.V. Hindman 1978:98B)

By the early 1950s, however, overseas trips, increasing contacts with astronomers, and a gradual accumulation of astronomical knowledge had caused a clearer picture to emerge of how the radio work fit into astronomy as a whole.[101]

The early 1950s represent a watershed from several other perspectives. For the first time research on solar noise was overtaken in quantity by that on "cosmic" (nonsolar) noise – the percentage of solar papers dropped from ~70% before 1951 to ~40% during 1952–54. At this time also, Pawsey and Bracewell wrote a masterful monograph, *Radio Astronomy* (mostly written in 1952–53, but not published until 1955). It formed a capstone to the first stage of the field's development and remained the definitive textbook for a decade. And of course the 1952 URSI meeting also happened at this juncture.

The most important change that began during the early 1950s was the move from a large number of relatively small experiments to a smaller number of projects on a large scale. This was the start of the transition

---

[99] Kerr:author, 6 April 1987. Piddington expressed a similar sentiment in 1959 in a planning document for the 210 ft diameter Parkes dish, then under construction [in "Proposed early investigations using 210 ft radio telescope," 16 September 1959, file A1/3/11/17, RPS].

[100] Pawsey, "Solar radiation at radio frequencies and its relationship to sunspots: interpretation of results," talk delivered to an ANZAAS meeting, 19 August 1946, file A1/3/1a, RPS.

[101] For further context regarding RP's postwar research, see *Explorers of the Southern Sky* by Haynes *et al.* (1996), a thorough history of all of Australian astronomy.

from Little Science to Big Science – or as Wild has pungently described it, moving from trailers "with a characteristic smell" to air-conditioned buildings.[102] Progress in the science now demanded huge antennas and arrays and many of these were beyond the capacity of RP to produce in-house. For example, not until 1951 were *outside* bids for antennas sought (for 50 ft and 80 ft dishes).[103] In 1953 RP funded its last major antenna from its own resources: a 1500 ft Mills Cross array at Fleurs for £2500 ($5500).[104] More costly ventures did not come easily, however, for the government and CSIRO were not willing to support large capital projects (Bowen 1978:44T, 1981:267; Bolton 1978:56T). Nascent thoughts about a "Giant Radio Telescope" and its funding began as early as 1948. Bowen (1984:98–9) at that time tried to convince the Royal Australian Air Force to build a huge radar antenna that could do radio astronomy on the side. Several designs were studied over the next few years, some as large as 500 ft in dimension, but the funding never materialized. By 1951 the search for funds shifted to non-military sources and eventually key money came from American foundations. But it took a decade: only in 1961 did RP finish the transition to Big Science with the commissioning of a 210 ft dish at Parkes.[105,106]

The maturation of the science was of course also mirrored in the growing seniority of the RP staff. Pawsey had had great success in scientifically rearing his junior colleagues, but RP was not like a university department with its steady stream of students – the RP "students" had no where to go and there were few positions for junior staff. Already in 1951 Pawsey was saying that the outstanding defect in the radio astronomy group was its lack of the "research student" type with which he worked so well.[107] Through the 1950s the various group leaders became strong-willed, confident individuals, arguing their own particular visions of how radio astronomy at RP should be done. Major disputes centered on two questions: (a) should the focus continue on small technique-oriented groups or shift to a single major facility? and (b) which types of antennas would pay off best? The first major figure to leave Pawsey's group was Bolton, who in 1953 switched to cloud physics (after denial of funding for a new type of interferometer) and then in 1955 founded a new radio observatory at the California Institute of Technology (Stanley 1994:511–3). Furthermore, Bowen and Pawsey began to work less well as a team and developed significant differences. The tale of the later 1950s was to turn out to be anything but a simple extension of the pioneering period here covered, but this later period is beyond the scope of the present study.

Australia's history had been linked with astronomy from the start – Captain James Cook's first voyage, which led to the discovery of Australia's eastern coast, was as much to observe the 1769 transit of Venus across the solar disk in Tahiti as it was to explore for *Terra Australis Incognita*.[108] But 175 years passed before Australians became part of the first rank of world astronomical research. And when this happened, from cliff edges only a few miles removed from Cook's landing site at Botany Bay, it was in a most unlikely manner, for they did their astronomy not with glass lenses, but with rods of metal.

---

[102] As in note 15.

[103] "Tentative specifications for 80-foot and 50-foot radio astronomy aerials," 8 August 1951, file A1/3/1h, RPS.

[104] B.Y. Mills, 31 March 1953, "Large aerial for meter wavelengths," file A1/3/10, RPS. The Mills Cross antenna (part of a small pilot version is visible in Fig. 7.13; Mills and Little 1953) was an economical invention of Mills that allowed efficient and sensitive all-sky surveys of radio sources. Starting in 1955, these surveys turned out to fundamentally disagree with those (especially the 2C survey) from Ryle's group in Cambridge, which led to a decade of acrimony (Edge and Mulkay 1976).

[105] For complete details of the fund-raising, design, construction, and research program of the Parkes dish, see Robertson (1992). This book discusses fascinating negotiations between Bowen and Lee DuBridge regarding Bowen founding a major new radio observatory at Caltech (pp. 117–21). This never came to pass, but it did lead (a) in 1954 to the Carnegie Foundation providing the first money for a giant Australian dish, and (b) in 1955 to Bolton moving to Caltech to set up radio astronomy in the Owens Valley.

[106] For another take on the development of the Giant Radio Telescope and competing ideas, including Bowen's fundraising, see Munns (2002:85–110).

---

[107] Pawsey:Oliphant, 10 September 1951, file A1/3/1n, RPS.

[108] Furthermore, Australia's first colonizing expedition in 1788 included William Dawes, an astronomer who set up an observatory immediately upon arrival.

TANGENT 7.1 VACUUM AND HIGH-ENERGY PHYSICS IN THE RADIOPHYSICS DIVISION, 1945–48

At war's end Bowen transferred some of RP's best researchers (including B.Y. Mills) into vacuum and high-energy physics, but then could not hold many of them from taking overseas appointments or scholarships. Nor was he ever able to attract to RP an outside expert in the field.[109] In 1945 plans were centered on developing a device that could produce millimeter-length waves as efficiently as the cavity magnetron did centimeter ones. There were also efforts in 1947 to develop a bolometer or receiver suitable for solar measurements in the 1–10 mm wavelength range, but these came to nought.[110] Vacuum physics research did, however, lead to a magnetron-type resonant cavity producing high-energy (~1 MeV) X-rays.[111] The next goal was to chain several of these devices together and accelerate electrons and protons to extremely high energies, but the costs needed to effectively compete with US groups rapidly escalated out of reach and the linear accelerator project was shut down.[112]

---

[109] Bowen, 24 May 1946, "Notes on future programme of the Radiophysics Laboratory," p. 5, file D1/2, RPS.

[110] Bowen:Woolley, 15 April 1947, file A1/3/1a; ca. July 1947, "1946/47 Annual Report of Radiophysics Division," p. 4, file D2/6 – both RPS.

[111] E. G. Bowen, O. O. Pulley and J. S. Gooden (1946), "Application of pulse technique to the acceleration of elementary particles," *Nature* **157**, 840.

[112] (Bowen 1984:106). More information is in Munns (1997:307–12) and Home (2006:227–31), "The rush to accelerate: early stages of nuclear physics research in Australia," *Historical Studies in Physical Sciences* **36**, 213–41.

8 • Ryle's group at the Cavendish

Alumni of the largest of England's wartime radar laboratories, the Telecommunications Research Establishment, were central to British pre-eminence in postwar radio astronomy. This chapter tells the story of Martin Ryle's group at the Cavendish Laboratory and the following chapter tells that of Bernard Lovell's at Jodrell Bank. Both groups were indelibly marked by the war. Both were successful despite a limited number of men and meagre financial resources. However, there were great differences in their research styles, techniques, scientific problems, and personalities.

After a period of uncertainty about his postwar research direction, Ryle jumped on to extraterrestrial noise and steadily built a talented group around him. He started off monitoring the sun, but soon became fascinated with the enigmatic radio stars. By 1950 fifty were listed in the "1C" catalog, including the strongest of all, Cas A. Along the way he developed clever instrumental techniques such as a noise-balancing receiver, the Michelson interferometer, and phase switching. By the early 1950s, with strong support from the Cavendish Lab, he was firmly on the road to developing a Fourier mapping technique that came to be called aperture synthesis and for which he would eventually win a Nobel Prize.

8.1 THE SETTING AT CAMBRIDGE AND TRE

Since 1874 the Cavendish Laboratory of Cambridge University had been Britain's, and arguably the world's, preeminent physics laboratory. The first Cavendish Professor of Experimental Physics was James Clerk Maxwell, followed by lights no less than Rayleigh, Thomson, and Rutherford. Ernest Rutherford's tenure, from 1919 until his death in 1937, culminated in the *annus mirabilis* of 1932 during which his group split the atom and discovered the neutron. Rutherford's style of science and management suffused the Cavendish.

Experiments should be cheap and simple in concept: the watchwords were "string and sealing wax" and "we have no money, so we shall have to think," sometimes rendered as "use your brains rather than your wallet." These Rutherfordian attitudes remained an important part of the laboratory's ethos despite the fundamental changes wrought by World War II and a redirection of the Cavendish program away from nuclear physics by his successor W. L. Bragg.[1]

Although the greatest fame of the pre-war Cavendish came from its nuclear physics, fields such as hydrodynamics, low-temperature physics, atmospheric electricity, and radio physics of the ionosphere were also first-rate. For instance, Edward Appleton's 1924 experiment demonstrating the existence of the ionosphere had been conducted at the Cavendish (Appleton and Barnett 1925), and was often later cited not just for its scientific import, but also as a prime example of proper experimental style: Appleton had used nothing more than a simple receiver, together with an existing broadcast transmitter, in contradistinction to a contemporaneous American experiment requiring a special pulse transmitter and receiver (Ratcliffe 1975:463). When Appleton moved to London in 1924, ionospheric work in the Cavendish became the province of his student J.A. "Jack" Ratcliffe (1902–1987),[2] who led a research group (two to five in size) primarily studying shortwave propagation. Ratcliffe's research was notable, but his real forte was in teaching and training. His lectures were models of clarity, and a worldwide survey

---

[1] Information about the pre-war Cavendish Laboratory can be found in *The Cavendish Laboratory, 1874–1974* by J. G. Crowther (New York: Science History Publications., 1974), and *Cambridge Physics in the Thirties* by J. Hendry (ed.) (Bristol: Adam Hilger, 1984).

[2] For Ratcliffe's career, including its interactions with Ryle's group and radio astronomy, see K. G. Budden in *Biog. Mem. Fellows Roy. Soc.* **34**, 671–711 (1988).

by Gillmor (1982:403) found that he trained "the largest number of students who have made significant contributions to ionospheric physics." These skills were put to good use during the war when Ratcliffe founded an Army school that gave six-week crash courses in radar principles and techniques to those with scientific background, but with no previous work in radio. Many principals in postwar radio physics went through this school, including Stanley Hey (Section 5.2). In 1941 he moved over to the Telecommunications Research Establishment (TRE), soon to be located at Malvern, in Worcestershire, and ran a "post-design" service. This service took brand-new radar models, installed them in the field and trained operators, and then recommended design changes based on this experience.

Since even before the war TRE had been the premier British radar laboratory, the direct descendant of the small group that Robert Watson-Watt assembled in 1935 to design the first operational radar system (Section 5.1). Run by the Ministry of Aircraft Production and the Royal Air Force (RAF), TRE under the leadership of A. P. Rowe developed a wide variety of radar systems for air raid warning, bombing, navigation, fighter interception, location of ships and submarines, and jamming of enemy radars. TRE's size (~8000 civilian and military staff by the end of the war) dominated its sister laboratories, the Royal Navy's Admiralty Signals Establishment, which concentrated on radar transmitters and vacuum tubes, and the Army's Air Defence Research and Development Establishment. Success of these radar laboratories hinged on technical expertise and on quickly turning raw ideas into usable operational systems. For instance, liaisons with the RAF, with industry, and with the Air Ministry in London were enhanced at TRE by Rowe's "Sunday soviets," freewheeling sessions where everyone from Members of Parliament to RAF pilots exchanged views. As Watson-Watt put it:

> The Air Marshal took advice from the junior scientific officer on how to make war, and the laboratory assistant was told by the Admiral why physics has sometimes to give way to psychology in the planning and conduct of operations.[3]

8.2 TRANSITION TO PEACETIME

In early 1945 Ratcliffe, then one of the top administrators at TRE, began to rebuild his Cavendish radio group by circulating among the best young men at TRE with a sheet of paper listing his planned topics for postwar research – these included his own area of radio propagation, but also research on electronic devices, radio properties of solid materials, and optical studies of the atmosphere. One of the topics he thought most likely to succeed *and* not to require "external help" (a desideratum) was a category curiously called "Investigation of radiation on radio wavelengths from 'non-radio' sources." This included cosmic, solar, and ionospheric noise – all "non-radio" in the sense of not involving a manmade signal. Ratcliffe appears to have been particularly interested in the possibilities of spectral features in this noise, either at the gyromagnetic frequency (Appendix A) or at frequencies characteristic of molecular excitation. With these tentative programs he recruited many radio physicists to join him back in Cambridge, including Martin Ryle, Derek D. Vonberg, Kenneth E. Machin, and F. Graham Smith.[4] Ratcliffe also became Bragg's right-hand man in effecting not just a restoration, but a transformation of the old Cavendish. For instance, the budget and administrative support per researcher doubled from prewar standards (one's papers could now be typed by a secretary!) and the laboratory's staff jumped from 40 to over 100 (Ratcliffe 1975).

Pioneers of Radar (1999), C. Latham and A. Stobbs (Thrupp, Gloucestershire: Sutton Publishing). For an overview of British military radar labs, see several chapters in *Design and Development of Weapons* by M. M. Postan, D. Hay, and J. D. Scott (London: HMSO, 1964). For those (including TRE) antecedent to the present Royal Signals and Radar Establishment, see "The RSRE: a brief history from earliest times to present day," by D. H. Tomlin (1988), *Inst. Elec. Eng. Review* **34**, 403–7. For an American view, see Guerlac (1987:731–56, 817–23), including (unsuccessful) American attempts to emulate the intimate ties in Britain between military and civilian radar people.

[4] For example, Ratcliffe:Machin, 11 March 1945, a letter offering support as a research student in the Cavendish [photocopy from Machin]. Ratcliffe's sheet of paper, with his own handwritten notes, was found in the Ryle papers ["Notes on radio research programme," 6 May 1945, file 18 (uncat.), RYL]. Ratcliffe's plans did not appeal to at least one person: "Ratcliffe is collecting bodies for what appears to me to be a deadly dull programme on propagation at Cambridge" [P. I. Dee:A. C. B. Lovell, 16 April 1945, file R22 (uncat.), LOV].

---

[3] "Radar in war and in peace," *Nature* **156**, 319–24 (1945), p. 322. For details of TRE's wartime operations, see "T. R. E." by "J. A. and A. D. S." in *Discovery* **9**, 205–11 (1948), and

By the autumn of 1945 only Ryle and Vonberg had made their way to Cambridge; Machin and Smith came a year later. Vonberg had been trained as an electrical engineer at Imperial College in London during the war, while Ryle (1918–1984) was a few years older.[5] Ryle had been raised in an upper-class family distinguished for generations in the professions, the church, and academia; indeed, both his father and uncle were Oxbridge professors, the former, John A. Ryle, an authority in social medicine and one of the King's physicians and the latter the noted philosopher Gilbert Ryle. He took a physics degree at Oxford in 1939 (and ran the university's amateur radio club), but postgraduate studies with Ratcliffe were summarily aborted by the war. At TRE he developed airborne antennas and headed a radio countermeasures group that endeavored first to deduce and then to thwart the enemy's radar capabilities:

> At TRE Ryle developed the keenness of his intellect in competing with the German radar development groups. ... I can remember him showing me a very sketchy little photograph, taken by Intelligence from an aircraft, of a trial radar installation in Germany. From this he scaled the aerial dimensions [to estimate the operating wavelength] and got his countermeasure system all ready so that the "Gerry" system only worked for a night or two before ours went into full-blast operation and rendered it useless. This kind of tremendous intellectual stimulus to outwit the Germans obviously had a great effect on him.
>
> (Vonberg 1971:7–8T)

Perhaps his greatest contribution during the war was to the design and execution of an electronic spoof, a fake invasion at Dover that successfully diverted the Germans' attention from the real D-Day invasion in Normandy.[6] But as much as this kind of work could be exhilarating, by the end of the war it had left Ryle (1971:11T) drained:

> It is difficult to explain ... what a demoralized state one felt in at the end of six years of day-to-day struggles with new problems every week – problems that had to be coped with in a great hurry and that involved many skills other than messing around with circuit diagrams and soldering irons, problems of [dealing with] Air Marshals, problems of actually getting into aeroplanes and making equipment work, problems of making sure it continued to work and that people knew how to use it. ... Anyway, damn it all, it was all concerned with killing people and one got fed up with that. So one just wanted to get out of the whole shoot, and had no positive idea at all of where to go. And Ratcliffe was certainly very important to me in suggesting something positive to do.[7]

In fact, during his closing months at TRE Ryle seriously considered abandoning altogether a scientific career, and it appears that it was largely the intervention of his father that persuaded him to accept Ratcliffe's offer.[8]

Ratcliffe first wanted Ryle to study the ionosphere using radio propagation, but Ryle saw little future in what he called "ionomy." Another possibility was to use captured V-2 rockets for *in situ* measurements of the upper atmosphere,[9] but this too didn't appeal:

---

[5] See the obituaries of Ryle by A. C. B. Lovell in *Quarterly J. RAS* **26**, 358–68 (1985) and by F. G. Smith in *Biog. Mem. Fellows Roy. Soc.* **32**, 497–524 (1986). The latter is especially detailed on Ryle's wartime work.

[6] Ryle gave his own account of one aspect of the D-Day preparations in "D-13: some personal memories of 24th–28th May 1944," *Inst. Elec. Eng. Proc.* **A132**, 438–40 (1985). The overall work in radio countermeasures at TRE is well documented in a reprint of a 1945 report by R. Cockburn, "The radio war" on pp. 423–34 in the same issue. Note especially the comments on pp. 430–1 regarding the continual "crash" basis with which the countermeasures group had to produce equipment. Ryle also recollected stories of his TRE years on pp. 133–7 (and see photo on p. 248) of *Pioneers of Radar* by C. Latham and A. Stobbs (1999, Thrupp, UK: Sutton Publishing).

[7] Ryle's postwar state of mind after his wartime successes can perhaps be gauged by these lines written in Germany during the war by Bertolt Brecht: The designers sit/ Hunched in the drawing offices:/ One wrong figure, and the enemy's cities/ Will remain undestroyed. [From the poem *1940* on p. 347 of *Bertolt Brecht Poems 1913–1956* (1976, ed. J. Willet) (New York: Methuen).

[8] Smith, p. 506 as in note 5; J. A. Ryle:A. P. Rowe, 24 May 1945, file 4 (uncat.), RYL.

[9] From the summer of 1945 onwards, Ratcliffe was a member of a committee charged to consider possible scientific experimentation using V-2 rockets [Ratcliffe:Lovell, 25 July 1945, file R13 (uncat.), LOV]. Although Ryle did not want to pursue rocket science, he wrote up his ideas in "A suggested method for using long-range projectiles for the investigation, at radio frequencies,

I can remember a great deal of pacing up and down in his office while Ryle churned over the possibilities of using these rockets. But I think he didn't like the idea because he was a very independent and highly original person, and the idea of being dependent on rockets sent up by someone else didn't appeal to him. ... It would be out of his control. ... He was rather looking for something *really* new, to get away from the old ionospheric studies.

(Vonberg 1981:140B:40)

During his first autumn and winter back at Cambridge Ryle was basically unhappy about his resumed postgraduate studies and continued to vacillate regarding the direction of his career. Advice from several senior physicists, as well as his father, was sought and received. To assuage his concern about whether he could ever make the transition to basic research, especially in the rarefied atmosphere of the Cavendish, his father wrote to him at the time:

This adaptation to the rigid traditions and minds of the old universities is a beastly uncomfortable process. I'm sure what one has to bank on is that time and chance will bump one up against a good problem of one's own and a congenial mind or two. ... You have experience, ideas and techniques besides which the mathematical physics of your associates is no more than a school exercise in value. ... Remember that we Ryles are unduly sensitive and feel "inferior" for unsound reasons and are rebels and not easy mixers. Being commonsensical and having the ability to take broad views puts us in a different category from the often rather vain specialists.

I should have thought that methods and techniques had been every bit as important to the advancement of physical science as the fundamental ideas and mathematical principles which precede, but sometimes develop out of, them. Therefore I think you have every right to consider yourself a real physicist.[10]

A match-up finally clicked when Ratcliffe shifted Ryle from the ionosphere to extraterrestrial radio noise, which had come to Ratcliffe's attention primarily through K. F. Sander's wartime studies of the effects of the galactic background on receiver performance (Tangent 6.2).[11] Ryle (1976:4–6T) thought this more exciting and liked its advantage of no possible connection with war. Once he started on it, he wrote:

I am settling more into the present problem of the structure of the interstellar gas, although it is difficult to see how the work can be of much value to anyone. At the same time, it is hard to see how it could be used as a weapon of war!. ...

At the moment I am making a device for measuring the temperature of outer space! Not of terrific practical significance, I am afraid, but at any rate more harmless than atom-splitting, the only other main line of physics.[12]

In November 1945 Ryle set forth a proposed program of research on both solar and "cosmic" (galactic) measurements, first at frequencies of ~200 and 3000 MHz and later on other bands in the 70 to 10,000 MHz range.[13] The solar studies were to be aimed at studying time variability, and the cosmic studies at finding the temperature and "blackness" (absorption coefficient) of the interstellar gas. Ryle emphasized that "absolute" measures of the received noise energy were important, by which he meant measures well calibrated with respect to standard noise sources. After preliminary data were in hand, he proposed to build an electrically steerable, broadside array of 24 × 24 dipoles.

This long-term program was fine with Ratcliffe, but the next hurdle was that he and Ryle disagreed on how to launch it. Ryle (1971:5T, 1976:12T) wanted to postpone observations until he had designed and built a really first-class receiver and antenna system, but Ratcliffe urged him to start off with something quick and dirty, a lash-up for jumping in with the competition.

---

of atmospheric refractive index, with particular reference to the measurement of electron density in the ionosphere," Rept. MRP 282 (8 April 1946) [box 1/1(S), RYL].

[10] J.A. Ryle:M. Ryle, 30 October and 11 November 1945, file 4 (uncat.), RYL.

[11] Interview with Ratcliffe by F. G. Smith, January 1985, side A of tape in file 1 (uncat.), RYL.

[12] Drafts of letters (recipient unknown), undated (about February 1946), file 4 (uncat.), RYL.

[13] Ryle, "Proposed programme for investigation of solar and cosmic noise," 21 November 1945, 3 pp., CAV.

Vonberg (1971:11T, 1981:140B:110) later recalleed this conflict:

> Ryle was a great one for planning out and he used to plan out a bit in the way I imagine Montgomery must have done in the Army. He really mobilized his thoughts and then could move ahead with tremendous confidence and precision. But Ratcliffe was pressing us to get some measurements done, saying "for goodness' sake don't spend all this time thinking."
>
> At one point Ratcliffe got hold of some old field radar set, and told Ryle, "Now get it out in the field, set it up, and see if you can't see *something*. See some noise, point it at the sun, look at the Milky Way – anything you like, but get some measurements going." But Ryle kept digging in his heels and saying, "No, no, this isn't the way to do it – what we should do is first work out every aspect of what we're going to do, the sensitivity, the calibration, and so on. Then when we do get a measurement it will mean something." As it turned out, we ended up in a field with me rushing around with this wretched bit of radar, pointing it around, and Ryle being thoroughly annoyed and disinterested. Of course we saw nothing, or if we did, we didn't know what it was.

Before better equipment could be built, however, Ryle and Vonberg devoted much of their first year to restocking the bare shelves of the Cavendish with everything from voltmeters and cable to giant antennas. This could not be done with purchases, for there was no money, and so they took advantage of the plethora of war surplus electronics available for a song (Ryle 1971:12). They were of course familiar with the British equipment, but Ryle also knew which *German* equipment was desirable. Moreover a German item was easier to acquire because it didn't have a serial number on it that needed processing by some bureaucrat. From his countermeasures work, Ryle had developed an intimate acquaintance with the thoroughness and precision of German engineering; in fact, during the war he could even tell whether an unknown transmitter was British or German because the Germans' frequency stability was markedly superior.[14] More direct information came immediately after V-E Day when he joined a TRE team participating in RAF Exercise "Post Mortem," wherein the British tested the performance of German radar defenses in Denmark by staging a mock attack while observers below watched German operators and evaluated equipment (Ryle 1981:142B:670).

Five truckloads of surplus gear were looted from a depot at the Royal Aircraft Establishment (RAE) at Farnborough. The most spectacular finds were several 3 m and 7.5 m diameter Würzburg dishes, which later played an important role in solar and radio source measurements (Tangent 4.2). As Ryle (1976:43–4T) later recalled:

> We went to RAE to arrange for the transport of the Würzburgs, but found that they'd unfortunately just been sold to a scrap metal merchant, which was sad. But we drove around to see the merchant and he was a very nice chap. When we told him we wanted them for scientific research, he said, "You can have them! I like science." We swapped them for some other stuff and we all parted happily. But then we were in a bit of a fix because gifts to the University of Cambridge have to be recited in the Senate House in Latin and we weren't quite sure how we'd translate all this stuff about German radar sets!

After six years of trying to outwit and defeat German engineering, Ryle could now profit from it.[15]

8.3 SOLAR OBSERVATIONS

Ryle and Vonberg started off with observations on meter wavelengths because the required equipment was more familiar, more easily available, and cheaper. More specifically, a wavelength of about 2 m was a good compromise between the abundant solar activity at longer wavelengths and the better angular resolution afforded by shorter wavelengths. But what for an antenna? They wanted a scheme that would allow good discrimination of the sun from the background

---

[14] R.V. Jones, *Most Secret War* (London: Hamish Hamilton, 1978), pp. 229–30. Jones relates fascinating case studies of British scientific intelligence during World War II, much of it involving radar.

[15] This account of the early development of Ryle's group has profited from Edge and Mulkay (1976), in particular pp. 20–28. Note also that all cited 1971 interviews were carried out by Edge and Mulkay (see Appendix B), and that their book carries the development of Ryle's group forward to the 1960s.

cosmic noise, yet not require an antenna too costly or cumbersome. Sometime in the winter of 1945–46 Ryle came up with the idea of an interferometer, two separated small antennas whose signals would be combined by cable. This idea was undoubtedly influenced by his wartime work on aircraft radars that determined direction of a target by combining signals from two separated antennas, one on each wing. Such an interferometer would produce a series of intensity oscillations ("fringes") caused by the varying difference in path length from the radio source to each of the antennas as the source moved across the antennas' beams (see Appendix A). In this way one could obtain an angular resolution measured by the wavelength of operation divided by the *spacing* of the two elements, a resolution that could far exceed even that from a giant antenna. And by increasing the spacing to the point where fringe amplitudes became weaker, one could estimate the radio source's size. Furthermore, since the beam of each individual element was broad, a source such as the sun could be monitored for several hours each day without any tracking. This scheme was analogous to the optical interferometer developed by Albert A. Michelson in the 1880s and used by him and Francis Pease in the 1920s for measuring the angular diameters of stars. T. L. Eckersley had even employed such an instrument for ionospheric work in 1938,[16] but Ryle (1976:4T) later recalled that the idea came to him from elsewhere:

> I don't think the idea arose by analogy with an optical Michelson interferometer because one had forgotten all one's physics by the end of the war. I think it arose from the idea that if you have a null by interference between two aerials, then this could fairly easily tell you something about a compact source. And then you could make that null narrower by separating the aerials. I suppose we were reinventing the Michelson interferometer, but I think it came from a rather simple-minded thinking about aerials rather than from saying, "Ah, I remember in my physics book – optics."

The receiver system was also novel, designed by Ryle to allow a continuous, accurate measurement no matter how noise power level might vary (such as during a solar burst, or during fringe oscillations). They called it the "cosmic radio pyrometer," but even today it is known as a Ryle–Vonberg receiver (Appendix A). The antenna signal was compared (at a rate of 23 Hz, using a capacitance switch from a German Junkers 88 nightfighter) with a reference noise source, whose magnitude was then continually varied so as to equal the signal (Fig. 8.1). The noise power emitted by this reference was thus a measure of the signal strength, and, so long as the reference source could track gain changes accurately and quickly enough, omnipresent variations in gain or other receiver parameters, for instance those arising from a varying mains voltage, had minimal effect. The function of the conventional receiver in the system, a modified air-to-surface-vessel radar, was thus reduced to producing an error voltage to control the noise source. But such a scheme required sophisticated servomechanism circuits to control the variable noise source, and it became Vonberg's lot to wrestle with the cantankerous "thyratron" vacuum tubes involved. The wideband noise source was a tungsten-filament diode operating under temperature-limited conditions,[17] which Ryle well knew from wartime "jamming" practices. Overall system temperature was ~3500 K and tests showed that a 0.5% change in system power could be detected.

Each interferometer element was an array of eight half-wave dipoles, mounted on a support of angle iron covered with a chicken-wire reflecting screen (Fig. 8.2). They were connected by low-loss, air-spaced coaxial cable whose nickname of "Jerry cable" revealed its origin. Ryle had managed to scrounge a kilometer of this captured German cable – nothing of comparable quality was available in Britain for many years – but it came only in relatively short lengths. This necessitated tedious hours making plug and socket connectors, which unhappily led to severe attenuation whenever rainwater leaked in (Ryle 1971:15). These cables and their troublesome connectors were used by the group for almost a decade (Baldwin 1972:14A:360).

They had a going system by June 1946, but interest in the Milky Way, which gave the same trace every day, quickly waned in favor of the changeable sun. For many months Ryle had been ruminating on the relation between solar and galactic noise, concluding that

---

[16] T. L. Eckersley, "A wireless interferometer," *Nature* **141**, 369–70 (1938).

[17] See Tangent 6.3 for more details.

Solar observations   161

RYLE-VONBERG RECEIVER

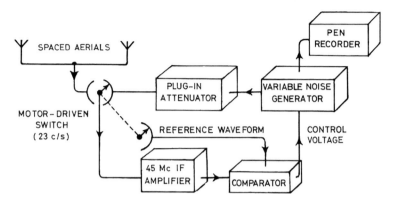

Figure 8.1. Block diagram of the noise-balancing receiver developed by Ryle and Vonberg in 1945–46.

Figure 8.2. First observing hut (leftmost) and early antennas of Ryle's group, circa 1948 at the "Rifle Range" field station off Grange Road, Cambridge (see Fig. 8.3). Several solar interferometers are shown. From the left, (1) (in front) 80 MHz Yagi; (2) (behind) 80 MHz 4-dipole broadside array with reflecting screen; (3) 175 MHz 8-dipole array; path; (4) antenna paired with 3; (5) (in front) pair with 1; (6) (behind, cubical structure higher than shack) 214 MHz 4-Yagi array (operated as a single antenna, based on a wartime "Lightweight Warning" set); (7) pair with 2; (8) alternate pair with 3 (with polarization crossed). Ryle's house is off to the left and the tower of the main University Library is visible in the background.

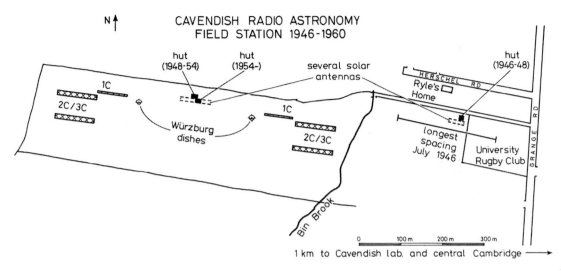

Figure 8.3. Map of the Grange Road field station of Ryle's group in the 1940s and 1950s, based on various contemporary sources.

for any progress solar data were first needed.[18] He now posed himself basic questions concerning the sizes of the active radio regions on the sun, whether or not a base level of solar emission existed at his frequency of 175 MHz, and correlations between radio and optical activity. In late July a giant sunspot group appeared and Ryle and Vonberg (1946) seized the opportunity to measure the angular size of the accompanying radio-emitting region. Within a week they hauled their dipole array to three ever-more-distant spacings, from 25 to 140 wavelengths (240 m). This all took place on the periphery of Cambridge, about a mile from the centrally located Cavendish Laboratory (Fig. 8.3). The site, just off Grange Road and known as the "Rifle Range" because it was shared with the Officers Training Corps, had been used by Ratcliffe's group since the 1920s and was still an active ionospheric field station. Until they moved to an adjacent plot in 1948 Ryle's group was confined to a wooden hut optimistically dubbed "Mon Repos" and measuring only fifteen feet on a side (Fig. 8.2).[19] And when they began needing antenna separations of over 200 m, they found themselves pushed onto the adjacent University rugby grounds.[20] The trouble was worth it, for the ratio of maximum to minimum intensity measured on the longest baseline, at which the fringes were spaced by 25′ (Fig. 8.4), enabled them to place an upper limit of 10′ on the emitting region's east–west extent and therefore a lower limit on its brightness temperature of 2×10^9 K. By crossing the polarity of the two elements, they also measured 100% circular polarization in the active region, leading them to argue that the strong magnetic field of the sunspot group (known from optical measurements) was important for the radio emission process. These results were similar to those in the same year

---

In some sense, therefore, it can be said of the early Cambridge group that they, like Reber, built their antennas in their backyard!

Ryle and Smith also became more than colleagues when in 1947 Ryle married the sister of Smith's wife (note 23). Likewise at Jodrell Bank, Hanbury Brown became linked to Lovell through more than science when he married the cousin of Lovell's wife.

[20] Once the rugby season started in the autumn, sport took precedence over science and Ryle's forays on to the rugby grounds quickly came to a halt. [p. 6 of the notes for a talk by Ryle on "Cosmic noise" to the Cambridge University Astronomical Society, 13 November 1946, file 8 (uncat.), RYL]

---

[18] Ryle, "Increase of radio frequency radiation from the sun during sun spot activity," 16 February 1946, 8 pp., file 17 (uncat.), RYL; Ryle, "Notes on the origin of extra-terrestrial noise," 9 April 1946, 6 pp., file 6 (uncat.), RYL.

[19] Within 100 m of the hut visible in Fig. 8.2 was a large house shared by Ryle, Vonberg, Smith, Henry G. Booker (note 40), and spouses; later, Ryle's family lived there alone (see Fig. 8.3).

In the autumn of 1946 Ryle's group doubled in size when Machin from TRE and Harold M. Stanier, who had done antenna work at RAE during the war, also began postgraduate study under Ratcliffe. Machin improved the circuitry (Fig. 8.5) for their basic receiver (for example, shortening response time to ~0.2 sec to better follow solar bursts) and Stanier developed antennas, in the first instance for a second frequency of 80 MHz. By December the group was able to commence a dual-frequency solar monitoring effort that led to a second paper in *Nature* and a summary paper in the *Proceedings of the Royal Society* (Ryle and Vonberg 1947, 1948). The latter included a particularly succinct explication of the problems and principles of measuring small radio noise levels.

Ryle and Vonberg found that solar radio bursts, lasting no longer than a few minutes, could become as strong as one hundred times the daily mean at 80 MHz and ten times at 175 MHz. These bursts were on top of variations of a similar magnitude in the daily mean itself. Furthermore, the levels never dropped below an equivalent brightness temperature of 1×10^6 K (assuming a source diameter of 30'), as Pawsey (1946) had earlier discovered. Although emission at the two frequencies tended to go up and down together, this was not uniformly true, leading to their suggestion that radio emission from active regions might be beamed differently at different frequencies. They searched for any correlations of the radio variations with visible sunspot regions, in particular for tie-ins with solar rotation, and gave graphical evidence for three distinct periodicities of 26.3, 27, and 28 days in the radio data. These, they thought, might be linked to the sun's differential rotation with latitude. But on the whole the monitoring program did not prove profitable and, although continued until 1953 (see Fig. 13.1), from 1948 onward it garnered less and less attention as Ryle was lured away by the radio stars, to which we now turn.

8.4 RADIO STARS

During the early months of 1948 Ryle became intrigued with the source of emission discovered by Hey's group in the constellation of Cygnus (Section 6.2), as well as in the many additional sources picked up by John Bolton (Section 7.4). The solar records taken over the course of 1947 had often shown tantalizing weak ripples, possible indications of discrete

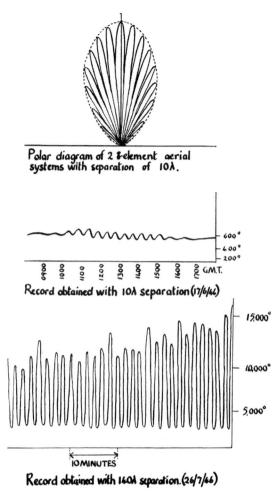

Figure 8.4. Calculated antenna response and observed fringes at 175 MHz from an active solar region in July 1946. These observations are the first with a Michelson interferometer and the Ryle–Vonberg receiver, but not yet with a phase-switch (note the large offsets). (Ryle and Vonberg 1946)

by Hey's group (Section 6.4) and by the Australians,[21] but the real significance was in the introduction of the Michelson interferometer. This turned out to be Ryle's first step in a thirty-year journey of developing novel techniques for interferometry.

---

[21] In particular, the deduction of a small size for the radio-emitting regions had been made in February 1946 by McCready, Pawsey and Payne-Scott (1947) using a sea-cliff interferometer (Section 7.3.2).

Figure 8.5. 214 MHz "Cosmic Radio Pyrometer," an improved version of the Ryle–Vonberg receiver built by Machin circa 1948. In the left unit is the motor-driven capacitance switch (large box), noise diode (smaller box), and RF amplifier and mixer. The middle unit contains the 45 MHz IF amplifier (a "Pye strip" taken from wartime radar) and the right unit contains power supplies.

sources, but of inconsistent strength from day to day (Vonberg 1981:140B:590). Interest was further piqued by the latest news from Sydney, borne by Joe Pawsey in the spring of 1948 when he paid an extended visit to the Cavendish Laboratory as part of an around-the-world tour (Section 7.5.1). Ryle decided to investigate the polarization of the radiation from the Cygnus source, in order to see whether it corresponded to the circularly polarized sunspot radiation or to the unpolarized quiet sun. His picture was of a star with a strong magnetic field, and so it seemed natural, especially when one considered the intensity variations shown by Cygnus, to expect a high degree of polarization. Chief collaborator in the radio star work was Graham Smith, who had joined the group in the summer of 1947 after finishing undergraduate physics at Cambridge, which had been interrupted by three years at TRE (primarily developing test equipment and procedures). Smith (1976:1–3T) carried out various solar observations during his first year, but then focused on radio stars.[22,23]

Ryle and Smith first observed Cygnus simply by tilting the solar dipole-array antennas up to the appropriate declination. But their 80 MHz interferometer (of ten wavelength spacing) gave only weak, unusable signals and the 175 MHz interferometer none at all. So they changed to a higher gain, four-Yagi array

---

[22] F. G. Smith, August 1948, "Radiations from the sun and stars," Rept. to the Dept. of Scientific and Industrial Research, 25 pp. and 9 figs., [file 6 (uncat.), RYL]. This report summarized Smith's activities on the sun and radio stars over the previous year and has been an invaluable source for this section.

[23] Smith's wife Elizabeth, another TRE veteran, was also a technician in the group during 1946–47. For a while she constructed solar recording equipment and kept it running, but left the group to raise a family. [F. G. Smith:author, 17 July 2007]

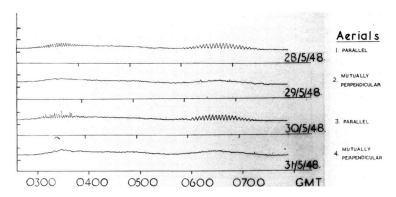

Figure 8.6. Interferometric traces obtained with 80 MHz four-Yagi elements and a spacing of 450 m (120 wavelengths) by Ryle and Smith (1948) in May 1948. Fringes centered at 0330 UT are from Cyg A and those at 0640 UT from newly discovered Cas A; note the longer periodicity of the fringes from Cas A. Comparison of the records on 28 and 30 May shows irregular intensity fluctuations associated with Cyg A, but none with Cas A. Records on 29 and 31 May, taken with the elements cross-polarized, show no detectable circular polarization from the sources, only a slowly varying level of galactic background noise. Tick marks at the left indicate intensity levels of 0 and 50,000 Jy.

and moved the spacing to 120 wavelengths in order to minimize signals from general galactic noise. Now they hit the jackpot, for in May 1948 on the first night in which they left this new 80 MHz system running (still with the noise-diode balancing scheme), not only did fringes from the Cygnus source show up, but three hours later even stronger ones (Fig. 8.6) – they had discovered Cassiopeia A, the strongest radio source in the sky:[24]

> We said, "Let's build an instrument to look for [this fluctuating region in Cygnus]." And it was actually very exciting because we built this instrument and left it running and overnight, damn it, there were *two* of them! (Ryle 1976:17T)
>
> I remember turning up at the laboratory after the thing had been set running, and there was an interferometer trace of Cygnus just as it should be – absolutely marvellous. And then three hours later, another [sinusoidal] trace, which subsequently turned out to be Cassiopeia A, … and with a *different* periodicity than Cygnus A. This was a complete surprise. We thought it must be something to do with the way things moved in the sky, but neither of us had any training in celestial coordinates. … I went away to think about it, and I decided the fringe rates would be different according to the distance from the pole. So that was my first introduction to spherical astronomy. … I remember then coming into the Cavendish Lab an hour or two later and saying to Ryle and Ratcliffe, "I think it's in a constellation called Cassiopeia."
>
> (Smith 1976:10–11T)

Ryle and Smith had accidentally discovered not only a new radio star, but a powerful method for determining positions. The source's right ascension came from measuring the sidereal time when it transited, that is, when the fringes reached greatest amplitude, and the declination from the periodicity of the fringes. Although their individual antennas had a ~60° beam and therefore very little discrimination in the north–south direction, the fact that stars nearer the celestial pole moved more slowly across the interferometer's pattern of lobes could be used as a neat measure of declination (see Appendix A). Encouraged by his initial foray, Smith (1976:11–12T) now dove into the geometry of measuring star positions:

> I was in fact invigilating in the Tripos examinations during that summer and I spent the hours of peace

---

[24] At Ryle and Smith's operating frequency of 80 MHz, Cas A was about 50% more intense than Cyg A and comparable to the quiet sun (Fig. A.2 and Table 14.1). See Tangent 6.4 for a discussion of why Cas A, despite its strength, had eluded earlier recognition by all observers.

one gets in an examination room working out spherical astronomy from scratch. I worked out how to correct for the lack of east–west [alignment] and the lack of level of our transit instrument, and I'm proud to say I got the right signs for these corrections. And then after I had completed my calculations and found the positions of these sources, I was told that a chap called Bessel had done it a few years earlier!

But amidst these successes a serious problem soon arose. Pawsey brought news of Bolton's best position for Cyg A and it turned out to disagree grossly with Ryle and Smith's. Long discussions over possible errors were held, but no one could see how either the sea-cliff or the Michelson interferometer could be off by as much as 2°. As Ryle wrote to Bernard Lovell at Jodrell Bank in June 1948:

> At the moment we are not entirely satisfied with the calibration of our aerial system – we can see no possibility of appreciable error, yet we find a serious discrepancy in the declination for the source in Cygnus as compared with the Australians. … We hope to have it cleared up in a few days.[25]

Ratcliffe and Pawsey persuaded Ryle to hold off publication of any Cyg A position until he and Bolton could sort out who was wrong and why.[26] Ryle wrote to Bolton about all this, and further discussed the possibility that they *both* might be right if the source were actually moving:

> It is clear that one of us must be wrong – unless the source has moved. … We are wondering if you will be able to make another measurement soon. If you still get your original figure, we will have to work out which of our systems is giving the wrong answer, but if you find it has moved – then the astrophysicists must think again![27]

Ryle and Smith (1948) sent a note to *Nature* announcing their new radio star in Cassiopeia and that the signals from Cas A and Cyg A were unpolarized and came from regions less than 6′ in size. A position was given for Cas A, but readers must have wondered, given the authors' touting of their method's positional accuracy, why none was proffered for Cyg A. And in Ursa Major they had also registered a *third* source – it was at the same declination and almost opposite in right ascension to Cas A, and about 20% as intense (although variable). This, however, turned out to be bogus, for later work showed that these fringes were in fact caused by circumpolar Cas A passing through the *back* sidelobes of the Yagi arrays as it daily scraped the northern horizon at lower culmination (Ryle 1976:23T).[28] Ryle and Smith were also impressed that the new Cassiopeia source remained steady in its signal level while Cyg A showed strong fluctuations. From these fluctuations they inferred that the source of emission was less than 20 light-sec in size, which meant that, even if it were as close as the nearest star, its brightness temperature, incredibly, was at least 10^{14} K. And yet because they had found the radiation to lack any polarization, they suggested that it still might be thermal, "possibly more nearly analogous to the radiation from the undisturbed sun."

Enticed by three strong sources and hints of many weaker ones, Ryle decided that the next step was to find more of these radio stars by building an interferometer with far greater sensitivity. Thus was born what the group called the "Long Michelson" (Figs. 8.7 and 8.3). Ryle shared strongly in the Cavendish tradition of simple apparatus applied to experiments of elegant design, and to him the east–west Michelson interferometer was a particularly economical device: its small, stationary antennas allowed one to neatly derive both coordinates for all the radio stars in the sky, and in only one day. Element spacing remained at about one hundred wavelengths (of 3.7 m each), but each element was now enlarged to a 10 × 1 wavelength broadside array of 40 half-wave dipoles made from brass tubing. This was designed to be more sensitive and to produce fewer fringes for each passage of a radio star through

---

[25] Ryle:Lovell, 10 June 1948. [photocopy from V. A. Hughes]
[26] Pawsey (in London):Bowen, 24 June 1948, file F1/4/PAW/1, RPS. In the end it turned out that they were *both* about 1° off in declination, straddling the correct position. See Section 14.1 and Fig. 14.1 for details.
[27] Ryle:Bolton, 22 June 1948, file F1/4/PAW/1, RPS.

[28] This incident of forgetting about an antenna's backlobes is reminiscent of a serious earlier mistake made in September 1939 by Chain Home (Section 5.1) operators. At that time they took radar reflections from British fighters *behind* the antennas to be approaching German airplanes *in front*, which led to other fighters shooting down their countrymen. [Fisher, D. E. (1988), *A Race on the Edge of Time: Radar – the Decisive Weapon of World War II* (New York: McGraw-Hill), pp. 84–91.

Figure 8.7. The eastern element of the Long Michelson (1C) interferometer (ca 1950), with Graham Smith standing alongside. It comprises an array of 2 × 40 81 MHz halfwave dipoles mounted 1/4 wavelength above a reflecting screen. In the distance can be seen the Library tower, Ryle's house (just below it and to the right), and the covered stands of the rugby grounds (see Fig. 8.3).

the primary beam of the antennas, thus allowing better recognition of the "essential maximum," that is, the fringe of greatest amplitude. Its £175 ($500) cost made it by far their largest project to date (Fig. 8.8):

> There was quite a pile of angle iron needed, so we ordered it up from the ironmonger down the road, and got him to chop them off the right length and drill the holes in it, because of course we didn't have a power drill. And then we put posts in the ground and poured the concrete around them. We mounted the dipoles on top of these posts using insulators that we'd got from an old German radar.
> (Ryle 1976:32T)

Observations began in December 1948 and within a few months Ryle, Smith, and Bruce Elsmore, who had joined the group as a technician in the autumn of 1948, had cataloged a couple of dozen sources. Elsmore had been a Cambridge undergraduate with a longtime interest in both radio and astronomy and he soon showed that he was capable of full participation in the group's research effort; as he put it, his status changed from dog's-body to experimentalist (Elsmore 1971:9T). In principle the chart record from just a single twenty-four-hour run was sufficient for a complete survey of the northern sky, but in practice it was found that the fringe patterns were inconstant from day to day (see

Figure 8.8. Graham Smith surveying for the Long Michelson interferometer (summer 1948).

Fig. 14.6). The weaker sources were especially bothersome – sometimes there, sometimes not. A method was needed for keeping track of the long-term behavior of all candidate sources and so Elsmore devised a planisphere-like device, inspired by ones he had known as an amateur astronomer:

> When we were obtaining these somewhat crude interferometer records, we were confronted with the problem of trying to find out where the sources were. But the observations were pretty inaccurate, so we thought that if we kept repeating them time after time we might get some sensible results. A graphical technique sprang to mind and I set about constructing a circular planisphere about three feet in diameter, with a scale round the outside which enabled us to adjust for the [solar] time and date of observation so that it produced the right ascension appropriately. And there was a radial scale, which in fact was marked out in [fringe] periodicity, to give us declination. And so one measured the times and periods on the records and then made a mark on this thing [for each candidate source]. And [after many weeks] when one got a conglomeration of points in one place on the planisphere, you'd say, "That must really be a genuine source there."
>
> Elsmore (1971:5–6T)

This analog device even had a curved cursor whose shape allowed for the Long Michelson's baseline being 7° offset from east–west. This offset had arisen because the group had found out the hard way that horses would trample cables laid in open fields; they therefore ran their baseline and its associated cables and posts along a convenient hedge.

First observations with the Long Michelson were almost *too* successful: the sky was so full of fainter sources that their fringe patterns on the charts often blended, making it difficult to discern their locations or even their correct number. In April 1949 Ryle gave a paper to the Royal Astronomical Society (RAS) showing an initial map with 21 new sources besides Cas A and Cyg A. But he was not at all confident of their positions, for he noted that "the distribution on the map is ... not directly related to that actually occurring in the sky as the detection of a source on the record is often limited by the confusion arising from adjacent sources."[29] He thus decided to double the east–west length of each element, so narrowing the beam that sources now exhibited only about five fringes on their daily transit, not ten, and signals were yet stronger. More importantly, he came up with an ingenious solution to difficulties stemming from the strong galactic background radiation (see Fig. 8.6). Because one was constrained to set the amplifier gain on the recorder such that the ink trace always stayed on the chart, sources could be found only if they registered more than a few percent of the maximum background signal recorded over the entire day. Ryle's fix came to be known as a *phase switch*, and its merit was such that it soon became the standard everywhere; for example, Bernard Mills (1952a:268) in Sydney installed it for his first survey of radio sources in 1950.[30]

Ryle originally invented the phase-switch in connection with advisory work on underwater sound detection that he was then doing for the Admiralty – for a while the group called it the "mermaid hunter."[31] In the underwater warfare case, one was trying to discern a submarine's signal against background noise caused by fish, currents and other vessels, but the principles were identical for detecting a radio star against the Milky Way background. Ryle laid out details of the technique in an October 1947 report,[32] but over a year passed before the phase switch was used at Cambridge.[33] The reason for

---

[29] Ryle, "Discrete celestial sources of radio-frequency radiation," presented to the Royal Astronomical Society on 8 April 1949, as recorded in *Observatory* **69**, 85–7 (1949).

[30] Mills learned of the phase-switch idea from Robert W. E. McNicol, who returned to ionospheric research in Brisbane after receiving his Ph.D. under Ratcliffe in 1949. Pawsey was careful, however, to insure that Mills gave Ryle priority for the technique. [Pawsey:Ryle, 1 May 1951, box 1/2, RYL; Mills (1984:149); F. F. Gardner:author, 12 November 1986]

[31] Smith (1984:240); Machin:author, December 1987.

[32] Ryle, "The detection of a small signal originating from a point source against a large background signal of diffuse origin," 29 October 1947, 8 pp., file 12 (uncat.), RYL. This report describes phase-switching in some detail, but only gives hints now and then that the intended application deals with underwater naval operations (for example, the use once of the term *transducer*). Also see note 35.

[33] The date of introduction of phase switching is uncertain. The written record of the RAS talk of April 1949 (note 29) suggests that the data discussed were not taken with phase-switching. In a letter of 7 June 1949 to Appleton [box 1/1, RYL], Ryle

this delay is not known, but Machin (1981:134B:570) later vividly recollected:

> Martin came in one day and said "Hey, I've just had a fantastic idea!" So he got out a piece of paper and drew on it on the floor. "Right," he said, "it's eleven o'clock, shall we go and try it?" And so we modified a standard set in an hour and a half – we simply rewired the RF switching so as to put an alternative half-wave in. And that showed it worked.

The phase switch was a capacitor switch (as in previous receivers), but now rapidly switching to a configuration that inserted an extra half-wavelength of cable into one arm of the interferometer. This phase reversal caused the maxima, or lobes, of the interferometer's antenna pattern to oscillate between two sky positions exactly one-half lobe spacing apart. And if one now synchronously recorded the *difference* of the signal strength in the two switch positions, any radiation coming from a region of sky larger than the size of a lobe spacing would be cancelled out, and the only remaining signals would be from discrete sources (see Appendix A and Fig. A.6). Furthermore, it turned out that this technique not only eliminated the deleterious effects of the galactic background (for baselines longer than one hundred wavelengths), but also minimized those from deficiencies such as manmade interference, receiver gain variations, and inequalities between arms of the interferometer (Ryle 1952; 1976:42–3T). Ryle's tutorial 1952 paper on the phase-switched Michelson interferometer (and its superiority over both the sea-cliff interferometer and dish antennas) became the essential primer for all his students and many others.

Investigations of radio sources by Ryle's group and others are covered in detail in Chapters 14 and 15. By mid-1950 the Long Michelson had spawned a list of 50 sources (later to be known as the 1C catalog) with quoted positional accuracies typically of 1° to 3° (Ryle, Smith, and Elsmore 1950). The group had also brought forward a series of arguments that these were *radio stars*, a "hitherto unobserved type of stellar body, distributed widely throughout the Galaxy," and that the general galactic noise arose from the integrated effect of these new stars. These stars were numerous ($\sim 3\,pc^{-3}$) and had a cool photosphere, but a hot corona that produced the radio emission. Repeated observations of the stronger radio stars with the Long Michelson revealed no effects of annual parallax, proper motion, or long-term intensity changes. Graham Smith and Antony Hewish also established, through a long series of observations both at Cambridge and in conjunction with Jodrell Bank, that the bulk of the short-term intensity fluctuations were caused by the ionosphere. Hewish had joined the group as a research student in 1948 after three years of wartime radio work at RAE and TRE (where he had known Ryle) and completion of his undergraduate physics degree. In his Ph.D. thesis he used the radio stars as handy background sources to study the ionosphere (Section 14.2).

Smith's thesis involved an all-out attack on getting highly accurate positions for the four strongest radio stars. He used the two 7.5 m Würzburg dishes that Ryle had scrounged many years before and with extreme care obtained positions accurate to an unprecedented 1′ (Smith 1951a). These positions first of all confirmed the Australian identifications of Tau A with the Crab nebula and of Vir A with the galaxy Messier 87 (Section 7.4). Secondly, such accuracy permitted Walter Baade and Rudolph Minkowski (1954a) within a few months to use precious time on the new 200 inch Palomar telescope to discover, at last, optical counterparts to the two strongest radio sources in the sky (Section 14.4). Cas A was found to correspond to a strange network of high-velocity filaments in our own Galaxy. And Cyg A, despite its location near the galactic plane, coincided with a faint (17^m) peculiar galaxy one hundred million light years distant. Its radio power dwarfed that at optical wavelengths and its existence implied that radio techniques could probe to far greater distances than optical ones. The notion of nearby radio *stars* suffered one of its greatest blows (although hardly fatal – see Chapters 14 and 15) and within three years Ryle's group instead found itself doing cosmology (Edge and Mulkay 1976; Sullivan 1990).

8.5 OVERVIEW

8.5.1 Ryle, Ratcliffe, Bragg

The success of Ryle's group was notable – within a few years he and his few students were in the same

---

mentioned experiments with a "new system for obtaining better discrimination against the galactic background." This would seem to imply that the technique was introduced in April or May 1949, but another document ("Solar noise programme," [file 6 (uncat.), RYL]) suggests that phase-switching went into operation as early as its date of February 1949.

league with the far larger groups at Jodrell Bank and in Sydney (Table 10.1, Fig. 17.2). The prime reason for this success was the man Martin Ryle himself. His scientific and engineering acumen, his ability to develop a loyal, cohesive group, his informality (to all he was "Martin"), his intensity and high-strung nature – these made for a powerful combination. The comments of three early colleagues are typical:

> Martin was such a fundamentally nice chap, and so clearly not interested in blowing the trumpet of Martin Ryle, but in actually doing some good work. We got swept along with him, trusted him absolutely, and would have followed him anywhere. And still he was one of the boys, a "dirty-hands" leader. ... Sometimes he would throw things at you, but then you just threw them back at him!
>
> He used to think so quickly that he would leave out two-thirds of the words and you had to assemble what he was thinking about from the one-third that did come out – there was usually just sufficient redundancy that you could do it.
>
> (Machin 1981:135A:0)

> It was almost impossible to think of something that Martin hadn't thought of an hour before. You'd come in in the morning and say, "I've had a good idea!", and he would say, "Shut up, shut up, let me tell you *my* good idea," which turned out to be the one you'd thought of the night before. It was very difficult and very inspiring.
>
> (Machin 1971:9T)

> Ryle really was brilliant at thinking out ways of getting enormous amounts of information with a couple of old bedsteads and some bits of copper tube. ... He was absolutely first-class. He was quite brilliant and scientifically uncompromising, but very, very approachable. It was quite difficult to keep one's end up in that he was so good and so clever that very few things which the ordinary mortal produced really contributed a great deal.
>
> (Vonberg 1971:5+9T)

> [The unity and success of our group] has come about very largely because of Martin Ryle. He is a great enthusiast and has his feet very firmly on the ground, and he has *not* become an administrator and committee man. He is very intimately concerned with the details of the hardware and with everything that goes on in the group. And I'm amazed that this has worked despite the fact that the group is so inbred – all but one of the present staff members have gone through the Cambridge system.
>
> (Peter Scheuer 1976:65B:700)

Ryle himself attributed much of his success to the remarkable amalgam of engineering and management skills that had poured forth from the wartime crucible:

> It was a fantastic occurrence to a young chap to get into radar research in England. ... You found yourself with the responsibility for really a rather surprisingly large amount for one's age. Two or three years later you'd even find young people with not even a proper degree (they'd only taken a one- or two-year course) running groups of twenty-five people. ... [You had the situation] whereby a young chap ... could come up with a brass tube [a radar gadget] and tell an Air Vice-Marshal, "That was a bloody silly thing to do – you must do it like this." ... I think [our success came from] this contact with life in the raw, as it were. Being able to talk to the people that were going to be flying and be killed in these machines made one grow up fast. ...
>
> This wartime work was an incredibly good training. The proportion of the small group of people at TRE that have gone to the tops of their professions in all sorts of fields is remarkably non-statistical.[34] And this has nothing to do with clever people being brought into the game – it was because the training, which lasted for six years, was very good. ... You had enormous responsibilities at an early age – there was nobody to ask about decisions because you knew more about it than anyone else. This is the sort of thing one ought to do in one's Ph.D. program, if one could think of how to do it without having another war.
>
> (Ryle 1976:28–9T, 2–3T)

---

[34] An example of a young TRE researcher rising to the very top of a field unrelated to radar is the physiologist Alan L. Hodgkin, a 1963 Nobelist for his work on the electrochemistry of nerve cells. See Hodgkin (1992), *Chance and Design: Reminiscences of Science in Peace and War* (Cambridge: Cambridge University Press).

Ryle's supervisor Jack Ratcliffe was also critical to his success. Ratcliffe took almost all of the administrative load of Ryle's group upon himself, saw to it that Ryle had minimal teaching duties, and tended to many of the necessary fund-raising chores, for example maintenance of support (by 1951 at a level of £4000 per year, excluding salaries) from the university and the government's Department of Scientific and Industrial Research (DSIR)[35]:

> Ratcliffe was very, very conventional, the chap who was always sitting in an office wearing a suit,[36] while Ryle was the chap who was out in a field with gum boots on. ... Ratcliffe was available as a sort of general grey eminence, ... a source of sane and quiet thoughts, one who would ask awkward questions.
>
> My impression is that Ratcliffe doesn't get enough credit for what he did. First of all, he spotted the right people to bring back after the war. Secondly, there must have been a tremendous organizational effort to make legal all of the looting of equipment we did – Ratcliffe was a terribly upright person. He was also a very good administrator, the ideal chap to be behind a research group. ... As Ryle grew more and more independent, Ratcliffe continued to do all of the lousy administrative work in the background, did a lot of the pushing for money and status ... meanwhile letting Martin run the group in his characteristic style. ... If Martin had had to do all the paperwork associated with running a group, he'd have gone mad, because he hated things like that.
> (Machin 1981:135A:210 and 1971:11–2T)

John W. Findlay, a colleague of Ratcliffe's during these years, also described how Bragg (see below) and Ratcliffe relieved Ryle of many of the usual administrative and teaching duties:

> As I watched Ryle grow, I also watched ... the great freedom he was given. I do not think that this was in any sense incorrect or unfair, but it was an excellent example of the way the British (in those days anyway) "picked winners." I don't know who did it for Ryle, but Bragg must have been in the forefront. Ratcliffe might not have been the leader, but he clearly should get enormous credit for not standing in Ryle's way; even more, for clearing the ground so that Ryle could exercise his talents as a scientific leader.[37]

Ratcliffe also played a role in the intellectual life of the group. He edited all proposed papers, and his extensive "suggestions" on matters of style, in addition to substantive comments, meant that most papers emerged in a house style referred to as "Ratcliffe English" (Machin 1971:40T; O'Brien 1981:141B:470).[38,39]

Another important aspect of Ratcliffe's influence was his teaching and research guidance on ionospheric topics and, especially, on physical applications of Fourier transforms. Gillmor's (1982:403) survey of the ionospheric community concluded that Ratcliffe is remembered as "the finest lecturer at the Cavendish in this century." Many of my interviewees have recalled the lucidity, insightfulness, breadth, and long-term influence of his series of thirteen lectures on physical applications of Fourier transforms.[40] A case in point is Ronald Bracewell, who earned his Ph.D. at the Cavendish in the late 1940s (Chapter 7) and who himself later made important contributions, as well as writing influential textbooks, on Fourier theory. Bracewell (1980:131A) later recalled how Ratcliffe never would simply state something like "the Fourier transform of a cosine is

---

[35] See Lovell (1977:166–8) and Section 17.3.5 on the greatly increased support for British science in the postwar decade.

[36] After reading this quotation, Ron Bracewell, a student of Ratcliffe's during these years, wrote: "Machin would not be aware that if you phoned Ratcliffe in the middle of the night to see something interesting in the ionosphere, he would come." [Bracewell:author, 11 December 1986]

[37] Findlay:author, 10 September 1986.

[38] A good example of Ratcliffe's style and clarity is his article on "Aerials for radar equipment" in *J. IEE* **93**, Pt. IIIA, 22–32 (1946). Page 24 of this review also discusses a wartime technique of "electrical beam-swinging" similar to Ryle's phase-switching technique.

[39] See Tangent 8.1 for Ken Machin's humorous frustrations with the internal refereeing of papers by Ratcliffe and Ryle.

[40] Ratcliffe's lectures on Fourier transforms were never published as such, but stenciled notes entitled "Fourier analysis and correlation functions in diffraction and circuit theory: a physical approach" were made in 1956. His most pertinent publication was "Some aspects of diffraction theory and their application to the ionosphere," *Rept. Prog. Phys.* **19**, 188–267 (1956).

a (double) delta function," but rather "a cosine variation in the ionosphere's density will cause an incident plane wave to be detected at the ground as two plane waves coming from either side of the vertical."

It was etched upon Ratcliffe's students' minds that a measurement of the pattern of signal phase and amplitude on the ground allowed one to map the sky brightness distribution and/or properties of any intervening medium. This approach was particularly profitable for Hewish's work on scattering and scintillation of radio waves by the ionosphere (Section 14.2) and Ryle's on principles of interferometry. As Hewish has put it:

> Ratcliffe was delighted at the link between radio astronomy and his interests and encouraged me to continue using radio stars, whatever they might be, to gain information about the ionosphere. ... I owe it to Jack Ratcliffe and his stimulating lectures that the right ideas were buzzing around in my head at the time.[41,42]

One step up the ladder from Ratcliffe was the Cavendish Professor, William Lawrence Bragg, and he too was strongly supportive of Ryle. From 1945 to 1953 (when Bragg left Cambridge) Ryle's group steadily grew to 7–12% (depending on one's measure) of the total Cavendish Lab effort and to about 60% of Ratcliffe's radio group.[43] Other large groups in the Cavendish were in nuclear physics (~35%) and in X-ray crystallography (~20%). The latter had long been Bragg's specialty – in fact in 1915 he had received (at age 25) the Nobel Prize with his father for pioneering the technique. Its application at the Cavendish Lab to biological molecules by F. H. C. Crick and J. D. Watson culminated in 1953 in one of the fundamental discoveries of twentieth century biology: the unravelling of the double helix structure of deoxyribonucleic acid. X-ray crystallography involves measuring the diffraction pattern of X-rays after their passage through a lattice-like arrangement of molecules. The Fourier transform of this pattern then yields the sought-for molecular structure. The technique is therefore directly analogous to the radio astronomer's analysis of extraterrestrial radio waves (see Sections 13.1.3 and A.11), albeit with scale changes in wavelength of $\sim 10^{10}$ and in structure sizes of $\sim 10^{31}$! Bragg was excited and delighted by the similarity of the two techniques despite their disparity in scale.[44] His career had been spent in devising ways to reveal structure through the interference of light. Techniques like Ryle's, which were the same thing in a new guise, naturally caught his attention.[45]

8.5.2 Group style and development

During these postwar years Ryle gradually grew independent of Ratcliffe and ran his group ever more autonomously; Ryle (1976:31T) drily recalled that "Ratcliffe perhaps saw that I was rather an awkward chap and should best be left alone." After two years in Cambridge, he was appointed a Lecturer and students such as Machin formally became his own, not Ratcliffe's. And even as he took on his own students and fended off outside offers,[46] he decided to skip his own Ph.D. (as had Ratcliffe in the 1920s). As he modestly put it:

> Around 1948 I finally abandoned irrevocably all attempts to get the Ph.D. because it was clear I wasn't going to be able to learn all that physics again.[47] ... When you go straight off [from

---

[41] Hewish:author, 5 September 1986.
[42] Budden (as in note 2, pp. 691–6) gives further information on Ratcliffe's skills in administration, scientific writing, and teaching.
[43] Ionospheric researchers under Ratcliffe during these years included Sidney A. Bowhill, Basil H. Briggs, Ronald N. Bracewell, Kenneth G. Budden, John W. Findlay, Frank F. Gardner, L. R. Owen Storey, Thomas W. Straker, and Kenneth Weekes; Henry G. Booker also worked closely with the group.
[44] Judson, H. F. (1979), *The Eighth Day of Creation* (New York: Simon & Schuster), pp. 104–7. In a biographical note for his 1974 Nobel Prize in Physics, Ryle also stated that "Bragg always showed a delighted interest in the way our work progressed."
[45] Another major postwar effort in the Cavendish Lab also benefited directly from World War II radar. A. Brian Pippard, who later himself became Cavendish Professor, used wartime-learned microwave techniques for his pioneering research on superconductivity. [Forman (1995:432–4)]
[46] These job offers were from the Solar Physics Observatory at Cambridge (in 1947), Lovell's group at Manchester University (1947), University of Western Ontario (1948), and University of British Columbia (1950). [box 1/1 and file 4 (uncat.), RYL]
[47] Note, however, that Lovell (as in note 5, p. 361) quotes Ratcliffe as saying: "Ryle often pretended to have lost all knowledge of fundamental physics, but that was all rubbish – just a pretence – he was a very sound classical physicist."

undergraduate study] and start putting boxes into aeroplanes, the physics evaporates. Furthermore, when you come back six years older, you find you can't learn things any more. This struck me so convincingly in 1947, '48 that I told Ratcliffe I wasn't going to do a Ph.D. He was of course upset and said it was all wrong.

(Ryle 1971:26T)

He had started off as a research student, but now was supervising his peers: Ph.D. theses were submitted by Stanier (1950b), Hewish (1951a), Smith (1951c), and Machin (1952). Vonberg, who always felt himself more an engineer than a scientist and who left in 1947 to work on cyclotrons for medical purposes, was the only member who departed without a degree. Stanier went into industry, Machin joined a biophysics group at Cambridge, and Hewish and Smith stayed on as Ryle's lieutenants. Three more of the seven students who began in 1950–52 also stayed with Ryle for decades: Peter A. G. Scheuer (1930–2001), John E. Baldwin, and John R. Shakeshaft (Fig. 8.9).

Two institutions within Ryle's group nicely illustrate how its remarkable cohesion and loyalty were fostered and maintained. The first were "Range Days," periodic events when everyone cut hawthorn bushes, repaired fences, and mowed the grass. The second were the rough-and-tumble Saturday morning sessions for the purposes of assessing what had happened, what needed to happen, and what was happening elsewhere. Two early participants later reconstructed the flavor of these sessions:

> These Saturday morning sessions were often not so much debates as monologues by Martin. ... He would go through the details of each program, say of Pat O'Brien and his solar measurements, and all we research students sat mutely as this went on. And although we could not then make any contributions, we did hear all of this and in the end I think it was quite useful, because Martin did a lot of thinking out loud at these sessions and we learned of the kind of considerations that mattered, the things that had to be dealt with.
>
> (Baldwin 1981:144A:390)

> We used to have a Saturday morning session, just sitting around on the floor saying "What the hell does it all mean?" ... And literature coverage was no problem in those days – about one *Nature* letter every two weeks or so.
>
> (Hewish 1976:10T)

Figure 8.9. The Cavendish radio astronomy group in front of their observing hut in 1954: (back row, l. to r.) John Thomson (joined group in 1952), John Baldwin (52), George Whitfield (53), Patrick O'Brien (50), Peter Scheuer (51), Robin Conway (53), Ronald Adgie (51), Charles Sharp (technician); (middle) Graham Smith (47), Martin Ryle (45), Antony Hewish (48); (front) Michael Turner (technician), John Shakeshaft (52), John Blythe (53), Bruce Elsmore (48).

Several interviewees also remarked that the building and installation of antennas and receivers was always considered a group effort, with each person "mucking in," chipping in his labor as needed. Publications were treated similarly, with Ryle just one of the team, often completely fading from a paper's authorship despite his central contributions.[48] Ratcliffe also fostered this brand of teamwork in his own research style.[49] Spare conditions also helped. Patrick A. O'Brien (1981:141A:460) later remembered his first impressions after joining the group from South Africa in 1950:

> I was a bit overawed by the Cavendish and its name in physics – it looked old, and even smelled old. Labs were very crowded and the group seemed to be operating on a shoestring, but it was a very nice atmosphere, with people being on friendly, first-name terms.

The same close-knit quality that led to marvellous camaraderie and scientific success often, however, came to be viewed by the other major radio groups as haughtiness and secrecy.[50] For example, Bolton (1978:111–2T) described the difficulties he encountered in 1950 when he requested a visit to Ryle's group. Ryle's program plowed ahead with minimal influence from others and results were seldom released prior to formal publication.[51] The group never recruited from outside,[52] nor did it ever host any conferences during the postwar decade. Indeed Cambridge University as a whole has had a reputation of being inbred and proceeding in relative isolation, with little reference to the work of others (Smith 1976:40–1T; Edge and Mulkay 1976:341). The kind of thing that annoyed non-Cantabrigians was exemplified in Ryle and Vonberg's long 1948 article on the sun. It contained a perfunctory paragraph at the start where all previous solar radio papers were cited, but none of their results (often of a similar nature and obtained contemporaneously) were ever again referred to. Smith later wrote:

> We took very little notice of any publications, either in journals or in textbooks, and relied on Ryle's insight. ... We were in the full flood of discovery, and we were self-propelled.[53]

Counterexamples, such as correspondence at times between Pawsey and Ratcliffe and between Ryle and Bolton, were infrequent and usually initiated from outside.

This is not to say that the Cavendish group felt no competition – indeed they did, for instance over measuring the best position for Cyg A (Smith 1976:5T). And by the late 1940s they closely guarded unpublished data, giving the appearance to outsiders of undue coyness. In 1950 John Bolton pressed Ryle to participate in a freer exchange of information, and that same year Robert Hanbury Brown of Jodrell Bank complained to Taffy Bowen:

> Ryle is concentrating on interferometers and has some very elegant stuff. He has a large collection of point sources (50) which he is soon publishing. ... I haven't even got a list myself as, of course, it is harder to find things out in Cambridge than in Sydney.[54]

---

[48] For example, complete copies *in Ryle's hand* of the *Nature* articles published *alone* by Machin (1951) and by Smith (1951a) are extant. [Undated manuscripts, files 16 and 17 (uncat.), RYL]

[49] Budden, as in note 2, p. 689 and p. 694.

[50] In contrast, minor radio groups, especially those just starting in radio astronomy, found their requests to Ryle for technical advice answered in great detail. [Correspondence in boxes 1/1 and 1/2, RYL]

[51] In October 1950, shortly after the Würzburg interferometer went into operation, Smith apparently sent Mills his latest position for Cyg A, but unfortunately with considerable error because the fringe periodicity had been inadvertently calculated on the basis of a solar (not sidereal) day. The Australians were never informed of this error, however, and the next communication on this from Cambridge to Sydney was almost a year later, a preprint of a *Nature* article (Smith 1951a) with a much improved position. This caused Pawsey to cry foul (to Ratcliffe) and Ryle to be even more convinced that informal communication of preliminary results only led to trouble.

[Undated draft of letter from Ryle to Pawsey, about August 1951, file 4 (uncat.), RYL]

[52] In 1951 Ryle considered hiring CSIRO Radiophysics researcher James A. Roberts for the last year of his Ph.D. work (Roberts had been working under Hoyle, but had run out of support). But Ryle was wary: "Do we want someone who is going back to Sydney so close to all our new ideas?" [Ryle:Ratcliffe (in US), 21 June 1951, box 1/2, RYL]

[53] Smith, as in note 5, p. 508.

[54] Bolton:Bowen, 31 May 1950, file F1/4/BOL/1; Hanbury Brown:Bowen, 30 April 1950, file A1/3/1a, both RPS. Also Jennison (1976:39–40T).

And it was not only observers – theorists too were not particularly welcome:

> We were very sensitive to the idea that we were the people that were producing results, and producing results does mean building apparatus and digging holes in fields. And what we wanted was a little bit of time to think about our results – we didn't particularly want to act merely as a fodder machine producing data fed to the theoreticians, who would then crack it all. ... Whenever there was a theoretician around, there was an air of tension, a "lock up the filing cabinets, chaps" attitude.
>
> (Machin 1981:134B:735)[55]

On the other hand, relations with optical astronomers, although hardly close, were different. Throughout the late 1940s Ryle's group kept contact with the local observatory under the directorships of Frank J. M. Stratton and (after 1947) Roderick O. Redman. Stratton had wide interests in solar physics and in solar influences on the earth's atmosphere, and was interested in the radio sun from the start. His staff even ran a joint experiment with Kenneth G. Budden of Ratcliffe's group to search at 13 MHz for long-delay echoes. He also encouraged comparisons of his own solar monitoring data with Ryle's.[56] Once Ryle's interests shifted to radio stars, the optical astronomers were often consulted about fundamentals of astrometry and information on optical objects that might correspond to the enigmatic radio stars. Early radio papers and my interviewees cited, besides the directors, Donald E. Blackwell, David W. Dewhirst, Michael W. Ovenden, and Harald von Klüber as being of particular assistance. Beginning in 1951 Dewhirst played an active role in seeking optical identifications for the radio stars (Section 14.4). The existence of this interaction is not, however, to say that it was on a peer basis. Despite attempts at communication such as listing radio star intensities in magnitudes in the 1C catalog, Elsmore (1976:60B:220) later remembered:

> The optical astronomers very much considered us an inferior race. In those days we were probably not even justified in calling ourselves radio *astronomers*. Looking back, I feel that they quite rightly viewed us with considerable scorn, although they were very helpful.

8.5.3 A turning point

Over the period 1949–52 Ryle and his group qualitatively changed in several aspects of the style and content of their science. In particular, they were moving from solar to extra-solar studies and from small, follow-your-nose studies to large, long-term programs. They began in an exploratory mode where newly discovered phenomena were followed up without any particular long-range plan:

> One's mental attitude at that time was quite different from what one [now] expects. One was quite old – I was 28 or so – so one wasn't in a learning frame of mind. One didn't automatically go and say, "Ah, we'll study this problem, so now we'll read all that's been done about it before." This was because (a) nothing had been done about this particular problem before, and (b) [even] if it had, virtually nobody had written about it. The first thing was probably just to look and see what was there. Only quite a bit later did one try to see what could make circular polarization or what could make the radiation at all. ... During the first two or three years we were so much finding new things all the time that just to collect facts was really a fulltime occupation.
>
> (Ryle 1976:9–10T)

This stage ended during 1948–49 when Ryle wrote four interpretative papers, including a major review of the entire field in *Reports on the Progress of Physics* for 1950. And as he largely achieved independence from Ratcliffe, his group's interests shifted in the main from the sun to radio stars. That these two latter things happened together was not a coincidence. Work on solar radio emission had been not unlike studying another, distant ionosphere; moreover, two decades had been spent studying tie-ins between solar activity and the ionosphere's response. But the leap from the sun to the world outside the solar system took one far from ionospheric radio physics – stars and galaxies were unequivocally

---

[55] For three theorist's views of their exclusion by Ryle's group, see T. Gold (1976:5–7T), G. R. Burbidge (1979:2–3T), and F. Hoyle (1994:268–9).

[56] Nothing much ever came of this; see Edge and Mulkay (1976:37) for details.

part of astronomy.⁵⁷ In turning down an invitation to a conference on the ionosphere in 1950, Ryle wrote that his group was moving away from the ionosphere and that "we tend to regard ourselves more as astronomers."⁵⁸

The success of the Long Michelson and its 1C catalog with a sky full of ever more and fainter radio stars naturally led Ryle to think of a new survey that would be yet more sensitive and accurate. He still felt that an interferometer was the economical and elegant approach, and that a large dish antenna would be too costly and obvious a solution (Ryle 1952), but he had to concede that even the interferometric approach was going to be expensive this time (Ryle 1976:36–7T). Instead of about £350 as supplied by the University for the 1C interferometer, he now sought £6400 ($18,000) from DSIR for four tiltable parabolic cylinders, each measuring 12×98 m (Fig. 8.3).⁵⁹ This interferometer, which was built in 1952–3 and which eventually would produce the 2C survey of 1936 sources, surely marked the transition of Ryle's group from Little to Big Science.⁶⁰

The 2C survey proved to be a watershed in another respect. In the period 1952–54 following the identification of Cyg A with a very faint galaxy, Ryle came to accept that radio sources represented a new way to investigate the structure of the universe (Sullivan 1990):

> Perhaps the greatest discontinuity [in my career] was with the identification of Cygnus A. That showed that we were in the cosmology game. ... To me that was the point where one said, "Well now, if other things like Cygnus exist, here is something which we can likely see, even with our little instrument, as far away as the 200 inch can see. This is something much more interesting than it might have been – much more interesting than [the radio sources] being galactic objects, much more interesting than M87's [fairly normal galaxies]."
>
> (Ryle 1976:37–8T)

He thus set out on a long range plan: to survey radio sources and, eventually, to map them individually. Chapters 14 and 15 include details of this Cambridge work through ~1953.

In addition to these surveys, the task of forming an *image* of a source led through the 1950s to the development of what came to be called aperture synthesis, for which Ryle (together with Hewish) was awarded the 1974 Nobel Prize in Physics. The basic idea and language came right out of Ratcliffe's lectures, namely that any single interferometer measurement of signal amplitude and phase provided one Fourier component of the sky's brightness distribution and that this distribution could be calculated with a certitude commensurate with the number of components gathered (see Appendix A). The first statement of these principles found in Ryle's notes dates from 1946.⁶¹ But although the principles were reasonably straightforward, the realization was devilishly difficult. How could one best obtain all of the needed interferometer spacings? How could phases be accurately measured and tied to each other over weeks of observations? How could one carry out the tedious Fourier transformations? First steps were taken on the sun, because of its strength and because one could assume it to be symmetric, obviating the need for phases. In 1949 Stanier made the first one-dimensional synthesis, using 17 spacings at 600 MHz, followed by Machin's at 81 MHz. O'Brien then did the first two-dimensional studies in 1951–52, some of which also used phase information. These experiments are detailed in Section 13.1.3. Synthesis mapping projects were to constitute a major portion of Ryle's program for the next twenty years.

---

⁵⁷ Machin (1971:14–5T), Hewish (1971:2T), Edge and Mulkay (1976:25).

⁵⁸ Ryle:A.H. Waynick (Pennsylvania State University), 22 March 1950, box 1/1, RYL.

⁵⁹ See note 9.45 on the support given to this proposal by Lovell and others.

⁶⁰ De Solla Price, D.J. (1963), *Little Science, Big Science*, (New York: Columbia University Press). John Baldwin points out that Bigness meant that in 1952 Ryle's group underwent another transition: from the Iron Age to the Alumin(i)um Age. For at that time an iron framework for a new 10×10 m array of dipoles, which had been designed to be a movable element for an interferometer, was in fact found to be *un*movable with the available muscle power, despite the group's recent augmentation by three new bodies.

⁶¹ "If a maximum/minimum ratio [of the fringe amplitudes from the sun] were obtained with a large number of equivalent aerial spacings, it should be possible, by carrying out a harmonic analysis, to derive a true line distribution of power intensity, and this would show up the contribution of multiple spots and the disc as a whole." [Ryle, "Notes on future noise programmes," 23 August 1946, 2 pp., file 18 (uncat.), RYL]

Ryle did not explicitly publish these ideas until much later (Ryle 1952:356), at which time he acknowledged that the Australians McCready, Pawsey and Payne-Scott (1947:367–8) had been the first to publish such a concept (Section 7.3.2).

Membership in the Royal Society is usually awarded on the basis of a long career of past achievements, but when Ryle was named a Fellow in 1952 at the age of 34, he had only just begun.

TANGENT 8.1 AN EXAMPLE OF CAMARADERIE IN RYLE'S GROUP

Ken Machin's frustrations with the refereeing of papers by Ratcliffe and Ryle are well documented by the following ditty [from a scrapbook, CAV], composed for a party in December 1953 (to be sung to the tune of *Villikins and Dinah* [UK] or *Sweet Betsy from Pike* [USA]):

A paper on sunspots I promised to write,
So I started one evening and stayed up all night,
And by the next morning that paper was done
Apart from the title and Figure twenty-one.

Three weeks passed before I looked at it again
And only 'cos Martin kept asking me "When?".
I read it and read it until I got drowsy –
The physics was good, but the English was…
   unsatisfactory.

I tore the thing up and then started once more,
The sight of that paper by now made me sore.
But I sat down to write without further delay
And gave it to Martin the very next day.

Said Martin "Oh Christmas, oh blimey, oh cor,
The whole of this stuff seems to me pretty poor,
There're quite a few things that I would like to change,
We'll discuss it next Monday when up at the range."

I knew then that Martin was going to revise it,
I was sure after that I would not recognize it,
But when he returned it, it looked pretty rough –
Ten pages of somewhat illegible scruff.

The next version Martin said, "That's quite O.K.,
Now show Mr. Ratcliffe this draft straightaway.
We'll send it straight off to the *J. I. E. E.*,
Perhaps it'll be printed by next year but three."

Said Ratcliffe, "This paper is now very good
And publish immediately you certainly should.
There are a few points I consider unsound,
On these ten sheets of foolscap I've written them down."

At last it was finished, but then sad to say,
I posted it off and then later that day
I looked into *Nature*, I felt such a clot,
The whole thing had been published by Pawsey, Mills,
   Yabsley, Bracewell, and Payne-Scott!

9 · Lovell at Jodrell Bank

The tale has often been told of a research group that starts off with nothing more than a scientist and an idea, and gradually grows into a full-scale institute of international renown. Indeed this happened in the years after World War II for the particular case of Bernard Lovell, the idea of detecting cosmic ray air showers using radar, and the Jodrell Bank radio research station. But in the present case the twist is that success came despite the original idea never panning out.

Instead, Lovell's group became the world leader in studying radar echoes from meteor trails, opening up views of daytime meteor showers and amassing detailed information on the orbits of meteoric particles and their interactions with the ionosphere. *Passive* radio astronomy began several years later with the completion of a giant, 218 ft diameter fixed dish and the arrival of Robert Hanbury Brown, who directed important work on radio sources, in particular with his study of the Milky Way's neighboring galaxy Messier 31. By 1950 Jodrell Bank's staff numbered twenty, and almost thirty scientific papers had been produced. At this juncture Lovell's group's scientific focus was steadily shifting from meteors to the greater universe, and Lovell took the first technical and political steps toward building a 250 ft steerable reflector.

9.1 COSMIC RAY SHOWERS

In 1937 Patrick M. S. Blackett (1897–1974) succeeded W. L. Bragg as the professor of physics at the University of Manchester and within a few months completely changed his new department's direction of research from X-ray crystallography to the cosmic ray particles impinging on the earth's atmosphere. Blackett had developed cloud chamber techniques at Cambridge a decade earlier (and would in 1948 receive the Nobel Prize in Physics for this work) and was now applying them to study cosmic rays. Of particular interest were the highest energy particles, which created the largest "air showers" of secondary particles and which by that time had led to the discovery of the "heavy electron" or "mesotron" (now called the muon). Blackett quickly established a first-rate laboratory of high-energy physics at Manchester, only to be cut short by the outbreak of war in 1939 and the immediate mobilization of all scientific manpower. During the war Blackett achieved fame as one of the wizards of operational research, especially for naval applications such as determining the optimum size for a convoy of merchant ships menaced by U-boats. After the war his socialist politics and anti-nuclear-bomb stand made him prominent in British politics.

A. C. Bernard Lovell obtained his Ph.D. at the University of Bristol in 1936, working on the properties of thin metallic films on glass surfaces. With abilities in music and cricket rivalling those in science, Lovell then became one of Blackett's lieutenants at Manchester. During the war he joined the Telecommunications Research Establishment (TRE) and developed a number of cm-wavelength radars, in particular the airborne 9 cm "H_2S" radar that allowed night bombers to pick out features on the ground below. As part of the radar work, Lovell soon learned of the frequent transitory echoes picked up by radar sets at meter wavelengths (Section 6.3.1) and, with his background, naturally wondered whether the implied temporary regions of ionization were caused by cosmic rays. He discussed the idea with Blackett (who himself had been a major advisor in the development of prewar radar) and together in 1940 they worked up a short paper suggesting radar as a possible probe of cosmic rays (Blackett and Lovell 1941).[1]

---

[1] Information about Blackett is from Lovell (1975) in *Biog. Mem. Fellows Roy. Soc.* **21**, 1–115, and the biography *Blackett: Physics, War and Politics in the Twentieth Century* by M. J. Nye (2004) (Cambridge: Harvard University Press). Lovell's wartime work is covered by Saward (1984:42–119), Lovell (1990:44–104), and Lovell (1991), *Echoes of War: The Story of H_2S Radar*

There the matter lay until the summer of 1945 when they both returned to Manchester and resurrected the cosmic ray/radar idea as their best way to jump back into scientific research.[2, 3] After considerable time spent restocking empty machine shops, teaching and research laboratories, and staff slots – Lovell even made recruitment overtures in July to Martin Ryle and Stanley Hey[4] – Lovell started by asking Hey to look for the cosmic ray echoes. In July he wrote to Blackett:

> AORG looked for eight hours with an upward pointing beam in the region 2 to 20 km and did not see anything ... [but] they should have seen about a dozen returns. The negative result does not surprise me, however, since with the display system as it is they would be very unlikely to notice a return of duration shorter than, say, 1/2 second. Where do we go next?[5]

In September Hey, who was already embarked on a meteor radar program (Chapter 6), loaned Lovell a suitable Army 4.2 m radar set.[6] Hey (1973:31) later recalled:

> I went with my colleague Parsons to the University of Manchester for two or three weeks to set up the radar in optimum working condition, to fit a brightness-modulated display on the cathode-ray tube for cinerecording of transient echoes, and to advise on the operation of the equipment. The radar was duly installed in a quadrangle near the Physics Laboratory. Our visit proved quite hectic, interspersed by numerous discussions ... Lovell was bubbling over with enthusiasm, and bombarded us with questions and ideas, so we were frequently engaged in an amiable battle of argument as to whether this, that or the other of Lovell's latest suggestions was or was not feasible.[7]

After a month of experimentation, echoes from both low and high altitudes seemed present, but in the city environment reliable results proved impossible, especially with interference from electric trams. Searching for rural quiet, Lovell learned of a ten-acre site operated by the Department of Horticultural Botany in rolling countryside 20 miles south of the city:

> [Our new site] is rather nice ... and I am looking forward to working in the clean Cheshire air instead of the grime of Manchester![8]
>
> I remember Lovell writing to me, when it had been fixed for me to come, saying that he'd found a very nice place ... where they had a field or two – they were doing things like growing brussel sprouts and mangel-wurzels. He had got permission to put his lorries in the corner of the field, at this place called Jodrell Bank. (Clegg 1971:5T)

In December 1945 the vegetables were invaded by a small convoy of lorries and trailers containing the radar transmitter and receiver and a diesel generator. These were installed in a muddy field and within a few days Lovell and a parttime technician were in business. But they only picked up echoes at ranges above 100 km (already known from wartime experience and then being studied by Hey's group as probably from meteors). Nothing was definitely recognizable as an echo from an air shower, which was expected to be of very short duration and to originate at altitudes of

---

(Bristol: Adam Hilger). A fascinating coda to the H_2S story is found in Lovell (2004), "The cavity magnetron in World War II: was the secrecy justified?," *Notes and Records Roy. Soc. London* **58**, 283–94.

[2] During the summer of 1945 Blackett and Lovell also seriously considered doing cosmic ray observations from V-2 rockets. [Correspondence between Blackett, Lovell, and J.A. Ratcliffe, 26 June to 5 August 1945, file CS7/19/4, LOV]

[3] After the war E. G. Bowen also made attempts in Sydney to study whether cosmic rays might have caused certain echoes that he had observed in 1939–40 when flight-testing 200 MHz radars. His postwar 200 MHz searches for these echoes were unsuccessful; in retrospect, if he had used a frequency more like 75 MHz, he might well have stumbled on to meteor echoes as did Lovell. The subsequent program of the Radiophysics Lab might then have turned out very differently. [Bowen:Blackett, 27 August 1946, file CS6/1/1, LOV; Gilbert (1975:24)]

[4] Ryle:Lovell, 13 July 1945; Hey:Lovell, 14 July 1945 – both file CS7/13/3, LOV.

[5] Lovell:Blackett, 4 July 1945, file CS7/13/3, LOV.

[6] This radar was of the type used for defense against the V-2 rockets, with a single Yagi antenna looking upwards at an angle of ~55°. This type in turn had been adapted from the original GL Mark II that had led to the detection of solar radio waves by Hey (1942) (Fig. 5.2; Section 5.2).

[7] A similar account has been given by Parsons (1978:20–23T).

[8] Lovell:J. A. Clegg, 9 December 1945, Clegg personnel file, LOV.

only ~10–20 km. There was still the possibility, however, that some of the short-duration echoes might have their origin in air showers viewed far from the zenith, where a long range was compatible with a small altitude.

While busy with the experimental side, Lovell was not neglecting the theory of echoes from cosmic ray showers. In fact, ever since returning to Manchester, he had been answering requests from Blackett to rework his predictions on their expected numbers, strengths, and durations. The impetus was a wartime letter to Blackett from the ionospheric expert Thomas L. Eckersley (1886–1959) (Tangent 11.2), a Marconi Co. engineer, in which Eckersley had pointed out that the calculations by Blackett and Lovell (1941) were invalid because they omitted an important effect.[9] This was the so-called "damping factor," which arose from the fact that electrons reflected a transmitted radar pulse far less efficiently if they were in a dense gas. In such a case frequent collisions with surrounding molecules caused the energy of the radar pulse to be absorbed as heat rather than re-radiated back to the antenna. This effect was particularly critical for the lower radio frequencies and the denser (lower) parts of the atmosphere, where energetic cosmic rays would be expected to create ions and electrons. Lovell, however, stuck by his calculations[10] ignoring damping, and remained sanguine in a letter to Hey just before the latter arrived:

> There is undoubtedly something quite major to be got out of this. We now feel quite confident that the showers will give sufficient reflections and it is merely a matter of getting the right conditions. There is one problem which you might be thinking about. One of the first alterations will be to adjust the apparatus so that it can detect very short duration echoes, e.g. they may be as short as 10^{-4} or 10^{-5} second.[11]

Over the next few months, Lovell learned of Hey and Stewart's (1946) demonstration of high-altitude meteoric echoes (Section 6.3.1) at the same time as he took account of the damping factor in new calculations. In a January 1946 document he came to the opposite conclusion: the experiment was unlikely to succeed.[12] Undaunted, he pointed out the large uncertainties in his various assumptions and wanted to press on anyway.

Progress was hindered by the fact that Lovell had no colleagues at Jodrell Bank for some time. But after six months of efforts to pry people loose from TRE, Lovell at last succeeded in early 1946 with John A. Clegg (1913–1987), an antenna expert who had taught secondary-level physics before the war and now wished to pursue a higher degree. A few months later another new recruit arrived: C. John Banwell, a New Zealander, TRE veteran, expert on receiver electronics, and contemporary of Lovell's.

Lovell, Clegg, and Banwell continually improved the sensitivity of their receivers over the next six months,[13] began construction of new antennas (see below), and conducted various observations of echoes, always seeing far more (~10–20/hour) at altitudes of ~120 km than at 10–30 km. Blackett's patience, however, was running out, and he demanded a detailed report of all activities to date. In the resulting report of June 1946 Lovell noted that these echoes were "popularly attributed to the passage of meteorites [sic]," but that his observations so far gave no evidence for or against this.[14] It nevertheless seemed more reasonable that the bulk of his echoes were from meteors rather than from cosmic rays for the simple reason that even his most optimistic predictions were for only one cosmic ray echo (originating in a particle of energy ~10^{21} eV) per *week* of observation. All he could point to was one unusual echo that had lasted for a long time (10 sec) and that he thought might have indeed been caused by a cosmic ray.

For any chance with the cosmic ray showers Lovell and Clegg could find no alternative to the need for a huge antenna. In the first half of 1946 they therefore

---

[9] Eckersley:Blackett, 12 March 1941 (not sighted). The first known mention of this letter and of the importance of damping is in Blackett:Lovell, 17 August 1945, file CS7/19/4, LOV.

[10] Lovell, untitled, July 1945, 54 pp. of calculations, LOV.

[11] Lovell:Hey, 21 August 1945, file CS7/19/4, LOV.

[12] Lovell, "Radio reflections from cosmic rays," January 1946, 17 pp., LOV.

[13] Lovell (1990:163) later stated that at this time Banwell told him about the effect of cosmic noise considerably adding to the intrinsic noise in radar receivers, and that this was the first time that he (Lovell) had ever heard of this. If this recollection is correct, it shows that at least one leading radar developer during the war, albeit one who worked primarily at microwavelengths (where cosmic noise is weak and was of no operational consequence), was unaware of Jansky's and Reber's observations.

[14] Lovell, "Report on the present state and results of radar experiments," 21 June 1946, 10 pp. + 5 figs., LOV.

decided to build a 100 × 100 ft broadside array of dipoles. Undeterred by their lack of experience, or even of a concrete mixer, they started in on the construction of this behemoth, but quickly realized they had bitten off more than they could chew: the project abruptly ceased when the first section of scaffolding reached a 50 ft height. Besides, Clegg had a better idea: why not build a vertically-facing paraboloid? This would not need to be so high off the ground, could be used at a variety of frequencies, and would be easier to feed with transmitter power. Thus was conceived the 218 ft paraboloid, built over the next year (next section). Upon completion of the big dish in 1948, air shower echoes were duly sought at 72 MHz ("our first serious attempt is now being made to locate these echoes")[15], but with indefinite results. Two years later, by which time the dish's role had completely changed, Lovell had still not given up and planned to look again at 36 MHz with improved equipment (although this was in fact never done).[16]

The cosmic ray experiment served as a peculiar kind of focus, perhaps one should say more as a catalyst, to the early research program at Jodrell Bank. It hung around for many years, had constantly changing expectations of signal levels, and never in fact succeeded, but at the same time it ironically led to successful techniques and fruitful science in wholly other domains. Despite few actual searches for the cosmic ray echoes, nor ever any publications reporting negative results,[17] the cosmic ray experiment for several years nevertheless remained the *raison d'être* for Lovell's program: it allowed him, notwithstanding his steady wandering away from cosmic rays, to fit to some degree into Blackett's research program. Indeed, even as late as the summer of 1947, Lovell's overriding aim was still to detect cosmic ray echoes, and only in the meantime and secondarily was he studying meteors.[18] Furthermore, the experiment motivated him to do what might be called "atmospheric radar" and thus led him directly into meteor radar astronomy and ionospheric physics. And it provided the justification for the 218 ft dish, which proved critical in Jodrell Bank's swing toward passive radio astronomy (Section 9.5) as meteor radar eventually played itself out (Section 11.5). Finally, the successful example of the 218 ft antenna in turn provided an enormous boost to the idea of a yet larger and fully steerable dish. Although the resulting 250 ft antenna (Section 9.5) never searched for cosmic ray air showers (despite this intention being mentioned in various proposals for funding), its very existence can fairly be traced back to the experiment that never worked.[19]

9.2 METEOR RADAR

Lovell's predictions for echo strengths from cosmic rays changed over the course of 1945–46 by a factor of a million or more downwards and indeed none had shown up, but echoes from high-altitude meteors were everywhere. He therefore steadily shifted attention toward what *was* doable, that is, the ubiquitous echoes on hand, which in any case had to be sorted out before cosmic ray echoes could be identified.[20] Thus began the meteor radar astronomy and ionospheric physics that was to dominate Jodrell Bank for the next decade.

To supply this program the Jodrell Bank crew became masters at scrounging radar items (as in Fig. 9.1) from disposal centers around the country:

---

[15] Lovell, "Financial report of Jodrell Bank, December 1945 – July 1948…," October 1948, p. 5, file R7 (uncat.), LOV.

[16] Lovell, "Memorandum on a 250 ft. aperture steerable radio telescope," February 1951, pp. 11–12, LOV; Lovell, "Radio astronomy at Jodrell Bank," April 1950, file 15 (uncat.), RYL.

[17] In 1964 radio pulses emitted by cosmic ray showers were in fact detected using, appropriately enough, equipment at Jodrell Bank – "Radio pulses from extensive cosmic-ray air showers," J. V. Jelley, *Nature* **205**, 327–8 (1965). Although seemingly similar to Blackett and Lovell's idea, this phenomenon is distinct in that it involves natural radio emission by cascading shower particles, not radar reflections from shower-induced ionized clouds.

[18] June 1947, "Note on the Manchester work for Sir Edward Appleton," unsigned report, 3 pp., file C.232, APP; Lovell:Brearley, 9 March 1947, file CS6/1/1, LOV; Lovell, "Physics at the Jodrell Bank Experimental Station, University of Manchester," 7 June 1947, file R7 (uncat.), LOV.

[19] Parts of this section rely on Lovell (1968:1–5, 16–18) and his autobiography *Astronomer by Chance* (1990:149–54). I am also grateful to Lovell for photocopies of many of the cited items in this section. Clegg's obituary by Lovell (1988) has also been helpful (*Quarterly J. Roy. Astron. Soc.* **29**, 403–9). Lovell has written a detailed account of the cosmic ray shower echo calculations and how they changed over the period 1941–6 ("The Blackett–Eckersley–Lovell correspondence of World War II and the origin of Jodrell Bank," *Notes and Records of Royal Soc. London* **47**, 119–31 [1993]).

[20] Lovell:Hey, 2 May 1946, file CS7/11/5, LOV.

Figure 9.1 John Clegg (left) and fellow student J. C. Parker working in a Jodrell Bank hut filled with surplus radar equipment (ca. late 1946).

You could go out and buy a hundred kilowatt transmitter for ten pounds [$40], provided you towed it back yourself. And the receiver that went with it was five pounds. I remember those figures quite clearly. It stopped them from being thrown down old mine shafts![21]

Items in better shape, however, could be dearer; by 1948, Lovell had put out a total of £4000 ($16,000) for various surplus antennas, transmitters, and power generators, but they were worth a million or more.

While trying to understand the radar properties of meteors, Lovell also began inquiring about their astronomical aspects.[22] He found little interest among most professional astronomers, since quantitative meteoric observations were virtually impossible and in any case certainly not compatible with the use of large telescopes, which are unable to observe a large portion of the sky such as a meteor trail covers. British amateur astronomers, on the other hand, *did* methodically observe meteors, and Lovell discovered that one J. P. Manning Prentice (1903–1981), a Suffolk lawyer, was a world-class expert observer. For two decades Prentice had been the director of the Meteor Section of an amateur society, the British Astronomical Association (BAA); in 1934 he had also discovered the famous Nova DQ Her. In the summer of 1945 he had done some brief, inconclusive visual/radar observations with R. A. Naismith (Section 6.3.2), and in June 1946 was immediately enthusiastic about Lovell's proposal for collaboration. Prentice suggested that the upcoming Perseid meteor shower[23] would be a propitious time, and thus he and his junior assistants Michael W. Ovenden and Gerald S. Hawkins (both of whom later became professional astronomers)

---

[21] J. G. Davies (1978:7–8T). Davies came to Jodrell Bank in early 1947.

[22] Lovell's lack of astronomical knowledge in May and June 1946 can be seen in his use of *meteorites* when referring to *meteors*, as in the documents cited in notes 14 and 20.

[23] See note 6.13 on meteor showers and their radiants.

were on watch outside Jodrell Bank's radar van for a two-week period in August 1946. Lovell later recalled Prentice's method:

> With a deck chair as near horizontal as possible, Prentice would settle himself with a piece of string, a dimmed torch [flashlight], and a writing board. When the meteor streaked across the sky he would raise the stretched string at arm's length along the line of the transient meteor and read off with comparative leisure the stars which defined the beginning and end points of the meteor's track. His writing board soon became covered with symbols which, in the beginning, were mere hieroglyphics to us.[24]

And Clegg (1971:6T) recalled another night during the hard winter of 1946–47:

> Lovell sat looking into the tube in a lorry while I was sitting in a deck chair outside... in two or three feet of snow. I was dressed in a flying suit and flying boots with a telephone head and breast set on.... If I saw a shooting star or he saw something on the tube, we gave a yell, and thus we got correlations.

Prentice subsequently visited Jodrell Bank many times over the next several years, and at one stage even wanted to run his own radar system at home, although this never panned out. He not only directly aided in the meteor research, but, as Lovell (1976:9T) later recalled, also provided the group with basic education in astronomy:

> This was a marvellous training – if you're an observer of visual meteors, you know a hell of a lot about the sky. And that is how we learned our fundamental astronomy – in the best possible way... sitting out in a deck chair in the depths of the night by the side of Prentice.

Later that year Prentice, Lovell, and Banwell (1947) wrote up their results, based on 1800 echoes observed over two months with their vertically-looking 72 MHz Yagi antenna (~40° beam) (Fig. 9.2) and a peak pulse power of 150 kW. When each echo appeared on the screen, best estimates were recorded for time, range, duration, and intensity. As Hey and Stewart (1947) had previously found (Section 6.3.1), the altitudes of the meteor trails were almost all ~100 km. The echo rate ranged from 5 to 30 per hour, being highest during the Perseid shower, but also during a two-week period in late June. Since Hey and Stewart had been able to recognize the Quadrantid and Lyrid showers from echo rates alone (Fig. 6.9), Lovell and company were puzzled why the (visually) much more prominent Perseids could not be so distinguished. The primary radio characteristic that *did* discriminate the Perseids was the much higher fraction of echoes lasting longer than 0.5 sec, some as long as a minute; this too had been noted by Hey and Stewart (1947:874) for the Perseids of 1945. Clouds prevented more than a total of ten hours of simultaneous visual/radar observations, during which only 17 coincidences (~10%) were tallied. This was not particularly impressive, although 65% of *long* duration echoes had optical counterparts.

During 1946 Lovell and Clegg and Nicolai Herlofson, a Norwegian meteorologist-physicist who worked at both Manchester and Oxford during these years, began developing physical theories for formation of the meteor trail and for the reflection process. Modifying the results of Blackett and Lovell (1941), they derived the expected echo power from a trail, and with this theory in hand estimated that meteors seen visually produced trails with $1-10 \times 10^{10}$ electrons per cm of length (further details are in Section 11.4.3). Given their wide antenna beam and uncertainties in the relative orientation of any given meteor trail (side-on trails yielded much stronger echoes than end-on), they were not uncomfortable with the lack of correlation between radar and visual sightings.

After the Perseids run and the full realization of how important a trail's aspect angle was to its echo strength, the constraint of a solely zenith-looking antenna could no longer be tolerated. But how would they make a fully steerable, yet reasonably large aerial? Clegg solved the problem by adapting a wartime "Elsie," a radar-controlled searchlight used to track aircraft for gun-laying purposes. It consisted of five Yagi antennas mounted around a 5 ft diameter searchlight (Fig. 9.3). This arrangement could be easily steered around the entire sky and gave a main beam of $13° \times 28°$ at a wavelength of 4.2 m. Prentice had told Lovell of a predicted gigantic meteor shower in

---
[24] Lovell (1968:7). Information on Prentice is contained in his obituary by Lovell in *Quarterly J. RAS* **23**, 452–60 (1982) and in file CS7/10/1, LOV.

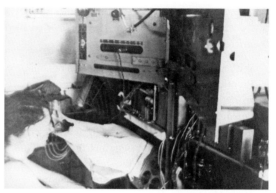

Figure 9.2 (a) The 72 MHz receiving Yagi antenna used for Perseid meteor radar observations in August 1946 at Jodrell Bank. John Banwell makes adjustments on top of the "Park Royal" trailer.
(b) Banwell looking (through a light-darkening funnel) for echoes on the cathode-ray-tube display inside the trailer. This was "the most sophisticated receiving and display equipment of the day" (Lovell 1990:124).

October (in fact Prentice in the 1920s had played an important role in the recognition of this Giacobinid shower), and so the Jodrell crew prepared for a continuous three-day watch with their new "searchlight aerial." As with other observers around the world, they were not disappointed. At the peak, which lasted for only a few minutes (Fig. 11.2), 170 echoes per minute were recorded on their film! To maximize the number of echoes they periodically moved the antenna to keep it at right angles to the visual shower radiant, but they purposely aimed it once *toward* the radiant and, indeed, the echo rate dropped dramatically. This was the first direct demonstration of the criticality of beam orientation with respect to a meteor trail (Lovell, Banwell, and Clegg 1947); moreover, Lovell now became convinced that *all* the echoes being received, even those of short duration, were from meteors, not cosmic rays.[25]

One advantage of studying meteors with the new radar technique was that one no longer had to suffer annoying interference from cloud, moonlight, or even daylight. For a year after the Perseids of 1946 Jodrell workers monitored every major nighttime shower. It turned out that typical echo rates obtained with their setup were fortuitously similar to the meteor counts that a single visual observer could log. This then gave a basis of comparison for interpreting the radio results when no visual observing was possible. In early May 1947 investigation of the nighttime η Aquarid shower began by nightly tracking the point on the sky 90° away from its radiant (in order to get the most echoes) until a few hours after dawn, at which time the radiant set and the echoes petered out as expected. But on one early morning's watch the observer overslept and thus it was discovered, long after the η Aquarids were supposedly gone, that the echo rate was again high.[26] Observations on subsequent mornings of this fishy result confirmed that they had stumbled upon a new major shower, with its radiant in Pisces. Further monitoring showed that this was only the beginning of a two-month span of daytime meteor showers stretching across the ecliptic,

---

[25] Lovell:author, 31 October 1985. Details of these and other Giacobinid observations are discussed in Section 11.2.

[26] Hughes:author, 9 January 1986; Hawkins:author, 10 September 1985.

Figure 9.3 The "searchlight aerial" adapted by John Clegg from a wartime gun-laying system. Five 4.2 m wavelength Yagi antennas (cross-elements do not show up in this photo) were mounted on a 5 ft diameter Army searchlight completely steerable in elevation and azimuth. For half a decade this antenna observed radar echoes from meteors, beginning with the Giacobinid shower of October 1946.

never too far from the sun's direction and therefore never observed by human eyes (Clegg, Hughes, and Lovell 1947). These showers were not minor – had they occurred at night, most would have been comparable to the Perseids. Hey and Stewart (1947:879) had identified one of these June daylight showers with their three-station triangulation method in 1945 (Section 6.3.1),[27] and in retrospect it could also be

---

[27] In 1949 Hey apparently reexamined his 1945 data and in retrospect found evidence for numerous daytime showers, similar to those later seen at Jodrell Bank. He proposed to write

seen that evidence of high daytime rates also appeared in the Jodrell Bank data for the summer of 1946. But now the full extent of the phenomenon was laid bare. Moreover, extensive study of these streams in following years showed them to differ significantly from nighttime showers (Section 11.4.1).

A further line of attack on the meteors was undertaken by students Clifton D. Ellyett (1915–2006), a New Zealander ionospheric physicist on leave from Canterbury University College, and John G. Davies (1924–1988), a Cambridge graduate in mechanical sciences who had worked on radio proximity fuses at the Royal Aircraft Establishment.[28] Because of the desire to calculate orbits for meteors and determine if they came from interstellar space (Section 11.4.2), it was becoming important to be able to measure their speeds in addition to their directions. Hey, Parsons, and Stuart (1947) had done so during the 1946 Giacobinids by directly measuring the changing range of a faint signal attributable to reflection off the meteor itself. But such signals occurred only in a few optimum cases, and Davies and Ellyett (1949), following a suggestion by Herlofson, devised a more general method. The idea was that one could quickly record successive intensity maxima and minima of the Fresnel diffraction pattern created by interference of waves reflected off different portions of the continuously lengthening meteor trail; the pattern was no different than that from a moving knife edge. Knowing the radar-measured distance R to the trail, then the time t between successive echo maxima numbered n_1 and n_2 would yield the speed at which the "reflective aperture" was growing, that is, the speed v of the meteor trail's head:

$$v = \sqrt{(R\lambda)(\sqrt{n_2} - \sqrt{n_1})/t},$$

where λ is the radio wavelength. The system hinged on an automatic recorder (Fig. 9.10b) with a camera that, triggered by an echo's onset, photographed amplitudes of 60 reflected pulses over the next 0.1 sec (Fig. 9.4). Ellyett and Davies's main problem in building this system was that it was constantly triggered by interference from auto ignitions, atmospherics, etc. This was solved by increasing the radar receiver's bandwidth by about three times, which caused interfering pulses to be considerably shorter than reflected meteor pulses. First successfully used during the Geminids of December 1947, this technique proved indispensable in ensuing years. The story of meteor radar research at Jodrell Bank (and elsewhere) is continued in Chapter 11.

9.3 THE 218 FOOT DISH AND THE ANDROMEDA NEBULA

The rationale in 1946 for a fixed paraboloid of large diameter was to acquire a large collecting area for the cosmic ray experiment at minimal cost – Lovell was able to finance it with about £1000 from university sources and another £1000 from the government. The diameter of 218 ft (66 m) was determined by the size of the available plot. Remembering their previous experience with the broadside array that quickly grew too high, Clegg's design this time had the basic constraint that the edges of the dish should be no higher than a ladder's reach. The shape of the surface was defined by 24 radial cables that connected a concrete ring at the center with the tops of 23 ft perimeter posts (steel tubes) supported by guy wires. Each of these cables was additionally strained to the ground at three intermediate radii so as to approximate a parabolic profile in four linear segments – the deviations from a paraboloid were less than 15 cm. The reflecting surface itself consisted of rings of wire running in an azimuthal direction with 20 cm spacing.

The focus of such a large, shallow paraboloid, however, was 126 ft above the ground and there was no way to avoid having the feed (typically two folded dipoles with reflectors) perched atop a tall mast (see Fig. 9.8). Access to the feed involved a ride in a bosun's chair or a climb up scaffolding, both not for the faint of heart. Construction of this dish, which remained the world's largest for a decade, commenced in 1946 and was finished by the end of 1947. At one stage, wives and children helped with the wiring of the surface in the inner part near the ground, while the men did the outer sections, sometimes standing on a platform on

---

up this new analysis for publication, but did not do so after Lovell in essence advised him that the results were of little interest in light of the recent Jodrell Bank work. [Lovell:Hey, 13 September 1949, file CS7/11/5, LOV; also see Hey in *Observatory* **71**, p. 14 + p. 80 (1951)]

[28] For an obituary of Davies by Lovell, see *Quarterly J RAS* **30**, 365–9 (1989).

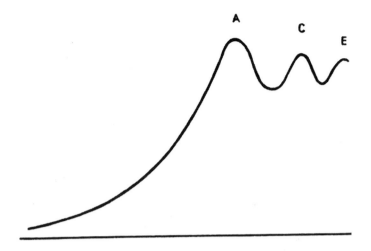

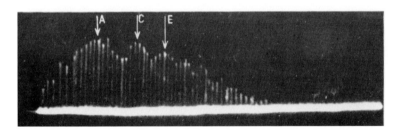

Figure 9.4 (Top) Curve of reflected intensity versus time expected on the basis of diffraction effects and a lengthening meteor trail; successive Fresnel-zone maxima are labelled A, C, and E. The radar beam and trail are perpendicular at a point about halfway up the rise leading to A. (Bottom) Photograph of the cathode-ray-tube display obtained with the automatic recorder system designed by Davies and Ellyett (1949). Each line represents a reflected radar pulse (1.3 msec spacing in this case, corresponding to ~50 m of meteor movement). The changing intensity of the train of pulses allowed calculation of the meteor's speed. This photograph was taken on 14 May 1949 during the summer daylight showers and exhibits a particularly well-defined Fresnel pattern.

a moving truck (Fig. 9.5). Sixteen miles of wire were needed in all.[29]

Besides the brief cosmic ray searches mentioned above, the chief user of the new dish was a postgraduate student, Victor A. Hughes (1925–2001), who had joined the group in December 1946 after an undergraduate degree in engineering physics and a couple of years at TRE. Even as the dish was being constructed, it became clear that *passive* observations of extraterrestrial radio noise would be a profitable new research direction. Hughes therefore set out to measure 72 MHz radio waves from the only part of the sky readily observed with the dish, namely the +53° declination strip that passed through the zenith. His first task was calibration of the antenna pattern, achieved through a standard wartime technique – an oscillator transmitting from a small RAF Avro Anson aircraft moving upwind at all of 40 miles per hour. Its position was followed and marked on the inside of a large *camera obscura* for over a hundred flights interspersed with periods waiting for the wind to change directions. The complete pattern exhibited a main beam width of 4.5°.[30]

---

[29] Information about the 218 ft dish comes from Lovell (1968:16–18), Lovell (1990:154–62), Clegg (1981:137B), and Hanbury Brown and Lovell (1955:547–51).

[30] Hughes obtained an M.Sc. in 1949 based on a detailed study of the zenith strip at 72 MHz; he then joined Hey in what

Figure 9.5 The 218 ft fixed paraboloid reflector under construction in the summer of 1947. The man on the truck is tying surface wires to the cables defining the surface shape.

At this point, in the autumn of 1949, fortune smiled on Lovell's group when out of the blue there appeared a man who since the 1930s had been one of the radar pioneers under Robert Watson Watt and at TRE. Robert Hanbury Brown (1916–2002; Fig. 7.3), with broad-based experience in virtually every aspect of radar systems, was a first-class radio engineer and the prototype *boffin*, a wartime neologism applied to scientists who could develop radar that worked not only in the lab, but also in real military operations.[31] He sought to join a university research group when Watson Watt's postwar consulting firm, of which he was a member, moved to Canada. Lovell had known Hanbury Brown at TRE and in his recruitment emphasized the research possible with the 218 ft antenna:

> When you come up next week, we must discuss what sort of work you might like to do. ... For example, there is a great harvest to be reaped in the cosmic noise field when we can bring some intelligence to bear on it.[32]

Hanbury Brown, who had also been an enthusiastic amateur astronomer as a boy, saw Jodrell Bank as an ideal environment for exciting research and so at the

---

turned out to be largely non-astronomical radio work at AORG (Section 6.5).

[31] See Hanbury Brown's (1991:57) autobiography *Boffin: A Personal Story of the Early Days of Radar, Radio Astronomy and Quantum Optics*, and his obituary by J. Davis and B. Lovell (2003) in *Biog. Mem. Fellows Roy. Soc.* **49**, 83–106.

[32] Lovell:Hanbury Brown, 20 June 1949, LOV (cited in Lovell [1973:9]); see Lovell (1973:8–10) for further details of Hanbury Brown's joining Jodrell Bank.

age of 33 signed on, with a 70% salary cut, to get a Ph.D.[33]

Hanbury Brown spent his first nine months designing and building an all new state-of-the-art receiver system. In this he was assisted by fellow student Cyril Hazard, freshly arrived with a physics degree from Manchester. In order to have superior angular resolution ($\sim$2.0°), he went to 158 MHz (1.9 m wavelength), the shortest wavelength tolerable given the leakage (40%) caused by the wide spacing of the wires making up the reflector. Although he replaced the lossy open-wire feeder lines running up the mast with polyethylene coaxial cable, almost half of the signal was still lost travelling the 300 ft to the amplifier on the ground. His receiver's gain stability was based on carefully constructed power supplies (good to one part in 2000) and the comparison scheme developed by Ryle and Vonberg (Fig. 8.1). System temperature was $\sim$1000 K. This fine receiver and huge dish led to unparalleled abilities: in a single drift curve one could now detect with some assurance a source of only 20–30 Jy (corresponding to an antenna temperature of 3–4 K), fully *fifty times* better than previously attained (see Table 14.1). Hanbury Brown proudly wrote at the time:

> I personally am fighting a grim battle against the cosmos with a 218′ aperture paraboloid.... I have spent many unhappy hours suspended at 126 feet and 4 inches above the ground with a hot bucket of black stuff trying to pour it into watertight joints.... [and] have got the equipment running sweetly. It will hold a level of $\pm$10° K for three weeks or so without being touched at all.[34]

Hanbury Brown also realized that the dish, with its long focal length, should behave tolerably well even with its central mast tilted significantly from the vertical. This opened new possibilities, for it allowed parts of the sky away from the zenith to be reached. Furthermore, by good fortune the nearest galaxy comparable to our own Milky Way turned out to lie just within the mast's range of maneuverability (passing 12° south of the zenith each day). Thus it was that Messier 31, the Great Nebula of Andromeda, became the first target of observation, and plans for a survey of the galactic noise were postponed. Although Reber's attempts at M31 had been inconclusive (Section 4.3), it looked like a good bet since calculations indicated that it should just be detectable if it radiated similarly to the Milky Way. On the other hand, not one of the discrete radio stars then known was reliably identified with a galaxy (Chapter 14), so perhaps the Milky Way emission was abnormal.

From July through November 1950 Hanbury Brown and Hazard took drift curve after drift curve on M31, a total of 90 at six different declinations. They were delighted to find that the Milky Way's neighbor was indeed detectable, but only one-third of the passes measured up to the theoretical sensitivity of their receiver – the exposed aerie of a feed aerial and the long length of cable made the system particularly susceptible to the vagaries of interference and weather. Most daytime scans were plagued by signals from the sun and from automobiles (especially since petrol rationing in England had just ended), and nighttime data, although usually better, often had to be discarded because of rain and manmade noise originating farther away.[35] A few days or even two weeks would be spent repeating a given declination in order to secure sufficiently high-quality data. And then when a new declination was called for, the mast had to be moved:

> It was extremely hard work indeed, because it took most of a morning by two people to move the antenna through half a beamwidth to take the next scan. The instrument had 18 steel guys, and you had to lift them off in the right order or otherwise the steel mast would kink, and you had to watch it with two theodolites at right angles.... We literally worked our fingers to the bone to do that survey. (Hanbury Brown 1976:52T)

---

[33] In the end Hanbury Brown never bothered to obtain a Ph.D. since he soon found himself in effect supervising several other Ph.D. candidates.

[34] Hanbury Brown:Bowen, 30 April 1950, file A1/3/1a, RPS. Other information comes from Hanbury Brown and Hazard (1951a:357–60); Hazard (1953); and Hanbury Brown (1976:5T, 1984:213–7).

[35] Over one several-week period, Hanbury Brown and Hazard became convinced that they had discovered a new, variable radio source; they were especially persuaded by the fact that the signal was arriving $\sim$4 min earlier each day, as it should for a sidereal source. But it turned out to be electrical interference from a nearby milking machine – the cows (and their farmer) were rising earlier each day with the springtime sun!

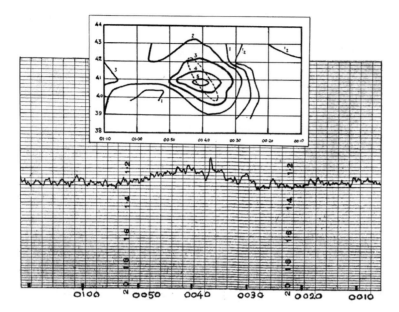

Figure 9.6 (Bottom trace) A transit of the Andromeda nebula through the beam of the 218 ft dish. Abscissa is right ascension and the r.m.s. noise is ~0.7 K in antenna temperature, corresponding to an r.m.s. flux density of 5 Jy for an unresolved source. At the top is a contour map (in right ascension and declination) of the Andromeda nebula source at a frequency of 158 MHz; one contour unit equals 10 Jy of radiation within the 2.0° beam. Background radiation from our own Galaxy has been largely removed. The dashed ellipse represents the galaxy M31 as seen "on a photograph." (Hanbury Brown and Hazard 1951a)

By the end of September they felt they had enough reliable data to write up a short note for *Nature* (Hanbury Brown and Hazard 1950), followed by a few more observations and a more detailed analysis in a paper submitted to the Royal Astronomical Society (RAS) the following March (Hanbury Brown and Hazard 1951a). In the first paper they did not properly appreciate how an extended source of emission affected the measured angular size and therefore underestimated both M31's angular size and its flux density. This was caught by the time of the second publication, aided by correspondence with Jack Piddington in Sydney, who pointed out various weaknesses in the *Nature* article.[36]

Despite months of data collection, the radio signal from M31 (Fig. 9.6) remained difficult to analyze in competition with terrestrial interference and gradients in emission from our own Galaxy. When these were accounted for, they derived an intensity of ~100 Jy, a position for the source within 30′ of the optical galaxy, and a total extent of ~2.5°. Much of their paper was occupied with calculating, based on various assumptions about the radio background of the Milky Way, the 158 MHz intensity a distant Andromedan would measure looking back at our own Galaxy. They arrived at values ranging from one to seven times their own reciprocal measurement and, considering the uncertainties, felt safe in concluding that M31 and the Milky Way were similar in their radio emission. It was exciting that at last a discrete radio source could be confidently associated with a "normal" optical object. The American Charles Seeger wrote at the time:

> I would have given my right arm to have had a hand [in the detection of M31]; while at Cornell I tried every possible trick to verify Reber's suspicion, but I could not overcome the handicap of inadequate equipment.[37]

While taking these data, it was convenient also to observe the region of Cygnus A, which is at almost the

---

[36] Correspondence between Piddington and Hanbury Brown and between Pawsey and Lovell, 4 July to 7 August 1951, file D5/4/X, RPS.

[37] C. L. Seeger:Lovell, 2 April 1951, file CS7/13/3, LOV.

same declination and only four hours earlier than M31. Hanbury Brown and Hazard also became intrigued with trying to detect other nearby galaxies and began a long program to that end. Finally, they decided to undertake a full survey of discrete sources in the accessible 30°-wide strip, and the resulting "HB" survey (Hanbury Brown and Hazard 1953b; Section 14.3.3) played an important role in the debates over the nature of radio stars. All of this work is discussed in Chapter 14. By this time the cosmic–ray radar origins of the 218 ft reflector had been long forgotten: the dish had instead become the centerpiece of *radio* astronomical research at Jodrell Bank. Joe Pawsey had been uncannily correct back in 1948 when he had visited and reported back to Sydney:

> They have built a 75 yard diameter parabolic aerial with the idea of detecting (a) the Blackett and Lovell cosmic ray bursts of ionisation, or (b) meteors. Both these seemed doomed to failure so I guess Lovell will simply discover something new. The possibility of drawing a blank I regard as both inartistic and unlikely.[38]

9.4 OTHER PROJECTS

During the early years at Jodrell Bank several other smaller investigations were mounted. The only solar publication arose when in July 1946 the sun burst so mightily that the antenna observing the Perseid meteor shower, although vertically directed, was overwhelmed (Lovell and Banwell 1946). Although the resulting report on the next two weeks of intense solar activity turned out to be the first scientific result to emerge from Jodrell Bank, Lovell and his colleagues never further studied the radio sun (in contrast to virtually every other postwar group over the subsequent decade).[39] In fact, it soon grew to be Lovell's deliberate policy not to overlap with topics (such as the sun) and/or techniques being covered by Ryle's group at Cambridge, in order that both groups could better justify their need for funding.[40]

Similarly to the intrusive solar bursts, on another occasion diffuse radar echoes unexpectedly began "frothing and bubbling" on the screen during the 1947 Perseids. Stepping outside, the group witnessed a magnificent aurora borealis streaming across the sky (Lovell, Clegg, and Ellyett 1947). This led in the early 1950s to radar study of aurorae as one of Jodrell Bank's several lines of atmospheric research (primarily by Hawkins, who joined the group as a student in 1949).

Lovell had also been interested in lunar radar ever since he and Banwell had made feasibility calculations at the end of the war.[41] The goal was more to study the nature of the ionosphere's influence on the reflected signals than to study the moon itself. The first students assigned the task of developing a suitably powerful transmitter and sensitive receiver were Douglas R. H. Forsyth and Frank Moran, both of whom joined the group in 1946. But Forsyth soon gave up for a career in law and Moran had no success until Ian A. Gatenby (1926–1963) arrived as a new student. He and Moran carried out observations at 72 MHz with Moran's millisecond-pulse equipment and various of the meteor antennas in 1949–50 and in the end did record lunar echoes, but only sporadically and never stronger than two to three times the receiver noise. This contrasted with calculations, based on his 20 kW of peak power and earlier American and Australian results (Chapter 12), that he should be getting signals at least twenty times stronger. In his M.Sc. thesis[42] Gatenby was unable to explain the discrepancy and these results were never further published. Not until late 1953 would Jodrell Bank have a properly working lunar radar, under the guidance of W. A. S. Murray and J. K. Hargreaves (Section 12.6).

Two further projects, fully discussed in Sections 14.2 and 14.5.1, deserve mention here. The first began in 1949 and was designed to answer the question as to whether the erratic intensity fluctuations of discrete radio sources were intrinsic to the sources themselves or

---

[38] Pawsey:Bowen, 11 June 1948, file F1/4/PAW/1, RPS.

[39] The major exceptions to Jodrell Bank's avoidance of the radio sun were test observations of the sun undertaken by R. C. Jennison and M. K. Das Gupta in 1951–52, as a prelude to radio star work (Section 14.5.1.2). Another exception was an offbeat project to detect extremely low frequency waves (7 Hz) from the sun (Tangent 9.1).

[40] Lovell:author, 31 October 1985; Edge and Mulkay (1976:237–9); Ryle:Lovell, 12 December 1949, file CS7/16/3, LOV.

[41] See note 12.18.

[42] I. A. Gatenby, 15 October 1951, "Radar reflections from the moon and the aurora," M.Sc. thesis, Dept. of Physics, Manchester University.

caused by the earth's ionosphere (Section 14.2). Lovell and his students C. Gordon Little and Alan Maxwell spent a year and a half monitoring Cyg A and Cas A with both single and spaced antennas and eventually established the ionospheric origin of the fluctuations (Little and Lovell 1950; Little and Maxwell 1951), the initial portion being done in coordination with the Cambridge group (Smith 1950). The second project began shortly after Hanbury Brown's arrival when he invented what became known as the intensity interferometer. As with the more common Michelson interferometer, the new scheme used two widely separated antennas to achieve much better angular resolution than possible with either alone, but now one searched for correlations in the small noise-like intensity fluctuations possessed by any natural signal, rather than correlations in the signal itself. With his student colleagues M. K. Das Gupta and R. C. Jennison he tested this idea first on the sun and then on strong discrete sources. It worked, and the results on the size and nature of Cyg A were particularly important (Hanbury Brown, Jennison and Das Gupta 1952; Jennison and Das Gupta 1953).

9.5 PLANS FOR A HUGE STEERABLE DISH

Lovell saw Jodrell Bank's first five years as only a prelude. The fixed 218 ft paraboloid and its successes under Hanbury Brown's direction whetted Lovell's appetite for a dish of similar size that could point anywhere in the sky – it would be able to detect radar echoes from four planets, from asteroids, and from extremely small meteors, and it would do passive radio astronomy with unprecedented sensitivity and flexibility. Sometime in 1948 Lovell first broached the subject with Blackett, where it struck a responsive chord, for Blackett firmly believed that leadership in postwar research would go to those with the large instruments. Great Britain and Jodrell Bank now had the largest paraboloid in the world (by a factor of ~50 in area), but Lovell had gotten wind of plans for ~200 ft diameter steerable dishes in the United States and in Canada. Since such a structure had to be mounted very high in order to be able to look near the horizon, it was clear from the start that this was an enormous project. Inquiries to firms in 1949 about the project's feasibility brought replies of incredulity or lack of interest, but eventually Lovell found H. C. Husband, head of an engineering consulting firm in Sheffield. Husband immediately took a liking to the project and stayed with it throughout the 1950s.

Meanwhile, a strong base of support for this dish was being cultivated in the astronomical and radio community at large. Blackett, even more a force in British science with his recent (1948) Nobel Prize, asked the RAS to establish a special Radio Astronomy Committee and its chairman, William M. H. Greaves, an astronomer at the University of Edinburgh and past RAS President, took its charge as:

> ... to keep radio workers who are exploring new methods in touch with astronomers of the classical type who might be able to help by suggesting problems for investigation. Apart from this it may be helpful for radio-astronomers to have the backing of the Society.[43]

In preparation for the first committee meeting, Hey, Ryle, and Lovell met in January 1950 to consider their priorities over the next five years. Ryle at first disagreed with the emphasis Lovell wanted to place on the big dish, but within a few weeks was persuaded to agree upon a document entitled "Radio Astronomy in Great Britain."[44] It covered all aspects of work in progress or foreseen for the near future, and its central theme was the need for a 200–250 ft paraboloid with potential for contributions to almost every sector of radar and radio astronomy. At a meeting in February the full committee gave its unanimous approval to Lovell's dream. Shortly thereafter the following resolution was passed:

> The RAS Council strongly endorses the proposals put forward ... for a steerable paraboloid aerial of 250 feet diameter. The Council considers that by the erection of this apparatus the prestige of science in Britain would be considerably enhanced ... The Council is impressed by the consideration that [this] would permit the continuation in the United Kingdom of new methods of astronomical research ... which are independent of climatic conditions.[45]

---

[43] W. M. H. Greaves:Ryle, 24 February 1950, box 1/1, RYL.
[44] February 1950, photocopy from Lovell, LOV; Lovell:Ryle, 23 December 1949, 9 and 31 January 1950, box 1/1, RYL.
[45] Lovell (1968:34). See Edge and Mulkay (1976:58–9) for more on the work of this committee. In particular, in late 1950 it also persuaded the RAS Council to pass a similar resolution

In the summer of 1950 the Department of Scientific and Industrial Research funded a design study by Husband. The diameter of the antenna eventually settled on 250 ft, with a deep bowl for easier access to the focus and for protection from interference. The antenna was to operate at wavelengths as short as one meter (where its beam would be ~0.5°) with a reflecting mesh of 5 cm spacing. It was to be mounted in an azimuth-elevation fashion, with tracking of radio sources controlled through a novel analog computer system. In elevation the dish would be driven by 15 inch gun-turret racks (obtained as surplus from battleships) mounted on trunnion bearings supported by two 190 ft steel towers. In azimuth the dish would turn on a circular double railway track.

The final antenna indeed ended up looking like the original proposal. But construction costs quickly soared far beyond the 1950 estimate of £50,000–100,000 ($150,000–300,000), and ensuing engineering problems contrasted with the statement in the submission to the RAS Committee that "both firms [consulted] consider the project fairly simple from an engineering point of view." In March 1951 a £259,000 ($700,000) request was made to the government and a year later the project was approved once the private Nuffield Foundation (money associated with the Morris Motors auto firm) agreed to split the cost. Construction in the Cheshire countryside began in September 1952, but not until five years later and an expenditure of ~£750,000 (~$2,000,000) would the dish first operate. Agar (1998) has made interesting sociological arguments regarding how the big dish eventually became an icon of British scientific and technological prowess, a "public spectacle of science." The trials of constructing and financing the big dish are beyond the scope of the present study, but the great radio telescope's origin as a natural outgrowth of the fixed 218 ft paraboloid is very much a part of the early years of Jodrell Bank.[46]

9.6 JODRELL BANK AFTER FIVE YEARS

By 1950 Lovell had established Jodrell Bank as the clear world leader in meteor radar astronomy (Chapter 11). As an offshoot this work had further led to productive research on the physics of the ionospheric regions affected by the meteors. Nearly thirty scientific papers had been produced, doctorate degrees had been awarded to Ellyett (1948) and Clegg (1948a), students and staff numbered around twenty (Fig. 9.7), and the annual budget (exclusive of salaries) had grown from £2000 in 1946–47 to £10,000 ($30,000). Many factors contributed to this success, including Lovell's effective leadership both at home and outside in the scientific community, Blackett's strong support, a well-trained staff of ex-radar workers, and the productive atmosphere they created at Jodrell Bank.

Lovell in those days was later described by Hawkins (1984:163A) as "stimulating, speculative (always throwing out hundreds of ideas), impetuous, driving, an excellent organizer and scientific politician and fund raiser." Much of this no doubt stemmed from his wartime experience as a group leader who had to get things done; it was there that he learned how to get the most out of people. As he wrote to Blackett just before war's end:

> [In your advice to me at the beginning of the war when I was getting frustrated with the lack of organization in the government laboratories] you told me that I would soon realize that 75% of the new work consisted of dealing with people. For the past two years I have spent 95% of my time dealing with people and I am filled with dismay at the thought of having to teach myself to deal with apparatus again.[47]

The lesson may have been painful, but it paid dividends for the rest of his career.

Several interviewees commented on the high group morale during these years. Exciting results rolled in, and all hands pitched in with a high-intensity team approach that many had learned in the war. Over three years after the war, Lovell himself pointed out the connection in a newspaper article: "The radio techniques which enabled our bombers to see the

---

strongly supporting Ryle's future funding needs for a large array of antennas for the 2C survey (Section 8.5.3).

[46] Lovell (1968:24–36 et seq.) and Lovell (1990:191–211 et seq.); "Memorandum on a 250 ft aperture steerable radio telescope" (the main proposal to the government), February 1951, photocopy from Lovell, LOV. Full details of the checkered course of construction problems and eventual triumph with the 250 ft dish during the 1950s are in Lovell (1968, 1990), for the protagonist's view, and Agar (1998), for a sociological analysis. The dish, after fifty years and several upgrades and being named

the Lovell Telescope, is still one of the world's leading radio telescopes.

[47] Lovell:Blackett, February 1945 (quoted by Saward [1984:103]).

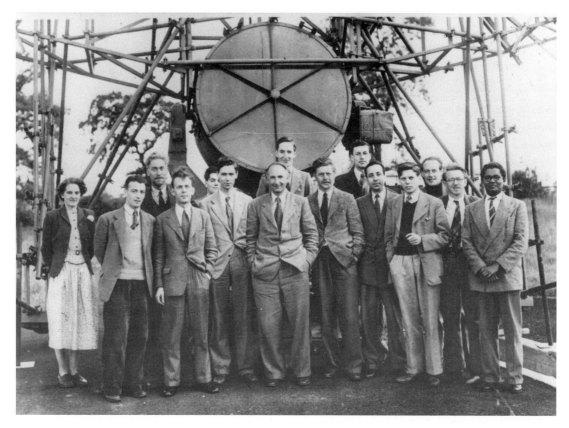

Figure 9.7 The Jodrell Bank group in 1951, photographed in front of the searchlight aerial on the occasion of Lovell assuming the new University of Manchester chair in radio astronomy. From left to right (with date of joining the group): Mary Almond (1949), Cyril Hazard (49), Roger Jennison (50), Stanley Greenhow (48), Gerald Hawkins (49, partially hidden), John Davies (47), Bernard Lovell (45), Gordon Little (48, behind Lovell), John Clegg (46), Alan Maxwell (49), Ismail W. B. Hazzaa (49), Tom Kaiser (50), Sandy Murray (50), Tim Closs (49), and "Das" Das Gupta (50). Missing are D. Greening (49) and Robert Hanbury Brown (49) (present in Fig. 7.3).

Tempelhof aerodrome [in Berlin] through ten thousand feet of mist and cloud have now been deployed in reverse ... for a new branch of astronomy."[48] In a 1949 letter to a former TRE colleague, Lovell commented: "You would be surprised if you could see our Experimental Station here. I am sure it would remind you of some of the early days of T.R.E."[49] And John Clegg (1971:34,5–6T) recalled:

We still had the wartime atmosphere with us. At TRE during the war ... you could be doing a piece of research to try to design something and you needed a 360 ft tower. So you just asked for a 360 ft tower and a lot of characters came and began to build it next morning. You just plunged in and did it.

[At Jodrell Bank] we were all in it together – still near enough to remember the old days of danger. You made tremendously good friendships there. You worked like mad. You had the feeling all the time that there was something round the corner.... When I joined the staff, you worked absolutely day and night. Sometimes you used to give a lecture in the morning [in Manchester], then you'd go out to Jodrell Bank, you designed an aerial system, you dug holes, you put concrete in them, ... you built the scaffolding up, you put the aerials on the scaffolding,

---

[48] Lovell, *Manchester Guardian*, 10 November 1948, "Radio astronomy – a new science in Manchester."

[49] Lovell:B. O'Kane, 27 January 1949, file CS6/1/1, LOV.

and then at night you tried to get correlations between meteors and things.

It's not hard to understand why one newspaper reporter called them "picked graduates and graduates with picks,... navvies, constructional engineers, steeplejacks and 'Boffins' all in one."[50]

Blackett liked the way Lovell did research, the way he ran Jodrell Bank, and the way his astronomical results complemented Blackett's own group's research on cosmic rays.[51] Lovell recalls how he was backed to the hilt:

> Blackett and I had our clashes, as you can well imagine, but in the broad canvas he was a very critical factor in the whole affair. After all, he allowed Jodrell Bank to grow twenty odd miles from his main department and he gave me money ... and manpower, which he might well have said he wanted for his own cosmic ray research. It is an interesting question as to whether Blackett would have had that attitude if the original incentive had not been in cosmic rays. I think the answer is: yes, indeed he would have. He was so fascinated by the whole of this, by the techniques and by the subject matter. If I now think of my own attitude to one of my own individuals doing that sort of thing, I would like to think that I would behave as Blackett behaved, but I'm damn sure that I wouldn't.[52]

Blackett's career, itself full of speculation and failures as well as successes, meant that he was not deterred by Lovell's failure to find cosmic ray echoes.[53]

Lovell in turn remained loyal to Blackett and the university and in 1947–48 turned down several salary-doubling offers for chairs in Canada and Australia.[54] In 1951 Lovell, then Director of the "Jodrell Bank Experimental Station" (Figs. 9.8 and 9.9) and Reader in Physics, was rewarded with the world's first chair in radio astronomy.

"Radio *astronomy*" was indeed an appropriate description of Lovell's endeavor, for although Jodrell Bank hardly looked the observatory, radio techniques there were viewed solely as a means to answering astronomical (and ionospheric) questions, not as an end in themselves or for practical applications. For example, by 1950 the surplus radar transmitters and receivers (Fig. 9.10) were becoming obsolete relative to the latest electronics, but they were kept because they were still doing good astronomy. Lovell (1971:18T) later recalled that "the transmitter was just the thing one turned on – the meteors and aurora were the exciting things." He also made strong efforts to integrate and legitimate his group within the mainstream of the astronomical community – his first talk at an RAS meeting was two months after the October 1946 Giacobinid shower.[55] His popular book with Clegg, *Radio Astronomy* (1952), was the first of its kind, and his monograph *Meteor Astronomy* (1954) became the premier reference on meteors. Three-quarters of Jodrell Bank's papers through 1950 were not in engineering journals, but in publications of the RAS and BAA or in *Nature*.[56]

Lovell also fostered friendships with leading astronomers such as Fred Whipple of Harvard University and in turn gained valuable advice from them about ripe research topics. Whipple was the keynote speaker at a Jodrell Bank conference on "Meteor Astronomy" held in September 1948 and at that time stressed the importance of determining meteor velocities, a program pursued by Jodrell workers for many

---

[50] R. Calder, London *News-Chronicle*, ca 1–15 May 1947, "Midday star study."

[51] Blackett's group's cosmic ray research in the postwar decade was also going very well and led to several advances, in particular discovery of new particles such as the K-meson. In fact, one of the first pieces of evidence for these was produced in Blackett's lab by a cosmic ray on 15 October 1946, the very same week as the great Giacobinid meteor shower observed by Lovell. [Rochester, G. D., "The early history of the strange particles," pp. 299–321; Elliot, H., "Cosmic ray intensity variations and the Manchester school of cosmic ray physics," pp. 375–84 – both in *Early History of Cosmic Ray Studies* (1985), Y. Sekido and H. Elliot (eds.) (Dordrecht: Reidel)]

[52] Lovell (1971:13T). Also see comments in Lovell (1968:1–36).

[53] M. J. Nye (2004), *Blackett: Physics, War and Politics in the Twentieth Century* (Cambridge: Harvard University Press), p. 178.

[54] Saward (1984:135–7). These offers were from McGill, Sydney and Adelaide Universities. Lovell's reasons for turning down these offers are set forth in Lovell:Ryle, 7 August 1948, file 4 (uncat.), RYL.

[55] Lovell (1990:131–2) has a nice description of his experience giving this talk.

[56] Before submission, all papers underwent a rigorous examination (and often heavy editing) by Lovell. [Hawkins:author, 10 September 1985; Jennison (1972:8T, 1976:12–13T)]

Figure 9.8 Aerial view of Jodrell Bank in 1948 from the southeast (see Fig. 9.9). The 23 ft perimeter poles and 126 ft mast of the 218 ft diameter fixed reflector dominate the scene. Behind the big dish is the "searchlight aerial" (Fig. 9.3) and to the upper right poles for part of the meteor radiant aerials (under construction; see Fig. 11.7). Behind the searchlight aerial is a lunar radar antenna and in front of it the "Park Royal" trailer (Fig. 9.2). It is easy to see why locals called this collection "the fairground."

years thereafter (Section 11.4.2). Besides a steady stream of individual visitors, Lovell arranged for the Physical Society to meet at Jodrell Bank in March 1947 and the RAS in July 1949; *Observatory* reported for the latter meeting that "The hissing sound of 'Cygnus calling' was made clearly audible to the visitors through a loudspeaker system."[57] Further important contact with traditional astronomy was effected through weekly lectures at Jodrell Bank by Zdenek Kopal, an astronomer who came to Manchester University in 1951.[58]

Jodrell Bank in 1950–51 was at an important juncture. Before 1951 only three of 28 papers dealt with passive radio astronomy. But in the years 1951–52 the radio astronomy fraction rose to 40%, led by the contributions from Hanbury Brown and his group.[59] Furthermore, Lovell became preoccupied with financing and managing the project to build the 250 ft diameter, fully steerable paraboloid; the philosophy of the big dish was in the ascendancy.

---

[57] *Observatory* **69**, p. 122 (1949).

[58] Hanbury Brown (1984:220–1); Davies (1971:6–7T + 38–9T, 1978:6T); Lovell (1976:9T). Another theoretical astronomer who interacted with the Jodrell Bank group after 1951 was Franz Kahn (Edge and Mulkay 1976:283–4, 318–19).

[59] These figures are based on a Jodrell Bank list of publications, excluding popular articles. Also see Edge and Mulkay (1976:81).

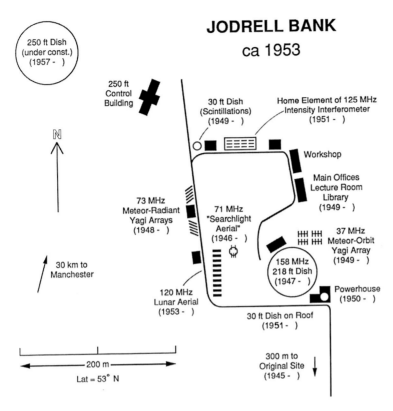

Figure 9.9 Map of Jodrell Bank as it was ca 1953, based on various contemporary sources. About six antennas primarily used for ionospheric research have been omitted. Antennas on the map not mentioned in this chapter are discussed in Chapters 11, 12, and 14.

Meteor radar, although hardly dead, was no longer the undisputed focus of research and in fact steadily declined in importance over the next five years. Clegg left the group in 1951 and joined Blackett for research in paleomagnetism. Active participation in the remaining meteor radar work passed from the hands of Lovell and Clegg to others, in particular Davies, Hawkins, and new staff members T. R. Kaiser and J. S. Greenhow. This led to the paradoxical situation of a site where one set of researchers was assiduously blasting out power (for meteor radar) at the same time that another set agonized over any signals interfering with their detection of faint cosmic noise. Although a system of blanking off the passive receivers in synchronism with the radar pulses was soon developed (and this usually worked reasonably well), it made radio astronomy much trickier at Jodrell Bank than elsewhere.[60] Finally, research came to be conducted in four or five small groups working on distinct problems with little overlap.[61]

Jodrell Bank was also maturing as an institution. Styles of research were necessarily changing as the group grew, equipment became more complex, and problem areas gained definition:

> In the early days if you wanted to do an experiment, or a student was going to write his thesis on a particular thing, you would spend the first year building the equipment, and then you'd

---

[60] Unattributed quotation in Edge and Mulkay (1976:293). Also Jennison:author, 8 October 1985; Davies (1978:113A).
[61] Gilbert (1975:106, 1977); Edge and Mulkay (1976:291–311); Section 11.5.

do an experiment and write it down. As time went on, of course, a person got more and more of the equipment given to him, which has both advantages and disadvantages. Some students never understand what the equipment does. (Davies 1978:7T)

Similarly, Hawkins (1973:3, 7T) later recalled:

When I turned up in 1949, Lovell said, "You're working in that hut there." It was the "Radiant Hut." He had obviously hand-picked me to crack the problem of the meteors from the point of view of survey measurements ... I was basically told I had to run this equipment, observe the showers as they came through, and write them up.

Just as in 1949 permanent buildings replaced the vans and lorries on site (although all power came from diesel generators even as late as ~1953), so Jodrell Bank's science was evolving from an exploratory phase to planned programs. Jodrell Bank joined Radiophysics in Sydney and the Cavendish Laboratory in a trio of groups that led worldwide radio astronomy for over a decade. And yet it all started with, and to a significant extent for many years was guided by, questionable calculations that led to the unsuccessful pursuit of bouncing radar off cosmic ray showers.

TANGENT 9.1 ATTEMPTS TO DETECT AUDIO-FREQUENCY SOLAR RADIO WAVES

An offbeat project was undertaken by Roger Jennison and W.A.S. Murray while still undergraduates in 1948 (both subsequently undertook postgraduate

(a)

Figure 9.10 (a) 50 kW ex-Army radar transmitter used for 72 MHz meteor studies at Jodrell Bank (ca 1951).

(b)

Figure 9.10 (b) Receiver and display associated with the transmitter in (a). Against the wall in back is an automatic recorder for meteor echoes, as developed by Davies and Ellyett (1949).

work under Lovell in 1950). In a note to *Nature* in early 1948 Donald H. Menzel of Harvard Observatory and Winfield W. Salisbury of Collins Radio Co. had suggested that the sun might commonly emit electromagnetic waves at extremely low frequencies, less than 500 Hz. These were presumed to be excited by sunspot regions and therefore to have wavelengths comparable to the size of a spot. The Americans cited some evidence for these, but in fact could say little about the contents of secret military reports detailing these peculiar low-frequency signals (Menzel 1976:57A). To follow this up, Jennison and Murray constructed a receiver based on an iron-core coil and sensitive to 7 Hz waves. Numerous times they detected a type of pulse that seemed of non-manmade origin, but they were unable to distinguish between direct radiation from the sun and variations in the ionosphere or magnetosphere.[62]

In Sydney, Lindsay McCready and Fred Lehany also attempted to confirm the scrappy American report, but soon gave up.[63]

---

[62] Jennison and Murray, "Report on a preliminary investigation of electromagnetic radiation at sub-audio frequencies," November 1948, Jodrell Bank (photocopy from Jennison); Jennison (1972:1–2T, 1976:2–5T).

[63] Pawsey:Bowen, 31 December 1947 and 13 February 1948; Bowen:Pawsey, 14 January 1948; McCready:G. Newstead, 18 July 1949 – all A1/3/1a, RPS. Also see note 7.93.

10 • Other radio astronomy groups before 1952

Postwar radio astronomy was dominated to an extraordinary degree by the British and Australian accomplishments, but important groups in Canada, France, Japan, the Soviet Union, and the United States also sought to understand the radio sky. This chapter tells the stories of their diverse origins and first investigations, as well as those of small groups in Germany, Sweden and Norway. Table 10.1 presents a concise summary of these groups, as well as those encountered in earlier chapters; Figure 1.2 graphically defines each of these groups and shows their relationship with each other. *Radar* astronomy groups are not included, but discussed in the following chapter and tabulated in Table 11.1. The origin and first decade of the important group at Leiden University in Holland is covered in Chapter 16.

10.1 UNITED STATES

Although postwar radio astronomy in the United States lagged significantly, there was some activity. Grote Reber's work in Illinois and at the National Bureau of Standards has been recounted in Chapter 4. The discovery of the 21 cm hydrogen line by Harold I. Ewen and Edward M. Purcell at Harvard University during 1949–51 is found in Chapter 16. In addition there were three other American efforts in the late 1940s: at the Radiation Laboratory of the Massachusetts Institute of Technology (MIT), at the US Naval Research Laboratory, and at Cornell University. An overall discussion contrasting this American work with that in England and Australia will be deferred to Chapter 17.

10.1.1 Dicke and the Radiation Laboratory

Although the first postwar American investigation did not develop into a long-term program, it was important from several perspectives. Robert H. Dicke (1916–1997) conducted these observations in 1945 at the Radiation Laboratory on the MIT campus in Cambridge. The "Rad Lab" was the largest American radar laboratory, created from whole cloth in late 1940 after consultation between the elite of American physics and a British scientific mission headed by Henry Tizard (and also including E. G. "Taffy" Bowen, future director of the Radiophysics Lab in Sydney; see Chapter 7). The visitors had come to share the secret of the cavity magnetron, a recently-invented source of microwave transmitter power some thousand times stronger than anything before. This "tube" (it was actually a hollowed-out block of copper) was a critical technical innovation that allowed the Allies to overtake the early German lead in short-wavelength radar. The magnetron had little direct effect on early radio and radar astronomy, however, since its value was in *transmission* (not reception), and early *radar* astronomy (Chapters 11 and 12) was not at microwavelengths.

Among American research and development efforts radar was second only to the Manhattan District (the atomic bomb project). Civilian radar development, primarily at the Rad Lab and Bell Labs, concentrated on microwaves and left longer wavelengths to military laboratories. Under the leadership of Lee A. DuBridge, the Rad Lab grew by the end of the war to a scientific and technical staff of 1200 (4000 total) spread over 15 acres of floor space. It participated in the development of about one-half of all American radar sets deployed during the war. A menagerie of radars was produced at wavelengths of 10, 3, and 1 cm (the so-called S, X, and K-bands), in addition to a system (LORAN) that was to become fundamental to world navigation for two decades.[1]

DuBridge, taking note of British experience as well as the fact that microwaves were more a laboratory curiosity than a practical technique, recruited key staff members from the ranks of physicists rather than radio engineers. Dicke, fresh from his Ph.D. in

---

[1] On the Radiation Lab see Guerlac (1987), Kevles (1977:302–23), Buderi (1996:chapters 1–11), Brown (1999:166–74) and Section C.1.

Table 10.1 *Radio astronomy groups before 1952*[a]

Dates[b]	Organization	Researchers[c]	Site	Antennas	Frequencies	Programs[c]	Papers[c]	Chapter
USA								
1939–47	–	G. Reber	Wheaton, Ill.	31 ft dish	160, 480 MHz	Gal., sun	7	4
1947–	Radio Div., US Naval Research Lab.	J.P. Hagen, F.T. Haddock, C.H. Mayer	Washington, DC	several 6–10 ft dishes	9400, 35000	sun	5	10.1.2
1947–51	Central Radio Propagation Lab., Natl. Bureau of Standards	G. Reber, H.V. Cottony, J.R. Johler, J.W. Herbstreit	Sterling, Va.	3 (7.5 m) 31 ft dish (Reber), dipole arrays	25 → 480	sun, Gal. (also meteor radar)	6	4.6.2
1948–	School of Elec. Eng., Cornell Univ.	C.L. Seeger, R.E. Williamson	Ithaca, N.Y.	dipole arrays, 17 ft dish	205	sun, Gal., source scintillations	6	10.1.3
ENGLAND								
1945–48	Army Operational Research Group	J.S. Hey, J.W. Phillips, S.J. Parsons	Richmond Park (near London)	Yagi arrays	64	Gal., sun (also meteor radar)	6	5, 6
1946–	Cavendish Lab., Cambridge Univ.	M. Ryle, F.G. Smith, A. Hewish, 6 others[d]	Cambridge	many	38 → 500	sun, sources	27	8
1946–	Physics Dept., Manchester Univ.	A.C.B. Lovell, R. Hanbury Brown, C. Hazard, M.K. Das Gupta, R.C. Jennison	Jodrell Bank	218 ft dish	72, 158	Gal., sources (mostly meteor radar before 1950)	15	9
AUSTRALIA								
1945–	Radiophysics Lab., CSIRO	J.L. Pawsey, E.G. Bowen, L.L. McCready, 17 others[e]	in and around Sydney	many	18 → 24000	sun, sources, Gal., moon (also lunar radar)	61	7

201

Table 10.1 (cont.)

Dates[b]	Organization	Researchers[e]	Site	Antennas	Frequencies	Programs[c]	Papers[c]	Chapter
CANADA								
1946–	Radio & Elec. Eng. Div., Natl. Res. Council	A.E. Covington	near Ottawa	4 ft dish	2800	sun	7	10.2
USSR								
1947–	Oscillations Lab., Lebedev Inst., Moscow	S.E. Khaykin, V.V. Vitkevich, B.M. Chikhachev	Crimea	many	$\sim$40$\rightarrow\sim$10000	sun, sources	7	10.3.2
1949–	Inst. of Tech. Physics, Gorky Univ.	V.S. Troitsky	Zimenki (on Volga River)	several	$\sim$75$\rightarrow\sim$10000	sun, moon,	1	10.3.3
FRANCE								
1948-54	Inst. d'Astrophysique, Paris	M. Laffineur	Meudon Obs.	Würzburg	250, 550	sun	6	10.4
1948-54	Ecole. Norm. Sup., Paris	Y. Rocard, J.-F. Denisse, J.-L. Steinberg, E.-J. Blum	Marcoussis	Würzburg	160, 1200	sun	25	10.4
NETHERLANDS								
1949–	Astron. Obs., Leiden Univ.	J.H. Oort, H.C. Van de Hulst, C.A. Muller	Kootwijk	Würzburg	1420	21 cm H line	14	16
JAPAN								
1949–	Tokyo Astron. Obs.	T. Hatanaka, F. Moriyama, S. Suzuki	Mitaka	Yagis, dipole array	60, 100, 200	sun	2	10.5
1950–54	Osaka City Univ.	M. Oda, T. Takakura	Osaka	horn, 1 m dish	3260	sun	2	10.5
1951–	Res. Inst. of Atmospherics, Nagoya Univ.	H. Tanaka, T. Kakinuma	Toyokawa	2.5 m dish	3750	sun	1	10.5

[a] This table does not include groups who did only meteor radar work; see Table 11.1 for those. Table includes only those observational groups that maintained a program over at least several years. Six other smaller groups that started in the early 1950s are briefly described in Section 10.6 and two others in the footnotes to Fig. 13.1.
[b] Initial dates refer to the first reported observations; closing dates are given only for those groups ending before 1955.
[c] Researchers and programs refer to the period before 1952. Papers are those published before 1953 on radio astronomy topics; the total for each group is tabulated only as a very rough indicator of quantity of research effort.
[d] B. Elsmore, K.E. Machin, P.A. O'Brien, P.A.G. Scheuer, H.M. Stanier, D.D. Vonberg.
[e] J.G. Bolton, R.N. Bracewell, W.N. Christiansen, J.V. Hindman, F.J. Kerr, F.J. Lehany, B.Y. Mills, H.C. Minnett, R. Payne-Scott, J.H. Piddington, C.A. Shain, O.B. Slee, S.F. Smerd, G.J. Stanley, K.C. Westfold, J.P. Wild, D.E. Yabsley.

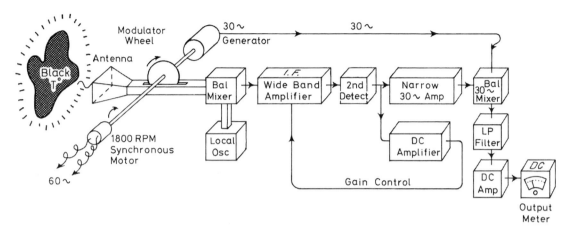

Figure 10.1 Block diagram of the original Dicke radiometer. The horn antenna is depicted looking at a blackbody emitter. The shaped modulator wheel dips in and out of the input waveguide at a 30 Hz rate. The signal is then mixed with a local oscillator, amplified at an intermediate frequency, detected, mixed with a 30 Hz oscillator (synchronized with the wheel) to bring it down to baseband, sent through a low-pass filter, and displayed. (Redrawn from Dicke 1946)

nuclear physics at Rochester University, joined the Rad Lab in 1941. He became a key member of Purcell's Fundamental Development Group (Chapter 16), who acted as pathfinders for the rest of the laboratory, continually developing techniques at higher and higher frequencies. Dicke worked on antennas and on microwave components such as crystal mixers and a "Magic Tee" waveguide junction, amassing nineteen patents before war's end. It was in the process of making measurements of the field pattern of a horn that he became impressed with the merits of the "lock-in" (switching) amplifier and realized that extraordinary sensitivity could be achieved if this were combined with a wide-band receiver. And so in 1944 he designed what soon became known as the Dicke radiometer (Fig. 10.1), the prototype for every microwave receiver in radio astronomy to this day.

Dicke (1946) masterfully introduced his receiver in an article in *Reviews of Scientific Instruments*, as well as setting forth the basic thermodynamic principles of microwave radiometry. The essential innovation was a (Dicke) switch, which caused the input signal to be rapidly alternated (say at 30 Hz) between the desired antenna signal and a reference signal of similar magnitude; Dicke also invented the term *antenna temperature*, although the concept had existed before (see Appendix A for details). Inevitable slight changes in a normal receiver's gain were disastrous for detecting any signal small compared to the receiver's own intrinsic noise; for instance, detecting the moon, which produced an antenna temperature of only ~20 K, required a gain stability of 0.1% from a conventional receiver, extremely difficult to achieve. In contrast, his novel 1.25 cm radiometer readily measured temperature differences as small as 1 K despite only modest gain stability. The reference signal was supplied by thermal emission from a shaped wheel rotating in a slot in the waveguide directly behind the input horn (Fig. 10.1). Another key component was a balanced mixer, which in essence handled its two inputs and one output in a much cleaner fashion than earlier mixers, preventing signals from propagating in undesired directions.[2]

Purcell suggested that Dicke should try some measurements of the earth's atmosphere with his radiometer, for it had been recently found that prototype K-band radars were not performing well on humid days. Although this was known to be due to absorption by water vapor, the magnitude of the effect was alarmingly higher than had been predicted when the frequency for K-band had been chosen.[3] In April 1945 a group led by

---

[2] See Forman (1995:435–41) for an insightful detailed discussion of the Dicke receiver (and especially the later, related lock-in amplifier) in a broader historical context.

[3] This natural feature of the atmosphere's spectrum, which turned out to be centered at 22 GHz (1.35 cm), unhappily

Figure 10.2 Robert Dicke (right) and colleagues (l. to r.) Robert Kyhl, Robert Beringer, and A. B. Vane with their 1.25 cm radiometer on the roof of the MIT Radiation Laboratory in 1945. The horn setup is the one used for measurements of the earth's atmosphere; Dicke is holding a "shaggy dog," a piece of absorbent material used for calibrations.

Dicke therefore conducted measurements in Florida at wavelengths of 1.00, 1.25, and 1.50 cm (Dicke, Beringer, Kyhl, and Vane 1946). Their procedure was to point a small horn first at the zenith and then at a series of angles away from the zenith. Each measured antenna temperature was a combination of the atmospheric temperature and the opacity of the atmosphere to microwaves (Appendix A) – their data indicated that radio waves would be attenuated by as much as a factor of three over a 100 km path on a muggy summer day. But Dicke was also curious to see if his setup (as in Fig. 10.2) could measure noise from "stars and other cosmic matter" at his operating frequency, fully one hundred times higher than Reber's or Jansky's. Any such signal was expected to show up as a relatively constant contribution on top of the atmospheric signal. None, or more exactly less than 20 K, was found.[4]

Back at MIT, Dicke and physicist E. Robert Beringer next made the first measurements anywhere of a radio solar eclipse and of the *absolute* intensity of extraterrestrial noise; this required the superb stability of their radiometer, as well as special attention to calibration of antenna and receiver gains. Using an 18 inch diameter dish mounted on a tripod on a Rad Lab roof, they observed a partial (~60%) solar eclipse that happened their way on the early morning of

---

was not appreciated early in the war by those who selected the various frequency bands for military radar (Purcell 1977:29T, Guerlac 1987:507–24).

[4] No attention was paid to the position of the Milky Way in the sky during Dicke's 1945 search for microwave background noise. This experiment became of particular interest twenty years later when Dicke was involved with the discovery of the 3 K cosmic microwave background. Dicke also made (unpublished) measurements at the Rad Lab in 1945 indicating that any background cosmic noise that might exist did not vary by more than 1 K around the sky (Dicke 1976:3–4T).

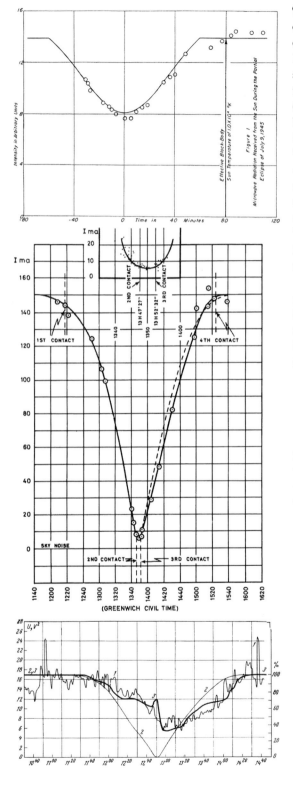

Figure 10.3 Three early radio eclipse curves. The top one was taken by Dicke and Beringer (1946) at 1.25 cm wavelength during a partial solar eclipse of July 1945; open circles are measured points and the solid line is the fraction of the optical disk exposed ("optical curve"). The other two curves are for a total eclipse of May 1947, the middle one at 3.2 cm by Hagen (1949, redrawn), and the bottom one at 1.5 m by Khaykin and Chickhachev (1947). In the middle plot, the symbols are the same as for the top, with the symmetrically-drawn dashed line designed to illustrate the asymmetry of the radio eclipse. In the bottom plot curve 1 shows the radio intensity, curve 2 the optical, and curve 3 is calculated based on exposed hydrogen prominences and filaments. Notice as one goes to longer wavelengths the decreasing depth of the radio curves (relative to their optical counterparts), as well as the increasing asymmetry and irregularity.

9 July 1945. The overall curve of radio intensity very closely followed the fraction of the optical sun being obscured (Fig. 10.3), leading to the conclusion that the 1.25 cm sun was the same size as the optical sun (Dicke and Beringer 1946).[5] They also measured an effective blackbody temperature for the sun of 10,000 K ($\pm$1000 K), but did not comment on why this might be different from the photospheric temperature of 6000 K (see Sections 5.6 and 13.1). Finally, another "first" was their October 1945 detection of the moon, with a reported effective temperature of 292 K. But no more was done other than this single shot at the moon and Dicke (1979:25T) later regretted that he hadn't thought of following the moon through all its phases. This would have beaten by three years Piddington and Minnett's (1949a) fruitful study (Section 12.5.2), which also used a Dicke radiometer at 1.25 cm wavelength.

Dicke took a job as an assistant professor in the physics department at Princeton University in 1946 (and Beringer took a similar position at Yale) and one might have expected that with such a bang-up start he would have continued research into extraterrestrial noise. But in fact no more was done. As he recalls:

> There were two things that made me shy away. First of all, I went to Princeton without a tenured position and I did have a family to worry about.... And also,

---

[5] If anyone had been doubting, these eclipse observations also showed that the detected radio energy, by exhibiting no time lag with respect to the optical eclipse, indeed travelled through space with the speed of light. (Reber and Greenstein 1947:20)

at that time the physics department was interested in cosmic rays and nuclear physics – what would they have thought of an assistant professor flying off in the wild of radio astronomy? That isn't even physics!...Years later, I realized how completely mistaken I was about the department – I hadn't realized that they really were a remarkably tolerant bunch.... But I think I'd still have done it if I'd had any support from the astronomy department. (Dicke 1979:21T; also see Edge and Mulkay [1976:424])

Dicke's last statement refers to contact he made with astronomer John Q. Stewart, who had some interest in meteor radar work (Section 11.3), but reported that no one in the department (then headed by the venerable Henry Norris Russell) was interested in collaboration. Dicke thus eschewed radio astronomy and embarked on a distinguished career in atomic physics, radiation processes, gravity, and cosmology, the last of which led in 1964 to renewed contact with radio astronomy in connection with the discovery of the 3 K microwave background radiation.[6]

10.1.2 US Naval Research Laboratory

The second American postwar group was part of the US Naval Research Laboratory (NRL) in Washington, DC, the flagship American laboratory for naval scientific research since its founding in 1923. During the 1930s NRL had developed a 200 MHz radar that by the time of Pearl Harbor was in service on about twenty ships. Wartime saw a great expansion of the radar work and the design of a multitude of naval radars.[7] Most of this was at frequencies less than 500 MHz, but the Centimeter-Wave Research Branch exploited higher frequencies. It was headed by John P. Hagen (1908–1990), a physicist who had worked on microwaves at NRL since 1935. Among his Branch's projects were a slotted-waveguide radar for submarine periscopes and a ground-speed indicator for aircraft, both at 3 cm wavelength. At war's end this kind of work continued, but now there was also room for some basic research. As Hagen (1976:4,7–8T) recalled:

> We had a large group trained in these centimeter techniques and so we cast about for some useful scientific thing to do and radio astronomy looked like the right thing.... [There was] interest in promoting basic scientific work that might at some time possibly benefit the Navy. A certain portion of the budget was set aside for such research, in which there was no pressure to show a direct use. We were able to convince them that radio astronomy was a good effort to get into, principally because it led to the early development of useful radio techniques.

These techniques involved large dishes, sensitive receivers, and methods of precision measurement. Hagen's group, renamed the Radio-Frequency Research Branch, in fact spent about three-quarters of its effort until 1954 on Navy matters such as developing radar systems and studying atmospheric propagation.

Hagen, who had wanted to become an astronomer even as a physics student in the 1930s, saw the opportunity to combine his group's radar expertise with his own love of astronomy. Over the next decade ten persons participated in the radio astronomy effort at one time or another. The sun was the focus, for it soon became clear to Hagen and his number-two man, Fred T. Haddock (1919–2009), a physicist who had worked under him since 1941, that nothing else was accessible with the available sensitivity of microwave systems. Moreover, the sun was important to the Navy in terms of its effects on the atmosphere and hence radio communications, radar propagation, and missile flight. The first solar observations came in the summer of 1946 using a 3.2 cm receiver and a 10 ft diameter dish. Over the next two years the group put two other similar dishes into service, one with a polar mount and a dual feed able to measure polarization at 3.2 cm, and the other working at the extremely short wavelength of 8.5 mm (35,000 MHz). Sporadic monitoring of the sun at 3.2 cm began in 1947. In July 1948, while using the sun to study the terrestrial atmosphere, the group accidentally picked up a strong burst (at the shortest wavelength at which one had ever

---

[6] This account of Dicke's postwar work is partly informed by Dicke (1976:1–10T, 1979:17–26T). In the mid-1950s Dicke also played an important advisory role in the founding of the US National Radio Astronomy Observatory.

[7] Good historical accounts of the NRL radar work are available in Allison (1981) and Swords (1986:101–112). See Hevly (1987:chapters 1, 2) for NRL's prewar and wartime history in radio communications, ionosphere, radar, and related work. Van Keuren (pp. 137–56 in Blumtritt et al. [1994]) and Brown (1999:64–7) also discuss prewar radar development. Useful photographs and bibliographies can be found in L. A. Gebhard (1979), "Evolution of naval radio-electronics and contributions of the Naval Research Laboratory," NRL Report 8300 (448 pp.).

been observed) and reported it in their first publication (Schulkin *et al.* 1948). Two years of subsequent monitoring of the 3.2 cm "quiet" sun revealed variations in apparent temperature from 10,000 to 16,000 K, the level correlating somewhat with sunspot activity. Lunar measurements at 3.2 cm were also made, but their low accuracy precluded any study of variations with lunar phase. Their 8.5 mm receiver, with its system temperature of 25,000 K, pushed the art of radio electronics to its limit. There were no vacuum tubes for such high frequencies, so the mixer was a silicon point-contact crystal detector and the local oscillator was a special klystron. But its lack of Dicke switching and instability meant that few results from it were ever published. It was noted that 8.5 mm lunar radiation varied neither during a lunar eclipse nor over a lunation. The average temperature of the 8.5 mm sun in early 1949 was reported to be 6700 K ± 10%, with only small variations from day to day; this datum provided the deepest radio penetration yet into the solar atmosphere (Chapter 13). On one occasion in September 1948 the 0.2° beam (the smallest then in existence) even allowed the tracking of a sunspot group across the face of the sun (Hagen 1949:77–89, 1951:551).[8]

The second approach to the sun was to travel to its eclipses, and in this Hagen was encouraged by NRL's Director of Research Edward O. Hulburt, a senior ionospheric theorist (see Section 12.1) with a particular interest in the sun and eclipses[9] (Haddock 1983:116). The first eclipse was that of 20 May 1947 across South America and, as Hagen wrote, the idea was to see exactly where the solar radiation came from:

> The sun's effective temperature [at 3.2 cm wavelength] is higher than the true temperature measured optically and is due to an additional radiation in the radio frequency part of the sun's spectrum.... This additional radiation could originate entirely from the photosphere or it may possibly come from the corona.[10]

Because of poor weather prospects in Brazil and a longer duration of totality at sea, Hagen decided to take advantage of his naval connections. Eclipse day therefore found him and three assistants on the destroyer escort USS *Spangenberg* near the equator in the South Atlantic, about 2500 km northeast of the Brazilian coast. They attached their 8 ft dish to a pedestal stabilized against the roll and pitch of the ship. This was high on a mast, 75 ft above the sea, and technician C. B. Strang was elected to perch there over the critical four hours, continually aligning the 0.8° beam by sighting the sun through an optical telescope. Photographs aligned with the dish were also taken to aid in post facto corrections, but it was cloudy through most of the eclipse and for 30 min the ship's gyroscope stabilization system failed completely (note the gap in data for 1310–1340 UT in Fig. 10.3). But the 3.2 cm Dicke radiometer worked fine and in the end they obtained a decent eclipse curve showing some asymmetry and a dip to only 4% of the sun's uneclipsed radio intensity. They concluded that the 3.2 cm sun was only barely larger than the optical sun.

In order to interpret these observations and in general understand the microwave sun Hagen (1949, 1951) carried out extensive calculations of solar models that then became the heart of his 1949 Ph.D. thesis under Francis J. Heyden, S.J. of the Astronomy Department at nearby Georgetown University. The eclipse data implied a strong brightening in microwaves along the outer 2% of the optical solar limb (Section 13.1).

Convinced of the value of eclipse observations for detailed solar studies, Hagen committed his group to further major expeditions. The next was in September 1950 on the dismal island of Attu at the very end of Alaska's Aleutian chain. On an abandoned air strip

---

[8] Primary source for these pre-1950 results from the NRL group is an unnumbered, undated (ca. 1950) 10-page report entitled "Extraterrestrial radiation research at NRL" by J. P. Hagen and F. T. Haddock (available from RPL). This was apparently written to document all the (mostly otherwise unpublished) 1945–50 NRL work. The 8.5 mm work was also discussed in Hagen's thesis (1949:29–39, 75–89) and in Hagen (1951:548–51).

[9] Hulburt also fostered NRL's postwar pioneering solar ultraviolet and X-ray work (using V-2 rockets) under Richard Tousey and Herbert Friedman. This was the opening to astronomers of yet another portion of the electromagnetic spectrum (Section 18.2.1). For information on this, on NRL's ionosphere research, and on the role of "basic research" at NRL, see Hevly (1987). For the beginnings of US space sciences see DeVorkin (1992).

[10] J. P. Hagen, R. B. Jackson, R. J. McEwan, and C. B. Strang, "Observations on the May 20, 1947 total eclipse of the sun," NRL Reprint (unnumbered) of March 1948, 5 pp.; this is almost identical to an article (classified confidential) in *Reports of NRL Progress* for March 1948. My eclipse account comes from this report and from Hagen (1949:90–113).

eleven men set up four dishes and an 18 ft long optical telescope for coronal photographs, but it all came largely to nought when they were hit on eclipse day with the tail end of a typhoon. They doggedly took data during high winds and driving rain, but the eclipse curves in the end looked ragged and untrustworthy (Mayer 1978:34–5T, Haddock 1983:117). Also along on this expedition was Grote Reber, then at the National Bureau of Standards – his data at the longer wavelength of 65 cm, although not so affected by the weather, suffered from ground reflections. Hagen, Haddock, and Reber (1951) worked up a few results, but only for *Sky and Telescope* magazine and talks to the American Astronomical Society.

There was no rest after Alaska, as the NRL group next prepared for their third eclipse, this one by contrast in the desert sands of Khartoum, Anglo-Egyptian Sudan (Hulburt 1952). They shipped seventy tons of equipment and joined over a hundred astronomers studying the eclipse (Fig. 10.4). Things went much better than in 1950 – aided by local police who stopped all auto, camel, and rail traffic for three hours during the eclipse – and nice 9.4 cm and 8.5 mm eclipse curves were procured. At 8.5 mm the sun fell to only 0.5% of its normal intensity, the closest to a total *radio* eclipse ever observed.[11]

Another major thrust was the design, acquisition, erection, and calibration of a 50 ft paraboloid reflector with a surface sufficiently accurate for operation at microwavelengths (Fig. 10.5). Ever since the end of the war, Hagen and Haddock had wanted a really large dish and by the late 1940s they persuaded the Navy to cover the price tag of $100,000. It had become obvious that with small dishes there was little to measure except the sun's total intensity, whereas a big dish would have sufficient resolution to study details of the solar disk without the bother of going to the ends of the earth, as well as enough sensitivity to pick up some of the new discrete radio sources and (probably) thermal emission from planets. And Haddock had wanted to do lunar radar even during the war (Section 12.2). A contract was let out to Collins Radio Co.[12] for fabrication of the reflector and its installation on the roof of NRL's main administration building.[13] Because an altitude-azimuth mounting was chosen for its simplicity and lower cost, special attention was paid to tracking in equatorial coordinates, which was effected by a unique analogue axis converter. The reflector comprised thirty aluminum sectors bolted together and adjusted to an accuracy of better than ±1 mm. Construction went smoothly over the period 1949–51, but serious troubles surfaced during the break-in phase. To save money the reflector had been placed on a twin five-inch anti-aircraft naval gun mount, but it was soon discovered that the fifteen-ton dish's huge moment of inertia caused unacceptable oscillations whenever the drive system came to a halt. In the end almost two years were spent working out this and other bugs. In the meantime Hagen tried 8.5 mm observations with a beamwidth of only 3′, allowing some mapping of the lunar and solar surfaces, but the data were never satisfactory enough to publish and did not fulfill his hope to see solar limb brightening (Hagen 1976:25T). After these years of frustration, in late 1953 it was decided that longer wavelength observations, where a larger beam would mean that pointing and tracking would not be so critical, would be more profitable. Thus a 9 cm receiver was installed on the dish by Haddock, Cornell H. Mayer (1921–2005), and Russell M. Sloanaker (the latter two also NRL veterans who had worked on radar during the war and on solar eclipses since the late 1940s). The results were spectacular – in just two weeks they picked up a dozen new sources (mostly ionized hydrogen regions such as the Orion nebula), the first discrete sources at such a short wavelength (Haddock, Mayer, and Sloanaker 1954). Moreover, for the first time, fully

---

[11] These results from eclipses are further considered in Section 13.1.3. Also see the comments on eclipse expeditions in Section 17.1.2.

[12] Over these years Collins Radio Co. (in Cedar Rapids, Iowa) was also developing a "radio sextant" for the US Navy, a small dish and microwave radiometer for fixing the sun's or moon's position despite clouds. By 1949 they had developed a solar radio sextant (4 ft dish and 1.9 cm receiver) that could automatically track the sun to within an error of 1′. In that same year D. O. McCoy and Winfield W. Salisbury used such a setup at 1.25 cm to observe two lunar eclipses. See *Sky and Telescope* for January 1950 (p. 50 and cover), as well as "Interim report [to the Office of Naval Research] on the microwave radiometry project of the Collins Radio Company" by C. M. Mepperle et al., August 1949, RPL. See note 17.122 on how this radio sextant concept was adapted to nuclear submarines later in the 1950s.

[13] Although not used since the late 1960s, this dish remains to this date on the same roof, an icon of NRL, and can be easily seen directly across the Potomac River when landing at Reagan National Airport, Washington DC.

Figure 10.4 The NRL eclipse expedition to Khartoum, Sudan in February 1952. The antenna is a 3 × 16 ft parabolic cylinder operating at 8.5 mm wavelength, in front of which stand (r. to l.) John Hagen, R. J. McEwan (a technician), and a native helper.

seven years since the discovery of Cygnus A, one had sources whose basic emission mechanism was *understood*, namely as free–free radiation from gas surrounding hot, young stars (Appendix A). For the 50 ft dish this turned out to be the beginning of a string of first-rate observations of neutral hydrogen clouds, discrete sources, and planets.

The year 1954 was when the NRL group came to full flower, able fully to profit from the decade of groundwork it had built up in microwave techniques. Its work had lagged behind the discoveries by the leading groups in England and Australia, for the lodes to be struck in that first decade of radio astronomy after the war were at *long* wavelengths, where signals from most sources were much stronger (Section 17.3.5). But now the incomparable 50 ft dish, which many foreign radio astronomers viewed as only a costly white elephant ("$100,000 for something that can only see the sun and moon?!"), at last began to pay off scientifically (Haddock 1984:4N). Furthermore, the branch was renamed "Radio Astronomy" and the Navy now allowed the NRL group to devote most of its resources to the science of the subject.[14] Publication

---

[14] Van Keuren (2001) skillfully describes how the NRL radio astronomy branch in 1954 got involved with a project to build a 600 ft diameter steerable dish, eventually sited at Sugar Grove, West Virginia. The project, finally cancelled in 1962 after huge expenditures and overruns, was in fact driven by James Trexler's NRL group (Section 12.6), who convinced the Navy of the value of such a dish for monitoring Soviet radio traffic via lunar reflections. Hagen's radio astronomers were to share time on the dish, but primarily served as a front for the secret eavesdropping project.

Figure 10.5 The NRL 50 ft diameter dish on the roof of the Lab's main administration building in the early 1950s.

10.1.3 Cornell University

A third American group was led by Charles L. Seeger (~1912–2002) and directed by Charles R. Burrows in Cornell University's School of Electrical Engineering, in central New York state. Burrows had spent most of his career at Bell Labs working on radio wave propagation and during the war was in charge of all US propagation research. At Cornell he initiated a large program of research in propagation[18] and as part of this in 1946 secured funding from the Office of Naval Research (ONR) for a "Microwave Astronomy Project" (the term *microwave* was used in anticipation of future observations at higher frequencies, which in fact never materialized). After one year the project had a payroll of five faculty (two fulltime and three consultants), one graduate student, five technicians, and a secretary.[19]

The seeds of the Cornell project extended back to before the war, for at that time Seeger, an ardent radio amateur who repaired radio sets to get through the Depression, learned of Jansky's discovery and was immediately taken:

> To me it seemed perfectly natural that there should be signals there. ... I suppose it was the way I was brought up – my father was a musician,[20] but he was interested in astronomy and he knew [Richard] Tolman and others and when I was a kid he used to tell me about things like the redshift of galaxies. (Seeger 1975:2T)

styles changed, too, for before 1954 the group, focusing on development, cared little to publish in the outside literature (Haddock 1984:5N, Mayer 1971:3T). Whereas before 1954 the *only* substantial NRL paper was Hagen's thesis results in the *Astrophysical Journal* (1951), in 1954 alone four such papers were published. This period, however, is beyond the scope of the present study.[15,16,17]

---

[15] For information about the early NRL years I have profited from interviews with Hagen (1976), Haddock (1984), Mayer (1971, 1979), McClain (1973), and Sloanaker (1973).

[16] A junior member of the NRL group from 1951 to 1955 was Nannielou Hepburn Dieter, one of the first women involved with radio astronomy. She studied astronomy as an undergraduate and obtained a Ph.D. at Harvard doing 21 cm hydrogen line research in B.J. Bok's group. For an autobiography see *Two Paths to Heaven's Gate* by Nan Dieter Conklin (2006, Charlottesville, VA: National Radio Astronomy Observatory).

[17] In 1955 Hagen became head of the Navy's Project Vanguard, the US program to launch an earth satellite that lost the race to the Soviet Union's Sputnik 1 in October 1957.

[18] Prominent among the ionospheric physicists in Burrow's Cornell program were Henry G. Booker and William E. Gordon, both from 1948 onward. Far more than radio astronomy happened under Burrows's grants – for instance, ONR was also amenable to support research on "electrostatic forces in the blood, the coagulation mechanism, and measuring red blood cell counts by the conductivity method."

[19] Burrows, 31 October 1947, "Microwave astronomy: Status report no. 1," 7 pp., Rept. EE2 of the School of Electrical Engineering, Cornell University (available from RPL).

[20] Charles Seeger's father (of the same name) was a noted musicologist at the University of California, Berkeley. Many of his family became professional musicians, including his younger brother, folk singer Pete Seeger. His uncle Alan Seeger was the World War I poet renowned for "I have a rendezvous with death."

Seeger became an engineering physics major at Cornell just before the war (at age 27) and came into contact with theoretical astronomers Ralph E. Williamson (1917–1982) and Donald A. MacRae. The three of them were enthusiastic about the potential of using radio to study the Galaxy, but the war soon intervened. In fact, as a graduate student Williamson had heard Reber's 1939 colloquium at Yerkes Observatory (Section 4.3). After the war Seeger was appointed an assistant professor in the Electrical Engineering Department, but as it turned out, Williamson and MacRae left Cornell in 1946, although the former remained for many years a consultant to the Cornell group from his post at David Dunlop Observatory in Toronto. In 1947 he wrote a particularly nice review of radio astronomy, one of the first (Williamson 1948).[21]

Seeger first built a 205 MHz Dicke radiometer and modified a US Army SCR-268 radar antenna into a single 48-dipole array (Fig. 10.6). Solar monitoring began in July 1948 and continued for many years, but little was done with the data until 1951 when Leif Owren (1954), a Norwegian on leave from the University of Oslo (Section 10.6.3), used it for his Ph.D. research.[22] Despite this lack of attention to the Navy data already in hand, Burrows in 1948 also contracted with the US Air Force to install a battery of solar radio telescopes at a new (optical) solar observatory on Sacramento Peak in New Mexico; first observations were in 1950 and full operation by 1952.

Another early project of Seeger and Williamson's (1951) involved an intensive week of observations in October 1948, followed by over a year of analysis. Their object was to determine the plane of symmetry of the 205 MHz galactic radiation and compare it with the numerous determinations already available from optical studies. To this end they made 320 precisely calibrated cuts across the galactic plane when it was perpendicular to the horizon; in this way data reduction was simplified and the constant-elevation scans were least subject to various instrumental effects. In the end the radio position of the north galactic pole was determined to an accuracy of 0.4° and found to agree well with the optical pole. These data were also used to argue strongly that the galactic noise could not be due to free–free radiation and that it was emanating from a typical distance of about 1 kpc (Chapter 15).

Seeger's third line of research involved the scintillations of Cyg A, which he took to arise in the ionosphere (Seeger 1951). He began observing in the summer of 1949 with a US Navy dish of 17 ft diameter. Cornell press releases often expressed the diameter as "204 inches," a jocular reference to the brand new Palomar 200 inch optical telescope, the world's largest. Its surface was accurate enough to use at a wavelength of 10 cm, it could be rotated around any of four axes, and it cost $30,000. One project planned as early as 1947 for this dish was to search for a predicted 21 cm line of hydrogen, although no record of such has survived (Section 16.2).

This phase of Cornell radio astronomy ended in 1950 when Burrows and Seeger developed irreconcilable differences and Seeger left to start up radio astronomy in Sweden under Olof Rydbeck (Section 10.6.2). After a year he moved on to Leiden Observatory, where he stayed for a decade. Seeger (1975:7–9T) had felt from the beginning that radio work should be part of astronomy, not radio engineering, and this attitude melded well with the Dutch (Chapter 16).

10.2 CANADA

Upon Canada's entry into the war in 1939, the National Research Council (NRC) in Ottawa began developing radar. The Tizard mission also visited NRC in 1940, causing the primary thrust of NRC's radar effort to be in microwaves; by war's end NRC supported a professional staff of sixty. One of the radio physicists in NRC's Radio and Electrical Engineering Division was Arthur E. Covington (1913–2001), who as a youth had had a strong interest in both amateur astronomy and radio and whose graduate studies in physics at the University of California at Berkeley had been curtailed by the war in 1942.

After the war NRC switched over to civilian pursuits and Covington, remembering a chance encounter with a Reber article before the war, decided to study Milky Way radiation at much shorter wavelengths.

---

[21] Another astronomer associated with the group was Martha E. Stahr (later Martha S. Carpenter), whose bibliographic work (indispensible to the research for this book – see Appendix C) was also supported by ONR.

[22] Owren's was the third Ph.D. thesis in the US to be awarded on the basis of radio astronomical work; the first two were by J. P. Hagen (1949) (Section 10.1.2) and H. I. Ewen (1951) (Section 16.3).

Figure 10.6 The 48-dipole, 205 MHz array at Cornell University's field station at an airport about five miles from campus (ca. 1949). The antenna was a modified Army radar. (Seeger and Williamson 1951)

During the war Covington had already constructed a Dicke switch at 10.7 cm (2800 MHz, or S band), at which many radar components were available, and so he started there. By the spring of 1946 he had an operating receiver connected to a small horn (30° beam) and put emphasis on securing a good calibration, a theme that was to permeate his career.

While trying to measure a background level for the zenith sky, Covington (1947a) was surprised now and then to detect bursts of "microwave sky noise." He switched to a 4 ft dish, but the bursts persisted and he ended up monitoring them for over a year. In the end he concluded that they were real and often correlated with geomagnetic activity such as aurorae (Covington 1950).[23] Concerned to separate variations in sky noise from those in solar noise, as well as to respond to the minority who felt that the claimed solar bursts actually originated in the ionosphere, Covington (1948:455, 1976:18T) used one antenna pointed at the sun and a second at the zenith as controls for each other. He explained the 10.7 cm sky bursts as resulting from heating and cooling of the upper atmosphere, whose undisturbed antenna temperature he measured as ~50 K – although this last figure was "only a first approximation, since a measurement so close to absolute zero is subject to considerable error" (Covington 1950:33).

Covington's primary task, to confirm Reber's galactic radiation, was foiled, however, as he measured no 10.7 cm signal (< 5 K in antenna temperature) from the center of the Galaxy, nor any from the moon, Mars or Jupiter. On the other hand, the sun, which he was at first using only as a handy reference source, gave a signal much higher than expected. After noticing a naked-eye sunspot, simple measurements of solar radiation in July 1946 showed that the microwave sun had not only a base level, but also enhanced emission apparently associated with sunspots. Covington next turned his equatorially-mounted 4 ft dish and receiver (Fig. 10.7), mostly built from scavenged radar parts, to a partial solar eclipse that fortuitously fell across his field station (just south of Ottawa) on 23 November 1946. Through a careful comparison of his radio eclipse curve with optical photographs taken at the nearby Dominion Observatory, Covington (1947b) found that when a major sunspot group was covered by the moon, the overall radio signal noticeably declined. Despite his dish's 6° beam, he thus effectively attained a resolution of order 1–2′ and established that the solar active region had a 10.7 cm brightness temperature of 1.5×10^6 K. In addition a region located just off the solar limb was at 3×10^6 K even though it showed no particular optical activity. These regions were in contrast to an average solar value of 60,000 K, three times George Southworth's (1945) value measured four years earlier (Section 5.6). It was not clear if this difference was a serious calibration

---

[23] The nature of these nonsolar microwave bursts remains unknown today, as no one appears to have carried out extensive follow-up observations. See also Section 14.2 for discussion of low-frequency bursts detected by Ryle's group in 1950.

Figure 10.7 Arthur Covington's field station (1950) at Goth Hill, south of Ottawa. Left to right are a 4 ft dish (10.7 cm wavelength), Covington standing at a wideband (10–30 cm) receiver and horn, and a Yagi (1.5 m) for monitoring solar bursts.

problem, or an effect of the approaching maximum in the eleven-year solar cycle.

Through conversations with solar astronomers and with NRC colleagues, Covington (1976:22–3T) also became intrigued with the puzzle of how the solar cycle controls ionospheric properties. Although the number of sunspots and prominences dramatically rose and fell every cycle, calculations indicated these to have little effect on the total power output of the sun. It was therefore not clear how the sun exerted such a strong influence on the earth's ionosphere that it could, for instance, cause the electron density of the F region (Appendix A) to vary by a factor of five or ten over a cycle. Most guessed that it had something to do with the unmeasured solar ultraviolet radiation. This problem, plus a general interest in the radio sun, led Covington in February 1947 to start a daily solar patrol that in fact continues to this day (see Figs. 13.1 and 13.2). The longevity of this patrol can be ascribed to the happenstance that a wavelength of about 10 cm turned out ideal as an index of solar activity, although its original choice was dictated strictly by radar technology. Indeed, after only six months Covington (1948) could see his 10.7 cm data points rise and fall in perfect synchronism with the sunspots carried along by solar rotation (Section 13.2.3).

All of Covington's contributions in the postwar years stemmed from measurements with nothing more than a 4 ft dish, but by 1950 he and his group of five began to see that something larger was needed to stay competitive. Right after the war there had been design studies for a 30 ft microwave dish (which was killed once the cost reached $40,000), and in 1949 Covington began to think even of a 150 ft microwave dish once he learned of Reber's plans for a giant dish (Section 4.6.1). But financial constraints, largely a result of the Korean War, meant that such a gigantic project could not be funded. Visions were pared back, first to a 24 × 150 ft parabolic cylinder, and finally to a 1.5 × 150 ft V-shaped trough with a slotted waveguide line feed (a technology well-known from NRC radar work). This last went into operation in late 1951 and provided the highest solar detail then available (about 8′) without an eclipse or an interferometer.[24,25]

---

[24] Primary sources for the early years of Covington's work are Covington (1967:314–20, 1976:1–26T, 1983, 1984:317–26, 1988). Useful background can be found in *Radar Development in Canada: the Radio Branch of the National Research Council of Canada 1939–1946* by W. E. K. Middleton (Waterloo: Wilfrid Laurier University Press, 1981). Jarrell (1997) gives many details of the institutional development of Canadian radio astronomy over the period 1945–65.

[25] See Section 11.4 for the start of meteor radar research (in 1946) at NRC under Peter M. Millman and Donald R. W. McKinley.

10.3 SOVIET UNION

10.3.1 Ginzburg and Shklovsky

The contributions of Soviet researchers to radio astronomy were far more in theory than in observation. This was true from the start in 1946, with seminal papers on solar radio emission from the two men who were to dominate the Soviet theoretical scene for decades.

The first was Vitaly L. Ginzburg (Fig. 10.8). His intellectual gifts were such that he had gained both a Ph.D. and a D.Sc. from Moscow State University by age 26, one of the youngest doctors of science in the entire nation. During World War II he turned from elementary particles and quantum electrodynamics to the more urgent concerns of ionospheric radio wave propagation. This began his long association with radio physics and led in 1945 to a Chair of Radio-wave Propagation at Gorky State University, as well as to his 1949 monograph *Theory of Radio-wave Propagation in the Ionosphere*. Most of his work in radio physics was concentrated during regular periods every month in Gorky (where his wife was politically exiled), while in Moscow at the P. N. Lebedev Physics Institute (FIAN) he made important contributions to an amazing variety of fields including solid-state physics, superconductivity (for which he shared the 2003 Nobel Prize in Physics), plasma physics, optics, general relativity, and the development of the Soviet hydrogen bomb (Ginzburg 1990).[26] In short, it should be recognized that Ginzburg's work relating to radio astronomy and cosmic rays was only about 15% of his entire corpus (by count of papers through 1952).

In late 1945 Nikolai D. Papaleksi (1880–1947) suggested to Ginzburg that he work out the feasibility of bouncing radio waves off the sun. This was an extension of earlier calculations on lunar radar (Section 12.2) by Papaleksi and Leonid I. Mandel'shtam, close colleagues and the leading Soviet radio physicists of their day. Over forty years Papaleksi had worked on development of radio tubes, frequency stabilization, nonlinear oscillation theory, and radio for navigation and geodesy.[27]

Ginzburg soon found that the solar atmosphere was really just a huge ionosphere and amenable to a similar theoretical approach.[28] In a paper submitted in March 1946 to the Soviet Academy of Sciences (Ginzburg 1946) he showed how waves longer than a meter would indeed be reflected by the solar corona (Section 12.2). But the more important consequence of his calculations was that metric and centimetric radio waves would in fact be absorbed by the corona, which then implied (from Kirchhoff's radiation law) that the corona must also *emit* these same waves. Ginzburg worked out the equation of transfer through the corona for various frequencies and produced much of the basic model that would be used over following years (Section 13.1). He then discussed two possible values for the coronal temperature: 6000 K, the same as the surface (photosphere), or 600,000 K, which came from what he called the "conjecture" or "hypothesis" recently put forward to explain the corona's high ionization levels (Section 13.3.1). Ginzburg's stance was that radio measurements could now test the hot-corona idea by making a "direct determination" of the coronal temperature. But in the end he found the existing radio data insufficient (in the 1944–45 papers of Southworth, Reber, and Appleton), and concluded that measurements at wavelengths longer than Reber's 1.9 m were needed to determine how hot the corona really was.

At about this same time, a few kilometers away, independently and published a few months later, Iosif S. Shklovsky (1916–1985) (Fig. 10.8) carried out his own study of solar radio emission. Shklovsky was born in the Ukraine, but spent much of his youth in the western Soviet Union where his step-father was an engineer on the Trans-Siberian Railroad. After helping build railroads for a couple of years, he studied physics in Vladivostok and at Moscow State University. He switched to astrophysics for postgraduate work and was awarded a Ph.D. in 1944 despite the university being forced to relocate in central Asia. His dissertation, "The

---

[26] Ginzburg (1990) has written a frank, post-*glasnost* autobiographical piece, especially nice to read in conjunction with Shklovsky (1991). Two other rambling, fascinating collections of articles by Ginzburg – covering his life, career, colleagues and Soviet times – are *The Physics of a Lifetime* (2001, Berlin: Springer Verlag) and *About Science, Myself and Others* (2005, Bristol: Institute of Physics Publishing). Article 18 (pp. 357–67) in the latter contains many personal details of Ginzburg's involvement with hydrogen-bomb research and of political persecution suffered by him and his wife.

[27] On Papaleksi see Salomonovich (1981) and V. J. Frenkel's article in the *Dictionary for Scientific Biography*, Vol. 18, pp. 705–7 (1990), ed. F. L. Holmes (New York: Scribners).

[28] Ginzburg (1976:5–7T, 1984:290).

Figure 10.8 Vitaly Ginzburg (ca. 1975) (l.) and Iosif Shklovsky (ca. 1960).

concept of electron temperature in astrophysics," largely dealt with the solar atmosphere, which remained his primary interest right through the appearance of his monograph *The Solar Corona* (Shklovsky 1951a). Throughout these years Shklovsky was a member of the Sternberg Astronomical Institute at the University, where he soon developed a reputation for working on unconventional topics. Shklovsky (1979:125B) described himself as very happy in his role of *wunderkind*: it was "as if I had open eyes and others were blind." He later wrote:

> As a scientist I was a strange mixture of artist and conquistador, a phenomenon possible only in eras when established conceptions are breaking down and new ones arising. That style of work is impossible today; now Voltaire's dictum that "God is always for the big battalions" is rigorously observed. (Shklovsky 1991:90)[29]

Always fascinated by the latest discoveries, he had been initially attracted to the solar corona by ideas from the West (which he readily adopted) that it had an incredible temperature of at least 300,000 K (Section 13.3.1). When at a lecture in mid-1946 he learned that the British and Americans had discovered radio emission from the sun and Galaxy, he sprang into this new field also and quickly produced one of its important early papers. It was published in *Astronomichesky Zhurnal* in late 1946, with a summary version in *Nature* the following year.

Shklovsky worked out the expected thermal or "free–free" radiation from not only the solar corona, but also from the ionized portion of the interstellar gas. But he made a basic mistake in modifying the already correct formulation of Henyey and Keenan (1940), with the result that all his optical depths were much too high.[30] For the case of the Galaxy this meant that, unlike all other early theoretical studies (Chapter 15), he obtained an expected intensity (for a 10,000 K gas) too *large* when compared to Reber's 1944 map. To "save

---

[29] *Five Billion Vodka Bottles to the Moon* (Shklovsky 1991) is a delightful anecdotal tour of Shklovsky's life and times, comparable in its insight and humor to Richard Feynman's *Surely You're Joking, Mr. Feynman!* (1985, New York: W. W. Norton).

[30] See Tangent 10.1 for details of Shklovsky's calculations.

the phenomenon," he suggested that perhaps the fraction of space occupied by ionized gas was a good bit smaller than the 10% value he had adopted.

For the solar emission Shklovsky argued for both a nonequilibrium component and a thermal component, the latter related to Reber's and Southworth's observations. Despite his error in the physics, he reached the correct conclusion that Southworth's microwave results referred to the chromosphere and Reber's meter-wave data to the corona, chiding both authors for talking about the photosphere's 6000 K temperature as relevant: "to connect solar radio emission with the photosphere makes as much sense as to compare its visible radiation with the inner regions of the sun" (Shklovsky 1946:337). But a combination of inaccuracies in the reported data and in Shklovsky's calculations meant that he derived an intensity about one hundred times greater than observed by Reber. With typical resourcefulness, however, he reconciled matters by pointing out that the optical evidence implying a hot corona actually only applied to the inner corona (out to 1.3 solar radii) and that the outermost corona could well be very cold. Defending an electron temperature there of only ~3500 K, he wrote:

> This supposition seems to be absurd at first, but remember that only a few years ago one could not have imagined that the electron temperature of the inner corona is of the order of hundreds of thousands of degrees. (Shklovsky 1946:339)

In the final section of his paper Shklovsky tackled the enhanced solar radiation reported by Appleton (1945a) (Section 5.5). Appleton had found that the sun had sudden bursts in the 10–40 MHz range that were as much as 10^4 times the intensity expected from the entire sun if thermally emitting at 6000 K. Shklovsky pointed out that the enhancement was probably much worse (more than 10^6) since the radiation probably originated in a large solar active region, which was likely to cover at most 1% of the entire sun. Then arguing that it was "physically unacceptable" to have an active region with temperature high enough to account for the sporadic emission, he proposed a nonthermal solution that he had previously developed for his earlier coronal studies. Plasma oscillations, longitudinal waves that physicists had shown could propagate at frequencies related to a plasma's critical frequency (Appendix A), would have roughly the observed radio frequencies. The idea was somehow to cause electrons in the solar atmosphere to oscillate in unison and thus efficiently radiate, but what could set them going? Shklovsky proposed that, just as supersonic motion in a gas causes shock waves, rapidly streaming (5000 km s^{-1}) charged particles ejected by a solar active area could set off plasma oscillations. He admitted that both the theory and his knowledge of it were uncertain,[31] and indeed later observations did not agree well with his predictions. In the early 1950s, however, plasma oscillations were again seriously discussed as the main agent for radio bursts, eventually becoming accepted in that role (Section 13.2.2).

Although these first papers of Ginzburg and Shklovsky had their problems, on the whole they were on the right track and they exemplify the kind of fruitful theory and interpretation that the two men were to produce over and over. By 1952 each had written about a dozen papers on radio astronomy, compared to three Soviet papers *in toto* reporting radio observations. Soviet theory launched in a splendid manner, but neither man succeeded during this period in instigating significant experimental work. Shklovsky (1982:15) in particular made efforts and often ended his papers exhorting Soviet radio engineers to get going. Ginzburg (1976:37T) also recalls:

> I was always in a very difficult position because I didn't observe myself. So I always had to persuade somebody to do it,...but to persuade anybody to do anything is very hard.

Shklovsky wrote pioneering papers on possible radio spectral lines (Section 16.2) and on radio source emission mechanisms (Chapter 15), while Ginzburg's most important work was in applying the theory of "magnetobremsstrahlung" (synchrotron emission) to the galactic background radiation (Section 15.4). They both also wrote numerous reviews of the field (Ginzburg's in 1947 and 1948 were among the earliest anywhere), and Shklovsky's (1953a) book *Radio Astronomy: a Popular Sketch* followed Lovell and Clegg's popular book by only a year. The irony is that their work, although totally based on Western data, had no influence in the West until the mid-1950s (Section 10.3.4).[32]

---

[31] Shklovsky (1982:9) later marvelled at his audacity in making the suggestion in 1946 of plasma oscillations as the origin of solar radio bursts, noting that "too much erudition can fetter scientific research."

[32] Information on Ginzburg and Shklovsky comes largely from 1976 and 1979 interviews, an obituary of Shklovsky (*Sov.*

10.3.2 Lebedev Institute observations

The first Soviet radio observations took place during the solar eclipse of 20 May 1947 in Brazil, the moon's umbra covering their antenna about one hour before it did Hagen's out at sea on a US Navy ship (Section 10.1.2). Papaleksi, who had previously organized ionospheric radio observations at the time of eclipses, was excited by the calculations of Ginzburg and Shklovsky indicating that the meter-wavelength sun was much larger than the optical sun. Since the angular resolution of feasible antennas was woefully inadequate to establish whether or not metric radiation came from the outer corona, the idea was simply to measure how much radio intensity remained despite the moon blocking the *visible* sun. This radio experiment was part of a large Soviet eclipse expedition sponsored by the Academy of Sciences, and as Ginzburg (1976:10–11T) remarks: "In Russia, you know, if a decision is made on the highest level, everything afterwards works." Papaleksi, however, suddenly died in February, and so the radio experiment was taken over by Semen E. Khaykin (1901–1968), a fellow FIAN radio physicist whose work before had been in nonlinear oscillation theory and radio properties of materials and during the war in radar development. He was assisted by Boris M. Chikhachev (1910–1971) (Fig. 10.9), an older postgraduate student of Papaleksi's who had been one of the leading engineers in the Soviet vacuum tube industry.

The expedition was given use of the freighter *Griboyedov* (Fig. 10.10), but their departure from Latvia was considerably delayed by unusual late-season ice – in the end only an icebreaker got them out. Later they hit a storm severe enough to force them to shelter in England. But in the end they made it to the Bay of Baía, just west of Salvador, Brazil. The bulk of the expedition, which included Ginzburg and Shklovsky, planned to do optical and ionospheric observations well inland, but the former were a complete loss due to rainy skies. The radio astronomical observations, in contrast, took place on board the ship. The idea was to use a "mattress-spring" array of 96 dipoles (a Lend-Lease American radar) operating at 200 MHz. But how to steer this large ~8 × 10 m array? This was accomplished in the first instance by mounting the antenna in a cargo "hoist cradle" that could be tilted up and down by windlasses. But to move it in azimuth the solution was to rotate the entire ship! Although they selected a site in the bay protected from winds, problems still cropped up with the required maneuvering because of strong currents and a sea bottom ill-suited for anchoring. Nevertheless, by raising and lowering special anchors and by adjusting hawsers thrown to shore, they were able to track the sun sufficiently well. Their chart recorders, however, had proven so unreliable that data were recorded by men writing down meter readings every minute.

Six months later Khaykin and Chikhachev (1947, 1948) published these observations, and the final eclipse curve turned out to be worth all of the effort, for it clearly showed that the metric-wavelength radiation came from a region much larger than the optical sun, the first direct demonstration of this fact (Fig. 10.3). Whereas the optical sun had been completely occulted, the radio emission fell to only 40% of its uneclipsed value, implying that the radio sun's radius was about 35% larger than the optical's. An asymmetry in the eclipse curve also indicated that a uniform radio disk was not a good model for the radio emission. The asymmetry was best explained by supposing that enhanced radiation originated in prominences and filaments high in the solar atmosphere (curve 3 in Fig. 10.3).[33]

Despite the impetus that might have been expected from this notable achievement, further radio observations over the following five years were meager. Given the economic realities of recovering from the war (observatories near Leningrad [St. Petersburg] and in

---

Astron. **29**, 364–5 [1985]), and articles about Ginzburg's career (Ginzburg 1984, 1985, 1990; *Sov. Phys. Uspekhi* **9**, 782–4 [1967] and **19**, 872–7 [1976]). Also see "The Shklovskii Phenomenon" by N. S. Kardashev and L. S. Marochnik (1992), pp. 7–23 in *Astrophysics on the Threshold of the 21st Century*, (ed. N. S. Kardashev) (Philadelphia: Gordon & Breach). Shklovsky (1982) has also written an informative and wonderfully candid booklet (63 pp.) containing his recollections of Soviet radio astronomy. I am deeply grateful to Julian Barbour for translating this booklet into English for me; it should be published in the West. Another 275-page volume available only in Russian, *Outline of the History of Radio Astronomy in the USSR* (Salomonovich 1985), contains a collection of 13 articles on the development of all Soviet groups.

---

[33] Information on the Soviet eclipse expedition comes from Ginzburg (1976:8–12T, 1984:291–2), Salomonovich (1984:270–3, 1981:627–8), Khaykin and Chikhachev (1947, 1948), and Shklovsky (1991:89–103) (a great story!).

Figure 10.9 (l. to r.) Boris Chikhachev, Martin Ryle, and Viktor Vitkevich at an IAU Symposium at Jodrell Bank in 1955.

the Crimea, for instance, were completely destroyed), as well as the developing Cold War, there was little money for peripheral sciences such as astronomy. Upon his return to FIAN in Moscow Khaykin determined to start up a program in radio astronomy, but the only viable way for many years was as a secondary part of more practical investigations, so-called "applied problems" (Salomonovich 1981:629, Kaydanovsky 1980:1N, Strelnitsky 1995). Since monitoring the signal from a strong radio source was a good way to study how the earth's atmosphere affected radio waves, programs on radio propagation often allowed some radio astronomy on the side. It is not surprising, then, that most Soviet results in the decade following the war dealt with the sun and the strongest discrete sources. For example, Khaykin devoted a solid year to a project important to the military, namely measuring radio atmospheric refraction using primarily solar signals.

Over the period 1948–53 Khaykin built up a group at FIAN's Oscillations Laboratory – its field station was on the Crimean coast of the Black Sea, in particular at Alupka, Alushta, and on Mt. Koshka above Simeiz. The first two sites had already been in use by Papaleksi for his work on applications of radio interferometry and the third was home to the (optical) Crimean Astrophysical Observatory. They used a wide variety of antennas and receivers at frequencies ranging from 40 to 10,000 MHz, mostly in the sea-cliff interferometer mode (Section 7.3.2). Notable among the early experimenters were Alexander E. Salomonovich (1916–1989) (head of the station after obtaining his Ph.D. in radio physics in 1949), Pavel D. Kalachev (a mechanical engineer who designed many of the large antennas), Naum L. Kaydanovsky (a radio engineer who had worked under Mandel'shtam and Papaleksi for over a decade), and Viktor V. Vitkevich (1917–1972) (Fig. 10.9). Vitkevich had been a student of Khaykin's and a radio engineer in the Soviet Navy for six years before joining the FIAN group in 1948. A strong-willed man, he soon distinguished himself as its leader, in particular by inventing and implementing new observing methods.

Their first antenna, installed in the Crimea in 1948–49, was the very same 200 MHz dipole array that had been used on the *Griboyedov*. Most of the other early antennas were also converted military radars, including one of the ubiquitous German Würzburg dishes (see Tangent 4.2). It was assembled from scraps located all over the country, but with an all-new solid-aluminum surface that allowed operation at wavelengths as short as 10 cm. But published radio astronomy results were few. Chikhachev (1950) obtained his Ph.D. based on observations of solar activity, but did not otherwise publish his results until 1956. Vitkevich published several papers in the early 1950s, mostly on interferometric techniques. The most significant was a 1951 article in which he suggested observing the radio source Taurus A (Section 7.4) in June when the outer solar corona by chance passes in front of it. Vitkevich showed that by tracking Tau A during the week in which it passed close to the sun one could gain information about many properties of

Soviet Union 219

Figure 10.10 The cargo ship *Griboyedov* and the "mattress spring" dipole array on its deck for the Soviet eclipse expedition of May 1947.

the outer solar corona: refractive index, electron temperature and density, magnetic field, etc. One practical problem foreseen was that the sun's emission was normally much stronger than that of Tau A, but Vitkevich's neat solution was to observe with an intereferometer whose lobe spacing would be such that the solar signal, being from an extended region, would be minimized while that from small-sized Tau A would remain. He even thought it possible that radio waves from Tau A might be observed after having *reflected* off the plasma of the corona. Vitkevich (1955) first tried this experiment in June 1951, but intense and variable signals from small solar regions were impossible to disentangle from Tau A. In 1952 he again had bad luck when torrential rains and thunderstorms caused enough uncertainty in his measurements that he never presented them. Finally, in June 1953 persistence paid off as all went well with 3.5 and 6m wavelength observations using large arrays of dipoles. These revealed the corona to exert an appreciable effect on radio waves as far out as *fifteen* solar radii – so much farther out than expected that Vitkevich called it the "supercorona" and it became the object of several more such experiments in succeeding Junes.[34,35]

---

[34] As it turned out, the main effect during the occultation of Tau A by the solar corona was that the apparent angular dimensions of the source widened as its radiation passed through denser portions of the corona and underwent scattering off density irregularities. Also note that a parallel, independent set of experiments along this line was being carried out in Cambridge by Kenneth Machin and Graham Smith (1951, 1952). In their case they were unsuccessful during the 1950 and 1951 occultations (due to solar activity), but succeeded in 1952 with results similar to Vitkevich's.

[35] Information on the beginnings of radio astronomy at FIAN comes primarily from Salomonovich (1981:628–30, 1984:272–7), Kaydanovsky (1980:1–2N), and obituaries of Khaykin (*Sov. Phys. Uspekhi* **12**, 149–52 [1970]) and Vitkevich (*Soviet Astron. – AJ* **16**, 562–3 [1972]).

10.3.3 Gorky

The other active early Soviet group was at Gorky State University. As we have seen, Ginzburg headed the Department of Radio-wave Propagation there, but the experimental work was in the Institute of Technical Physics under the general leadership of Gabriel S. Gorelik. In 1947–48 Vsevolod S. Troitsky (1913–1996) built the first of a series of Dicke-switched radiometers (beam switched) at 4 m wavelength, and as the years passed he moved to wavelengths ever shorter, reaching 3.2 cm by 1952. As with the FIAN group in Moscow, Troitsky (1976:9T) and his colleagues were strongly influenced and encouraged by Khaykin, and atmospheric propagation studies had priority over radio astronomy. Troitsky, who obtained his radio physics Ph.D. in 1949 on the theory and practice of techniques for solar observations, set up a field station in 1949 at Zimenki, 40 km outside the city on the banks of the Volga, and by 1952 headed a group of about a dozen workers. By that same year microwave observations of the moon and discrete sources were also underway, but no reports of these emerged from Gorky until the mid-1950s. The group was concerned with developing accurate radiometry at the shortest wavelengths obtainable. In this regard the Gorky group presents a similar case to that of NRL in the United States: both situations led to few astronomical publications until the mid-50s, but then paid off when microwave technology finally became good enough to produce workable results on the weak signals the sky offers at these wavelengths (see Section 17.3.5).[36]

10.3.4 Overview

In a lecture just one month before his death Papaleksi said:

> There is every ground for thinking that the application of radio methods for astronomy opens a new era, which we can compare in its significance with the discovery of the Fraunhofer lines and the application of spectroscopy in astrophysics.
>
> (quoted in Salomonovich [1981:630–1])

And yet, despite this enthusiasm and that of his successor Khaykin and despite the technical leadership from Vitkevich and Troitsky, observations from Soviet radio astronomers contributed little to the development of worldwide radio astronomy in the decade following the war. Because governmental policy was such that almost all available support for radio physics came from agencies interested in practical information, there was little opportunity for experimenters to produce results of a more scientific nature. Strong ideological control of science during the Stalin era also meant that experimental research was often discouraged in favor of theoretical work, which it was thought could be more easily shaped. For example, the theory of synchrotron radiation as an explanation for the radio galactic background was extensively developed in the Soviet Union (Section 15.4.2). Moreover, in an economy still recovering from the devastation of war, theorists had the advantage of needing little more support than pencil and paper.[37] Men like Ginzburg and Shklovsky were thus hindered less by the Soviet system[38] and able to reach the ranks of world leaders in radio astrophysics.

The isolation between East and West science during the Cold War was marked. It led to many duplications and parallel developments out of ignorance: for example, Vitkevich and R. L. Sorochenko published an account in 1953 of their idea of a grating array of many individual antennas, whereas they could have seen it in action the year before if they had attended the Sydney URSI meeting and toured Christiansen's array.

---

[36] Sources for the early years of the Gorky group include Troitsky (1976:1–6T) and Salomonovich (1984:272–7).

[37] Many characteristics of Soviet science, including its much greater strength in theory than in experiment, are analyzed by T. Gustafson in pp. 31–67 of *The Social Context of Soviet Science* (1980), (eds. L. L. Lubrano and S. G. Solomon) (Boulder: Westview Press). Also see L. R. Graham (1993), *Science in Russia and the Soviet Union* (Cambridge: Cambridge University Press).

[38] Although a theorist's lot may have been easier than an experimenter's in the postwar decade, Ginzburg and Shklovsky, both Jews, frequently suffered from Stalin's anti-Semitic campaigns (Ginzburg 1990, Shklovsky 1991; also see the two Ginzburg books cited in note 26 and the Shklovsky story related in Section 15.4.2). Khaykin apparently also suffered from anti-Semitism, which led in 1953 to his leaving FIAN to set up a radio astronomy department at Pulkovo Observatory near Leningrad (V. Abalakin [1993], "On some aspects of the history of Russian astronomy"; abstract in *Bull AAS* **25**, 1335).

Or again, Shklovsky (1954:486) complained mightily about Walter Baade and Rudolph Minkowski delaying formal publication of their results on identifications of radio sources, as well as not acknowledging his published ideas:

> Soviet astronomers always attentively and conscientiously study the investigations of their foreign colleagues. We rejoice in each new exploit of the science, whether from West or East. We hope that an atmosphere of mutual respect and understanding will be established between Soviet and American astronomers.

As soon as he learned of Shklovsky's article, Minkowski wrote that it was all unintentional and due to delays of 1–2 years in learning of Soviet work.[39]

The mutual isolation worked greatly to the disadvantage of Soviet observers, for they would have greatly profited from contact with Western groups. On the other hand, if Western radio astronomers had been exposed to the ideas of Shklovsky and Ginzburg, they too would have been notably enriched. Publication of Soviet results, when it did happen, was only in Soviet journals, in Russian, and little read in the West;[40] and only after Sputnik in 1957 did the West worry about translating these journals. Although the major Western journals were available in the Soviet Union during this period, not until about 1953 were Western articles and books on radio astronomy translated into Russian. Personal contacts, too, were absolutely nil before 1955, which was the first time for any significant Soviet participation in radio astronomy matters in either the IAU or URSI (Fig. 10.9). In sum, the Soviet and Western stories of radio astronomy for the 1945–55 decade developed in separate spheres, although each side would have benefited from interaction.

10.4 FRANCE

The French situation in radio technology after five years of occupation was hardly favorable economically or technically when compared to the English-speaking world, but neither was it barren. Radar development had been undertaken in France before its fall in 1940 and some radio physicists and engineers even managed to carry out clandestine research afterwards (Swords 1986:120–6).

Two groups made radio observations in France during the late 1940s. The primary one was started by Yves Rocard (1903–1992), an electronics expert and head of the physics laboratory at the Ecole Normale Supérieure (ENS) in Paris, France's leading scientific and engineering institute. Rocard had been a particularly effective spy on German radars in the Resistance and later was an officer in the Free French Navy. His laboratory specialized in those areas of physics that could particularly profit from electronic instrumentation and thus in 1946 he started a small effort to study solar noise, with a particular interest in the connection between the radio sun and the ionosphere. The initial team of researchers comprised physicists Jean-François Denisse and Jean-Louis Steinberg (Fig. 7.3), who had been working together on plasma physics. During the war Denisse had been teaching in Africa, while Steinberg had been in the French underground and had survived a year in concentration camps. They were joined in 1949 by Emile-Jacques Blum, a radio signal officer during the war who afterwards obtained his physics degree and worked in electronics research for the Navy. Denisse brought to the group his expertise in physical interpretation of the radio emission, and Steinberg and Blum theirs in building equipment and carrying out experiments. These skills were well reflected in their respective Ph.D. theses for degrees awarded in 1949, 1950, and 1952.[41]

At war's end Rocard was in charge of the disposition of German scientific facilities and personnel encountered in the French zone of occupation. His excellent connections with the Navy meant a ready availability of captured German and surplus American radar equipment, particularly valuable since French industry could supply little of use to them. Among

---

[39] Minkowski:Shklovsky, 17 February 1955, boxes 4–5, MIN.

[40] The only exception to publishing in Soviet journals before 1955 was a 1947 note in *Nature* by Shklovsky. Yet even for those Western scientists who knew Russian, the journals were often unavailable. As one consequence, for instance, there are many issues from the postwar decade of *Uspekhi Fizicheskikh Nauk* (Achievements of Physical Science), one of the premier Soviet journals of physics, that are not held by any of the major American or British libraries.

[41] Denisse's (1949b) thesis was published in two parts (1950a, b) and also translated into English by Haddock and Sloanaker (NRL Translation No. 257, NRL Library, Washington, DC). Steinberg's (1950) thesis was presented in a series of three articles in *L'Onde Électrique* (Steinberg 1952).

these was one of the German Army's giant Würzburg dishes[42] acquired with the aid of Edward Appleton, who as a favor diverted its shipping destination from Britain to Paris. In 1947 this 7.5 m diameter reflector was set up 25 km south of Paris at Marcoussis on the site of the French Navy's Research Laboratory (Laboratoire du Service des Études de la Marine), which had been recently augmented not only by German equipment, but also by twenty indentured German engineers and scientists. The ENS group profited from this expertise, in particular from the antenna engineering of Siegfried Zisler. They began monitoring the sun at 158 MHz in June 1948, but this was primarily for the Navy's Ionospheric Prediction Service and little of it was ever analyzed for its astronomy. The first significant results came from a partial solar eclipse observed in April 1949 both with the dish at Marcoussis and a smaller 3 m Würzburg dish on ENS's rooftop for which Steinberg had built a 1200 MHz receiver. These observations were combined with Marius Laffineur's (see below) and those of optical astronomers and published in *Annales d'Astrophysique* (Laffineur *et al.* 1950). The radio eclipse curve gave evidence that it was faculae (regions bright in the H_α line), and not sunspots or a bright annulus, that were producing 25 and 54 cm radiation. This eclipse turned out to be the first of several that the French groups studied (Orchiston and Steinberg 2007). In particular the French Navy transported six tons of equipment to Markou, near Ségou on the Niger River in French Sudan (present Mali), for an annular eclipse in September 1951. Observing at 1.8 m wavelength with a converted US Army radar and at 3.2 cm with a 1.5 m converted military searchlight (Fig. 10.11), Blum, Denisse and Steinberg (1952a, b) confidently deduced an elliptical distribution of solar radiation (see Section 13.1.3).

While this was going on, Denisse was producing a stream of papers dealing with the general physics of radio emission and the conditions giving rise to various types of solar radiation (Section 13.1.2). He also authored two of the earliest reviews of extraterrestrial noise (Steinberg and Denisse 1946, Denisse 1947c), and published articles dealing with plasma physics and the ionosphere (for example, he made ionospheric observations during the 1947 eclipse in Brazil). One of his most fruitful investigations arose from two years spent with the radio physicists of the National Bureau of Standards in Washington, DC. Denisse became well acquainted with North American work,[43] and in fact Covington turned over to him data from the first fifteen months of his 10.7 cm solar monitoring. With these data Denisse was able to considerably sharpen the intimate relation, first pointed out by Covington (1948), between the 10.7 cm solar intensity and sunspots (see Fig. 13.2). He called the week-to-week semi-periodic variations *une composante lentement variable* (Denisse 1949a) and soon this "slowly varying component" was recognized as a third major category of solar emission in addition to the quiet sun and the sudden bursts (see Section 13.2.3 for further details). While in America Denisse in fact wrote his Ph.D. thesis on the theory of solar emission and officially received his degree the following January upon his return to Paris (Denisse 1949b) – the first doctorate awarded for a radio astronomical topic.

As with other groups in the early 1950s, the ENS group grew to the stage where it felt the need for a much larger instrument to provide high angular resolution without the inconvenience of infrequent, short eclipses on other continents. Steinberg attended the 1952 URSI meeting in Sydney (see Fig. 7.3) and was particularly impressed with Christiansen's 21 cm wavelength grating array of 32 small dishes. He decided that French radio astronomy could only compete with something big of its own and convinced Rocard that they should seek funding for a large antenna system.[44] They found a sympathetic ear in

---

[42] On Würzburg dishes see Tangent 4.2. Rocard and the French Resistance also played a vital role in the 1942 Bruneval raid on a Würzburg installation, described in note 5.8.

[43] Fred Haddock (1983:116–17) told a delightful anecdote about the theorist Denisse and the engineer Reber (then both at NBS) when they came to visit NRL one day in 1948:

> Reber brought down this short Frenchman in handmade shoes and rolling his own cigarettes, and said, "This is a Frenchman! He won't wire power supplies, but he just sits around and calculates. He wants to see your antenna".... That began my friendship with Denisse, who instead of wiring power supplies was writing up his thesis on a full general theory of the quiet sun.

[44] Steinberg (1976:49B:550); Steinberg:Christiansen, 20 October 1952, file A1/3/1a, RPS. In 1949 the ENS group had independently come up with the idea of a solar grating array and had built a prototype at Marcoussis that consisted of four small cylindrical paraboloids operating at 450 MHz, but no

Figure 10.11 (top) A modified US Army radar (on an equatorial mount) about to be used for 1.8 m wavelength observations of an annular solar eclipse in French Sudan (present Mali) in September 1951. (bottom) A 1.5 m diameter US-surplus searchlight converted by Steinberg into a 3 cm wavelength radio telescope; note the waveguide feeding the dish from the top and the optical finding telescope. Photo taken on the ENS roof in Paris (note the dome of the Pantheon in the background) before this dish was used for the 1951 African eclipse.

the government and soon had a grant of 25,000,000 francs (~$70,000). A search for a suitable site eventually turned up Nançay 200 km to the south and from 1953 onwards this became the focus of French radio

---

observations were ever published. [Steinberg:Covington, 29 April 1950 (photocopy from Covington); Steinberg (1976:49B:500); Steinberg:author, 23 October 1986]

astronomy. In that same year Denisse returned from two years teaching in Dakar and the entire group shifted organizationally and physically from ENS to nearby Meudon Observatory.

The second French group was headed by Marius Laffineur (1904–1987) (Fig. 7.3), an electrical engineer who had also been an avid amateur astronomer. He was hired as an instrument builder at the Institut d'Astrophysique in Paris after the war, and soon convinced his supervisor, Daniel Chalonge,[45] to let him pursue the new radio noise reported in *Nature*. In 1947 he experimented with 64 MHz Yagi antennas in the gardens of Paris Observatory, but found the interference hopeless in the heart of the city. He then set up an interferometer (modelled after Ryle and Vonberg's at Cambridge) 10 km away at Meudon, where the Observatory's Astrophysical Section (Section 2.5) operated several optical telescopes in a park setting dating back to Louis XIV. Limited results in late 1947 and early 1948, however, led Laffineur to the conclusion that he needed a much larger antenna. He had already made a foray to the Zeppelin factory in Germany, where the Würzburg reflectors were manufactured during the war, and had returned with two of them. Mounting one dish on a concrete pedestal (Fig. 10.12) near the solar telescopes, he boasted that the whole thing cost him only 300,000 francs ($900).[46] From the start Laffineur envisioned his program as profiting from Meudon's strong optical solar group (which, for example, included the famed instrumentalist Bernard Lyot), and indeed over subsequent years he engaged in many collaborations. At one point he even installed a system whereby he could ring a bell at the optical telescopes whenever he picked up a healthy radio burst.

Laffineur's receiver was also based on war surplus, mostly from a German airborne radar, and this determined the operating frequency of 555 MHz (54 cm wavelength). Because his 7.5 m diameter dish had no tracking ability, readings were made by watching the needle of a voltmeter rise and fall as the sun drifted through the stationary beam. Laffineur (1976:2T) later recalled the excitement of the first solar observations in 1948, carried out with Jakob Houtgast (1908–1982), a Dutchman from Utrecht then visiting the Observatory for six months:

I said to him, "Tell me if you see something on the voltmeter as we try to cross the sun." Then I turned the handle [to move the dish] and he said, "Ça monte! Ça monte!" It was the sun coming into the beam. And then a few seconds later, "Ça descend!", as it went out of the beam. And we were absolutely glad and kissed each other.

Laffineur and Houtgast (1949) recorded and analyzed solar bursts in this manner in the autumn of 1948, but the tedium soon led to installation of a strip-chart recorder and an "equatorial pilot" that allowed the dish to smoothly track an astronomical source despite its altitude-azimuth mount. The pilot was a small equatorially-aligned device, run by a sidereal clock, that generated appropriate electrical signals for servomechanisms to control the reflector's position. This proved quite adequate for tracking the sun all day long with the 10° × 5° beam and in February 1949 Laffineur (1954) began regular monitoring for radio bursts, a program that eventually earned him a Ph.D. in 1951.

By 1952 the French had placed themselves on the radio astronomy map despite their considerably handicapped start compared to the British, Australian, and American groups chock full of radar veterans. They confined themselves to solar work partly as a strategy to help them compete, partly because of the relative ease of solar observations, and partly as a result of Rocard's and Denisse's strong interests in plasma physics and ionospheric effects. Meudon Observatory's long tradition of solar work also had a strong influence, and in fact Laffineur's case is particularly interesting in that as early as 1947 here was a radio engineer studying solar noise as a staff member of an astronomical observatory – the first such situation anywhere.[47] His general outlook was indeed more astronomical than that of other radio workers. More typical was Steinberg's (1976:49A:590) attitude at that time:

I reasoned that if we had no data, we could not start anything. Therefore I was more interested in developing instruments than in learning astronomy.

---

[45] The astronomer Chalonge first became excited about the new extraterrestrial radio noise from reading Reber's articles and hearing a lecture by Appleton in Paris in 1945. As he wrote: "The trail is being blazed for a new field of research, full of promise for astrophysicists." ["Parasites cosmiques," *l'Astronomie* **59**, 167–8 (1945)]

[46] Laffineur's Würzburg dish was moved to Bordeaux Observatory in 1962, but sixty years later its concrete pedestal remains, just outside the present cafeteria of Meudon Observatory.

[47] Pestre (1997:195) argues that Laffineur's hiring was part of a broader scientific "turf battle," and was designed to ensure that Rocard's ENS group did not monopolize the new radio field.

Figure 10.12 The 7.5 m Würzburg reflector operated at 54 cm wavelength by Marius Laffineur at Meudon Observatory from 1949 onward.

Steinberg (1976:49A:500) also later emphasized how the ENS group started from scratch, with little knowledge of even the basics such as building stable amplifiers and power supplies. For example, he recalled a two-week visit to England in 1948 after which he imported back to France several precious volumes on radar techniques from the MIT Radiation Laboratory (Ridenour 1947–53). But in general the French had little contact with the leading foreign groups during this period (nor did they ever publish in English) and their progress, although significant, was undoubtedly thereby hindered.[48]

10.5 JAPAN

Radar was also intensively developed in Japan during World War II, although it was not as technically advanced as that of the Allies.[49] Once the defeated nation began to recover in the late 1940s, however, radio physicists could draw not only on domestic stores, but also on American radar parts, readily available from war surplus dealers. Japan's long tradition of research on the ionosphere and radio communications (Section 11.2.1), natural for an island nation, also aided the development of radio astronomy. By 1951 no less than three separate groups were studying the radio sun, despite problems such as a general power system that could only supply frequencies fluctuating between 54 and 61 Hz and whose voltages varied between 85 and 105 V (Tanaka 1984:335).

The first group was at the Tokyo Astronomical Observatory (TAO) whose director Yusuke Hagihara (1897–1979), a specialist in celestial mechanics, had

---

[48] Information on the early years of the French groups comes primarily from 1976 interviews with Denisse, Steinberg, Blum, and Laffineur, as well as from the group's publications; also see Steinberg's "Les débuts de la radioastronomie en France," *l'Astronomie* **99**, 479–86 (1985) and "The scientific career of a team leader," *Planetary and Space Sciences* **49**, 511–22 (2001). The wartime exploits of Rocard are treated on pp. 260–3 of *Most Secret War* by R.V. Jones (London: Hamish Hamilton, 1978). Pestre (1997) provides a useful historical analysis of the interactions between French ionosphere and radio astronomy research efforts during these years. Orchiston *et al.* (2007:221–3, 233–7) provide further details on the early years of the ENS and Laffineur groups.

[49] Brown (1999:135–40, 371–3), Swords (1986:130–5), Guerlac (1987:917–24); also see S. Nakajima (1992), "Japanese radar development prior to 1945," *IEEE Antennas and Propagation Mag.* **34**, No. 6, 17–22.

worked on radio propagation theory during the war. In 1944 he had advised M. Huruhata on radar observations of meteors (Section 11.2.1) and after the war he set up an Ionosphere Research Committee that provided a monthly forum for groups from a variety of institutions to discuss their work in the broad field of solar-terrestrial relations. Hagihara thought it important to study how the new phenomenon of solar noise fitted in with ionospheric activity and with the familiar optical sun long monitored at TAO. In 1949 one of his staff members, Takeo Hatanaka (1914–1963), chief of the Spectroscopy Section, started a radio project after being intrigued by discussions of the radio sun by visitor Richard Woolley (Suzuki 1978:94A:690). As assistants Hatanaka took on Fumio Moriyama, who had just received an undergraduate degree in astronomy, and Shigemasa Suzuki, a radio engineer with several years' experience in industry. Observations began in September 1949 among the domes at TAO's Mitaka site, about 25 km west of downtown Tokyo, with a 200 MHz receiver and dipole array. This was soon followed in 1950 by solar monitoring at three frequencies, including new 60 and 100 MHz double-Yagi systems constructed by Suzuki. Polarization measurements were also started up, a technique that later in the 1950s became a Japanese specialty.

Japanese observations also took place in the Physics Department of Osaka University. Minoru Oda, who had developed microwave radar during the war and whose primary interest was in cosmic rays, carried them out with graduate student Tatsuo Takakura. A large cone-shaped horn, scaled in size from typical Japanese wartime radars, was placed on a naval searchlight mount and connected to a radar receiver modified by Takakura into a radiometer operating at 3300 MHz (Fig. 10.13). Solar observations began with this unique antenna in November 1949, but after a move to the new Osaka City University, the horn was replaced with a 1.5 m diameter dish. Solar monitoring then continued until 1954 when Takakura moved to TAO.

The third Japanese group was part of the Research Institute of Atmospherics run by Nagoya University. As its name indicates, its program was dominated by the study of atmospherics (mostly from lightning), but because its director Atsushi Kimpara also felt that a program of solar noise observations could be helpful for understanding ionospheric disturbances, in 1949 he hired Haruo Tanaka (1922–1985), a fresh electrical engineering graduate from Tokyo University. Over the next two years Tanaka built a mesh-surface 2.5 m diameter dish and a receiver whose frequency of 3750 MHz was chosen because the telephone company was developing various components for microwave links at a similar frequency. These were located at a field station in Toyokawa, 60 km southeast of Nagoya. In April 1951 Tanaka commenced a program of accurate solar monitoring that has in fact continued to this day and from the beginning has provided a valuable comparison to Covington's data at a similar frequency. As with Covington, Tanaka was particularly concerned about absolute calibration. In pursuit of the best possible flux measurements, for example, he was able to report in 1951 that the background sky temperature at 8 cm wavelength was no more than 5 K (Tanaka et al. 1951; Tanaka 1984:341).

Although three groups were in full operation by the early 1950s, their work was little known outside Japan. Publications at the start were few and often in Japanese; even when in English, periodicals such as the *Proc. of the Research Inst. of Atmospherics* were not read and had little influence. Nevertheless, the solar monitoring programs, unglamorous though they were, eventually contributed notably to studies on the long-term behavior of the sun.[50]

10.6 OTHER SMALL EARLY GROUPS

For completeness brief mention will be made of a few other groups who began work before 1952 (this chapter's cutoff date), but who did not start observations or produce publications until later. Several of these groups became important in the later 1950s.

10.6.1 Germany

The wartime observations in Germany of Kurt Fränz have already been discussed in Tangent 6.1, but this did not lead to anything after the war, nor could it, for up until the establishment of the Federal Republic of Germany in 1949 the Allied occupation authorities forbade any kind of research that looked like radar,

---

[50] Primary sources for the early Japanese years are Tanaka (1984) and 1978–81 interviews with Tanaka, Moriyama, Takakura, and Suzuki.

Figure 10.13 Minoru Oda (l.) and Tatsuo Takakura operating their 3300 MHz radiometer at Osaka University in late 1949. Japanese microwave systems of this period followed wartime practice in using circular waveguides and conical horns.

including radio astronomy. But by 1949 the Allies had also confiscated all of the useful radar booty, as has been recounted in several earlier instances, and so those Germans who wanted to pursue radio astronomy were still strapped. Theoretical research, however, was still possible. At Kiel University, the distinguished astrophysicist Albrecht Unsöld (1905–1995), Professor of Astronomy and Theoretical Physics and known in particular for his work on stellar atmospheres, took a strong theoretical interest in the first observations of radio noise. He was a close friend of Otto Struve and had been infected with the latter's enthusiasm for Reber's and Jansky's results when he spent time at Yerkes Observatory in 1939. During the war Unsöld also came into contact with Fränz (and Kiepenheuer, see below) when in 1944 they were all enlisted for a crash radar

project after Hitler realized that his 1940 decision to de-emphasize radar research had been a mistake. In 1944 Unsöld wrote a remarkable review entitled "The cosmic shortwave radiation," eventually published in 1946 in *Die Naturwissenschaften*. After reviewing in this article Jansky's, Reber's, and Fränz's observations and covering the free–free theory of Henyey and Keenan (1940) in detail (Section 4.3), Unsöld concluded that the longer-wavelength data forced one to accept that an electron temperature of ~100,000 K existed in interstellar space, much more than standard ideas of ~10,000 K. But he pointed out that such a high temperature was no longer so absurd when one considered that recent (optical) evidence indicated that the solar corona itself was several hundred thousand degrees (Section 13.3.1). Unsöld urged that astronomers search for links between the radio radiation and the locations of hot stars and interstellar H_α emission. In addition, although at the time he wrote he did not know of any radio detection of solar emission, he pointed out that the sun's signal level should correspond to blackbody emission at a temperature of 6000 K.[51] Finally, Unsöld pointed out that radio investigations might shed light on the question as to whether or not a tenuous medium existed between the *galaxies*. He calculated that the lack of a "noticeable effect" (meaning that the sky was not uniformly bright in all directions) implied that the density of any intergalactic medium had to be less than 10^{-4} electrons cm^{-3}. His closing paragraph reveals a fascination with these new signals from space:

> The old dream of wireless communication through space has now been realized in an entirely different manner than many had expected. The cosmic shortwaves bring us neither the stockmarket nor jazz from distant worlds. With gentle murmurs they rather tell the physicist of the endless love play between electrons and protons.
> (Unsöld 1946:40)

Unsöld continued his interest in interpreting the galactic background radiation over the next

---

[51] Unsöld of course did not know about the wartime solar radio detections. He also missed out on a nice prediction: if he had combined his free–free theory with the high temperature of the corona and its estimated density, he might have predicted that the sun's meter-wavelength signal would be indicative of the hot corona rather than of the photosphere.

decade (see Chapter 15), but he also wanted to get radio observations going at Kiel. He had good relations with the British military officer in charge of overseeing scientific research at the university, and, in cooperation with W. Kröbel, Professor of Applied Physics, simple equipment was built in the late 1940s and even some observations undertaken, but never very successfully until 1953. At that point Kröbel put his student Franz Dröge onto the project and Unsöld hired Wolfgang Priester, fresh from a doctorate in astronomy at Göttingen. By the end of the year these two had started a 200 MHz all-sky survey with an array of dipoles.[52]

Karl-Otto Kiepenheuer (1910–1975) was another prominent astrophysicist who unsuccessfully tried to get radio astronomy going after the war, not succeeding until 1955. Kiepenheuer's entire career focused on the sun and in the late 1930s he had tried to open up the ultraviolet portion of the spectrum to astronomy with solar observations from the 3600 m Jungfraujoch in Switzerland, as well as from balloons. In 1939 he founded the Fraunhofer Institute for solar studies at Freiburg University and during the war was responsible for optical solar observations at several observatories (including Meudon in occupied France) needed for prediction of ionospheric conditions for radio communication (Hufbauer 1991:120–3). After the war Yves Rocard in Paris (Section 10.4) helped Kiepenheuer's institute, which was in the French occupation zone, get back on its feet, producing ionospheric predictions for the French military. Nevertheless, Horst Wille, who was a postgraduate student under Kiepenheuer during these years, recalls that a large array of dipoles that he had built and was using for solar observations was confiscated in about 1947–48 by the French Navy. Wille then built a steerable Yagi setup and, because all of the Würzburg reflectors had been hauled away, started to build his own large paraboloid in about 1949. Funds proved insufficient, however, and the French occupation authorities were still uncooperative;[53] in the end Wille turned to optical observations for his doctoral thesis. Despite Kiepenheuer's efforts to become involved with the radio sun, it was ironically *non*solar radio astronomy where, just at this time, he made his greatest contribution. In 1950, during a visit to several American universities, he proposed that the Milky Way's radio emission originated in synchrotron radiation from relativistic electrons. This was to be one of the seminal papers of radio astronomy (Kiepenheuer 1950; Section 15.4.1).[54]

10.6.2 Sweden

Early solar observations in Sweden were conducted under the direction of Olof E. H. Rydbeck (1911–1999), a radio physicist with interests in both ionospheric observations and theory, as well as in development of microwave amplifiers. Rydbeck worked at the Chalmers University of Technology in Göteborg, and in 1949 set up a coastal field station south of the city at Onsala. There 150 MHz solar monitoring was done, as well as meteor radar observations instigated by Bertil-Anders Lindblad of Lund Observatory as early as 1950. By 1952 several Würzburg reflectors (Tangent 4.2) from remote Norwegian mountain sites had also been imported by barge. Hein Hvatum, who worked under Rydbeck as both an undergraduate and postgraduate student, built many of the receivers and antennas involved during these years. But the group was much more interested in instrumental developments than in solar astrophysics and not until the mid-1950s did any of their work lead to a scientific publication.[55]

10.6.3 Norway

A small radio group arose at the Solar Observatory, Institute of Theoretical Astrophysics, University of

---

[52] For the work at Kiel I have profited from interviews with Unsöld (1976) and Priester (1973).

[53] Kiepenheuer:A. C. B. Lovell, 6 April 1949, file R6 (uncat.), LOV.

[54] The primary source for the Freiburg story is Wille:author, 25 April 1979, and an obituary of Kiepenheuer by A. Bruzek in *Solar Physics* **43**, 3–7 (1975). Pestre (1997) details the vicissitudes over the postwar decade of the Freiburg institute's work vis-à-vis similar ionospheric predictions being done in France under P. Lejay.

[55] Interviews with Rydbeck (1978) and Hvatum (1985) have been helpful. See also "Technological Drift in Science: Swedish radio astronomy in the making, 1942–1976" by M. Hård (1993), pp. 378–97 in *Center on the Periphery*, S. Lindquist (ed.) (Canton, Mass.: Science History Publications). There also exists a well-illustrated 800-page biography of Rydbeck: *Femtio år Som Rymdforskare och Ingenjörsutbildare: Olof Rydbeck* (1991, Göteborg: Chalmers Tek. Hög.).

Oslo (at Blindern, near Oslo). From 1949 onwards Gunnar Eriksen, an engineer, led an effort to acquire a 7.5 m Würzburg dish, mounted it equatorially, and built a 1.5 m wavelength receiver for solar observations. This group's first publication reported observations of a solar eclipse in June 1954.[56]

10.6.4 Soviet Union

Two other early Soviet groups deserve mention. One was at the Byurakan Astrophysical Observatory, founded by Viktor A. Ambartsumian after the war in the mountains 30 km outside of Yerevan, Armenia. In 1950 he invited Vagarshak A. Sanamyan, who had had some wartime radar experience, to join his staff and get radio astronomy going. Ambartsumian's work on stellar associations had led to a strong interest in what might be called "energetic locales," places where objects or matter seemed to be ejected from small volumes, and to him the new radio stars seemed like good candidates. Over 1950–52 Sanamyan (1980:1–2N) built several meter-wavelength interferometers and observed the sun and the strongest four discrete sources, but no results were published until 1954–55.

The second Soviet "group" was Andrei P. Molchanov (1980:1–2N), a radio physicist who, starting in 1949, carried out 3.2 cm solar observations with a small dish at Leningrad University, but did not publish until the mid-1950s after he had made observations in central Asia (along with groups from Gorky and FIAN) at the total eclipse of February 1952 (Molchanov 1956).

10.6.5 United States

Finally, there were two additional groups in the United States. The first was part of the Department of Terrestrial Magnetism (DTM) of the Carnegie Institution of Washington, DC. Its director Merle A. Tuve (1901–1982) had eclectic interests: he had made important radio measurements of the ionosphere in the 1920s (Section 3.2.1), but was also known for his prewar work in nuclear physics and his wartime work on the proximity fuse for antiaircraft shells. After the war Tuve modified a 200 MHz radar for some preliminary solar measurements in 1947, with the idea of tying them into both ionospheric and cosmic ray work then going on at DTM. But this was abandoned, and not until several years later did Tuve again get fired up, this time after the 1951 discovery of the 21 cm spectral line of neutral hydrogen (Chapter 16). By late 1952 his group was on the air with a 21 cm receiver installed in a Würzburg dish (Tangent 4.2), the first of a variety of radio astronomical antennas and observations at DTM over succeeding years.[57]

The second American group was led by John D. Kraus (1910–2004) in the Electrical Engineering Department of Ohio State University. Kraus had had an interest in extraterrestrial radio noise since the early 1930s (Section 5.4.1), but his primary efforts were directed toward antenna design. After his invention of the helix antenna, in 1951 he began testing its promise as a radio telescope. By the end of 1952 a grant of $2000 from a private foundation had allowed him to build a steerable array of twelve helices, soon expanded to 48. Observations of the galactic background and of discrete sources over 1952–53 then led to the group's first publications in radio astronomy in 1953.[58]

TANGENT 10.1 SHKLOVSKY'S FREE-FREE RADIATION CALCULATIONS

In working out the total absorption coefficient at radio wavelengths for an ionized gas, Shklovsky reasoned that not only was there free–free absorption as in the optical case, but also absorption due to collisional damping,

---

[56] G. Eriksen, "Solar radio noise recording equipment built for the solar observatory at Harestua," Rept. No. 3, Inst. Theoretical Astrophysics, May 1953.

[57] Information on the start at DTM comes from a 1974 interview with Tuve, DTM Annual Reports published in the Carnegie Institution's Yearbooks (especially those for 1946–7 and 1972–3), and Needell (1991). T. D. Cornell (1990) has written the entry for Tuve in the *Dictionary for Scientific Biography*, Vol. 18, pp. 436–41 (1990), F. L. Holmes (ed.) (New York: Scribner); P. H. Abelson (1996) has a memoir in *Biog. Mem. NAS* **70**, 406–23. Also see Chapter 22 in L. Brown (2004), *The Department of Terrestrial Magnetism* (Cambridge: Cambridge University Press).

[58] For the start at Ohio State I have used Kraus (1975:5–7T, 1976:97–108); also see Kraus's article on his first radio telescope in *Sky and Telescope* **12**, 157–9 (1953). The all-sky map of Fig. 1.1 is based on observations with the 48-helix array.

that is, oscillating electrons giving up their energy in collisions with protons. But in fact these are one and the same process (Shklovsky 1982:9–10, Ginzburg 1984:291, Denisse 1950b:168). Not until Shklovsky (1952a) was this error corrected.

In a related incident, Shklovsky (1952a:421) stated that Unsöld (1949a) made a "gross arithmetic error, which it is astonishing that nobody abroad ever noticed." This claimed error made a given amount of gas (in emission measure) produce 1000 times less radio free–free emission than Shklovsky calculated, and this Shklovsky took as the reason why Unsöld calculated that there was no significant contribution from free–free emission to the observed galactic background. My own analysis, however, leads to the conclusion that both investigators had correct coefficients in their equations, but that Shklovsky probably did not account for Unsöld's expressing specific intensities in mks units (as opposed to Shklovsky's cgs units). Such an oversight amounts to exactly the required factor of 1000.

11 · Meteor radar

Through the 1930s continual variations of the ionosphere – with location on earth, with the 11-year solar cycle, with time of day, and in apparently random ways – were at once puzzles and sources of insight. Edward Appleton's terminology of D, E, and F layers (or regions) was applied to the atmospheric heights of approximately 70, 120, and 200–400 km that produced distinct radio reflections (the E layer being equivalent to the original Kennelly–Heaviside layer) (see Appendix A). Despite the success of Appleton's magneto-ionic theory in describing the basic laws of propagation for radio waves in the ionospheric plasma, it was still not agreed whether the electrons or the ions were more effective in reflections. A theory of how the various layers might be maintained by an external ionizing agent was developed by British theorist Sidney Chapman. It was clear that the sun was the main source, but debates swirled over whether the solar agent was corpuscular (electrons? cosmic rays? neutral atoms?) or radiative (ultraviolet? X-rays? cosmic rays?). And the fact that the ionization maintained a healthy *nighttime* level was difficult to understand unless the processes of recombination (of electrons with ions) and attachment (of electrons to neutral molecules and atoms) acted much more slowly than expected. Or was it that charged particles from the sun curved around to the nightside of the earth, or that some other mechanism such as thunderstorms or meteors operated?[1]

Other puzzling phenomena arose. One example was the sudden ionospheric disturbance or shortwave fadeout (Section 5.4.3). Another was the occurrence of *sudden* increases in ion density even *during the night* when solar radiation could not be directly active. Various mechanisms were suggested for this impulsive type of ionization, which eventually became known as "abnormal E" or "sporadic E," but no consensus developed until after World War II, when meteors became accepted as one principal cause. These bombarding bits of rock from outer space were then used by some researchers as probes of the ionosphere and by others as the basis for a novel system of communications. Yet others studied astronomical aspects of the meteors themselves. This chapter concentrates on the last aspect and examines three phases of what came to be called meteor radar astronomy: prewar intimations, postwar denouement, and development by 1950 into a mature, albeit short-lived, field that for a time represented a significant fraction of early radio astronomy (Figs. 1.2 and 1.3).

After a brief review of prewar indications that meteors might well be causing ionization and affecting radio propagation in the upper atmosphere, we examine the postwar beginnings of meteor radar astronomy that followed Hey's initial end-of-war discoveries (covered in Section 6.3). The spectacular October 1946 Giacobinid meteor shower was the key event that excited many researchers and convinced them to invest in what became meteor radar astronomy. The leading groups were at Jodrell Bank (Chapter 9), Canada's National Research Council, and Stanford University. These groups developed the antennas, receivers, and analysis methodology to study tens of thousands of meteors. They discovered daytime meteor showers stronger than anything at night, as well as played the principal role in settling a contemporary astronomical feud: did meteoric particles originate in interstellar space or within our solar system? Unlike (passive) radio astronomy, however, meteor radar astronomy did not continue to grow through the 1950s; by 1960, a

---

[1] There exists no general history of ionospheric research. A useful brief overview of various aspects is given by C. S. Gillmor (pp. 101–114 in *Space Science Comes of Age* [eds. P. A. Hanle and V. D. Chamberlain, Washington, DC.: National Air and Space Museum, 1981]). Lovell (1948:415–28) provides a detailed summary of prewar work on abnormal effects in the E region and their relation to meteors, as well as the first comprehensive review of meteor radar.

few of its researchers were involved in planetary radar (Section 12.6), but most either left the field altogether, or used the meteor radar technique to study the ionosphere or develop novel communications systems.

11.1 PRE-1945 INTIMATIONS OF RADIO AND METEORS

11.1.1 Nagaoka

The first suggestion that meteors might affect the ionosphere was published in 1929 by Hantaro Nagaoka (1865–1950), chief physicist at the Institute of Physical and Chemical Research, Tokyo, in the *Proc. of the Imperial Academy of Tokyo*, with additional comments two years later (Nagaoka 1931:183–4). Nagaoka was impressed with the coincidence of height (~100 km) between the ionospheric layer and where visual meteors occur. Not only might meteors directly cause quick ionization, but he argued that long-lived fine dust resulting from disintegration of the original meteor particles would act as nuclei for collecting ions or electrons from the surroundings, thus causing a region of *less* ionization, or a "disturbance":

> Irregular and diffuse reflections from the ionised layer are the general result. The effect is somewhat analogous to the reflection of light from a mirror with numerous scratches made in a disorderly confused manner.
>
> (Nagaoka 1929:235)

Based on this idea, communications would be affected during a meteor shower, perhaps preferentially in the direction of the shower's radiant. These ideas led to a monitoring program by Eitaro Yokoyama (1930) during the time of four major showers during 1928–29, but the "grinder" and "click" atmospherics that he found on 30 kHz were judged not abnormal in any way.

11.1.2 Skellett

A. Melvin Skellett (1901–~1991) studied physics at Washington University in St. Louis and joined the Bell Telephone Laboratories in 1929 at their Deal, New Jersey field station, only 15 miles from Holmdel where Karl Jansky was then working. Deal dealt with transmitters and Holmdel homed in on reception. Although research topics in those Depression years had to link closely with communications, Skellett had long had a strong interest in astronomy and often managed to work on projects exploiting his knowledge of extraterrestrial matters. In fact, over the years 1930–33 he frequently commuted forty miles to Princeton University to work on a Ph.D. in astronomy, all the while working fulltime at Bell Labs. This rare combination of skills in radio physics and in astronomy not only served him well, but also aided Jansky in his astronomical interpretations (Section 3.3). Skellett's astronomical projects included studies of solar activity and the invention of a television-based "coronaviser" to see structural details in the solar corona.[2]

Another astronomical tack of Skellett's was to search for a connection between meteors and the ionosphere. In early 1931, unaware of Nagaoka's work,[3] he noticed that contemporary ideas on meteors nicely meshed with recently discovered properties of the Kennelly–Heaviside layer (the ionosphere). First of all, an altitude of ~100 km was not only where the ionosphere (or more exactly, its E region) caused radio reflections, but also where visual observations indicated that meteors typically burned out. Second, standard theory said that 98% of a meteor's kinetic energy went into ionizing the air, and this allowed Skellett to estimate that meteors as a class could contribute, along with cosmic rays, the bulk of the energy needed to maintain charged particles in the ionosphere. And for the case of a particularly large meteor, Skellett calculated that as many as 10^6 electrons and ions cm^{-3} could be produced, more than enough to affect shortwave signals. Third, shortwave signals often varied in strength ("fading"), indicating turbulence in the ionosphere; this could be caused by the constant mixing action of the meteors. For instance, fellow Bell Labs researcher Raymond A. Heising had just found that the reflecting layer sometimes suddenly dropped in altitude at night as though "great masses of electrons are tossed into

---

[2] A.M. Skellett (1940), "The coronaviser, an instrument for observing the solar corona in full sunlight," *Proc. Natl. Acad. Sci.* **26**, 430–3.

[3] Skellett was unaware of Nagaoka's paper when he published his own in 1931. When Nagaoka saw Skellett's paper, he decided to reprint his own article, with a bit of further commentary, in 1932 in the more widely disseminated *Reports of Radio Research in Japan* (Vol. **2**, 49–53).

the atmosphere rather quickly."[4] Furthermore, it was known that the layer usually lowered after midnight. Skellett reasoned that this might be a consequence of the well-known fact that the number of visible meteors substantially rises after midnight when the observer is situated on the *leading* side of the orbiting earth and therefore collides with more interplanetary debris.[5]

In a one-page note to *Physical Review* in 1931 Skellett summarized these ideas and concluded by saying that he planned observational tests during future meteor showers. He went into further details in an article in the *Proceedings of the Institute of Radio Engineers* (1932a) and in a Bell Labs memorandum, the latter closing with words clearly indicating that his motivation was as much astronomy as communications:

> From the astronomical standpoint it appears that suitable radio equipment may be valuable in obtaining meteoric data. Such equipment is not affected by cloudy weather and, although in present form would give little more than relative numbers, it is possible of a great deal of elaboration.[6]

Skellett tested these ideas over a two-year period. From the evidence of his laboratory notebook[7] he was originally expecting meteors to cause a whistling sound in his headphones, for he built a 1.7 MHz "whistler amplifier" in the summer of 1931 and used it for observations of the δ Aquarid, Perseid, and Leonid meteor showers[8] later that year. The idea was to have one person outside looking for meteors and the other inside searching for associated effects. Skellett heard mostly "tweeks," that is, static impulses followed by tones rapidly descending in frequency, but there were also musical "swishes." His notebook is full of vividly described sounds – "a sky rocket in reverse...a xylophone key being struck...someone whistling far off...wind blowing through the trees...several people swishing reeds or whips in the air" – but their relation to meteors was uncertain. Cloudy weather harried visual observations, but when it cleared there was no definite correlation of radio sound and meteoric light.[9]

In addition to his own setup, Skellett enlisted the aid during the same showers of two other colleagues at Deal, J. Peter Schafer and William M. Goodall. They had just developed a fancy 2.4–4.8 MHz pulsed transmitter and receiver in order to study ionospheric reflections. This operated simultaneously on two frequencies and had a cathode-ray-tube display for monitoring. Schafer and Goodall (1932) too had frequently observed nighttime increases in ionization and puzzled over their origin. Now, particularly for the Leonids of November 1931, they recorded many intermittent reflections "not normally found," as well as a very large deduced electron density for the reflecting layer. But an unfortunate lack of any clear weather and therefore any detailed meteor correlations, as well as the presence of a magnetic disturbance (they wondered whether it might itself be associated with the meteor shower), brought caution to their conclusions about any meteor–ionosphere link.

Skellett's next big push was for the Leonids in November 1932. This shower had been closely watched ever since its fantastic display in 1833, when meteors were described as falling from the sky like snowflakes at estimated rates as high as ten thousand per hour. The Leonids were known to be extremely strong every 33 years, and although the expected 1899 event had

---

[4] R. A. Heising (1928), "Experiments and observations concerning the ionized regions of the atmosphere," *Proc. IRE* **16**, 75–99. [Quotation from p. 84]

[5] Thus does the windshield of a speeding automobile collect more bugs than the rear window.

[6] Skellett, 20 May 1931, "The effect of meteors on short-wave transmission through the Kennelly–Heaviside layer," (Memorandum 327-112-MM-5/20/31-GZ), p. 17, Case 16916 (Vol. N), BTL. Skellett also presented his ideas and early results to astronomers, at a December 1931 meeting in Washington, DC of the American Astronomical Society (abstract in *Pub. AAS* **7**, 101 [1933]).

[7] Skellett, 27 July 1931 to 27 November 1933, Laboratory Notebook No. 10788, pp. 72–91, BTL.

[8] See note 6.13 for a brief explanation of meteor showers and their radiants.

[9] Many of these sounds correspond to what later were called *whistlers*, which had been picked up on radio circuits since the turn of the century. In the 1920s these too were speculated to be the result of meteors. For example, at a meeting of the Institution of Electrical Engineers in London in 1929, J. E. Tayler and T. L. Eckersley briefly discussed the possibility [*Journal IEE* **67**, pp. 1030–2 (1929)]. Whistlers are now understood as lightning-induced atmospherics causing 0–15 kHz radio waves to propagate along the earth's magnetic field lines from one hemisphere to the other; see L. R. O. Storey (1953), *Phil. Trans. Roy. Soc.* **A246**, 113–41.

proved a bust, 1932 was still a year to watch.[10] Skellett (1932b:434) sounded a call to arms in the form of a short note in *Science* urging radio researchers to observe during the shower – it afforded their best chance to test "the hypothesis of a meteoric effect on radio transmission since the advent of radio, or that is likely to occur again until 1965."

Skellett, Schafer, and Goodall observed the Leonids on two nights: despite a full moon and a shower that turned out to be unspectacular, the cold clear weather allowed 50 meteors to be visually noted. Concentrating on meteors passing within 20° of the zenith (toward which the transmitter beamed), Skellett recorded the approximate magnitude, location, and direction of travel of each meteor. Eleven meteors passed near the zenith on the two nights and indeed at the time of each the amount of ionization (presumably inferred from the height of reflections) increased or remained at a relatively high value (Fig. 11.1). Furthermore, all times when the ionization level jumped rapidly were accompanied by a meteor overhead. An event at 12:58 a.m. on 16 November was particularly impressive: "When this bright meteor was sighted, the meteor observer and the operator at the receiver called out simultaneously" (Skellett 1935:145). They immediately began running through a sequence of transmitted frequencies and determined that the ionization was of the order of 10^6 electrons cm^{-3}, an extremely high value even for noontime.

Skellett's last observations were of the Leonids a year later. He visually observed almost two hundred meteors on two nights, but there was a disappointing "lack of coincidences of ionization and meteors overhead." This he attributed to the lack of bright meteors and the generally low background level of ionization, making it more difficult for a meteor to increase the ionization level to the point where a radio echo would result.[11] For some reason these 1933 data were never mentioned in either Skellett's Princeton thesis (1933) or in his essentially identical summarizing paper in the *Proc. IRE* (1935). Both were titled "The ionizing effects of meteors" and stated that the radio studies were "strongly suggestive of meteoric ionization" (Skellett 1935:132).

Thus ended the most detailed experiments before the war involving visual meteor watches combined with radio data. The case for an association was hardly airtight, but Skellett's pursuit of this notion produced significant progress.

11.1.3 The situation before 1945

Besides Nagaoka's and Skellett's studies, there were many other hints before and during World War II that meteors impinging on the upper atmosphere affected its state of ionization and the propagation of radio waves. For example, two follow-up experiments testing Nagaoka's ideas were conducted in Japan. In India two radio engineers studied a close correspondence between visual meteors and whistles[12] that they heard on shortwave channels, arguing that the whistles were a Doppler effect in the beat tone between a directly received wave and a wave reflected off a rapidly changing ionization region created by the meteor (Chamanlal and Venkataraman 1941). In an article featuring this work, entitled "Radio-astronomy traps music of the stars," the *New York Times* said:

> Now, it seems, the astronomer and the radio man are about to pioneer again – side by side – in a branch of science from which may spring a new conception of ionospheric mechanics.[13]

Details of these and other Japanese, Indian, and American studies up until 1945 are given in Tangent 11.1.

Besides Skellett, the only prewar work in America on radio and meteors was by John A. Pierce (1907–1996), a radio physicist at the Cruft Laboratory, Harvard University.[14] The theory that he developed turned out

---

[10] This 33-year periodicity of the Leonid shower arises because the associated stream of interplanetary debris is not spread uniformly along its orbital path about the sun, but highly bunched. The earth thus intersects the orbit of the meteors every year, but only rarely at the point where the debris stream is located.

[11] Skellett, 15–16 November 1933, Lab. Notebook No. 10788, p. 90, BTL.

[12] These *whistles*, as well as all others discussed henceforth in this chapter, are not the same as the *whistlers* described in note 9.

[13] *New York Times*, 10 May 1942, p. X10. The headline contains one of the first uses of the term "radio astronomy," although the meaning here varies from later usage (see Section 17.1.1).

[14] John A. Pierce should not be confused with (a) John R. Pierce (aka J.J. Coupling to science fiction readers) of Bell Labs and Caltech, famous for his work in electronics and communications, nor with (b) Joseph A. Pearce, a Canadian astronomer of this era.

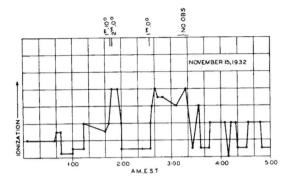

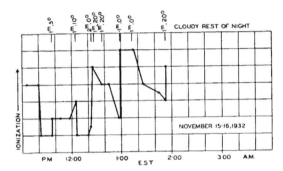

Figure 11.1 Amount of ionization (presumably inferred from the height of the reflecting layer) measured over two nights of the Leonid meteor shower in November 1932. The times and brightness (in magnitudes) of meteors visually observed are also indicated. (Skellett 1935)

to be important for postwar studies. In 1937 Pierce was investigating abnormal ionization in the ionosphere by monitoring a 10 MHz signal from a distance of 30 km. During the nighttime, this placed him well within the so-called skip zone, the region near a transmitter where signals reflected from the ionosphere usually are not possible to receive, since a wave leaving the transmitter at the appropriate (high) elevation angle penetrates the ionosphere. Yet sometimes he (and others) observed anomalous "bursts of signal" when for a period of 5–15 minutes the signal came in loud and clear. The bursts were taken to be associated with transient regions of high ion density and Pierce favored Skellett's idea that meteors were the agent.

Pierce (1938) described these data and more importantly worked out the first theory for the nature of the reflection process. Passage of a meteor was considered to create a cylinder of high ion density, which would quickly expand in radius as the ions diffused and then weaken as the electrons recombined with the ions. Reflection of radio waves was taken to occur only if the electron density stayed above the critical value (Appendix A) over a region larger than a wavelength in size. Pierce stressed that backward-directed reflections occur only when observing along a normal to the meteor trail and showed that the expected signal strength for a trail of radius 250 m was comparable to those observed. He worked out that a -2^m meteor (similar in intensity to a bright planet) was required to produce the radio effect and that current ideas on the total meteor flux striking the earth yielded such meteors frequently enough to explain the observed rate of anomalous signals. Pierce also calculated that the E-layer ionization through the night could possibly be *entirely* maintained by meteoric bombardment. Postwar studies later indicated that some aspects of Pierce's theory required substantial revision (Section 11.4.3), but his basic model of a cylinder of ionization remained unchanged.

The British ionospheric community also contributed several theoretical and experimental studies before 1946 about the possible role of meteors. Tangent 11.2 details these studies.

The situation at war's end thus included evidence and suggestions, dating back as far as 1929 and from both experimental and theoretical sides, that meteors could have a significant influence on the E region and therefore also on radio propagation. Only a fraction of investigators, however, was convinced of the importance or reality of meteoric phenomena. The importance of meteors was hardly gospel for the ionospheric community and even believers were uncertain about basic questions such as: Did all visual meteors produce a radio effect? Were meteor showers of any special importance for radio propagation? Which abnormal E effects were a result of meteoric bombardment? Could the *entire* nighttime ionization of the E region be supported by meteors?

This then was the state of knowledge encountered by J. Stanley Hey and Gordon S. Stewart when they conducted their 1944–46 investigations at the Army Operational Research Group in England (fully recounted in Section 6.3.1). Their seminal work definitively established that radar could detect meteor trails and usefully study the meteors themselves. Hey and Stewart (1946, 1947) carried out multi-station observations, monitored both sporadic and shower echoes, determined shower radiants, and discovered daytime showers. After fifteen years of intimations, meteors and radar were at last firmly linked.

11.2 THE 1946 GIACOBINIDS

In 1900 Michel Giacobini discovered a comet that was subsequently lost and rediscovered by Ernst Zinner in 1913. An associated Giacobinid (or Draconid) meteor shower with a 6.5-year period was first picked out in 1926, and 1933 provided a splendid display, one of the best showers of the century. This was thought to be because the earth in 1933 crossed the comet's orbit only 80 days after the comet's passage. Circumstances looked even better thirteen years later: predictions were that the earth would be late by only 15 days. Meteor watchers around the world, a fraternity now including radar workers, harbored great hopes that this might be the display of a lifetime, whether to the eye (despite a full moon) or on a cathode-ray tube. As it turned out, few were disappointed.

At Jodrell Bank, Manning Prentice had advised Bernard Lovell of the Giacobinids and in preparation John Clegg had built the "searchlight aerial," five 72 MHz Yagis mounted for steerability on a surplus Army searchlight (Fig. 9.2). Lovell (1968:8–9) later vividly described the night predicted for the shower's peak, 9–10 October 1946:

> We were [at first] seeing two echoes an hour, which was quite a normal background rate.... Then in a flash, it seemed, everything was transformed. Just after midnight our echo rate began to rise dramatically and simultaneously meteors streaked across the sky.... Soon they were coming so fast that we were unable to write down any details and by 2:30 a.m. there were so many that we could not even count them.... By 3 a.m. the sky was streaked with trails and looked like the drawings of the great meteor showers of the eighteenth century which we had always thought to be imaginative. That was the peak. Our echo rate [then] decreased and by 6 a.m. all was normal, except we ... were in a state of great excitement. Moreover, term had begun, and both Clegg and I had to give lecture courses in Manchester at 10 a.m.[15]

---

[15] The times in this recollection are somewhat at variance with the curves of Fig. 11.2.

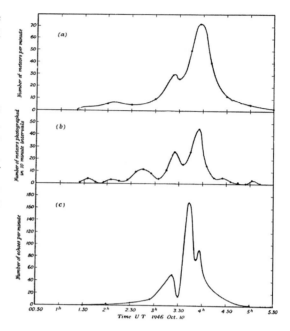

Figure 11.2 Activity of the 10 October 1946 Giacobinid meteor shower. The Jodrell Bank radar echo rate on the bottom was compared with visual counts in the United States (top) and photographic records from Millman in Canada. (Lovell 1954:330)

The novel feature in this experiment was the steerability of the antenna, which was kept looking perpendicularly to the radiant's direction, that is, perpendicular to the apparent direction of arrival of the meteor shower. This led to an echo rate that soared to 3 per second at 0340 UT, although it stayed at those ethereal heights for only ten minutes (Fig. 11.2). Moreover, just before the peak the antenna was turned to point directly *toward* the radiant, and, sure enough, the echo rate dropped by a factor of 25, a nice confirmation of Pierce's (1938) point that one obtained echoes only when looking normal to a meteor trail. They had also rigged up a movie camera on the cathode-ray tube and took three minutes of film at the shower's peak. This allowed accurate measurements of the intensities of individual echoes and detailed analysis of the decay of the longer-lived ones. Using Nicolai Herlofson's theory (Section 9.3), Lovell, C. J. Banwell, and Clegg (1947) also were able to say something about the size distribution of the meteors, finding an exponential distribution and a typical size of $\sim 10^{-2}$ cm. Planned visual observations, however, were thwarted by typical English weather allowing less than two clear hours.

Similar excitement reigned in Richmond Park where Hey's group also geared up for a Giacobinid watch (which turned out to be their swan song in meteor radar research). They invited the press out for a tour on the day before and received considerable publicity therefrom.[16] With a smaller antenna and constrained to look only toward the zenith, Hey's group recorded a peak of "only" 300 echoes per hour at 60 MHz, but their new item was a high-speed movie camera that John Parsons had built. Combined with a brighter display, this innovation allowed wholly new details of the strongest traces to be discerned, in particular a much sharper and weaker feature that in some cases preceded the main echo (as in Fig. 11.3). Hey, Parsons, and Stewart (1947) interpreted this precursor as a faint reflection off a ball of ionization (a "head") surrounding the meteor itself. Its quickly changing range was then a direct measure of the velocity of the meteor along the line of sight, and, with the assumption of a straight path, even the total velocity of the meteor was calculable. Less than 10% of the meteors yielded signs of a head, but 22 cases produced a remarkably uniform result. The Giacobinid stream was moving at a speed of 22.9 ± 1.3 km s^{-1}, a value in excellent agreement with theoretical calculations based on the associated comet's orbit.

Appleton and Naismith (1947) also made special efforts for the Giacobinids at the Radio Research Station, Slough. Operating at the much lower frequency of 27 MHz meant that they found an enhancement in the "ionization burst" rate of merely a factor 20 above normal and a duration of the shower of ~90 min, although still unlike anything previously seen. Appleton now strongly opined that the short-lived bursts were all meteoric and that a large fraction of the longer-lived, sporadic E bursts were probably likewise.

In the United States, four different groups observed the Giacobinids, although with fewer useful scientific results. A radio physics group at Stanford University (Section 11.3) monitored signals from two transmitters during the shower (Manning *et al.* 1946). They were particularly interested in the plainly audible whistles. At the height of the shower, bursts of signal strength overlapped so completely that the meteor-reflected signal was nearly continuous. Laurence Manning later recalled that "the visual display was nothing less than awe-inspiring, as the sky streaked with flashes of light, uncountably many per second at the peak."[17] At Harvard, John Pierce (1947), continuing his prewar work (Section 11.1.3), made observations similar to those at Stanford and also found abnormally high levels of ionization for over four hours. Recording over 3500 meteor whistles through the night, he explained them as the result of reflections off an "ionic bow wave" caused by the expanding (due to diffusion) girth of the meteor's ionization trail.

At Sterling, Virginia, Ross Bateman, Alvin G. McNish, and Victor C. Pineo (1946) of the Central Radio Propagation Laboratory of the National Bureau of Standards, used an Army SCR-270 radar. Although they had synchronized two outside cameras with a cathode-ray-tube camera, this went for nought as rain and fog prevailed. Their high operating frequency of 107 MHz meant that only the densest meteor trails produced echoes, but nevertheless a peak in the echo rate of 70 per hour showed up just before 0400 UT.[18] One last American experiment was a case of "too much producing too little." John Q. Stewart, the Princeton astronomer who had been Skellett's advisor fifteen years earlier, persuaded the US Army to stage an operation using no less than twenty-one radars in New Mexico, Idaho, Texas, and New Jersey. Although operating at five frequencies from 100 to 10,000 MHz, the only definite signals were at 100 MHz – nothing more than a *Sky & Telescope* article emerged (Stewart *et al.* 1947).[19]

---

[16] Hey (1973:32–3) later recounted an incident with respect to the planning of this press visit in which Appleton yet again attempted to gain credit for work in which he was minimally involved (see Section 6.3.2). Furthermore, immediately after the Giacobinids, Appleton issued a press release stating that "the Slough Radio Research Station first detected these echoes in 1932 and has continued to investigate them ever since. A theory was evolved that they were caused by the reflection of radio pulses from meteor trails, and evidence was progressively built up to support the theory" (*Science* **104**, 435 [1946]). The evidence on prewar ideas presented in Section 11.1, however, indicates that things did not proceed in so straightforward a manner.

[17] Manning:author, 31 August 1987.

[18] After a hiatus, this group resumed meteor radar research in 1948, directed toward the practical concerns of communications.

[19] Another part of this effort was to use the Project Diana radar system, which had successfully reached the moon in January 1946 (Section 12.3), to search for echoes from Comet Giacobini–Zinner itself. None were received, however, much to the disappointment

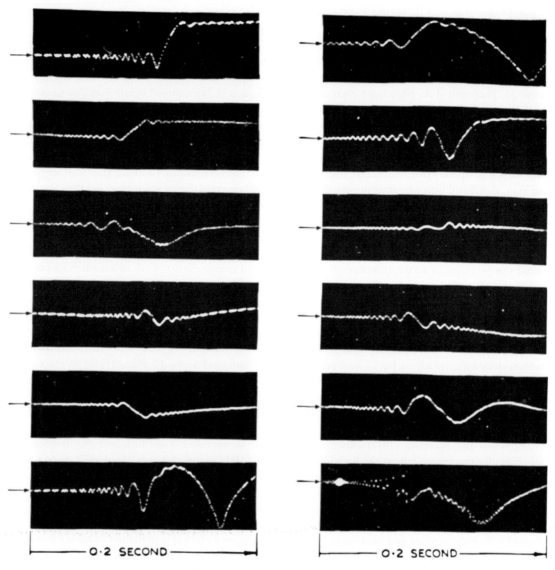

Figure 11.3 Examples of amplitude-time records of meteor echoes or "Doppler whistles," recorded at 30 MHz at Ottawa. Time increases to the right. Note the variations in frequency and amplitude of the oscillations before and after closest passage of the meteor (at mid-graph). Slow variations after passage arose from a "body Doppler" drift of the meteor trail caused by ionospheric winds. (McKinley 1951)

The eighth and final effort on the 1946 Giacobinids was the first radar or radio astronomy observation of any kind in the Soviet Union.[20] B. Yu. Levin, a longtime leading meteor researcher at the Geophysical Institute in Moscow, and P. O. Checkick found a maximum radar echo rate at 0350 UT, at least twenty times more than

---

of a national radio audience listening in on a live broadcast. [H.A. Zahl (1968), *Electrons Away* (New York: Vantage Press)]

[20] An earlier passive radio experiment involving meteors was carried out in the Technical Physics Institute, Ashkhabad, Turkmenistan. During the 1944 Leonids and 1945 Perseids, I. S. Astapovich, encouraged by Skellett's papers, searched for correlations (but found none) between visual meteors and sounds from a loudspeaker connected to a 375 kHz receiver. ["Radio observations of meteors in Turkmenistan in 1944–1945," *Izv. Turkmenistan Filial Akad. Nauk SSR*, No. 3–4, 163–9 (1946)]

that earlier the same night. Although Levin (1946) commented that the radar data were very interesting, there seems to have been no further meteor radar research in the Soviet Union until the 1950s.

Skellett had said that if radio people missed the opportunity of the 1932 Leonids, they wouldn't have another such chance until 1965, but he didn't reckon with the 1946 Giacobinids. Even the strongest doubters now had to admit that meteors and radar echoes went together. Hey and Stewart's (1946) note in *Nature* had already been published (Section 6.3.1), but the 1946 Giacobinids were *the* galvanizing event that forced astronomers to pay attention to the unconventional radar techniques. For instance, a Royal Astronomical Society (RAS) meeting two months later featured talks by Prentice, Lovell, and Hey, and even today the excitement in the reported discussion is palpable.[21] Whereas it had taken Prentice over twenty years to observe 25,000 meteors, Lovell had now chalked up 17,000 in less than a year, and with unprecedented accuracies. As Prentice himself put it: "We are like the first possessors of the telescope, unexpectedly armed with new powers of observation."

11.3 STANFORD AND OTTAWA

Harvard University and the Massachusetts Institute of Technology have long been rivals in everything from hiring faculty to winning crew races on the Charles River. Thus it was fitting that the principal wartime government laboratory for the development of microwave radar, the Radiation Lab at MIT (Section 10.1.1), was "opposed" by the principal locale for radar countermeasures work, Harvard's Radio Radiation Laboratory (RRL). Frederick E. Terman headed RRL and had recruited many of his staff from his academic home, Stanford University in California.[22] One of these was Oswald G. (Mike) Villard, Jr. (1916–2004), a young electrical engineer and amateur radio enthusiast who in 1945, while listening to shortwave broadcasts from New York, sometimes noticed unusual brief whistle-like sounds superimposed on the programming. Many of the whistles were accompanied by bursts of signal strength and Villard recalled the paper by Pierce (1938) proposing columns of ionization formed by a meteor. It soon all made sense: he was receiving the beat frequency between a direct ground wave and a wave reflected off the region of ionization surrounding the head of a fast moving meteor. The changes in whistle pitch resulted simply from the geometry of reflection off a trail passing an observer. The meteor's total velocity remained constant, but the component of velocity along the line of sight (which is what matters for a Doppler shift) changed smoothly from a negative (approaching) value through zero (at the point of the trail closest to the observer) to a positive value. This produced the usually oberved pattern of a first decreasing audio frequency, or descending whistle, sometimes followed by an ascending whistle. Villard discussed these ideas with meteor expert Fred Whipple (1906–2004) (also then on the RRL staff), who told him about the earlier reports of whistles by Chamanlal and Venkataraman (Section 11.1.3 and Tangent 11.1). But Villard's explanation of the changing Doppler shift was different, as he wrote in an article for the amateur radio magazine *QST* in 1946.

Villard returned for postgraduate study in electrical engineering at Stanford and, as time allowed, conducted experiments on the meteor whistles. When he and Laurence A. Manning (whose career had followed a similar path to Villard's) learned of the expected great Giacobinid meteor shower in October 1946, they naturally wanted to make an effort. Villard commandeered the Stanford Amateur Radio Club transmitter and set it up for continuous-wave (no pulses) operation at 29 MHz together with a half-wave dipole and receiver about a mile away. The experiment, with further assistance from Robert A. Helliwell and his student William E. Evans, Jr., came off well (as described in the previous section). Both during and after the brief shower peak, they were particularly impressed that whistles could be recorded with as little transmitter power as 1 kW (Manning *et al.* 1946).

After the Giacobinids, Manning (1948) concentrated on the theoretical side and began an analysis

---

[21] Report of the RAS meeting of 13 December 1946, *Observatory* **67**, 2–8 (quotation on p. 3) (1947).

[22] RRL was only about one-fifth the size of the Radiation Lab, and was staffed primarily by engineers, rather than the physicists of the Radiation Lab. Further details about Terman and RRL can be found in A. M. McMahon (1984), *The Making of a Profession: a Century of Electrical Engineering in America* (New York: IEEE Press), pp. 183–206; and in C. S. Gillmor (2004), *Fred Terman at Stanford: Building a Discipline, a University, and Silicon Valley* (Stanford: Stanford University Press).

of whistles, using a model that they resulted from a Doppler shift off a moving head of ionization. He published details of how to determine meteor speed from a single station, how to use three-station data to get the full vectors of meteor position and velocity, and how to correct observed meteor burst rates for aspect effects. The determination of speed v was straightforward if one had measured the rate of change of whistle frequency v' (as in Fig. 11.3) near the point of closest approach: $v' = -2\, v^2/(\lambda R)$, where λ is the wavelength and R is the range to the meteor. The relationship between this whistle Doppler method for obtaining speeds and the Fresnel diffraction method developed at Jodrell Bank (Section 9.2) was at first not clear. Eventually, though, it was realized that the two approaches were essentially equivalent near the point of closest approach, but that the Fresnel theory was a more complete description.[23] Furthermore, "head echoes" off a region of ionization surrounding the meteor itself were found to be rare.

Villard (1948) pursued the meteor whistle work for his doctorate.[24] He built three transmitters for a range of frequencies from 6 to 50 MHz and developed various recording methods to study the relationship between ionization bursts and details of the whistle waveform. His observations culminated on one night of the August 1948 Perseid shower, in concert with Manning and Allen M. Peterson (1922–1994), a new student in the group (Manning *et al.* 1949). Two continuous-wave transmitters for obtaining whistles were operated at differing frequencies and a third in pulse mode was for ranges; the antennas were no more than half-wave dipoles. A team of visual observers, stationed in the nearby hills, relayed their sightings by radio. Speeds were derived from meteor whistles, but for only six of these were unambiguous visual sightings also in hand (this was necessary to establish identity as a Perseid because of the lack of any directivity in the transmitter antennas). Nevertheless, these six cases produced a remarkably concise mean speed of 62.3 ± 1.6 km s^{-1} for the shower. The only point of comparison was one Perseid that Whipple had captured photographically in 1937, and it gave 61.2 km s^{-1}, so the method seemed reliable – the whistles gave excellent accuracy and derived speeds really did refer to the meteor head, not to something else such as an expanding region of ionization.[25]

Manning, Villard, and Peterson, joined later by many others, notably Von R. Eshleman as a student in 1949, continued their work on meteors throughout the 1950s, but the 1948 Perseids represented their last foray into the astronomical side of things. Meteors were viewed as a marvellous probe of the ionosphere, but there was no interest in designing research programs for astronomical ends. They were in a department of electrical engineering and the group's focus was on communications and ionospheric propagation effects. They also followed their department head Terman's philosophy in concentrating on developing techniques and exploiting them only for a short while, leaving long-term follow-up to other groups.[26] Furthermore, from the beginning support came from substantial contracts with the Air Force, the Army Signal Corps, and the Office of Naval Research – one case of many at Stanford in which Terman built up a large research group based on military funding.[27] Meteor radar research was thus pushed toward practical ends and – as it became apparent in the early 1950s that a viable communications system could be based on reflections off meteoric ionization

---

[23] See Tangent 11.3 for further information.

[24] This 1948 Ph.D. in meteor *radar* astronomy was a year earlier than the first Ph.D. awarded for *radio* astronomy research, to Jean-François Denisse (1949b) (Section 10.4).

[25] Primary sources for the early years of the Stanford group are an interview with Manning (1979), Manning:author (31 August 1987), and Chapter 1 ("Historical Summary") of Villard's 1948 thesis. This chapter outlines the "student's story" of how the topic developed from start to end. Would that all theses contained such a chapter.

[26] Manning:author, 31 August 1987.

[27] On the build-up of electrical engineering and physics research at Stanford during these years, based almost entirely on military funds and largely engineered by Terman, see S. W. Leslie and B. Hevly in "Steeple building at Stanford: electrical engineering, physics and microwave research," *Proc. IEEE* **73**, 1169–80 (1985); Leslie, "Playing the education game to win: the military and interdisciplinary research at Stanford," *Hist. Stud. Phys. & Biol. Sci.* **18**, 55–88 (1987); Leslie (1993), *The Cold War and American Science* (New York: Columbia University Press); R. S. Lowen (1997), *Creating the Cold War University: the Transformation of Stanford* (Berkeley: University California Press); and C. S. Gillmor (2004), as in note 22. Much research in later decades by the Stanford meteor radar group was directed to military needs – for example, Villard later was a key developer of over-the-horizon radar systems.

Table 11.1 *Principal early groups in meteor radar*

Initial year	Location	No. publications in meteor radar before 1950	Principal researchers before 1950
1945	Army Operational Research Group, Richmond Park, England	6	Hey, Stewart
1945	Jodrell Bank, Dept. Physics, U. Manchester, England	25	Lovell, Clegg, Prentice, Davies, Ellyett, Herlofson
1945	Electrical Eng. Dept., Stanford U., California, USA	7	Villard, Manning, Eshleman, Peterson
1947	Radio & Electr. Div., Natl Research Council/ Dominion Observatory, Ottawa, Canada	7	Millman, McKinley

These four groups accounted for about 75% of all meteor radar papers during 1946–50 (based on a modified version of a bibliography assembled by Gilbert [1975]).

bursts (Section 11.5) – also toward a perfect meshing of the interests of the military and the Stanford group (see Section 17.3.5).

In Canada the last of the principal early groups in meteor radar (Table 11.1) launched also as a result of the 1946 Giacobinids. Impetus came from news of the British observations in a talk given in Ottawa by Appleton in late 1946. Peter M. Millman (1906–1990) and Donald R. W. McKinley (1912–1984) both heard the talk, became enthused with the possibilities, and started a major program that lasted over a decade. Millman had obtained his doctorate in astronomy at Harvard in 1932, analyzing the few optical spectra of meteors then existing. He subsequently made significant contributions to identifying the elements causing emission lines observed from meteors, and during the 1930s obtained many more spectra at the David Dunlap Observatory, Toronto. During the war Millman was involved in operational research as a Canadian liaison officer in England, and thus came to meet Hey and learn of his and Stewart's early meteor echo work. He returned to a post at the Dominion Observatory, Ottawa, and after Appleton's talk teamed up with McKinley, whom he had known as a student at Toronto. McKinley, with a Ph.D. in experimental physics, had managed several microwave radar projects during the war for the National Research Council, and was now part of its Radio and Electrical Engineering Division (Section 10.2). The matchup of McKinley's radar expertise and Millman's astronomical knowledge seemed ideal.

Millman felt that a major weakness of the early radar studies was a lack of thorough work on visual/radar correlations (McKinley and Millman 1949:364). Thus the first observations in Ottawa, at the time of the August 1947 Perseids, included six visual observers as well as direct and spectrographic cameras. On the radio side McKinley suitably modified a surplus 33 MHz radar, operating simply with a half-wave dipole for antenna. Altogether, 3700 echoes and 1100 visual meteors were obtained over five nights (Millman, McKinley, and Burland 1948). As the Jodrell Bank workers had found, the longer-lasting echoes correlated well with the brightest meteors, but less than 3% of the echoes correlated at all and many bright meteors had no corresponding echo. With such a wealth of visual data, in particular each meteor's distance from the shower radiant, they were also able to establish that a good fraction of the long-lived radar echoes originated when the meteor was travelling other than perpendicular to the line of sight. These echoes were postulated to come not from a continuous column of ionization, but from a string of ionic clouds.

Since their operating wavelength of ~10 m and transmitter powers of up to 400 kW appeared to yield no pronounced aspect effect for the long-lived echoes, they decided to try radar triangulation. In August 1948 a three-station radar setup spread over a 50 mile region went into operation. They had been impressed with the Stanford work on whistles and using that technique acquired a very large sample of meteor velocities in order to search for any possible interstellar component (Section 11.4.2). By the end of 1948 Millman and McKinley (1949) had toted up 100 meteor photographs, 5000 visual meteor sightings, 50,000 photographic records of whistles (Fig. 11.3), and 1,500,000 radar echoes! With this plethora they formulated a detailed phenomenological scheme classifying the filmed appearances of echoes on the radar screen (McKinley and Millman 1949). They were struck by the speed and regularity with which a wide variety of observed morphologies were formed. Since ionospheric winds should not behave so regularly, they instead postulated the existence of an M ("meteor") region in which meteor ionization took place, slightly below but overlapping the E region. The M region contained striae or patches, ~1 km in size and spaced by ~5 km, that for some unspecified reason were more easily ionized by a passing meteor. This created trails with fine structure along their length and, depending on viewing angle and how the sectors of the trail diffused with time, could produce many types of echo.[28]

11.4 SCIENTIFIC RESULTS BEFORE 1952

11.4.1 Daytime showers

With the meteor radar technique in hand it was of course apparent that for the first time the *daytime* half of the sky could be explored. Indeed, as discussed earlier, Hey and Stewart (1947) in their initial three-station monitoring had found a daytime shower in June 1945 (Section 6.3.1) and Lovell's group extended this with the discovery of an entire complex of daytime showers in mid-1947 (Section 9.2). The delineation of these showers over succeeding summers became one of the major programs at Jodrell Bank and by 1951 the previously unknown streams of debris bombarding the sunlit side of the earth had been thoroughly mapped.

In May and June 1947 Lovell's group found that the (already known) nighttime η Aquarid shower was followed by a remarkable procession of daytime showers stretching until early August. These showers seemed to "follow" the sun in the sense that their radiants were always 1–2 hours preceding the sun's shifting position with season. The basic observing method had been devised and tested by Clegg (1948b) the previous winter. He had shown that one could derive radiants to an accuracy of a few degrees with only a single antenna of modest directivity – they used the "searchlight aerial" (Fig. 9.2) with its 13° × 28° beam. The geometry of the method was complicated, but its main principle was that an antenna responded only to those trails perpendicular to its beam axis and occurring at the E region height of ~100 km. It was possible to calculate the expected (relative) echo rates and radar ranges of echoes associated with a given radiant position. Turning this around, the Jodrell Bank group could derive a radiant by studying the changing echo rates and ranges over a few hours at any one antenna position, and then repeating this for other antenna positions.

In practice the antenna was pointed just above the eastern horizon, where shower echoes would have a maximum range (and be first observed) when the radiant passed due south. From this the radiant's right ascension was known. Ranges from a given shower would then steadily decrease until no more were detected. Standard procedure was then to move the antenna to a southwest position and again watch the radiant move through the "collecting area" perpendicular to the beam's direction. The lag in time between these two instances of maximum range then depended in a known way on the declination of the radiant. That was Clegg's method in a nutshell, although there were many corrections needed for effects such as refraction and height of the meteor zone (faster showers produced meteors at higher altitudes). The main uncertainties arose from the large beamwidth, which meant that at

---

[28] Sources for the beginnings of the Ottawa group are a 1979 interview with Millman, as well as Millman and McKinley (1967). Also see I. Halliday's (1991) obituary of Millman in *JRASC* 85, 67–78.

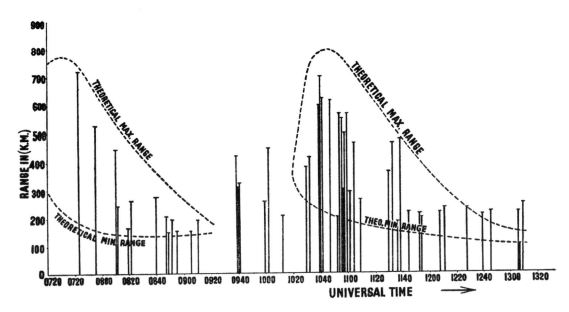

Figure 11.4 Ranges and times of radar echoes observed with the "searchlight aerial" (Fig. 9.2) over a six-hour period on 7 May 1947. The η Aquarid shower was represented by the group extending to 0900 and the newly discovered daytime shower (then called the Piscids) came in at 1030. Theoretical curves such as those indicated allowed determination of the positions of shower radiants (see text). (Clegg 1948a)

any given time one expected in fact a range of ranges. Furthermore, shower echoes always had to compete with the background echo rate. Figure 11.4 illustrates the data and the kinds of curves that were fit.

During May, June, and July of 1947, for seven hours daily, ranges of the daytime echoes were read off a cathode ray tube, until at last the meteors petered out. Over this period the echo rate seemed stuck in the 10–50 per hour range (versus a normal 1–2 per hour). And because it was known from nighttime shower work that the Jodrell Bank equipment fortuitously gave similar echo rates to those seen by a single visual observer, it was argued that these new showers far surpassed anything seen after dark (Fig. 11.5). But mapping out the showers by fitting curves such as those in Fig. 11.4 to the complex summer daytime showers turned out to be no easy task, fraught with ambiguities. Clegg, Hughes, and Lovell (1947) in the end produced the map shown in Fig. 11.6, where a total of six new showers beyond the η Aquarids were displayed with radiants following a line just a bit north of the ecliptic plane and dominated by the Piscids, the Arietids, and the Taurids.

Their interest piqued and now joined by new student Arnold Aspinall, Lovell and Clegg similarly monitored the 1948 daytime showers and found that they had the same general pattern, although now a total of *ten* daytime radiants could be distinguished from Aquarius to Gemini (Aspinall *et al.* 1949). But because something better was needed to untangle properly the complex structure, Clegg and Aspinall had the Jodrell Bank workshop build for them what became known as the "radiant aerials" (Fig. 11.7) – two 73 MHz Yagi arrays, one facing 28° south of the western horizon and the other 22° north.[29] Each antenna was fed from the same transmitter and the resulting echoes separately received and photographically recorded side by side. Several improvements over the old system now allowed radiants to be determined to ~1.5° accuracy: the two antennas operated simultaneously and automatically,

---

[29] It was intended that the two aerials be symmetric about an east–west line, but the 3° offset arose because of a surveying error in which the 9 minute difference between true noon at Greenwich and at Jodrell Bank was neglected. [G. S Hawkins:author, December 1985]

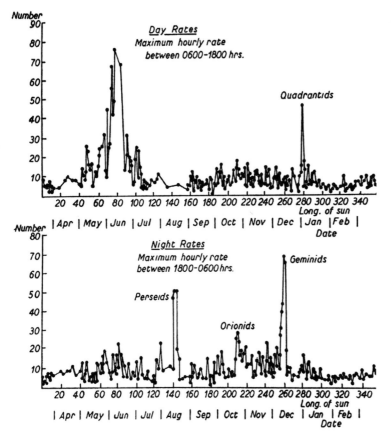

Figure 11.5 Distribution of meteoric activity over the year 1949–50 as observed at Jodrell Bank. The unique character and strength of the summer daytime showers (top plot) are apparent; the main peak comprised the Arietids and ζ Perseids, while the secondary peaks on either side were the η Aquarids and β Taurids. Although visible in the darkness, the Quadrantids appear as a daytime shower here because their peak occurs shortly after 6 a.m. Note the strength of the summer daytime showers compared to the familiar largest nighttime showers. (Lovell and Clegg 1952:98)

beamwidths were ~10°, the system recorded a higher echo rate, and a double-pulse technique allowed much better discrimination between noise blips and real echoes (Aspinall, Clegg and Hawkins 1951).

From September 1949 onwards Jodrell Bank was thus in business for a comprehensive, sensitive, 365-day, 24-hour survey. Also at this time Gerald S. Hawkins (1928–2003), who as an eighteen-year-old had been one of Prentice's apprentices at Jodrell Bank for the 1946 Perseids (Section 9.2), joined the group as its first student with a strong astronomical background. He took up residence in the "radiant hut" (visible in Fig. 11.7) and for several years was in charge of the interpretation of the general survey. Aspinall and Hawkins (1951) mapped the summer daytime radiants with much greater accuracy, as well as reexamining data from previous summers in light of the new information. Only three daytime streams could be confidently identified as having occurred each summer; they concluded that many previously published radiants arose from a mixture of ambiguities, errors, and real changes from year to year. The Arietids dominated the scene for the entire first half of June, followed by the strong ζ Perseids with a radiant only 14° to the east. But perhaps the most interesting daytime meteors were the β Taurids of early July, for

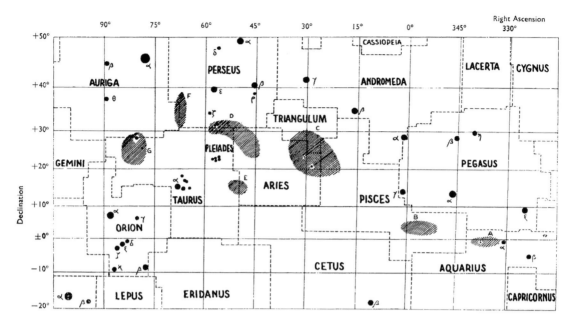

Figure 11.6 Daytime shower radiants (hatched blobs labelled with capital letters) as delineated in the summer of 1947 at Jodrell Bank. The map also shows official boundaries of the constellations and the brightest stars. The radiants were mainly slightly north of the ecliptic plane and situated about two hours (30°) before the sun. Named showers and periods of activity in 1947: A – η Aquarids (1–10 May), C – Piscids (6–15 May), D + E – mixture of Arietids and ζ Perseids (30 May – 17 June), and G – β Taurids (20 June onward). This first map of radiants was later substantially revised (see text). (Clegg, Hughes, and Lovell 1947)

their existence had in fact been predicted. As Lovell related at a March 1948 meeting of the RAS:

> Whipple and Hoffmeister have independently sent me a list of daylight radiants to be expected from the known orbits of short-period comets of small inclination, the other sides of whose orbits produce radiants in the night sky at the opposite season of the year. Both predict that Encke's comet should give a daylight shower with maximum at June 22 and their calculated radiants agree remarkably well with the active centre near β Tauri.[30]

Encke's comet, with a period of only 3.3 years, had been recognized for many years to be the source of interplanetary debris in a large sun-girdling torus. The Taurids of October and November emanated from this, and Fred Whipple (1940) had written that in late June and early July the earth should again intersect this stream as it moved away from the sun. At that time Whipple's only hope for verification was that perhaps a few daytime fireballs might be observed, but now meteor radar had neatly done the job.

Stronger evidence for the association of the daytime streams with nighttime counterparts came after accurate velocities were measured in the summer of 1950 by research students John G. Davies and J. Stanley Greenhow (1951). Furthermore, student Mary Almond (1951) produced orbits for each of the main summer streams after computational advice from J.G. Porter of the Royal Greenwich Observatory (Fig. 11.8). With periods of 1.5–3.2 years and paths that kept them inside Jupiter's orbit, these orbits were on average a good bit smaller than those for the nighttime meteor streams, suggesting a different origin for the daytime particles. And it turned out that, in addition to the β Taurids, two more of the daytime streams had low enough inclinations to the plane of the earth's orbit that they too intersected it twice and thus also produced nighttime showers.

11.4.2 Meteors from interstellar space?

Radar turned out to be ideal for investigating one of the key controversies in meteor astronomy of the 1940s: are

---

[30] *Observatory* **68**, p. 52 (1948). Also see C. Hoffmeister, *Observatory* **70**, p. 75 (1950).

Figure 11.7 The "radiant aerials" used for determination of the radiants of meteor showers and for studies of sporadic meteors at Jodrell Bank (ca. 1950). In the front and back of the hut are two arrays of six horizontally directed Yagi antennas, operating at a wavelength of 4.1 m. One radar array is directed slightly south of due west and the other slightly north.

there any meteors whose orbits are hyperbolic in shape and which therefore must have originated in interstellar regions? The possibility of an interstellar component excited astronomers and had therefore received frequent attention. Most studies had concluded that the constituent particles of meteor showers followed bound, elliptic orbits, and some showers could be associated with specific recurrent comets, but what of the so-called sporadic meteors comprising the bulk of objects hitting the earth every night?

In 1925 astronomer Cuno Hoffmeister (1892–1968) of Sonneberg, Germany published a catalog of the most reliably determined meteor speeds. This compilation of visual data on track lengths and times of duration for 600 fireballs led to the conclusion that ~80% of the sample had geocentric speeds greater than $72 \,\mathrm{km\,s^{-1}}$ and were therefore of interstellar origin.[31] This became the accepted view in astronomy, and

meant that studying meteors was possibly a direct way to learn about the composition of interstellar regions of our Galaxy. Harlow Shapley of Harvard College Observatory became excited by the idea of meteors as a branch of galactic research (his specialty) and in 1930 mounted a major meteor expedition to Arizona to study the "contents of interstellar space" (Doel 1996:22–33). Leading the group was the Estonian astronomer Ernst J. Öpik (1893–1985), who invented an ingenious device called a rocking mirror for observations of meteor speeds. The observer looked down onto a mirror driven so that its normal described a cone on the sky at a 10 Hz rate. Stars appeared as 0.5°–sized ellipses and a meteor shooting across the field of view traced out a cycloid-like curve. The meteor's time of flight could be figured from counting the number of loops seen in the mirror. Öpik made a complex statistical analysis of his Arizona data, finding ~50–70% of the meteoric orbits to be hyperbolic. He strongly argued that sporadic meteors originated outside the solar system and that meteoric particles populated most of the space between stars. Further confirming observations in the late 1930s at

---

[31] This critical speed of $72 \,\mathrm{km\,s^{-1}}$ is $(30 + \sqrt{2} \times 30) \,\mathrm{km\,s^{-1}}$, where $30 \,\mathrm{km\,s^{-1}}$ is the earth's orbital speed. Further discussion is in Tangent 11.4.

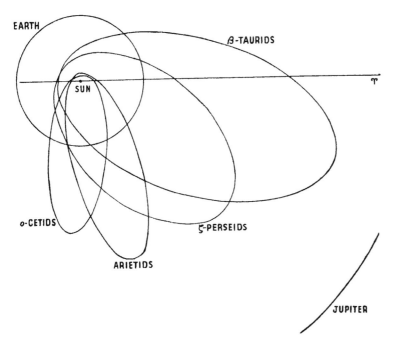

Figure 11.8 Orbits of four summer daytime meteor showers projected on to the ecliptic plane, as calculated from 1950 observations by Mary Almond (1951). Inclinations to the plane ranged from 4° to 34°. Note earth's orbit and a short section of Jupiter's orbit for scale.

Tartu, Estonia increased his sample size to almost 800 meteors.

During World War II Porter severely criticized Öpik's results, saying that many key assumptions in the statistical anaylsis were unfounded and that the two stations used in Arizona were too closely spaced (40 km). Furthermore, he could discern no definite hyberbolic velocities in his own compilation of all available (1200) meteors observed by at least two stations. Öpik parried by arguing that Porter's data set, derived mostly from amateurs' records and extending back to the nineteenth century, was too heterogeneous to yield reliable results.

The final pre-radar approach to meteor velocities, also at Harvard in the 1930s, was first developed by E. Dorrit Hoffleit and Millman and then extensively by Whipple. Much of their interest was fomented by Öpik's presence at Harvard and the prospect of studying actual interstellar particles falling to the earth.[32] They photographed meteor trails from two different stations with cameras having rotating vanes in front, creating periodically interrupted lines. Accuracy was unparalled, but limited to only the brightest meteors (at least 0^m) and in practice required a hundred hours of film exposure to secure a single meteor velocity. Whipple nevertheless persevered and managed over the period 1936–50 to nab seven sporadic meteors, only one of which (possibly) had a hyperbolic orbit. In reply Öpik pointed out that Whipple's results applied only to the very brightest and largest meteors, which might not be a fair sample of meteors as a whole, and that the photographic method was inherently biased against faster meteors, which exposed each bit of film for a shorter time and thus left a fainter track.

This then was the scene that the radar meteor people happened upon.[33] The problem looked ripe for solution by using radar's capability to amass a huge number of reasonably accurate speeds. Would a significant number of speeds, or indeed even *one*, be above $72\,km\,s^{-1}$? At Jodrell Bank, where Clifton Ellyett and Davies had

---

[32] Whipple (1979:122B); F. L. Whipple (1972), "The incentive of a bold hypothesis: hyperbolic meteors and comets," *Proc. New York Acad. Sci.* **198**, 219–24; Whipple:Lovell, 12 December 1949, file R17 (uncat.), LOV; Doel (1996:22–33).

[33] In this summary of pre-radar data and arguments about meteor velocities I have drawn heavily on Chapters 8–12 of Lovell (1954).

developed the Fresnel method for measuring speeds, Lovell in the end invested four years of data-taking in the controversy. At a conference held at Manchester University in September 1948 (Section 9.6) Whipple urged the radar workers to continue their contributions to meteoric astronomy, in particular "by verifying the existence of hyperbolic meteors in the fainter visual range or setting an upper limit."[34] Hoffmeister was forbidden to leave East Germany to attend, but Öpik was there, sanguine that radar techniques would break the impasse since his own theory of meteor interaction with the atmosphere predicted that the ratio of ionization to light increased sharply for higher velocities. This would account for the lack of hyperbolic meteors in Whipple's bright-meteor sample, and it gave him hope that radar would be able to investigate much fainter, and presumably faster-moving, objects.

Only ten days after this conference Lovell's group commenced their program to settle the issue. The experimental design suggested by the astronomers was clever: the "searchlight aerial" (Fig. 9.2) would be pointed such that it would most likely pick up meteors of the highest speed, namely those approaching from the earth's apex, the direction of its orbital motion. This configuration was most favorable for obtaining meteors with speeds greater than $72\,\mathrm{km\,s^{-1}}$, that is, unambiguously of interstellar origin. In practice this "apex experiment" involved autumnal observations with an antenna pointing at sunrise toward the eastern horizon. Such observations were sensitive to meteor trails with radiants along the north–south meridian, which near the autumnal equinox includes the apex. Thus it was that for 56 days in the fall of 1948 Almond, Davies, and Lovell (1951) took sunrise data and acquired over 1400 echoes. A first analysis of about 15% of the total sample led to the surprising conclusion that as many as one-third of the meteors indeed were of interstellar origin. Lovell wrote to Öpik in January 1949:

> You will no doubt remember that we were fairly convinced that all the meteors we observed here had less than parabolic velocities, and we regarded the experiment, which Whipple and yourself urged us to do, as in the nature of a "negative results" experiment. However, it has turned out far from being in this category, since we have measured many hyperbolic velocities.... The results are so amazing that we wish to avoid any premature publicity until we have completed the analysis.[35]

And when reporting the same results to Whipple, he pressed him as to whether speeds $>72\,\mathrm{km\,s^{-1}}$ *had* to be interstellar:

> If our measurements show that some of these meteors have a velocity greater than $72\,\mathrm{km/s}$, is it quite certain that these are interstellar?... We have no one here sufficiently experienced in astronomical thinking to work out the implications.[36]

But by March Lovell's group changed its stance and came to distrust their putative hyperbolic meteors because they found that higher-speed meteors produced shorter echoes, which yielded unsatsifactory pulse patterns for recognition of a Fresnel pattern. This effect, plus echoes that were too weak or had no measured range, meant that only about 5% of all echoes turned out to be reliable for measuring speeds – and of those the final census had only five speeds (of 67 total) greater than $72\,\mathrm{km\,s^{-1}}$. But before any publication, the bias against high speeds had to be dealt with, and the simplest way around it was to switch to a longer wavelength where echoes lasted longer. In the autumn of 1949 they thus moved from 4.2 to 8.1 m and now figured that they could reliably detect meteors even as fast as $140\,\mathrm{km\,s^{-1}}$. Overall results, however, did not change – only ~10% of the speeds were greater than $72\,\mathrm{km\,s^{-1}}$ and ~2% greater than $80\,\mathrm{km\,s^{-1}}$.

Öpik and Hoffmeister refused to give up, and criticized these experiments (even before the results were formally published) as being subject to effects selecting against velocities above ~$70\,\mathrm{km\,s^{-1}}$ and thus eliminating the strong interstellar component that they felt sure was there.[37] In response, the Jodrell Bank team mounted a third effort, an *ant*apex experiment that observed

---

[34] *Observatory* **68**, 226–32 (1948) (quotation from p. 227). See Lovell (1990:140–2) for a personal view of this conference and its disputes and disputants; for example, Öpik is described as "entirely absorbed and tenacious."

[35] Lovell:Öpik, 12 January 1949, file R6 (uncat.), LOV.

[36] Lovell:Whipple, 31 December 1948, file R17 (uncat.), LOV.

[37] To follow the debate blow-by-blow at its peak, see Hoffmeister (1950), *Observatory* **70**, 70–6; Almond et al. (1950), *Observatory* **70**, 112–3; Millman and McKinley (1950), *Observatory* **70**, 156–8; Öpik (1950), *Irish Astron. J.* **1**, 80–96; and Lovell (1954:234).

eastwards again, but now in the spring and at sunset. This was expected to yield the *lowest* possible speeds. If a large proportion of meteors at speeds well above 72 km s^{-1} existed *and* if a marked ~65 km s^{-1} cutoff of observed speeds that had shown up in the apex experiment was indeed of instrumental origin, then antapex observations should likewise show a 65 km s^{-1} cutoff. But if the cutoff was a real property of the meteor speeds, then the antapex distribution of speeds should be well shifted to lower values. In the event, the antapex distribution did shift to the 20–40 km s^{-1} range (similar to that shown in Fig. 11.9) and Almond *et al.* (1951) therefore all the more confidently concluded that there was no evidence for a significant hyperbolic component.

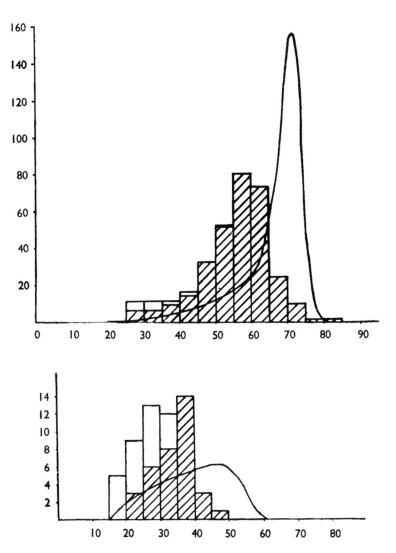

Figure 11.9 Distributions of meteor speeds (km s^{-1}) obtained at Jodrell Bank for the autumn 1950 apex (top) and spring 1951 antapex (bottom) experiments. The non-hatched areas represent meteors whose ranges of less than 250 km implied that they originated from an antenna sidelobe, and therefore not from the apex or antapex. The curves are the distributions expected for uniformly distributed meteors in parabolic orbits. (Almond, Davies, and Lovell 1952)

Still, there was the possibility that yet fainter meteors would behave differently and so another year's worth of apex and antapex data were taken with a radar system fully ten times more powerful. The resulting distributions (Fig. 11.9) were close repeats of the first set, but several results were modified. First, calculations by Greenhow and Hawkins (1952) had revamped the conversion from echo strength to optical magnitude (see below) and accordingly Almond *et al.* (1952) now estimated that their sample included meteors as faint as 6.0–7.5^m. Second, their comparison of observed and expected distributions was revised. Clegg (1952) worked out, assuming the meteor orbits were all parabolic (on the elliptic/hyperbolic borderline) and randomly oriented, how their observed distribution would be affected by the broad beam, by aspect effects of reflection off the trails, and by errors in measured speeds. Although a significant error in these calculations was not caught until the second paper (Almond *et al.* 1952), the new comparison of observed versus expected made it even clearer that hyperbolic meteors were, at the most, insignificant in number.

The final step was to push to meteors as faint as eighth magnitude by tilting the antenna elements upwards in the autumn of 1951, considerably reducing the typical range of observed meteors. Similar conclusions again ensued (Almond, Davies, and Lovell 1953). Altogether about a thousand speeds had been measured by now and only seven cases exceeded 80 km s^{-1}. These could not be ruled out as a possible interstellar component, but their numbers profoundly disagreed with the expectations of Hoffmeister and Öpik. Furthermore, the observed distributions indicated that the detected meteors travelled on average in elliptic orbits of relatively short period (~2 years).

Over this same period the hyperbolic orbit question had also impressed McKinley, who responded with measurements at Ottawa of no less than 11,000 speeds! He used the continuous-wave technique at a frequency of 30 MHz for monthly periods of 1–3 days from December 1948 to March 1950. At this lower frequency than Jodrell Bank, McKinley encountered more manmade interference and a loss of sensitivity whenever the Milky Way was above the horizon, but on the other hand he had a higher echo rate and as many as 10% of his echoes were suitable for measuring speeds. Antenna directivity was also sacrificed – simple half-wave dipoles mounted 1/4 wavelength above the ground were used for both transmitting and receiving (8 km apart). This meant that interpretation of the observations required more assumptions than for the Jodrell Bank experiments, but also that more echoes could be picked up.

McKinley (1951) employed photographs of the "whistle" oscillations in signal amplitude (Fig. 11.3) and estimated that resultant speeds were accurate to 5% and that speeds as high as 150 km s^{-1}, if present, could be satisfactorily measured. Yet he found only 0.3% of his sample with speeds as high as 75–84 km s^{-1}. Since even these few candidates might well have undergone planetary perturbations, he concluded that he too had no definite evidence for interstellar meteors. McKinley also found that his observed meteors were best explained as having heliocentric velocities close to the parabolic limit and a preponderance of direct orbits over retrograde (Fig. 11.10).

One last approach to the question of the origin of sporadic meteors culminated in the Ph.D. thesis of Hawkins (1952), although not published in the regular literature (Hawkins 1956) until after he had emigrated to Harvard to work with Whipple. The history of this approach extended back to 1866 work in Milan by Giovanni V. Schiaparelli, who worked out the expected diurnal and annual variation in meteor rates under the assumption that meteors impinged upon the earth with uniform speeds from all directions. Since the slower the mean heliocentric velocities of the meteors, the more efficacious the earth's orbital motion in causing a modulation of meteor rates, observations of hourly rates alone in principle could provide information on the mean velocities of meteors. Over the years this method had been refined and applied to ever more data, usually indicating that meteor orbits were hyperbolic. Hoffmeister too had applied this type of analysis to his own extensive observations in South Africa, and had concluded by the late 1930s that nearly 70% of the meteors were interstellar in origin. Hawkins's doctoral project was to use the radiant aerials (Fig. 11.7) in 1949–51 to record over 200,000 echoes from sporadic meteors. The yearly average of his measured hourly rates showed daily peaks at 2, 6, and 10 a.m., that is, fixed with respect to the sun, not to interstellar space. Such peaks implied that meteors came preferentially from the ecliptic plane (therefore not of interstellar origin) and overwhelmingly had direct elliptic orbits of high eccentricity.

Neither Hoffmeister nor Öpik, however, accepted the radar verdict. Öpik vehemently argued that the

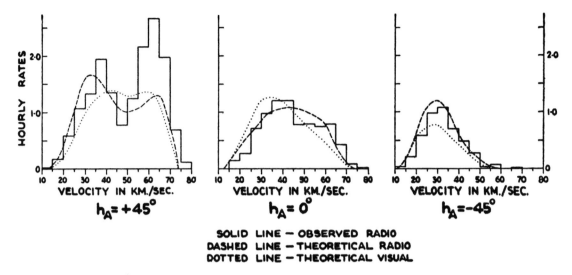

Figure 11.10 Distributions of meteor speeds obtained at Ottawa from 1948 to 1951. The histogram in each case corresponds to the measured radar echoes (with a halfwave dipole antenna) and the dashed-line curve is the best-fit distribution expected on the assumption of near-parabolic orbits and a uniform distribution of meteor radiants except for a concentration at 135° from the apex (i.e., more direct orbits than retrograde); the dotted curve refers to a prediction for visually-observed meteors. The meteors selected for each plot occurred at times corresponding to the indicated value for h_A, the elevation angle of the earth's apex. (McKinley 1951)

highest velocity meteors were still being discriminated against (either by producing echoes too weak or too short-lived to show clean diffraction patterns), and fought on for many more years, in particular in a scathing review of Lovell's 1954 book *Meteor Astronomy* (Öpik 1955). Hoffmeister for his part maintained that the apparent speed of the ionization trail could well be slower than that of the meteor itself. Nevertheless, most meteor experts accepted the radar results and agreed that classical analyses had suffered from various biases such as the assumption of meteors approaching uniformly from all directions and incompleteness of data sets derived from visual observing.[38] As Whipple wrote to Lovell in 1950:

> This completes beautifully a chapter in the history of meteor astronomy. Thank heavens that we can now proceed without this continual uncertainty

concerning this interstellar contribution to the meteors in the visual range.[39,40]

11.4.3 Ionosphere physics

Meteor radar, a child of prewar radio studies of the ionosphere, did not forget its origins and throughout its postwar heyday made important contributions to knowledge of the ionosphere. In this section some of this research is outlined, with emphasis on those aspects affecting astronomical interpretations.

Evidence for winds in the upper atmosphere had accumulated over the prewar years from the visible motions of aurorae, noctilucent clouds, and long-lived trails of visual meteors. But meteor radar provided the best *quantitative* studies of these winds. All the early groups noticed that after a few seconds the longest-lived trails often exhibited amplitude fluctuations, which were taken as a natural consequence of E region winds distorting the initial trail. Furthermore, small drifts in

---

[38] Although Öpik never acquiesced to the radar results, much later he did accept a 1961 optical study that found less than 1% of all meteors to have hyperbolic orbits. This led him to admit in a remarkable article (scientists seldom discuss their serious mistakes!) that analysis of the Arizona Expedition's meteor velocities must have been badly in error, and to discuss the probable reasons why this had occurred. [Öpik, "The failures: interstellar meteors," *Irish Astron. J.* **9**, 156–9 (1969)]

[39] Whipple:Lovell, 18 May 1950, file R17 (uncat.), LOV.
[40] See Gunn (2005:111–116) for another account of Jodrell Bank's role in the interstellar meteor controversy, in particular with more details of the heated year of 1950 (as in the papers cited in note 36).

range could sometimes be detected, apparently arising from a bulk drift of the entire trail. The Canadian group also found evidence for complex distributions of scattering ions in the varied forms of recorded echoes (Fig. 11.3).

Ionospheric winds were studied mainly at Stanford and Jodrell Bank. Manning, Villard, and Peterson (1950) measured what they called "body Doppler," that is, few-hertz shifts occurring well after the whistle, corresponding to bulk motions of ~40 m s^{-1}. Wind vectors were determined by measuring the variation of this Doppler shift at different azimuths. Meanwhile, Ellyett (1950) and Greenhow (1950) studied intensity fluctuations in long-lived echoes at Jodrell Bank, using the same pulse radar technique as for the Fresnel velocity measures. Fluctuations with periods of 0.01–0.1 s were taken to indicate strong windspeed gradients or turbulence of order 20 m s^{-1}. Continuation of this type of work through the early 1950s eventually enabled measurement of distinct diurnal and seasonal patterns in the wind vectors.

The other principal theme in meteor ionosphere studies was the physics of formation of a trail and its interaction with incident radio waves. The first theories of Pierce (1938) and Blackett and Lovell (1941) pioneered the way, but required substantial revision when tested against new data and theoretical ideas. Lovell and Clegg (1948) corrected the 1941 work (Section 9.1) for the case of a column of ions and electrons with radius much smaller than a radio wavelength. The incident radio wave was considered to cause each electron first to oscillate freely and then to re-radiate. Knowing the transmitted power, the power in the returned echo, and the range to the trail, one could calculate q, the number of electrons per unit length along the trail. Combining this with visual observations during 1946–47, Lovell's group associated a fifth magnitude meteor, at the eye's limit, with $q \sim 10^{10}$ cm^{-1}, and a zero magnitude meteor with a line density 100 times larger. One nice confirmation of the theory was that, as predicted, echo duration was found to be proportional to the square of the observing wavelength. Meanwhile, Nicolai Herlofson (1948) worked out that a meteor's kinetic energy would be converted into heat, light, and ionization energy in the proportions of $10^4:10^2:1$ as it passed through the atmosphere. This was an extremely difficult problem involving knowledge of how an object of millimeter to centimeter size moving tens of times faster than a bullet interacted with molecules and atoms and gradually either slowed down or ablated away. But it was heartening that Herlofson's theory, building largely on work by Öpik in the 1930s, produced a ratio of visible light to ionization energy corresponding to the observations.

A key theoretical discrepancy, however, came in explaining the high proportion of echoes of long duration, especially at longer wavelengths. Rapid diffusion of the trail's electrons should have caused the reflected waves from the various electrons to be out of phase, greatly weakening the received echo within a few tenths of a second. Millman (1950) also pointed out that his spectra of once-ionized calcium implied that calcium alone produced more electrons in the trail than theory predicted. Gradually it became accepted that the long-lived echoes resulted from trails where the electron volume density was sufficient to cause mirror-like reflection. The trail could be considered as a metallic cylinder of varying radius (at first expanding and then contracting) inside which the density exceeded the critical density for the wavelength of operation. This picture went back to Pierce's original notion, but with a far more slender column of a few meters radius at most. The new theory's upshot was that line densities deduced from bright meteor trails became some one hundred times larger (Greenhow and Hawkins 1952). In addition, four times greater line density, for example, produced an echo lasting four times longer in the new theory (versus a prediction of the same lifetime in the low-density theory) and with power twice as strong (versus sixteen times stronger). The borderline between the two cases was at $q \sim 10^{12}$ cm^{-1} (roughly the inverse of the classical radius of the electron). Jodrell Bank researchers Thomas R. Kaiser and R. L. "Tim" Closs (1952) worked out this detailed theory and Greenhow (1952) applied it to meteor trails. Simultaneously at Stanford, Eshleman (1952, 1953) derived similar results in his thesis. But although theorists began to assume some confidence, notions about the mechanisms by which a meteor interacted with the atmosphere remained uncertain throughout the 1950s.[41]

---

[41] To give one example of the uncertainties at that time in knowledge of the meteoric process, research later in the 1950s established that most meteors were not at all composed of solid stone or iron, as are the meteorites that survive the trip

11.5 THE METEORIC RISE AND RAPID DECLINE OF A FIELD

Meteor radar astronomy was an important part of early extraterrestrial radio research and yet it stands apart from (passive) radio astronomy in many respects. Those who studied meteors (with the exception of Hey) were distinct from those studying the sun and Milky Way. Jodrell Bank would seem to be a counterexample, although there too we find a demarcation, although now in time (see below) rather than in place. And even as both types of research coexisted at Jodrell Bank in the early 1950s, research groups fit either in one mold or the other and persons did not switch sides (Edge and Mulkay 1976:293–4). Millman (1979:125B) also later recalled that meteor radar researchers at URSI meetings and other gatherings found little in common with the radio astronomers. Meteor radar directly led to results of interest to ionospheric researchers, whereas the connection of passive radio astronomy to ionospheric work was less intimate.

The meteor radar community also differed from the radio astronomers in being more in touch with (optical) astronomers. Visual observations of meteors did not require fancy equipment and from the start they were considered essential for establishing ties between shooting stars and radar blips. Even before the war there was the example of Skellett, who himself was as much an astronomer as a radio physicist, and afterwards there was Millman and the strong encouragement and assistance from men like Whipple, Prentice, and Öpik. Radio astronomers, on the other hand, found themselves often observing phenomena with little more than a tenuous connection with the optical universe. The string of serendipitous discoveries made by radio astronomers led to an exciting life, but certainly not a methodical one. In contrast, meteor radar programs more often proceeded in an orderly manner: a question was posed and an experiment carried out to secure an answer. (Although note that only the Ottawa group *started* in a purposeful fashion; the other three in Table 11.1 all first detected meteor echoes by accident.)

From the very start, too, the meteor radar groups participated in astronomical conferences in both Europe and North America. Besides the RAS meetings and Jodrell Bank conferences already mentioned, the American Astronomical Society featured meteor radar work in a special symposium held at its June 1949 meeting in Ottawa. In 1948 the International Astronomical Union's Commission 22 (on meteors and the zodiacal light), with Whipple as President, passed a resolution proposed by Öpik that urged workers to carry out simultaneous radar and visual observations because these were "essential to the proper advance of meteoric astronomy." The Commission's report further stated:

> Progress in the application of radar and other electronic techniques to the observation of meteor ionization trails has recently become of outstanding importance. Contributions of many workers ... are revolutionizing the methods of meteor observation; Hey ... and Lovell report the actual measurement of meteor velocities electronically!
>
> (Whipple 1950:240–4)

One participant noted that the field had been "invaded with shattering effect by the physicists."

In the United States, however, the meteor radar groups, led by Stanford, were more closely linked to the ionospheric and engineering communities than to astronomers. Gilbert (1975:32–6) has shown that American papers in meteor radar were more often in journals for ionospheric research whereas British work was published in astronomical journals. The differences between research goals and sources of funding at Stanford and at the other major centers have already been pointed out; this distinction between postwar developments in the United States and those elsewhere is further analyzed in Section 17.3. Even Whipple, whom we have seen in this chapter in an astronomical context, was intimately involved in ionospheric studies funded by the military (DeVorkin 1992:273–300).

What of meteor radar's contribution to an understanding of the solar system? It was of course a wholly new technique, but did it do more than simply add to the number of meteors observed? The ability to observe at all hours, independent of daylight and clouds, led to a better knowledge of meteor orbits as a whole, but in itself produced no insights of a fundamental nature. Rather, the main result from the masses of data was the distribution of *speeds* of a large number of meteors, extending to magnitudes much fainter than the visual limit. The observed distribution indicated that there were very few, if any, meteors from interstellar regions

---

to the ground. Meteors are now taken to be more of an easily fragmented complex of particles, like a clump of cigarette ash.

and that the birthplaces of meteors would have to be sought in places like comets and asteroids. This was a major contribution to astronomy. Only a few years of work had overturned the prevailing opinion of the previous half century – Michael Ovenden (1951:97) was one astronomer impressed with "the relative ease with which the radar methods have obtained a definitive solution to one of the outstanding controversies of a generation of visual observation." Another classical astronomer and RAS President, W. M. Smart, remarked that he was "sure that it must be about a hundred years since our Society has devoted so much attention to meteors."[42] In fact the radar technique was only one component of a general revival of work on meteors and comets at this time. Along with the key experimental studies on meteor speeds by McKinley (1951) and Almond, Davies, and Lovell (1951, 1952, 1953), three theoretical proposals in 1950–51 injected new life into the field. These were Jan Oort's model of a cloud of comets surrounding the solar system at a distance of ~50,000 astronomical units, Whipple's model of a comet's nucleus as a dirty snowball, and Ludwig F. Biermann's suggestion that comet tails pointed away from the sun because of a fast-moving plasma shed by the sun (now called the solar wind). Solar system astronomy did not again see such a fruitful harvest of ideas and data until the space age in the early 1960s.

In the period 1950–55, the groups at Jodrell Bank, Stanford and Ottawa continued as leaders in the field, now joined by a larger effort at the US National Bureau of Standards and small antipodal groups at Adelaide and Canterbury. As remarked in Section 9.6, before 1950 Jodrell Bank did little but meteor work, but afterwards its focus progressively shifted to passive radio astronomy, culminating with the completion of the 250 ft diameter dish in 1957, by which time the fraction of meteor radar papers had fallen to about 20% (Edge and Mulkay 1976:81). Worldwide, the number of published papers dealing with reflections off meteor trails continued to increase until the late 1950s (Gilbert 1975:47, 1977:107), but by 1955 these dealt mostly with communications and the ionosphere, no longer with astronomy. A strong influence on this change was that throughout the 1950s military forces in various NATO nations were intensely interested in the development of a communications system based on reflections off the intermittent trails of meteors. Its advantages were primarily in immunity to eavesdropping, freedom from many classes of disturbances such as polar blackouts, and operation at lower powers and higher frequencies (30–300 MHz) than normally possible over thousands of kilometers.[43] Project Janet in Canada was one of the chief efforts. Meteor radar researchers found the military also interested in their expertise on the problems of detecting incoming missiles and of studying the ballistics of any object entering the atmosphere at high speeds. NASA also funded efforts to map the distribution throughout the solar system of what astronauts viewed as dangerous space debris.

Gilbert (1975, 1977) has treated in detail the sociological development of the field of meteor radar astronomy. He characterizes a research specialty as having three stages over its lifetime: exploration, rapid growth, and decline. In the present case the exploratory phase corresponded to roughly the first three years after the war, when participant groups acted in isolation and had not yet drawn the boundaries of the nascent specialty. Duplication of results, often unintentional, was frequent at first, but gradually groups tended to avoid direct competition through differentiation of their research topics. In Gilbert's growth phase the number of researchers rapidly expands, groups learn of each other, and research programs become better defined. This corresponded to the years 1948–52 for meteor radar astronomy. The third phase of decline is identified by Gilbert with the late 1950s, on the basis that not until then did the total number of papers and researchers decrease in numbers. But, as discussed above, a decline in the late 1950s can only refer to meteor radar *astronomy*, not to meteor radar as a whole. One can also examine the field in yet another fashion, compared to the total effort in (passive) radio astronomy. Under this criterion meteor radar astronomy

---

[42] *Observatory* **69**, p. 126 (1949).

[43] The modus operandi of a meteor-trail communications system was to have one antenna always sending a continuous-wave signal (typically at ~50 MHz) toward the second. When a meteor trail appeared that caused the second antenna to receive the signal, it relayed a start signal back to the first, which then, in the second or so available, dumped its message. With the paper tape and tape recorder technology then available, it was no mean feat to transmit information in this way, but it worked. See the December 1957 special issue of *Proc. IRE* for further information. This technique, now called "meteor burst" communications, is still widely used.

by 1947–48 had already peaked at 25% and declined thereafter, to 7% by 1953.[44]

Why did meteor radar astronomy have a relatively short lifetime? It started with a bang with Hey's discoveries and with the 1946 Giacobinids and only six or seven years later Lovell's monograph *Meteor Astronomy* (written in 1952 and published in 1954) can be considered as the almost-final review of the field. In Section 9.6 we have already learned of how Jodrell Bank in the early 1950s was shifting to the use of large dish antennas for passive radio astronomy; given Jodrell Bank's dominance of the entire field, this change by itself was enough to put a significant dent in meteor radar work. Davies (1978:8–9T) later emphasized that it was not just the shift of resources and psyche toward passive work, but also the technical problems of trying to operate sensitive radiometers in the middle of radars continually blasting away.[45]

Meteor radar astronomy had a limited longevity fundamentally because those in the field ran out of interesting and significant experiments to do. Once one had surveyed the entire sky over several years and done an analysis of echo rates and speeds, what was left? One could push to more powerful transmitters and thereby study smaller meteors, or study meteors in the Southern Hemisphere. Indeed, both these were done in the 1950s and beyond, but no breakthroughs emerged. Gilbert (1975:39–49) discusses several other possible research topics, such as study of the masses or ages of meteors, but none of these seemed practicable with existing knowledge and techniques. Meteor radar astronomy had run its course. New information about meteors would have to come from other approaches:

> We had cracked the method of measuring velocities and of meteor distribution. It then seemed, I think, apparent to a lot of us that the main issues before meteor astronomy had been solved. One could go on and on and get heaps of Ph.D.s out of it, but it would be rather a matter of cataloguing, and not profitable compared to the potentialities of what our steerable dish could do.
>
> (Lovell 1976:14T)

It is a remarkable coincidence that initial discoveries leading eventually both to meteor radar astronomy and to radio astronomy occurred at the same place and at the same time. Jansky's first paper was immediately followed by Skellett's in the *Proc. IRE* for 1932, and Skellett's May 1933 thesis coincided with the public announcement of Jansky's discovery. Moreover, these discoveries similarly languished until after World War II when each was enthusiastically pursued by ex-radar researchers and each quickly produced a major impact on astronomy. These early parallels, however, only serve to make more striking the contrast in the 1950s between the demise of the one field and the continued vitality of the other.

TANGENT 11.1 PRE-1945 IONOSPHERE/METEOR STUDIES IN JAPAN, INDIA, AND THE UNITED STATES

Japan A few years after Nagaoka's suggestion of meteors affecting the ionosphere, Tsutomu Minohara and Yogi Ito (1933) were inspired to follow up in Tokyo. Operating at 2 MHz, they determined the height of the Kennelly–Heaviside layer by observing interference between waves that arrived at a distant receiver directly versus those that had bounced once off the layer.[46] On a dozen different nights in 1932 Minohara and Ito monitored the reflecting layer's height, including once during the Perseids, but they found a peculiarity only on the night of 16–17 November, which was the expected peak of the anticipated strong Leonid shower. On that night they recorded a large number of anomalous echoes indicating, if taken at face value, heights above 2000 km. They also observed the sky visually, but

---

[44] These percentages are based on Gilbert's (1975:47) counts of meteor radar papers and my own (Section 17.1.2) for radio astronomy papers.

[45] A sensitive radio astronomy receiver cannot gather data while transmitters are emitting power. The only way that several meteor radar transmitters and radio astronomy receivers could efficiently operate at Jodrell Bank was via a time-sharing scheme. All operating radars sent their pulses out at the same instant, simultaneously producing suppression pulses that temporarily turned off the radio astronomy receivers for the pulse duration.

[46] This was Appleton and Barnett's (1925) method of determining the existence and height of the layer, although the pulsed radar method of Breit and Tuve (1926) gradually became the standard.

presented no analysis of times of occurrence of meteors and anomalous echoes. The evidence of this one night was taken as definite proof that meteors affected radio wave propagation, even to the point of discussing in some detail the behavior of "ion clouds" created by meteors.[47]

Another Japanese episode occurred during World War II. Masaki Huruhata of Tokyo University's Astronomy Department collaborated with Y. Hagihara of the Physical Institute for Radio Waves to observe four meteor showers during late 1944. The experiments were not written up until four years later and the published article (Huruhata 1949) gives no clue as to his original motivation, but it is remarkable that such experiments were carried out in Japan in the closing year of the war. The article concentrated on the Geminids of 14 December 1944, for which a total of 250 meteors were sighted by several observers and a similar number of echoes picked up by a 25 MHz radar. From ~50 correlations in time Huruhata tried to understand which properties of meteor paths led to radar reflections, but the brightest meteors often yielded no echoes, in particular a −7^m fireball, and strong echoes often had no associated meteor. Huruhata also observed the 1948 Perseids, but that was to mark the end of Japanese meteor radar.

India Responding to Skellett's early work, radio physicists in Calcutta searched for an ionospheric effect during the 1933 Leonids. S. K. Mitra, P. Syam, and B. N. Ghose (1934) measured the reflecting layer's critical frequency throughout three nights and found a marked increase in electron content for the two nights nearest the predicted peak of the shower; this "strongly suggested that the effect was due to the impact of meteors on the upper atmosphere." J. N. Bhar (1937) of the same group followed up with a short study on the 1936 Leonids in which he worked at somewhat higher frequencies (6–10 MHz) in order further to check on any meteoric effect on the upper (F) reflecting layer. Bhar confirmed that the lower (E) layer was considerably enhanced, but found no unusual behavior for the upper layer, which on the meteor hypothesis seemed reasonable since visual meteors were seldom observed above 200 km. Neither of these studies employed any visual watches.

During the war in Delhi, two engineers of All-India Radio, C. Chamanlal and K. Venkataraman (1941), heard peculiar whistling effects while monitoring nearby shortwave (5–15 MHz) stations – they were described as like the "ping" of a bullet ricocheting off a rock. The whistles lasted anywhere from a fifth of a second to several seconds, came mainly in the early morning, almost always descended in pitch (typically from a ~3 kHz tone to zero frequency), and sometimes were as frequent as 5–10 per minute. They were explained as representing a beat tone between the directly received wave and one which was somehow being bounced off a postulated rapidly decelerating surface. In such a case the Doppler effect would cause the reflected wave to be continuously decreased in frequency, which, when mixed with the direct wave, would produce the decreasing tone. Working out the magnitude of the effect, they found they needed a deceleration from ~60 km s^{-1} to zero in a few seconds or less, and proposed they were observing the retardation of a meteor and its associated ionized gas as it hit the earth's atmosphere.[48] Meteors were further fingered as the culprits when visual observations showed that visible meteors always produced simultaneous whistles (although most whistles were unaccompanied by meteors). These observations demonstrated, perhaps even more convincingly than Skellett's work, that meteors could be detected with radio waves.[49]

---

[47] The same group also briefly described 1934 Leonid observations, but the evidence for the claimed meteor effect in this case too seems weak. [T. Minohara, Y. Ito, H. Shinkawa, and M. Yamamoto (1936), "A continuous recorder for ionospheric heights and its recent results," *Nippon Elect. and Communications Eng.*, pp. 505–11]

[48] Later work showed that Chamanlal and Venkataraman's explanation for the phenomenon could not be correct. Their whistles were undoubtedly produced by meteors, but meteors are in fact very little retarded as they travel through the atmosphere.

[49] In England, whistles (called "tadpoles") had been noticed and studied by BBC engineers headed by H.V. Griffiths as early as 1936–37, but no reports were then published. By 1940, this group had acquired shortwave direction-finding equipment and proved to their satisfaction that the whistles originated with meteors, but were then anticipated by Chamanlal and Venkataraman (1941) and still did not publish anything. In 1947 Griffiths put together a letter intended for *Nature* explaining all this and asked Edward Appleton for advice – it is not known what happened thereafter, but no such results ever appeared in print. [Griffiths:Appleton, 4 December 1947, file C.227, APP]

United States Inspired by Nagaoka's suggestion, Greenleaf W. Pickard (1931) of RCA Victor in Boston examined the correlation over five years between meteor rates and signal strength from a Chicago longwave station. He claimed "strong indication" of a meteoric effect, especially with the Perseid shower, but his results did not influence others and indeed the data today seem statistically weak.[50,51]

John Pierce of Harvard also attempted some correlated visual meteor watches in connection with a 1940 solar eclipse expedition to South Africa for ionospheric observations. On the morning of 14 November 1940 he used a standard pulse transmitter at frequencies ranging from 3 to 7 MHz and his own eyes in the 0.4–0.6 μm wavelength band. Several Leonid meteors were indeed followed shortly by bursts of radio reflections from heights of 100–200 km, but not in every case. Pierce (1941) emphasized that this new observational technique could prove valuable to study the development of a meteor trail's region of ionization, as well as to make meteor counts at any time of day and in any weather.

In September 1945 Oliver P. Ferrell (1946) sent a short report to *Physical Review* saying that he had frequently observed radar echoes of 0.5 to 3 sec duration and 30 to 125 km range while stationed with the US Army in India during the war. Ferrell had already made propagation tests in 1941 that led him to advocate the idea of meteors causing short-duration scattering, and he therefore was eager to investigate "in an unofficial capacity" using a 105 MHz SCR-270 radar, a stacked array of 32 halfwave dipoles. He found a strong maximum in the occurrence of anomalous echoes at 0400 local time and a minimum in late afternoon, as well as echoes often coincident with bright meteors.

Another American study took place during the war under the auspices of E. W. Allen, Jr. of the Federal Communications Commission (FCC) in Washington, DC. For over a decade it had been known that signals could sometimes be received in brief "bursts" at great distances on frequencies so high that they should not at all be reflected from the usual ionospheric layers. The FCC was concerned whether these had any practical consequences for broadcasting with the new FM and television techniques and in 1943–44 conducted experiments with a 44 MHz station in Massachusetts, monitored from sites as far away as 2200 km. They revealed not only "sporadic E" ionization events, but also much faster burst phenomena, when a signal would be enhanced for no more than a few tenths of a second. Surprisingly, the signal was usually not distorted during these bursts and speech and music came through with clarity. Extensive tests revealed that the height of the reflecting agency was ~100 km. Strong evidence that meteors were the cause came from comparisons of burst rates with published annual and diurnal visual meteor rates, although agreement was not by any means exact. In addition, for a year starting in the summer of 1944 occasional meteor watches were maintained near one of the monitoring sites in Maryland and correlations were indeed found between bright meteors and bursts. However, until many years later (Allen 1948) these observations were available only as a government report (FCC 1944).

TANGENT 11.2 PRE-1945 IONOSPHERE/METEOR STUDIES IN BRITAIN

One theme of British research was what came to be called "sporadic E" or "abnormal E," that is, transient disturbances in the reflective properties of the E region. For example, Thomas L. Eckersley (1886–1959), a Marconi Co. engineer, spent much of his career elucidating both theory and data on what he called scattering, namely signals propagating through means other than reflection off smooth layers. Eckersley (1940) studied in detail a class of 10 MHz anomalous signals that were common if transmitter power was large enough, that lasted 0.5–10 sec, and whose existence he attributed to temporary clouds of high electron density in the E region. Eckersley (1937) at one time speculated in *Nature* that "this unceasing supply of scatter clouds" required a steady external source of ionization energy such as meteors or cosmic rays emanating from the stars.[52] After all, as he also pointed out: "Disturbances

---

[50] Pickard (1931:1166–7) also stated that at the time when the Earth passed through the tail of Halley's comet in 1910, he had conducted a "series of audibility meter measurements," but found no comet effect.

[51] E. Quäck of Transradio in Germany apparently conducted a similar study in 1931 and came to similar conclusions as Pickard (Lovell 1954:26).

[52] Two months after sending his 1937 note to *Nature*, Eckersley apparently tried, on at least one night of the Leonids

(high-frequency noise) have been shown by Jansky to originate in the Milky Way." This note, which contained not a single citation, prompted Skellett (1938) to send off his own letter to *Nature* pointing out his earlier work and strongly arguing for a meteoric origin for Eckersley's clouds.[53]

Since the late 1920s Appleton too had been puzzled as to how the nighttime E region not only refused to decline much in ionization level, but sometimes even showed a general increase. During the 1932–33 International Polar Year his group ran an ionospheric observatory at Tromsö in arctic Norway and his 2–4 MHz data were full of abnormal E reflections, sometimes lasting for as long as minutes. In a summary as to the cause of this phenomenon, Appleton, Robert Naismith, and L. J. Ingram (1937:259) stated that "no general conclusions are reached." A likely candidate was charged particles from the sun, largely because of correlations with auroral phenomena (which were also known to occur at an altitude of ~100 km). But in the end they stated (1937:254) that "the relative influences of extra-terrestrial sources (charged particles and meteorites) and terrestrial sources (thunderstorms) are not yet clear." In addition, Appleton *et al.* noted E reflections at frequencies of 5 to 10 MHz that lasted only a few seconds. These, in retrospect, almost certainly were caused by individual meteors, as were similar reflections observed in 1935–36 by Robert A. Watson Watt, A. F. Wilkins and E. G. (Taffy) Bowen. Their 1937 paper is notable in that it gave the first photographic record of such a reflection, a 5 sec burst of intense ionization at 80 km altitude.[54]

Another British study was conducted by Eric Eastwood (1910–1981) and K. A. Mercer from January 1945 to July 1946. Appleton initially prompted them to collect data primarily using the 23 MHz Chain Home radar equipment at Bawdsey, Suffolk. Eastwood was the RAF's expert on calibration of coastal radars and in particular worked on propagation problems, for example reflections off nearby hills and flocks of geese (eventually leading to a 1967 book called *Radar Ornithology*!). For two full days each week over 18 months the horizon-looking 1 MW radar recorded ionospheric bursts. Eastwood and Mercer (1948) then plotted annual and diurnal variations of the hourly rate of bursts and found them to agree in a qualitative sense with similar curves for meteors, that is, with well-defined maxima in the autumn and in the early morning hours. Furthermore, they found from their range measurements that the heights of origin of the bursts were very narrowly distributed about 86 km, again agreeing with optical meteor data. They also took advantage of a solar eclipse whose path happened to pass over northern Scotland in July 1945. Because the eclipse produced no perceptible influence on eight different radar equipments, they concluded that the sun was not a source of burst-producing particles. In the end, they argued that the evidence supported the idea that a large proportion of the bursts arose from meteors.[55]

TANGENT 11.3 THE DOPPLER METHOD AND FRESNEL THEORY OF METEOR ECHOES

Identical oscillation frequencies for the ascending-in-pitch whistle are predicted by Fresnel diffraction theory as by the Doppler approach. In one case the signal providing a reference phase (against which the signal from the newest trail sector is being compared) is supplied by reflection from the already existing trail and in the other case by the direct ground wave from the

---

(15 November 1937), to correlate 9 MHz "scatter cloud" events and visual meteors. He found no correlation and did not publish these results until well after the war. [*Nature* **162**, 24–5 (1948)]

[53] Later work showed that a large fraction of Eckersley's anomalous signals were indeed due to irregular scattering, but from the ground, not from the E region. The point was controversial even before the war; for example, see the report of 10 MHz experiments by C. F. Edwards and K. G. Jansky (1941), "Measurements of the delay and direction of arrival of echoes from near-by short-wave transmitters," *Proc. IRE* **29**, 322–9.

[54] The data reported in this paper were obtained at Orford Ness, Suffolk, as part of the first year of development of British radar (Swords 1986:434–8; Bowen (1987), *Radar Days* [Bristol: Adam Hilger]). Bowen later recalled that on the night of 12 December 1935, and only on that night, a huge number of "sporadic E"

echoes were detected [talk in Sydney, 4 September 1980]. In retrospect, it seems probable that they were unwittingly observing the Geminid meteor shower.

[55] As mentioned in Section 6.3.2, Appleton and Naismith (1947) also conducted a monitoring program of the E region events, with a 27 MHz RAF radar at Slough from January 1944 to mid-1946. They published annual and diurnal rate curves similar to those of Eastwood and Mercer and likewise concluded that the ionization bursts arose from meteors.

transmitter. But the Fresnel theory is more complete, for instance in predicting the amplitudes of the waveforms. Because recording at Jodrell Bank was only initiated *after* detection of a strong echo, the oscillations observed were only those immediately *following* the ionization burst at the point of perpendicular reflection. The descending whistle, which occurs before the ionization burst, was therefore not used in the Fresnel method. The reasons for much more common *descending* whistles were probably (a) geometrical effects that tended to produce more and stronger echoes from *approaching* meteors, and (b) distortion of the meteor trail with time, tending to mask more often the later-occurring ascending whistles. The clearest explication of this entire question was given by McKinley (1951:226–32).

TANGENT 11.4 THE VELOCITY CUTOFF FOR INTERSTELLAR METEORS

Basic dynamics teaches that any particle measured at the earth with a speed (relative to the sun) exceeding $\sqrt{2}$ times the earth's circular motion of $30 \,\mathrm{km\,s^{-1}}$ cannot be in a bound orbit and therefore, barring an earlier energy gain from an (unlikely) close encounter with another planet, must have come from interstellar space. This is true no matter what the *direction* of the meteor's flight – if the speed of a meteor as measured from the moving earth exceeds $72 \,\mathrm{km\,s^{-1}}$ ($30 + \sqrt{2} \times 30$), then its kinetic energy assures a hyperbolic orbit. On the other hand it is possible for a meteor whose geocentric speed is less than $72 \,\mathrm{km\,s^{-1}}$ to nevertheless have a hyperbolic orbit if its orbit is such as to overtake the earth – to establish this requires knowledge of not only speed, but also direction of motion. Such a velocity vector then allows one to work out complete details of its orbit, but generally can be garnered only by two-station observations or by working with shower meteors whose radiant is known.[56]

---

[56] In this discussion I have been careful to distinguish between the terms *velocity*, the full vector associated with an object's motion, and *speed*, the magnitude of that vector. Other places, however, follow the usage prevalent in the literature of the time, namely that *velocity* could refer to either speed or velocity proper. Note that both the continuous-wave Doppler and Fresnel diffraction methods measured the *transverse component of velocity* at a meteor's closest approach, which in fact is its speed.

12 • Reaching for the moon

From radio's earliest days the possibility of radio contact between earth and other solar system bodies has been tantalizing. At first people thought of communications with whomever might be out there, but nothing was found. But as the art of radio and the capabilities of radar continually developed, the goal switched to that of bouncing signals off the moon or planets. A few made brief tries both before and during World War II, but it was not until 1946 that the first successful lunar radar was mounted. This chapter describes these early experiments on the moon, which were primarily either engineering feats or investigations in ionospheric physics. They included a well-funded US Army project, an amazing experiment in war-torn Hungary, and in 1948 a study at the Radiophysics Lab of the ionosphere through its effects on lunar echoes. In addition, a pioneering study in Sydney of the thermal radio radiation from the moon is described. By the early 1950s lunar radar survived in two forms: in the open at Jodrell Bank and in secret at military laboratories such as the US Naval Research Laboratory.

12.1 PREWAR THINKING

Most early speculations about planetary contact involved communication with extraterrestrial inhabitants, particularly those hailing from Mars. In 1900 the eccentric Serbian–American electrical pioneer Nikola Tesla heard unusual signals with a definite one-two-three pattern and attributed them to Martians. In the early 1920s several radio experimenters, including Guglielmo Marconi, picked up unexplained signals at ~2 kHz, setting off extensive debate about signals from Mars.[1] During the particularly close opposition of Mars in 1924, astronomer David Todd persuaded the US Army Signal Corps to listen for Martian signals at hourly intervals over a three-day period.[2]

In 1929 Edward Hulburt (Section 10.1.2) of the US Naval Research Laboratory (NRL) published an investigation into which types of radio waves Martians themselves might use. Calculating the expected ionosphere, he showed that wavelengths greater than 50 m were most suitable for Martian radio circuits. After then noting that such long waves do not penetrate the Earth's ionosphere, Hulburt (1929:1527) concluded:

> It may be that there are no short-wave receiving stations on Mars, except possibly those for experimental or research purposes. … From the present calculations, quite apart from other considerations, it is concluded that only a very optimistic experimenter would look for successful wireless communication between the Earth and Mars.[3]

As radio techniques developed and transmitters became more powerful, many persons thought of the

---

[1] Marconi's claims as to the nature and origin of the signals he received varied over the period 1919–22, but in 1922, ten years before Jansky's discovery, he did state that the signals "might have come from any region in the universe where electrons are in vibration." For his part, Jansky had to field questions as to whether he might be detecting "interstellar signalling" from intelligent creatures (Section 3.3.3).

[2] Interplanetary communications in this period are covered in Dick (1996:401–14). Dick discusses in detail a 1937 popular article by the Harvard astrophysicist Donald H. Menzel, in which he suggested possible messages and modulation schemes for radio communication between intelligent beings on earth and on Mars. This article is accompanied by another calculating required power levels, etc.; G. C. Southworth (Section 5.6) contributed extensively to this latter analysis ["Can we signal Mars by shortwave?," *Short Wave and Television*, December 1937, pp. 406–7 + 450–4].

[3] Bruce Hevly and Hulburt's sometime colleague William A. Baum have told me that Hulburt's consideration of Martian radio stations was probably in part a joke, done with his tongue firmly in his cheek; he was known for a puckish sense of humor, even within official writings.

more definite possibility of simply bouncing a radio beam off our moon and detecting its echo 2.6 sec later after it had travelled ~800,000 km. The first (and only prewar) publication of this idea appears to have been in 1927 by radio popularizer and science fiction writer Hugo Gernsback in his *Radio News* magazine. Gernsback felt confident that radio waves shorter than 2 m could penetrate the Heaviside layer (ionosphere) and proposed a gigantic "beam system" that would direct no less than 100 MW of power toward the moon, Venus, or Mars. It was noted that the feed of the transmitting antenna "would naturally become white-hot due to the titanic amount of energy radiated" and would have to be "a pretty heavy bar of silver or copper." The echo would be picked up by a conventional antenna, perhaps on the other side of the earth. Gernsback hoped to rope a millionaire philanthropist to support this radio venture, as had been done so well for the major American optical telescopes of previous decades.

In a more sober treatment, the ionospheric radio physicist David F. Martyn (Chapter 7) contemplated lunar radar in a two-page, unpublished write-up entitled "The reflection of wireless signals from the surface of the moon." Writing in May 1930 shortly after his arrival in Australia, Martyn noted recent European reports by C. Størmer and B. Van der Pol of anomalous radio echoes with delays of 1–30 sec, thought to be reflected from temporary ionized regions far beyond the earth's ionosphere. Martyn was struck by this evidence that manmade radio waves had left the earth and then returned, and saw in it a new tool:

> Thus a wireless beam may be used as a probe to investigate regions of the solar system, in much the same way that X-rays have been used to examine the structure of matter. Observations on the time taken by the signal to return, the attenuation of the signal, its state of polarization, frequency, intensity and waveform, will all yield information regarding the size and shape of the object encountered, its constitution, and its velocity.[4]

He calculated that a 1 kW mean-power[5] transmitter operating with a large, shortwave antenna would suffice to detect lunar reflections. Besides scientific experiments, he was also enthusiastic about the possibilities of using the moon as an intermediary for communications between distant points. Martyn closed by recommending that the Radio Research Board should support such an experiment, but nothing more was ever heard of the idea.

Through the 1930s many researchers undoubtedly actually tried for the moon, although no one bothered to publish negative results. At NRL an attempt at lunar reflections on a frequency of 32 kHz was carried out in the 1920s. Later, Leo C. Young and his colleagues at NRL probed for the moon in 1935 with a meter-wavelength radar set, and this experiment was repeated with each more powerful NRL radar over the next decade.[6] In the end, however, detecting lunar reflections required the technology of World War II as well as the time to try it out during peacetime. But before relating the first success, we examine some ideas prevalent during the war.

12.2 WARTIME CALCULATIONS AND OBSERVATIONS

The feasibility of lunar radar was considered in a 1940 three-page report by Jack Piddington (see Section 12.5.2) of the CSIR Radiophysics Laboratory in Sydney (Chapter 7). He began:

> With the enormous powers available from R.D.F. [radio direction finding] equipment, the possibility of obtaining echoes from the Moon appears worthy of investigation. It would then be possible to measure its distance from the earth by R.D.F. technique and also to send ultra-short

---

[4] D. F. Martyn, "Progress report to Committee of Radio Research Board. Period 17th February to 24th May, 1930" (Appendix 2), series 4, item CZ/231(7), CSIRO Archives, Canberra. This report is mentioned in Evans (1973:280); I thank Rod Home for bringing it to my attention.

[5] *Mean power* refers to the average power (over a period of minutes, say) that a transmitter can generate. A pulsed radar, however, sends out energy only a small fraction of the time, dependent on the pulse length and the pulse repetition rate. *Peak power* refers to the output *during* the pulse and is much larger than mean power. Unless otherwise specified, all powers quoted in this chapter are *peak* power.

[6] Taylor, A. H. (1948), *Radio Reminiscences* (Washington, DC: NRL), p. 133, cited in Butrica (1996:12); Trexler (1958:286–7). Also see Van Keuren (1997:12).

wave signals to the other side of the earth via a reflection at the moon.[7]

Piddington worked out that a transmitter with a peak power of 100 kW connected to a 4 m wavelength, 16-dipole array, quite within capabilities at that time, should be able to do the job. But no experiments were undertaken in Sydney during the war.

In Germany too, radio workers thought about the moon. In 1943 the rocket development center at Peenemünde apparently attempted to get radar echoes from the moon.[8] Another second-hand account relates that a powerful 15 cm experimental radar near Berlin *did* (accidentally) detect lunar echoes during the war.[9]

The most probable wartime detection of lunar echoes was by Wilhelm Stepp and Willi Thiel in the winter of 1943/44 using an experimental radar set near Göhren on the island of Rügen, ~200 km northeast of Hamburg (Brown 1999:437–8). Stepp was a leading radar engineer working for Telefunken and Thiel his radar technician. Seeking ranges of up to 300 km for coastal radars searching for bombers, Stepp's group had combined elements of the Würzburg radar system (Tangent 4.2) with a lower-frequency radar called Wassermann. The hybrid, dubbed Würzmann, consisted of a very tall (~35 m) array of 640 dipoles operating at ~560 MHz with a power of 120 kW and receiver system temperature of ~350 K (Fig. 12.1); its vertical lobes were only 0.9° wide. While testing Würzmann, Thiel noticed strange enhanced noise (or interference) from the radar for one or two hours at certain times. Further investigation by Thiel and Stepp revealed that this occurred only when the array was facing eastwards over the Baltic Sea – was this distant enemy activity towards Finland? Furthermore, the noise did not commence until a few seconds after the transmitter was turned on, and it persisted for ~2.5 sec after the transmitter was turned off. The final clue came when they noticed one time that the moon was rising above the sea during this behavior. The moon had apparently been reached.

Stepp did not write a report at the time, but well after the war he described this incident in several publications. The earliest record is an enigmatic footnote (see Brown 1999:438) inserted at the bottom of a page of Stepp's 1946 Ph.D. thesis: "We hit an extraterrestrial target for the first time ever when we observed the rising moon at the start of 1944 with the Würzmann on Rügen."[10]

In the Soviet Union, Nikolai Papaleksi of the Lebedev Physics Institute in Moscow (Section 10.3.1) had long been interested in the idea, ever since the anomalous radio echoes of Størmer and van der Pol. It was natural to think of the moon as the cause, but his calculations with Leonid Mandel'shtam showed that the strength of any lunar echoes would be minuscule compared to those reported. In 1943 the same two re-did their calculations in light of new radar capabilities and concluded that lunar reflections had become doable. On 1 February 1946, within a week after hearing of a successful lunar radar in the United States (see below), Papaleksi (1946a, b) outlined his work in a lecture at the Lebedev Institute. His focus was on the ability of either a light pulse or a radar pulse to measure the distance to the moon with accuracy far superior to the ±10 km obtainable by traditional methods. Papaleksi worked out the expected echo strengths for various pulse lengths, beam sizes, and scattering laws at the lunar surface. Concluding that optical pulses would not work because of bright reflected sunlight,[11] he found on the other hand that radar technique was completely practicable. But despite his urging, no Soviet lunar radar experiments in the 1940s are known to have been conducted. Further, it turned out that it was not until the late 1950s that pulse lengths could be made short enough to achieve the accuracies he envisioned. His

---

[7] J. H. Piddington, 10 December 1940, "Radio echoes from the moon," Radiophysics Lab Technical report RP60.1, 3 pp., RPL.

[8] W. Bopp, G. Paul and W. Taeger (ca 1960), *Radar: Grundlagen und Anwendungen* (Berlin: Schiele & Schön), p. 187.

[9] H. Gartmann (1960), *Science as History* (London: Hodder & Stoughton), p. 170.

[10] W. Stepp (1946), p. 95 in "Über die Reichweite drahtloser Anlagen im Wellengebiet von 1 cm bis 20 m," Ph.D. thesis, Fäkultat für Electrotechnik, Technischen Hochschule, Darmstadt. See also H. Mogk (1992), "Die Mondentfernung 1943 funktechnisch vermessen," *Funkgeschichte*, No. 87, 323–4. I thank the late Louis Brown for sharing copies of all his sources regarding this story and Karl-Heinz Böhm for help with translations.

[11] Papaleksi (1946a, b) did hold out hope, however, that the great flash of light said to accompany an atomic bomb explosion would be visibly reflected from the moon, and suggested such observations when atomic bombs were in the future exploded for scientific purposes.

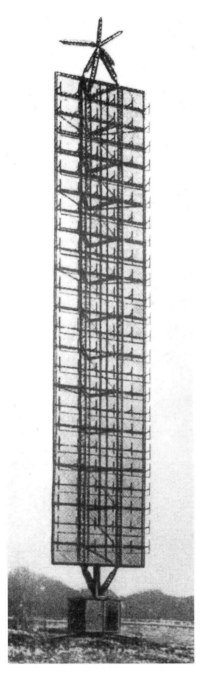

Figure 12.1 The German Wassermann M1 aircraft-warning radar antenna, a 36 m-high array of 250 MHz dipoles. Two of these towers, modified to have 640 560 MHz dipoles, were combined with a Würzburg transmitter/receiver system to make the experimental "Würzmann" system with which the moon was probably detected by radar in early 1944.

study's greatest effect on radio astronomy, however, was that it brought into the field the talents of young Vitaly Ginzburg (1976:5–7T, 1984:290), for at this same time Papaleksi asked Ginzburg to look into the possibility of bouncing radio waves off the *sun*. Ginzburg had been developing the theory of ionospheric radio wave propagation and found this well adaptable to the solar corona. Ginzburg (1946) ended up producing a pioneering paper on the thermal radio *emission* of the sun (Section 10.3.1), but was pessimistic about the chances for solar radar. The basic problem was that only low frequencies were reflected from the sun's atmosphere (others were absorbed), and the required antenna size at these frequencies was just too large.

Other examples of wartime thoughts about lunar radar and communications have come to light. In 1940 Grote Reber and John Kraus, both then working at the US Naval Ordnance Lab, discussed the possibilities of using Reber's dish for lunar radar.[12] In 1941 Donald Menzel and Winfield Salisbury calculated that using the moon for emergency naval communications was feasible, but the concept was never tested. Menzel remained interested throughout the war and in 1945 even met formally with astronomers at the US Naval Observatory to discuss possible scientific payoffs from a lunar radar.[13] Edward McClain (1973:1–2T), while developing microwave radar equipment at NRL, unsuccessfully tried in 1942 to bounce radio waves off the moon with a 3 cm radar having a 100 kW transmitter and 2 ft diameter dishes. In 1944–45 Fred Haddock (1983:116, 1984:2N) and others at NRL became interested in the idea of building a large dish to search for moon echoes. Haddock calculated that existing 10 cm-wavelength radars needed a 30 ft diameter dish, but the Navy never came through with funding for such a project. The notion of a big dish at NRL never faded for Haddock and John Hagen, however, eventually leading to a 50 ft microwave dish in the early 1950s (Section 10.1.2).

A remarkable Italian paper on the feasibility of lunar radar was published by Pietro Lombardini in 1944 in the *Commentationes* of the Pontifical Academy of Sciences. Nothing is known about Lombardini or

---

[12] J. D. Kraus (1988), "Grote Reber, founder of radio astronomy," *JRASC* **82**, p. 109.

[13] D. H. Menzel (1947), "Radar and astronomy," *Astron. Soc. of the Pacific Leaflets* **5**, No. 217, p. 139.

the circumstances of this paper, but he obviously had a decent knowledge of radio techniques and radio propagation and he certainly produced the only paper cited in this book with an abstract in Latin![14] Perhaps he was part of the small Italian effort in wartime radar. In any case, he argued that frequencies of 100 to 200 MHz were needed to penetrate the ionosphere as well as any more distant electron clouds that might surround the earth. He then calculated that a 60 kW transmitter connected to an antenna with a 10° beam would produce a signal detectable with the best receivers. Although his calculation contained several errors and did not consider that one needed an exceptionally long pulse (see below), the paper was still prescient. Lombardini closed by noting that for "only a few thousand lire" one could do this experiment and get very good distances and radial velocities for the moon. A version of the paper was apparently submitted for publication to *Nature* in the autumn of 1945, but was rejected.[15] Lombardini's work therefore remained obscure and in fact his paper was never to be cited.

In England, Edward Appleton (1945b:349) in an April 1945 lecture also calculated that "radiolocating" the moon should be possible with the very largest equipment. In the same year his countryman Thomas Gold (1976:2–3T), working on radar at the Admiralty Signal Establishment, wrote a report proposing that a large, fixed spherical reflector could be used not only for lunar and planetary radar, but also as a home base for communication with ships via lunar reflection.[16]

Finally, Bernard Lovell was not only thinking about cosmic ray radar during the war (Section 9.1), but also lunar and eventually planetary radar.[17] By 1946 he and C. J. Banwell had worked out in detail the expected echo intensities from the moon, sun and planets for various antennas and wavelengths.[18] Despite optimistic conclusions that the moon was immediately reachable and that Mars and Venus were not too far off, it took seven years of intermittent efforts before Jodrell Bank had successful lunar radar gear (Section 12.6).

When radar was declassified in August 1945, *Time* magazine enthused about both its role in winning the war and its future promise: "[Scientists] already dream of beaming a radar echo off the moon."[19] It was clear that lunar echoes were going to happen, and were going to happen quickly once the war ended. The only question was where. As it turned out, there were two locales, the one as completely unsurprising as the other was unpredictable: prosperous America and war-torn Hungary.

12.3 PROJECT DIANA

John H. DeWitt, Jr. (1906–1999) studied engineering at Vanderbilt University and worked at Bell Labs 1929–32 before becoming chief engineer at WSM, Nashville, Tennessee, one of the largest radio stations

---

[14] …Auctor censet posse radio-echos e corporibus extra terram positis.…Scilicet, simplicibus suppositis suppuntans, ostendit ordinem magnitudinis campi a Luna remissi, cum in ipsam fasciculus transmittentis undarum ultra-brevium mediae potentiae dirigatur, eum esse ut revera echus signum recipi sinat. (Lombardini 1944) [Translation: The author believes that a radio echo can be produced from bodies placed outside the earth.…Using a simplified calculation, he shows that the order of magnitude of an [electric] field reflected from the moon is such that it permits the radio echo actually to be received [detected] when a pulse of ultra-short waves from a transmitter of moderate power is directed toward the moon.]

I thank Larry Sandler for a translation of the Italian paper; Jim Naiden helped with the Latin.

[15] R. L. Smith Rose:E. V. Appleton, 27 October 1945, file C.351, APP.

[16] T. Gold, "Radio communications via the moon," report of the Admiralty Signal Establishment, Witley, ca. 1944–45 [not sighted]; also see Hoyle (1994:268) in *Home is Where the Wind Blows* (Mill Valley, CA: University Science Books). In the late 1950s Gold, then at Cornell University, played an important role in the development of the Arecibo 1000 ft diameter reflector, of a similar design.

[17] For instance, on 28 February 1945 P. M. S. Blackett wrote to Lovell that "the radio detection of big showers and the distance to the moon must be fully thought out." [Lovell 1990:177–8]

[18] A. C. B. Lovell, undated (May 1946 from other evidence), "Note on immediate prospects of obtaining radar responses from the moon and the possibility of responses from the sun and planets," 5 pp., LOV (photocopy courtesy of Lovell); C. J. Banwell, ca. 1946–47, "Planetary radar possibilities" [not sighted]. For another British calculation at this time see Arthur C. Clarke, "Astronomical radar: some future possibilities," *Wireless World* **52**, 321–3 (October 1946). Also see Clarke's earlier paper, where he speculated that radar might make it possible to determine if the surface of Venus is largely aquatic or not: "The astronomers' new weapons: Electronic aids to astronomy," *J. British Astron. Assn.* **55**, 143–7 (August 1945).

[19] *Time* **46**, 20 August 1945, p. 82.

of its day. Every Saturday night he dusted off the transmitter tubes so that the Grand Ole Opry could broadcast its fifty kilowatts of country music throughout the heartland of America.[20] DeWitt was also an amateur astronomer, at one time building with his brother a 12 inch optical telescope and observatory. This unusual combination of radio and astronomy led to an attempt in 1935 to detect the Milky Way radiation reported by Jansky.[21]

In 1940 DeWitt resumed astronomical observations with radio equipment. His notebook testifies:

> It occurred to me that it might be possible to reflect ultrashort waves from the moon. If this could be done, it would open up wide possibilities for the study of the upper atmosphere. So far as I know, no one has ever sent waves off the earth and measured their return through the entire atmosphere of the earth. In addition, this may open up a new method of world communication.[22]

His original scheme was to use radio pioneer Edwin Armstrong's experimental 44 MHz FM station in New York City as the transmitter. This station broadcast CBS network programming, which was also available to DeWitt over the regular land lines. Thus he listened on one or two nights for an echo of ~2.5 sec delay between the land lines and an antenna pointed at the moon, but heard nothing. This technique was then abandoned in favor of using an 80 W transmitter that he had developed for a 138 MHz studio-transmitter link. This transmitter was beamed toward the moon with the aid of a rhombic antenna (gain of 25) on a single night. No echoes could be heard, which DeWitt ascribed to probable losses in his transmission line and in the ionosphere. But he was determined to go after the moon in much grander style and immediately began drawing up plans for a larger transmitter and a giant pyramidal horn antenna. The transmitter was built, but the horn was later judged too cumbersome to steer around the sky. Instead, over the summer of 1940 DeWitt built a 30 ft tower on which were supported a pair of steerable rhombic antennas with a combined gain of ~100. After several evenings of successful reception of 111 MHz Milky Way radiation (Tangent 6.1), for a few hours before sunrise on 26 September he transmitted with a similar setup and listened for moon echoes. The transmitter delivered 500 W to the antenna in keyed, short pulses, but no echoes forthcame.

At this point DeWitt ceased his experiments despite calculations indicating that success was close. These calculations, however, were in error, for they took the expected echo's strength to be proportional to the inverse *square* of the distance to the moon, as opposed to the correct inverse fourth power (see below). He thus never had a chance with this attempt, but it nevertheless planted a seed.[23]

In 1942 DeWitt joined in the development of radar and by two years later was a Lieutenant Colonel in the Army and director of Evans Signal Laboratory at Belmar, New Jersey. This was a part of the US Army Signal Corps Engineering Laboratories at nearby Ft. Monmouth, one of the principal American laboratories for radar research and development.[24] He directed the development of a wide variety of systems, for instance a radar able to detect mortar shells and direct guns to their source, but he never forgot his prewar attempts to bounce radar off the moon. When V-J day came in August 1945 and DeWitt found himself with nine months still to serve in the Army, he determined to give it another shot. He set up (unofficially) Project Diana, after the Roman goddess of the moon, and legitimized its existence on the basis of a Pentagon directive to look into the uses of radar in detecting and

---

[20] WSM's *Grand Ole Opry* show happened to preempt the regular network broadcast of the NBC Symphony Orchestra conducted by Arturo Toscanini. Because DeWitt preferred the latter, he built a link between the WSM studio and his home whereby he and his friends were the only ones in Nashville hearing Beethoven rather than Roy Acuff on Saturday nights.

[21] See note 3.25.

[22] J. H. DeWitt, Jr., 21 May 1940, Lab. Notebook no. 3, p. 139 (from a photocopy courtesy of DeWitt). This notebook and related materials for Project Diana are now in file HL Diana 46 of the Historical Office, US Army Communications-Electronics Command, Ft. Monmouth, New Jersey.

[23] Information on DeWitt's pre-war attempts comes from his notebook (pp. 139–60 and 200–06) and an interview (1978:107B).

[24] For an outline of prewar radar development in the US Army Signal Corps, see Swords (1986:112–18). An official history is contained within the three volumes on the Signal Corps in the series *United States Army in World War II*; the lunar radar is described on pp. 628–30 of the third volume, *The Signal Corps: the Outcome* (1966) by G. R. Thompson and D. R. Harris (GPO: Washington, DC).

Figure 12.2 The Project Diana team at the Evans Signal Laboratory in early 1946 after the first successful lunar radar experiment: (l. to r.) Jacob Mofenson (radio engineer), Harold Webb (physicist), John DeWitt (leader and radio engineer), King Stodola (scientific leader and radio engineer), and Herbert Kauffman (technician).

controlling guided missiles such as the Germans' deadly V-2. DeWitt took advantage of the postwar surplus of men and equipment and chose an ideal crew (Fig. 12.2), one that had been working together on radar for the anticipated invasion of Japan. Its scientific leaders were E. King Stodola, chief of the Lab's Research Section, and Harold D. Webb, a former physics professor.[25]

The team's mission was to assemble as quickly as possible a radar system that could do the job, scrounging parts here, building others there. They started with the so-called radar equation:

$$P_r/P_t = A\,G\,\sigma/(4\pi R^2)^2$$

which expresses the ratio of power P_r in the received echo to that transmitted (peak power P_t) in terms of the effective area A of the receiving antenna, the gain G of the transmitting antenna (Appendix A), the cross section σ of the target,[26] and the distance R to the target. For the strongest echo, one of course wanted the largest possible transmitter, antenna, and target, but note that the signal depends on the inverse *fourth* power of distance (pulse energy diverges on *both* legs of its round trip). The moon may be a big target compared to an aircraft ($\sigma \sim 40\,\text{m}^2$), but its astronomical distance (compared to, say, 100 miles) meant that one still needed a radar about one thousand times better than the most powerful then operational. The best candidate in the American arsenal was the Army's SCR-271, a workhorse throughout the war for early warning of aircraft attack.[27] This radar's normal antenna was a "mattress-spring" array of 32

---

[25] Walter S. McAfee (1914–1995) was an African-American physicist and expert in radar theoretical calculations who contributed to the team by calculating the moon's effective cross section and expected signal levels. He was acknowledged at the end of the paper by DeWitt and Stodola (1949:241), but was "invisible" during the extensive 1946 publicity and in publications. [Oral history (1994) of McAfee at 161.58.53.125/mcafee.html]

[26] The effective cross section of a spherical target such as the moon can be considered as its geometric cross section multiplied by reflectivity and directivity factors. The reflectivity factor accounts for what fraction of the incident radiation is not absorbed by the surface material (for the moon, this is ~10%) and the directivity factor for what fraction of the reflected radiation scatters back toward the observer.

[27] The SCR-270 mobile radar, and its fixed-tower version SCR-271, were among the first American radar systems developed in the late 1930s and were put into operation, for instance, in time to pick up Japanese aircraft approaching Pearl Harbor on 7 December 1941. During the war about 800 sets were

Figure 12.3 The Project Diana antenna overlooking Shark River Bay in northern New Jersey. The US Army SCR-271 radar had an array of 64 dipoles, twice the usual number. The antenna tower was 100 ft high and stood on a cliff 75 ft above the ocean as it looked toward the rising moon.

dipoles mounted on a 100 ft high tower. For Project Diana the tower was reinforced and a second identical array was linked to the first (Fig. 12.3), creating a 40 ft square array with a gain of 250.

The greatest efforts went into the electronics. The SCR-270 normally operated with a 20 μsec pulse length and a pulse rate of 625 per second, but these

---

manufactured. An informative account of this radar is given by R. B. Colton in *Proc. IRE* **33**, 740–53 (1945).

would not at all suit the moon. In order unambiguously to identify any echo as coming from the moon, the interval between pulses was set at 4 sec, considerably greater than the ~2.5 sec roundtrip to the moon. The "radar depth" of the moon, that is, the time difference between reflections from its nearest and farthest points, is twice the lunar radius divided by the speed of light, or 12 msec. In order to have any chance for an echo, they wanted reflected waves from all parts of the moon to be received together. The Diana team produced a long

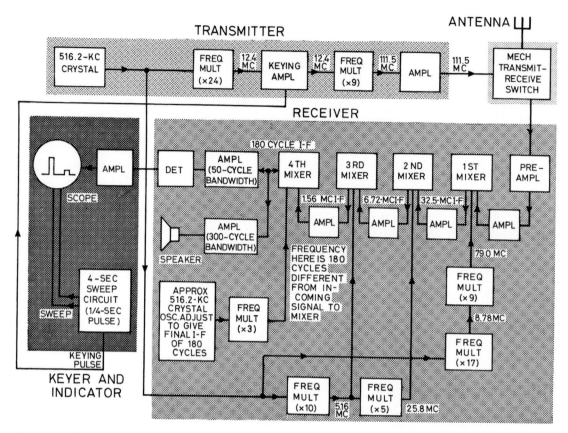

Figure 12.4 A block diagram of the 111.5 MHz receiver and transmitter of Project Diana. Note that all frequencies were controlled by two precision oscillators and that four heterodyne stages were necessary to bring the signal down to a frequency suitable for the narrow bandwidth required by the 250 msec pulse. (Redrawn, with slight corrections, from Mofenson 1946)

250 msec pulse that easily satisfied this requirement, and was also advantageous in that noise could be minimized through use of a narrow receiver bandwidth (see Appendix A); the chosen bandwidth of 50 Hz was larger than the optimum of 4 Hz = $(250\text{ msec})^{-1}$, but was the smallest practicable.

The narrow bandwidth in turn necessitated exceptional frequency stability and tunability, for which purpose they were fortunate to have a state-of-the-art system designed by none other than Edwin Armstrong, the developer of FM radio and a wartime consultant to Evans Lab. Armstrong's transmitter could deliver more than 3 kW throughout the long pulse, for a pulse energy of about one kilojoule, six hundred times the usual for the SCR-271! The receiver employed no less than four superheterodyne stages (Fig. 12.4) to bring the signal down to the audio range (180 Hz), where it could be filtered and impressed on a loudspeaker. Frequency stability was insured by quartz-crystal oscillators. Furthermore, the oscillator for the receiver's last stage was tunable to allow reception of echoes offset in frequency. This facility was needed because the expected Doppler shift arising from the rotation of the earth and the orbital motion of the moon, could be as much as 300 Hz at their frequency of 111.5 MHz, significantly greater than the receiver bandwidth. Each observation of the moon therefore required calculation and careful tuning to the anticipated echo frequency.

First experiments were in September 1945. Ten days of failure, however, convinced DeWitt that major changes were needed (Kauffman 1946:66). This led to the larger antenna and a major re-design of the T-R (transmit–receive) switch. This switch, which protected the receiver during the period when the transmitter was blasting away, is an essential part of any radar system, but did not work well because of the extraordinarily long pulses needed for the moon. Instead of the usual gas-discharge tubes, the group in the end relied on a mechanical T-R switch that with a great clatter lay down a bar to short the receiver input during transmission. A low-noise pre-amplifier was also added (Mofenson 1946:97), making the system noise five times better than standard.

Daily tries commenced with this improved system in December 1945. Since the antenna was fixed at the horizon and could be moved only in azimuth, the moon was available for only about a half hour as it rose each day. When pointed eastwards, the antenna looked out from a 75 ft high cliff (Fig. 12.3), thus gaining from water reflections similar to those of the sea-cliff interferometer (Section 7.3.2), but which here strengthened *both* transmitted pulse and returned signal (equivalent to increasing the transmitter power by a factor of sixteen). Despite these advantages, observations on the rising moon for a month conjured no echoes. This was especially discouraging because the group expected, based on calculations of the moon's radar cross section, signals fully one hundred times stronger than the receiver noise. This was assuming a rough surface of volcanic rock with a reflectivity of 0.17. Many of the precious half-hour observing periods were in fact lost, however, because of manmade interference from amateur operators, auto ignitions, other equipment on site, and neon signs; near new moon, even "cosmic interference" from the sun pestered them (Webb 1946:3). Furthermore, there were numerous breakdowns arising from the jury-rigged nature of their setup and from system components being pushed to their limit.

Diana dragged on into the new year and DeWitt began to wonder for how much longer he could justify such a major effort on an unofficial project, but perseverance finally paid off at 11:58 a.m. on 10 January 1946 when, as *Time* magazine put it, "man finally reached beyond his own planet."[28] Webb and technician Herbert Kauffman were on duty in the late morning when they heard the beep come through above the loudspeaker noise; on later days bumps on the oscilloscope trace occurring ~2.5 sec after transmitted pulses gave further confidence that a celebration was in order (Fig. 12.5). Suddenly their radar's range had jumped by a factor of a thousand from previous typical "long ranges." The echoes also always came with the right frequency shifts, with roughly correct time lapses,[29] and only when the moon was in the beam. From the start, echo intensities varied greatly from pulse to pulse and day to day, and never did reach expected levels. But although DeWitt was convinced of his team's success, as the news went up the chain of command it struck a skeptic in the person of one General George Van Deusen, head of research and development at Signal Corps Headquarters (DeWitt 1978:108A). First he wanted a notarized affidavit from the entire group and then he decided that outside experts should check the results before any public announcement was made. In due time the appointed experts from the MIT Radiation Lab and the War Department, as well as the general, arrived to witness the tests and were soon won over. As one of them later wrote:

> Today, a third of a century later, I recall vividly the excitement I felt as the moon rose over the horizon. For a few minutes, nothing happened; the expected echoes were not observed. Then, suddenly, I saw the blip on the 'scope and heard the audio tone. It was real!.... The actual event carried an emotion that mere figures could not.
>
> (D. G. Fink, quoted in Clark 1980:47)

Reassured, General Van Deusen now planned to announce successful contact with another world at the end of a banquet put on by the Institute of Radio Engineers in New York City. A press statement was given out earlier that day with a release time of 11 p.m., but the after-dinner speaker went on interminably, with the result that Van Deusen was scooped by boys in the lobby of the Astor Hotel hawking newspapers headlining the Army's *tour de force*. According to the press release, the significance of this success included "the possibility of radio control of long-range jet or rocket-controlled missiles, circling the earth above the stratosphere," the potential for using similar techniques to study the ionosphere and planets, and the establishment with certainty

---

[28] *Time*, 4 February 1946, "Diana," p. 84. A similar account in *Newsweek* for the same date (pp. 76–7) refers to several earlier radar successes for which I can find no other evidence.

[29] The Army team never did accurately measure delays for the echoes, but did accurately measure frequency shifts. [DeWitt:author, 21 January 1986]

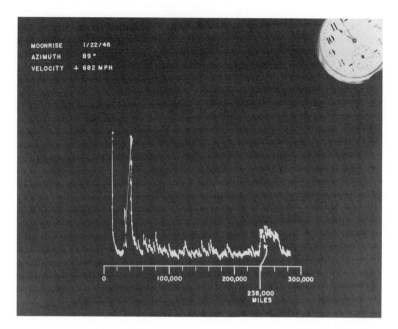

Figure 12.5 Cathode-ray tube trace of a moon echo obtained with the Diana system on 22 January 1946. The receiver was offset by 230 Hz to accommodate a relative radial velocity of the moon and antenna of 682 miles per hour (0.3 km s^{-1}, mostly due to the earth's rotation). Pulse transmission began at "0" on the scale, the high intensity shortly thereafter was an artefact caused by the T-R box, and the reflected echo came back 2.6 sec later, corresponding to the moon's distance of 238, 000 miles (383, 000 km).

that radio waves could penetrate the ionosphere.[30] It is curious that the Signal Corps, of all organizations, did not mention the possibility of using the moon for a communications link. It would also seem that the ability of radio waves to penetrate the ionosphere had been sufficiently established by Jansky's and Reber's previous observations of Milky Way radiation. Certainly the Diana team never had any doubts, although they did not know what propagation effects might be encountered. But apparently many needed the psychological impact of an extraterrestrial "bounce" to bring the point home. Indeed, the *New York Times* commented that "somehow... the moon and all the heavenly bodies become more real... more than a guide to navigators and an inspiration to poets... tangible objects to which we can reach out."[31]

Only two days after the press release Americans were able to hear the moon echoes in their own living rooms on a live national radio program.[32] Not all, however, were enthusiastic – in *Harper's Magazine* one writer cited the feat as further evidence of the last stage of the general disintegration of civilization: "The penetration of the ionosphere... in making radar contact with the moon is a magnificent coda to the invalidation of all that once insulated and protected human life."[33] An Army Air Forces General suggested that eventually a radar code might reach the planets (although DeWitt emphatically stated that this was not possible with

---

[30] 25 January 1946, *New York Times*, "Contact with moon achieved by radar in test by the Army," p. 1 and 19. Also see Clark (1980:48).

[31] 26 January 1946, *New York Times*, "Reaching for the moon," editorial, p. 12.

[32] Mutual Broadcasting System broadcast, 27 January 1946, 15 min in length, Tape 166A (copy courtesy of J. H. DeWitt); also see *New York Times*, 28 January 1946, p. 21, where the reporter was much more impressed by the display of signal strength on the oscilloscope from a nearby car's interference than by the moon's echo, which was "so short and so faint as not to rival a baby chick's first attempt at barnyard articulation." In contrast to the state-of-the-art electronics, he was also struck by the crude shack housing it: an 18 by 20 ft room with unfinished planking on the floor and 2 by 4 inch beams overhead.

[33] J. H. Spigelman (July 1946), *Harper's Magazine* 193, No. 1154, p. 1.

present capabilities) and that any beings present might respond. Another writer, harking back to America's attempts to remain isolated from Europe's problems at war's start, opined that "from now on no doubt even the planets will not be able to pursue a policy of splendid isolation."[34]

After the initial success, several improvements were undertaken to make a scientific study of the lunar echoes – until that point it had been "better classified as an engineering achievement" (Mofenson 1946:92). The power of the transmitter was quadrupled and provision was made for photographic recording of all echoes and quantitative measure of signal levels.[35] But although DeWitt (1978:108A) wanted a full-scale program, he could not enlist Army support, and observations were carried on only intermittently and at low priority. Neither the Army nor apparently any other American research sponsor reckoned that any possible payoffs justified the great expense of a sustained program.[36] DeWitt and Stodola (1949), both of whom shortly moved to industry, did not complete a final paper until two years later. Their paper pointed out how echo strengths could change by as much as a factor of 100 from one echo to the next, although never getting higher than 15% of the expected strength. Two suggested causes for this "fading" were rapid changes in atmospheric absorption and refraction and lunar librations (see Section 12.5), but their data were insufficient to pursue the issue in any detail.[37]

Project Diana had been a rousing engineering success for the Army and "contact with the moon"

had excited the popular imagination. But lunar radar in the United States in fact was soon to become a Navy domain and be done in secret (Section 12.6).

12.4 BAY IN HUNGARY

It is a remarkable fact that a small country like Hungary produced so many scientists of world renown in the first third of the twentieth century. Most of these men emigrated to the United States, and the names of Eugene P. Wigner, John Von Neumann, Leo Szilárd and Edward Teller are well known.[38] Another man in this tradition was Zoltán Bay (1900–1992), educated in physics at Budapest University where he obtained his Ph.D. in 1926. After four years in Berlin, Bay returned to his homeland as one of its leading experimental physicists. By the late 1930s he held both a chair of atomic physics at the Technical University of Budapest and the directorship of the research lab of Tungsram, a large manufacturer of lamps and radio tubes. His outstanding work during the prewar years was the development of the electron multiplier tube as a particle detector and high-speed counter.

When Germany attacked the Soviet Union in 1941, Hungary became an Axis ally and soon thereafter Bay was asked to develop an early-warning radar. For this he rounded up a group of about ten scientists and thirty technicians and started from scratch, since he was denied access to German radar findings (Bay 1980:134A). Despite this handicap, by early 1944 his laboratory had produced a 2.5 m wavelength radar that soon went into service for the Hungarian Army.[39]

In March 1944 Bay decided that his team should devote time to fundamental research at a shorter wavelength: his chosen vehicle was an attempt to bounce 50 cm radio waves off the moon. He had long been interested in astronomy and cosmology and felt that here was a chance to take a major step in both technology and science: as he told his group at the time, they could be the first to do an *experiment* in space, not just an observation. From the beginning his chief colleagues

---

[34] 26 January 1946, *New York Times*, "Radar code to the planets envisioned for the future," p.1; A. O. McCormick, "New frontiers and old in the light of the radar beam," p. 12.

[35] [Project Diana group], 28 February 1946, "Radar echoes from the moon," Report 5/4, Evans Signal Laboratory, Belmar, New Jersey, 13 pp. and 11 figs, RPL; also cited as Report PB-13742 of Publication Board, US Dept. of Commerce (available from the Library of Congress).

[36] J. J. Slattery (Chief, Radar Branch):Director, Evans Signal Lab, 7 April 1948, Project 1041–3 memo. on "Moon radar," (photocopy from DeWitt); also available in file HL Diana 46, Ft. Monmouth, N. J.

[37] The story of Project Diana has been pieced together primarily from the contemporary accounts by Kauffman (1946), Mofenson (1946) and Webb (1946), as well as my DeWitt interview (1978) and papers he kindly supplied to me. Clark (1980) has written a nice historical article, as well as assembled the file cited in note 21.

[38] This situation was discussed in a paper entitled "A case from the peripheries: the background of the 'Hungarian phenomenon'" by G. Pallo at the International Congress for the History of Science, Berkeley (Session Xe, 5 August 1985).

[39] A few more details of the development of this Hungarian radar system are given by Swords (1986:144–7).

Figure 12.6 György Papp (left) and Zoltán Bay conducting their lunar radar experiment in February 1946.

in this venture were György Papp (Fig. 12.6), Karoly Simonyi and Edvin Istvanfy. Initial calculations indicated that it should be just doable, but things were soon interrupted when in the summer the entire radar lab was moved to the countryside as a precaution against Allied air raids. Equipment for the moon experiment was then constructed, but again the effort came to a halt when in September the lab fled back to the city in the face of the approaching Red Army. The siege and fall of Budapest that autumn and winter made further efforts impossible until March 1945, and even then only under most trying circumstances. These too came to nought, however, when the Russians dismantled the entire Tungsram Factory and shipped it all back home. Not to be deterred, Bay (1946:2) drove relentlessly on because he felt certain that lunar radar projects must be going on in other countries, although his isolation meant that he knew of no specific competitors. He thus started the project up for the fourth time in August, cannibalizing Army radar sets and using tools, machines, and radio equipment that had been hidden in the countryside as a precaution against bombing. He also switched at this time to the longer wavelength of 2.5 m because of a greater availability of parts and ease of operation. At last he was able to work continuously for many months, although still contending with the hardships of a devastated economy.

From the first calculations in 1944, it had been clear that success, although within reach, would be tricky. As it turned out, their transmitter's mean power of 3 kW and frequency of 120 MHz were very similar to Project Diana's. Their pulse length was shorter at 60 msec and the effective gain of their specially-built antenna, an 8 × 6 m array of 36 dipoles on an altitude–azimuth mount, was much smaller. But where the Hungarians were really at a disadvantage was that they could not achieve the precise frequency stability demanded for a narrowband pulse. Their receiver's bandwidth had to be 200 kHz, compared to an ideal width more like $(60\ \text{msec})^{-1} = 17\ \text{Hz}$. Bay was able to impose a filtering of ~20 Hz *after* detection of the received pulse, but this resulted in an improvement of signal-to-noise ratio some 100 times less than if filtering before detection had been possible.[40]

The net result of all this was that the expected signal, assuming that the moon's reflectivity was 0.1, was still no better than one-tenth the receiver noise. Something more was needed and Bay's answer was "a suitable method of cumulation." If one could somehow add the effects of hundreds or thousands of echoes, then their feeble strength would increase with time in a linear fashion whereas the accompanying noise would only increase as the square-root of the time. In this way the ratio of signal to noise could be arbitrarily improved. Bay needed at least a half hour of integration, but how could he do this? The glow on a cathode-ray tube persisted for only a few seconds and quickly saturated. Storing charge on capacitors was another electronic approach, but those available were far too leaky for such a long period. Bay's ingenious solution, as strange in the context of radio electronics as it was sound in principle, was borrowed from chemistry and involved the coulometer, a device that measures an amount of electrical charge through chemical reactions. In Bay's setup (Fig. 12.7) each of ten coulometers was connected to the receiver's output with a common anode (in the reservoir above) and a cathode at the bottom in a solution of potassium hydroxide. The amount of hydrogen gas formed at any cathode, and then bubbling upwards through the capillary tube, was directly proportional to the charge flowing to that cathode, that is, the cumulative signal. The length of the column of gas was then a direct measure of electrical signal as integrated over a long period of time. In practice Bay's runs lasted about

---

[40] If one has a broad bandwidth containing noise and somewhere within it a weak narrowband signal, it is much better to filter the signal before detection than afterwards.

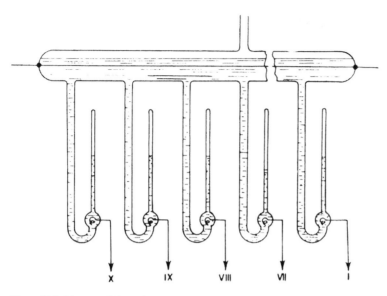

Figure 12.7 Diagram of the coulometers connected to the output of Bay's receiver; successive coulometers corresponded to echo delays spaced by 90 msec. The amount of hydrogen gas measured at the top of each tube was proportional to the integrated echo signal at that delay. (Bay 1946)

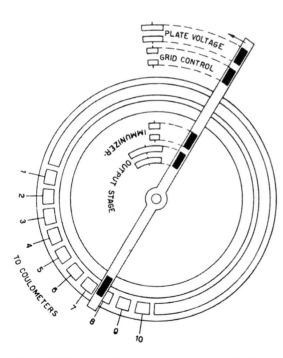

Figure 12.8 Diagram of the master rotating switch of Bay's lunar radar setup. The switch arm rotated once every 3.0 sec and its various contacts controlled the actions of the transmitter, receiver, and coulometers. (Bay 1946)

30 min, or 600 transmitted pulses, and thus he expected an improvement in signal-to-noise ratio of about a factor 25, calculated to be enough to raise the echo to a discernible level.

Bay's setup is nicely summarized by the diagram of Fig. 12.8 (Bay 1946:16). A master switch arm rotated once every 3.0 sec. The pulse was sent out after the switch turned on the plate voltage of 8000 V for the high-power transmitter tubes and grounded their grids for 60 msec, thus defining the pulse length. During this transmitter activity the receiver and the postfilter were "immunized" against damage from the transmitter. The receiver output was then successively connected to each of the ten coulometers, the first being 2.1 sec after the pulse and the last 2.9 sec. Any lunar echo would then cause an excess reading in the coulometer corresponding to the moon's distance.

Experimental runs began in early January 1946, with Papp in charge of the observing crews each night. Standard procedure was to follow the moon by manually moving the antenna, and to integrate the receiver's signal for about 30 min. Control runs were also taken while pointed at empty sky and while the receiver was disconnected from the coulometers. News of the American success in the middle of these attempts must have disheartened the team, but they pressed on and claimed their own contact on 6 February, instigating much excitement in the Hungarian press.[41] Bay's paper, published in English in Volume 1, page 1 (1946) of *Hungarica Acta Physica* (a journal that Bay helped establish after the war), showed results for only five coulometers for the 6 February run; after a half hour, each accumulated a volume of hydrogen of ~100 units, and their offsets (relative to a run where the antenna was pointed off the moon) were 0, +4, −1, −1, and −1 unit. The second coulometer was thus ~4% higher than the mean with an error "under 1%." Almost daily runs were then made for the next three months, and the basic result (an excess in the expected coulometer) was confirmed many times,[42] but the only other published run was for 8 May. With all ten coulometers in operation on that day, number six gave a reading ~4% higher than the mean, a signal some 10–15 times the estimated noise. Using a method fashioned by Papp for injecting a signal of known strength into one of the coulometers, Bay worked out that the received signal was indeed at about the expected level. It is curious, however, that Bay did not greatly strengthen his case for lunar echoes, especially since he could not display individual pulses, by presenting evidence that the coulometers with high signals indeed corresponded to the travel time for a pulse to the moon and back.[43]

In his article Bay (1946:21) pointed out that a combination of advanced electronics and the method of cumulation might in the future allow even planetary radar experiments. He also emphasized that this feat added "the most direct evidence to that which confirms the validity of the Copernican view." By this he meant not that any twentieth-century scientists doubted that the moon was another ball of rock like the earth, but only that here was a more tangible demonstration. As he said, "we can 'touch' the moon by a pencil of electromagnetic waves whenever we wish."

12.5 AUSTRALIA

12.5.1 Lunar radar

Lunar radar experiments also took place at the CSIRO Radiophysics Laboratory (RP; Chapter 7), but there the motivation was different – to study the ionosphere. Frank J. Kerr (1918–2000) had worked during the war primarily on radar systems and propagation effects (Sullivan 1988b) and C. Alexander Shain (1922–1960) on radar countermeasures. While most of their RP colleagues were exploring solar and galactic radiation, Kerr and Shain decided to do radar work at ~20 MHz, following up on unexplained echoes reported during the war. They set up in a sheltered valley at Hornsby Valley in the bush north of Sydney (Figs. 7.4 and 12.9) and went fishing with a variety of experiments. For instance, they searched without success for the still-unexplained, long-delay echoes that had puzzled Martyn fifteen years before. They also tried unsuccessfully for a radar reflection from the *gegenschein*, a faint reflection of sunlight sometimes visible at night and caused by particles outside the earth. In April 1947 they switched to studying echoes from meteor trails for a couple of months, but then gave up on those in favor of something yet farther away – the moon.

With the example of Project Diana before them, Kerr and Shain were particularly curious about why the echoes recorded by DeWitt's team had behaved so erratically, often coming in strong and then fading away after only a few seconds. This, they thought, might well be an ionospheric effect, or even caused in interplanetary space by streams of charged particles shot from the sun. But to get strong echoes from the moon and best study the fading, they needed a powerful transmitter and a large antenna, an expensive and time-consuming proposition. This problem was neatly skirted, however, by enlisting Radio Australia to provide one of its transmitters when not broadcasting overseas. The arrangement as it finally evolved was that for three hours early every morning they had use of a 50 kW, 20 MHz transmitter located in Shepparton,

---

[41] A listing of Hungarian popular articles in early 1946 is given on p. 46 of the book *Zoltán Bay: Atomic Physicist* by F. S. Wagner (Budapest: Akademiai Kiado, 1985), which can also be consulted for aspects of Bay's personal life. More context is supplied by a fascinating memoir by Bay of the 1941–48 period in Hungary (just before he emigrated to the US), which includes his dealings with first Nazis and then Communists, all the time trying to run a factory. [Z. Bay, *Life is Stronger* (1991, Budapest: Püski)]

According to a footnote in Bay (1946:2), the Hungarian press also carried accounts at that time of a sucessful *Soviet* lunar radar experiment in January 1946, but neither Bay nor I have ever found any evidence for such a lunar radar. Note also that in Papaleksi's May 1946 article (Section 12.2) there is mention only of the American success.

[42] Z. Bay:author, 4 December 1985.

[43] See Tangent 12.1 for an analysis of Bay's data.

Figure 12.9 Rhombic antennas, transmission lines and equipment huts at RP's Hornsby Valley field station. In 1947–48 Kerr and Shain studied lunar echoes received here and transmitted from Radio Australia 600 km distant.

Victoria, 600 km southwest of their receiving array of two rhombic antennas (Fig. 12.9). Pulses were initiated from the Hornsby Valley site by signals sent over a land line, and returning echoes were picked up with a communications receiver modified to have a narrow bandwidth, as well as photographed on a cathode-ray tube (Fig. 12.10). But because the moon rarely rose above the northeastern horizon during the requisite daily block of time, only 28 days of observations could be logged over the year beginning in November 1947. These data nevertheless represented the first systematic investigation of lunar echoes and allowed several facets of the problem to be illuminated.[44]

Kerr and Shain (1949 [with technician Charles S. Higgins]; 1951) focussed on the fading phenomenon and found that it occurred on time scales of seconds, minutes, and days. No explanation could be found for the day-long variations, but the other two were interpreted as due to the moon and ionosphere, respectively. The fastest variations turned out to correlate well with the apparent rocking motions of the moon, known as librations.[45] Assuming that irregularities on

---

[44] Note that any number of powerful shortwave transmitters *could* have been used, as much as a decade earlier, to obtain lunar echoes by this same method, but apparently no one tried it before Kerr and Shain in 1947.

[45] Lunar librations are apparent oscillations about (variable) axes that can amount to as much as 4° per day, causing the lunar limbs to move at ~1 m s^{-1} either toward or away from the earth. The three principal contributions to librations are (a) the shifting position from moonrise to moonset of the observer on the rotating earth, (b) the moon's variable orbital angular velocity sometimes leading and sometimes lagging its uniform rotation rate, and (c) a north-south tipping motion every month caused by the moon's rotation about an axis inclined 6° to its orbital

Figure 12.10 Equipment used by Kerr, Shain, and Higgins (1949) to detect and analyze 20 MHz lunar echoes. On the right a camera records the individual pulses as displayed on a cathode-ray tube.

the moon's surface were such that the received echo contained components reflected from many portions of the visible hemisphere (as is the case with sunlight reflected off the moon), then differential motions of the various parts of the moon would cause phases of these components to constantly change, resulting in a highly variable total echo intensity. The correlation between librations and echo fading also provided a demonstration that the moon indeed had a rough surface as "seen" by 15 m waves.

The minute-to-minute fading was taken to result from "roughness" caused by randomly moving irregularities in the ionosphere's F region. Furthermore, the elevation angle of the moon when first detected, as well as the average echo strength, ranged far from predicted values and the deviation depended on the critical frequency of the F_2 region. For instance, the echoes would usually be first detected a full 10° higher above the horizon than expected and were always 10 to 300 times weaker than predicted.

---

axis. The net result of the librations is that from earth we can see (at one time or another) more than just the near hemisphere of the moon, in fact 59% of its surface.

Although Kerr spent some time working out the more general possibilities for solar and planetary radar (see below), these experiments were the beginning and the end of astronomical radar at RP. Kerr was particularly enthusiastic about solar radar and in 1948 pushed for construction of a large (perhaps 600 ft diameter) array of dipoles.[46] But such follow-up projects were viewed as too costly to mount and too limited in scope. So Shain stayed at Hornsby Valley and mapped out the galactic noise at frequencies around 20 MHz, while Kerr went to Harvard for a year of learning astronomy (Section 16.5).

12.5.2 Passive lunar observations

In 1948, also at RP, Jack Piddington (1910–1997) and Harry Minnett (1917–2003; Fig. 7.3) studied the thermal radiation emitted by the moon. Piddington was an Australian who had taken his Ph.D. in 1938 at Cambridge under Appleton doing ionospheric radio research and who then joined RP from Sydney University's Electrical Engineering Department in 1940. As a senior staff member during the war, he worked on many aspects of radar system design and field testing. Minnett, as with Piddington and many others at RP, had emerged from a combined B.Sc./B.E. program at Sydney University and then immediately joined RP in 1940. During the war he worked on antennas and switches, and then became expert on microwave receivers after a six-month, technique-gathering visit to the MIT Radiation Lab in 1945. After the war they both worked on radar aids to civil air navigation, but in 1948 they gained exclusive rights (at RP) to extraterrestrial noise research at the highest technically feasible radio frequencies – their mandate was to explore the sky at the shortest wavelengths (less than 25 cm) and see what turned up.[47]

The foray began with Minnett building a superheterodyne receiver at 1.25 cm (24,000 MHz or "K band") patterned after the one Robert H. Dicke (1946) had developed at the Radiation Lab (Section 10.1.1). This employed a balanced crystal mixer that combined the incoming sky signal with a klystron local oscillator.

The beauty of Dicke's scheme was in its use of a disk of absorbent material that was periodically (25 Hz) rotated into the input waveguide and that in effect acted as a continual monitor of receiver gain. With this "Dicke-switching," accurate measurements were possible where they otherwise would have been unthinkable. Minnett achieved fluctuations in antenna temperature over 6 sec as small as 1.2 K (corresponding to a system temperature of ~5700 K). These electronics were connected to a 3.7 ft diameter paraboloidal dish formerly used as a searchlight, the largest available collecting surface of the accuracy required for such short wavelengths. Observations with this apparatus (Fig. 12.11) were from the top of a hut known as the "Eagle's Nest" perched on the roof of the main Laboratory building. At first they had nothing but a small feed horn at the dish's focus (in order to minimize blockage of incoming signal) at the end of a long run of waveguide to the following electronics, but this setup plagued them with badly drifting baselines, even with the Dicke system. When the rotating disk, mixer, and IF preamplifier were all brought close to the focus, however, the resultant arrangement, although ungainly looking, worked beautifully.

The obvious first things to look at were precisely those that Dicke had very briefly checked at the same wavelength in 1945 (Section 10.1.1) – the sun and the moon. The sun yielded an apparent disk temperature of 10,000 ± 500 K, but disappointingly exhibited only small intensity enhancements (< 5%) over six months of monitoring (Piddington and Minnett 1949b). But monitoring the moon from April to June 1948 produced a pioneering study on the nature of the lunar surface (Piddington and Minnett 1949a). Note that this was simply measuring the moon's natural radio emission, *not* a lunar radar experiment. It was expected that lunar emission would simply vary along with the mean temperature of the observed moon, which depends on its phase, that is, the fraction of the lunar surface we see lit. This had certainly been the experience of the best earlier investigations along this line, those by Edison Pettit and Seth Nicholson in the 1930s using the Mt. Wilson 100 inch telescope at infrared wavelengths of ~8–14 μm.[48] But after one cycle of phases they were

---

[46] Memoranda and correspondence, October 1947 to December 1948, file A1/12, RPS.

[47] Minnett (1978:16T); another motivation for work at short wavelengths was the possible existence of spectral lines there. [Bowen: F. G. W. White, 21 March 1949, file A1/3/1a, RPS]

[48] Unlike at optical wavelengths, where the reflected solar spectrum dominates over any intrinsic lunar emission, solar infrared and radio waves *reflected* off the lunar surface are far weaker than those thermally *emitted* from the surface.

Figure 12.11 3.7 ft microwave dish on top of the Radiophysics Laboratory building used by Piddington and Minnett for a variety of investigations over the period 1948–50, in particular of the moon, sun, and galactic background. Note the optical telescopes for accurate alignment of the dish. The box at the focus contains the input waveguide with its Dicke-type rotating disk, a 1.25 cm mixer, and an intermediate-frequency preamplifier.

excited and puzzled to find that the expected warming and cooling was *not* in phase with the moon, but seemed to lag it by about three days, or one-eighth of a lunar month. Two more months confirmed this trend, as shown in Fig. 12.12 where each datum represents the maximum signal recorded in a series of drift scans across the moon. Their dish was equatorially mounted, but pointing of the 45′ beam was uncertain to the extent that the moon had always first to be found in an attached optical telescope. Strong winds or thick clouds often caused observations to be discarded. They finally concluded that the microwave equatorial temperature had a sinusoidal variation of $\pm 50\,\mathrm{K}$ superimposed on a mean value of 250 K (measured to an accuracy of $\pm 5\%$) and shifted in phase by 45° from the usual lunar phase. This contrasted strongly with the Mt. Wilson infrared results showing an in-phase cycle from ~120 K to 390 K.

The phase shift was the key anomaly calling for explanation, and Minnett (1978:19–20T) later recalled:

This phase lag of three days was a strange thing and got us excited straightaway. We went on

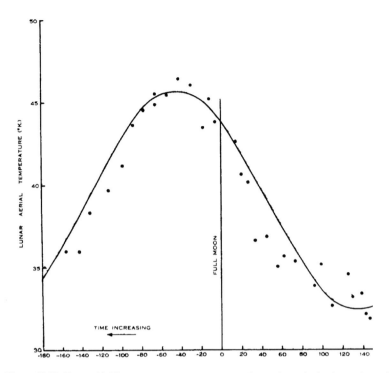

Figure 12.12 Curve of 1.25 cm antenna temperature versus lunar phase obtained over the period April–June 1948. Note the 45° (three day) lag of the microwave curve compared to optical phase. Lunar equatorial temperatures deduced from these data corresponded to about six times the plotted antenna (aerial) temperatures. (Piddington and Minnett 1949a)

for many months and kept asking, "Why is this happening?" And I remember Jack Piddington one day saying in the cafeteria, "I think I've got it. It's the thermal inertia of the moon's surface. We can work it out."

The obvious source to consult was *Conduction of Heat in Solids* by Horatio S. Carslaw and John C. Jaeger (1907–1979). Jaeger was a first-class applied mathematician who had left his University of Tasmania post to work at RP during the war and still worked at RP parttime.[49] With this text in hand Piddington and Minnett worked out an initial model for the moon's surface, but still could not make it match the monthly curve. At that point Jaeger and Alan F. A. Harper (1950) suggested that maybe a sprinkling of dust on top of their model surface would fix things, and sure enough it did.[50]

The full calculations were complex, but basically involved working out, for an assumed composition of the lunar surface, how the varying solar illumination of the moon affected the temperature at the surface and below. Knowing this, one could then calculate the contribution from each layer to the escaping radiation at infrared or microwavelengths. As mentioned, a single-component model could not be made to produce the observed amplitudes and phase shifts of both the 1.25 cm and infrared measures, but a two-component model worked well. The idea was that a

---

[49] During the war Jaeger worked on calculations dealing with antenna patterns and propagation (see note 7.41). But there were also non-radio problems such as calculating the temperature rise within a person's retina when looking directly at the sun. This issue arose because anti-aircraft gunners' eyes were suffering serious damage while tracking Japanese dive bombers attacking from the sun's direction. CSIR was asked to look at the problem, and in the end special goggles were designed based on Jaeger's calculations and on experiments with monkeys and rabbits. For the eye story see Mellor (1958:271–2) and p. 175 of Paterson, M. S. (1982), *Biog. Mem. Fellows Roy. Soc.* **28**, 163–203 (obituary of Jaeger).

[50] Piddington and Minnett (1949a:73); Jaeger:Bowen, 5 May 1948, file A1/3/1a and Z1/5, RPS.

poorly conducting layer of dust, no more than a few millimeters thick, overlay rock (perhaps broken up) made of something like basalt. The thermal inertia of this dust, that is, its resistance to a change in temperature, was some 25 times less than that of the rock, and the infrared radiation came primarily from the postulated dust, where temperature responded rapidly to changes in sunlight. The microwave radiation, on the other hand, was not as heavily absorbed by these types of materials, and thus came from an average depth of 40 cm, where the monthly variations in temperature were a good bit smaller and lagged behind those on the surface. The amount of phase lag, one-eighth of a period, also neatly fell out from the theory.

These lunar microwave observations remained alone until the mid-1950s and the accompanying theory, although certainly refined, also endured. The idea of a dust layer was not new – earlier optical evidence from polarization and monthly cycling in the optical and infrared also leaned that way – but Piddington and Minnett's precise radiometric data gave new strength to the thin-layer model and were still being cited during 1955–65 debates within NASA over whether spacecraft landing on the moon would be swallowed by a deep ocean of dust.

There being nothing beside the moon and sun detectable at 1.25 cm with their apparatus, Piddington and Minnett embarked on a two-year program of further microwave observations at wavelengths of 3, 10, and 25 cm, finding more and more to study as they approached the longer wavelengths where their colleagues were working. These studies of the sun, Milky Way, and a few discrete sources are covered in later chapters. But the weak, albeit important, signals presented at the short wavelengths cried out for an antenna larger than a few feet in diameter. They proposed in 1950 that RP build a semi-steerable dish of ~50 ft diameter, but Pawsey felt that the required resources of manpower and shoptime were too much and the project died on the vine.[51] This led to Minnett going back to problems of air navigation and Piddington diverting away from observing and toward theorizing about the radio emission from the sun and the Galaxy (Section 15.2.2).

12.6 THE 1950s

Progress in lunar and planetary radar went slowly after these initial studies. Powerful transmitters and large antennas were not only beyond the budgets of most research groups, but the potential scientific payoffs were seen as limited in comparison to passive radio astronomy. This rationale, plus the increasing realization that the moon could be useful as part of a communications or eavesdropping circuit, meant that most efforts in the early 1950s were connected with the military and carried out in secret. The following is an extremely brief overview of this period.

D. D. Grieg, S. Metzger, and R. Waer (1948) of the Federal Telecommunications Laboratory in New York City authored the first public, detailed analysis of what they called moon-relay communications. They pointed to possible use of frequencies above 100 MHz over long distances, but also noted the operational disadvantages that usable bandwidths were likely to be small[52] and that the moon had to be visible to a large antenna at each end. They concluded that basic experimental data, such as the roughness of the moon's surface and the effects of librations, were needed before a proper assessment could be made of the moon circuit.

Inspired by this article radio engineer James H. Trexler (1918–2005) of NRL led a group that set out to get such data in a series of steps over the 1950s. He sold the Navy on a scheme of using the moon as a means for intercepting Soviet radio transmissions, as well as for moon-relay communications: Project PAMOR (Passive Moon Relay) was born. For $50,000 a giant hole-in-the-ground antenna of 220 × 260 ft was built at Stump Neck, Maryland, with a clever offset feed arrangement that allowed daily tracking of the moon for almost an hour at frequencies around 200 MHz (Fig. 12.13). The hole was paved with asphalt embedded with a mesh of galvanized iron. First lunar echoes were obtained in October 1951, although not made public until 1957 when Trexler convinced the classifiers that NRL should receive credit for its prior achievements in the face of later Jodrell Bank results then being published.[53] With 750 kW peak power feeding the huge dish in 10 μsec

---

[51] Minnett (1978:32T), Piddington (1978:26T); Pawsey:Bowen, 19 February 1951, "Notes on cosmic noise programme in Radiophysics," file A1/3/1a, RPS.

[52] The usable band of frequencies on the moon circuit depends on the size of the lunar region causing the bulk of the reflections. If the region is too large, only a narrow range of frequencies have their relative phases properly preserved after reflection.

[53] Trexler (1975:34B); Trexler:author, 8 January 1986.

Figure 12.13 The Naval Research Lab's hole-in-the-ground dish at Stump Neck, Maryland, run as a lunar radar from 1951 onward under the direction of James Trexler. The reflecting surface is a 220 × 260 ft offset paraboloidal section, paved with asphalt embedded with wire mesh and usable up to a frequency of 300 MHz. The 198 MHz feed horn (90 ft above the dish) allowed tracking of the moon through a system of cables for about one hour per day.

pulses, Trexler (1958) quickly demonstrated that, unlike the indications and assumptions of earlier workers that the lunar signal originated from a large portion of the surface, fully one-half of echo power came back in the first 50 μsec, meaning that a central area of only ~1% of the lunar disk was responsible for much of the echo. The radar moon had more the glint of a billiard ball than the uniform appearance of the optical moon. This then implied that broader bandwidths, say up to a few kHz, could be used, and in fact the US Navy did use the moon link for communications from the mid-1950s until the era of artificial, geostationary satellites dawned in the early 1960s.[54]

In the Soviet Union also, public word of lunar radar experiments was slow to appear. For instance, the first microwave echoes were obtained in July 1954 at a wavelength of 10 cm by a group at Gorky State University led by M. M. Kobrin, but this was not published until 1959. In the West, Benjamin S. Yaplee of NRL led a group that conducted a microwave lunar radar in 1957, also at 10 cm wavelength.

Interest in lunar radar never abated at Jodrell Bank. After slow progress in the late 1940s, culminating in a scant detection of echoes by Ian A. Gatenby (Section 9.4), the next major effort centered around students William A. S. "Sandy" Murray and John K. Hargreaves, each of whom obtained a Ph.D. in 1954. Unsuccessful tries were first made with a 30 ft dish in 1952, after

---

[54] For the Trexler story I have profited from van Keuren (1997; 2001:211–5), where many more details can be found. The first demonstration of long-distance communication over the moon circuit was in October 1951 at a frequency of 418 MHz from the Collins Radio Co. in Iowa to the National Bureau of Standards 800 miles away in Virginia. The receiving antenna was none other than Grote Reber's 31 ft dish (Tangent 4.3). See P.G. Sulzer *et al.* (1952), *Proc. IRE* **40**, 361. Unknown until many years later was the fact that NRL had achieved a similar feat exactly one week earlier.

which it became clear that a much larger antenna was needed. This led to a north–south row of ten tiltable elements, each having eight full-wave 120 MHz dipoles backed by a flat reflector that allowed the moon to be observed for about an hour each day (location shown in Fig. 9.9). First echoes were picked up in the autumn of 1953. Their intensity variations over periods up to an hour showed the same fading behavior as Kerr and Shain had seen, but they proposed a different cause, namely that the combination of an ionospheric plasma and magnetic field was causing so-called Faraday rotation of the plane of polarization of both outgoing and incoming waves. B. Burrell (1952) had suggested to the Jodrell Bank workers that future radar experiments might exploit the expected Faraday rotation in planetary atmospheres,[55] but Murray and Hargreaves (1954) applied this to our own atmosphere and neatly proved that Faraday rotation was the culprit causing fading. They did this by transmitting linearly polarized pulses from the dipole array and receiving echoes with the 30 ft dish[56] equipped with a dual-polarization feed: when the received signal went away for one sense of linear polarization, then it invariably could be found with the orthogonal sense. This immediately suggested that fading could be avoided by transmitting and receiving *circularly* polarized waves, and this eventually became standard procedure. After 1954 lunar radar at Jodrell Bank was continued and refined for many years by John V. Evans.

After the moon had been conquered, those with the big radars naturally thought of going for the planets, but the harsh realities of the radar equation were sobering. The leap from the moon to Venus (the best case for a planet) required a radar *ten million* times more powerful – recall that the factor in going from an aircraft to the moon was only one thousand! Kerr (1952) published the first major study of the feasibility of planetary and solar radar, arguing that the sun was a much better bet than any planet in terms both of signal-to-noise ratio and scientific interest. He showed that 30 MHz was the optimum frequency for a radar to study the outer solar corona, and, as it turned out, a group at Stanford University finally did the job at about that frequency in 1959.[57] Success on the three inner planets soon followed. After some questionable claims in 1958 and 1959, Venus fell in 1961 to an onslaught by groups in the United States, in the Soviet Union and at Jodrell Bank, with Mercury and Mars close behind during the next two years.[58]

In summary, lunar radar experiments during the postwar decade – Project Diana, Bay in Hungary, Radio Australia, Jodrell Bank, and NRL – constitute a fascinating mixture of engineering feats and ionospheric studies, but any new *astronomical* knowledge from them was minimal. Radar systems and methods of data analysis were simply not powerful enough to yield the quality and quantity of signals necessary to carry out envisioned astronomical projects such as mapping the moon and refining its orbit. These early investigations, however, set the stage for the emergence in the late 1950s of a sufficient combination of high-gain antennas, powerful transmitters, and digital electronics. Only then could information on precise distances, rotation rates, and surface features begin to pour in, and only then could one fairly talk about lunar and planetary radar *astronomy*.

TANGENT 12.1 LUNAR RANGES IMPLIED BY BAY'S PUBLISHED DATA

A detailed analysis has been carried out based on timings derived from the switch positions shown in Fig. 12.8 (Bay 1946:16) and on the distances to the moon on 6 February and 8 May 1946 (BPS-1). I find that the respective signals should have occurred 2.5 ± 0.4 and 0.8 ± 0.6 coulometer "widths" from where they were in fact reported (the quoted uncertainties arise largely because of the lack of a specific time [only a date is given] for each detection). The latter discrepancy is not

---

[55] Correspondence between B. Burrell and A.C.B. Lovell, February to April 1952, file R5 (uncat.), LOV.

[56] This 30 ft diameter dish began its life as a display at the 1951 Festival of Britain in London. Mounted on top of the ~150 ft-tall Shot Tower, the goal was to display real-time cosmic signals to Festival visitors below. First plans were for a lunar radar, but this was scaled back to solar observations. Jodrell Bank was involved in the planning and execution of all this, and after the Festival acquired the dish for research. [Agar 1997:20, 1998:99; *Times* (of London) for 17 November 1949, p. 4; 26 October 1950, p. 2; 10 April 1951, p. 7; 17 May 1951, p. 4; 22 May 1951, p. 4; 24 May 1951, p. 4; 5 June 1951, p. 4]

[57] "Radar echoes from the sun," Eshleman, V.R, Barthle, R.C. and Gallagher, P.B. (1960), *Science* 131, 329–32.

[58] Details of these and later planetary radar experiments are given by Butrica (1996) and Buderi (1996).

of major concern, but the claimed signal for 6 February, whose intensity in any case was only four or five times the estimated noise level, appears to be well removed from its expected time. There is some evidence in Bay's paper, however, that the switch shown in Fig. 12.8 was not yet in use for the run of 6 February. If the five coulometers of the 6 February run did not in fact correspond to coulometers number 6 though 10 of the 8 May run, despite what is suggested by the structure of Bay's (1946:19) Table 1, then one does not have sufficient information to calculate the expected location of the 6 February echo. In summary, the evidence given in Bay (1946) is unfortunately lacking sufficient detail to give one confidence in the 6 February detection.

I thank E. K. Stodola, J. H. DeWitt and Z. Bay for correspondence regarding these calculations.

13 · The radio sun

The radio sun presented an attractive source to early radio astronomers for a variety of reasons. Solar signals were strong[1] and therefore required a minimum of equipment. Furthermore, they could be readily correlated with data from a long tradition of visual observations of the sun. The radio sun also seemed amenable to theoretical investigation using principles straight from ionospheric radio physics. Finally, studies of the radio sun promised a better understanding of various solar influences on the earth's atmosphere and on radio communications.

Almost every postwar group began with solar observations and for most the sun long remained their research focus (see Figs. 1.2 and 1.3). Study of solar radio emission dominated early radio astronomy: fully two-thirds of all papers during 1946–53 (excluding meteor radar) were on the sun, with the remainder covering the rest of the universe (Fig. 17.1).

To understand the radio sun it seemed essential to monitor its daily vicissitudes, which led to many such projects around the world. Figure 13.1 summarizes these monitoring programs, the long-lived ones of which began in April 1946 with Clabon Allen at Mt. Stromlo in Australia, followed within a year by Martin Ryle and Stanley Hey in England and Arthur Covington in Canada. From early 1947 onwards radio data were also regularly sent via Allen and Stefan Smerd to the IAU's *Quarterly Bulletin on Solar Activity*, where they joined longstanding runs of optical and geophysical data (Fig. 13.2 illustrates typical data). Furthermore, from 1950 onwards a group headed by A. H. DeVoogt of the Netherlands (see Chapter 16) within URSI's Radio Astronomy Commission V strove to establish a worldwide chain of solar monitoring stations. Their goal was that the radio sun would be continuously monitored with identical methodology at a variety of frequencies, in particular at 200 and 3000 MHz, but this was not attained until the late 1950s.

Monitoring in the pre-1953 period was conducted with simple antennas, ranging from a few dipoles or Yagis at meter wavelengths to small dishes for microwaves. But because these antennas were unable to resolve the solar disk, information was provided only on short- and long-term changes in *total* solar emission. Special interferometric or eclipse observations were required to study the emission in further detail.

This chapter gives an overview of early observational and theoretical work on the radio sun up until ~1952. By this time the major components of solar radio emission had been identified and detailed comparisons with optical solar phenomena were being conducted. I first consider the so-called quiet sun, that is, when emitting only thermal radiation, followed by the messier topic of the "disturbed" radio sun, which took much longer to untangle. In order to obtain higher angular resolution on the sun, early investigators travelled to solar eclipses and developed clever interferometric schemes. These allowed measurement of the sizes and positions of ever-varying radio-bright regions, as well as detailed checks of theorists' predictions of how radio emission originated and propagated through the complex solar atmosphere. The radio data pointed to a million-degree corona, far hotter than the surface 6000 K and a startling result at first, but in accord with other optical evidence. They also led to the concept of a "slowly varying component" of emission at shorter wavelengths, which turned out to be extremely useful (and complementary to sunspot numbers) for long-term solar activity indexes. In Australia, special attention was paid to the detailed radio spectra of outbursts, leading to a first taxonomy of radio events.

---

[1] Although usually the strongest source in the sky (Fig. A.2), the radio sun, especially at higher frequencies, does not dominate the radio sky to the degree that the optical sun dominates its sky; this is why radio observations can be carried out during the daytime.

The quiet sun 285

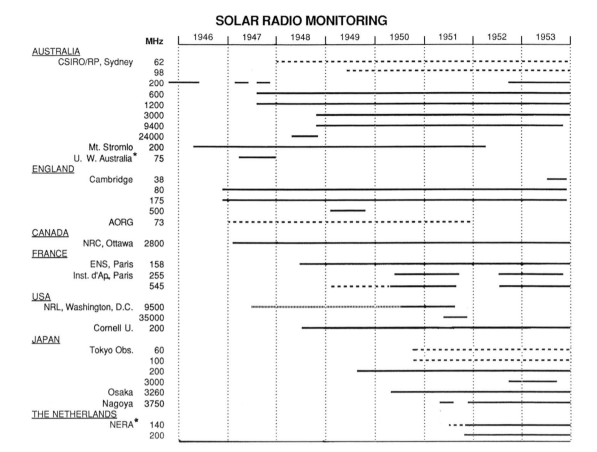

Figure 13.1 Radio monitoring of the sun, 1945–53. Solid lines – continuous monitoring; dashed lines – strong bursts only; dotted lines – intermittent. Diagram largely based on reports in the IAU *Quarterly Bulletin on Solar Activity*. The two asterisked groups are described in a footnote[2].

Solar radio astronomy had stronger ties to traditional astronomers than did other subfields of radio work – each side could understand the value of the other's data and models, and also found comfort in the knowledge that they were indeed working on the same object! But although significant, the contributions to solar physics and astronomy from the radio side were not as revolutionary as in other parts of early radio astronomy.[3]

13.1 THE QUIET SUN

13.1.1 Observed spectrum

As recounted in Section 7.3, Joseph Pawsey of the Radiophysics Laboratory (RP) in Sydney first recognized that the 200 MHz solar flux density, although variable by factors of up to a hundred from day to day, seemed never to drop below a certain floor level

---

[2] Sydney E. Williams carried out solar monitoring with a 75 MHz Yagi antenna in the Physics Department of the University of Western Australia, Perth. He had little contact with other groups before his effort ended in early 1948. See *Nature* **162**, 108 (1948) and *J. Roy. Soc. W. Australia* **34**, 17–33 (1948), as well as an account in Orchiston *et al.* (2006:50–1). NERA (Nederhorst den Berg Radio) was a reception station of the Dutch PTT. Under the direction of A. H. De Voogt (Chapter 16) a 7.5 m Würzburg dish was used to monitor the sun. See *Proc. Of the General Assembly of URSI* **9**, Sec. 6 (Comm. V), 57 (1952), *Hemel en Dampkring* **50**, 209–14 (1952), and Strom (2005, 2007).

[3] See Hufbauer (1991) for an excellent history of solar research in general.

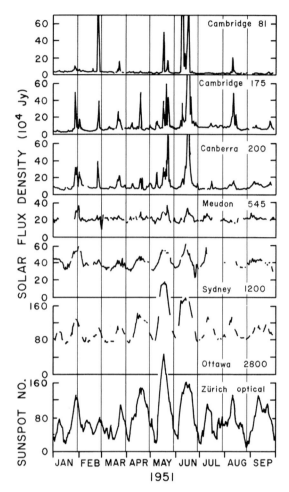

Figure 13.2 Examples of solar intensity as measured daily during 1951 at the indicated frequencies (MHz) and sites – note the change in character as frequency increases. The bottom curve is of sunspot number. (Adapted from Waldmeier [1954:20–21])

Pawsey was naturally interested in seeing how the quiet sun behaved at frequencies other than 200 MHz, for this would give insight into the emission mechanisms, temperatures, and densities of the sun's atmosphere. To this end he and Donald Yabsley (1949) collated and synthesized data from several RP and overseas programs. After establishing that at each wavelength there indeed existed a base level, they produced a spectrum of the quiet sun (Fig. 13.3) in which T_a increased from $\sim 1 \times 10^4$ K to $\sim 1 \times 10^6$ K in a more or less linear fashion (on a log–log plot) as wavelengths increased from 1 cm to 2 m. But creating this plot was tricky, for one could not simply adopt values of T_a found in the literature. Instead, Pawsey and Yabsley went through an involved statistical analysis for the long-wavelength data in order to ferret out the relatively weak constant component buried amidst the extremely strong bursts. Bursts at *short* wavelengths, on the other hand, were manageable since they were infrequent and weak, but here the problem was that several studies had shown that the quiet-sun level itself gradually varied (by as much as a factor of three), roughly following the number of sunspots (Section 13.2.3). Although a truly unblemished condition for the sun was extremely rare, measurements made with various numbers of sunspots allowed extrapolation to the desired base value of T_a.

Pawsey and Yabsley's second goal was to establish that this base level was indeed of thermal origin, independently of whether its intensity and spectrum fit any solar model. This they found undoubtedly to be the case, based on three criteria: (1) thermal emission should exhibit a random, noiselike waveform (Fig. A.3), and indeed qualitative observations, both by listening to "rushing" sounds on loudspeakers and by observing "grass" on cathode-ray tubes, indicated that the quiet sun did so; (2) thermal emission should be unpolarized, as indeed was the quiet sun; and (3) thermal radiation should emanate from the entire solar disk in a roughly uniform fashion, and indeed available eclipse and interferometric observations of the base-level radiation indicated this to be so.

(Fig. 7.10). If this minimum emission came from the entire visible disk of 30′ diameter, a corresponding mean brightness temperature was $\sim 1 \times 10^6$ K. Pawsey later called any such solar brightness temperature an "apparent" solar temperature T_a.[4]

13.1.2 Theory

Theory of the undisturbed sun's radio emission quickly achieved a state widely considered satisfactory, in contrast to a lack of understanding of most other extraterrestrial radio radiation. The key ideas were put forward

---

[4] T_a is a rough indicator of the temperature of the emitting region if it is not too different from 30′ in size and emits optically thick, thermal radiation, provisos that turned out to be crudely correct for the quiet sun. For radio bursts, however, which came from small fractions of the solar disk and which most researchers took to emit nonthermally, T_a was no more than a handy measure of intensity.

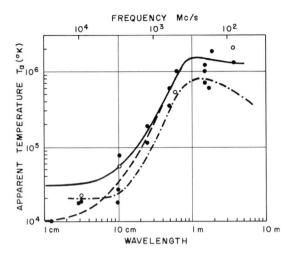

Figure 13.3 The spectrum of solar apparent temperature (see text) as measured in the years before 1951, similar to that published by Pawsey and Yabsley (1949). Observational points are shown, as well as theoretical curves (Smerd 1950a) for chromospheric temperatures of 1×10^4 K (dashed line) and 3×10^4 K (solid line), both with a coronal temperature of 1×10^6 K. Also shown is the theoretical curve (dot-dash line) of Martyn (1948). (Adapted from Pawsey and Smerd 1953:485)

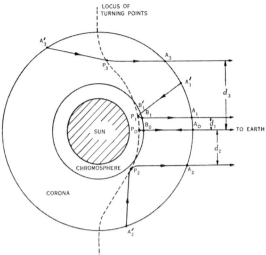

Figure 13.4 Sample trajectories through the solar atmosphere for radio waves observed from earth (note that the size of the chromosphere is exaggerated). The sample frequency illustrated occurs in the steeply rising portion of the spectrum seen in Fig. 13.3. (Smerd 1950a)

independently by Vitaly Ginzburg (1946) and Iosif Shklovsky (1946) in the Soviet Union (Section 10.3.1) and by David Martyn (1946a) in Australia (Section 7.3.3), but in the West they became solely identified with Martyn once he had fleshed them out in a 1948 paper in the *Proceedings of the Royal Society*. Martyn transferred his expertise on ionospheric propagation to conditions in the solar atmosphere and tackled the question of the observed rays' origins by turning the question around: if a wave were transmitted from the earth to a given portion of the solar disk, where in the solar atmosphere would it be absorbed? Kirchhoff's law stated that any such region of absorption was then also precisely a region of emission for a sun-to-earth wave.

Martyn found that radio waves of lower frequencies originated primarily from the corona, for which he adopted an electron temperature of 900,000 K (see Section 13.3.1), while higher frequencies were representative of deeper layers, in particular the chromosphere (at 30,000 K). Using a fall-off of electron density with height from prewar German work, Martyn calculated the minimum coronal height that various observed waves encountered along their trajectories (Fig. 13.4). For lower frequencies and on the limb (the disk's periphery), this ray tracing showed that waves were "turned back." This turning happened as the refractive index first became smaller and eventually imaginary (allowing no propagation) when the critical or plasma frequency (proportional to the square root of the electron density) became greater than the wave frequency (see Appendix A). Then, knowing that most absorption took place in the vicinity of this turnaround point, he calculated the observed brightness temperature by estimating the temperature and amount of absorption there. In contrast, higher frequencies and ray trajectories nearer to the disk center were little refracted and encountered (before being turned back) regions dense enough that their energy was mostly absorbed by the action of particle collisions. In such cases the observed brightness temperature was taken to be the electron temperature where the optical depth along the ray path was about one.

Martyn also figured in further complications wrought by the presence of a general solar magnetic field. Based on observations thirty years earlier at Mt. Wilson by George Ellery Hale, the accepted value for this field was about 50 gauss (at the polar surfaces). Standard magneto-ionic theory then said that an

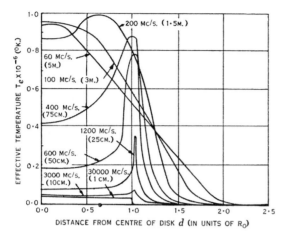

Figure 13.5 Calculated radial profiles of brightness temperature for various frequencies. Note the marked limb brightening at the higher frequencies. (Smerd 1950a)

initially unpolarized wave split into two circular components with significantly different interactions with the plasma. In particular, their critical frequencies, and therefore their turnaround points, differed and this led Martyn to predict that left- and right-hand-circularly polarized antennas would observe noticeably different spectra and brightness distributions. Such polarization effects were searched for in vain in subsequent years, until it became clear that in fact the sun's general magnetic field had been grossly overestimated.[5]

Martyn's predicted spectrum, which showed a peak value of $T_a \sim 800{,}000$ K at wavelengths of ~ 1–2 m and which fell off rapidly at shorter wavelengths and slowly at longer ones (Fig. 13.3), agreed roughly with the two or three measured values of T_a then available. Even more interesting, however, were his predictions concerning the brightness profiles expected across the solar disk, for these indicated substantial limb darkening (relative to the disk center) at the longest wavelengths and limb brightening at wavelengths shorter than 1 m (similar to that shown in Fig. 13.5). Limb

---

[5] Smerd (1950c) analyzed 1948 radio eclipse data and obtained an upper limit of ~10 gauss for any general solar magnetic field. Optical work at this time by G. Thiessen in Germany and Harold D. Babcock at Mt. Wilson also was calling into question Hale's canonical value. See Chapter 7 of N. S. Hetherington's *Science and Objectivity* (1988, Ames: Iowa State University Press) for an analysis of Hale's measurements and their eventual demise.

darkening arose because of the increased reflection (or, equivalently, decreased absorption) of long waves glancing off the upper corona, while brightening occurred at the limb as one essentially sampled a higher (and hotter) region of the solar atmosphere than near the center. This limb brightening was taken to be a solid observational test of Martyn's theory, if only the angular resolution required to discern the narrow limb could be achieved, and it became a major concern of solar observers over ensuing years.

In the late 1940s Martyn returned exclusively to ionospheric theory, abruptly curtailing his short but successful incursion into radio astronomy. However, theory of the quiet sun was not dropped in Australia. Richard Woolley and Cla Allen (1950) at Mt. Stromlo Observatory and Jack Piddington (1950) at RP used the Martyn-type theory as well as optical eclipse data to assemble models of the run of temperature and density in the solar chromosphere. Understanding the chromosphere was considered critical because it was this ~15,000 km-deep region within which the temperature somehow shot up from the photosphere's 6000 K to the corona's 10^6 K. But detailed models of the chromosphere from these two studies and their contemporaries (such as Hagen [1951]) were discordant, arising from uncertainties in the optical and radio data and different assumptions in the physics.

Stefan F. Smerd[6] (1916–1978) was the man who took over Martyn's mantle in Australia. As a Viennese university student, he had fled the Nazis in 1938 and finished his undergraduate instruction in England. He then spent several years at Birmingham University and the Admiralty Signal Establishment working on theory of circuits and tubes, in particular of the cavity magnetron. In 1946 he answered an advertisement in *Nature* looking for adventurous researchers willing to travel to Australia, and subsequently joined Pawsey's solar noise group at RP after a year spent designing millimeter-wave tubes. Together with Kevin C. Westfold (1921–2001), a recent graduate of Melbourne University with a specialty in mathematical physics, Smerd worked to improve and extend Martyn's theory. Their first paper

---

[6] J. P. Wild has written a memorial article on Smerd's life and career: "The sun of Stefan Smerd," in *Radio Physics of the Sun* by M. R. Kundu and T. E. Gergley (eds.) (1980, Dordrecht: Reidel). I have also profited from interviews with Smerd (1978) and Westfold (1978).

(Smerd and Westfold 1949) was highly theoretical, but Smerd's two solo papers the following year were more accessible to most radio astronomers. The first (Smerd 1950b) was a handbook on all aspects of the sun relevant to radio astronomy and had evolved from an RP report in 1948. That report in turn was based on his first assignment from Pawsey, namely to work out tables and graphs of all the solar parameters relating to radio observations. This also led Westfold and Smerd to give a series of tutorial lectures to the solar noise group on radio propagation in the solar atmosphere and on solar astrophysics in general.[7]

Smerd's other paper (1950a) was the culmination of the early theoretical work in Sydney and for a long time remained the standard against which observations were compared. His assumed physical conditions on the sun differed little from those of Martyn, but the numbers were worked out much more accurately and there were many improvements in the physics used to calculate brightness temperatures expected from a given ray and from the integrated effect of the entire disk. In particular, the "solar radius" shown in Martyn's plots was an ill-defined quantity, varying with frequency, whereas in Smerd's plots (Fig. 13.5) "solar radius" always referred to the photosphere. Furthermore, Smerd's industry with a Brunsviga calculating machine allowed him to work out expected signals for a variety of possible electron temperatures in the corona and chromosphere. Qualitative results on limb brightening and darkening were similar to Martyn's, but there were many differences in detail.

There was other theoretical work on the quiet radio sun during this period, but it did not have as much influence as the Australian work because (a) much of it was seriously in error or incomplete by comparison, and/or (b) it came from institutions less central to the radio work. The best was by Jean-François Denisse (Section 10.4), a French theoretical physicist who worked closely with solar observers. His first calculations on the quiet sun (Denisse 1947a:11–12) erred in matters such as ignoring the non-unity radio refractive index in the corona (see below). Also, despite accepting a corona with a temperature of $\sim 5 \times 10^5$ K (from optical evidence), Denisse missed the opportunity of simultaneously with Martyn predicting a million-degree *radio* corona, and instead took the quiet sun's radio intensity to be that of a 6000 K blackbody. Later, however, Denisse nicely laid out an improved theory of radio radiative transfer in his Ph.D. thesis (1949b; published in 1950a,b), the first in radio astronomy anywhere. His thesis included ray-tracing, detailed comparisons of the physics of emission and absorption (considered either classically or with quantum mechanics), and numerous model calculations for frequencies ranging from 25 to 750 MHz. Here also the concept of the "slowly varying component" was first put forth (Section 13.2.3). Denisse's work was independent of and essentially simultaneous with that of Smerd, although he did not calculate as complete a range of models. Moreover, an observer wanting to compare brightness profile data with a model found Smerd's published curves much handier to use.

In Germany Albrecht Unsöld (1947) (Section 10.6) also published calculations of expected radio intensity and brightness profiles at a wide range of wavelengths, including predictions of microwave limb brightening. He built on the work of Martyn (1946a), but being an astrophysicist couched the problem in terms of free–free radiation rather then the theory of ionospheric radio propagation. But this work, as well as that of Max Waldmeier and H. Müller (1948) at Zürich Observatory, was seriously in error in that it too ignored the fact that the refractive index, especially at the longer radio wavelengths, was often much less than 1 in the solar atmosphere (something that never happened at optical wavelengths). This oversight produced incorrect (straight-line) ray paths (see Denisse [1949c] and Smerd [1950a:54–6]). The Germans Gerd Burkhardt and Arnulf Schlüter (1949) did pick up this error and produced a more accurate model, but only for wavelengths longer than 5 m. At Cambridge University in 1947–48, the polymath theorist Fred Hoyle (1915–2001) (1949:110–14) also carried out calculations on the radio sun, but he too ignored the low values of refractive index. Finally, John Hagen (1949 [the second earliest Ph.D. thesis in radio astronomy];1951) worked out a model for radio emission at microwavelengths as then being used at the US Naval Research Lab (Section 10.1.2). Hagen was aware of the importance of the small radio refractive index, but nevertheless argued that even for wavelengths as long as 50 cm it did not matter for ray tracing. He concluded that limb brightening

---

[7] Memorandum from F. J. Lehany announcing upcoming series of lectures by Smerd and Westfold, 18 November 1947, file A1/3/1a, RPS.

290   The radio sun

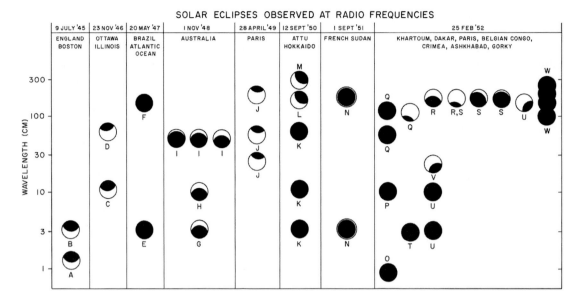

Figure 13.6 The dates, wavelengths, and circumstances of solar eclipses observed by radio techniques over the period 1945–52. Letters by each disk refer to the following papers: **1945** - (A) Dicke and Beringer (1946), (B) Sander (1947); **1946** - (C) Covington (1947b), (D) Reber (1946); **1947** - (E) Hagen (1949), (F) Khaykin and Chikhachev (1947, 1948); **1948** – (G) Minnett and Labrum (1950), (H) Piddington and Hindman (1949), (I) Christiansen, Yabsley, and Mills (1949); **1949** - (J) Laffineur et al. (1950); **1950** - (K) Hagen, Haddock, and Reber (1951), (L) Kawakami (1951), (M) Hatanaka, Suzuki, and Moriyama (1951); **1951** - (N) Blum, Denisse, and Steinberg (1952a, b), Denisse, Blum, and Steinberg (1952); **1952** - (O) Hulburt (1952), (P) Haddock (1954), (Q) Laffineur et al. (1952, 1954), (R) Blum, Denisse, and Steinberg (1952b), Denisse, Blum, and Steinberg (1952), (S) Coutrez, Koeckelenbergh and Pourbaix (1953), (T) Molchanov (1956), (U) Troitsky et al. (1956), (V) Vitkevich (1956), and (W) Vitkevich and Chikhachev (1956). The April 1949 eclipse was also observed at Cambridge at 50 cm wavelength, with results published in a thesis (Stanier 1950b:29–33). (Corrected and modified version of a diagram by Haddock [1954])

should be a noticeable effect, amounting to a factor of four at a wavelength of 11 cm.

13.1.3 Eclipses and interferometers

The angular resolution available with single antennas at this time was usually much worse than the entire 30′ disk of the sun, hardly adequate to discern the details plotted by Smerd in Fig. 13.5. But because brightness profiles from the center to the edge of the solar disk were viewed as a key observable quantity to check theoretical predictions, herculean efforts were undertaken around the world to improve effective resolution through either solar eclipses or interferometers. The former were usually costly, inconvenient, and limited to a few minutes of data-taking, but allowed investigation of details as small as a few arcminutes.[8] On the other hand, Michelson interferometer data (at least for the first several years of the technique) did not produce accurate results unless the sun's radio brightness was (uncharacteristically) symmetric and unchanging. Sea-cliff interferometry was able to handle a rapidly varying sun, but was inflexible (e.g., limited to observations near sunrise or sunset) (Section 7.3.2) and of insufficient resolution for most purposes. A third method, scanning the sun with the beam of an array of antennas, was introduced by 1952. This quickly became established as superior in most regards, at least at the decimeter wavelengths where it was practicable.

Solar eclipse expeditions have been discussed in several previous chapters and are summarized in Fig. 13.6.[9] As can be seen, the February 1952 eclipse in

---

[8] See Tangent 13.1 for early discussion of another technique, observing diffraction fringes as the moon occulted small radio regions on the sun or (nonsolar) discrete radio sources. This technique, however, was not used until much later.

[9] See Section 17.1.2 for a discussion of broader geopolitical aspects of eclipse expeditions.

Africa was by far the most heavily observed during this period. This was a consequence of the general growth of radio astronomy, of the favorable path of the line of totality, and of the fact that the sun was approaching minimum phase in its 11-year activity cycle (Fig. 13.11), making observations of a "quiet" sun more possible. Altogether, nine radio groups observed this eclipse at eighteen different frequencies. The center of action for both optical and radio groups was at Khartoum, although French and Soviet groups observed at other sites.

The attraction of an eclipse was that the predictable motion of the moon's sharp edge across the sun allowed one to pinpoint the locations and measure the sizes of both active "hotspots" and the quiet solar background. The earliest radio eclipse curves have been discussed in Chapter 10 (see Fig. 10.3) and already revealed much of what gradually became established trends. Robert Dicke and Robert Beringer (1946) found the 1.25 cm sun to have a size indistinguishable from the optical (Section 10.1.1). Covington (1947b) got a similar result at 10.7 cm for another partial eclipse, but in addition noticed a sudden drop in intensity as a sunspot group was being covered up, indicating that enhanced emission was confined to a 1–2′ region and had a brightness temperature of 1.5×10^6 K (Section 10.2). Finally, the first radio observations of a *total* solar eclipse were in May 1947 from Soviet and American ships in the South Atlantic (Sections 10.3.1 and Section 10.1.2). Semen Khaykin and Boris Chikhachev (1947, 1948), although obtaining a rather ragged, asymmetric 200 MHz curve because of an active sun, found that the radio intensity at mid-eclipse fell to only 40% of its normal value, which was the first direct evidence that the meter-wavelength radio sun, as expected, was much larger than the optical sun. Meanwhile, at the other end of the radio spectrum, Hagen (1949) measured an eclipse curve in which only 4% of the 3.2 cm solar intensity remained at mid-eclipse, and which he interpreted as also indicating substantial limb brightening.

Further major eclipse results followed in November 1948 when a partial eclipse was observed by RP staff at 50, 10, and 3.2 cm wavelengths.[10] "Chris" Christiansen (1913–2007; Chapter 7) organized the 50 cm effort, for which ten-foot portable dishes were taken to Tasmania and a suburb of Melbourne, while at the Potts Hill field station near Sydney a rectangular 16 × 18 ft dish was used (visible in Fig. 7.14). Observing from three separate sites meant that a given active region was covered and uncovered by the moon at three different cut-angles and times. This removed many of the uncertainties of single-site observations, but the simultaneous presence of many radio-bright regions still often made the analysis ambiguous. Detailed work by Christiansen, Yabsley, and Mills (1949) on abrupt changes in slope evident in the three eclipse curves allowed good measures of the positions and sizes of radio-bright regions, most of which coincided with optical features such as current or recent sunspots. Subtracting this radiation from the active sun (20% of the total), the Australians could estimate the quiet-sun brightness profile. They were reasonably sure that almost half of the radio radiation came from outside the solar disk, but were unable to decide between a uniform disk extending out to 1.3 solar radii and something akin to Smerd's new limb-brightened distribution as in Fig. 13.5. Harry Minnett and Norman R. Labrum (1950) were similarly inconclusive based on their 3.2 cm observations, but Piddington and James V. Hindman (1949) argued that their 10 cm curve definitely showed that ~30% of the solar signal came from a bright limb.

During this era most other eclipse observations at wavelengths less than 30 cm were put forth as evidence for limb brightening, but in practice even an eclipse curve with 1% errors made it difficult to distinguish between various models. This stemmed from ambiguities caused by the way successive solar annular regions were covered and uncovered by the arc of the moon's edge, the presence of numerous active regions, and the technique's still-insufficient effective resolution (relative to the thinness of the predicted bright limb). Furthermore, one could usually do no better than to assume a circularly symmetric brightness distribution, despite the fact that the optical corona, especially at solar minimum, was known often to exhibit equatorial broadening. This restriction was relaxed for the first time when Emile-Jacques Blum, Denisse, and Jean-Louis Steinberg (1952a, b), using 1.8 m wavelength eclipse data from an exceptionally quiet sun observed in 1951 in French Sudan (Section 10.4), argued that their measured radio corona was more extended along the sun's equator than its polar axis. Further data of theirs from the 1952 eclipse, plus those of Marius Laffineur *et al.*

---

[10] Orchiston *et al.* (2006:45–8) give a detailed description of these eclipse observations. Also, Wendt *et al.* (2008) describe (1) plans by RP (eventually cancelled) to observe the 1947 eclipse in Brazil, and (2) unpublished observations by RP of a partial eclipse in Australia on 22 October 1949.

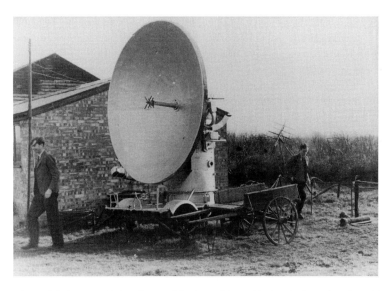

Figure 13.7 The movable 3 m diameter Würzburg dish used by Stanier in conjunction with a similar fixed dish for multi-spacing interferometry of the sun (Fig.13.8). Also seen are Martin Ryle (left) and Ken Machin.

(1954) observing at 0.6 and 1.2 m at Khartoum, also strongly favored an elliptical corona.

The second approach to measuring a detailed solar brightness profile, and the province solely of Martin Ryle's group in the Cavendish Laboratory, was Michelson interferometry. Three students included results using this technique in their doctoral dissertations: Harold M. Stanier (Ph.D. granted in 1950b), Kenneth E. Machin (in 1952), and Patrick A. O'Brien (in 1954). As discussed at the end of Chapter 8, the primary historical importance of this interferometric work was that it marked the first stage of Ryle's developing the principles of assembling a radio image from interferometric data. Most of these principles were resident at the Cavendish and at RP from early on, but the problem was turning them into practice with available technology. It was known that a measurement of the amplitude and phase of the fringes obtained with an interferometer of baseline n (measured in wavelengths), the so-called visibility, gave information about the Fourier component of the sky brightness distribution having a periodicity of $1/n$ radians and a sky orientation dependent on that of the baseline (Sections 7.3.2 and Section A.11). If one then made measurements at a series of baselines, a Fourier transform of the fringe visibility function would yield the sun's brightness "strip profile" along a certain direction. But it often took as long as a week to gather the needed data, during which time the sun, especially at longer wavelengths, was inevitably variable. The solar profile that resulted was then an ill-defined average. Furthermore, receivers and cables of that day caused electrical phases to wander around so badly that measuring an accurate fringe phase was a *tour de force*. In most cases, one gave up and forfeited any knowledge of the phases, which then forced any answer into the procrustean mold of a symmetric strip profile. One still did not have, however, the desired profile along a solar radius, but rather a strip profile oriented on the sun parallel to the interferometer baseline and having no resolution at all in the perpendicular direction. To go from this strip profile to a radial profile required the further assumption that the solar emission was circularly symmetric.

The calculations involved in this last step and in the Fourier transform process were long and tedious. Days, weeks, and even months (when working on two-dimensional transforms) were required even though the labor was considerably reduced by borrowing a technique from the X-ray crystallographers resident in the Cavendish Lab at that time (Section 8.5.1). Ryle's group used so-called Lipson–Beevers strips,[11] cardboard strips printed with, for a given value of A and n,

---

[11] H. Lipson and C. A. Beevers, "An improved numerical method of two-dimensional Fourier synthesis for crystals," *Proc. Phys. Soc.* **48**, 772–80 (1936); also in *Nature* **137**, 825–6 (1936).

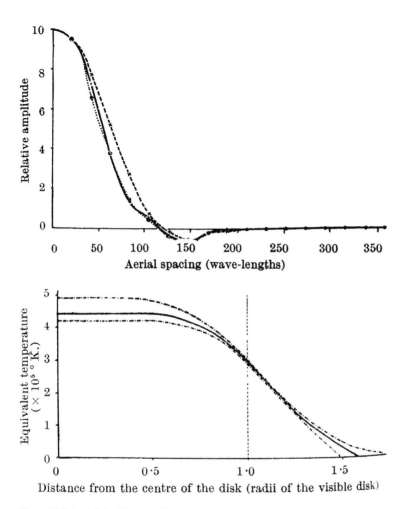

Figure 13.8 (top) Solar fringe amplitudes (normalized to 10) observed by Stanier (1950a) at 60 cm wavelength on three separate days in 1949 (negative amplitudes refer to fringes 180° out of phase). (bottom) Derived solar brightness temperature radial profile (solid line); the dashed lines indicate the estimated uncertainty.

the values of $A \cos nx$ in (say) 3° increments of x. By arranging these strips appropriately side by side, one could form the sums needed for the terms approximating the Fourier integral. Despite all of the above difficulties, the interferometry seemed worthwhile: it allowed one to choose times and places of observation more ideal than what an eclipse usually dealt out and it could better distinguish between different theoretical models. Moreover, the sun provided a nice strong signal to map – ideal for Ryle to test out his nascent techniques of interferometry.

Stanier monitored the sun at 60 cm wavelength throughout 1949 and on three separate occasions, when the sun seemed spot-free, used two 3 m diameter Würzburg dishes, one mounted on a trolley (Fig. 13.7), to measure fringes at many spacings. There were 17 east–west spacings in all, out to a maximum of 220 m; no one had done interferometry at such a short wavelength before and the technical problems were myriad, such as long lengths of cable causing signals to drop by a factor of 3. He used a Hollerith punched-card machine to assist in calculating the required Fourier integral and a graphical technique to derive a radial profile from the measured strip profile. The result, which represents the first time in radio astronomy that a fringe visibility function was turned into a source profile, is shown in Fig. 13.8, taken from a note to *Nature* published in March 1950 (Stanier 1950a).

Stanier's 60 cm profile was unexpectedly considerably limb-*darkened*, in stark contrast to eclipse results and the limb brightening predicted by Martyn (1948) and everyone else. As Ryle wrote at the time: "They [the Australians and the French] seem to find it very hard to believe."[12] Yet Stanier (1950b) cogently argued in his thesis that it was no artifact introduced, say, by the presence of a few minor sunspots or by his particuliar choice of interferometer spacings.[13] Although he realized that the apparent limb-darkening might be the result of a radio sun that was elliptically shaped despite his assumpton of circularity, he thought this unlikely (Stanier 1950b:93).

A year later, Machin followed Stanier, but now at a wavelength of 3.7 m, using four fixed and two movable antennas (each a broadside array of dipoles) and Ryle's new phase-switching scheme (Section 8.4 and Section A.11). He likewise obtained a one-dimensional set of fringe visibilities, but improved on Stanier's mathematics by noting that, assuming circular symmetry, the radial brightness profile was more directly calculated using Bessel functions. At this longer wavelength, though, the sun was sensibly quiet on only a few days, so Machin (1951, 1952) could publish only one visibility function, obtained over a five-day period in August 1950. The resulting brightness profile agreed reasonably well with Smerd's theoretical curve (Fig. 13.5) out to 1.5 solar radii, but farther out fell off much more slowly. Machin proposed that the combination of his and Stanier's profiles argued either for a lower temperature in the outer corona (causing a higher radio opacity) or for the presence of density irregularities in the corona, which would cause multiple scattering and a higher effective optical depth.

The most complete and general interferometric observations were by O'Brien. In 1951–53 he conducted no fewer than fifteen separate runs at three different wavelengths, including 7.9 m where large, fixed corner-reflectors fed by dipoles were employed. His 7.9 m brightness profile was about as expected and his 3.7 m results, although differing in detail, basically agreed with those of Machin. Most of his effort went into the 1.4 m experiments since at this wavelength the radio sun's activity was better behaved, the antennas were smaller (only two persons were needed to move them!), and the resolution was best for the longest baselines afforded by the constricted Grange Road site (Fig. 8.3). O'Brien's main new contribution was to do the first *two-dimensional* Fourier synthesis by measuring one-dimensional visibility functions at a variety of angles across the sun. He achieved the various angles by using either (1) baseline orientations differing by as much as 60° from east to west, or (2) repeated observations on the same baseline as the earth's rotation caused the sun to move over a range of hour angles. Although he started out gathering a data set over a week or two, he soon learned that the sun was too fickle for such leisure. In the end he recruited fellow students several times to help him with experiments, each involving a long single day of hauling and installing antennas all over the pastures. Once at a position, data were quickly taken, often over no more than one-half a fringe period, and then it was on to the next position:

> What we were doing was really extremely primitive – picking up an aerial, walking along, and then plunking it down somewhere else. It was the string-and-sealing-wax tradition, except ours was angle iron and concrete.
>
> (O'Brien 1981:141A:620)

O'Brien's busiest day, for nine hours on 1 April 1952, included 43 spacings over five different baseline orientations cutting across the sun at angles spread over a range of 64°. And on this day he even measured fringe *phases* and found them to change slowly and smoothly with increasing antenna spacing, indicating a single region of enhanced emission located a few arcminutes from the sun's center. After excising this source, a contour map of the sun was made that for the first time superseded previous assumptions of circular symmetry with one of quadrilateral symmetry about the sun's equator and axis of rotation. And indeed, after all the Fourier transforming, O'Brien (1953,1954) found that the polar width of the sun was about 20% smaller than the equatorial (Fig. 13.9). He considered

---

[12] Ryle:Stanier, ~1 April 1950, box 1/1, RYL.

[13] Stanier (1950b:57) measured fringe amplitudes at spacings extending from 0 to 365 wavelengths in uniform steps of 21.5 wavelengths. His maximum spacing determined the angular resolution, i.e., the smallest discernible detail (in this case, ~1/365 radians = 9'); he could go no farther because of attenuation in his connecting cables. Similarly, his step size determined the *largest* scale of features that the resultant image could contain (~1/21.5 radians = 2.7°). The step size was chosen based on a conservative guess as to how large the radio corona might be at 60 cm wavelength.

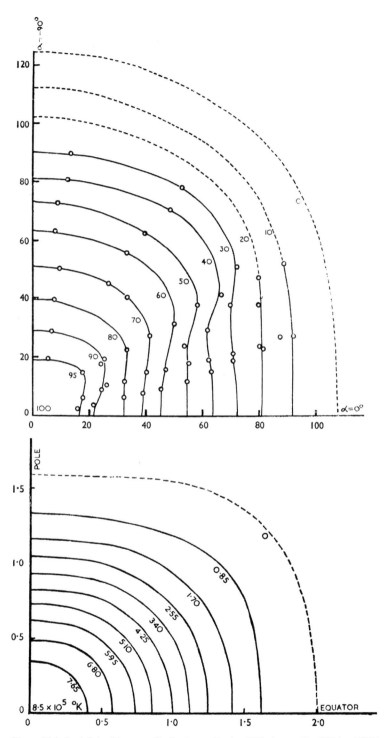

Figure 13.9 (top) Solar fringe amplitudes (normalized to 100) observed by O'Brien (1953) at 1.4 m wavelength on one day in May 1952. The axes represent antenna spacings in wavelengths in the north–south and east–west dimensions. (bottom) Derived two-dimensional solar brightness map of a quadrant; axes are in units of solar photospheric radii.

Figure 13.10 "Chris" Christiansen and his 32-dish east–west array at Potts Hill Reservoir field station in Sydney in 1953.

these interferometric data as sounder evidence for an elliptically-shaped corona than the earlier French eclipse observations (mentioned above), and also noted that over the 1950–52 period the radio corona seemed to be shrinking, perhaps as a result of the approaching minimum of the solar activity cycle.

By 1952 the toils and vagaries of eclipse expeditions and interferometry began to give way to the more direct approach of using a single antenna with one dimension large enough to supply the required resolution. This work only started at the end of the period being discussed here, but deserves brief mention. The outstanding instrument in this class was an east-west array of 32 six-foot diameter dishes that Christiansen lined up along one border of the Potts Hill reservoir (Fig. 13.10) (Christiansen 1953; Christiansen and Warburton 1953; Christiansen 1984:118–22). Inspired by somewhat analogous effects of an optical system designed by the French solar astronomer Bernard Lyot, Christiansen realized that such an array would produce a single narrow beam on the sun. He was allotted all of £10 ($30) per dish, but wooden-post mountings and cheap aluminum reflectors made it possible. Aiming the dishes was a manual operation – to track the sun, every 15 minutes one person had to cover the 200 m length of the array in 3 minutes, moving each of the 32 dishes one notch along on its east–west axis.[14] Operating at 21 cm, the dishes were all connected with a branching, open-wire transmission line system to a single receiver. The overall effect was that of a diffraction grating, producing a series of widely separated $3' \times 10°$ fan beams. From 1952 onwards the sun daily drifted through this array's pattern, yielding strip profiles with $3'$ resolution. Active regions could now be easily traced as they rotated with the solar surface from day to day. The

---

[14] Christiansen (1984:117–22); Haynes *et al.* (1996:209); Christiansen interview in Bhathal (1996:38–9).

lower envelope of all these cuts also nicely limned the quiet sun and revealed substantial limb brightening. Christiansen argued that Stanier's method had washed this out because of its assumption of a symmetric disk despite the likely presence of strong asymmetries, which indeed were now daily observed.

Mapping ability at Potts Hill grew even better a year later when Christiansen added a similar north–south array. By tracking the sun for several hours each day, he used the earth's rotation to change the effective orientations of his baselines and their resultant cuts across the sun. A two-dimensional solar map could be laboriously reconstructed over months using Lipson–Beevers strips and a huge sheet of graph paper covered with many long lines of numbers running across it at various angles (Christiansen and Warburton 1955). This earth-rotation technique was much easier than hauling antennas around a Cantabrigian field and later, when digital computers allowed for faster two-dimensional Fourier transforms, became an important part of imaging interferometry.[15]

13.2 THE ACTIVE SUN

13.2.1 Meter wavelength bursts

The radio sun on meter wavelengths was spectacular in its ability to increase within a minute to (routinely) hundreds of times or (rarely) a million times its normal intensity (Figs. A.2, 13.2 and 13.12). This was what had brought the sun to the attention of Hey in 1942 and what caught the imagination of many studying extraterrestrial noise in the postwar era. These giant bursts inspired awe:

> The first time I recorded a solar storm, I just sat there and listened to it roar away. You could hear[16] it sighing up and down, like being on a beach with occasional big waves crashing among the rocks. It completely overpowered the receiver noise and the recorder was going wild. Boy, it was quite a thing!
> (Seeger 1975:11T)

One researcher, Karl-Otto Kiepenheuer, was even concerned with the effects of such intense radio fluxes on the earth's atmosphere and on life.[17]

The postwar period was ideal for study of the active sun, for 1947–49 turned out to coincide with a maximum phase in the eleven-year cycle of solar activity (Fig. 13.11). And not only was the cycle at its peak, but that particular cycle was the most active since the start of reliable counts of sunspots in the eighteenth century. Indeed, the sunspot groups that spawned the great radio events studied by the Australian and British groups in February and July 1946 were record-breaking, and the one of March 1947 has still never been equalled.

There were many groups studying the active sun at meter wavelengths in the late 1940s (Fig. 13.1), but progress in the field was dominated to an extraordinary degree by the Radiophysics Laboratory (RP) in Sydney. Beginning in late 1945 with the work of McCready, Pawsey, and Payne-Scott (1947) in establishing the association of enhanced radiation with sunspots (Chapter 7) and culminating in the spectrographic burst categories of Paul Wild in 1950, RP researchers led the way in systematizing and measuring the sun's sporadic events. Besides the RP work, there were only two other meter-wavelength studies of note. The first was by Allen (1947) at Mt. Stromlo, using 200 MHz equipment supplied by RP (Section 7.3.4). After a year of monitoring, he found an association of radio activity with sunspots, flares, and geomagnetism, but it was not by any means a close one-to-one correspondence, for there were often radio events with no accompanying optical phenomena, and vice versa. He also coined the terms "outburst" for the most intense and longer bursts (usually of minutes duration, often preceded by a flare), and "noise storm" for the enhanced levels that could last for hours or days, usually associated with sunspots. The second study was at Cambridge by Ryle and Derek Vonberg (1948), who, as described in Section 8.3, interferometrically

---

[15] A second early high-resolution antenna was built outside of Ottawa, Canada by Covington and Norman Broten (1954). This was a 150 ft-long slotted-waveguide array that began 10.3 cm observations in late 1951, and that also clearly indicated limb brightening with its 8′-wide fan beam (Covington 1984:325).

[16] Standard practice was not only to measure the solar signal on a strip-chart recorder, but also to monitor it as impressed on a loudspeaker. Atmospherics and manmade interference, which on the recorder trace could easily mimic a solar burst, sounded very different from the characteristic rushing or whooshing sound from the sun.

[17] See Tangent 13.2.

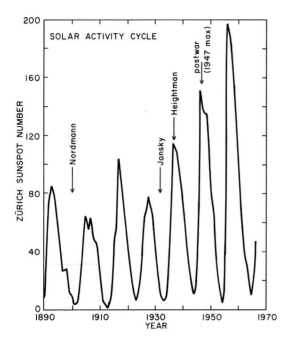

Figure 13.11 The solar activity cycle. Events in the history of radio astronomy affected by the solar cycle are indicated – solar minimum was unfortunate for Nordmann (Section 2.5), but greatly aided Jansky (Chapter 3). Maximum phase was important for Heightman in the 1930s (Section 5.4.2) and for all of the postwar solar work.

measured the size of emitting regions as smaller than the solar disk, and found that the enhanced radiation was almost always 100% circularly polarized.

13.2.1.1 Payne-Scott's work

Ruby Payne-Scott, with active direction from Joe Pawsey, was the central person for the early meter-wavelength effort in Sydney. On the heels of her seminal work in 1945–46 together with Lindsay McCready and Pawsey (Section 7.3.2), she organized an intensive fortnight of monitoring at four frequencies in July–August 1946. This showed that bursts often appeared on only a portion of the meter-wavelength band and that variability was much greater at the longer wavelengths (see Fig. 13.2). In many cases, however, it was possible, particularly on adjacent bands such as 60 and 75 MHz, confidently to associate events occurring on different frequencies and thus to study their relative timing. A histogram of these timings showed that bursts definitely arrived earlier on 75 than on 60 MHz, the most common delay being ~2 sec. But this result was published only in an RP report[18] until the impetus of the arrival at 0431 UT on 8 March 1947 of *the* most intense extraterrestrial radio signal ever received – greater than 10^{11} Jy on 60 MHz![19]

Payne-Scott, Yabsley, and John Bolton (1947) (see also Section 7.4) sent *Nature* a note about this behemoth in which they emphasized the above results from the previous year's monitoring and pointed out that the new outburst clearly showed a sequence of 200–100–60 MHz arrival times over six minutes (Fig. 13.12). Suggesting that the outburst was related to "some physical agency passing from high-frequency to lower-frequency levels," they then adopted Martyn's (1946a) idea that radiation at any frequency originated at the coronal height where the observed frequency equalled the plasma frequency. With a standard model of the coronal electron density fall-off with height, this led to the "physical agency" moving 0.3 solar radii in 6 minutes, or at a speed of ~600 km s^{-1}. This seemed reasonable when compared with the velocities of prominence material (hundreds of km s^{-1}) and particles causing aurorae on earth (~1500 km s^{-1}), such as the one in Australia the day after the outburst. The authors, however, were not willing to tell the reader the speeds of ascent implied by the ~2 sec delays that Payne-Scott had commonly seen, probably because they came out so much higher (~25,000 km s^{-1}) than anything ever measured on the sun.[20] As an alternative interpretation, they suggested that these short delays might be due to a dispersion of propagation speeds with frequency.

In England too, people had been thinking about the frequency/time relation of bursts as a way of probing

---

[18] R. Payne-Scott, "A study of solar radio frequency radiation on several frequencies during the sunspot of July–August, 1946," Radiophysics Lab Rept. RPL 9 (33 pp. + 15 figs.), August 1947.

[19] Paul Wild later stated that the calibration of this extremely large value of intensity was deemed uncertain by Payne-Scott and himself (Wild:author, 4 August 1987). The sunspot associated with this radio burst remains today one of the largest ever recorded (covering 0.6% of the sun's disk).

[20] In a write-up for an RP talk, Payne-Scott *did* work out speeds for the bursts of ~60,000 km s^{-1} (20% of that of light), but then dismissed them as "apparently unlikely." [R. Payne-Scott, "Differences between times of arrival of bursts of solar radiation emitted at different frequencies – preliminary measurements," 9 February 1948, file A1/3/1b, RPS]

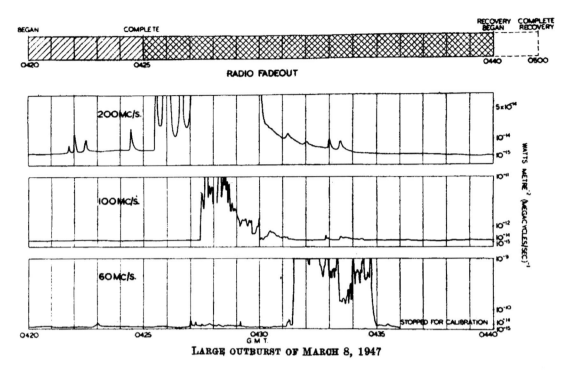

Figure 13.12 Traces of the great outburst observed by Payne-Scott, Yabsley, and Bolton (1947) at three different frequencies; notice the sequence in frequency of onset. The 60 MHz intensities (at least 10^{11} Jy) were the highest ever recorded (and still are to this day). At the top, information on an accompanying radio fadeout in shortwave communications is given.

motions on the sun. In October 1946 RP's Chief Taffy Bowen had discussions with Edward Appleton and Jack Ratcliffe in which this basic idea came up, and he wrote home to Pawsey for RP to hop on it.[21] Only a month after the Australian report in *Nature*, Hey, Parsons, and Phillips (1947) deduced the presence of an outward-moving "solar disturbance" from the fact that a large burst occurred a few minutes after a radio fade-out on earth (for instance, as in Fig. 13.12). The fade-out (a loss of long-distance shortwave communications signals) was taken to be caused by increased ionization in the ionosphere arising from a sudden high ultraviolet flux from the photosphere, while the burst originated in the corona. But in general Hey (1949:201) and Ryle (1950:211) could not see reliable patterns in the arrival times of bursts in their own data and so were inclined to think that a case had not been established. Ryle wrote to Pawsey expressing this sentiment and from the latter's reply it is clear that even the Sydney side had some doubts:

> With regard to the "seconds delay" cases, I am not entirely happy about the evidence. I believe that the tendency exists but do not know how often relative to zero delay, which is common, or to unco-ordinated effects.[22]

All of this interest in time delays of bursts led to a major effort in 1948 at RP's Hornsby field station in which for a nine-month period Payne-Scott (1949) monitored the sun at four frequencies and accumulated hundreds of events. For those cases where the same event could be unambiguously recognized on different frequencies, once again the higher frequency always arrived first. But the delays were always of order seconds – throughout 1948 there was not a single case of a minutes delay, as for the huge March 1947 event. And now, rather than work out implied speeds in the corona

---

[21] Bowen:Pawsey (from London), 13 October 1946, file A 1/3 1a, RPS.

[22] Pawsey:Ryle, 3 July 1947, file A 1/3 1a, RPS.

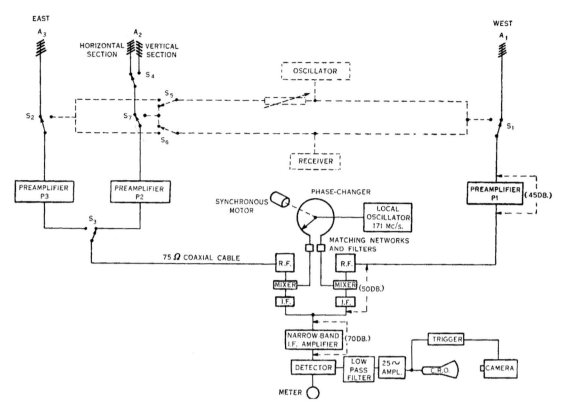

Figure 13.13 Schematic diagram of the 97 MHz swept-lobe interferometer at Potts Hill of Little and Payne-Scott (1951). The Yagi antenna separations were 790 and 130 ft.

as she had before, she chose to interpret the delays in terms of ideas being put forward by RP's applied mathematicians John C. Jaeger (Section 12.5.2) and Westfold (1949). Their picture was that all frequencies from a burst originated at a common level in the upper corona and that the higher frequencies arrived first because of their higher propagation velocity arising from a larger index of refraction. This also explained why many bursts were observed to be double-humped (that is, show two peaks in intensity separated by 2–10 sec), for radiation in the first hump had come directly to the earth and that in the second was an "echo" that travelled to the inner corona before being reflected back at the point where the refractive index became zero.

RP next tried a more direct attack on motions of the bursts, involving a clever, unprecedented "swept-lobe" interferometer, one that had good enough angular and time resolution to track the location of a burst from second to second. This Payne-Scott did with Alec G. Little (1925–1985), a technical assistant who was well on his way to becoming a full-fledged researcher. Little had begun as a messenger in 1940, but his technical talents were soon recognized and, after a spell working on X-ray vacuum tubes, he transferred to the radio astronomy group. His first major project was to build this interferometer along 1000 ft of the reservoir bank at Potts Hill. The paper describing it (Little and Payne-Scott 1951) served as an excellent primer on interferometry of the day (Fig. 13.13).

The interferometer's key innovation was to reduce the time needed to measure solar fringes – the normal pace supplied by the earth's rotation (about one fringe every three minutes) was far too slow. The heart of the system was a phase changer rotating at 1500 rpm that in effect inserted extra lengths of cable into one of the interferometer's arms (although the phase change was actually applied to the local oscillator). This had the effect of sweeping the interferometer's lobes back and

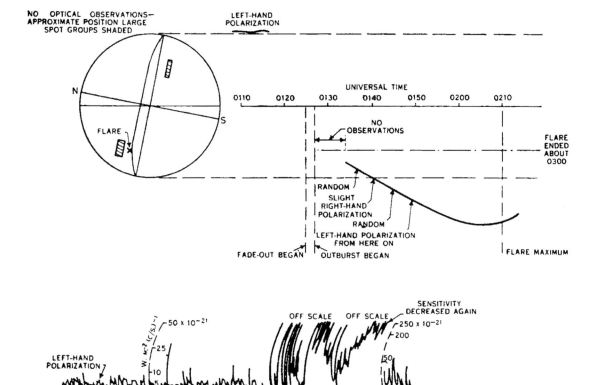

Figure 13.14 An outburst observed with the swept-lobe interferometer by Payne-Scott and Little (1952). The bottom record is of the intensity received with another antenna at 97 MHz, with right- and left-hand polarization alternating every 30 sec. Downward kicks are 5 min time marks. The top diagram indicates the optical sun and the changing east–west position and polarization of the radio emission (displaced to the right for convenience; in 30 min the emission region moved from 0.8 to 1.7 solar radii).

forth by two lobewidths every 1/25 sec, and, when synchronously displayed on a cathode-ray tube, produced a sine wave corresponding to the desired fringe. This fringe's phase gave the source's position, or, if more than one source were present, their weighted mean position. There remained the ambiguity, however, of *which* lobe's phase was being measured and this was removed through a second interferometer of slightly different baseline and lobe spacing. The system automatically switched back and forth between the two interferometers every third of a second and in this way gave a source's east–west position to an accuracy of ~2′. Polarization was measured by yet another fast switching between two orthogonal Yagi antennas. The final touch was to have a movie camera automatically recording the cathode-ray tube when the solar signal exceeded a predetermined threshold (of the order of 0.5 MJy). In this way the position and polarization of any 97 MHz burst that occurred on the noontime sun was measured every second.

Starting in May 1949 Payne-Scott and Little (1951, 1952) observed for over a year with this swept-lobe interferometer, harvesting 30 noise storms, 25 bursts, and six outbursts. The noise storms were tracked across the face of the sun from day to day and often went well outside the optical disk. Assuming that the noise storm radiation originated directly above the sunspot group usually also present, they concluded that the radiation came from 0.3 to 1.0 solar radii above the photosphere, well beyond the level where the plasma frequency equalled 97 MHz. And yet the handedness of the circular polarization they measured almost always agreed with the magnetic polarity of the single largest sunspot in the group, implying that the noise storm, despite its lofty perch, in some manner was associated with this spot alone.

Each of the six outbursts they examined, such as the one shown in Fig. 13.14, shifted positions over an hour or so. They were not able to make much of the outbursts' complex polarization situations, but these first measures of position changes were amenable to a fairly direct interpretation. After an optical flare and terrestrial radio fadeout, the outburst appeared and usually moved in an apparently outward direction relative to the flare. Payne-Scott and Little calculated a velocity for whatever agency initiated an outburst by using (a) the delay between times of flare and of outburst, and (b) the distance from the photosphere to the initial height where the outburst was observed, typically 0.3–0.6 solar radii. In this way typical velocities of 500 to 3000 km s^{-1} were derived. Many of these velocities were a good bit larger than those being inferred in similar outbursts by Wild (see below), and it was suggested that he might have gone astray by using incorrect values of coronal electron density.

In 1951 Payne-Scott decided to have a family and that summarily was the end of her short and fruitful research career (Chapter 7). It was also the end of the swept-lobe interferometer, for Little moved on to help Bernie Mills with the latter's new idea of a cross antenna. Little did try to analyze the 25 bursts that they had caught, but the positions never yielded consistent results and he gave it up (Little 1978:94B:580).[23]

13.2.1.2 Wild's work

Payne-Scott and Little had exited the RP solar-burst scene, but a new character had earlier entered and was by now conducting his own research with no less success and technical innovation. J. Paul Wild (1923–2008) (Figs. 7.4 and 16.9) had come to RP in early 1947 after a degree in mathematical physics at Cambridge University and several years as a radar officer in the Royal Navy. Although an Englishman, his frequent visits to Sydney on the battleship *King George V* led to meeting his future wife and settling in Australia. At first he languished in the RP equipment test room for a year, but then Pawsey gave him his "release" in the form of an option to work either with Bolton on radio stars or with McCready on a new solar instrument. The latter seemed to offer more independence and so in 1948 he found himself trying to put into practice McCready and Pawsey's design ideas for a device that would instantaneously measure a wide portion of the solar spectrum, allowing time delays in bursts to be studied in a much cleaner fashion than Payne-Scott's separate receivers and antennas. During the war there had been wideband intercept receivers for locating the frequencies of enemy radio transmissions, but the sensitivity and calibration requirements were far severer for this new application (Wild 1987:98). As Wild (1978:6–7T) later recalled

> It was a question of getting a good noise figure over something approaching a two-to-one frequency band – anything you did at one frequency tended to upset the other frequencies. ... We also had trouble getting enough power out of the local oscillator over the frequency range, using an extraordinary tube called a "doorknob tube" (because of its shape). ... Another task was getting rid of all the parasitic oscillations which seemed to want to creep into the system.

Wild and McCready (1950) set to work, greatly aided by electrical engineer John D. Murray, who contributed the display apparatus, and by their technician William C. Rowe. By early 1949 they had a system that, by tuning with capacitors attached to a fast-rotating shaft, in less than 0.1 sec could display a solar spectrum from 70 to 130 MHz as a trace on a cathode-ray tube. To lessen manmade interference (although it remained a serious problem) Wild set up a new field station on a cattle stud farm in Penrith, about 30 miles from Sydney. His choice for antenna was the rhombic, a workhorse for wideband shortwave communications originally developed by Edwin Bruce at the same time and place where Jansky made his original discovery with another Bruce design. It was a diamond-shaped affair that offered a reasonable gain over a broad range of frequencies and whose high sidelobes could be tolerated for solar observations. The rhombic (Fig. 13.15) was mounted on a long pole movable across the sky (for five hours daily) by cranking a winch and adjusting ropes – hard work in the summer sun, but the large ~25° beam meant that observers only needed to go through these calisthenics about once every twenty minutes.

---

[23] Much more detail about Payne-Scott's research and career is given by Goss and McGee (2009).

Figure 13.15 The rhombic antenna built by Wild and McCready at Penrith, with technician W. C. Rowe standing at the base and another person on top of the pole (on left) adjusting a guy wire. During 1949 this antenna operated with a swept-frequency spectrograph over a 70–130 MHz range to study solar bursts. The antenna was daily steered east–west using a winch at the base, with seasonal north–south adjustments also possible. (RPP)

Commencing in February 1949, standard observing procedure was for Wild and Rowe to alternate shifts staring at the cathode-ray tube, waiting for something to happen, whereupon they would start hand-cranking a movie camera that photographed the successive spectra:

> Sometimes a whole week would go by with absolutely nothing. ... One would wait for long periods and then all of a sudden these things rose out of the cathode-ray tube's scan. It was tremendously exciting and whoever was observing would try and turn on the camera as quickly as possible and shout or scream to get the other one to come along and have a look. (Wild 1978:8–9+13T)

By June they had fifty days of good observations and had exposed enough film that it seemed a good time to stop and analyze the data. Wild first checked out narrowband signals that occurred at certain fixed frequencies – could they be spectral lines from hydrogen, the most abundant element in the sun? Although this led to interesting astrophysics in an RP report and (much later, after the discovery of the 21 cm hydrogen line in 1951) a paper in the *Astrophysical Journal* (Sections

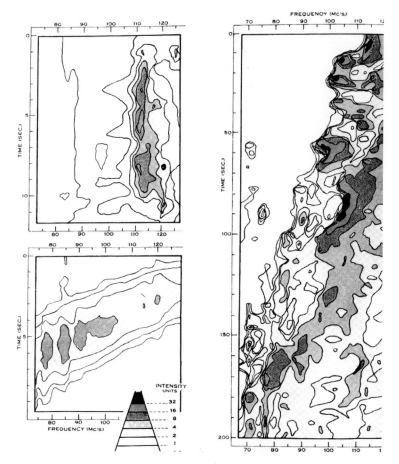

Figure 13.16 Frequency–time contour diagrams illustrating the three spectral types defined by Wild based on his swept-frequency spectrograph observations in 1949: (upper left) Type I, (right) Type II, and (lower left) Type III. The unit of intensity is 1 MJy. Note the different rates of frequency drift of the major "ridge line" in each case. This illustration was originally published with colored shading for the contour intervals (Wild and McCready 1950); the present version is almost identical and comes from a review by Pawsey and Smerd (1953).

7.2.3 and Section 16.6.2), it turned out that the putative spectral lines were spurious, arising from standing waves in the line between receiver and antenna.

Other observed patterns, however, were not artifacts and in fact Wild's analysis became the foundation for all later study of solar bursts. He set forth what he called Type I, Type II, and Type III bursts,[24] classified solely on the basis of their dynamic spectra, that is, on how their spectra changed rapidly over time (Fig. 13.16). Wild's plots of frequency versus time dramatically illustrated the character of three distinct spectral types into which almost all bursts seemed to fall, and proved so useful for studying the radio sun that over the next decade they became a standard method of presentation.

---

[24] Wild's first scheme for naming solar bursts called for spectral types α, β, and γ (corresponding to, respectively, Types III, II, and I), but for some reason he changed nomenclature in the last months before submitting his article. [see *J. Geophys. Res.* 55, 191 (1950) for the abstract of a talk delivered by Wild in Sydney in January 1950]

Within a six-month period Wild (1950a,b;1951) followed up his overview paper with three more delving into the details of each type of burst. The great majority were of Type I, also called by him "storm bursts" since each lasted for a couple of seconds on top of the noise storms. Wild's major new result was that the bandwidth of any particular storm burst was only ~4 MHz and the center frequency, although sometimes rapidly fluctuating, showed no characteristic drifts with time (Fig. 13.16).[25]

Wild (1950a) also caught five outbursts, called Type II in his new classification scheme, and found that their dynamic spectra (Fig. 13.16) had a well-defined low-frequency cutoff that slowly drifted to lower frequencies over the few minutes of the outburst – a typical drift rate was ~0.2 MHz s^{-1}. The complex frequency-time structure now seen in the Type IIs made it clear why Payne-Scott had not been able to find them despite long observations during 1948 with separate receivers at spaced frequencies. Wild assumed that their drift in frequency indicated a disturbance moving outwards through the solar corona, somehow exciting the emission of radio waves in regions of lower and lower density. In each region a broad band of frequencies might be generated, but only those with frequency greater than the critical frequency at that location could escape. The disturbances' frequency drifts typically led to inferred velocities that increased from 300 to 700 km s^{-1} over a five-minute period. This order of velocities he took to tie in reasonably well with either the faster motions observed in prominences or the slower ones of particles that caused geomagnetic storms.

Finally, there were the Type III bursts. These were of broader bandwidths (50 MHz or larger) and showed a very fast drift rate to lower frequencies (~20 MHz s^{-1} – see Fig. 13.16). Wild (1950b) found them to constitute over half of the "isolated bursts" (those occurring at times other than during noise storms), the rest being complex mixtures whose dynamic spectra defied categorization. But although these Type IIIs were no less a distinct class than Types I and II, their origin was more puzzling since there were no known optical solar phenomena with which they correlated – the radio data were on their own. Payne-Scott (1949) had found that Jaeger and Westfold's (1949) theory of different frequencies propagating at different speeds (mentioned above) seemed to work well, so Wild set out to test it against his more detailed observations. He found the theory wanting because of one test, namely the slight curvature of the "ridge lines" along the bursts as displayed in Fig. 13.16. Wild argued[26] that the theory was unacceptable because in all cases it predicted a degree of curvature much larger than the slight amount observed. Instead, he liked the same scheme as for the Type II bursts, namely an outward-moving disturbance somehow setting off radio waves at the local critical frequency of each level. Now, however, the much greater rate of frequency drift in the Type III bursts implied that the disturbance was moving at speeds ranging from 20,000 to 100,000 km s^{-1}, the latter being one-third of the speed of light. Nothing had ever been seen to move on the sun at faster than ~1000 km s^{-1} and this interpretation inevitably met with considerable skepticism. Wild himself was not sure of it, but pointed out that ionized material moving at such speeds might well give no optical emission at all, and that even here on earth lightning flashes created such fast particles. Finally, he noted that these speeds all depended on a "normal" fall-off of electron density with height in the corona; if there existed regions with substantially higher densities and density gradients, deduced speeds would be reduced.

Despite great success in sorting out the solar bursts, or perhaps more because of it, Penrith was closed down after only a half year of observations. The manmade interference had been abominable and now Wild's team knew what they really needed to study these fascinating

---

[25] A statistical analysis of almost 700 of these bursts also revealed the curious effect that the bursts strongly preferred ~115 MHz and avoided 89 MHz. Wild (1951) argued that this effect seemed real, but that it should be verified with a larger sample (and indeed later studies failed to confirm it).

[26] Wild has recalled how the proof published in Wild (1950b:556–7) came about:

About this time [1949–50] Professor Jaeger used to come and visit us frequently, as a kind of theoretical consultant and father confessor.... I can remember proudly showing him my proof [that the theory of Jaeger and Westfold (1949) made a prediction in disagreement with the data on Type III bursts] which covered six closely written and closely argued foolscap pieces of paper; and he looked at it and said, "Yes, that is very interesting." He came back next morning with a very small sheet of paper and the whole thing elegantly proved in five lines.

[Wild (1974), *Records Austral. Acad. Sci.* 3, No. 1, 93 (quotation from p. 103)]

Figure 13.17 Paul Wild with one of his three rhombic antennas to study solar bursts. Photo taken in 1952 at the Dapto field station south of Sydney.

bursts – more frequency range, greater sensitivity, and a better recording setup. Regarding the last, Wild had become convinced of the merits of instantly being able to see the nature of the bursts:

> Wild had a tremendous emphasis on the need for a real-time display, so that one could actually *see* the solar event and have the excitement of watching it right in front of you and get ideas, to some extent. You could say, "Ah, yes, this is a slow-drift burst and that's a fast-drift one, this one is continuum and that one's narrowband." Later refined analysis was of course important, but Paul realized that the immediate results were important so that one didn't accumulate miles and miles of film and never get anything out of it.... The "cream-skimming" basic discoveries came right during the observations.
>
> (Smerd 1978:92A:425)

Wild, Murray, and Rowe (1954) thus built a completely new system on a dairy farm at Dapto, about 60 miles south of Sydney (Fig. 7.4) and shielded from the city by a steep escarpment. Covering a tripled range (from 40 to 240 MHz) and using three separate rhombic antennas and receivers (Figs. 13.17 and 13.18), it went

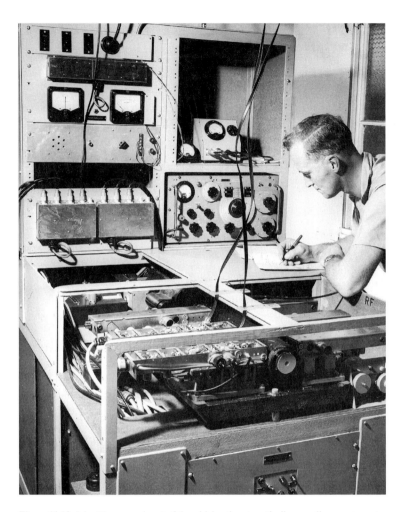

Figure 13.18 John Murray and part of the wideband, automatically recording spectrometer receiver at Dapto in 1952.

into operation just in time to be one of RP's prides put on display for the August 1952 URSI General Assembly in Sydney (Section 7.5.1). Its work is beyond the scope of this study, but one early result deserves mention. Although things were not as lively on the sun as they had been three years previously at Penrith (minimum phase was approaching in the solar cycle), the first strong outburst they caught, on 21 November 1952, made it all worthwhile. Their increased frequency coverage revealed that this Type II burst showed the usual slow drift in frequency, but now the signal was concentrated at any time into *two* distinct bands with a frequency ratio close to 2.0 (Wild, Murray, and Rowe 1953). The natural interpretation was that the higher frequency was a harmonic of the lower, and this helped establish the idea that radio emission in Type II and III bursts (for which harmonics were also measured) was involved with a resonant process such as plasma oscillations (see below). Wild's group was now firmly launched on a program of detailed study of the active sun that was to lead the world for another three decades.

13.2.2 Burst radiation theory

Although by 1952 theories of emission for the quiet sun and the slowly varying component (see below) were beginning to gel, the situation with radio bursts remained confused. Since a certain fraction of bursts were associated with sunspot groups and optical flares, attention focused on how these special regions could somehow produce brightness temperatures of 10^{10} K or more. But despite years of gathered data and attention from some of the best theorists of the day, there was still no agreement as to how the radio sun frequently managed to brighten to thousands of times its quiescent level. As Ryle (1950:223) put it: "The resolution of these difficulties presents one of the most important problems not only in radio astronomy but in astrophysics generally." Ideas during this period fell into three main categories: (a) gyromagnetic emission, (b) emission from high-energy electrons, and (c) emission induced by plasma oscillations.[27]

Kiepenheuer (1946) initially suggested that radiation at the resonant frequency of electrons spiralling in the strong magnetic field of a sunspot could account for the intense emission from radio bursts. In this scheme of gyromagnetic (or cyclotron) emission (see Section A.5), the electrons had only the thermal energy associated with each particle in a 6000 K gas, but Kiepenheuer calculated that one could nevertheless obtain apparent solar temperatures of $\sim 10^{10}$ K. Although first out of the starting gate, this suggestion did not get beyond the first turn because of criticism on the fundamental grounds that emission from a gas in thermal equilibrium, with or without a magnetic field, can never produce a brightness temperature higher than the gas's own temperature (Unsöld 1947:200, Martyn 1947:27). Kiepenheuer had gone astray by summing the expected emission from each parcel of gas without considering the counteracting absorption that also took place. Denisse (1946, 1947a) attempted to get around the absorption problem by arguing that a steep fall-off in the sunspot's magnetic field would allow the radiation to escape, but this was argued by Ryle (1948a:86–7) to be incompatible with the magneto-ionic theory of propagation, for which no escape seemed possible in a direction toward a region of lower magnetic field. Hoyle felt that the waves could escape, however, and came into sharp conflict with Ryle over this point at several Cambridge seminars.[28]

A second way to generate copious radio waves was somehow to create high-energy electrons that would radiate either by the high-energy analog to cyclotron radiation, namely synchrotron radiation (see Section A.5), or by decelerations caused by collisions. The former suggestion arose soon after synchrotron radiation was first proposed as an explanation for radio emission from discrete sources and the galactic background (Alfvén and Herlofson 1950; Kiepenheuer 1950 – Section 15.4). The astrophysical applications of synchrotron radiation were being fully worked out at this time by Ginzburg and his student German G. Getmantsev primarily in the context of the Galaxy, but they also favored it for solar bursts (Getmantsev 1951:539–40, Getmantsev and Ginzburg 1952). This work was unknown in the West, but Ryle (1951:110+133) also talked about synchrotron radiation in solar bursts (after abandoning his original ideas described below), although he never developed this idea beyond a conference presentation. Synchrotron radiation in essence represented high harmonics of the cyclotron frequency, and so there was no problem with it escaping from the solar atmosphere. It also seemed to be able to supply sufficiently high brightness temperatures, although no one discussed exactly where the high-energy electrons came from. But although the synchrotron idea was to prove bountiful in other contexts (Chapter 15), it was eventually laid to rest for solar bursts. This was largely the work of Piddington (1953), who argued that the characteristics of solar bursts and outbursts (circular or no polarization, narrowband spectrum) did not match those of synchrotron radiation (linear polarization, broadband spectrum).

A common hypothesis was that electrons could gain high speeds through acceleration in solar electric fields and then radiate a goodly portion of their energy away (via bremsstrahlung [free–free radiation] – see Section A.5) when they collided with other particles. One popular idea for the source of these electric fields was a fast-changing magnetic field associated with an active region, perhaps during the growth of a sunspot or during a flare (Denisse 1947b; Giovanelli 1948). On

---

[27] An informal précis of the main arguments concerning the origin of solar noise can be found in an account of the Royal Astronomical Society meeting of 23 April 1948, at which Appleton, Hey, Ryle, Hoyle, Pawsey, and Martyn all spoke. [*Observatory* **68**, 178–83 (1948)]

[28] Ryle:Hoyle, 26 November 1949; Ryle/K.C. Westfold (at Oxford) correspondence, November 1949 – April 1950; both box 1/1, RYL.

a larger scale, Ryle (1948a, 1949a, 1950:222–36) took up a prewar suggestion of Hannes Alfvén that electric fields would be produced by the combination of solar rotation varying with distance from the solar equator and the presence of a tilted, general solar magnetic field. These fields accelerated electrons, which then radiated in the general corona at an effective temperature of $\sim 10^6$ K and in the vicinity of sunspots even up to $\sim 10^{10}$ K. Ryle thus economically invoked the same mechanism for the puzzling high temperature of the corona (see below) as he did for the radio bursts. Both were considered to be of *thermal* origin since the electrons radiated independently of one another and were moving randomly with a distribution of velocities appropriate to a high-temperature gas. The great bursts were analogous to electrical discharges and their circular polarization arose naturally from the differing intensities of the emerging ordinary and extraordinary components. Within a few years, however, Ryle's theory was found to have a fatal flaw: he had incorrectly taken the collision cross sections (and therefore rates of emission) for high temperatures to be the same as for the earth's ionosphere at $T \sim 300$ K (Piddington and Minnett 1951a:140, Ryle 1951:109). But at high temperatures particles zip past each other faster and interact less, with the effect that the amount of emission (or absorption) is strongly reduced (Section A.5).

Even before this error was discovered, however, many were bothered by the presence of what Ryle called "genuine" temperatures as high as 10^{10} K. As Ryle (1976:13T) later recalled:

> We realized that if this was going to be what you might call a thermal mechanism that you'd have to have a gas with particles of [very high energy]. There was considerable opposition to this – that it was daft, that you couldn't talk about temperatures like that.

To overcome this and other problems, many researchers came to prefer a mechanism known as plasma oscillations, orderly motion of large groups of particles of like charge (Section A.5.4). Such oscillations had been studied in the laboratory since the 1920s and in fact were a key ingredient in the operation of many vacuum tubes, including the cavity magnetron of World War II. But the key challenge for a theorist was to make a jump of nine orders of magnitude from plasma in a bottle to the sun. He had the double task of figuring out how to create and maintain the plasma oscillations *and* of how to get them to transfer energy into radio waves able to escape. As Ryle (1976:13T) put it:

> The only other way of doing it was as in broadcast aerials, to unloose more than one electron at a time. ... This sort of oscillation was known to exist in gas-discharge plasmas. [But for the sun] there were great problems in starting it up and great problems in getting it to escape because it always happens at frequencies that can't get out through the boundaries. These don't matter for a plasma in a bottle because its gradients [of density and fields] are so fast at the edge, and one has bits of wire sticking in [allowing radio energy to escape].

As discussed in Section 10.3.1, Shklovsky (1946, 1947) made the first suggestion of plasma oscillations in connection with radio bursts. Martyn (1947) independently made the same suggestion, pointing out that the oscillations would cause electromagnetic waves when there were sufficient collisions between the oscillating charges and the ambient gas. At the US Naval Research Laboratory Andrew V. Haeff (1948, 1949), who learned of the solar noise problem from Hagen's group in the next building (Section 10.1.2), worked on adapting his own theory of amplification in microwave "electron wave" tubes to the solar situation. His scheme required intermingled streams of electrons of differing velocity that interacted with initially weak plasma oscillations so as to amplify them and cause intense radio emission. In effect, the kinetic energy of the streams was converted into radio waves. Such conversion was the basic principle behind several new microwave tubes being developed in the late 1940s, such as the klystron and the travelling-wave tube, and the possible relation between these and solar bursts was nicely set forth in an article by John R. Pierce (1949) of Bell Labs.

Hoyle (1949:116) and Ryle (1949a), however, strongly argued that plasma oscillations simply would not work in the solar atmosphere. They could not see how electrons could oscillate in united phase over a region many wavelengths in size. Ryle in particular argued that the maintenance of Haeff's ordered streams of particles was unlikely. Furthermore, he pointed out that even if the radio waves were generated, they would be unable to escape because the gradients of solar density and magnetic field along the line of sight were far too small.

And so the arguments went back and forth about which processes caused radiation from solar bursts. It was not until the late 1950s that further observational and theoretical work led to the general acceptance of plasma oscillations as the main agent. As mentioned above, perhaps the most important event leading to this consensus was Wild's observations in 1952–53 of bursts where both a fundamental and a first harmonic frequency were observed. But although this pointed indubitably to a resonance, with plasma oscillations as the best candidate, the details of a satisfactory theory took many more years to work out.

13.2.3 Slowly varying component

In 1948 the activity of the microwave sun was recognized as qualitatively different from that at longer wavelengths: (a) short microwave bursts were rarer and less intense, and (b) over months one observed a quasi-periodic kind of variation (see Fig. 13.2). This behavior could be found in 480 MHz monitoring over 1946–47 (Reber 1948a), but Reber focused on the few bursts he picked up and did not discuss the month-to-month variations. It was Covington (1948) who first showed that the microwave sun's intensity neatly followed the number of visible sunspots (as in Fig. 13.2). In February 1947 he began monitoring the sun at 2800 MHz with a 4 ft dish in Canada (Section 10.2), and his first nine months of data showed a marked 27-day periodicity corresponding to the sun's rotation carrying sunspot groups in and out of view. Although Covington chose sunspots for his correlation simply "as one measure of solar activity" and made no suggestions as to the mechanism of origin of the radio emission, Denisse, then visiting the US for a year, soon took things further. Denisse (1949a,b; 1950a:193–9; 1976:3–4T) proposed that microwave solar emission should be broken up into three components: quiet sun, sudden bursts, and a new "slowly varying component." He showed that microwave intensity in fact correlated much better with the total *area* of all visible sunspots than with their number, and also that Covington's 10.7 cm wavelength of observation was ideal for uncovering this new component. At 62 cm (Reber's data) and 3 cm (unpublished data from Hagen and Fred Haddock at NRL), radio intensity was far less correlated with sunspots.

Covington was fortuitously operating on a wavelength for which the sun's overall intensity served as a remarkably accurate indicator of solar activity, in many ways as good as the sunspots themselves and yet observable in every type of weather. This encouraged him to continue and to improve the accuracy and consistency of his monitoring as the solar cycle wound down following its 1947–49 peak. After a few years, it became apparent that the mean intensity of the monthly variations was also declining with the cycle and thus that the microwave sun accurately charted not only daily and monthly sunspot variations, but also yearly ones. By the mid-1950s his Ottawa 10.7 cm solar intensities had become so useful that they were accepted by astronomers, ionospheric physicists, and radio communications services as one of the standard indexes of solar activity.

Denisse (1949b, 1950a:191–3) considered the slowly varying component to be gyromagnetic emission from electrons in the enhanced magnetic fields above sunspots. In order to produce frequencies of order 3000 MHz, fields of ~1000 Gauss were required and he argued that observers were in effect seeing emission from the surface of a bubble where the field centered on a sunspot had fallen to this level – emission from within the bubble could not escape outwards. Since the electrons were powered by no more than their thermal kinetic energy, he calculated that a temperature of at least 4×10^6 K was required at the base of the corona to produce the inferred high radio brightness temperatures. This picture also explained why the slowly varying component was absent at other frequencies: at lower frequencies, the optical depth of the corona prevented reception of any emission from near sunspots, while higher-frequency radiation could not be generated even within a sunspot magnetic field.

During this time Waldmeier was also championing special regions in the lower corona. Several years earlier he had argued that enhancements of the coronal green line (from thirteen-times-ionized iron atoms) were evidence for 100,000 km-sized "coronal condensations" twenty times denser and perhaps hotter than the surrounding medium. Waldmeier (1912–2000) was a prominent solar astronomer who, as director of Zürich Observatory, maintained the most widely used series of sunspot numbers, extending back two centuries. He had taken an early interest in interpreting the radio sun (Section 13.1.2) and wrote several review articles, as well as one of the first popular books on radio astronomy (Waldmeier 1954). He was intrigued by the close correlation between sunspots and Covington's intensities and, similarly to Denisse, split the microwave radiation into three components (Waldmeier 1987:114A). In fact, Waldmeier (1948) had originally proposed this type of scheme for metric wavelengths, but now he saw that it all worked better with Covington's microwave data and a tie-in with his coronal condensations. The microwave intensity was explained as thermal emission from coronal condensations, ignoring any effects of magnetic fields (Waldmeier and Müller 1950).

The Australians were also studying the slow microwave variations during this same period. Fred J. Lehany and Yabsley (1949) monitored the sun at 25 and 50 cm wavelengths for three months at the end of 1947 and pointed out an excellent correlation with sunspot area. A year later Minnett and Labrum (1950) found a similar correlation at 3.2 cm, although with considerably less amplitude than at longer wavelengths. This work plus Covington's led Piddington and Minnett (1951a) to synthesize a detailed physical model of the slowly varying component, called by them the "S component." They modelled it as thermal emission from hot ($\sim 10^7$ K), high-density (~ 20 times normal) regions extending into the lower corona, usually above sunspots. Although the idea of these regions in many respects resembled Waldmeier's coronal condensations and was published a year later, it appears that it was completely independent of the European work (Piddington and Minnett 1951a:132), another example of how Australian isolation sometimes led to duplication of results (Section 7.5.1). Piddington and Minnett, however, did include the effects of the sunspot's strong magnetic field extending up into this dense gas. By using standard magneto-ionic theory, they were able to produce a spectrum, polarization, and intensity consistent with the microwave data. Altogether, by the early 1950s these studies led to a developing consensus about the general origin and significance of the slowly varying component.

13.3 OVERVIEW

13.3.1 Acceptance of a hot corona

One of the main influences of radio observations on solar astrophysics of the mid-1940s was assisting in the general acceptance of a startlingly high coronal temperature of $\sim 1 \times 10^6$ K. But the exact way in which the radio observers and their data interacted with the astronomical establishment and their optical data is complex and appears to have varied from one person to another.

Since the nineteenth century it had been known that the corona (observed at the time of total eclipses) had a dozen or so spectral lines whose wavelengths did not match any of the known terrestrial elements. The hope at first was that entirely new elements might be discovered (as indeed turned out to be the case for helium, first observed at an 1868 eclipse in the chromosphere and not identified on this planet until a quarter century later), but as the periodic table filled up, prospects faded for a postulated "coronium." But if not a wholly new element, what *was* the explanation for the coronal lines? The breakthrough came in 1939–41 when Walter Grotian and Bengt Edlén explained the lines as forbidden transitions of highly ionized atoms of iron, nickel, and calcium. The very existence of these ions, representing atoms stripped of 10–15 electrons, required extremely energetic collisions or other special processes. Indeed, *if* the coronal gas were in thermal equilibrium, an astounding temperature of $\sim 10^6$ K was needed to supply collisions of sufficient energy.

These line identifications quickly gained acceptance over the wartime world – for example, the Royal Astronomical Society in 1945 awarded Edlén its Gold Medal for this work.[29] Acceptance of the identifications, however, did not lead to unanimity on the nature of the corona and why it possessed such energetic ions. The mystery of coronium had now been replaced by the mystery of the unprecedented conditions of excitation in the corona. Many astronomers argued that the most straightforward interpretation, although at first hard to believe, was that the corona indeed was a gas in thermal equilibrium at a temperature of $\sim 10^6$ K. This camp could also point to corroborating evidence in the form of the breadth of the coronal lines (caused by high random velocities of the ions), the relatively great extent of the corona (which on hydrostatic theories implied a high temperature gas), and chromospheric spectral

---

[29] K. Hufbauer gives an excellent account of Edlén's work and its reception in "Breakthrough on the periphery: Bengt Edlén and the identification of the coronal lines, 1939–1945," pp. 199–237 in *Center on the Periphery* (ed. S. Lindqvist, Canton, MA: Science History Publications).

lines indicating a temperature well above that of the photosphere. But the problem with a hot corona was that no one could satisfactorily explain *how* the solar atmosphere managed to change from a temperature of 6000 K in the photosphere to $1-2 \times 10^6$ K only a short distance above. What indeed was the corona's source of energy? Theories were not hard to find, but their very diversity and multitude spoke to a lack of understanding: heating by the accretion of interstellar dust and gas, or by shock waves emanating from the solar surface, or by magnetohydrodynamic waves, or by the sun's rotating magnetic fields. Some theorists took the stance that the corona might not be at all in thermal equilibrium (it was only *as if* the temperature were 10^6 K) and appealed to ejection of particles from below the photosphere or from nuclear reactions in active sunspot regions. Still others wrote papers showing that the ions could indeed result from a thermal equilibrium, but at a more reasonable temperature of $\sim 10^4 - 10^5$ K. Finally, a significant faction of astronomers took a wait-and-see attitude and remained noncommittal.

This then was the astrophysical cauldron into which Pawsey and Martyn in late 1946 dumped their *Nature* papers giving observational and theoretical evidence for a corona emitting radio waves equivalent to a blackbody of $\sim 1 \times 10^6$ K (Section 7.3.3). Was this line of evidence the clincher for a hot thermal corona, or did it merely aid its universal acceptance by the late 1940s? There is evidence on both sides. Hendrik van de Hulst (1984:389–90), who during this period was starting his astrophysical career (Chapter 16), later argued that the radio observations were critical for acceptance of this "impossible temperature" and concluded:

> I think the full credit goes to radio astronomy. Until then we only knew that the corona looked quite different from one eclipse to the next, but that spectacular changes *during* the five minutes or so of an eclipse were absent. Radio astronomy not only coined the term "quiet sun" but also for the first time revealed the corona as a constantly present, quietly radiating gas.

Michael W. Ovenden (1976:75B:540), a young astrophysicist at Cambridge Observatories at this time, and Machin (1981:134B:380) also recalled that the radio evidence played an important role. On the other hand Donald H. Menzel (1949:208), a leading solar astrophysicist of the day and well acquainted with radio work (Section 16.3.1), gave five "independent analyses" of the high temperature of the corona in his popular book *Our Sun*, but never mentioned radio. Similarly, Waldmeier (1978:114A:220) recalled that the radio data were confirmatory, not a vital step. Woolley (1947a:10) too gave the radio evidence no pride of place, saying instead that the coronal lines "lead very directly" to a high temperature while the quiet solar noise "is consistent" with it. But degree of "directness" is often in the eye (or antenna) of the beholder – it has already been mentioned in Section 10.3.1 that Ginzburg (1946) felt that radio measurements hold great promise to sort out the "conjecture" of a hot corona by allowing a "direct determination" of the temperature.

Overall, the radio data were deemed at most confirmatory by the solar experts of the day (and sometimes not even considered important), but for many others they were impressive evidence in building the picture of a hot corona.

13.3.2 Imaging

We have seen in Section 13.1.3 how Ryle's students Stanier, Machin, and O'Brien at Cambridge carried out successively more complex observations of the sun, culminating in a two-dimensional image computed from Fourier transforms of data gathered over a busy day hauling a small antenna all over a field. At the same time in Australia Christiansen, using his east–west and north–south arrays of small dishes, achieved the same aim by letting the earth effectively change the array orientations as the day passed. In both cases the sun was the object of astrophysical interest, but the two groups were equally intensely interested in the mathematical techniques required to reconstruct an image from imperfect data.[30]

As mentioned at the end of Chapter 8, through the 1950s and beyond, Ryle's group developed imaging techniques that came to be called aperture synthesis, for which Ryle (together with Antony Hewish) was awarded the 1974 Nobel Prize in Physics. On the Australian side, Christiansen also further developed imaging techniques, but a greater impact was made by his colleague Ronald N. Bracewell (1921–2007), an Australian who obtained a Ph.D. in 1949 in Jack Ratcliffe's ionospheric

---

[30] The importance of creating a radio *image* to aid communication with the optical astronomers is discussed in Section 17.2.5.

radiophysics group at Cambridge (Section 8.5.1),[31] co-authored the first monograph in radio astronomy (Pawsey and Bracewell 1955), and emigrated to Stanford University in 1955. There he set up his own array of dishes to study the sun, but more importantly produced several seminal papers on the mathematics of assembling an image based on a set of individual crosscuts. These techniques turned out to have broad applications outside of astronomy, in particular in medical imaging. They played an important role in the development of X-ray computed tomography ("CT scan") in the 1960s and 70s, for which the 1979 Nobel Prize in Physiology or Medicine was awarded to Godfrey Hounsfield and Allan Cormack, who worked more directly on the medical problem than did Bracewell.[32]

13.3.3 The radio and optical suns

The radio sun, especially at meter wavelengths, turned out to be notably different from the optical one. The coronal plasma's high opacity, its curved ray paths, and its radio changeability meant that in contrast to the optical corona, which supplied about one-millionth as much light as the photosphere, the radio corona was *all* that one observed at meter wavelengths. The corona had previously been of interest to solar astrophysicists, but its study was stymied by the fact that it could only be seen at total eclipses or to a limited extent with highly-specialized coronagraphs. As Ovenden (1976:75B:540) remarked:

> Essentially, you saw the corona only at the time of an eclipse – most of the time you could forget about it! But when you stuck your radio telescope up, a two-degree-wide sun was there *all the time*. The corona became something that people were interested in, instead of being a marginal thing. When radio astronomers looked at the sun, the corona was their interest – they couldn't care a jot about the photosphere.

---

[31] Note the influence of Ratcliffe (Section 8.5.1) with regard to these two "Fourier-thinking" disciples, Ryle and Bracewell.

[32] Kevles, B. H. (2003), "The physical sciences and the physician's eye: dissolving disciplinary boundaries," pp. 615–33 in *The Modern Physical and Mathematical Sciences* (ed. M. J. Nye) (Cambridge: Cambridge University Press). Also see Bracewell (1984:183–5, 2005:80–1).

The optical sun showed all kinds of variability, but its total output never wavered by more than 0.1%; on the other hand, the radio sun varied by factors of thousands and millions. The robust postwar maximum in the eleven-year solar activity cycle could not have been better timed from the point of view of providing good "material" on which to work. The resulting coronal radio bursts, however, turned out to be difficult to explain because of the complexities of plasma physics, as well as the lack of optical correlates and counterparts to the radio events. But once some understanding emerged, the radio data themselves began to contribute to knowledge of conditions such as the run of temperature and density in the quiescent corona and chromosphere, as well as the nature of energetic particles.

Solar radio research blended with traditional astronomy more easily and quickly than did the other branches of early radio astronomy. From the start people knew basically what they were dealing with, in contrast to studies of the galactic background or of discrete sources. In this way optical and radio astronomers had common grounds for discussion and this chapter has described the involvement of many conventional solar astronomers in the radio results. Some were theorists interested in the physics of the radio bursts, while others were "phenomenologists" who liked to monitor the sun and earth's atmosphere in every way imaginable and then search for correlations between measured quantities. They embraced the radio data and happily admitted them to the tabulations in the *Quarterly Bulletin on Solar Activity*, where they provided several more quantities with which correlations could be sought.

Although the solar radio work was integrated more fully with the optical than for other radio topics, it was never more than a small portion of the total field. One gauge is the fraction of pages devoted to radio results in contemporary books on the sun: 7% in Hoyle (1949), 3% in Menzel (1949), 12% in Kuiper (editor, 1953; see Pawsey and Smerd 1953), and 6% in Waldmeier (1955). And these figures are despite the fact that solar radio papers over the 1946 to 1953 period were always a high percentage of all radio papers, ranging from 85% to 55% (Fig. 17.1).

Solar radio work had less influence on solar astronomy as a whole than did the other major parts of radio astronomy on their respective fields. To be sure, it pointed to new phenomena and assisted in an improved understanding of the solar atmosphere and its

314 The radio sun

variegated activity, but there was not the frontal assault on basic ideas and approaches that occurred with, for example, meteor radar (Chapter 11), discrete radio sources (Chapter 14) or galactic radiation (Chapters 15 and 16). Rather, it allowed detailed exploration of regions and regimes previously only dimly reconnoitered. Solar radio astronomy's contribution was more in the nature of a revelation than a revolution.

TANGENT 13.1 PROPOSALS TO OBSERVE LUNAR OCCULTATIONS OF RADIO SOURCES

As a further refinement on standard solar eclipse observations, G.G. Getmansev and V.L. Ginzburg (1950) described how in some cases one could discern details even as small as 1–10″, but their paper in a Russian journal went unread in the West. The method, as it turned out, was not used until much later, when for instance it played an important role in the discovery of quasars in the early 1960s. Their technique was to observe the *diffraction* fringes produced by any small radio-bright region if it happened to be occulted by the "knife edge" of the moon moving across it. They also pointed out the possibility of improving even this angular resolution by another factor of ~100 in the event that a radio source were aligned *exactly* behind the moon's center. In such a case, by analogy with François Arago's early nineteenth-century experiment demonstrating the wave nature of light (often called "Poisson's spot"), one expected a strong oscillating signal centered on the moment of alignment. No one has ever tried this latter experiment.

In addition, lunar occultation was envisaged at an early time for also studying *non*-solar emission, although it was not successful until 1955–56 (on IC 443 and Tau A). In 1946 Ryle considered this technique to explore fine structure in galactic noise (as well as solar noise) ["Notes on future noise programmes," 23 August 1946, file 18 (uncat.), RYL]. Furthermore, in 1950 D.H. Menzel and W.W. Salisbury submitted a note to *Nature* entitled "Occultation of radio stars by the moon," but publication was denied [A.C.B. Lovell: editors of *Nature*, 7 March 1950, file R12 (uncat.), LOV]. Also in 1950, J.L. Greenstein suggested in a letter that an airplane could be used rather than the moon: "A large plane (say a B-29) could be used to occult point [radio] sources completely, and by optical triangulation might give very accurate positions, and perhaps sizes" [Greenstein:Menzel, 17 February 1950, box 4, GRE. I thank David DeVorkin for pointing this letter out to me.].

TANGENT 13.2 SOLAR RADIO BURSTS AND PLANT GROWTH

At the 1948 IAU meeting in Zürich K.-O. Kiepenheuer gave a paper on possible effects of radio bursts on the weather and on life. Apparently his interest was aroused by what he described as the common wartime experience that trees grew better when irradiated by radar beams. He initiated experiments with a botanist in which plant roots were exposed to 200 MHz radio waves, and cited preliminary results that at intensities of ~0.1 V m^{-1} the cell division rate was significantly increased. He then said that this level was eight times less than the strongest solar bursts, but this number seems in error. I find that even the "record-holding" burst of 8 March 1947 (reported as >10^{11} Jy at 60 MHz) was in fact at a level of >0.003 V m^{-1} (assuming a bandwidth of 60 MHz, as he did). Along these same lines, Waldmeier (1949:10) wrote that it was known that 5 m radiation in small doses could assist cell growth, and therefore that the sun was "continually giving us a high-frequency therapy." ["Sur quelques effets terrestres du rayonnement solaire dans la gamme des ondes métriques," paper given 10 August 1948 to the Joint Commission for Solar-Terrestrial Phenomena, IAU General Assembly, Zürich, file 11 (uncat.), RYL; the report of this in *Observatory* **68**, 171 (1948) is only partially correct.]

14 • Radio stars

Although the "radio stars," or discrete radio sources as they came to be known, at first played a minor role in radio astronomy, by the mid-1950s they were central to the plot. Hundreds were listed, and a handful revealed optical counterparts with exciting potential for new astrophysics – distant colliding galaxies, abnormal galaxies, supernova remnants, and new types of gaseous nebulae. But it was discouraging that the vast majority remained anonymous and mysterious, devoid of optical tie-ins, unable to be placed even inside or outside the Milky Way. Moreover, inconsistency among the various surveys led to frequent controversy among the three major groups and doubts about which radio stars were really there. This chapter follows these observational developments to the point in 1953 when the first round of major surveys had been completed.[1] I discuss theoretical efforts to understand the nature of the radio stars, also in a confused and unsettled state, in the following chapter.

We begin with the earliest efforts by Bolton's group in Sydney and Ryle's in Cambridge to understand the first handful of radio stars, in particular Cyg A. This also led to the first suggested optical identifications, in particular of Tau A with the Crab nebula. The second section traces early studies of source intensity scintillations and the arguments back and forth as to whether they were intrinsic to the radio stars or caused by the ionosphere; by 1950–51 consensus had been achieved for the latter.

Accounts of the first four major surveys (1950–54), as well as the discovery of a source at the galactic center, are next given; Table 14.1 lists properties of ten sources important for the early history. Cambridge's 1C survey found a discouraging lack of correspondence of its sources with stars or galaxies and concluded that its 50 "radio stars" were a nearby, common type of dark star.

Mills's interferometric survey of 77 "discrete sources" was interpreted in terms of two types of sources: strong and near the galactic plane (and within the Galaxy) and weak and uniformly distributed, either very close or very distant. Hanbury Brown and Hazard tabulated 23 sources near the zenith of their large fixed dish and found similar statistical results to Mills, but no agreement in detail. Finally, Bolton's group turned up 104 sources over 1951–53, based on sea-cliff interferometry, and likewise mostly agreed with Mills's statistics, but not the details of his individual sources.

Optical identifications of the two strongest radio stars, Cyg A and Cas A, are next discussed. In both cases, precise positions obtained by Smith at Cambridge (supported by similar work by Mills) proved the key that allowed optical astronomers Baade and Minkowski to search with the 200 inch telescope and pin down Cyg A as a distant, faint, weird extragalactic object (taken to be two galaxies in collision), and Cas A as an unusual network of filaments of uncertain origin. Baade and Minkowski became the key persons for optical identifications, working with all the radio groups and producing definitive papers by 1953 on over a dozen reliable identifications. Such an optical identification, with its seemingly more direct and rich information, conferred on any radio source a stamp of approval and a reality that it otherwise would not have had.

Angular sizes of sources were another important parameter to measure. Hanbury Brown and Twiss developed the concept of the intensity interferometer, and Hanbury Brown's students Jennison and Das Gupta built the equipment and carried out the observations that soon showed that Cyg A in fact was a double source, with all of the radio emission coming from well outside the optical object. The other major groups also measured angular sizes for a half-dozen radio stars, indicating that they were not at all as small as stars. This provoked vigorous debates on whether most sources were "radio stars or radio nebulae" and on the

---

[1] Edge and Mulkay (1976:87–111) have also provided an excellent account of portions of this history.

Table 14.1 *Radio sources important in the early history of radio astronomy**

Recognized								Flux density (Jy^x)	
Year	Group	Source	Optical ID†	α 1950	δ	b'	Angular size	178 MHz	1400 MHz
1946	Hey	Cyg A	G	19^h 57^m 44^s	+40° 36'	+05°	2'	8700	1600
1946	Reber	Cyg X	H II	20 30	+41	~00	10°^{a}	~10,000	~10,000
1947	Bolton	Tau A (Crab nebula)	SNR	05 31 30	+21 59	−04	5'	1500	930
1947	Bolton	Vir A (M87)	G	12 28 18	+12 40	+74	5'	1000	210
1947	Bolton	Cen A (NGC 5128)	G	13 22 24	−42 45	+19	7' (2°^{b})	1000^b	300^b
1948	Ryle	Cas A	(SNR)	23 21 11	+58 33	−02	5'	11,000^c	2100^c
1949	Bolton	Pup A	(SNR)	08 20 30	−42 50	−03	1°	350	120
1950	Hanbury Brown	Andromeda nebula (M31)	G	00 40 00	+41 00	−21	2°	200	70
1950	Piddington & Minnett	Sgr A (the galactic center)		17 42 29	−28 59	−01	1'	~150^d	~500^d
1951	Hanbury Brown	Per A (NGC 1275)	G	03 16 30	+41 20	−12	6'	60^e	10^e

* All quantities in this table are modern values, although note that galactic latitude b' is in the old (pre-1958) system.

† G = galaxy; SNR = supernova remnant; H II = region of ionized hydrogen; parentheses indicate interpretations later than 1953.

^{x} Jy = jansky = 10^{-26} W m^{-2} Hz^{-1}

^{a} Boundaries rather arbitrary

^{b} Consists of inner and outer double sources and an extended region of ~10° size; flux densities refer to only the outer pair

^{c} Epoch 1950; flux density decreases ~1% per year (not known until ca 1960)

^{d} Radio emission distribution is complex and varies with frequency; values only indicative

^{e} Most of flux density comes from core region of < 1' in size

biases inherent in the various measurement techniques and (sometimes) in the measurers, too.

14.1 FIRST STEPS BY BOLTON AND RYLE: 1948–49

We have earlier seen how John Bolton's group in 1947–48 made an extensive survey of the sky with their 100 MHz sea-cliff interferometer at Dover Heights, Sydney (Section 7.4). Bolton and Gordon Stanley (1948b) found that Cygnus A was no larger than 8′ in extent and that their best position for it corresponded to nothing in particular on photographs. Bolton (1948) then discovered six other discrete sources with sizes less than a degree and became convinced that distinct "radio-type stars," a new class of astronomical object, existed. Realizing the shakiness of his positions based only on observations of rising sources, he next sought westward-facing cliffs. Moreover, he wanted coordinated observations of Cyg A's fluctuations between Sydney and a distant site – if the fluctuations were intrinsic to the source, both sites should record similar data, but if they were caused by local ionospheric conditions, little correlation should occur. The required cliffs were to be found in New Zealand and thus it was that from June to August 1948 Bolton and Stanley went on what they called the New Zealand Cosmic Noise Expedition.[2]

Most of their time was spent near the fishing village of Leigh, 50 miles north of Auckland. There they had a cliff much higher than at Dover Heights (920 versus 260 ft, leading to narrower interferometer lobes and the promise of greater accuracy), a clear view both to the east and north, and far less manmade interference than in Sydney – even daytime observations were clean. They lived in a cramped trailer with their equipment, which was connected above to a 100 MHz dual-Yagi antenna. Isolation and wintry storms made life tough from the start, as Bolton wrote back to Taffy Bowen:

When I arrived at Leigh a week last Sunday, I found the trailer on site but with no power, Stanley with a very bad cold, myself with an incipient one, and a public holiday the following day. Since then, I am pleased to say, things have gone much better. ...

Cooperation both official and unofficial has been magnificent. The farmer on whose land we are sited has raised no objection to us using his timber, digging holes in his paddocks, etc. – in fact he has done everything to assist. They have even brought us tea and sandwiches at five o'clock in the morning on the last two nights – for which we are very grateful. Nine hours at a stretch without Dover Heights' comforts is just a little tough.[3]

About three weeks into the expedition the importance of the observations was emphasized by a letter from Martin Ryle. As recounted in Section 8.4, Ryle had learned of the best Australian position for Cyg A (from Joseph Pawsey, then visiting England) and was alarmed to find that his own position, based on Michelson interferometry at 80 MHz, was fully 2° different (compare points R48.5 and BS48.0 in Fig. 14.1, which is a compilation of all positions for Cyg A over 1947–52). Pawsey and Jack Ratcliffe had considered this dirty laundry unacceptable for general display and so had persuaded Ryle to hold off publication of any Cyg A position until things could be sorted out. Hence we find (a) the curious lack of any position in Ryle and Graham Smith's (1948) note in *Nature*, where they only reported that Cyg A was less than 6′ in size, and (b) Ryle's letter of 22 June to Bolton explaining all of the above, including the very real possibility that the source might have moved. Things were getting very interesting indeed – a reporter for the *New Zealand Herald* quoted Bolton at the time: "Some scientists look on the [Cyg A] phenomenon in the same way as the small boy reacted to his first sight of a giraffe. They do not believe it."[4]

Even cursory looks at the fringes on their New Zealand chart recordings, however, told Bolton and Stanley that their previous positions were badly in error. As Cyg A daily traced out an arc low in the northern sky (never rising farther than 13° above the horizon),

---

[2] See Orchiston (1994) for some further details of this expedition and early solar radio astronomy in New Zealand.

[3] Bolton:Bowen, 15 June 1948, file A1/3/1m, RPS.

[4] "Cosmic 'noise' from region of the Milky Way," 26 June 1948, *New Zealand Herald*, RPS; Ryle:Bolton, 22 June 1948, file F1/4/PAW/1, RPS.

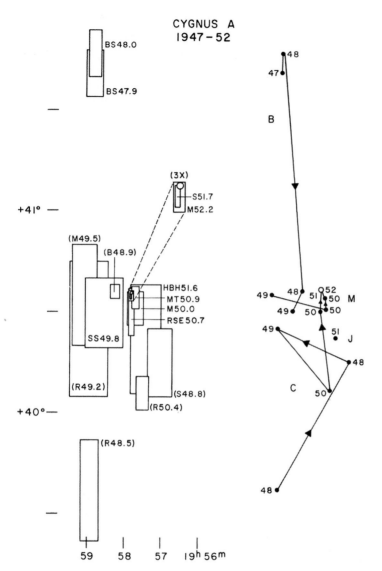

Figure 14.1 Fifteen measured positions (epoch ~1950.0) for Cygnus A over the period 1947–52. Each box represents a position and its cited error and is labelled with author and date of submission (in the case of an article) or date of a report or private communication (if in parentheses). Measurements were made on frequencies ranging from 80 to 215 MHz with a variety of interferometers and (for HBH51.6) a single dish. The circle of 1′ diameter (most easily seen in the 3× blowup) indicates the modern position of the source and that of the optical identification by Baade and Minkowski. On the right, the same measurements have been shown in a different manner, shifted and simplified in order to show more clearly the sequence of results obtained by each group: C = Cambridge, B = Bolton, M = Mills, and J = Jodrell Bank. Key: BS47.9 (Bolton and Stanley 1948a), BS48.0 (Bolton and Stanley 1948b), S48.8 (Smith 1951c:17), SS49.8 (Stanley and Slee 1950), RSE50.7 (Ryle, Smith, and Elsmore 1950), MT50.9 (Mills and Thomas 1951), HBH51.6 (Hanbury Brown and Hazard 1951b), S51.7 (Smith 1951a), M52.2 (Mills 1952b); sources for unpublished values, such as R48.5, are given in a footnote.[5] See Fig. 14.10 for a different display of these data.

---

[5] R48.5 = Ryle [note 4]; very similar to position in an August 1948 report to the DSIR by Smith [note 8.22]. B48.9 = Bolton [note 12]; also the report cited in note 10. R49.2 = Ryle [note 13]. M49.5 = (Mills) Pawsey:Ratcliffe, 28 June 1949, file A1/3/1a, RPS. M50.0 = Mills [note 59]; also cited in the abstract of a January 1950 talk by Mills [*J. Geophys. Res.* **55**, 199 (1950)]. R50.4 = Ryle:Appleton, 29 May 1950, box 1/1, RYL; also in report by Bolton [note 40].

it exhibited many more fringes than expected, which implied that it rose higher in the sky than expected and that its position had to be at a more southerly declination. Moreover, after six weeks they moved to the west coast, at Piha 20 miles west of Auckland,[6] and there found that their radio star Coma Berenices A was setting fully a half-hour later than expected and that now it must needs be renamed *Virgo* A![7] In mid-August they returned home laden with these greatly improved data. Analysis of all this cost Bolton three months of drudgery, doing things such as correcting for the earth's curvature and tidal heights, as well as calculating accurate times for the strip chart recordings,[8] which ran at erratic speeds because of poor power service. When it was finished, the New Zealand data showed that the source positions (other than Cyg A) that Bolton (1948) had published based on Dover Heights observations were horrendous, off by 5–15°.[9]

Bolton's focus was on Cyg A, but even at its new position it disappointingly remained a mystery in terms of coinciding with any optical object. In a Radiophysics Lab report[10] he set forth its "mean apparent position" (point B48.9 in Fig. 14.1), that is, a grand average of the somewhat discrepant values from each of the 25 days of observation. But then in November a second letter from Ryle complicated matters, for Ryle had found Cyg A apparently variable in position and tending to be at either of two distinct positions, and/or perhaps having a nearby companion source.[11] Bolton replied that he had no evidence for two favored positions (although his declination was half-way between Ryle's two positions), but that over a night the source often wandered off its mean track by as much as a degree, far more than his formal statistical error of $\pm\,3'$. Bolton was also worried that Cyg A's apparent intensity changes might be a product solely of the source changing its position among an interferometer's lobes, and so he suggested that Ryle check with a single antenna whether or not changes in intensity still occurred. He also told Ryle of their results from simultaneously observing Cyg A's intensity fluctuations from New Zealand and Sydney (done by Bruce Slee):

> A further result of the N.Z. Expedition is that the variations are "local" – there was no correspondence between Dover Heights and N.Z. records. I am beginning to wonder whether there might be some hard radiation from the source in addition to R.F. [radio frequency radiation], which affects the ionosphere particularly along the line of sight.[12]

Bolton closed by suggesting that Ryle and he should write a joint note to *Nature* describing the variations in apparent position, along with the "controls" they had in other sources that did not move around. Pawsey also suggested a joint paper to Ratcliffe, but the Cambridge side was initially not interested. Pawsey then pointed out that they should guard against the possibility of someone

---

[6] Piha had been one of the wartime radar stations where Elizabeth Alexander investigated solar radio emission (then called "the Norfolk Island effect") in 1945 (Section 5.3, Alexander [1945]).

[7] Coma Berenices A was transformed into Virgo A when it moved 9° southeast between Bolton (1948) and Bolton, Stanley, and Slee (1949). Moreover, the early Cambridge position (12^h28^m, $+14°$) also put the source in Coma Berenices and caused Ryle's group to be so much in the habit of labelling it Coma Berenices that they continued with this name long after the time of Bolton *et al.* (1949) and of Cambridge's own 1C position in Virgo in 1950 (for example, see Hewish [1951a:21]).

[8] See Tangent 14.1 for the basic methodology of the sea-cliff interferometer.

[9] Beside difficulties with refraction, the original Dover Heights source positions reported by Bolton (1948) were in error because of (a) an incorrect value supplied by the Army for the cliff height, and (b) a tendency to use records in which fringes could be more easily distinguished among the bothersome scintillations. Later, it became clear that these more tractable fringes tended to be the ones "stretched" by ionospheric effects, and therefore constituted a biased sample. [Bolton:author, 29 April 1986]

[10] Bolton, "Cygnus-A: Determination of position," New Zealand Cosmic Noise Expedition Report No. 4, undated (about late October 1948), 18 pp. + 6 figs., file A1/3/1m, RPS.

[11] This letter from Ryle, which arrived in Sydney on 11 November 1948, is referred to in Bolton's letter cited in the following note, but unfortunately the letter itself has not been found on either side of the world. The evidence for two positions and/or sources, some of which is contained in an August 1948 DSIR report by Smith (note 8.22) apparently was based on (1) records showing a beat effect indicating two close but different fringe periodicities, and (2) differing positions measured for Cyg A while at the same time Cas A remained unchanged (in the manner of Fig. 14.2, although this plot was drawn three months later).

[12] Bolton:Ryle, 12 November 1948, file A1/3/1a, RPS. Bolton's idea of "hard radiation" from Cyg A was that very energetic particles or photons emitted by the source might somehow cause different changes in the ionosphere's structure at different locations.

else, perhaps an American, independently discovering the same effect, and this prompted the Cavendish group into agreeing to publish adjacent notes. Drafts were written, but the whole situation suddenly turned around in December 1948 when Ryle's group started up their "Long Michelson" interferometer (Fig. 8.7). Its much greater sensitivity and superior timing system indicated that the position of Cyg A was now *not* changing (at least not more than 20′) and, as Ryle wrote in February, "the question now arises as to whether our earlier measurements were as accurate as we thought – whether in fact we can account for all our observations by a single stationary source." Ryle argued that a more judicious editing of bad records, due primarily to variable chart recorder speeds and infrequent timing marks, as well as a recalibration of his first position from May (point R48.5 in Fig. 14.1), now produced the conclusion that the source had not changed over a period of nine months by more than 20′ in declination or 5′ in right ascension (point R49.2).[13] Ryle's evidence at this time is contained in Fig. 14.2, which has been traced from a plot found in the Ryle papers.[14] It can be seen how the position of Cassiopeia A stayed relatively constant while that of Cygnus A was highly variable and uncertain, especially prior to October 1948. This confused situation led Ryle and Bolton to scrap plans for coordinated papers and to press forward with further observations. In the end, Bolton's group's best position for Cyg A (point SS49.8 from Stanley and Slee [1950:241]) never was able to be reconciled with those from other groups.[15]

Bolton and Stanley's positions of unprecedented accuracy naturally led them to search for optical counterparts. Cyg A proved fruitless, but the second strongest source, Taurus A, was another matter. In a report written shortly after returning to Sydney, Bolton derived its east–west size limit as 6′ and its position as $05^h30^m37^s \pm 30^s$, $+22°06′ \pm 15′$. This allowed him to say:

> The Source is close to the Star ζ Taurus as far as identification from a star map is concerned, but the limits of position enclose the most remarkable object in this region – NGC 1952 or the Crab nebula.[16]

This in fact was a gross understatement, for the Crab nebula was one of the most remarkable objects *in the entire sky* (Fig. 14.3).[17,18] Known since the eighteenth-century work of Charles Messier, where it held a place of honor (M1) in his catalog, it was observed to be expanding at an unprecedented speed of 1000 km s^{-1}. Its nature had remained a mystery, however, until detailed studies at Mt. Wilson by Walter Baade and Rudolph Minkowski built on the accumulating evidence that the nebula was in fact the remains of what Chinese chroniclers called a "guest star" when it suddenly appeared on 4 July 1054. It had been visible in the daytime for three weeks and at night for nearly two years before finally leaving the eleventh-century sky, and now in the twentieth century it seemed that this must have been a supernova, a rare type of exploding star normally only observed at great distances in other galaxies. One can therefore

---

[13] Ryle:Pawsey, 24 February 1949, file A1/3/1a, RPS.

[14] Smith and Ryle, undated plot and table entitled "Positions of Cygnus and Cassiopeia on 214 Mc/s and 80 Mc/s up to February 20, 1949," file 7 (uncat.), RYL.

[15] Bolton (1978:38–9T) later felt that the Cyg A position from the New Zealand expedition was thrown off by ionospheric irregularities known as "spread-F" (see Section 14.2), which were worse near midnight, when Cyg A was observable.

[16] Bolton, "Taurus-A," New Zealand Cosmic Noise Expedition Report No. 3, 8 September 1948, 9 pp. + 3 figs., file A1/3/1m, RPS. The Crab nebula (Table 14.1) is in fact 53^s east of Bolton's cited ± 30^s position for Tau A. Bolton and Stanley (1949) later gave a revised position for Tau A (using, for example, better values for the location and altitude of the Leigh site), and a new error box that did include the Crab nebula.

[17] The Crab nebula was remarkable enough that Ivan Atanasijević (1919–1998), a Yugoslav who later worked with Laffineur at Meudon in 1952, noted in October 1948 that the position of Tau A given by Bolton (1948) "is situated in the immediate vicinity of the Crab nebula" (even though the Tau A position was 7° away and quoted as accurate to 1°). This suggestion, however, remained unknown to the principal players, for Atanasijević's paper was written in Serb! ["Solar and galactic radiation on short radio wavelengths," [in Serb], *Bull. Soc. Mathematiciens et Physiciens et la Rep. Pop. de Serbie*, No. 2, 23–39 (1949) (quotation from p. 29). Also Atanasijević:Ryle, 26 January 1951, box 1/1, RYL]

[18] Shklovsky (1982:27) recalled how in early 1949 he enlisted S. E. Khaykin (Section 10.3.2) to search for 200 MHz thermal radiation from the Crab nebula. The project failed, however, for the available sea-cliff interferometer in the Crimea was blocked by mountains from a view of the Crab when rising or setting.

First steps by Bolton and Ryle: 1948–49 321

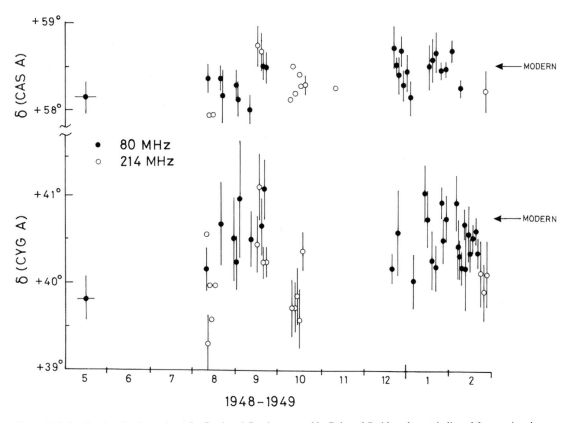

Figure 14.2 Declinations (and error bars) for Cas A and Cyg A measured by Ryle and Smith at the two indicated frequencies; the modern values are indicated on the right. Cyg A appeared to be changing in its position far more than Cas A. Data up to October 1948 were from their earliest interferometers (Fig. 8.2) and after December 1948 from the much more sensitive and stable "Long Michelson" interferometer (Fig. 8.7). (Adapted from an unpublished plot in the Ryle archives, note 14)

imagine Baade and Minkowski's interest when they received the following from Bolton in April 1949:

> The most interesting source is the one in Taurus whose position corresponds very closely to that of the Crab nebula. ... The intensity of the radiation at 100 Mc/s gives an equivalent [brightness] temperature of about a million degrees for an angular width of 5′. From your results on temperature and density in the Crab nebula it seems unlikely that this equivalent temperature could be due to strictly thermal processes in the nebula. I thought perhaps friction resulting from differential expansion within the nebula or perhaps general expansion into interstellar matter might provide an explanation. I would be interested to hear your opinion on this.
>
> [Centaurus A] corresponds closely to N.G.C. 5128, an extragalactic nebula which Shapley describes as a "pathological specimen" [in his popular book *Galaxies*], and [Virgo A] corresponds to Messier 87. If these two nebulae are responsible for high radio frequency emission it would be hard to imagine them outside our own Galaxy.
>
> We have not detected any radiation from the close galaxies – the Magellan clouds and Andromeda – or from the other recent galactic supernovae – Tycho's and Kepler's.[19]

---

[19] Bolton:Minkowski, 4 April 1949, cartons 4–5, Minkowski papers, MIN. At this same time Bolton (1976:15–16T) also sent the news to Jan Oort in Leiden and received a long reply, especially with regard to the Crab nebula. Oort had earlier worked not only on modern astrophysics of the Crab, but also with the Chinese historian J. J. L. Duyvendak on the eleventh-century evidence for a corresponding guest star.

Figure 14.3 Photograph of the Crab nebula (see Table 14.1), taken in a narrow band centered on the Hα line with the Palomar 200 inch telescope. The photo is ~7′ wide. North is up and east to left (as in all other photographs in this chapter). (Baade 1956)

These ideas about Tau A and the Crab were soon published (Bolton and Stanley 1949; Bolton, Stanley, and Slee 1949). In the latter note (in *Nature*), they also proposed the two other correspondences mentioned above between optical and radio positions, that is, what came to be known as "optical identifications" (Section 14.4). M87 was a bright (9ᵐ) spherical nebula in the Virgo cluster of galaxies – this put its distance at six *million* light-years and made its radio luminosity (intrinsic power output) incomprehensibly gargantuan. Likewise, NGC 5128, called a "most wonderful object" by John Herschel over a century before, was usually taken to be a strange, nearby galaxy (Fig. 14.4), although its spectrum of emission lines was very unusual for a galaxy and there was a recent study proposing it as a planetary nebula within our own Galaxy.[20] Bolton was clearly more comfortable with the idea of *galactic* noise originating in *galactic* sources:

> If the identification of these objects [M87 and NGC 5128] with the discrete sources of radio-frequency energy can be accepted, it would tend to favour [that they are diffuse nebulosities in the Galaxy], for the possibility of an unusual object in our own galaxy seems greater than a large accumulation of such objects at a great distance.
>
> (Bolton, Stanley, and Slee 1949:102)

Minkowski replied immediately to Bolton's letter with further information on the three objects and suggestions for other nebulae that might be radio emitters. He favored an extragalactic location for both M87 and NGC 5128 and sent a short-exposure photograph of the former that revealed a unique "linear jet" feature shooting out of the nucleus. This had been known for thirty years as an oddity of M87, but nothing had ever been made of it. Bolton replied that his Radiophysics Lab colleagues were fascinated with the jet and had even suggested an analogy with large solar flares.[21]

These identifications, however, were in fact not viewed as definitive, for the radio positions were still uncertain by 10–15′ and the wording was of "*possible*

---

[20] D. S. Evans (1949), "Photometry of NGC 5128," *Mon. Not. RAS* **109**, 94–102.

[21] Bolton:Minkowski, 20 May 1949, cartons 4–5, MIN. The letter of Minkowski to Bolton of 14 April has not been found.

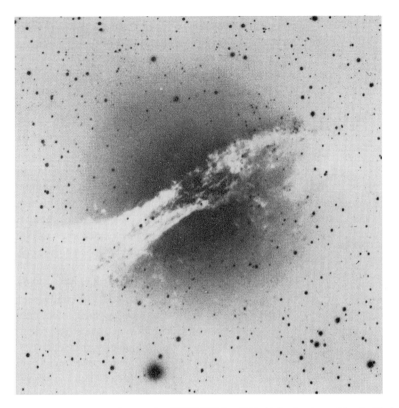

Figure 14.4 (Negative) photograph of NGC 5128 (see Table 14.1) taken in a red band with the Palomar 200 inch telescope. Diameter of the nebula is ~7′. (Baade and Minkowski 1954b)

associated visible objects" (my italics).[22] Over the next two years contemporary sources indicate that most people indeed viewed these three identifications as tentative; at a May 1950 Royal Astronomical Society (RAS) meeting, Bolton himself stated that "[our] accurate positions ... *suggest* their identification with ... "[23] (my italics). A mid-1951 administrative report of the Sydney group used the words "provisionally identified."[24] Why should these identifications have been accepted as gospel when their positions had changed by 5–10° from Bolton's earlier values, and when no one could see any optical object as obvious for the much stronger sources Cyg A and Cas A (Ryle 1976:34T)? Of the three suggestions, greater credence was generally given to the Tau A/Crab connection because of the patently energetic nature of the expanding nebula,[25] but then why could one not detect the remnants of *other* "recent" supernovae such as those of Tycho and Kepler (Ryle, Smith, and Elsmore 1950:521)? As late as 1951 Ryle was skeptical about even the Crab:

> I do not think there is really enough evidence yet to be sure of [identifying Tau A with the Crab nebula]. ... There were [none of our 50] radio stars

---

[22] Bolton later recalled, however, that he had "total faith in the identifications," and that Joe Pawsey (who had final say on the claims of all RP publications in radio astronomy) "always wanted to water down the claims" [Bolton:author, 4 March 1989]. Stanley (1994:510) also recalled that, although he doubted the M87 identification, Bolton never hesitated in accepting it.

[23] Bolton, part of the report of the 12 May 1950 RAS meeting, in *Observatory* **70**, 135 (1950).

[24] 1950–51 Annual Report of the Division of Radiophysics, p. 14, file D2/8, RPS.

[25] For instance, the eminent astronomer Jan Oort (note 19) remarked in April 1950, "I have been impressed by Bolton's suggestion of the Crab nebula as a possible source." [Oort:Ryle, 18 April 1950, box 1/1, RYL]

near the positions of other supernovae, similar to the Crab nebula, and the conditions likely to exist in the shell of such a body do not seem very favourable theoretically to the emission of radio waves.[26]

As he later recalled:

> Not being optical astronomers, we didn't know how remarkable the Crab was. ... At that stage we didn't know anything about it except that it was a little fuzzy thing on a photograph.
>
> (Ryle 1976:27T)

And regarding the other two identifications, Ryle wrote in February 1950:

> I do not think that there is any real evidence yet to suggest that any of the small sources of radio waves which have been detected are outside our own galaxy.[27]

Smith (1976:6T) later recalled:

> I think it's fair to say that we were pretty blinkered and busy on our work and we didn't take an awful lot of notice of what was happening elsewhere. And so it was some time before it dawned on us that the Australian positions were leading to real identifications.

The Cavendish group changed their minds only in mid-1951 when Smith obtained far more accurate positions that confirmed the identifications (Section 14.4.1).

The short paper by Bolton, Stanley, and Slee (1949) was one of the most important in early radio astronomy, presenting a first plausible link between "galactic noise" and traditional astronomy. And what an exciting link it was, too, for this handful of intense radio stars was being associated with objects that were much fainter than any of the five thousand objects visible to the naked eye, yet still unusual enough to be included in manuals such as *Norton's Star Atlas*, the amateur astronomer's *vade mecum* that was frequently consulted by Bolton's group. Although the source of radio energy from these objects was far from clear, what *was* clear was that if these proffered identifications held up, fields of astronomy other than the sun and meteors were going to be strongly affected by radio results.

14.2 SCINTILLATIONS

The fluctuations in intensity of Cyg A that had led to its original discovery by Stanley Hey's group (Section 6.2) were a remarkable phenomenon (see Figs. 6.4, 7.11 and 8.6). If progress were to be made in understanding radio stars, one had to establish how such dramatic changes could be produced over a few seconds or minutes. Most observers of the late 1940s took these fluctuations, or *scintillations* as they came to be known, as an intrinsic property of radio stars, while others felt it more likely that the ionosphere was modulating the radio waves in some fashion.[28] Of course it was also possible that both mechanisms were operative, perhaps to varying degrees at different times and for different sources. C. Gordon Little (1952), a postgraduate student at Jodrell Bank who carried out important investigations of the scintillations, listed in his Ph.D. thesis the reasons why before 1950 the scintillations were generally taken to be intrinsic. First of all, from Hey onwards people had discussed the striking analogy between solar bursts and radio star scintillations – both were much stronger at lower frequencies and associated with stellar-like objects. There were, however, differences: for instance, Ryle and Smith (1948) found the scintillating component not to be circularly polarized. Second, several results suggested that radio stars scintillated to differing degrees, some even not at all, whereas if the ionosphere were the culprit, then all radio stars should scintillate. For instance, Bolton (1948) classified his first radio stars as either (a) constant, (b) variable over a long-period, or (c) "Cygnus-type variable." Moreover, Ryle and Smith (1948) and Ryle (1949b:493) reported Cas A to fluctuate considerably less than either Cyg A or their Ursa Major source, even though all three sources passed close to the zenith. As visiting Pawsey reported to Sydney in June 1948: "[Ryle's] 'wobbles' on Cygnus vary from day to day and on a 'wobbly' day on Cygnus, Cassiopeia is quite steady – strong evidence against ionospheric

---

[26] Ryle:Atanasijević, 8 February 1951, box 1/1, RYL.

[27] Ryle:E. Tandberg-Hanssen, 10 February 1950, box 1/1, RYL.

[28] For instance, see the report of the August 1948 IAU meeting in Zürich which states that the radio observers present were emphatic that the Cygnus fluctuations "could not be of ionospheric origin since they can be observed at all seasons of the year and at different solar times" [*Observatory* **68**, 170 (1948)]. One person who felt all along that the fluctuations were caused by the ionosphere was Ratcliffe [Machin (1971:11–12T), Hewish (1976:5–6T)].

origin."[29] Ryle also was unable to find any correlation between the times of ionospheric storms and those of scintillation activity.[30]

The third reason for favoring intrinsic source variations was that Bolton and Stanley (1948b) had found that Cyg A's intensity always *increased* above a base level, as did the sun's, whereas an ionospheric effect should show as many decreases as increases. Fourth, the *mean* intensities of several sources seemed variable on a long time scale. Ryle and Smith reported a source in Ursa Major in 1948 that could not be later found, as did Stanley and Slee (1950:241) with several other possible "temporary sources." Finally, it was difficult to understand how the ionosphere could so devastatingly affect radio waves at frequencies as high as 160 MHz, 20–30 times the critical frequency (Appendix A) at the zenith.

Together with Stanley and Slee from 1947 to 1951, Bolton undertook a series of monitoring and spaced-receiver experiments in and around Sydney. By the end of 1949 the group was confident of an ionospheric origin for the scintillations,[31] and these results were first published by Stanley and Slee (1950), and later in detail by Bolton, Slee and Stanley (1953). Stanley and Slee found that, contrary to their previous conclusion, Cyg A's time-averaged, mean intensity did not vary by more than 10%. Furthermore, two years of monitoring showed that the degree of 100 MHz fluctuations varied from 10 to 60% and exhibited an annual pattern. This effect, combined with the complete lack of correlation between scintillations observed simultaneously in Sydney and in New Zealand, led them to "suggest, contrary to previous views, that these fluctuations are of terrestrial rather than of extra-terrestrial origin." They also pointed out that, if this were true, Cyg A was likely such a good scintillator because it had a small angular size.[32] The situation was directly analogous to that at optical wavelengths where stars "twinkled" and extended objects such as planets did not.[33]

Meanwhile in Cambridge, Ryle felt sure that the Cyg A fluctuations he was observing in 1948–49 were intrinsic to the source, even after Bolton informed him of the New Zealand/Sydney experiments. Ryle wrote to Pawsey in January 1949:

> I am very anxious to know your views on the New Zealand results.... The results, if true, are most significant. The possibility that fluctuations are introduced by tropospheric or ionospheric refraction seems so important ... [and] we feel so sure that the fluctuations that we see are genuine [intrinsic to the source] that we would like to repeat the experiment using normal incidence, and I have suggested to Lovell that we might do some experiment jointly.[34]

Ryle was suspicious that Bolton's scintillations with the sea-cliff interferometer, necessarily taken at very low elevation angles, were indicative not of "genuine" (intrinsic) variability, such as one could observe with Cyg A and Cas A's high elevation angles in England, but of another effect. For example, Bolton's apparent positions might dance around due to severe variable refraction when observing through the long atmospheric path near the horizon. Such movements would then lead to apparent intensity fluctuations, and, indeed, as mentioned above, Bolton too had once been concerned about this possibility. The Australians therefore held off publication of their results until corroborative results could be sought in England.

The existence in Cyg A of "genuine" scintillations was also important to Ryle because it constituted one of his key reasons for favoring the idea that radio stars were stellar-like objects (next section). Such variations

---

[29] Pawsey:Bowen, 11 June 1948, file F1/4/PAW/1, RPS.

[30] Ryle:Lovell, 10 August 1949, box 1/1, RYL. Also Fig. 3 of Smith (1950).

[31] The change in confidence can be gauged by the wording in two letters to Minkowski: on 20 May 1949 Bolton wrote that the Sydney–New Zealand observations "shewed the variations were probably local," while on 4 October 1949 new data and analyses led him to be "quite convinced that the variations in signal strength are terrestrial in origin." [Cartons 4–5, MIN]

[32] Stanley and Slee (1950:246) stated that "recent measurements indicate that the source in Cygnus has an angular width of less than 1.5′," but gave no further details. It is remarkable that this result was not substantiated or emphasized, for it was far superior to other limits at that time.

[33] See Tangent 6.5.

[34] Ryle:Pawsey, 12 January 1949, A1/3/1a, RPS. In his letter to Lovell proposing a joint experiment, Ryle flatly stated: "I do not believe [the Australian] result" [Ryle:Lovell, 22 December 1948, file R6 (uncat.), LOV].

(lasting as short as 20 sec) allowed the important deduction that the source of emission could not be larger than ~20 light-seconds in size, only three times the diameter of the sun (Ryle and Smith 1948:463; Ryle 1949b:493–4). This result was based on the standard astrophysical argument that an object could not physically change in a time shorter than it took light to travel across its dimensions. The argument would, however, be invalid if the scintillations arose not within the source, but in the intervening medium between observer and source.[35]

Lovell took Ryle up on his proposal for a joint experiment on the scintillations and, together with their students Little and Smith, they coordinated 81 MHz observations over a six-month period in 1949. They found that although the general degree of scintillations from Cyg A and Cas A on any given day was correlated over the Cambridge–Jodrell Bank distance of 210 km, the detailed minute-to-minute variations were not at all in common. The scintillations also were as often positive as negative (relative to the mean), which strongly suggested some kind of ionospheric (or near-earth) diffraction process, although it was puzzling that there was no correspondence between scintillation activity and ionospheric or magnetic storms. Each group also conducted its own spaced-antenna experiments at shorter baselines and found that the amount of correlation between the scintillations steadily declined as the separation between antennas increased, all correlation disappearing when baselines reached ~5–20 km. This was strong evidence that the scintillations (or at least the bulk of them) were not intrinsic to the radio stars.[36] On this the participants agreed, but they nevertheless published separate, adjacent notes in *Nature* (Smith 1950; Little and Lovell 1950),[37] largely because of a phenomenon that the Cambridge side accepted as valid, but which Jodrell Bank had not observed and thought too uncertain (Lovell 1984:203).

The phenomenon in question was the observation by Ryle's group of simultaneous, 15 sec-long, strong bursts on 45 MHz at sites as far as 160 km apart. Although the antenna beams were so large that one could not determine the direction of the bursts as from either Cas A or Cyg A, the long antenna separation and similarity of amplitudes seemed to preclude manmade interference as the cause, and so Smith (1950) concluded that "the observations suggest ... genuine variations of the emission from the galactic sources." The Cavendish group's monitoring for over a year also showed many days when only Cyg A or only Cas A (usually the former) was active.[38] This was interpreted as indicating the existence of two kinds of fluctuations, one ionospheric and one intrinsic – the latter's rapidity, as discussed above, implying objects of stellar dimensions. The point was forcefully made by Ryle (1950:221–2) in a review paper of the time. Many were skeptical:

> Cambridge have bursts which they think come from the sources. We think that this may be true but that it would take a hell of a lot of proving! In radio astronomy it is only too easy to ascribe cosmical significance to what is, in effect, activity in the local tramway system.
>
> (Hanbury Brown)[39]

> The bursts are remarkably like solar "isolated bursts" and there is no evidence that they come from Cygnus.
>
> (Bolton)[40]

These isolated, strong bursts had been rare and were in fact never again observed. Although not regarded by Ryle's group as proven,[41] they nevertheless were frequently cited over the next two years as supporting evidence that most sources were indeed radio *stars* (Chapter 15; Edge and Mulkay [1976:100]).

---

[35] See Tangent 14.2 for another possible effect of the intervening medium (Shklovsky 1950).

[36] Once Ryle became convinced of the local nature of the radio star scintillations, he even began to question the origin of *solar* radio bursts. Spaced-antenna observations of the sun were coordinated between Jodrell Bank and Cambridge in 1950, but nothing was ever published. [Ryle:Lovell, 4 August 1949 and other correspondence in file R6 (uncat.), LOV]

[37] When the Jodrell Bank and Cambridge results appeared in *Nature* in March 1950, the Australians were furious at the lack of any acknowledgment about their unpublished work, as well as the lack of any opportunity for them to publish alongside after they had expressly delayed publication. Their own paper (Stanley and Slee 1950) appeared several months later (Slee 1994:522–3).

[38] Ryle:Lovell, 24 September 1949, box 1/1, RYL.

[39] Hanbury Brown:Bowen, 30 April 1950, file A1/3/1a, RPS.

[40] Bolton, 31 May 1950, "Visit to the Cavendish Laboratory (Radio Astronomy)", file F1/4/BOL/1, RPS.

[41] For example, see "Report to IAU Commission 40," Ryle:Woolley, 30 October 1950, box 1/1, RYL.

Over the following two years, intensive efforts on the scintillations quickly turned them into a tool of ionospheric physics more than a concern of radio astronomy. Radio sources could now be treated as handy test transmitters whose signals probed the *entire* depth of the ionosphere, not just that below the F region. At Cambridge work was centered on Antony Hewish whose dissertation (1951a) and first papers (Ryle and Hewish 1950; Hewish 1951b, 1952) set him on a career focused on how an intervening medium can influence a radio signal. For fifteen months he used the Long Michelson interferometer elements together and singly to monitor scintillations from the four strongest radio stars. He found that each source had its greatest scintillations when it transited between 2200 and 0400 local time, no matter what the season of year. Ryle and Hewish also concluded that the degree of scintillation on any given night was well correlated with the ionospheric condition called "spread-F," thought to be associated with an irregular electron distribution in the F region (at an altitude of ~400 km – see Appendix A). A similar association with spread-F in the Southern Hemisphere was also reported by Mills and Thomas (1951), who in 1949 had conducted monitoring and spaced-receiver experiments. Ryle and Hewish suggested that the density irregularities might be caused by interstellar particles intercepting the earth's atmosphere as they fell toward the sun, analogously to the effect of meteors on the E region (Chapter 11). Such infalling particles, whose maximum effect on the ionosphere was indeed predicted for ~0200 each day, had been hypothesized before the war by Fred Hoyle and Raymond A. Lyttleton as a way to heat the solar atmosphere and incidentally cause climatic changes on the earth.

Hewish also built up an extensive theory of the scintillation phenomenon, founded on earlier ionospheric theory from Ratcliffe's group. In particular, Henry G. Booker, Ratcliffe, and Douglas H. Shinn (1950) had developed an elegant theory of the effects of an irregular screen such as the ionosphere upon incident radio waves, whether from below and reflected back (the original motivation) or from above and transmitted through. The ionosphere acted on radio waves in the same way as a sheet of glass of variable thickness distorts an image. Significantly extending this theory in a way that allowed quantitative measurements of ionospheric properties, Hewish showed that the fluctuations observed on the ground arose from movements ("winds") of the F region irregularities, which created diffraction patterns that sped across one's antenna as fast as 0.5 km s^{-1}. For instance, two receivers spaced by 1 km usually registered identical patterns of fluctuation, but with the one delayed by 2 to 15 sec with respect to the other. When this fact was combined with fluctuation time scales ranging from 1 to 100 sec, it was straightforward to infer that the ionospheric irregularities had a size of ~5 km.[42]

At Jodrell Bank, Little (1951) worked on similar theory. Little (1952) and New Zealander and fellow research student Alan Maxwell[43] (Little and Maxwell 1951, Maxwell and Little 1952) carried out observations parallel to those of Hewish. A war-surplus 30 ft diameter dish was used together with a mobile dual-Yagi array to come to conclusions very similar to those of Hewish, with the addition that their ability to track sources all day allowed a better study of the strong dependence of the scintillations on elevation angle. They also set up three antennas in a triangular arrangement 4 km on a side and thereby determined the full velocity vectors of the ionospheric winds.

Thus it was that by 1950–51 the cause of the scintillations had been solidly ascribed to the ionosphere. Most radio astronomers, whose focus lay increasingly on the radio stars far beyond, therefore lost interest in any further pursuit of the phenomenon, except in understanding its deleterious effects on their data. Moreover, observations on the whole were steadily moving to higher frequencies where the scintillations were weak and of no practical importance. The very scintillations that had given birth to the radio star phenomenon now ironically left the scene.

14.3 NEW DISCRETE SOURCES

After each group made its initial findings on the few brightest radio stars, it became clear that methodical

---

[42] This result relied on a key theorem proved by Booker, Ratcliffe, and Shinn (1950), namely that (for relatively weak scattering) the characteristic size of the variations in the ground pattern was equal to that of the irregularities in the ionosphere. The theorem actually stated that the autocorrelation function of a phase-modulating screen is identical to that of the diffraction pattern it produces on a distant parallel plane.

[43] Maxwell had earlier (1948) obtained an M.Sc. in physics at Auckland University College, New Zealand; his thesis was titled "Enhanced solar radiation at 3 metre wavelengths."

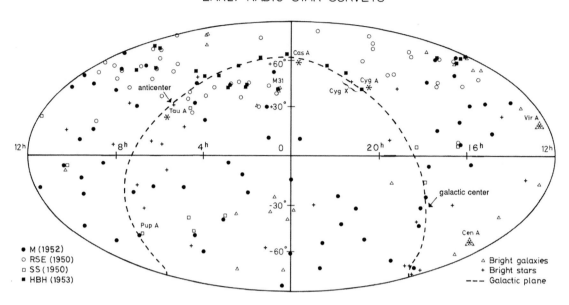

Figure 14.5 An all-sky plot (equatorial coordinates) of the positions of the discrete radio sources listed in the first four surveys: Stanley and Slee (1950); Ryle, Smith, and Elsmore (1950); Mills (1952a); and Hanbury Brown and Hazard (1953a). Sources of particular interest are labelled (large asterisks indicate the five strongest), and the galactic plane is shown as a dashed line. The 27 (optically) brightest galaxies and 25 brightest stars are also shown. Error boxes for these sources varied greatly, but in general were 1–5° (no more than three times the width of the symbols in this diagram). Note the general lack of agreement of surveys with each other and with the optical objects.

surveying of the entire sky was a prerequisite to any proper understanding of the phenomenon. Over the period 1950–54 five major surveys were published, but glaring inconsistencies between the surveys (Fig. 14.5) meant that little consensus would emerge. In addition, in 1954 a unique new source (later called Sgr A) was established to exist at the center of our Galaxy.

14.3.1 The 1C survey at Cambridge

In Section 8.4 we have described the events leading up to the 1C survey, "A preliminary survey of the radio stars in the Northern Hemisphere" by Ryle, Smith, and Elsmore (1950), as well as details of the antenna used and the data analysis. The "Long Michelson" interferometer (Fig. 8.7), operating at 81 MHz with its fixed beam of 3° × 90°, daily swept the 40% of the sky passing near Cambridge's zenith. Observations were made over almost a year in order to secure as many reliable records as possible, free from manmade interference and from ionospheric scintillations. Good recordings such as that shown in Fig. 14.6 were used to derive source positions and intensities, with their repeatability from day to day giving estimates of measurement errors. Errors were quoted typically as 1–2°, but Ryle's group understood that positional accuracy was not being limited by the sensitivity of their apparatus or by the number of days over which they gathered data. Rather it was, as they said, the "confusion of the interference patterns produced by adjacent radio stars." In other words, the sinusoidal signature of any particular source (Fig. 14.6) could well be modified by those of other sources simultaneously in the antenna beam. Although the effects of stronger sources could usually be recognized, more insidious was the combined influence of many sources, most of which were probably too weak to be themselves cataloged. Because of this, it was pointed out that error estimates could be too small for some of their weakest sources, where the position given might in fact refer rather to a kind of "centre of gravity" of the positions of several sources.

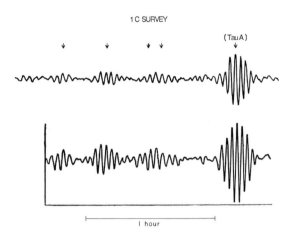

Figure 14.6 Traces of the same portion of sky on two different days of observation with the "Long Michelson" interferometer (Fig. 8.7) for the 1C survey in 1949–50. The bottom trace comes from Ryle, Smith, and Elsmore (1950) and the top from Smith (1951c). I have drawn arrows at the approximate positions of listed 1C sources – the four sources on the left were typical, whereas Tau A was the third strongest radio star known.

Fifty radio stars were tabulated in what later became known as the 1C ("First Cambridge") survey. The distribution on the sky was found to be isotropic, showing no preference whatsoever for the galactic plane (Fig. 14.5). Correlations with every conceivable type of optical object were attempted: the brightest normal stars, the nearest stars, stars with high proper motion, novae and supernovae, planetary nebulae and diffuse nebulosities, star clusters, and nearby galaxies. Only the last yielded any modicum of success. Although comparisons of their 50 radio star positions with those of the nearest 300 galaxies yielded only nine coincidences (which was not significantly more than expected given the sizes of each source's error box), another approach worked better. By restricting attention to the five brightest galaxies, they found that two (M31 and M51) fell within estimated error boxes and two others (M33 and M101) just outside. These four sources were among the weakest in the survey, but the strength of each was nevertheless of the same order as would be expected if our own Galaxy were placed at its distance. They emphasized, however, that these bright optical galaxies could just barely be detected in the radio, whereas much stronger sources tied in not at all with nearby galaxies. This fact, together with theoretical difficulties in explaining huge areas of high brightness temperature ($>10^8$ K) in the sources if they were distant, was put forward as the chief argument against an extragalactic location. Instead, the sources were truly radio *stars*, a nearby and common (~ 1 pc^{-3}), but "hitherto unobserved type of stellar body, in which a very intense radio emission is associated with a very small visual intensity" (Ryle, Smith, and Elsmore 1950:522). For over a year Ryle had already been arguing for this point of view, and the development of these ideas will be discussed in Chapter 15.

The Cavendish team also undertook two other radio star projects at this time. In the first Smith looked for shifts in the positions of radio stars attributable either to a proper motion (velocity transverse to the line of sight) or a parallactic motion (apparent shift back and forth over a year arising from the earth's changing orbital position). Detection of either effect would indicate that a radio star was relatively nearby, and a good value of parallax would immediately yield its distance. With suggestions on the nature of the sources still ranging from comets[44] to galaxies, even negative results would be of use. For a full year Smith (1951b) weekly used the Long Michelson and Würzburg dish interferometers (next section) to search for drifts in the relative positions of any of the four strongest radio stars. Because of ionospheric effects on the longer wavelength of the Long Michelson,[45] better results were obtained at 1.4 m wavelength with the dishes. In the end he found no detectable shifts and placed upper limits of 5 to 20″ on any proper motions or parallax, implying that the sources were at least 0.2 pc away. This was only part way to the nearest visible star, but at least it was well outside the solar system!

---

[44] D. H. Menzel and D. J. Crowley (1950), "Point sources of radio noise" [considered as comets], *Nature* **165**, 443. Also note Martyn's 1947 suggestion of same, mentioned in Section 7.3.4. In early 1950 Robert d'E. Atkinson of the Royal Greenwich Observatory also looked in detail at possible correlations in the positions of comets and Ryle's radio stars [Atkinson:Ryle, 13 March 1950, box 1/1, RYL].

[45] Smith discovered that at 3.7 m wavelength there were systematic shifts in the radio star positions whenever observed at dawn or dusk. These he tied in with horizontal gradients in the electron density of the F-region as it adjusted to the rising or setting sun. [Smith (1952), "Ionospheric refraction of 81.5 Mc/s radio waves from radio stars," *J. Atm. Terr. Physics* **2**, 350–55]

In the second monitoring project Ryle and Elsmore (1951) reasoned that many types of visible stars varied in their brightness on scales ranging from hours to years, and so searched for any similar variations in the intensity of the radio stars. Comparing the daily strip chart recordings from the Long Michelson over a period of eighteen months, they found no long-term intensity changes at all and concluded that the radio stars were steady emitters to within 5–10%, or, as they alternatively expressed it in a nod to astronomers, to within 0.05–0.10 magnitudes. This behavior seemed in contrast both to visible stars and to the radio sun.

14.3.2 The Mills survey

After detailed work on the position of Cyg A (described in the following section), Bernard Mills (1952a) undertook a general attack on the "discrete sources" (as he called them) at a new field station located on Badgery's Creek, a CSIRO cattle research station about 30 miles southwest of Sydney (Fig. 7.4). He and the RP workshop built three antennas configured into two phase-switched 101 MHz interferometers with baselines of 60 and 270 m, yielding fringes spaced by 3° and 40′, respectively. The original purpose of having two interferometers was to aid in identifying the central fringe of each source, but an unanticipated bonus turned out to be even more important: relative strengths of fringes on the two baselines allowed a rough estimate to be made of the (east–west) angular extent of sources larger than 10′. With each element being a large, tiltable flat reflector holding an array of 24 dipoles (Fig. 14.7), all of the visible sky (90% of the entire sky) could be sampled with his $14° \times 24°$ beam. For most of the year of 1950 Mills would set the antennas for a given declination, sample the sky around the clock until he obtained two or more good-quality records (free from receiver problems and interference from lightning, sun, or man), and then move 10° to the next declination strip. He took pains to operationally define a source "in terms of a particular pattern produced on the pen recorder" (Mills 1952a:271), a pattern having a definite periodicity and length dictated by antenna beamwidth, source declination, and source intensity. Like the Cambridge survey team, he was concerned about the effects of confusion and endeavored to minimize them through careful comparison of scans taken at adjacent declinations; nevertheless, he warned that "the position and intensity of the weaker sources (in fact even their existence in a few cases) is rather doubtful." In the end he estimated that the positions of his 77 sources were typically good to 1–2°.

The sources showed an isotropic sky distribution when taken as a whole (Fig. 14.5), but Mills pointed out that the strongest sources behaved differently and were heavily concentrated to the galactic plane. To be exact, 8 of 11 of his sources stronger than 300 Jy occurred at galactic latitudes less than 12° (which only includes 20% of the sky). This suggested that there existed two kinds of sources, which he called Class I (near the galactic plane) and Class II. This division seemed yet surer when he found that plots of log (flux density S) versus log (number N of detected sources per steradian of S or greater)[46] were remarkably different for the two Classes (Fig. 14.8). Mills found that the Class I (low latitude) sources had a slope of 0.75, as expected for a population of objects closely confined to the galactic plane; he estimated their typical distance as ~1000 pc. On the other hand the Class II sources had a slope of 1.5, consistent with a uniform, isotropic distribution. Moreover, his dual-spacing interferometer had allowed him to deduce that three sources were extended, with sizes in the range 20–35′. Two of these (Puppis A and a source in the direction of the galactic center) were Class I, but the third (Cen A) was not. Cyg A and Cas A were also both at low galactic latitudes and therefore perhaps Class I, but their angular sizes were known to be very small. As with the 1C sample, the isotropic distribution of the Class II sources could be interpreted as putting them either very near or very far away and Mills discussed both options in detail (see Chapter 15). Although in the end he leaned toward an extragalactic solution, he was unable to come to a definite conclusion (Mills 1952a:279–81, 1976:28T).

For various 1C samples Ryle (1950:241) (Fig. 14.8) and Smith (1951c:95) had earlier made plots that were equivalent to log N–log S plots, obtaining slopes of 1.5. Ryle argued that Mills's two classes of radio star could well be artifacts of the statistics of small numbers and the effects of a wide dispersion in the intrinsic powers of radio stars.[47]

---

[46] See Tangent 14.3 for the basics of log N–log S plots, as well as my analysis of this plot for the 1C sample.

[47] Ryle:Minkowski, 30 March 1953, cartons 4/5, MIN.

Figure 14.7 One of three 101 MHz antennas, at Badgery's Creek near Sydney, used by Mills (1952a) as elements for his interferometric survey. The array consists of 24 end-fed halfwave dipoles (not visible) and could be tilted about the horizontal axis in order to observe the entire sky. The man working on the antenna is likely Adin Thomas.

14.3.3 The Jodrell Bank survey

Besides their studies of various individual sources, Robert Hanbury Brown and Cyril Hazard (1953a) compiled a list of 23 "localized radio sources" observed at 158 MHz with the giant 218 ft reflector at Jodrell Bank (Fig. 9.7, Section 9.3). These were all within a very restricted range of declination near their zenith (only 16% of the sky), but their sensitivity was 10–40 times superior to the other surveys of the day, especially for those cases where they searched hard at the position of an optical object by averaging records over several days. The other unique aspect of their survey was that it did *not* use an interferometer, which led to major differences with other contemporary surveys. They found reasonable agreement when comparing their own list with the strongest sources from other surveys, but the reverse was not true, that is, only four of the 13 strongest Jodrell Bank sources showed up elsewhere. Furthermore, they pointed out the utter discrepancy between the other two surveys (see Fig. 14.5) – in a common area of sky where each survey had ~35 sources there were only four that agreed.

The notable result from Hanbury Brown and Hazard was, as with Mills, that the strongest sources clustered near the galactic plane; ten of their thirteen strongest fell within 5° of the plane (less than 25% of their surveyed area). They were happy to adopt Mills's nomenclature of Class I and Class II sources and argued that the Class I sources were a rare type of galactic object, at typical distances of ~1 kpc. They too favored the Class II sources to be extragalactic (see Chapter 15 for details), but in the end had to say that the "evidence is weak and a definite conclusion must await further observations."

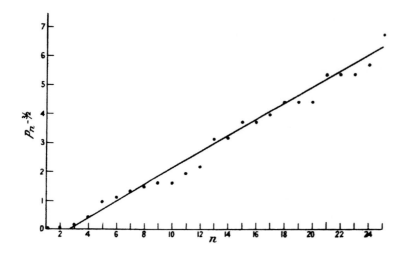

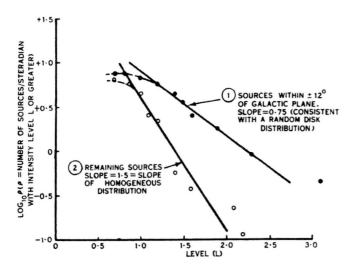

Figure 14.8 The first "log N–log S" plots of radio sources. The top plot was made by Smith and is from Ryle (1950), based on a preliminary sample of 25 from the 1C survey. P_n is proportional to the source's intensity and n is its ordinal ranking by intensity; a straight slope indicates a uniform distribution of sources. The bottom plot (Mills 1952a), based on a survey of 77 sources, shows a different behavior for sources near to or away from the galactic plane. The level L is equal to log (flux density), where flux density is measured in units of 10 Jy.

14.3.4 Bolton's group's surveys

Stanley and Slee (1950) published results on discrete sources based on sea-cliff interferometer observations made with Bolton over the period 1947–49 (see Section 7.4).[48] Their modus operandi was to point their antenna and Slee (1950): "The work described in this paper was inspired by Bolton and he also took a large part in the experimental work. He quixotically excluded himself from authorship on the grounds that papers should be shared with his assistants" [file D5/4/91, RPS]. Bolton (1978:35–6,40T) also commented in interview on his policy of encouraging authorship by his assistants – he had learned during the war to give proper credit to "the people who kept your plane together."

---

[48] In the archives of the Radiophysics Lab is the following note, probably by Pawsey or Bowen, attached to a reprint of Stanley

(a 100 MHz array of nine Yagis with a 17° beam) at a spot on the horizon, observe any fringes at that position for three nights or so, and then move on to the next spot. In this way they tallied 18 sources altogether, although three of these proved so problematic that in the end no positions were tabulated. They warned that, except for the three strongest sources, "no great [positional] accuracy is claimed"; error boxes were typically 1° by 7°. Further cause for concern was the possible existence of "temporary sources," as evidenced in several instances by detections of sources that could not later be found. Eighteen sources were too few to say much about their distribution on the sky, although they noted a "general tendency to crowd towards the galactic equator." They also presented the first spectra of any radio sources, comprising measurements at five frequencies from 40 to 160 MHz. Cyg A, Vir A, and Cen A all showed intensities varying inversely with frequency, whereas Tau A's strength proved independent of frequency, consistent with thermal radiation from an optically thin shell.[49] Throughout their paper Stanley and Slee used the term "discrete source" rather than "radio star," but they had little to say about the nature of the sources.

The next major effort by Bolton's group took place over 1951–53 using the same Dover Heights cliff and the same frequency of 100 MHz, but with a larger antenna and improved electronics that better discerned source fringes. The going was tedious, largely because ionospheric activity and thunderstorms drowned out most of their observing time (Bolton 1978:59–60T). Bolton wrote in 1952:

> It is a slow process covering the sky. We only do about 10° declination × 10 hrs RA per week under optimum conditions and modifications have to be made to the equipment to cope with different parts of the sky. Working across the [celestial] poles and away from major sources is easy, but distinguishing sources in the regions near the [galactic] plane where the background noise changes rapidly introduces difficulties which we can cope with, but take time.[50]

These surveys would be the last hurrah of sea-cliff interferometry in Sydney; Bolton and Slee (1953) wrote up many details of the technique, including several new types of interferometer, but the greater flexibility and power of Michelson interferometry were winning the day. Stanley (1994:509) later recalled that "custom had corrupted our thinking; had we constructed the simplest of east–west interferometers, we could have detected Cyg A [and sources in general] much more easily at Dover Heights." Moreover, sea-cliff interferometry lost its chief proponents when Bolton's group largely disbanded in 1953. This followed denial of funding for a proposed, very large 400 MHz antenna and Bolton's resultant shift to cloud physics research.[51]

Bolton, Stanley, and Slee (1954) were able to list 104 discrete sources over the 70% of the sky available to them. All sources were smaller than their fringe spacing of 1° and positional error boxes ranged from less than a degree in size to typically 1° × 3° or more (a second survey for extended sources is described in Section 14.5.2). They realized that confusion was a serious problem – as early as 1949, Bolton (1982:353) and Westfold had discussed this at RP under the rubric of "Detectability and Discernability" and Bolton, Stanley, and Slee (1954:111) stated:

> Where confusion exists on the records, the observer naturally adopts the simplest possible explanation of the complex patterns and trusts that his interpretation is correct. If incorrect, it can result in extreme cases in the assignations of positions and flux densities to sources that do not in fact exist at all.

They placed much more confidence in sources that occurred not only in their survey, but also in others (agreeing in position to within, say, 3°) – they reasoned that other surveys also suffered from confusion, but entirely different antennas should produce a different set of spurious sources. Their list of sources was therefore broken up into five sub-lists depending on optical identification, agreement with other catalogs, and degree of confusion evident on the records. From these criteria they considered that about 70% of their sources "definitely exist."

Bolton's group suggested that seven of their sources could be identified with optical galaxies called "abnormal radio emitters," that is, with radio luminosities

---

[49] At the same time, Ryle (1950:218) very briefly reported spectra for Cyg A and Cas A that showed *flat* slopes from 1.4 to 6.7 m wavelength (in contrast to Stanley and Slee [1950]).

[50] Bolton:Minkowski, 20 September 1952, cartons 4–5, MIN.

[51] Bolton (1978:63–9T); Bolton (1982:357); Stanley (1994:509–13); Bolton, "A new aerial for source survey work," undated (~1952) ms., 9 pp., file A1/3/10, RPS.

exceeding those of a normal galaxy by a factor of one hundred. But as with the other surveys, the positional accuracy of the great majority of their sources was such that unambiguous optical correlates could not be found. Examination of the sky distribution of the 104 sources showed that, like the Mills survey, the strongest sources distinctly crowded toward the galactic plane. Their log N–log S plot also agreed with Mills's findings in the galactic plane (a slope of −0.6), but outside of the plane it had a slope much steeper than the −1.5 expected for a uniform distribution. This was still of no avail in deciding on the nature of the Class II sources, for they ended their paper with:

> A plausible explanation is that the Sun (if these sources are galactic) or the Galaxy (if the sources are extragalactic) is in a local region of low source density and that somewhere towards the limit of the survey we reach a region of much higher density. ... However, there is not much point in speculating too far on this result as it could also be produced by a large dispersion in absolute magnitudes amongst the sources of the survey.
>
> (Bolton, Stanley, and Slee 1954:129)

These were the first surveys of radio stars. The next major survey was the 2C catalog of 1936 sources from Ryle's group, conducted in 1953–54 and published in 1955. A steep log N–log S slope was found and interpreted to have far-reaching cosmological implications, which led to a decade of quarrels. That story[52] is fortunately beyond the scope of this book, which is already too long.

14.3.5 A galactic center source

From the beginning, Jansky's and Reber's all-sky maps showed that cosmic noise was strongest in the general direction of Sagittarius, which astronomers took (on other grounds) to be towards the center of the Galaxy (Chapters 3 and 4). As angular resolution improved, however, a further effect was teased out, namely that within that broad maximum there existed a much smaller-sized discrete radio source. This source, later named Sagittarius A (Sgr A in Table 14.1), would prove very important as both an anchor for a new system of galactic coordinates adopted in 1958 and a means to study the otherwise dust-obscured region of the galactic center.[53]

The first significant discussion of such a discrete source was by Jack Piddington and Harry Minnett (1951b), who in 1948–50 in Sydney conducted surveys of the galactic radiation at the high frequencies of 1210 and 3000 MHz, affording them beamwidths as small as 2.8° (using a small dish). Pushing their receivers to the limit and employing techniques such as Dicke-switching between loads and switching between sky positions, they pointed out "a new, and remarkably powerful, discrete source ... on the border between the constellations of Sagittarius and Scorpius" (Piddington and Minnett 1951b:465, 467). At 1210 MHz its size was no more than 1.5° with an estimated intensity (after removal of the general background) of 2600 Jy. It lay very close to the galactic center direction, and since it was in the galactic plane it was likely far away. At an assumed distance of 10 kpc, an estimated radio luminosity was "about 100 times greater than the *total* power radiated by the sun" (italics in original). A possible optical identification of the source with a star cluster (NGC 6451) was also proffered.

In retrospect we can see that at least two other Australian catalogs of the early 1950s also included Sgr A in their listings, but it was weak at lower frequencies and no particular notice was paid. Major international attention *was* garnered, however, when a *Nature* letter by Richard X. McGee and Bolton appeared in 1954 entitled "Probable observation of the galactic nucleus at 400 Mc./s." McGee was a young RP staff member who had just been assigned to work with a new "hole-in-the-ground" 80 ft diameter dish at Dover Heights field station. Bolton had been inspired by the 218 ft behemoth of a fixed dish at Jodrell Bank (Fig. 9.7) during his 1950 European tour.[54] At Dover Heights a parabolic hole was painstakingly dug out shovelful by shovelful on the sandy edge of a cliff,[55] and its surface

---

[52] See Edge and Mulkay (1976) for a major treatment of this story and Sullivan (1990) for a small part of its beginning.

[53] Goss and McGee (1996) and Orchiston and Slee (2002) have analyzed in detail the sequence of events and publications surrounding the recognition of Sgr A; my account heavily relies on their research.

[54] Bolton, J. G., "Visit to Jodrell Bank Experimental Station," 27 June 1950, file F1/4/BOL/1, RPS.

[55] Bolton (1982:355) and Orchiston and Slee (2002:23–6) recall how Bolton and Slee did this digging secretly during three

covered in concrete overlain with wire mesh. A 40 ft long mast carried the feed element and a preamplifier (Fig. 14.9). At 400 MHz the beamwidth was 2°. The mast could be tipped by as much as 16°, allowing mapping of the galactic center region (which passed nearly overhead every day).

As Fig. 14.9 shows, Piddington and Minnett's source was now very striking at a lower frequency and with a much larger dish. Joe Pawsey, leader of the RP radio astronomers, sent a pre-publication version of this diagram to Baade, who then showed it to other astronomical experts such as Jan Oort and Henk van de Hulst in Leiden, all of whom were excited by the agreement between this strong, discrete source's position and other, indirect estimates for the position of the galactic center.[56] Baade wrote "Frankly, I jumped out of my chair the moment I saw what it meant. I have not the slightest doubt that you finally got the nucleus of the center of our galaxy!!"[57] In the previous decade Baade had established the whole notion of Populations I and II in galaxies (see Chapter 15), and felt that the general background radio radiation (taken to be the integrated effect of dark radio stars) was part of Pop. II and that therefore a galaxy nucleus (also Pop. II) should be a strong radio emitter. But the RP internal reviewers (Section 7.5.3) were more skeptical than Pawsey about whether this *was* indeed the galactic center/nucleus and sent McGee's drafts through the wringer. The final compromise was the word "probable" in the title.

14.4 OPTICAL IDENTIFICATIONS

Even if one accepted the remarkable optical identifications suggested by Bolton, Stanley, and Slee (and many did not), there remained the central puzzle of why the two strongest radio stars in the sky, Cyg A and Cas A, could not be associated with any optical object. As it turned out, they were very hard nuts to crack. And even after they had been cracked in 1952, progress was excruciatingly slow on other sources – within the next two years only about seven more identifications had been made (Table 14.1). This was a tiny fraction of all cataloged radio sources, but it nevertheless began to shed light on which kinds of objects could generate such intense radio emission.

14.4.1 Cygnus A

We have seen that the measured positions of Cyg A had a tortured history in 1947–48. This continued for at least three more years – Figs. 14.1 and 14.10, charting Cyg A positions (both published and from letters or reports) over the period 1947–52, give insight into how the complex situation evolved.[58] It is clear from these diagrams that before 1950 errors of positions were grossly underestimated. During the next few years, however, the elimination of instrumental and methodological errors steadily progressed.

Two major radio astrometric efforts spawned the eventual optical identification of Cyg A. The first was by Mills, who decided in 1949 to abandon the sun (Section 7.2.3) and dove into the radio star problem; coaxed and coached by Pawsey, he was leery of the intrinsic problems of the sea-cliff interferometer. As he wrote in December 1949:

> We have been concentrating particularly on obtaining an accurate position [for Cyg A] and for that reason have abandoned the type of interferometer used by Mr. Bolton. The horizontal base-line interferometer which we use is much less dependent on unpredictable ionospheric effects.[59]

---

months' worth of lunchtimes in 1951. They feared that their boss Joe Pawsey would not approve of such an unassigned project, but in fact he was delighted with the first 72 ft version and authorized it to be enlarged and improved the next year.

[56] The direction to the galactic center could be estimated from the planes of symmetry derived from star counts over the sky, from velocities and proper motions of stars, and from velocities of the newly discovered 21 cm hydrogen line (Section 16.6). The last turned out to be the most accurate and would become the basis for a new definition of galactic coordinates in 1958.

[57] Baade:Pawsey, 16 February 1954, file A1/3/1, RPS.

[58] Figs. 14.1 and 14.10 contain an arbitrary element in that the positions plotted for Cyg A represent only those for which reports, articles, or letters are extant. Furthermore, in the case of letters the authors never intended that the positions would become public and often gave various qualifications to their correspondents. Nevertheless, the figures summarize the picture that existed at any given time, and give insight into how systematic and random errors of the various observers evolved over five years.

[59] Mills:Minkowski, 16 December 1949, file A1/3/1a, RPS; Mills (2006:3).

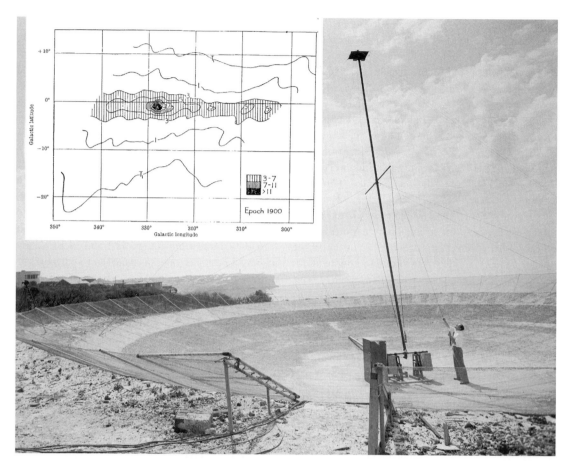

Figure 14.9 The 80 ft (24 m) diameter "hole-in-the-ground" parabolic dish at Dover Heights, Sydney, in 1953. Richard McGee adjusts nylon guy ropes to move the mast (as much as 16° from the zenith) to a new declination. The large box at the base of the mast contained an inner box with the 400 MHz receiver's back-end electronics. This waterproof box was designed to float up and down on those occasions when great amounts of rain and unreliable drains created a large pond! Inset: the 400 MHz map (in old-style galactic coordinates) produced by McGee and Bolton (1954), showing the strong source at the galactic center. Contour units are 30 K of brightness temperature above the background temperature T_1 of 55 K (outer contours).

Assisted by Adin B. Thomas, an English engineer who briefly worked at RP, Mills observed with a modified form of the interferometer at Potts Hill (two 97 MHz Yagi antennas separated by ~300 m) that had been carefully calibrated by Alec Little and Ruby Payne-Scott for precisely locating solar bursts (Section 13.2.1.1). Over the second half of 1949 Mills and Thomas (1951) measured the fringes from Cyg A as it again and again made its brief, daily appearance above the northern horizon. Realizing that their errors were exactly analogous to those of an (optical) transit telescope, they took advantage of centuries of experience at squeezing out accurate stellar positions. For example, just as one switched a transit telescope around in its bearings in order to cancel errors in collimation due to any mechanical asymmetry, in the radio case one could interchange the antennas and cabling. Other analogous errors arose from insufficiently knowing clock time, baseline length and orientation, and atmospheric effects. The worst offender, however, was wholly unknown to any optical astronomer, namely unequal ground reflections at the two ends of the interferometer. Especially at Cyg A's low elevation as seen from Australia, source radiation could reach an antenna both directly and after reflection off the ground, causing spurious phase shifts in the recorded fringes.

Optical identifications 337

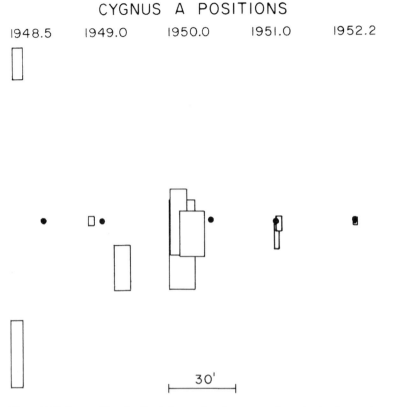

Figure 14.10 Best positions for Cyg A from each major group as they existed at five epochs over 1948–52. The modern position is indicated by dots. Data sources are as in the caption for Fig. 14.1.

Despite these difficulties, by December 1949 Mills felt that he had a position for Cyg A good to ~5′ (point M50.0 in Fig. 14.1), which he communicated to Minkowski and argued to be valid despite its evident disagreement with the latest from Bolton's and Ryle's groups. He also immediately checked out a print of a photographic plate (taken on the Mt. Wilson 100 inch telescope) that Minkowski had earlier sent Bolton (Fig. 14.11) and found a coincident object within 3′ of the radio position:

> We find that the source can quite reasonably be identified with an object which you have marked as an extra-galactic nebula. In view of your interest in the subject I have taken the liberty of writing this letter in the hope that you may be able to furnish some particulars about the nebula in question.... You will probably be familiar with the argument that indicates a small size for the Cygnus source because of the rapidity of the amplitude fluctuations. This, of course, is at variance with the identification with an extra-galactic nebula. How certain is it that the object in question is extra-galactic?[60]

Minkowski quickly put a damper on Mills's enthusiasm for this faint nebula:

> It is very gratifying to see the gradual improvement of the determination of the position. Ultimately, this should make it possible to investigate the spectra of the brighter stars within the area outlined by the probable errors. But in a field as rich as the Cygnus field this does not yet seem to be a reasonably promising method of attack. From the astronomical side, the most useful bit of work

---

[60] As in previous note.

Figure 14.11 Photograph of the field of Cyg A, taken by Minkowski with the Mt. Wilson 100 inch telescope. The error box for Cyg A is from an overlay published with this photo by Mills and Thomas (1951) (point MT50.9 in Fig. 14.1). The extraglactic nebula of interest to Mills in 1949 (not discernible in the present reproduction) lies in the upper-left (northeast) portion of the error box.

would be the search for a star with high proper motion. ... I do not think that it is permissible to identify the source with one of the faint extragalactic nebulae in the area. These nebulae are undoubtedly extragalactic, at a distance of the order 10^7 parsecs or more.[61]

---

[61] Minkowski:Mills, 29 December 1949, file A1/3/1a, RPS.

Publication of these results was delayed by work on their new interferometer (Section 14.3.2), but Mills and Thomas (1951) eventually refined their position to 3′ accuracy (point MT50.9), printed Minkowski's plate along with an aligned overlay sheet showing their error box (Fig. 14.11), and presented Minkowski's and their own arguments as to why the extragalactic nebula (galaxy) in question could not be associated with Cyg A. Its brightness temperature would have to be at least 10^{10} K and it was unlikely that an entire object as large as a galaxy could be abnormal, particularly since it was not distinguished in appearance from other nearby faint galaxies. They concluded that Cyg A was most likely a faint nearby star.

Strenuous efforts on Cyg A were also underway in England. Hanbury Brown and Hazard (1951b) used the giant reflector at Jodrell Bank to observe the source, which by good fortune was within their zenithal field of view and at almost the same declination as the Andromeda nebula, which they were also studying (Section 9.3). Their position for Cyg A was useful in that it was the only one of that era measured with a single antenna (or with a "pencil beam," as they put it), but their 2° beamwidth only allowed a determination good to 16′ (60× HBH51.6 in Fig. 14.1).

The key step in identifying Cyg A, however, was taken by Smith, who buried himself in an all-out effort at Cambridge to determine the positions of the four strongest northern radio stars:

> It was obvious that we could get good positions on the sources. We knew of the Australian "attempts," as we called them, to identify and we thought they were not good enough. We thought that we had to get better positions and so we built an accurate interferometer.... The Australians and we were both measuring positions, for Cygnus A in particular, and it really very quickly became a competition as to who got an accurate enough answer to be meaningful.
>
> (Smith 1976:12T,5T)

Smith used several different interferometers over the 1948–51 period, but the most accurate one, which came into operation in the summer of 1950, consisted of the two 7.5 m diameter Würzburg reflectors[62] (Fig. 14.12) that the group had salvaged from a junk dealer at the end of the war (Section 8.2). For a cost of £300 ($800) these were erected and mounted as transit instruments, movable in elevation by hand, at a spacing of 280 m (200 wavelengths). These dishes were superior to arrays of dipoles or Yagis for a host of reasons: their electrical (phase) centers were better defined, they suffered less from ground reflections, and they could be readily directed to different parts of the sky. Another important decision was to go to a relatively high frequency: Smith installed 214 MHz receivers with a system temperature of ~2000 K. Although sources were weaker at the higher frequency, it also meant less trouble from the ionosphere, narrower interferometer lobes, and a smaller beam for the dish, leading to less confusion with other sources. Moreover, with aids such as a precision Shortt clock borrowed from the Royal Greenwich Observatory and strip charts that raced under the recording pen at six inches per minute, Smith was able to measure a time of transit good to about one second. As with Mills, the site was carefully surveyed (Fig. 8.8) and the methodology of a transit (optical) telescope was applied to the radio case.

Smith (1951c, 1952a) also developed many variants on the basic interferometer that allowed vital cross-checks to be made on source positions; his 1952 paper was an especially nice tutorial exposition. For instance, because the Grange Road site dictated a baseline that was 7° different from east–west (Fig. 8.3), one could, for circumpolar sources such as Cas A, observe a source at both upper and lower culminations (due south and due north) and thereby solve for its position in a different way. Another clever technique was what he called the "displaced collimation plane" method, which obviated knowledge of one's precise operating frequency. For this he inserted a cable of known electrical length first into one arm of the interferometer and then (on the next day) into the other. Such insertions shifted the apparent time of transit of a radio star, and by measuring the induced difference between transit times one could derive the source's angular velocity (and hence its declination).

By the summer of 1951 Smith (1951a) had pinned down the locations of Cyg A and Cas A to ~0.3′ × 1.5′ error boxes and Vir A and Tau A to ~1.5′ × 15′ boxes (less accurate because of their weaker intensities and equatorial locations). All doubts about the Australian identifications of Vir A and Tau A were now removed, for the new positions still lined up with M87 and the Crab nebula, now with error boxes smaller by a factor

---

[62] Tangent 4.2 describes these dishes.

Figure 14.12 Graham Smith standing below one of the 7.5 m diameter Würzburg dishes used in the interferometer at Cambridge for accurate determinations of radio source positions (Smith 1951a).

of ~10 over those of Bolton, Stanley, and Slee (1949). But it was the positions for Cas A and Cyg A (point S51.7) that were pregnant with possibility and Smith immediately passed them on to Roderick Redman, director of the Cambridge Observatories. Redman set in motion various optical work in England (see below) and recommended a letter to Baade asking that the 200 inch Palomar Mountain telescope, which had only been in useful operation for a year or so at that point, be brought to bear on the optical identification problem. Smith did so in August 1951 and soon received an enthusiastic reply from Baade:

> We [have] decided to undertake the search [for Cyg A and Cas A] along two different lines. Minkowski will take spectrograms of all the stars falling in the areas defined by three times the errors of the present coordinates since it is possible that the radio stars show spectral peculiarities (emission lines, etc.) in the visible range. ... I myself intend to go after the colors of the stars in the same areas, particularly their behaviour in the infrared. ... Both lines of attack are of course straight gambling ... but one has to start somewhere. ... There are good reasons to suspect that [the radio stars] are nearby objects and therefore should have considerable "proper motion."[63]

Baade also reaffirmed his faith in the identification of Tau A with the Crab nebula and said that he now thought the coincidence of Vir A and M87 not merely accidental. This caused Smith to reply that "it would be most remarkable if the 'stars' are in fact extra-galactic – even more remarkable than the present hypothesis of extremely common 'dark stars.'"

After several exchanges between Baade and Smith concerning Cas A (see below), the first news on Cyg A from the optical realm came in a letter of 29 April 1952,

---

[63] Smith:Baade, 22 August 1951 and Baade:Smith, 3 September 1951, LOV. I thank Graham Smith for photocopies of this correspondence, which is also partially reproduced by Edge and Mulkay (1976:105–8) and almost in its entirety by Smith and Lovell (1983). The Smith:Baade correspondence between August 1951 and May 1952 is also found in folder 37, box 19, Baade papers, Huntington Library, San Marino, California.

for some reason not until seven months after Baade's initial observations:

> I photographed the [Cygnus] field at the 200 inch last fall after you had sent me your accurate position. The result was very puzzling. At the place you gave me there is a rich cluster of galaxies and the radio source coincides closely with one of the brightest members of the cluster. ... This galaxy, which has a diameter of about 0.6′, is a queer object. In fact the 200 inch picture suggests strongly that we are dealing with two galaxies which are in actual collision. ... As soon as the Cygnus region comes up again, we intend to shoot the spectrum. If no [high-excitation emission lines] show up,[64] we certainly can forget about the coincidence of this galaxy with the Cygnus source since there are thousands and ten thousands of galaxies of similar brightness scattered all over the sky.

Baade's queer 18th magnitude galaxy in the northwest corner of Smith's error box (Fig. 14.1) was none other than the one discussed at the start of this section: Mills's suspect (Mills and Thomas 1951) and Minkowski's reject of two years previous![65] Now, however, its credibility was vastly heightened, for the 200 inch plate (Fig. 14.13) revealed a bizarre appearance even as it fell within Smith's tens-of-times smaller error box. A month later Baade sent more exciting news:

> Last week Minkowski obtained the spectrum of the [candidate] nebula with the new grating spectrograph at the 100 inch. Its outstanding feature is a strong emission spectrum with [Ne V] present!! That we are dealing with an extragalactic object is convincingly shown by the large redshifts of the emission lines, which amount to about 15,000 km/sec. Altogether it seems to be well established now that the object coinciding with the Cygnus source represents two galaxies in actual collision.[66]

---

[64] Baade's expectation that colliding galaxies would exhibit high-excitation emission lines was largely based on NGC 1275, which was then taken to be such a case, even before its identification with Perseus A (Section 14.4.4). [Baade:Ryle, 5 June 1952, box 1/3, RYL]

[65] Much later, Mills (2006:3) remarked that his deference to Minkowski in 1949 was "the beginning of my development of a healthy skepticism toward authoritative pronouncements and the confidence to rely on my own judgment."

[66] Smith:Baade, 26 September 1951; Baade:Smith, 29 April 1952; Baade:Smith, 26 May 1952, LOV.

If anyone had any doubts about Smith's positions, these were quelled six months later when Mills (1952b) produced ones of equal quality. For the six strongest sources visible from Sydney, Mills carried out special, precisely calibrated observations with his 101 MHz interferometer (Section 14.3.2), achieving error boxes typically $2′ \times 8′$ in size. There was excellent agreement for three sources in common (including point M52.2 for Cyg A in Fig. 14.1), and a better position for Cen A strengthened its identification with NGC 5128. New positions for Hydra A and Fornax A, however, yielded no optical identifications.

The implications of the emerging identifications, especially for Cyg A, were astounding. The *New York Times* talked about Cyg A as a radio station at a distance of 600 million million million miles and with a power of 400,000,000,000,000,000,000,000,000,000 kilowatts.[67] William H. McCrea (1984:372), a leading British cosmologist, later recalled:

> Cygnus A was identified with a faint double galaxy ... for which a radio luminosity about a million times greater than that of our Galaxy was inferred. This was most astonishing and astronomers had to practise hard before breakfast at believing it – that two objects should have comparable emission in one part of the electromagnetic spectrum but emissions differing by a factor of a million in another part.

Not only did acceptance of this identification raise numerous astrophysical questions about the nature of the beast, it also led one to wonder how many of the other weaker sources might be similar specimens. If this were so, radio techniques were probing to distances in the universe where even the 200 inch telescope had no hope of finding optical counterparts (Baade and Minkowski 1954b:225).

14.4.2 Cassiopeia A

When Redman was given the new radio positions by Smith in mid-1951, he requested that both the Royal Greenwich and University of London Observatories take plates in order to search for high proper motions that would reveal candidate radio stars. This they did

---

[67] *New York Times*, 23 August 1953, p. E9.

Figure 14.13 Photograph of the optical object associated by Baade and Minkowski (1954a) with Cyg A in 1951–52. This was taken on the Palomar 200 inch telescope with a filter centered at a wavelength of ~5200 Å, emphasizing emission from a line of twice-ionized oxygen. The size of the photo is ~1′, comparable to the small circle in Fig. 14.1.

with negative results,[68] but it was an effort by his own young staff member David W. Dewhirst that meanwhile paid off. Although primarily a solar (optical) astronomer at that time, Dewhirst became fascinated with the radio star problem and searched tomes of astronomical catalogs for any possible coincidences in position with the radio stars, finding nothing brighter than 14th magnitude. After Minkowski sent him a print of the same plate of Cyg A's field that Mills had worked with in 1949 (Fig. 14.11), he confirmed that the faint galaxy that Mills had pointed out was now an even better bet with Smith's new, much smaller error box. Reasoning that infrared wavelengths were adjacent to radio ones, Dewhirst scanned a wide area around the Cyg A position with a lead sulphide photometer, but to no avail. Lacking any photographs for Cas A, he next took several in the visual and near-infrared bands with a 36 inch reflector. On a blue plate exposed on 23 September 1951 his eye first caught a "globular nebula similar in size to that in Cygnus, close to the radio source" (about 1.5′ from Smith's position, as shown in Fig. 14.14). Dewhirst (1951) reported these results to an RAS meeting on 12 October, where he emphasized the similar appearance of the nebulae he provisionally associated with Cyg A and Cas A and suggested that they might represent a new type of nebulosity within the Galaxy.

---

[68] See *Nature* **168**, 418–19 (1951) and *Mon. Not. RAS* **111**, 195–6 (1951). Another photographic and spectroscopic investigation of the Cyg A region was undertaken in the spring of 1950 by Donald A. MacRae using a 24 inch Schmidt telescope at the Warner and Swasey Observatory, Cleveland, Ohio. In consultation with Charles Seeger and Ralph Williamson at Cornell, MacRae searched for nearby dwarf stars of very low luminosity. These results were never published, although MacRae had one letter to *Nature* rejected. [MacRae:Ryle, 6 May 1950; Ryle:MacRae, 26 May 1950; 3 page ms. entitled "Radio stars" by MacRae – all box 1/1, RYL; undated excerpt from a MacRae letter, file 7 (uncat.), RYL. Also Lovell:editors of *Nature*, 5 July 1950, file R12 (uncat.), LOV]

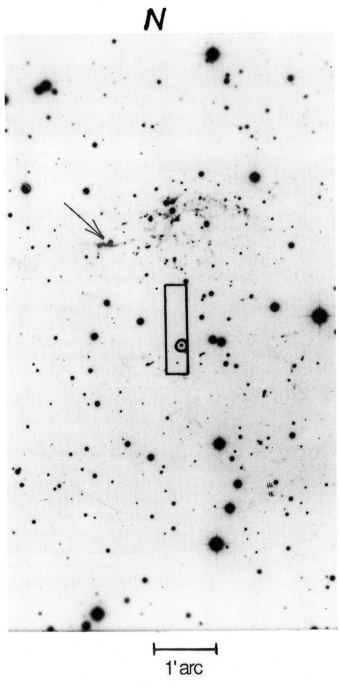

Figure 14.14 Photograph (negative) of the gaseous filaments associated by Baade and Minkowski (1954a) with Cas A in 1951–52. This was taken on the Palomar 200 inch telescope with a filter centered at a wavelength of ~6550 Å, emphasizing emission from a line of hydrogen. The main filaments are ~1′ to the north and northeast of the indicated radio position of Smith (1951a). The small circle with central dot marks the center of the 6′-diameter region of (mostly faint) optical nebulosity. The nebulosity independently discovered by David Dewhirst (1951) is marked with an arrow.

Valiant as Dewhirst's work was with a small telescope in a poor climate, it was soon engulfed by news from southern California. Baade wrote to Smith two weeks after Dewhirst's talk:

> My search for the Cassiopeia radio source at the 200 inch has turned up [on 8 September] an exceedingly interesting object close to your measured position. It is an emission nebulosity 2.8 minutes of arc long of a most abnormal type. In fact the only nebulosity with which it can be compared in its intricate structure is the well-known Crab nebula! ...Although the present data do not yet establish the identity of radio source and nebulosity, the coincidence of the radio source with a very abnormal astronomical object appears certainly suggestive.

Baade noted that he was not concerned that this object was 1.9′ north of Smith's position, for there were strong indications on his blue plate of heavy obscuration by dust in the region and he suspected that a red plate would show more nebulous structure to the south. A month later, after another observing run at Palomar, Baade again wrote, this time convinced of the identification. His new red plates had indeed shown a wider network of nebulous patches, in a circular area of 5′ diameter, and the center of this circle exactly coincided with Smith's position (Fig. 14.14). Baade told of his plans to search for the high expansion velocities expected if Cas A were indeed a supernova remnant, and also sent prints of the region, causing Smith to remark:

> The red exposure shows an amount of detail which amazes us; we are not experts on telescopes, so we showed the two photographs to Professor Redman and others at the Observatory here. They were equally impressed.[69]

Cyg A and Cas A, at long last, had succumbed.

14.4.3 Baade and Minkowski

Walter Baade (1893–1960) and Rudolph Minkowski (1895–1976) were two of the leading astronomers of the twentieth century (Fig. 14.15). Their careers largely paralleled one another, beginning about 1920 with doctorates in Germany and shared time at Göttingen and Hamburg Universities. Baade at that time was more the astronomer and Minkowski the atomic physicist and laboratory spectroscopist. In 1931 Baade took up a staff position at Mt. Wilson and in 1935 Minkowski followed him there (as a refugee from Nazi anti-Semitism) and forged what turned into an ideal scientific partnership. They were superb observers and, given their access to great quantities of time on the Mt. Wilson 100 inch telescope and (after 1950) the Palomar 200 inch reflector and 48 inch Schmidt camera, their stocks of high-quality data steadily accumulated. Minkowski's specialty was spectroscopy of supernovae and gaseous nebulae, Baade's direct photography of everything from asteroids (he discovered Icarus) to galaxies. Before radio sources entered their lives, Minkowski was best known for his work on the classification of supernovae, while Baade was acclaimed for his work showing that galaxies contained two basic populations of stars. With effectively no competition and with a tendency to publish only when they considered a subject as reasonably "closed" (Minkowski 1974:11T), their style was to share findings with the astronomical community more by word of mouth and handwritten correspondence than by the printed page. They were famous for enlightening visitors to their offices on almost any topic of the day, often by pulling a plate or unpublished plot from a desk drawer or an unkempt pile. By admitting the new radio researchers to their network, they forged an early and vital link between the optical and radio worlds (Section 17.1.3). In fact, the flow of unpublished information to Pasadena from Sydney, Cambridge, and Jodrell Bank over the years grew to a point where it far exceeded that shared between the rival radio groups themselves. The optical astronomers were viewed as safe neutral territory, although on occasion they could also act as a conduit![70]

---

[69] Baade:Smith, 23 October and 25 November 1951; Smith:Baade, 14 December 1951, LOV.

[70] Alan Needell (1991) has discovered an instance that nicely illustrates how much Baade and Minkowski valued their relationships with the overseas radio astronomers. When the possibility was put forward to them in 1952 of a major new radio telescope being associated with Mt. Wilson Observatory, they balked because they feared that the flow of unpublished radio data would be jeopardized if they were perceived to be part of a competing group. [Ira Bowen:V. Bush, 25 September 1952, Carnegie Institute of Washington archives]

Figure 14.15 Walter Baade (above) and Rudolph Minkowski.

Their enthusiasm for astronomical puzzle-solving and challenging new directions in research is well illustrated by their plunge into studying the enigmatic radio stars despite being in their late 50s.[71]

We have seen how the story of Baade's and Minkowski's work on Cyg A and Cas A appeared from Smith's viewpoint in England, but more is known, based largely on Baade's recollection four years later.[72] The day after Baade replied to Smith's first letter (4 September 1951), he began an observing run at Palomar and squeezed in the first 200 inch plates of Smith's Cyg A position. (Four nights later he took his first Cas A exposure, as described above.) Immediately upon developing the Cyg A plates, he knew he had something extraordinary. As quoted by science writer John Pfeiffer (1956:108):

> There were galaxies all over the plate, more than two hundred of them, and the brightest was at the center. I couldn't make head nor tail of it. It had a double nucleus and showed signs of tidal distortion, gravitational pull between the two nuclei. I had never seen anything like it before. It was so much on my mind that while I was driving home for supper, I had to stop the car and think.

What Baade's roadside cogitation produced was the notion of witnessing two galaxies in the process of collision, a cosmically rare event that must somehow be producing the huge radio power implied by the great distance to the optical object(s). The idea of colliding galaxies was hardly fabricated from whole cloth, for Baade was at that time what he later described as "collision conscious." Astrophysicist Lyman Spitzer, Jr. of Princeton University and Baade (1951) had recently collaborated on how galaxian collisions, which would take place especially within crowded clusters of galaxies, would affect the collidees. They had concluded that the stars of the two systems would pass through each other relatively unscathed, but that the interstellar gas masses would strongly interact, in the process heating up (with accompanying high-excitation spectral lines) and denuding each galaxy of its gas. The idea was attractive because such gas-free galaxies (S0 types) were indeed found mainly in

---

[71] This sketch of Baade's and Minkowski's careers and styles has been largely taken from obituaries of the former by A. Sandage (*Year Book of the Amer. Phil. Soc.*, 108–113 [1960]) and of the latter by D. E. Osterbrock (*Biog. Mem. Natl. Acad. Sci.* **54**, 271–98 [1983]); also see Osterbrock (2001) for a biography of Baade. I have also profited from Greenstein (1984:73–8) and interviews with Minkowski (1974) and Greenstein (1980).

[72] Baade:John Pfeiffer, 22 August 1954 and 11 April 1955, PFE. These letters were placed on deposit by Pfeiffer along with others generated during the research for his popular book *The Changing Universe* (1956).

clusters of galaxies, while gas-rich spirals (such as the Milky Way) were outside.

When Baade tried out his colliding-galaxies idea on his colleagues in Pasadena, however, he encountered uniform skepticism that a radio star could be so distant and, even if it were, that a collision could produce so much radio noise. Baade's story went like this:

> In 1951 at a seminar talk that Minkowski gave in Pasadena ... he reviewed all the other theories [of radio stars] first and then, at the end, as if he were lifting a hideous bug with a pair of pincers, he presented my theory. He said something like: "We all know this situation: people make a theory and then, astonishingly, they find the evidence for it...." I was angry and said to him, "I bet a thousand dollars that Cygnus A is a collision." Minkowski said he could not afford that – he had just bought a house. Then I suggested a case of whiskey, but he would not agree to that either. We finally settled for a bottle of whiskey and agreed on the evidence for collision – emission lines of high excitation....
>
> Several months later Minkowski walked into my office and asked "Which brand?" He showed me the spectrum of Cygnus A – it had neon-five in emission.... I said that I'd like a bottle of Hudson Bay's Best Procurable, the strong stuff the fur hunters drink in Labrador.
>
> But that was not everything. For me, a bottle is a quart, but what Minkowski brought was a hip flask. I did not drink it, but took it home as a trophy.
>
> But that's still not the end of the story. Two days later Minkowski visited me, saw the bottle and promptly emptied it.[73]

No matter who drank the whiskey, the important point for the history of radio astronomy is that these two expert optical astronomers were willing to spend considerable time searching for and studying visual objects associated with radio sources. They spent much of 1952 and 1953 on this and in June 1953 finally submitted a pair of papers for publication. The only previous mention in print of their seminal work was in an annual report for 1952.[74] Word nevertheless got around through their usual informal channels and also via a talk Baade gave in September 1952 to the Commission for Radio Astronomy at the IAU General Assembly in Rome. The impact of this talk on at least one person can still be gauged, for extant notes taken by Martin Ryle convey the excitement:

> Cas A – remarkable nebulosity (he hasn't seen one like it); situation [of large velocities] never before observed; one filament contains H, the other contains no H!!!
>
> Cyg A – of 800 galaxies so far observed [with spectra], only one similar emission spectrum.
>
> Perseus A – dying octopus.[75]

Baade and Minkowski's papers in the *Astrophysical Journal* for January 1954 became the bible on optical identifications. They were authoritatively written and filled with photographs, spectra, historical details, and copious notes gained from circulating drafts to the three major radio groups in March 1953.

Four major categories of optically-identified radio source were defined (also see Table 14.1). The first was unimaginatively named "peculiar emission nebulosities." Its archetype was Cassiopeia A, presented as a new type of galactic nebulosity consisting of broken bits of emission scattered over a region of ~5′ (Fig. 14.14). Shifts of the measured spectral lines occurred over an unprecedented range of −1000 to +2000 km s^{-1}, and it was thought that the intense radio emission might be connected with these turbulent motions. Because there was no apparent pattern of expansion, either in the observed velocities or in the changes detected between plates taken in 1951 and 1953, it was argued not to be the remnant of either a nova or a supernova. This point of view was vigorously opposed by Iosif Shklovsky (1954), who had earlier suggested in the Soviet Union

---

[73] Related by I. Robinson, A. Schild, and E. L. Schucking (1965, eds.) in *Quasi-stellar Sources and Gravitational Collapse* (Chicago: University of Chicago Press), p. xi. Slightly varying versions of the Baade/Minkowski bet story are available in Baade's first letter to Pfeiffer [previous note] and in Baade:Lovell of 26 May 1952, reprinted in Smith and Lovell (1983:161–3). The version I have cited, already a decade after the events described, undoubtedly is to some degree apocryphal, but the embroidery bears Baade's style.

[74] (Baade and Minkowski), "Identification of extended radio sources," *Carnegie Institution of Washington Yearbook for 1951–52*, No. 51, pp. 14–15.

[75] Notebook for September 1952 IAU General Assembly, file 27 (uncat.), RYL.

that Cas A's energetics and morphology pointed to an explosion and who had even located a candidate historical supernova, from AD 369 (Shklovsky 1953c:26, Section 15.4.2).

Grouped with Cas A was Puppis A, which had been first cataloged by Stanley and Slee (1950) and then found by Bolton *et al.* (1954) to be an extended source. After Bolton notified Baade and Minkowski of this in 1951, they discovered at its position an assemblage of filaments similar to those of Cas A, but spread over a full degree on the sky and exhibiting much smaller velocity dispersions and no measurable expansion. Another tentative member of this category was the source that Piddington and Minnett (1952) had named Cygnus X (Table 14.1), a large region only 6° removed from Cyg A, but with a flat radio spectrum. The Australians had suggested that this spectrum implied thermal emission from ionized gas known to exist in that part of the Milky Way, but Baade and Minkowski were more struck by the similarity of the unusual filaments found in Cyg X to those in Cas A.

Baade and Minkowski's second category of radio source contained "peculiar extragalactic nebulae," including Vir A, Perseus A, Cen A, and Cyg A. Vir A was still taken to be associated with the bright elliptical galaxy Messier 87, but how its unique nuclear jet could account for such radio power was inexplicable. Moreover, the latest measurements of the size of the emitting region (see below) showed it to be far larger than the jet, more like the size of the entire galaxy. In the same class was Per A, a radio source that Hanbury Brown and Hazard (1952a) had suggested to originate from the integrated effect of the entire membership of the Perseus cluster of galaxies. Although there were no measurements of angular size to discriminate between this and the possibility that only a single galaxy might be the cause, Baade and Minkowski argued for the latter. Other clusters (such as that in Coma) had not shown themselves to be strong sources and the Perseus cluster indeed contained a suspicious lone culprit. This was NGC 1275, long known to be peculiar and which Baade now interpreted as two spiral galaxies in collision, based on its morphology and spectrum (linewidths of 4500 km s^{-1}). A third example was Cen A, identified as NGC 5128 (Fig. 14.4) and now with a new interpretation, that one was again witnessing a collision, this time between a spiral galaxy (seen edge-on) and an elliptical galaxy. Finally, there was Cyg A. The redshifts of its emission lines yielded a recession of $17,000 \text{ km s}^{-1}$ and a distance of over 30 Mpc,[76] which then meant that its radio luminosity was fully 3×10^5 times that of our own Galaxy. This was the third example of colliding galaxies producing a strong radio source, an attractive idea that was to dominate thinking for many years.

The third category of radio sources comprised "normal galaxies," primarily as investigated by Hanbury Brown and Hazard. Ever since their initial detection of the Andromeda nebula in 1950 (Section 9.3), they had been studying the relation between optical and radio emission from galaxies by exploiting the superior sensitivity of Jodrell Bank's giant reflector to the fullest – a source as weak as 5 Jy (compare the source intensities in Table 14.1) was capable of detection, albeit with one month's effort! By 1953 they had reasonably secure detections of half a dozen spiral galaxies and possible detections of two clusters of galaxies, and it was clear that the radio luminosities of most galaxies were far lower than those of the peculiar ones discussed above. In order to compare radio and optical intensities, Hanbury Brown and Hazard (1952a) even introduced a system of *radio* magnitudes that survived for almost a decade. Besides individual galaxies, they also endeavored to discern small changes in the general background radiation that might be attributable to varying counts of galaxies across the sky.

The last of Baade and Minkowski's categories contained supernova remnants, for which the Crab nebula, alias Tau A, was of course the outstanding example (Section 15.4.2 covers later work, primarily in the Soviet Union, to understand its strong emission). A tentative second example was the weak source Cas B that Hanbury Brown and Hazard (1952b) had reported to coincide with Tycho Brahe's supernova of 1572. This source coincided with the sixteenth-century position (no modern optical remnant was known) to within its ~20–30′ error bars. On the other hand, Johannes Kepler's supernova of 1604, for which Baade *had* long ago discovered a faint optical remnant, had not been detected in the radio.[77]

---

[76] Throughout these papers Baade and Minkowski used a distance scale corresponding to a Hubble constant of $540 \text{ km s}^{-1} \text{ Mpc}^{-1}$, despite Baade's notable recent work showing that all extragalactic distances needed to be doubled. (Modern extragalactic distances are ~7 times farther than in these papers.)

[77] Kepler's supernova was located at −21° declination, too far south to be reached by the sensitive Jodrell Bank dish.

Discounting normal galaxies, the scorecard of galactic versus extragalactic identifications stood at a very tight 5 to 4. But what of the other two hundred unidentified sources?

14.4.4 The nature of optical identifications

14.4.4.1 *Techniques*

Finding optical counterparts to discrete radio sources in the early 1950s was a difficult and frustrating business. In the first place, optical atlases, catalogs, and photographic surveys covered only the brighter objects (for example, the Shapley–Ames catalog of galaxies of 1932 [to 13^m] and the Franklin Adams all-sky photographic survey of 1911 [to 15^m]), and therefore optical information about any specific position on the sky was generally not available. This situation was different, however, for at least one man in the world. Throughout this period Minkowski was supervising the National Geographic Palomar Sky Survey, a collection of photographs to twenty-first magnitude of the entire northern sky. When in the late 1950s this survey was reproduced for observatories around the world, it became a standard research tool, but before that time it was only available to Pasadena visitors under Minkowski's watchful eye.

Secondly, even if one had a photograph, the early radio positions were so poor (compared to the sub-arcsecond accuracy of optical positions) that a typical error box encompassed a region of sky containing hundreds, if not thousands, of stars and nebulae (for example, see Fig. 14.11). How was one to choose? One had to select an object that somehow stood out from the rest, either because it appeared peculiar in some respect (such as its brightness or morphology) or because it fit in with one's ideas of what a radio source should be. But what does one call "peculiar"? And what expectations does one have? Obviously, the answers to these questions varied with each researcher and with the state of knowledge in the field (Edge and Mulkay 1976:103). As it turned out, even the strongest radio sources were finally successfully associated only with very faint optical objects, and this of course meant that there were even more plausible candidates within a given error box. For instance, the Shapley–Ames catalog of galaxies listed one galaxy (of 13^m or brighter) for every 30 deg^2 of the sky. If one's error box for a radio source had an area of x, then one would expect an accidental coincidence of galaxy and source in roughly $x/30$ of all cases. For $x \geq 3$ deg^2 (which was common), the expected number of accidental coincidences was therefore bothersome. For yet fainter galaxies, the situation quickly grew hopeless, for there are, for each magnitude fainter, about four times as many galaxies in each square degree. And remember that Cyg A had been found to correspond to an *eighteenth* magnitude galaxy – this had only been possible because the radio error box had been reduced to 0.0002 deg^2. Mills found out firsthand about the trials of attempting optical identifications when he visited Pasadena for several months in 1953:

> My study of the Schmidt plates has not yielded any very startling results, as the number of "peculiar" objects in the sky is extraordinarily large, and I can sympathise with the astronomers in demanding more accurate positions.[78]

A further facet of the astronomical sky is that there are far more stars (to a given magnitude) than extended objects such as galaxies and emission nebulae. In the example given above, the number of *stars* brighter than 13^m ranges from 50 to 400 per deg^2. Only a tiny fraction had ever been observed and the amount of telescope time to obtain, say, the spectra and proper motions of all stars within the error box of even one radio source was prohibitive. Once it became clear that the strongest sources did not correspond with the brightest stars, it also became clear that positions of extremely high precision were going to be required to reliably identify any given star with a radio source. This led Baade and Minkowski to prefer working on sources known to be extended (next section), for they could then concentrate on the *non*stellar optical candidates, on the supposition that an extended radio source implied an extended optical counterpart. It also gave them a second criterion for the goodness of an identification, namely the match between angular sizes (and sometimes rough shapes), as well as that between positions.[79]

14.4.4.2 *Identity and reality*

Great importance was ascribed to optical identifications during this period. Indeed, some radio astronomers

---

[78] Mills (in Pasadena):Bowen, 16 November 1953, file F1/4/MIL/1, RPS.

[79] Minkowski:Ryle, 16 January 1952, box 1/3; Baade:Ryle, 1 July 1952, file 4 (uncat.) – both RYL.

studied radio sources only as a means to the end of learning their optical nature. For example, the ultimate purpose of Bolton, Stanley, and Slee's (1954:122) survey was stated to be the identification of its sources, and in another instance we find in a Cambridge thesis that "the most direct way of investigating the nature of radio sources is by observing individuals optically."[80] Again, from a Cambridge student: "If a radio star can be identified with a visible object, many powerful methods become available for determining its distance directly."[81] In the Soviet Union, Shklovsky (1954:483) was a strong supporter of early radio astronomy (Section 10.3.1), but still went so far as to say: "Baade and Minkowski's work shows in a most striking manner that it is only through the methods of optical astronomy that we can resolve the problem of the nature of cosmic radio radiation." An optical identification and spectrum promised rich information (e.g., temperature, density, chemical composition, motions). A single photograph also contained far more information than a strip chart trace.[82]

Optical identifications lent a sense of reality and reliability to the very existence of radio sources. Note the language employed.[83] First, were these radio sources so enigmatic that they acquired an "identity" only upon the discovery of an optical counterpart? Although an interesting strong conjecture, this is not supported by the contemporary evidence. But the word *identify* can carry several connotations: "to regard as the same," "to prove the sameness of," "to link inseparably," or simply "to associate." A taxonomist *identifies* (in the second sense) a plant by finding its place in a prearranged system, but in (optical) astronomy the word had been used, even before radio astronomy, in the last two senses; for example, Oort wrote in 1942 that "the guest star of 1054 has an identification with the Crab." This usage of "identification" was taken over into radio astronomy by Bolton, Stanley, and Slee (1949) right from their first article suggesting an association of radio sources with optical objects. An "optical identification" was no more than a strong link between radio and optical emissions from a certain patch of sky, although note that the linkage was not symmetrical. Even though one occasionally finds a usage such as "optical objects identified as radio sources" (Baade and Minkowski 1954b:215), the reverse was the norm. Moreover, the term "optical source" was only ever used once (in the thesis of M. K. Das Gupta [1954:113]), and the term "radio identification" never did appear.

Although optical identification did not confer an identity per se, a source nevertheless did thereby become more *real*. With a photograph in hand, one could now appeal to the accumulated authority of modern astronomy and not have to argue a source's existence based solely on somewhat suspect and arcane interferometry. Edge and Mulkay (1976:101) have also noted how the optical astronomers, from their position of consensus in both theory and technique, in effect served as arbiters to the radio star controversies. A photograph was something that one could immediately grasp and admire, as Bolton wrote to Minkowski in 1952:

> I hope that you will send a photograph of the Puppis object when convenient. Photographs are much more satisfying than the evidence we get out of our machinery.[84]

To get something more palatable out of his machine, Mills (1953:Plate 2) accordingly turned his derived elliptical models (Section 14.5.2) for the radio brightness distribution of three sources into "radio pictures," astronomy-photograph-like white ellipses with fuzzy edges (added for verisimilitude) that he directly compared with photographs of the optical identifications (Fig. 14.16). Now he was doing *real* astronomy!

As we have seen, different radio surveys disheartingly differed in common regions of the sky (Fig. 14.5) – the situation was so desperate that in 1952 an URSI Committee took on the duty of

---

[80] J. R. Shakeshaft, December 1956, "The distribution of radio stars," Ph.D. thesis, p. 72, Dept. of Physics, Cambridge University.

[81] P. A. G. Scheuer, "The origin of galactic radio radiation," Hamilton Prize Essay, 1 October 1953, file 34 (uncat.), RYL.

[82] For instance, a photograph of the Crab nebula might contain (in modern terminology) a million pixels (independent values of intensity), whereas the total radio information about Tau A from the early 1950s could have been summarized in fewer than one hundred numbers.

[83] "It is impossible to dissociate language from science or science from language.... [Phenomena, concepts, and words] all mirror one and the same reality." A. L. Lavoisier (1789), *Traité Elémentaire de Chimie*.

[84] Bolton:Minkowski, 11 January 1952, cartons 4–5, MIN.

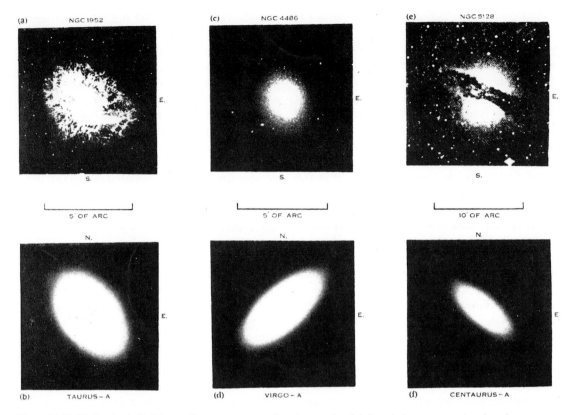

Figure 14.16 "Radio pictures" of three radio sources, compared to photographs of their optical counterparts (east is to the *right*). Each radio picture is a representation of three derived properties of the radio source (lengths of major and minor axis and orientation). Relative brightnesses are not shown – the "fuzziness" along the edge of each ellipse has been added for verisimilitude. (Mills 1953)

advising IAU astronomers "concerning the reality of each reported object."[85] Under these circumstances an optical identification was a prime means of distinguishing the radio wheat from the radio chaff. For instance, in order to designate those sources "whose existence has been definitely established" in an official list of radio sources assembled in early 1954, Joseph Pawsey (1955) put heavy weight on their having an optical identification, as well as having coincident positions in two or more surveys. Only eight sources qualified (the list was similar to Table 14.1).

The most telling instance I have found of the heightened sense of reality and persuasiveness brought on by optical identifications concerns Ryle and his colleagues. After discussions with Baade at the 1952 IAU meeting in Rome, Ryle returned home to Cambridge shorn of doubts he had had about the validity of some of the claimed identifications. In a colloquium he described these developments as "dramatic – a turning point in radio astronomy – the completion of the first stage of radio star observations." And before an evening discussion club of physicists he triumphantly declared:

> Radio stars have provided the Kapitza Club on several previous occasions with real opportunities for speculation – [there was] so very little information available concerning their constitution. There were, I think, even those who were not really convinced of their reality. This stage, I am afraid, is now over, as a number of

---

[85] "Report of sub-Commission Vb on Terminology and Units" (Chair: M. Laffineur), *Proc. General Assembly* [of URSI] **9**, p. 75 (1952, Brussels: URSI).

these have recently been identified beyond any reasonable doubt with visible objects.[86]

Four months later, in another talk, Ryle said:

> Radio astronomy makes use of this second "window" in the atmosphere, and our telescopes and photographic plates are replaced by large aerial systems and sensitive radio receivers.
>
> There are some who think that because we work with wavelengths which the senses cannot <u>perceive directly</u>, our information about the nature of the universe is second-hand – {or even not quite proper}. I do not think this view is right, and although there are many <u>practical limitations</u> in our observations as compared with optical ones, I think there is no difference <u>in principle</u> to the extension of one's senses in this way than by the use of photographic plates, or even more so, photoelectric cells. In fact very few observations are now actually made by looking through telescopes – the <u>advantages</u> obtained by taking a <u>photograph</u>, possibly with a <u>long exposure</u>, and examining and measuring it afterwards are obvious; when observations are made in the UV or IR direct observation is in any case impossible.
>
> {As a final argument, we can, if we want, provide an audible signal from our receivers. This would certainly be no less direct than a BBC broadcast of a concert. Later I hope that I can play a recording we made recently – so that you will at least have heard (if not seen) a radio star!}
>
> We can therefore regard radio astronomy as a straightforward extension of optical astronomy.[87]

The situation is strikingly similar to that of nuclear physics in the 1910–50 era, where Galison has argued that the "visualizability" of photographs produced by cloud chambers caused the particles of the subatomic world to become real in a compelling way: "C. T. R. Wilson's cloud chamber [ca 1911] had rendered the sub-atomic world visualizable – and consequently it took on a reality for physicists that it never could have obtained from the [earlier] chain of inferences." Galison cites a long list of philosophers of science and physicists, including Percy Bridgman, Max Born, W. V. O. Quine, and Mary Hesse, who have epistemically privileged the sense of *vision*. Recall the use of the word *direct* in the statements at the start of this section (e.g., "the most direct way of investigating the nature of radio sources is by observing individuals optically"). In effect these opinions acknowledged that optical data, even though they too had undergone much processing (i.e., the chemistry of the photographic darkroom), were epistemically superior to radio data (Fig. 14.17). Early radio astronomy, as in the case of cloud chambers and particle physics, also privileged optical data when assessing the reality of any given radio source.[88] This predilection was a key aspect of what I discuss as radio astronomy's "visual culture" in Section 17.2.5.

14.5 ANGULAR SIZES

In addition to accurate positions for the radio stars, it was early held that measurements of their angular extent were vital for deducing their nature. Baade and Minkowski in particular continually emphasized that optical identifications were only sure if there existed agreement between the radio and the optical in angular size as well as in position.

These measurements of angular size, however, turned out to require extraordinary technical dexterity by the three major groups.

14.5.1 The intensity interferometer

14.5.1.1 Theory

As related in Chapter 9, Hanbury Brown joined Lovell's Jodrell Bank group in 1949 after more than a decade of experience in developing radar techniques.

---

[86] Notes for talks given to the Cavendish radio group and to the Kapitza Club ("Recent results in radio astronomy"), 16 October 1952 and undated (ca October 1952) [file 8 (uncat.), RYL]. John Baldwin (1972:14A:380, 1981:144A:660), then a new postgraduate student, also recalls Ryle's excitement after returning from Rome and discussions with Baade. Ryle's doubts before the Rome meeting were expressed, for example, in Ryle:Lovell, 23 July 1952 and Ryle:Oort, 15 August 1952, both box 1/2, RYL.

[87] Talk given by Ryle to the Cambridge Philosophical Society, 16 February 1953 and to the Midlands branch of the Institute of Physics four days later. The phrases in braces { } were crossed out in the extant handwritten manuscript and may represent revisions to the first presentation. [File 8 (uncat.), RYL]

[88] P. Galison and A. Assmus, "Artificial clouds, real particles," pp. 225–74 (quote from pp. 268–9) in *The Uses of Experiment* (1989, eds. D. Gooding, T. Pinch and S. Schaffer) (Cambridge: Cambridge University Press); also see Galison (1997: 65–73).

"Just checking."

Figure 14.17

Besides his observations using the giant reflector (discussed above and in Chapter 9), he also invented and exploited, together with his students, a new technique called *intensity interferometry*.[89] The motivation for its invention was to find out if radio stars, which Hanbury Brown basically accepted as true (optical) stars, indeed had angular sizes as small as expected. At that time it had only been established that the few strongest radio stars had an upper limit of 5–10′ in size, thousands of times larger than optical stars. Hanbury Brown knew that Albert Michelson had been able to measure the sizes of a few of the closest and largest optical stars in the 1920s with a conventional optical interferometer of only 3 m baseline, but his central problem was to achieve comparable angular resolution at *radio* wavelengths, which then implied a baseline extending over thousands of kilometers. There was no hope that Michelson's (and Ryle's) brand of interferometry could work over such intercontinental distances – 20 to 50 km was a maximum with the techniques of the day, which over longer distances could not directly link two antennas or record signals separately and later bring them together.

During the winter of 1949–50 Hanbury Brown cooked up a variety of tentative solutions, most of which turned out to have fatal flaws. One scheme involved combining a radio star's signal (as received directly) with the same signal after it had bounced off the moon. Another involved recording a signal on magnetic tape over a period of time and then correlating this signal (at time t) with itself (at time $t + \tau$); the effective baseline of the "interferometer" would then be the distance travelled by the earth during the interval τ.[90] In November he wrote to Ryle about yet another idea, involving multiplication of the received signal to much higher frequencies:

> I would like to measure the diameter of these point sources! No doubt many other people would like to also. I should be obliged if you would chew over the following simple suggestion and shoot it down if possible.... I am worried about the passage of the incoherent noises from various parts of the star's disk through the non-linear frequency multipliers together with the incoherent receiver noises....[91]

His best idea, the key to the intensity interferometer, came along in February 1950:[92]

> I thought to myself... "Is there any other parameter [besides the usual signal amplitude and phase] that we could look at?" And into my mind came quite clearly the idea of a man sitting with a radio receiver looking at the noise on a cathode ray tube. I had an absolutely clear vision of the noise on two cathode ray tubes, one at each end of a baseline, and I thought to myself, "Ah, those two noises are the same, or ought to be!"
>
> (Hanbury Brown 1983:134)

---

[89] I use the term *intensity interferometer* for the Hanbury Brown–Twiss scheme, but during the early 1950s there were a variety of terms used, including "autoheterodyne interferometer," "post-detection interferometer," "post-detector correlation interferometer," and "video interferometer."

[90] Twiss (1981:145A:530, 640). The moon-bounce idea was much later worked out in detail; see "Radio interferometry by lunar reflections" by T. Hagfors et al. (1990), *Ap. J.* **362**, 308–17.

[91] Hanbury Brown:Ryle, 28 November 1949, box 1/1, RYL.

[92] Hanbury Brown:author, 5 March 1989.

Although preliminary calculations convinced him that a scheme based on this insight should work in principle, he was still uncertain about it in practice and therefore sought analytical help. This he found, in March through a mutual friend, in the person of Richard Q. Twiss (1920–2005), who was then a mathematical physicist on the staff of an Admiralty Lab in Hertfordshire, and had worked at the Telecommunications Research Establishment during the war and at MIT afterwards on various aspects of the theory of receivers. He and Hanbury Brown hit it off well, despite Twiss's alarming habit of bursting into Wagnerian song, and soon there existed pages and pages of mathematics showing that the proposed interferometer should indeed work. One no longer had to worry about phase as in a conventional interferometer; the measured correlation was now essentially of the noise-like modulations of the source signal rather than of the signal itself. As before, there would be less correlation for a source of given angular size as one went to longer baselines.[93]

By March 1950 they had convinced Lovell, who wrote to Ryle: "After much discussion it begins to appear that Brown's new scheme for measuring diameters is correct in principle.... We intend to carry it out."[94] In June Bolton visited Jodrell Bank and, as he later recalled:

> Lovell handed me a dog-eared school exercise book containing the mathematical formulation of the Hanbury Brown/Twiss intensity interferometer and asked me to see if I could find any errors. After a week of very long evenings I reported back that I could find no errors, but was at a loss for any physical understanding.[95]

The plan was to test this novel concept first on an intense object of known angular diameter, namely the sun, and then to move on to the radio stars.

The technique obviated a conventional interferometer's need for extreme accuracy in controlling instrumental phases at its two ends and in bringing the two signals together; extremely long baselines could therefore be realized. Its main disadvantages, which in fact led to its limited usefulness in radio astronomy, were that (a) its lack of phase information severely limited one's ability to model a source's brightness distribution (Appendix A), and (b) it required very strong signals to achieve high accuracy. The latter meant in practice that it could only be used on the two strongest radio sources, Cas A and Cyg A, and then only with relatively large antennas.

The intensity interferometer is historically important by virtue of more than its meteoric career in radio astronomy. An unexpected dividend of the technique was discovered when it became apparent that it worked perfectly well despite strong ionospheric scintillations. This showed that it could operate even within a turbulent medium and pointed to the idea of trying it at *optical* wavelengths, since the greatest obstacles to Michelson's measurements had been the similar stellar scintillations (twinkling) caused by the earth's atmosphere. Hanbury Brown and Twiss followed up this idea and by 1955 had demonstrated the optical technique by measuring 0.007″ as the angular size of Sirius, the brightest star in the sky. Later, fundamental (optical) stellar-diameter measurements under Hanbury Brown followed in Australia, but these episodes are beyond the scope of the present study, as are arguments by physicists of the day that such an interferometer, according to the laws of quantum mechanics, could not in principle work.[96]

14.5.1.2 Equipment

In the autumn of 1950 two postgraduate research students began at Jodrell Bank and soon were enmeshed in testing the intensity interferometer scheme (Fig. 14.18). Roger C. Jennison (1922–2006), who had been a navigator in the RAF for five years before taking his undergraduate degree at Manchester, had already worked several summers in Lovell's group and demonstrated his considerable skill in radio electronics and experimentation. Mrinal Kumar Das Gupta (1923–2005) had trained at Dacca University and done ionospheric research under S. K. Mitra in Calcutta. Jennison and Das Gupta were thus older, experienced students (ages 28 and 27) who became a close, productive team while working on their doctoral degrees over the next four years:

---

[93] See Tangent 14.4 for more details.
[94] Lovell:Ryle, 29 March 1950, box 1/1, RYL.
[95] Bolton (1982:353). Also see Bolton (from England):Bowen, 27 June 1950, file F1/4/BOL/1, RPS.

[96] See Tangent 14.4.

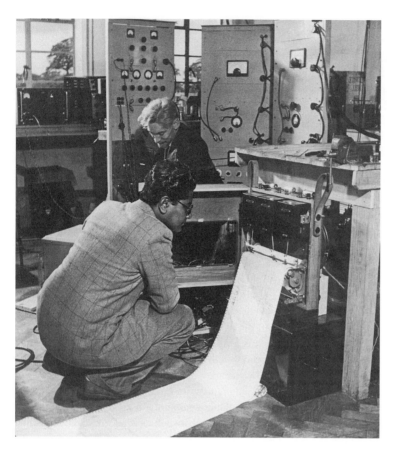

Figure 14.18 M. K. "Das" Das Gupta (foreground) and Roger Jennison working with their intensity interferometer at Jodrell Bank in late 1951. The tall racks in the back contain (l. to r.) the correlator and the two receiver back ends.

> "Das," coming from India, loved to go out and work on the aerial systems when it was snowing or freezing cold. But I used to love to go out into the sunshine and work on them then, and spend the wintertime fiddling with the electronics inside. And that's the way we subdivided the operations!
> (Jennison 1976:10T)

> To [Das Gupta] only praise; his stolid and continual support has been beyond all human ken. The erection of large aerial systems at wet, wintry, and windswept sites, in the face of ferocious farmers, alone promotes him to the ranks of the few.
> (Jennison 1954)

Jennison took the basic idea of the intensity interferometer and designed the needed hardware under Hanbury Brown's general supervision, while he and Das Gupta together built the apparatus and made observations.

It was felt initially that baselines of at least 100 km, and perhaps more,[97] were going to be needed and so the entire system was designed for such a (previously unapproached) scale. This meant that the 125 MHz receivers' bandwidths had to be narrow (200 kHz) and well-matched. Anticipated long baselines also led to the development of a radio link system for bringing the

---

[97] Hanbury Brown (1983:135, 1984:227; Hanbury Brown and Twiss 1954:681) originally was even considering the possibility of trans-Atlantic baselines, if needed. In that case, signals at each end would have been recorded on magnetic tape, with suitable timing marks, and then brought together days later. A technique related to this, Very Long Baseline Interferometry (VLBI), was in fact developed in the 1960s over intercontinental distances.

Figure 14.19 Lorry loaded with collapsed portions of Jennison's mobile 125 MHz antenna for the intensity interferometer (ca 1952).

detected signal (with a bandwidth of only 6 kHz) from a remote site to home base at Jodrell Bank, as well as to a delay system to compensate for the signal travel time between the two antennas. Jennison effected the required multiplication of the two antennas' signals in a novel device called a "correlator." His technique turned out to be analogous to Ryle's phase switch (Section 8.3), although at the time he knew only of the Dicke switch (Section 10.1.1 and Appendix A) and was attacking a wholly different problem from that of Ryle. If a and b are the signals from each arm of the interferometer, switching allowed a sensitive and accurate measurement of the difference between $(a+b)^2$ and $(a-b)^2$, which is proportional to ab. Jennison also developed a means of injecting 100%-correlated artificial noise (from a neon tube) into the two inputs of the correlator so as to provide a neat calibration and system check after each observation.

The final key ingredients were the antennas. The relative insensitivity of the technique meant that one needed (even for Cyg A) ~500 m² of collecting area in each antenna, yet one antenna had to be portable! Jennison's solution involved a tiltable array of 320 half-wave elements broken up into 16 bays that spanned 39×13 m. These were supported on a framework of pine and silver spruce. The dipoles in each bay folded down in unison "like a line of tin soldiers" and the entire assemblage collapsed into 32 units, each only a few inches thick (Fig. 14.19). The antenna, associated electronics, and a shed filled three truckloads and took one to two days to set up at a new site.

14.5.1.3 Observations

A race was on – Das Gupta later recalled Lovell daily urging him and Jennison to get the angular size prize before Cambridge or Sydney.[98] By the middle of 1951 he and Jennison had built a prototype system and tests began on the sun with short baselines and antennas one-quarter the final size. No reliable results were forthcoming, however, until Jennison, after trying numerous designs for the correlator, hit upon the switch idea. The switching was in fact first done at ~0.01 Hz (once per two minutes), that is, with Jennison himself tediously switching the equipment back and forth during the two hours that the sun took to drift through the antenna beam. On a baseline of about 50 m, the detected solar signals were indeed found to be less correlated than with the two antennas adjacent – the theory had passed

---

[98] M. K. Das Gupta (1988), "On sharing the excitement of a discovery," p. 4, Annual Souvenir, Dept. of Radiophysics and Electronics, University of Calcutta.

its first experimental hurdle. When Hanbury Brown saw this, Das Gupta recalled him excitedly exclaiming "Oh, Michelson, Michelson!"[99]

The next step was to measure a complete visibility curve and compare it with Machin's (1951) results from Cambridge. This stage, however, dragged on and on because the sun on almost every day exhibited sunspots, whereas they needed measures of the *quiet* sun. As winter came on and the noontime sun sank in the dank British sky, a further complication arose when their antennas were unable to dip far enough toward the southern horizon – they ended up erecting scaffolding in back and digging six-foot-deep trenches in front. Despite almost a year of trying, only six days of sufficiently quiet solar signal were observed. These data, however, formed a visibility curve (measured on baselines as long as 240 m) that steadily fell off in a manner similar enough to Machin's (resembling Fig. 13.7) that they felt confident in moving on to radio stars.

After building the radio link and increasing the antennas' size, observations of radio stars commenced in the summer of 1952. Standard procedure was to observe the transit of Cyg A with each flat array tilted at the appropriate angle, then shift each antenna 18° northwards to pick up Cas A three hours later. The first spacing was 0.3 km in order to test things out at an expected level of 100% correlation, but at this distance Cas A seemed to be already slightly decorrelated, intimating that its angular size might be far larger than anticipated. Plans had been to go next to a 20 km spacing, but at about that time Baade and Minkowski wrote to Hanbury Brown about their optical identifications of Cyg A and Cas A, which suggested that the radio sizes might well be of the order of 0.5′ to 5′, not sub-arcseconds. The first remote site was thus chosen to be nearer, only 4.0 km south of Jodrell in a corn field next to Lovell's home (Fig. 14.20, which should be consulted frequently during the present discussion). There they found that Cas A had indeed lost almost all correlation, implying a size larger than 2′. Cyg A's signal, however, was still 80% correlated, indicating a north–south extent of ~0.5′. It was decided that the next step should be to check the angular sizes at two other cut-angles across the sources, about 60° displaced on each side of north–south. By October this had been done at Goostrey and Lower Withington, and the results were truly intriguing: although Cas A was proving circularly symmetric in its radio brightness distribution, Cyg A would not at all fit into this mold. Cas A had a size of ~4′, agreeing well with the extent of the optical filaments found by Baade, but the three cross-cuts of Cyg A yielded equivalent sizes ranging from 0.6 to 2.2′ (Fig. 14.20) and strong indications of a complex brightness distribution (Hanbury Brown, Jennison, and Das Gupta 1952).

Although the details of Cyg A were hardly understood, it was clear by this time, with baselines of only a few kilometers, that the angular sizes of at least these two sources, as well as that of Tau A measured by Mills (1952c),[100] had nothing to do with the sizes of optical stars. This was an extremely important result, but ironically the Jodrell Bank team were at the same time keenly disappointed, for the intensity interferometer's ability to operate at far longer baselines had become superfluous:

> There was really no need to have developed the intensity interferometer; we could have done the same job with a conventional interferometer in half the time and with half the effort. We had built a steam-roller to crack a nut.
>
> (Hanbury Brown 1984:228)

The stage was now set for an all-out attack on the structure of Cyg A:

> Cygnus didn't make sense – there was something peculiar about the first results. The three measurements did not fit together, but we had confidence in them – we were sure we'd done them right, so the fault wasn't ours.
>
> I know this sounds like Archimedes, but it's perfectly true: one day I took a long bath and I was laying back in the tub thinking about this distribution. And then all of a sudden it clicked that if Cygnus were *two* blobs instead of one, then I could get this peculiar difference between the projections in different directions.
>
> (Jennison 1976:18–9T)[101]

---

[99] As in previous note, p. 3.

[100] Hanbury Brown had learned of this result directly from Mills at the 1952 Sydney URSI meeting (Lovell 1973:12).

[101] Jennison was also bothered by the discrepancy between his 2.2′ size for Cyg A and a value from Mills of about 1′.

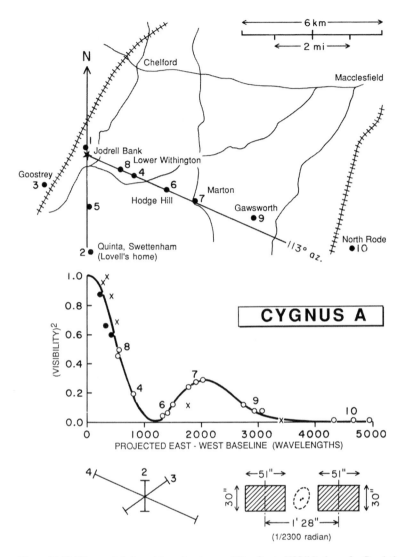

Figure 14.20 The graph (adapted from Jennison and Das Gupta [1956b]) shows for Cyg A the square of fringe visibility (or in the case of the intensity interferometer the equivalent correlation) versus projected east–west baseline, measured in wavelengths. The 125 MHz Jodrell Bank data (open circles) were taken with baselines between Jodrell Bank and the remote sites indicated on the map, numbered in the chronological order in which they were used over the period July 1952 to early 1954. Data at 210 MHz by Smith (1952b,c) (filled circles) and 101 MHz data by Mills (1952c, 1953) (crosses) are also shown. At the bottom (left) are the initial angular sizes derived from each of three crosscuts (site numbers indicated) measured by Hanbury Brown, Jennison and Das Gupta (1952), and (right) a model of radio brightness distribution based on more complete data (Jennison and Das Gupta 1953); also sketched is an outline of the faintest optical emission reported by Baade and Minkowski (1954a) from their colliding galaxies (with two dots representing nuclei as seen in Fig. 14.13). The Fourier transform of this model radio distribution is the solid line in the plot.

The double source model, however, also cleared this up, for Mills's value was then based on an incorrect interpolation of the points in his measured visibility curve (shown as crosses in Fig. 14.20).

Jennison's insight led to a hunch that Cyg A was a "binary" source with components roughly on an east–west axis. To confirm the position angle of the axis, site 5 (Fig. 14.20) was chosen to be at the same radial distance as sites 3 and 4; the value of correlation at this site would be very sensitive to the angle of the source's axis. In fact, it did confirm that the axis was very close to 113°. It was therefore decided to continue defining the visibility curve along a position angle of 113°;[102] the Fourier transform of this curve would then yield the source's detailed brightness distribution. To accomplish this, they drove their trucks over the next year to sites with names like Hodge Hill, Marton, Gawsworth, and North Rode (Fig. 14.20). This required both navigational ability and savvy in dealing with Cheshire farmers:

> The great problem in putting up a big aerial system at the distance and direction that you wanted was that you'd land right on top of some cows in a farmer's field, where he might not particularly want you.
> (Jennison 1976:17–18T)

Their setup in fact often had to be protected by electric fences. Jennison also came up with an ingenious idea that allowed a single site, with no additional move of equipment and consequent real estate negotiation, to provide a *triplet* of points on the visibility curve. This was to change the effective baseline (that is, as measured in wavelengths) of any given site by adjusting the equipment's operating wavelength ($\pm$ 7% was possible) – hence the triplet groupings of points on the visibility curve in Fig. 14.20.

Each site was chosen with careful regard to how Cyg A's visibility curve was shaping up with respect to the hypothesis of a double source. The most important baseline was at ~1500 wavelengths, where they wished to check on whether the curve actually had passed a minimum and was rising, as one would expect if Cyg A had at least one strong second component. By varying their operating wavelength, they measured the gradient of the curve at this point and after six months were convinced that it was indeed positive. They estimated the secondary maximum to be at ~2000 wavelengths and therefore took observations at this spacing, as well as confirming that the gradient reversed farther out. The longest baseline, at 12 km or ~5000 wavelengths, was designed to search for a possible *third* maximum, but none was found (<3% correlation).

In November 1953 Jennison and Das Gupta (1953) sent a note to *Nature* in which they presented the evidence that Cyg A was a double source. The details of the visibility curve and their data with north–south baselines allowed them to present reasonable estimates of the component sizes (largely derived from the exact depth of the first minimum), as well as their separation (from the position of the first maximum) (Fig. 14.20). There remained one important ambiguity, which arose from the intensity interferometer's forfeiture of any phase information: they could not know whether the phase of the secondary maximum was the same as that of the primary. If it were, then Cyg A had in fact *three* components (with a total extent of ~5'), whereas if the phase reversed, then the double model (2.3' size) was valid. They argued for the latter on the basis of Occam's Razor, the lack of a third maximum, hints in data taken at other position angles, and the feeling that the radio source's extent should be minimized, since it was already too large with respect to the optical identification – recall that Baade and Minkowski gave a size of only 0.5' for their colliding galaxies. But the size problem was less of a quandary than the assertion that the radio emission came wholly from outside the optical object. As Jennison (1976:28–9T) recalled:

> It was remarkable that there was absolutely nothing [optically] where the real signals were coming from. Baade and Minkowski's thing was coinciding beautifully with our "centre of gravity," but [the radio radiation] was not coming from that.

The possibility that the Milky Way's interstellar medium was somehow distorting the optical image of Cyg A (which does have a low galactic latitude) was seriously considered (Jennison 1954:128–32), but in the end this seemed unlikely.

One might have expected the *anti*correlation of radio and optical radiation to have shaken confidence in Cyg A's optical identification, but this did not happen. The authority of Baade and Minkowski and the fascination with their case of a galaxian collision was such that it survived unscathed.[103,104]

---

[102] Modern measurements of Cyg A indicate that its position angle is 112°, coincidentally very close to the orientation of Jennison and Das Gupta's baselines. Note, however, that all of Jennison and Das Gupta's models (as in Fig. 14.20) were based on an assumed position angle for Cyg A of 90°.

[103] By the early 1960s it would become apparent that Cyg A's surprising double structure was in fact not unusual for galaxies with excessive radio emission.

[104] The theses of Jennison (1954) and Das Gupta (1954), the belated, full publication of their results (Jennison and Das Gupta 1956a,b), and interviews with Hanbury Brown

14.5.2 Results from Cambridge and Sydney

At the Cavendish Laboratory in 1952 Smith, having measured accurate positions for Cyg A and Cas A the previous year and finished off his thesis, tackled the problem of finding accurate angular sizes for these same sources. He had originally written to Baade that the Cambridge data showed that the major sources were all less than 1′ in size, but now, after reconsideration, felt that 4′ was a better upper limit and that observations specifically designed for angular sizes were needed.[105] As before, he worked at the relatively high frequency of 210 MHz in order to increase angular resolution. The instability of electronics at such a frequency, however, meant that there was a severe problem in calibrating data measured days or weeks apart on different baselines. Despite this, he also knew that high accuracy was paramount for each measured point on the visibility curve, which, because of the Grange Road site, was limited to baselines of 700 m (Fig. 8.3). His colleague John Baldwin (1972:14A:470) recalled the situation from a new postgraduate's point of view:

> The extreme difficulty of trying to get our first source distributions was that on the site you could only get [400] wavelengths at the higher frequencies, such as 200 MHz. But making the first-stage pre-amplifiers function at 200 MHz and producing a reasonable noise factor was incredibly difficult. It only *just* lent itself to a rational procedure. I remember spending many hours sitting and watching Martin [Ryle] fiddling with these, trying to make them go with the large number of parameters to adjust, and occasionally with extreme irritation, even throwing them against the wall!

The final solution to these problems was a scheme involving a single, portable antenna (an array of dipoles) that was positioned at the desired distance, then quickly hooked up in succession to each of two fixed antennas (the two Würzburg dishes of Fig. 14.12) as a radio star transited through their beams. In this manner one could accurately measure the *ratio* of two fringe visibilities. A series of such ratios that included at least one very short baseline could then be used to construct a visibility curve.

Smith took such data beginning in the spring of 1952 for both Cas A and Cyg A and in August sent a detailed paper, especially on his techniques, to the *Proc. of the Physical Society* (Smith 1952b). For Cas A the visibility curve steadily fell to an amplitude of 0.5 at a baseline of 400 wavelengths and implied an east–west diameter of 5.5′ (for a uniformly bright disk), agreeing well with the extent of the optical filaments found by Baade and Minkowski (Fig. 14.14). On the other hand Cyg A, for which the points fell to an amplitude of only 0.8 (filled circles on the visibility curve of Fig. 14.20), exhibited an east–west diameter of 3.6′, much larger than that of the colliding galaxies. To Smith and Ryle this disagreement weakened the likelihood of the proposed galaxies being the correct identification, and Baade too was puzzled when he received the news.[106] These results also contradicted Australian inferences (based on the high degree of ionospheric scintillations) that Cyg A was <1.5′ (Stanley and Slee 1950)[107] and "much less than 3′" (Mills and Thomas 1951).

Meanwhile in Sydney, Bolton's group carried out a survey over 1951–53 for extended sources using their sea-cliff interferometer in conjunction with a Michelson interferometer on the cliff top (Bolton, Westfold, Stanley, and Slee 1954). This was called an "azimuth interferometer" and simultaneously produced two sets of fringes from a source's movement through lobes perpendicular to each other. The key novelty of the survey, however, was that very small baselines were used, yielding fringes with azimuth spacings from 3° to 14°. In this way Bolton's group discovered about ten sources of width greater than 1°. These included a source close to the accepted position of the galactic center, Puppis A (Section 14.4.3), and a 2° wide source that overlapped Cen A's position ("it would be a remarkable coincidence if these two bright sources had no physical connection"). This type of large source was virtually undetectable by most other interferometers, with their fringe spacings of 1° or less.

(1976) and Jennison (1976) have been invaluable in this section. Jennison has also written an account of these and following experiments in "High resolution imaging forty years ago," pp. 337–41 in *Very High Angular Resolution Imaging*, J. G Robertson and W. J. Tango (eds.) (1994, Dordrecht:Kluwer).

[105] Smith:Baade, 22 August and 14 December 1951, LOV.

[106] Ryle:Oort, 15 August 1952, box 1/2, RYL; Baade:Smith, 6 August 1952, file 4 (uncat.), RYL.

[107] See note 32.

Mills (1952c, 1953) was also carrying out detailed experiments to measure small angular sizes. He decided to build an interferometer that could distinguish whether or not any sources were smaller than 1′, which he viewed as a convenient demarcation between stellar and nebular objects. This dictated 10 km baselines, too long to lay cable connections, so he used a radio link. His otherwise conventional interferometer thus consisted of a battery-operated, portable receiver and small antenna (dual Yagi) connected by radio link back to one of the large antennas at Badgery's Creek (Fig. 14.7). The 101 MHz signal received at the remote station was converted to 195 MHz before transmission on the radio link, application of a compensating delay, and mixing with the home-station signal. Mills used mercury-filled acoustic delay lines, a familiar technology in the Radiophysics Laboratory because of ongoing work there on one of the first digital computers.[108] Special procedures, such as a very fast (2 kHz) phase switch, were necessitated by fringes with periods as short as five seconds.

For over a year Mills hauled his equipment around to a total of eleven different sites, mostly to the east or west, but with a few baselines close to north–south in order to check on the shapes of the radio brightness distributions. For three strong sources Mills (1953) derived elliptically-shaped models, presented in the form of "radio pictures" alongside photographs of their optical identifications (Fig. 14.16). Tau A's dimensions of $5.5 \times 3.5′$ and specific orientation agreed very well with that of the Crab nebula. Vir A's size of $5 \times 2.5′$ agreed approximately with that of the galaxy M87, but the radio shape was not the same, nor did the radio orientation agree with the inner optical jet. Cen A had the most problematic visibility curve, which did not fall off in a monotonic manner but had at least two bumps. The Fourier transform of this produced a brightness distribution with at least one minimum, but Mills cautioned that his inability to measure fringe phases (because of the radio link) led to serious ambiguities in any derived model. All he could be sure about was that about half of the radiation came from a region of ~5′ diameter and the rest from a much larger ~90′ region (as Bolton had also found). The final source was Cyg A, for which useful north–south data were precluded by the extreme northerly location of the source in Mills's sky. His east–west visibility curve (points shown in Fig. 14.20) yielded an effective size of 0.8′, not much larger than the colliding galaxies. In retrospect we can see that he missed the double nature of Cyg A because he had the bad luck not to choose a baseline in the critical interval from 500 to 1800 wavelengths where the curve minimum occurs.

As it happened, the fresh results on angular sizes from the three major groups converged in 1952 on the URSI General Assembly in Sydney (Section 7.6.1), which then served as a spur to further efforts. Smith reported his 3.6′ east–west size for Cyg A during the same session on 19 August in which Mills showed preliminary results that it was only about 1′ across. Hanbury Brown then reported Cyg A to be < 0.6′ in north–south extent, based on the first intensity interferometer observations (which had taken place only after he had left England and were communicated to him via cable). The apparent discrepancies were alarming and the principals tried to sort them out – could they originate in a spectrum varying across the source (since the data were taken at frequencies ranging over a factor of two), or in a complex source brightness distribution? Various checks were agreed upon and the idea arose of simultaneously sending short notes to *Nature*. Ryle and Smith were at first reluctant since their data had already been submitted for publication. But in the end, after a flurry of correspondence around the world, the Cambridge group did participate in a trilogy that was published at Christmastime – a unique event in the history of these three major groups. The adjacent notes (Hanbury Brown, Jennison, and Das Gupta 1952, Mills 1952c, Smith 1952c) disagreed in detail, but their unmistakable message to the reader was that the strongest radio stars had finite angular sizes and did not seem starlike.

14.6 RADIO STARS OR RADIO NEBULAE?

The above was the provocative title of a short note in *Observatory* by Australian radio physicist Ronald Bracewell (1952). Bracewell had spent three years working on ionospheric problems in Ratcliffe's group at Cambridge and was therefore well familiar with Ryle's work. Upon his return to the Radiophysics Laboratory he was astounded to find how little each group knew about the other's work (Section 7.6.1). He was struck

---

[108] See note 7.18.

in particular by major differences in interpretation of radio sources, which in Cambridge were taken as true, albeit strange stars and in Sydney as a mixture of galactic and extragalactic sources, of small and large angular extent. Indeed, in his note Bracewell raised the question as to whether *all* the sources might become "radio nebulae" if examined with enough resolution.

How did the major groups come to hold such disparate ideas about the nature of the radio stars? The answer lies in an amalgam of effects that include instrumentation, quirks in the northern and southern skies, and theoretical predilections. First, the antennas of each group differed in important ways. Cambridge's "Long Michelson" interferometer, with its baseline of 110 wavelengths, was the longest of any of the early survey instruments. This was a result of Ryle's (1950, 1952) emphasis on the advantages of a long-baseline, phase-switched interferometer in eliminating both the large and small angular structure in the intense galactic background radiation. Ryle favored the idea of true radio *stars* and therefore designed an instrument optimized to find entities unambiguously small in extent. The 1C list did not catalog large objects, which were suspected to represent nothing more than structure in the general Milky Way background. As Ryle explained at the time:

> It is very easy [with large dishes] to obtain maxima in the received flux which appear to be due to a source of small angular diameter, but which in reality are due to the angular variations of the general structure of the Galaxy. For this reason, we have always used interference systems of considerable resolving power to discriminate between "point" sources (i.e., sources having a diameter of less than 5–10 minutes of arc) and the general background radiation from the Galaxy.[109]

There was also no need to observe at more than one interferometric baseline because such a "point" source would have the same fringe visibility at all baselines. This is an excellent example of what Pickering[110] and Galison (1987:252, 256) have separately studied in other contexts, namely how methodology and instruments can be shaped by a researcher to "tune in" on phenomena consistent with his scientific commitments. Galison has also emphasized how the task of removing the background (in his case, in high energy physics experiments) is one and the same with identifying the foreground.[111]

In contrast, baselines in Australia tended to be shorter (50 wavelengths for Stanley and Slee [1950] and for Bolton, Stanley, and Slee [1954]) and two major surveys involved very short baselines (20 wavelengths for Mills [1952a] and 4 to 20 wavelengths for Bolton *et al.* [1954]). The short baselines, and in particular comparisons of a source's record on different baselines, gave new information:

> In a sense, as soon as we switched on our close-spaced interferometer, we realized that we had [sources of large angular extent]. It was the first thing that hit us in the eye. We'd look at the same part of the sky, first on a close spacing and see an enormous interference pattern and then on a wide spacing a small one. ... This I think was a source of a lot of the disagreement between Cambridge and us.
>
> (Mills 1976:31,33T)

Note, however, that especially away from the galactic plane, apparent extended sources in some cases could also be misleading – a blend of several discrete sources or a feature in galactic structure.

At Jodrell Bank, the situation was yet different, for surveys there were done with the giant paraboloid, whose response to the sky was roughly equivalent to that from a small dish plus a sum of interferometers having all spacings less than 35 wavelengths. Thus it too was sensitive to larger-size discrete sources, which turned out to be found mainly in the galactic plane. Hanbury Brown's later comments mixed up technical and personal issues:

> When you laid these beautiful traces through the Galaxy out in a long line, you could see these big sources. ... And so we noticed that there was a concentration of the more powerful sources into

---

[109] Ryle:E.A. Tandberg-Hanssen, 10 February 1950, box 1/1, RYL; also Elsmore (1976:62A:390) and Smith (1984:244).

[110] A. Pickering, "The hunting of the quark," *Isis* **72**, 216–36 (1981).

[111] Galison relates an apt apocryphal story. When Michelangelo was asked how he had carved his masterpiece *David*, he replied that it was simple – all one needed to do was to remove everything that was not *David*. Art meets science.

the galactic plane. I remember a visit to Jodrell Bank by a Cambridge team once in the early days when we pointed this out.... They looked at these things and said, "Well, this phenomenon simply doesn't exist! These sources must be sidelobes or something."... Much to my astonishment and dismay Ryle flatly refused to believe that the sources on our records were real. Apparently his interferometer in Cambridge didn't show them and as far as he was concerned that was that.... From [wartime] days... I knew Martin to be a capable radio engineer, but I hadn't realized how difficult he could be if you challenged one of his pet ideas.

(Hanbury Brown 1976:53T, 1991:101)

I shall never forget the strenuous arguments we had with those dedicated interferometrophiles.... It was part of the conventional wisdom at Jodrell Bank that Cambridge had only three standard reactions to our work: (1) "It is wrong," (2) "We have done it before," or (3) "It is irrelevant."

(Hanbury Brown 1984:220)

A peculiarly astronomical reason also contributed to some of the differences between the Australians and English, namely that they were observing from different hemispheres. The single most important source, Cyg A, was much more easily observed from England than from Australia, where it barely rose. On the other hand, the most varied and central parts of the Milky Way crossed near Sydney's zenith, but were too far south from England for either the Jodrell Bank dish or the Cambridge interferometer, which were confined to a relatively small part of the outer Galaxy. This enabled Mills to sample sources much better all along the galactic plane and to be better able to contrast their properties with those far from the plane. It also turned out that the Southern Hemisphere by chance had more extended strong sources away from the galactic plane (for example, Cen A and For A) and two relatively easy (bright) extragalactic optical identifications (Cen A and Vir A). In contrast, the two strongest northern sources were located in the galactic plane and had small angular sizes (Cas A and Cyg A). The conspiracy of these circumstances tended to make a northern observer less conscious of extended sources and of extragalactic possibilities, although in the case of Jodrell Bank the instrumental factor discussed above weighed heavier.

Other factors leading to controversy were the great physical separation between the Australian and English groups (discussed in Section 7.6.1) and the psychological isolation of the Cambridge group (Section 8.5.2). Discrepant results could not be negotiated by frequent personal contact at scientific meetings (in the case of the Australians) or through site visits designed to dig into the peculiarities of apparatus and the data produced. As Collins has emphasized, new scientific results do not derive solely from equipment, but also from the investigator.[112] Especially for controversial results, the degree to which one knows and/or trusts the investigator producing them (largely based on his previous results) plays an enormous role in gauging their credibility. The idea that results in these circumstances could have been checked by simply "duplicating" the experiment utterly fails in the present case for three reasons. The first was that the rival investigator, a vital part of the experiment, could not be (nor was wanted to be!) duplicated. Collins has called this "the experimenter's regress," namely that the integrity and significance of an experiment (or observation) can only be evaluated by examining the results of that experiment itself. A "good" experiment is one that produces "good" results, but these are only known to be good by virtue of their emergence from a good experiment.

The second reason was that rivals often had little respect for the data issuing forth from each other's types of antennas – the Michelson interferometer, the sea-cliff interferometer, the intensity interferometer, and the single reflector. There was no agreement on the criteria for evaluations of instruments. A strikingly similar episode over a century earlier (involving Michael Faraday and William Sturgeon arguing over the meanings of electrical measurements made by various instruments) has been studied by Morus.[113]

The third reason for the lack of closure on the radio source question was that areas of tacit technical knowledge, often vital, were not shared because of secrecy or lack of personal contact. The radio equipment built by

---

[112] H. M. Collins, "When do scientists prefer to vary their experiments?," *Stud. Hist. Phil. Sci.* **15**, 169–74 (1984); *Changing Order: Replication and Induction in Scientific Practice* (1985, London: SAGE Publications), chapter 3.

[113] Morus, I. R. (1988), "The sociology of sparks: an episode in the history and meaning of electricity," *Soc. Stud. Sci.* **18**, 387–417.

these groups was highly sophisticated, constantly being changed, and often obstreperous. Any interpretation of data produced by such equipment required an intimate sensibility to all of its problems:

> There is a major difficulty in interpretation of records where several sources produce interfering patterns. We at present use common sense and low cunning, but there may be better methods.
>
> (Pawsey)[114]

> An interference pattern is something which is terribly real and terribly false. The sum of n sine waves is still a sine wave.
>
> (Bolton 1978:44T)

Terminology was another interesting aspect of the debate over the nature of the radio sources. "Radio star" was used almost universally until about 1950, but over the next three years all of the leading researchers except the Cavendish group switched to other terms, usually "radio source" or "discrete source." By 1953, with still *none* of the optically identified sources coming up as a star, many such as Pawsey (1953:150) were calling "radio star" a misnomer: "Radio astronomy has thereby lost a graphic term; 'radio star' must be replaced by 'radio nebula'."[115] The Cambridge group's continued use of "radio star" was a reflection of their preferred interpretation for the bulk of the sources, and may well have contributed to their holding stalwart views longer than otherwise. In this respect the situation reinforces conclusions that Lesch reached regarding the discovery of the alkaloids by early nineteenth century chemists:

> Words and things are not so neatly separable, and the process of naming and classifying plays a more integral and dynamic role in research than is implied by the three-step progress from facts through taxonomy to theory. The positing of names and definitions cannot be understood as the passive and mechanical affixing of labels to things and the arrangement of the things in groups after the fact of their discovery. It is much more

fundamental in the sense that it constitutes the categories through which nature is perceived, and thereby tends to exclude from perception that for which no categories exist.[116]

Note, however, that Ryle's group continued with the term "radio star" even well beyond the time in 1954–55 when they dramatically changed their stance and forcefully argued that the great bulk of the radio sources were distant Cyg A-like objects (Sullivan 1990). Ryle's student Peter Scheuer explained his view of this in 1953:

> The continued use of the term "radio star" perhaps needs some justification. The term is used here in the same sense as the ancients used the word "star" – merely meaning a bright dot on the heavens – and its use does not pre-judge the nature of the objects. ... For everyday use among practicing radio astronomers (or sourcerers?) "radio stars" is much less cumbersome than phrases such as "discrete sources of radio radiation."[117,118]

14.7 STATUS OF RADIO SOURCES IN 1953

The radio sky of 1953, if not in turmoil, was hardly in a settled state. Over two hundred sources had been listed by one person or another, but agreement between surveys was scant. It was understood that the chief problem in assigning accurate positions and intensities to all but the strongest sources arose from "confusion," that is, the blending effects of sources, and yet everyone vastly underestimated how confounding this could be. Other problems arose from ionospheric scintillations and inadequate signal levels. Despite

---

[114] Pawsey:Reber, 20 March 1950, file A1/3/1a, RPS.

[115] Also see Bolton in *Observatory* **73**, p. 25 (1953); Jennison (1954:101); G. C. McVittie in *Observatory* **72**, p. 185 (1952); E. V. Appleton in *Science* **119**, p.103 (1954). Shklovsky (1953b) also emphasized this point (Section 15.4.2).

[116] J. E. Lesch (1981), "Conceptual change in an empirical science: the discovery of the first alkaloids," *Hist. Stud. Phys. Sci.* **11**, 305–28.

[117] P. A. G. Scheuer, as in note 81. Also Ryle (1976:35T).

[118] In the Ryle papers is a draft of the 1C survey paper (Ryle, Smith, and Elsmore 1950) with various suggestions in the hand of J. A. Ratcliffe (Ryle's supervisor). Ryle's draft used the term "radio star" far more than "radio source," but Ratcliffe (for some reason) favored the former even more. Ryle therefore cut down yet further on his use of the term "radio source" in the finally published paper. [File 30 (uncat.), RYL]

attempts to be as objective as possible in establishing criteria as to what should be called a source and what should be passed by, the early surveys were at best a crude approximation to the radio sky. Evidence in the Ryle papers regarding the formulation of the finally published 1C list, for example, is instructive. There are four successive working lists over a period of four months, and their mutual agreement on the positions and intensities of the radio stars is woeful except for the most intense one-third of the sources. Finally, in August 1950 the last source list crystallized into a table sent to the Royal Astronomical Society – official 1C radio stars had been declared.[119]

Study of the discrete sources fast became the hottest arena in radio astronomy. Despite the many uncertainties, investigators' research strategies could now be more clearly conceived (Edge and Mulkay 1976:110). In particular, the ten or so sources with optical identifications were a fascinating, heterogeneous mix of objects (Table 14.1) and raised all sorts of questions. The lack of trends was so bothersome that in 1952 Ryle set down the following desideratum in his notebook: "We must find 20–30 weak radio star positions with ~3 sec RA, 10′ dec. accuracy, to see if we can find two radio stars tying up with same type of visible object."[120] Indeed, why was there such a mix? As late as 1954 only eight radio sources on an official list could be said to have unassailable optical identifications and therefore to "definitely exist": why were the ranks of optical identifications not enlarging after each new survey appeared? Were most radio sources located in the Galaxy or outside? What was the relationship of the galactic background to the radio sources? Could Cyg A possibly be typical of most sources, implying that radio techniques could probe to distances untouched by even the 200 inch telescope? Whose statistics on source counts should be believed? Although it was reasonably established that at least the brightest radio sources were not stars and that a few of them were peculiar galaxies and peculiar emission nebulae, why were the bulk of peculiar optical objects nevertheless radio-silent? The strongest sources had measured intensities at low frequencies even greater than that of the sun and yet were at incomparably greater distances: how could they emit so much power?

First views on these questions are discussed in the following chapter.

TANGENT 14.1 THE SEA-CLIFF INTERFEROMETER

The basic methodology of the sea-cliff interferometer was to use sequential timing of fringes after rising and before setting to define two (horizon) circles on the celestial sphere at whose intersection the source was located. The mean of the sidereal times of rising t_r and setting t_s gave the source's right ascension. The declination δ came from the source's semi-diurnal arc, that is, the time interval between t_r and t_s: $\tan \delta = \cos (t_s - t_r) \cot \varphi$, where φ is the site latitude. Having observations of both rising and setting was also vital in that errors in one's assumed refraction could be minimized via an iterative technique. This contrasted with the situation at Dover Heights where one had only rising sources and therefore needed another intersecting (declination) circle, obtainable only by observing the *rate* of increase in elevation angle as the source rose. The anticipated rate (for a given source declination) was easily calculable if one accurately knew the experimental parameters such as operating frequency, antenna height above the water (including tides), and refraction at different elevations. This last factor was particularly critical, as incorrect values for the differential refraction between the elevations of successive lobes caused large errors.

Also see Section A.11 and Fig. A.6b.

TANGENT 14.2 INTERSTELLAR DISPERSION AND BURSTS IN 1950

Another study of the time that became moot when it was established that all of the scintillations arose in the ionosphere is nevertheless remarkable in light of what happened almost two decades later. Iosif Shklovsky (1950) pointed out that if a burst from a radio star emitted all frequencies simultaneously, the variable speed of propagation in the interstellar plasma would cause the high frequencies to arrive before the lower ones, allowing one to estimate a distance to the source. Using a published plot by Smith (1950) and an

---

[119] Files 30 and 7 (uncat.), RYL. In *How Experiments End* Galison (1987) has analyzed this problem in great detail – when does the experimenter say "this is enough – time to publish"?

[120] Notebook for September 1952 IAU General Assembly, file 27 (uncat.), RYL.

assumed interstellar electron density of $0.1\,\mathrm{cm^{-3}}$, he then derived an upper limit on the distance to Cyg A of 35 pc (although it appears that he mistook Smith's published data, which were for Cas A, to be for Cyg A). When pulsars were discovered in 1967 by Hewish and collaborators, the group soon realized (independently of Shklovsky) that their distances could best be estimated by just this method.

TANGENT 14.3 EARLY LOG N–LOG S ANALYSIS

For a sample of identical sources uniformly distributed in a certain volume, it can be shown that $\log N \propto -k \log S$, where the slope $k = 1.5$ for a sample in an isotropic distribution, $k = 1.0$ for a distribution in a thin disk with the observer unable to detect the edge, and $k < 1.0$ for a thin disk with the observer able to detect the edge. The value of k is determined by the manner in which the number of detectable sources N increases with distance r from the observer: $k = p/q$, where the value of p comes from $N \propto r^p$ for the region in which sources are detectable, and q is 2 because of the r^{-2} dependence of intensity received from a source.

My own analysis of the published 1C sample (Ryle, Smith, and Elsmore 1950) differs from that of the authors. I find a log N–log S slope of 2.1, with a demarcation similar to Mills's (1952a) for sources at low and high galactic latitudes. The 1C sample also shows an excess of strong sources near the galactic plane (six of seven of the strongest sources are at latitudes $\leqslant 12°$). These kinds of analysis were not presented in the 1C paper, although Ryle wrote once that both intense and weak radio stars in their survey were consistent with an isotropic distribution.[121]

---

[121] Ryle:Oort, 7 June 1951, box 1/2, RYL.

TANGENT 14.4 THE INTENSITY INTERFEROMETER

The intensity interferometer operated with two separated antennas and receivers as with a Michelson interferometer. The latter, however, preserved the phase of each incoming signal's waveform before combining them (Section A.11). The striking departure of the intensity interferometer from standard practice was to forget about the incoming phases and simply detect signal power levels (with a square-law detector). These detected signals, which exhibited their own (relatively slow changing) fluctuations (as in Fig. A.3d), were then the key to extracting information about the source. If two such signals, measured by separated receivers, were multiplied and suitably normalized, Hanbury Brown and Twiss (1954) showed that, despite the lack of any interferometric fringes, the resultant was in fact the square of the fringe visibility's amplitude as measured with a conventional interferometer. One correlated not the source signal itself, but essentially its noise-like modulations. As with a Michelson interferometer, there would be less correlation for a source of given angular size as one went to longer baselines.

A full account of the theory and astronomical results of the intensity interferometer, especially as used at optical wavelengths, can be found in Hanbury Brown's *The Intensity Interferometer* (1974, London: Taylor and Francis). Also see Hanbury Brown's (1991) autobiography. Today the field of quantum optics views Hanbury Brown and Twiss's work as fundamental and the "Hanbury Brown–Twiss effect" is regularly exploited in physics experiments using not only photons, but also particles undergoing high-energy collisions[122].

---

[122] For an interesting discussion of both the history and the physics see D. Kleppner (2008), "Hanbury Brown's steamroller," *Physics Today* **61**, 8–9.

15 • Theories of galactic noise

Radio astronomy began with Jansky's discovery in the early 1930s of radio noise from the Milky Way, but even twenty years later there was no consensus as to its cause. Around 1950 most people favored the idea that hot interstellar gas was ubiquitously emitting free–free radiation, but this raised several basic problems. The required temperatures of 100,000 K or more were far higher than admitted by other lines of evidence; nor did the observed spectrum well match expectations. Moreover, although galactic noise followed the general outlines of the Galaxy, it did not show extreme concentration to the galactic plane as did interstellar gas known from optical observations.

These problems led to proposals that the galactic background radiation was not at all an interstellar phenomenon, but rather the integrated effect of countless radio stars. This approach conflated the separate puzzles of radio stars and galactic noise, and discernible radio stars simply became the most intense examples of a vastly greater number. The situation was analogous to that at visual wavelengths where one saw bright stars superimposed on a general haze of emission from the combined effect of faint stars. But at optical wavelengths the Milky Way was negligibly faint compared to the nearest star (the sun), while at the longest radio wavelengths its intensity dwarfed that of the sun (Figs. 1.1 and A.5). As Martin Ryle put it in 1949:

> We are therefore faced with the remarkable fact that at radio wavelengths starlight [galactic noise] is normally far more intense than sunlight.[1]

Despite the importance of radio stars, their character remained mysterious. Moreover, what little *was* known lent no credence to their stellar nature – in particular, none of them could be associated with optical stars and the few successful optical identifications were with various peculiar nebulae and external galaxies (Section 14.4). It was perhaps understandable that models of galactic noise based on these radio stars were of a bewildering variety.

In the Soviet Union, meanwhile, an entirely different tack was followed. The radiation was taken to arise from high-energy cosmic ray electrons rather than quiescent gas heated by stars. Although this idea had been first proposed in 1950 in the West, it lay dormant there for years even as its ramifications were eagerly worked out in the Soviet Union. The resultant theoretical papers in Russian were little read or heeded by Western radio astronomers until it became increasingly apparent by the mid-1950s that this so-called synchrotron radiation worked very well for most discrete sources, as well as for the galactic background. Radio astronomy and cosmic rays became intimately linked, and high-energy processes became more important in astronomy.

This chapter discusses the three main explanations for galactic noise during the pre-1954 era: hot gas, the combined effect of innumerable radio stars, and synchrotron radiation. Debates over the nature of the radio stars are also covered, as well as the introduction of the synchrotron mechanism and the marked USSR–West dichotomy in its development and acceptance. Models of the Galaxy also led to the first applications of radio data to testing models of the cosmos as a whole.

15.1 EARLY SURVEYS

The first major surveys have been discussed at length in previous chapters – the most important were Jansky's (1935) at 20 MHz (Fig. 3.12), Reber's (1944; 1948b) at 160 and at 480 MHz (Fig. 4.6), and that by Hey's group at 64 MHz (Fig. 6.4; Hey, Parsons, and Phillips 1948a).[2]

---

[1] *Observatory* **69**, p. 86 (RAS meeting of 8 April 1949).

[2] Other useful but less complete data were added by Jansky (1937; Section 3.3.4), Friis and Feldman (1937; Section 3.5.1),

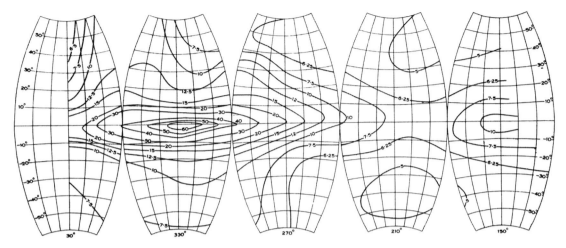

Figure 15.1 The 100 MHz survey of Bolton and Westfold (1950a) plotted in (old-style) galactic coordinates (the galactic center is at a longitude of ~330°); the polar caps are not shown. This was the survey most used as a basis for early models of galactic noise. The contours are labelled in units of 100 K in brightness temperature and the effective resolution, after beam restoration, was considered to be markedly improved over the antenna beam size of 17°.

These studies established that the intensity of the background radiation was much stronger at the lower frequencies, although calibration difficulties led to large uncertainties. In a relation of the form brightness temperature $T_b \propto \nu^{-\alpha}$ (where ν is frequency), reported values for the spectral index α ranged from 2 to 3. Many analyses adopted the value of 2.4 obtained at the US National Bureau of Standards by Jack W. Herbstreit and Joseph R. Johler (1948), who were the only ones at that time to observe the galactic noise at multiple frequencies (25 and 110 MHz) with well-calibrated broadbeam antennas.

The crude resolution (5–20°) of the early surveys could not hide a concentration to the galactic plane and to the galactic center, nor several secondary maxima, in particular toward Cygnus. Reber (1944, 1948b; also see Fig. 4.9) had interpreted these secondary maxima as indicating "projections" from the Galaxy, perhaps similar to the spiral arms seen in other galaxies and suspected to exist in our own. This notion was also used by John Bolton and Kevin Westfold (1950a, b) for their own 100 MHz survey of the southern sky (Fig. 15.1), a survey that became the standard for several years. Their nine-Yagi antenna had a beamwidth of 17°, but angular resolution was enhanced in their final map through an iterative procedure to mitigate beam-smoothing, similar to that used earlier by Hey, Parsons, and Phillips (1948a; Section 6.1).

15.2 FIRST THEORIES

15.2.1 Hot interstellar gas

As Jansky had pointed out from the beginning, one had the choice of ascribing the galactic noise to processes taking place either between the stars or at their surfaces. Up until 1950 most persons favored the interstellar explanation for, despite its evident problems, it dealt with a known astronomical process. Reber (1940a; Section 4.3) was the first to suggest that *bremsstrahlung*, or free–free radiation, might well explain his "cosmic static." The electrons and positive ions in an ionized gas constantly underwent near-collisions that led to emission of photons with a broad range of frequencies; the physics of this emission had been largely worked out by the time of World War II and its intensity at any given frequency depended on the temperature, density, and thickness of the gas (see Section A.5.1). We discuss elsewhere the first extensions of free–free theory from optical to radio wavelengths by Henyey and Keenan (1940; Section 4.3), Hendrik C. Van de Hulst (1945; Section 16.1), Unsöld (1946; Section 10.6.1), and Shklovsky (1946, 1947; Section 10.3.1). All but the last

---

Fränz (1942; Fig. 6.10), and Sander (1946) and Moxon (1946) (Tangent 6.2).

368 Theories of galactic noise

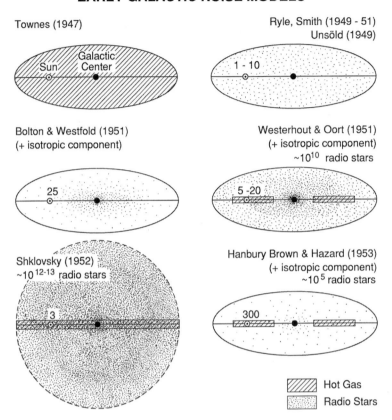

Figure 15.2 Schematic diagrams of early models of galactic noise as arising from hot interstellar gas by free–free emission and/or the integrated effect of a distribution of radio stars. In each diagram the Galaxy is viewed from the exterior and the indicated Galaxy "boundary" is arbitrary. The surface density of drawn dots is roughly proportional to the logarithm of each model's space density of radio stars (on a scale common to all diagrams); the number written near the sun's position in each diagram is the average spacing of radio stars (pc) in the solar neighborhood. The Westerhout and Oort model and the Shklovsky model have a range of dot sizes to indicate a dispersion in radio star luminosities.

of these studies agreed that Reber's observed intensities at 160 MHz jibed well with calculations for the interstellar medium of the day, which had a 10,000 K temperature and a density of 1 electron cm^{-3}. But they also agreed that Jansky's 20 MHz intensity corresponded to a brightness temperature of at least 100,000 K, unacceptable since it was thermodynamically impossible to observe a brightness temperature higher than a gas's (kinetic) temperature. Theorists had three options regarding such a serious discrepancy: ignore it, discard at least some of the data, or discard the theory. No one chose the last alternative; Henyey and Keenan offered no solution; Van de Hulst opined that Jansky's data must be somehow wrong; and Unsöld concluded

that interstellar regions must indeed be at temperatures of ~100,000 K (after all, he reasoned, had not the solar corona recently been found to be at an even higher temperature (Section 13.3.1), previously thought absurd?). Shklovsky did not even mention Jansky's data in his first studies, and also made an important error in the physical theory of free–free emission,[3] which led him to argue that only a small fraction of space contains free electrons.

After the war there were a few further studies along these lines. Richard Woolley (1947b), also never citing Jansky's data, found the radio emission most

---

[3] See Tangent 10.1.

likely to come from dense 10,000 K regions near hot stars. Albrecht Unsöld, noting the similarity in general shape of the galactic noise contours available at several frequencies, concluded that the interstellar gas was optically thin for all frequencies >40 MHz, which then implied that its temperature was at least 200,000 K (Unsöld 1947). Ralph Williamson (1950) continued this line of reasoning, finding that the width in galactic latitude of the radio emission varied only slowly from frequency to frequency. This then implied that the gas, assumed to be in a layer coincident with the galactic plane, had an optical depth varying as $\nu^{-0.7}$, not $\nu^{-2.0}$ as predicted by free–free theory. Finally, the American physicist Charles H. Townes had started thinking about galactic noise already during the war during breaks from developing radar bomb sights at Bell Labs. Immediately after the war he went to great pains, largely through correspondence and personal consultation, to secure the best possible "post hoc calibrations" for all available data. He ended up with six brightness temperatures defining a spectrum for the galactic center region. These varied in a monotonic manner from 120,000 K at 10 MHz (Friis and Feldman 1937) to <30 K at 30,000 MHz (Dicke *et al.* 1946), yielding a spectral index of 2.0. Townes (1947) then inferred from free–free radiation theory that the interstellar gas had a temperature of 150,000 K and a density of 1 cm^3, uniformly distributed in an ellipsoidal Galaxy (Fig. 15.2).

15.2.2 Combined effect of radio stars

Even at the time of their first solar burst observations in late 1945, the Australians had suggested that galactic noise might well represent the combined effect of countless "similar bursts of radiation from the stars" (Pawsey, Payne-Scott, and McCready 1946). But not until the discovery of numerous radio stars did Unsöld and Ryle become the first (in 1949, independently) to abandon the interstellar gas altogether and instead develop the idea that galactic noise (as well as cosmic rays) arose from the integrated effect of all radio stars.[4]

Within two or three years this picture was to dominate in the West.

In late 1948 Unsöld (1949a, b) completely changed his mind on the origin of galactic noise because (1) free–free radiation obviously failed to explain the new-found radio stars, (2) he could not understand how an interstellar gas as hot as 100,000 K would not quickly cool off, (3) photographs showed that ionized interstellar gas (unlike galactic noise) was very closely confined to the galactic plane, and (4) free–free theory, which allowed the spectral index to range only from 0 to 2.0, could not be reconciled with the observed spectral index. But from the time of Reber (1944:287) it had also been appreciated that the ratio of radio to optical intensity received from the Milky Way was fully 10^9–10^{13} times that same ratio for the (quiet) sun. This ruled out any explanation of galactic noise as simply the summation of innumerable *sun-like* stars. Unsöld therefore supposed that radio stars had extraordinary radio output and negligible optical brightness. It would suffice if the *entire* surface of a star could somehow emit radio waves with an efficiency comparable to that observed from sunspot regions during the most powerful of solar bursts (brightness temperatures of ~10^{13} K). Indeed, the radio spectrum of solar bursts (but not that of the quiet sun – see Figs. A.2 and A.5) agreed well with that of galactic noise. He further noted that there were several types of common, low mass stars known to exhibit much greater changes in their optical spectra than did the sun – could not these be accompanied by copious amounts of radio? Moreover, the distribution on the sky of these active stars reasonably matched that of the galactic noise.

Unsöld also liked his scheme (Fig. 15.2) because these same radio stars could also solve the puzzle of the origin of the high-energy cosmic ray particles that rained down upon the earth (Section 15.4). It had recently been discovered that events associated with sunspots could produce cosmic rays (as well as radio waves), and Unsöld pointed out that these temporary increases in cosmic rays from the sun, when compared to the steady flux from the entire sky, were similar in magnitude to radio solar events compared to galactic radio emission. Moreover, the (everyday) ratio of solar cosmic rays to all cosmic rays was within a factor of 100 of the same ratio for radio waves, a difference that could be explained by the confining effects on cosmic ray particles of an interstellar magnetic field. Amidst the usual

---

[4] At a meeting, Fred Hoyle also briefly discussed his "emotional belief" (as he put it) that galactic noise originated in enormous flares from a total of 20–30 stars, but he never further pursued this idea in print. [*Observatory* **68**, 16 (1948), report of RAS meeting of 9 January 1948]

dry prose of a scientific article Unsöld (1949b:194) revealed his delight with this new-found connection between galactic radio noise and cosmic rays:

> It does not appear to us to be correct, as one has always unconsciously done, to connect cosmic rays with the *visual* radiation from the heavens. Rather, one should imagine how a creature with "radio eyes" would observe the world!

Meanwhile at Cambridge, Ryle too was attracted by the economy of one mechanism tying together three phenomena. He was mainly motivated by the spectrum problems, as well as the existence of radio stars (which clearly could not be due to free–free radiation). It was also suggestive that the intensity ratio of the strongest radio stars to the total galactic noise was ~1%, very similar to the brightest visual stars as a fraction of the Milky Way's light. Although Ryle's notes indicate that he favored this general approach from as early as 1946,[5] his first relevant article was not until three years later when he proposed that radio stars could well account for both cosmic rays and galactic noise:

> It therefore appears that a type of star which has so far not been classified in astronomy at visual wavelengths may be responsible for two major astrophysical phenomena – the origin of cosmic rays and of the intense radio waves from the Galaxy.
>
> (Ryle 1949b:498)

It had earlier been speculated that cosmic rays could be accelerated to their high energies in the atmospheres of certain known types of stars, especially those with strong and variable magnetic fields; Ryle now suggested that the new radio stars also looked ideal for the job. Using intensity variations of Cyg A on a timescale of 20 sec to argue that radio stars were truly stellar in size (Section 14.2), Ryle proposed that this new class of star must emit free–free radio radiation from an envelope at the astounding temperature of ~10^{14} K. Escaped thermal particles from such a hot gas might well represent the genesis of observed cosmic rays.

Ryle did not carry the cosmic ray work further,[6] but once he and Graham Smith had preliminary results in hand for what was to become the 1C survey of radio stars (Section 14.3.1), they persuasively extended their model for galactic noise. If the galactic noise arose from a population of sources (both those observed discretely and those too weak to observe individually), one could derive their space distribution and intrinsic radio luminosities. This was done through an assumption for the depth of the emitting region combined with a comparison of the numbers of radio stars at various intensity levels (the so-called $\log N - \log S$ plot – see Section 14.3.2) with the observed brightness of the galactic noise.[7]

Calculations along this line indicated, for an assumed average extent of the Galaxy of 1000 pc (in the limited northern sky covered by the 1C survey), that the radio stars were separated by distances not unlike those of ordinary (visual) stars. Ryle (1950:241) and Smith originally found a typical separation of ~0.7 pc using a set of 25 sources from an early version of the 1C survey, but Smith (1951c:96) later derived ~10 pc using the final survey of 50 sources and a more detailed model[8] (see Fig. 15.2 for a schematic drawing of their model, as well as all others discussed). Because none of the 1C sources coincided with a bright star, radio stars, although far more radio powerful than the sun, were somehow visually dark. Ryle was particularly impressed that the radio stars now became common, albeit mysterious objects. As he wrote to Jan Oort in Holland:

> We feel the result to be of importance since the only previous published remarks (by Bolton and by Pawsey) have suggested that the sources are very unusual bodies, and that "a few thousands of such sources suitably distributed in space could account for all cosmic noise."[9]

---

[5] Ryle worked out that the galactic noise could arise from the integrated effect of free–free radiation from large (~10^4 astronomical units), hot (~10^8 K) regions surrounding ordinary stars. ["Notes on the origin of extra-terrestrial noise," 9 April 1946 (vws 29), 6 pp., file 6 (uncat.), RYL]

[6] At one point, however, Ryle was working with a student, T. E. Cranshaw, on the experimental possibilities of searching for discrete sources of cosmic rays in the sky (say with an angular resolution of 5°) and with the hope of finding coincidences with radio stars. [Ryle:Royal Society (regarding a fellowship application), 9 June 1950, box 1/1, RYL]

[7] See Tangent 15.1 for details.

[8] Although the value of ~0.7 pc for typical radio star spacing was also quoted by Ryle, Smith, and Elsmore (1950), it was soon changed to ~10 pc when a major error in scaling was discovered. [Ryle:Mills, cited by Mills (1952a:276)]

[9] Ryle:Oort, 9 May 1950 (BPS#4), box 1/1, RYL.

A different sort of analysis was carried out by Bolton and Westfold (1951) based on their 100 MHz survey of galactic noise (Fig. 15.1). Their main assumption was explicit in their first paragraph:

> We adopt the hypothesis that the background galactic radiation is the aggregate emission from the distribution of a new type of star, which is typified by the discrete sources. ... It [is] difficult to speculate profitably on the optical type of the discrete sources and their means of energy generation. At this stage we shall refer to the new type stars as *radio stars*, endowing them with the property of high radio and low optical emission.
> (Bolton and Westfold 1951:476, italics in original)

After showing that interstellar free–free radiation was a negligible contributor to their survey's emission, Bolton and Westfold set up a mathematical model to predict the run of intensity along the galactic plane. Assuming that the radio stars were arranged axisymmetrically about the galactic center, they found their numbers to be peaked at the center and steadily falling off to the 8 kpc radius of the sun (Fig. 15.2). In a similar manner to Ryle and Smith, they then derived that the radio stars probably had typical separations of ~25 pc and radio luminosities about 10^{14} times that of the sun. Bolton and Westfold also found that the intensity measured at the galactic poles was simply too high to fit comfortably into any of their models. Their solution was to propose that a second component existed, an essentially isotropic background of unknown origin and distance, from anywhere just outside the solar system to the farthest extragalactic regions!

Shortly after their analysis, another theoretical study bloomed in the completely different soil of Leiden Observatory in Holland. Since the end of World War II its director, the distinguished astronomer Jan H. Oort, had been excited by radio astronomy and trying to get radio observations going in a nation with essentially no radio researchers (Chapter 16). His first personal contact with a major radio group came in September 1949 when he discussed the problems of radio stars and galactic noise during a visit to Cambridge. Six months later Ryle sent Oort his analysis concluding that the radio stars were nearby and represented a new type of star. The Dutchman was intrigued by these postulated radio stars, but still felt that galactic noise was best explained by interstellar gas. Oort, who had been *the* world expert on Milky Way structure and dynamics for almost thirty years,[10] was also dismayed by the simplicity of Ryle and Smith's approach. Responding to Ryle's request for comments, he wrote:

> If the radio sources are comparable to ordinary dwarf stars, a [more complex and better] model like [I have] proposed might give a somewhat adequate picture of their distribution. The computational work involved would not be large. In case you would feel hesitation to venture into this astronomical domain I would be glad to ask one of our students to undertake it for you.[11]

Ryle declined the offer, but Oort's interest was further piqued by a visit to Leiden in July 1950 from Bolton and Westfold, who were doing a grand tour of European radio and optical observatories. Gart Westerhout (1973:2T), a young Leiden student who had majored in astronomy as an undergraduate, recalled a presentation on their radio results and model of the Galaxy:

> I still vividly remember that colloquium. There were stages when both of them were standing in front of the blackboard at the same time – jumping up and down and correcting each other, making remarks here and there.

Bolton (1978:114–15T) recalled intense intellectual intercourse during their extended stay:

> [Our] teaching at Leiden was based on a lot of work that Kevin and I had done: spherical integration methods, the concept of antenna temperature, and so on. These were really devoured at Leiden. ... [Besides three formal lectures over a month] Oort would call us in on two or three days a week and we'd just sit down and have a two or three hour discussion. ... Oort was a great seeker after the truth.

---

[10] Oort came from the tradition of countryman Jacobus Kapteyn (Section 16.7), who in the early twentieth century had produced (through detailed calculations and models based on exhaustive star counts) the then-standard model of our Galaxy ("the Kapteyn Universe").

[11] Oort:Ryle, 18 April 1950; also H.C. Van de Hulst:Ryle, 14 April 1950; both box 1/1, RYL.

Finally, later that summer, on another visit to England, Oort determined to do something. As he recounted a year later:

> I was started on this when I visited Dr. Hanbury Brown [at Jodrell Bank] and saw him struggling with problems that could be so much more easily tackled and judged by an astronomer specialized in that branch.[12]

Oort assigned Westerhout to look into the details of doing a proper analysis of the distribution of the galactic noise, that is to say, an *astronomical* one (see Section 17.1.1 for use of this term). Bolton and Westfold's 100 MHz survey was the one chosen for detailed modelling, for it was adjudged "by far the most complete and homogeneous, ... measured and reduced with special care." By May 1951 Westerhout and Oort (1951) had written the first article on radio astronomy to appear in a prestigious observatory periodical, namely the *Bulletin of the Astronomical Institutes of the Netherlands*. They began by citing optical evidence that ionized gas in interstellar regions occurred only within 100 pc of the galactic plane; their model then showed that this hot gas could produce only a narrow strip of free–free radiation, extending no more than 1° away from the plane. Although this would cause a brightness temperature of as much as several thousand kelvins if observed with sufficient angular resolution, Bolton and Westfold's 17° beam had severely diluted this and thus it was of no consequence.

Oort and Westerhout argued that the general shape of the galactic noise distribution immediately ruled out association with optical constituents either confined closely to the galactic plane (such as interstellar gas and hot, young stars) or those spherically distributed (globular clusters). Lacking any indication to the contrary, they adopted as a working hypothesis that the radio stars producing galactic noise were distributed like common stars such as the sun. Taking these same stars to comprise most of the mass of the Galaxy, one could then relate radio stars to models of mass distribution earlier developed by Oort. They obtained satisfactory matches between their models and 100 MHz radio intensities around the sky, but only if they added a further radio component. This was a 600 K isotropic "residual" that accounted for over 90% of the signal at the galactic poles – Westerhout and Oort thus found the same need to invoke an isotropic component as had Bolton and Westfold. The source of this signal was unknown. Although they opined that it was perhaps from a spherical distribution of sources in the far outer reaches of the Galaxy, they seemed to favor the "distinct possibility" that it represented the combined effect of distant galaxies: "this would give an entirely new datum regarding the structure of the universe" (Westerhout and Oort 1951:327).[13]

If they assumed that all radio stars had identical luminosities, Westerhout and Oort derived luminosities and separations for the radio stars similar to those found by Bolton and Westfold (Fig. 15.2). But they then introduced a new refinement, namely allowing for an arbitrary amount of dispersion in the luminosities. This problem had been long ago solved in optical astronomy and they showed mathematically that, for instance, if the luminosities were spread over a range of a factor 100 that this would greatly alter the results: radio stars would become 50 times more common and on average 700 times less luminous (although the *most* luminous among them would be detectable to much greater distances). Our Galaxy would then contain $\sim 10^{10}$ radio stars! This improved treatment was at once more realistic and more discouraging, for it demonstrated how little one was going to learn about radio stars based only on the supposition that they caused the galactic background.

At the Radiophysics Lab in Sydney, theorist Jack Piddington (1951) built a Galaxy model that took into account all frequencies observed. He argued that galactic noise originated in interstellar gas as well as in special stellar atmospheres that emitted nonthermally. At the highest frequencies where the spectrum was flat, free–free radiation from (optically thin) gas dominated; at mid-frequencies, emission from special stars reigned; and at the lowest frequencies where interstellar gas was optically thick, absorption caused a turndown in the spectrum.[14]

As a check on these models' descriptions of a narrow galactic layer of ionized hydrogen "invisible" to existing surveys because of poor resolution, Ryle and

---

[12] Oort:Ryle, 22 May 1951, box 1/2, RYL.

[13] See Section 16.1 for a discussion of Leiden astronomer H.C. Van de Hulst's (1945) analysis of radio data and cosmology, a study that influenced the work of Westerhout and Oort (1951).

[14] See Tangent 15.2 for details.

his student Peter A. G. Scheuer (1930–2001) looked for such a narrow, bright band with a new interferometer technique at Cambridge. They indeed found it, but it required a temperature of at least 18,000 K (Scheuer and Ryle 1953).[15]

At Jodrell Bank Robert Hanbury Brown and Cyril Hazard (1953b) incorporated data from many surveys and concluded that the radiation's apparent increase in galactic-latitude width at the lowest frequencies was completely ascribable to the much broader beams used at those frequencies. They also added isotropic radiation as a third component to (1) a distribution of ionized gas identical to that of Westerhout and Oort, and (2) an assumed elliptical distribution around the galactic center of "localized sources" (whose means of energy generation was not addressed). Their sources were mostly within 300 pc of the galactic plane and within 2 kpc of the galactic center; indeed, source density at the sun's position was only 1% of that at the center (see Fig. 15.2). The gas's contribution (at 100 MHz) steadily increased as one moved away from the center, but amounted to only 5% of the total galactic noise.

Having constructed this model based on only the observed background radiation, Hanbury Brown and Hazard (1953a) next inquired as to where the cataloged discrete sources fit in. Earlier, interpreting their own survey of sources, they had agreed with Bernard Mills's (1952a) scheme of splitting sources into Class I, the strongest ones found preferentially near the galactic plane, and Class II, the weak, isotropically distributed sources (Sections 14.3.2 and 14.3.3). If one located the Class I sources typically 50 pc out of the plane (analogous to the hot, young stars), then their typical distances were 1000 pc and separations 300 pc. Could the detected Class I sources now be identified as simply the local members of the model's (mostly) distant sources? Yes, Hanbury Brown and Hazard answered, the total signal expected from Class I sources on their model was consistent with that received from galactic noise.[16] This

source distribution was radically different from previous ones – a much smaller number of more powerful radio stars were now heavily concentrated toward the galactic center. It was now also unnecessary to invoke any contribution from Class II sources to the Milky Way's radiation, leaving the tantalizing possibility that Class II sources were all extragalactic. Their isotropic component, amounting to 60% of the total radiation at 100 MHz, might well then represent the integrated effect of all the weaker, undetected Class II sources.

Ever since their detection of the Andromeda nebula in 1950 (Section 9.3), Hanbury Brown and Hazard had been exploiting the unmatched sensitivity of their 218 ft dish to investigate radio emission from "normal" galaxies (Section 14.4.3). By 1953 they had detected half a dozen such galaxies and possibly two clusters of galaxies, and had developed a "radio index" for spiral galaxies, that is, the ratio of radio to optical intensity. But it was also clear that there existed anomalous objects such as Cyg A and Cen A with enormous radio output compared to optical. Thus attuned to the idea of both normal and peculiar galaxies as radio emitters, Hanbury Brown and Hazard naturally wanted to test whether their isotropic background component could represent the ensemble of all the universe's galaxies. In their published papers they only qualitatively discussed this possibility, but Hazard (1953:chapter 10) in his Ph.D. thesis made detailed calculations.[17] Combining the radio index with the optical intensities of galaxies, neglecting any effects from the redshift of galaxies, and assuming a radius for the universe of 550 Mpc, he found that the expected integrated radiation amounted to 50 K at 100 MHz, only one-tenth of the observed isotropic component. This discrepancy Hazard explained by (a) saying that types of galaxies other than the spirals they had investigated must have higher radio indices, and (b) citing their earlier work indicating that galaxies gathered into clusters emitted more radio power than an equal number of isolated galaxies. Hazard then went on and asked whether the intensities of Mills's collection of Class II sources were consistent with the (optical) distribution of galaxies and the observed radio indices. Mills (1952a:280–1) had earlier calculated that the total radiation at the

---

[15] See Tangent 15.3 for details.

[16] Mills (1952a:277–8) calculated, in contrast to Hanbury Brown and Hazard (1953b:957), that the Class I sources fell far short of accounting for background radiation near the galactic plane. This difference arose because Mills assumed a uniform density of sources throughout the Galaxy, whereas Hanbury Brown and Hazard postulated a distribution heavily concentrated toward the galactic center.

[17] Hazard's arguments that the Class II sources could well be extragalactic were also presented to a meeting held at Jodrell Bank in July 1953. [*Observatory* 73, pp. 193–4 (1953)]

galactic poles could be accounted for (within a factor of two) by assuming that his Class II sources were uniformly distributed throughout the universe; he thus leaned toward interpreting the Class II sources as extragalactic. Hazard's more sophisticated analysis, including the effects of dispersion in both optical and radio luminosities, led him to a more confident decision. The cataloged Class II sources were simply the brightest of a class of extragalactic objects with typical separations of 0.5–1.0 Mpc and optical luminosities only one-hundredth that of the Milky Way. The final sentence in Hazard's dissertation reads:

> It is therefore considered, in opposition to the ideas current at the commencement [1950] of the work discussed in this thesis, that the majority of the Class II sources so far observed are probably extra-galactic objects.

15.3 WHAT *ARE* RADIO STARS AND HOW DO THEY EMIT?

The basic problem with radio stars was that no one knew how far away they were. As Leiden astronomer Hendrik Van de Hulst lamented in a popular talk in 1951:

> There exists no distance criterion for the point sources. Thus it's very easy to make a gnat into an elephant, or vice-versa. Suggestions have ranged from a comet (0.1 pc) to a strange kind of extragalactic nebula (100,000 pc or more).[18]

Distances were fundamental not only in terms of locating objects in space and thus understanding their distribution, but also for understanding the means by which they emitted their radio power. The farther away an object, the larger its linear dimensions (for a given angular size) and the more power it emitted (for a given received intensity). In the absence of conclusive evidence from optical identifications, most people chose to minimize the size and power requirements by imagining a radio star as indeed a star-like object. A single weird star seemed a better application of Occam's Razor than an entire weird galaxy consisting of 10^{11} stars! Moreover, strange stars of many types were well known, but peculiar galaxies of that time were few in number and ill known. Mills (1976:34T) recalled:

> It was more a qualitative argument that a galaxy was a very, very large object consisting of very many stars, and it just seemed inconceivable that a large object could emit a great deal. It seemed in those days that everyone was thinking in terms of "local" physics, fairly small objects like stars.

By 1952, however, several radio stars had been identified as galaxies and we have seen above that the idea became viable, at least to some, that most radio stars were extragalactic. But in either case one was dealing with postulated objects that were unlike anything previously known – the ratio of radio to optical power utterly dwarfed that of either normal stars such as the sun, or normal galaxies such as the Milky Way. To explain the enormous radio brightness temperatures in either case, researchers attempted to use many of the same mechanisms and ran into many of the same problems that bedevilled theorists working on solar bursts (Section 13.2.2).

Ryle's (1949b) model, requiring a new type of star that produced both cosmic rays and radio emission, has been discussed in the previous section. The radio star had to produce a radio brightness temperature of 10^{14} K and cosmic rays of energy 10^{10} eV, and so Ryle posited a special stellar atmosphere that was an adaptation of his earlier model for solar radio emission (Ryle 1948; Section 13.2.2). A general electric field of ~1 volt cm^{-1} was required to accelerate particles to high enough energy, and for this he adapted, as he had before, a solar theory developed by Hannes Alfvén (whom he had just visited in Stockholm). Alfvén's theory generated such an electric field on a rotating star with a magnetic field. Although the speed of rotation times the strength of magnetic field had to be 10^8 times that for the sun, Horace W. Babcock of Mt. Wilson Observatory had recently reported just this sort of remarkable magnetic star. Lastly, the radio star had to accomplish all of this without emitting any significant visual radiation – it had to have a hot, thin envelope (a revved-up solar corona) surrounding a very cool, conventional photosphere. This picture of a radio star remained the standard at the Cavendish Laboratory until 1953–54.

No one other than Ryle specified in as much detail what went on in a radio star, but others made general comments. For example, one can feel the frustration and incredulity as Van de Hulst (1951:170) lectured to

---

[18] H.C. Van de Hulst, "De continue straling van het Melkwegstelsel," *Nederlands Tijd. Natuurkunde* 18, 145–50 (1952) (quotation from p. 146).

his students: "They may be stars, very cool or very hot or very old or very young, but at any rate very large." Unsöld came to favor flare stars, faint stars that had been recently discovered to brighten optically by factors of 10–100 within hours. The most influential comments were made by Walter Baade and Rudolph Minkowski. Over the years they steadfastly maintained that the radio stars must belong to Baade's so-called Population II, that is, the older component of our Galaxy, which tended to have a more spherical distribution (as did the radio noise) and represented the bulk of the mass of a galaxy such as ours or the Andromeda nebula. In 1950, heartened by the detection at Jodrell Bank of radio emission from the Andromeda nebula, Baade wrote:

> This new result strengthens very much the view that the radio stars are members – probably dwarf stars – of the stellar Population II. ... The tantalizing question now is: which stars of the Population II are the culprits? The astronomical identification of one of the isolated radio sources – leaving aside the Crab nebula which evidently is a special case – would solve the riddle, but I suspect that it will be a hard nut to crack.[19]

As we have seen, Baade's hope that a single optical identification would solve the riddle was dashed as identifications over the next two years turned out to be not with stars, but rather with nebulae in our Galaxy or with peculiar galaxies (Section 14.4). Still, in 1953 Baade and Minkowski (1954b:229) were again emphasizing that the general galactic noise most probably arose from the integrated effect of Population II stars, although "the concept of stars as [discrete] radio sources has dropped into the background." They were particularly unenthusiastic about prevalent quantitative models:

> While it is thus highly probable that radio stars exist, no estimate of their type or luminosity can be made. The attempts by several authors to estimate the number of radio stars on the basis of statistical analysis are completely invalidated by the fact that the strongest sources are now known not to be stellar and that it is completely unknown which, if any, of the known sources belong to the group of objects which emit the general galactic radiation.
> (Baade and Minkowski 1954b:231)

Lively direct confrontations over the nature of radio stars occurred between the Cambridge theorists Thomas Gold (1920–2004) and Fred Hoyle (1915–2001) on the one hand and Ryle on the other. Gold and Hoyle had worked on radar applications for the Royal Navy during World War II[20] and at war's end were intrigued by Hey's discoveries. They had considered the design of a large fixed spherical reflector for lunar and planetary radar (Section 12.2) and were interested as well in other sorts of radio physics, but were unable to find a patron at the Cavendish Laboratory, where Ryle was exclusively supported for such research (Chapter 8).[21] Gold and Hoyle nevertheless established themselves at Cambridge, followed radio astronomy closely, and made important theoretical contributions at Cambridge seminars and evening discussion clubs, as well as at national conferences. The best extant record of the spirited debates during this era is in the mimeographed proceedings (Boyd 1951) of a conference on "The Dynamics of Ionized Media" held in April 1951 at University College, London. Most of the meeting was devoted to the complex mathematical physics of discharge tubes and the ionosphere, but one session on the last day was entitled "Cosmical Applications" (!). Ryle (1951) there presented his case that the radio stars were nearby, dark, common stars in the Galaxy, not at all extragalactic. This was immediately countered by Gold and Hoyle, who insisted that an extragalactic location was just as well admitted by the available evidence (which consisted essentially of a few optical identifications and the observed isotropic distribution of the fifty 1C radio stars).

Gold (1951) then went on and considered both possibilities in his own presentation. If the radio stars were local, he argued that that each one must somehow act like a giant radar transmitter tube. Since ionized gases in space were normally extremely conductive, it was

---

[19] Baade:Lovell, 27 November 1950, file R14 (uncat.), LOV.

[20] Hoyle's most important contribution to radar technique was to work out how shipborne radars could determine not only the range, but also the height of an approaching aircraft. The method, taking advantage of the Lloyd's mirror interference (also independently studied at the Radiophysics Lab in Sydney and the basis of the later sea-cliff interferometer [Section 7.3.2]), became standard in the Royal Navy. [C. Domb (2003), "Fred Hoyle and naval radar 1941–5," *Astrophys. & Space Sci.* **285**, 293–302]

[21] Gold (1976:4–5T), Hoyle (1988:171A:470), Hoyle (1994:268–9).

difficult to generate the strong electric fields needed to accelerate particles to high energies. Gold's solution was to posit a very strong magnetic field surrounded by a low-density region; these together would play the role of the insulating glass of a tube, as well as allow high accelerations. Since this combination was not found in normal stellar atmospheres, he speculated that it might occur in the neighborhood of collapsed, dense stars (such as white dwarfs). As Gold (1976:12T) later recalled:

> Without understanding in detail how you jiggle the electrons, I knew that a vacuum tube will produce radiation in radio bands if you have a lot of electrons that suffer a high acceleration and if there is a large gradient of electron density. ... The nearest to a vacuum tube that I could think of was a strong [magnetic] field that has electrons in it. That was my picture.

But Gold also pointed out a problem: on this interpretation one would expect substantial fluctuations like those associated with solar flares, yet intensities of the radio stars were steady.[22,23] His talk next moved to the extragalactic hypothesis and pointed out that it predicted a ratio of radio to optical power (relative to a normal galaxy) much less extreme than the same ratio on the star hypothesis: perhaps 10^2 versus 10^8. He also mentioned the existence of a class of galaxies with "greatly broadened emission lines, suggesting that far more violent motions occur there than here [in our Galaxy]."[24]

This suggestion of Gold's did not fall on sympathetic ears. Hoyle (1988:171A:520, 1994:269–70) later recalled the absolute disbelief by almost all present. Another recollection comes from Geoffrey R. Burbidge, a postgraduate physics student in attendance:

> Tommy Gold got up and suggested that if the radio stars were so isotropic, wasn't it entirely possible, rather than being very close by, that they were very far away? And Tommy was then absolutely attacked from all directions for the next half an hour. That was the first indication that I saw that personalities entered into this thing very strongly. In those days I was naive and assumed that people really thought about science in an abstract and scientific way.[25]

In the closing discussion (Boyd 1951:143–5) Ryle and George. C. McVittie, a leading cosmologist, led the forces arguing that the radio stars were not extragalactic. Ryle opened his remarks with the biting comment that "I think the theoreticians have misunderstood the experimental data" (Fig. 15.3). He reiterated how difficult it would be for a galaxy to be a radio star – its entire interstellar medium would have to be at the unthinkable temperature of 10^8 K. Hoyle replied:

> From the remarks of Prof. McVittie and Mr. Ryle it might be thought that Gold and I have once more been riding a dogmatic hobby-horse – in this instance to the effect that the discrete radio sources are extragalactic nebulae. What we said was that the extragalactic nebulae hypothesis must be kept actively in mind. The boot is really on the other foot, for Prof. McVittie and Mr. Ryle have dogmatically asserted that the discrete sources cannot be of extragalactic origin, although of the half dozen or so discrete sources that have indeed been identified, five have been found to correspond with nearby extragalactic nebulae. Presumably a discrete source ceases to be a discrete source as soon as it is identifiable as a galaxy.

As to how the radio stars emitted, whether close or distant, the best minds in the field were baffled. As with

---

[22] Gold's suggestion of collapsed, dense objects with large magnetic fields that produce variable radio intensity bears a remarkable resemblance to the modern understanding of pulsars, discovered by Antony Hewish and Jocelyn Bell in Ryle's group in 1967. Indeed, shortly after their discovery Gold made a vital contribution to the theoretical understanding of pulsars as fast-rotating neutron stars.

[23] A suggestion similar to Gold's about the nature of radio stars was described in an unpublished note (intended for *Nature*?) by Jack Piddington in 1948. Piddington proposed that "the discrete sources are in very old, almost dead stars which have collapsed to small size and radiate very little light." He was interested in the extremely strong magnetic fields thought to be associated with white dwarfs, and reasoned they would probably generate powerful radio emission. ["Radio frequency energy from stars," 8 March 1948, 6 pp., file A1/3/1b, RPS]

[24] These galaxies were not further identified by Gold, but undoubtedly refer to those discovered by Carl Seyfert in the early 1940s, a class now known as Seyfert galaxies. At one time

Oort also suggested these galaxies to Ryle as possible radio sources [Oort:Ryle, 18 April 1950, box 1/1, RYL].

[25] G. R. Burbidge, 8 January 1980, excerpt (Tape 130A:50) from "Radio astronomy's historical role in astrophysics," talk given at AAAS annual meeting, San Francisco.

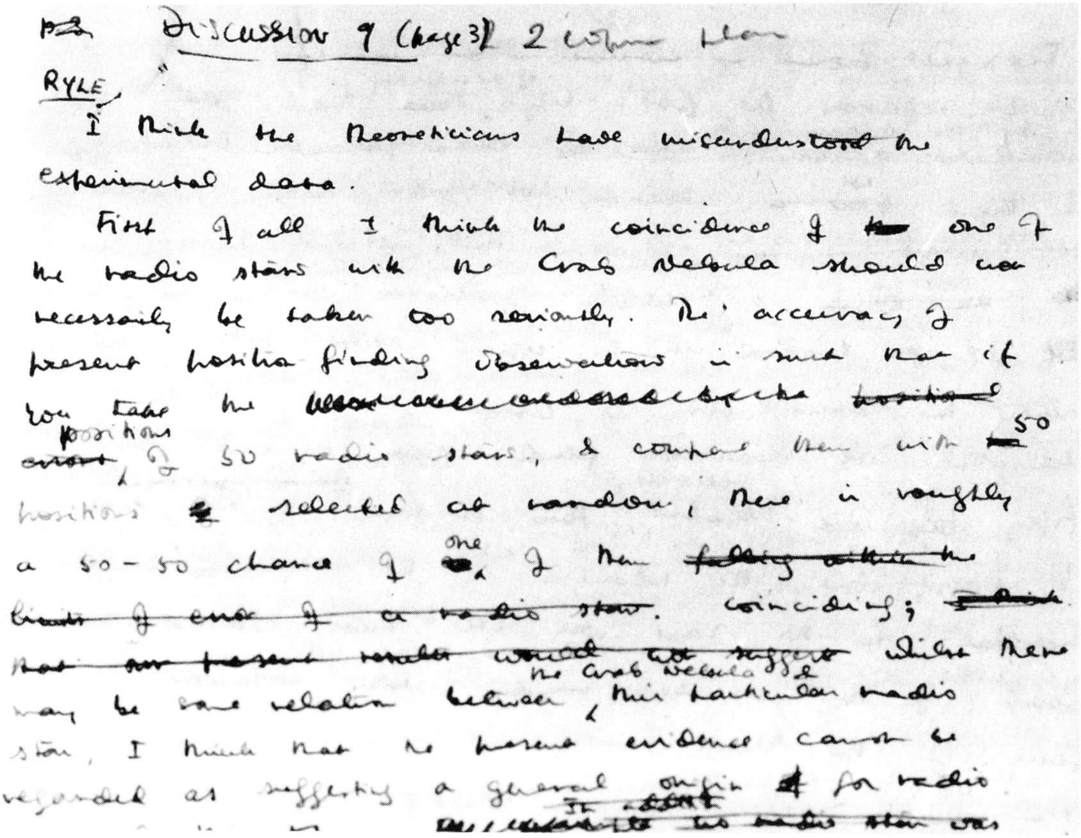
Discussion 9 (Page 3) 2 copies then
RYLE,
I think the theoreticians have misunderstood the experimental data.
First of all I think the coincidence of one of the radio stars with the Crab Nebula should not necessarily be taken too seriously. The accuracy of present position finding observations is such that if you take the positions of 50 radio stars, & compare them with 50 positions selected at random, there is roughly a 50-50 chance of one of them coinciding; while there may be some relation between the Crab Nebula & this particular radio star, I think that the present evidence cannot be regarded as suggesting a general origin for radio

Figure 15.3 Ryle's manuscript version of discussion remarks that he submitted to the editor of the proceedings of the April 1951 conference on "Dynamics of Ionized Media" (Boyd 1951:144). His opening remark that "I think the theoreticians have misunderstood the experimental data" illustrates the liveliness of the discussion with Gold and Hoyle.

solar burst theory, both cyclotron radiation and plasma oscillations were investigated in detail, but the same fundamental problem arose: even if one could generate ~100 MHz radio waves, reasonable configurations for the source regions would not allow their escape (Section 13.2.2). When it became apparent that objects associated with radio sources exhibited high velocities (Section 14.4.3), a possible new source of energy could be tapped: the kinetic energy of gas clouds colliding at high speeds. Cyg A was the exemplar for this, with its two Baadean colliding galaxies; moreover, large motions were also found in Cen A, Per A, Tau A and Cas A. Only a very low efficiency of energy conversion (1% or less) was required, but it was not enough simply to say that the colliding clouds would heat up and radiate thermally, for the free–free optical depth at radio wavelengths for the postulated $\sim 10^8$ K gas was extremely small. In fact, no one quite knew how cloud energy could be converted into radio waves – for example, Rudolph Minkowski and Jesse Greenstein (1954:238) could only vaguely appeal to "conversion by some type of gaseous plasma oscillation." In his thesis Roger Jennison (1954:113–8) considered possibilities for Cyg A including the effects of spiral arms, tidal distortions, jets, and dust particles, but in the end could not understand how the the two radio lobes that he had discovered (Jennison and Das Gupta 1953; Section 14.5.1.3) could be emitting so powerfully and yet be located well away from the optical image:

> Out there in apparently empty space the radiation seemed to be coming from these two enormous, symmetrical blobs. What was going on?
>
> (Jennison 1976:28T)

At one point he was even tempted to postulate a "Principle of Avoidance" between radio and optical emission (Jennison 1954:3). Yet even this was not as extreme as Ryle had once speculated during a talk in 1949:

> I suppose we should consider the possibility that intelligent beings on other satellites could cause electrons to move coherently in radio transmitters ... but it seems unlikely. [The radiation] extends over the whole range of wavelengths from 1.4 to 7 m and the powers are very great.
> If we now restrict ourselves to natural processes ... [26]

By 1954, fully eight years after their discovery, the answers to the questions posed in this section's title – what *are* radio stars and how do they emit? – evidently remained unanswered.

15.4 SYNCHROTRON RADIATION AND COSMIC RAYS

Ever since 1912 when cosmic rays were first recognized as a form of energy falling upon the earth's atmosphere, their study had occupied the mainstream of both the earth sciences and nuclear physics. They were important to geophysicists, for their trajectories, severely modified by the earth's magnetic field, gave important clues to that field's strength and shape in outer space. Researchers were also able to study the structure of the upper atmosphere through analysis of the complex chain of reactions set off when a primary cosmic ray particle collided with atmospheric particles and created showers of energetic secondary particles. On the other side, physicists treated these collisions as high-energy experiments unrealizable in the laboratory. Using a wide variety of detectors, at elevations ranging from sea level to the highest that balloons could fly, much had been learned by the late 1940s. Amongst the *secondary* cosmic rays were discovered particles whose existence provided critical evidence for fundamental theories of the structure of matter:

the positron, the muon, the pion, and strange particles such as the kaon. In contrast, the *primary* cosmic rays were almost all protons, with a small admixture of heavier nuclei. They arrived at earth equally from all directions with extraordinary energies, ranging as high as 10^{17} eV; the number of particles at any given energy E fell off approximately as $E^{-2.5}$. Although a small fraction of cosmic rays were associated with solar eruptions, the general origin and distribution of cosmic rays in space were complete unknowns. Arguments about their origin spanned no less a range than (1) the entire universe, (2) the Galaxy, and (3) the sun.[27]

About 1950 an idea emerged that, if correct, meant that cosmic rays and radio astronomy were intimately related. The idea was that radio radiation from both radio stars and the galactic background might be the result of cosmic ray electrons spiralling in a magnetic field, much as they did in a manmade synchrotron accelerator. After an initial pair of short papers, however, no person of influence in the major radio groups seriously took up this suggestion. In fact, this notion of synchrotron radiation[28] lay dormant in the West for a full five years. In the Soviet Union, on the other hand, first Ginzburg and then Shklovsky latched on to the theory and intensively developed it. In the end, however, Western acceptance of the theory ironically came about through *optical* observations, of the Crab nebula in 1955.

15.4.1 In the West

Classical physics indicates that charged particles circling in a magnetic field radiate and that the frequency and amount of radiation depends on the field strength and the particle's energy (Appendix A). About the time of World War II, this effect became important for particle accelerators of ever higher energy.

---

[26] Ryle, 12 March 1949, notes for a talk given to the Cavendish Physical Society, file 8 (uncat.), RYL.

[27] For the history of cosmic ray research, see contributions to Sekido and Elliot (1985) and to L. M. Brown and L. Hoddeson (eds.), *The Birth of Particle Physics* (Cambridge: Cambridge University Press, 1983). A brief history and reprinted key papers are available in A.M. Hillas (1972), *Cosmic Rays* (Oxford: Pergamon).

[28] The term "synchrotron radiation" was not coined until 1955 (by Jan Oort), but I use it for developments before that time.

The new synchrotrons could reach energies of hundreds of MeV, far more than previous cyclotrons and betatrons. Designs now had to compensate for the fact that circular electron beams of such high energy were expected quickly to decay due to loss of energy through radiation. This radiation was first actually observed (at visual wavelengths) in 1947. The theory was completely worked out at this time by Julian Schwinger of Harvard and thus in some quarters, such as in Ryle's group, the effect was called "Schwinger radiation."[29]

In Sweden, the physicist Hannes Alfvén (1908–1995) had been working since the early 1930s on plasma theory as applied to a wide variety of situations in astrophysics and geophysics. While investigating the theory of sunspots in 1942, he made perhaps his greatest contribution to the field by showing that transverse hydromagnetic waves (later known as Alfvén waves) could propagate along a magnetic field much like ordinary waves on a string. Another of Alfvén's interests, even before the war, was the origin of cosmic rays. By the late 1940s he took the stance that cosmic rays came not from extragalactic or interstellar space, but from the sun after many reflections off magnetic fields throughout a ~10,000 a.u.-wide circumsolar region.[30]

In 1949–50 Alfvén married his cosmic ray scheme to the basic idea of Unsöld and of Ryle (with whom he had extensive discussions) that radio stars produced cosmic rays. Alfvén collaborated at the Royal Institute of Technology in Stockholm with Nicolai Herlofson, an ionospheric theorist who had just returned from three years of research in England (including a collaboration with Lovell's meteor radar group [Section 9.3]). Alfvén and Herlofson (1950) found that extensive synchrotron radio emission was a natural byproduct of their cosmic-ray picture. As Herlofson wrote to Ryle:

We are looking into the question of how cosmic rays are generated, and whatever we do about it, we find radio noise at the same time.[31]

Radio stars were taken as stars surrounded by cosmic ray electrons emitting copious radio radiation as they moved within the confinement of a strong "trapping" magnetic field. Using Schwinger's theory, they showed that, for example, 150 MeV electrons in a 3×10^{-4} G field would suffice to produce 100 MHz radio waves. The strong magnetic field was best created by a relatively fast motion of the star with respect to an imbedding cloud of gas and dust. Moreover, they noted that one required a central star without strong optical emission in order to minimize loss of the electrons through collisions with optical photons. It was therefore no wonder that optical identifications were rare, for a radio star was an already dark object hidden in a dust cloud, not unlike Churchill's characterization of Russia as a "riddle wrapped in a mystery inside an enigma."

Alfvén and Herlofson's paper in the *Physical Review* was acted upon within months by the German solar astrophysicist Karl-Otto Kiepenheuer (Section 10.6.1), then on leave in America at Yerkes Observatory. The picture of electrons spiralling in a magnetic field and producing radio waves was attractive to him; in fact in 1946 he had (unsuccessfully) invoked a similar mechanism for solar radio bursts (Section 13.2.2). Kiepenheuer had also written about connections between solar activity and cosmic rays, but he believed that most cosmic rays came from interstellar regions. In his own *Physical Review* note, Kiepenheuer (1950) thus applied the synchrotron idea to the general galactic radio emission. If the Milky Way's magnetic field had a typical value of $\sim 10^{-6}$ G[32] and if hypothesized electrons comprised 1% of all cosmic rays, his rough calculations yielded

---

[29] For the first laboratory detection of synchrotron radiation, see F. R. Elder *et al.* (1948), *Phys. Rev.* **74**, 52–6; for a history (1873 to 1947) see J. P. Blewett (1988), *Nuclear Instruments and Methods in Physics Research* **A266**, 1–9; for the theory see J. Schwinger (1949), *Phys. Rev.* **75**, 1912–25. Theoretical studies of synchrotron radiation in accelerators and in the earth's magnetosphere were also carried out during this period in the Soviet Union. The principal study tapped by Ginzburg and Shklovsky was by V. V. Vladimirsky (1948), "Influence of the terrestrial magnetic field on large Auger showers" [in Russian], *Zh. Eksp. Teor. Fiz.* **18**, 392–401.

[30] H. Alfvén, "On the solar origin of cosmic radiation. I and II.," *Phys. Rev.* **75**, 1732–5 (1949) and **77**, 375–9 (1950).

[31] Herlofson:Ryle, 13 April 1950, box 1/1, RYL.

[32] The value of the general interstellar magnetic field was uncertain. Measurements of polarization of starlight suggested a field of order 10^{-6} G, and theoretical arguments regarding cosmic-ray confinement in the Galaxy and energy equipartition also indicated $\sim 10^{-6}$ G.

an intensity of synchrotron radiation approximately agreeing with radio observations; moreover, he pointed out that the spectral index of galactic noise qualitatively agreed with the synchrotron idea. But since no one had ever actually detected electrons as a component of cosmic rays (see next section), Kiepenheuer also briefly discussed why it was nevertheless reasonable that electrons might comprise as much as 1% of primary cosmic rays. Moreover, because of collisions with solar optical photons, cosmic ray electrons might well be more abundant in interstellar regions than near the earth.

These two short papers (and related talks) by Alfvén and Herlofson and by Kiepenheuer soon became known among radio astronomers, but synchrotron radiation was not seriously taken up by anyone in the major groups. There were only five isolated follow-up studies in the West,[33] the most noteworthy from Bernard Kwal (1951), a physicist at the Institut Henri Poincaré in Paris. Kwal, who was working on problems of radiation losses in particle accelerators, talked about radiation from *protons* (which is far less intense per particle than that from electrons), but nevertheless derived an important relationship between the energy spectrum of cosmic rays and the resultant radio spectrum of the synchrotron radiation they produced: if the number of cosmic rays at an energy E varied as $E^{-\gamma}$, then the radio spectral index α (in brightness temperature) was equal to $(\gamma+3)/2$.[34] Observed values for γ and α were indeed consistent with this relation.

15.4.2 In the Soviet Union

The two 1950 notes in *Physical Review* triggered one more follow-up study, by far the most fruitful. Vitaly Ginzburg (Section 10.3.1) of the P. N. Lebedev Physics Institute had wrestled with explaining galactic noise as free–free radiation, but could only match theory and observations by saying that Jansky's long-wavelength intensities must be too high and in any case needed confirmation (Ginzburg 1947, 1948). But in a short paper written in October 1950, "Cosmic rays as the source of galactic radio emission," Ginzburg (1951) switched and argued that Kiepenheuer's concept of interstellar synchrotron radiation provided a natural explanation for galactic radio emission, with no need to throw out the low-frequency data or to manufacture some unknown type of radio star. Ginzburg had always enjoyed problems dealing with radiation effects from moving particles; in particular, during the 1940s he made fundamental contributions to Cherenkov radiation theory. This background, together with his extensive research on ionospheric radio propagation (Ginzburg 1949) and interest in matters astrophysical, led to his enthusiasm for what he called "bremsstrahlung from relativistic electrons." He showed that the observed radio intensity could be explained by cosmic ray electrons if (1) there existed a general magnetic field of 10^{-6} G over a distance of 3 kpc, and (2) cosmic ray electrons in interstellar regions had a density of 10^{-10} cm^{-3} (a much higher value ["but not inadmissible"] than the upper limit then known for the top of earth's atmosphere).

A postgraduate student of Ginzburg's at Gorky State University, German G. Getmantsev (1926–1980), next developed the relation (independently of Kwal) between the electron-energy spectral index γ and the resultant radio spectral index α, and showed its consistency with available observations (Getmantsev 1951:551–3, 1952). Moreover, because the galactic center was a strong radio source at high frequencies and because the highest energy electrons emitted best at these frequencies, he argued that cosmic ray electrons probably originated in the center and were greatly depleted at the highest energies by the time they reached the sun's outer position.

Ginzburg forcefully developed the synchrotron idea, culminating in several papers prepared for a national conference in Moscow in May 1953 on the origin of cosmic rays.[35] Together with M. I. Fradkin, Ginzburg reexamined the question of the number of interstellar cosmic ray electrons required to account for observed radio intensities. Now using a path length through the Galaxy of 17 kpc, an average magnetic field of 10^{-5} G, and a more detailed theory than earlier, he found that a smaller cosmic ray density (and an electron

---

[33] Details of these papers are given in Tangent 15.4.

[34] The more usual formulation of this relation is in terms of specific intensity I rather than brightness temperature. If α is defined such that I varies as $\nu^{-\alpha}$, then $\alpha = (\gamma-1)/2$.

[35] *Proc. of the Third Conference on Questions of Cosmogony (Origin of Cosmic Rays)* = *Trudy Soveshch. Vopr. Kosmog.* 3 (1954, Moscow: Academy of Sciences).

component of 0.5%) would do the trick (Ginzburg and Fradkin 1953).[36]

By the end of 1953 Ginzburg had developed a confident grand synthesis:

> With the development of radio astronomy and of cosmic electrodynamics, the problem of the origin of cosmic rays has truly become an astrophysical problem and has progressed from a stage of mainly hypothetical conjectures that were impossible to check by observations.
>
> (Ginzburg 1953c:391)

This synthesis, set out fully in a long review paper entitled "The origin of cosmic rays and radio astronomy" (Ginzburg 1953c), revolved around the ability of radio observations to reveal *where* cosmic rays were generated and where they spent their subsequent lifetime. One now had a means of overcoming the limitation of sampling cosmic rays only at the earth and the frustration of finding no preferential directions of arrival. Ginzburg (1953b) agreed with Shklovsky's recent proposal that cosmic rays originated in nova and supernova explosions (see below). He also concluded that a popular acceleration mechanism due to Fermi (1949; see below) was not effective in the way it had been applied to interstellar regions, but that it worked fine in the expanding shells of novae and supernovae, where scale lengths were smaller and magnetic fields higher (Ginzburg 1953a). Ginzburg (1953c) also analyzed the various means by which cosmic ray electrons would gain and lose energy as they travelled for an expected lifetime of several hundred million years through the Galaxy.

Ginzburg developed these ideas in detail from late 1950 onwards, but until late 1952 they were not accepted by Iosif Shklovsky and the Soviet astronomical community (and were unknown in the West). As Ginzburg (1984:295–6) recalled:

> I at once believed that the synchrotron mechanism was responsible for non-thermal cosmic radio emission. I ascribe this not to any keen insight, but to the fact that I was closer to physics and rather far from classical astronomy.[37] In this situation the synchrotron mechanism seemed clear and realistic, whereas hypothetical, strange "radio stars" remained purely speculative. ... The reaction of astronomers was quite the opposite, i.e., the synchrotron mechanism seemed mysterious and speculative, whereas "radio stars," although posing riddles, were more acceptable – for what kinds of stars cannot exist?

Iosif Shklovsky[38] (Section 10.2.1) had first advocated free–free radiation to explain Milky Way radio emission, and then switched to a mixture of free–free radiation and radio stars. In contradistinction to Ginzburg, he was trained as an astronomer and drew on a vast knowledge of the sun, stars, and the Galaxy in assembling his theories. Shklovsky (1948) showed how secondary maxima in the galactic noise coincided with hot interstellar gas emitting optical H_α radiation and associated clusterings of hot stars. To account for the fluctuations observed in the radio star Cyg A he appealed to the so-called Wolf–Rayet stars, which had hot gaseous winds. Shklovsky proposed, as he had for solar bursts, that plasma oscillations within these flows set up radio star emission that was both fluctuating and of extraordinarily high brightness temperature.

Later, Shklovsky (1951b) adopted much the same model of radio stars that he found set forth in Ryle's (1950) review of radio astronomy. Shklovsky's picture, however, was even more extreme than Ryle's: radio stars were dark, cool objects of mass <0.05 solar masses, intermediate in size between stars and planets. Optical identifications over the next year, however, caused him to split radio stars into two classes. In this scheme the first class – the aforementioned, isotropically distributed majority of radio stars – created cosmic rays that emitted synchrotron radiation in high stellar magnetic fields (~1000 G). Shklovsky (1952c) was here building

---

[36] The 0.5% electron fraction applied only to energies >10^9 eV; for lower energy cosmic rays Ginzburg calculated that electrons, created by the collisions of cosmic ray protons with interstellar atoms, would in fact be comparable in number to protons.

[37] For example, all of Ginzburg's papers cited in this book are in journals of physics, not astronomy.

[38] In the remainder of this section, I describe all of Shklovsky's work in radio astronomy except that dealing with the sun (Chapter 13) and spectral lines (Chapter 16). Some of this does not deal specifically with synchrotron radiation (for example, his Milky Way models), but it is best included here, rather than with the comparable Western work described in Sections 15.2 and 15.3, because the majority of Shklovsky's work was unknown in the West until years later.

on the ideas of Alfvén and Herlofson and of Ginzburg, but with a major difference: the synchrotron radiation came from relativistic *protons*, a notion earlier discussed (for the interstellar medium) by his Moscow State University colleagues A. A. Korchak and Yakov P. Terletsky (1952). Shklovsky's second class of radio star contained the strongest few sources, occurring preferentially in the galactic plane. Inspired in particular by the identification of Tau A with the Crab nebula, he suggested that these strong sources had all been supernovae of a kind recently modelled by Soviet theorist E. R. Mustel. Such supernovae had a mass of $\sim 10^6$ solar masses, so large that immense gravitational fields caused emission from a 10^9 K gas to be redshifted all the way from gamma rays to the radio band![39]

Shklovsky (1952a, b) also developed his ideas of galactic radio emission in the context of current astronomical models of the Galaxy. He rejected the models of Bolton and Westfold (1951) and Westerhout and Oort (1951) because (1) they required an isotropic (probably extragalactic) component that he felt inadmissible (he calculated that galaxies could produce at best only 10% of this 600 K isotropic component), and (2) they considered free–free emission from the galactic plane of negligible import. He agreed with Piddington (1951) that the radio spectrum in various directions was a critical clue to a proper understanding, and that there existed two basic radio components. Through a detailed analysis of all existing background surveys he concluded that the first component was free–free radiation from ionized gas, revealed through spectra of the secondary maxima in the galactic plane, as well as toward the galactic center. The second component, dominating at long wavelengths, was manifest in the smooth background at high latitudes. It could not be due to a Ginzburg-type emission from particles in a general magnetic field, for magnetic fields required the presence of interstellar material and it was well known that such material stayed closely confined to the galactic plane. Shklovsky called this high-latitude component the Galaxy's "radio corona," spherical in shape and on a huge scale compared to the distance between the sun and galactic center (Fig. 15.2). Such a large feature in our own Galaxy also seemed plausible to Shklovsky because of an apparently analogous feature in the radio contours of the Andromeda nebula recently mapped by Hanbury Brown and Hazard (1951a; Fig. 9.5). This radio corona shared characteristics with other features of galactic structure known from optical astronomy (for example, globular clusters), but was unique in its combination of sphericity and only slight concentration toward the center. Shklovsky's corona was filled with 10^{12} to 10^{13} radio stars, far greater than the number of ordinary stars.

In 1952, however, Shklovsky, never one to be scientifically dogmatic, did an about-face, jettisoned his 10^{13} radio stars, and embraced Ginzburg's synchrotron mechanism. This happened sometime during the few months between when he served as an outside examiner for Getmantsev's 1952 dissertation and when he submitted a major paper to *Astronomichesky Zhurnal* in November.[40] The key new influence on Shklovsky was a mid-1952 theoretical study by Solomon B. Pikel'ner (1921–1975) of the Crimean Observatory.[41] Pikel'ner was concerned with explaining the isotropy of cosmic rays and concluded that the Galaxy must possess a huge halo of rarefied gas with a magnetic field of $\sim 10^{-5}$ G. Shklovsky had dismissed the idea of any significant magnetic field well out of the galactic plane because of the absence of any material there, but now he saw that a hot, rarefied medium was possible. Not only did Pikel'ner's halo provide the necessary magnetic field for synchrotron radiation, but in size it was neatly comparable to Shklovsky's radio corona. Rejecting that this corona consisted of radio stars, Shklovsky (1953b) now found it "absurd" and "increasingly less real" that there could exist so many strange radio stars in the Milky Way, far outnumbering normal stars. If for no other reason, so many stars emitting via (proton) synchrotron radiation would also imply a flux of cosmic rays far higher than observed. Moreover, he was impressed that not one of the half-dozen existing reliable optical identifications had been indeed with a *star* – all were nebulae of some sort. Thus was he led to the "fundamental, radical conclusion" that not only were radio stars unwanted in the galactic corona, but that in *any* capacity they must needs be discarded and

---

[39] This was in essence radiation from the vicinity of what is now called a black hole.

[40] Getmantsev (1978:8T).

[41] S. B. Pikel'ner (1953), "The kinematic properties of interstellar gas as related to the isotropy of cosmic rays" [in Russian], *Dokl. Akad. Nauk SSSR* **88**, 229–32.

replaced by the term "radio nebulae."[42] It now seemed to him "natural and perfectly justifiable" that the Galaxy's corona was the result of "electron-magnetic" (synchrotron) radiation from cosmic ray electrons.

But what of the radio nebulae? In a similar manner (but independently) to Baade and Minkowski's (1954b) work at about this same time (Section 14.4.3), Shklovsky put sources first into extragalactic (all those outside of the galactic plane) and galactic categories. The former were then split into normal galaxies (such as M31) that were relatively weak in their radio emission and the few that gave off tremendous radio power – for such an object Shklovsky's succinct new term of "radio galaxy" was to outlast Baade and Minkowski's clumsy "peculiar extragalactic nebula." Shklovsky discussed each of them in detail, in particular Cyg A, and marvelled at their tremendous range in the ratio of optical to radio power. He considered it the "best probability" that radio galaxies too emitted via the synchrotron mechanism, with the various radio outputs depending on cosmic-ray content.

Shklovsky's *galactic* radio nebulae also comprised two categories, the first of which included gaseous nebulae (such as the Orion nebula) that emitted by free–free processes. All of the *detected* galactic sources, however, he took to be supernova remnants emitting synchrotron radiation. Known radio-emitting supernova remnants included the Crab nebula (Tau A) and Tycho's remnant (Cas B; Section 14.4.3), but Shklovsky had also done his own search of compilations of historical novae and supernovae, coming up with an event that coincided reasonably well with the position of Cas A. For six months in AD 369 both Byzantine and Chinese observers had seen a very bright object in Cassiopeia, and the case for an association looked to him every bit as good as that for the Crab nebula with the Chinese event of 1054.[43]

Shklovsky (1953d, f) was enthused about supernovae not only as radio emitters, but also as the source of cosmic rays.[44] He worked out that electrons and protons could both be accelerated to high energies in the aftermath of a supernova explosion, the electrons efficiently radiating (at least for a few thousand years) and the protons being injected into the Galaxy at large. The estimated rate of injection roughly matched that needed to account for the observed flux of cosmic rays. Novae, although far less energetic in their explosions, occurred in the Galaxy ten thousand times more frequently than supernovae, and therefore they too might be significant contributors to the cosmic rays. Like Ginzburg, Shklovsky (1953f:475) was impressed with how radio astronomy had fundamentally altered cosmic ray studies; he went so far as to proclaim that "radio astronomy has made the study of the nature and origin of primary cosmic rays a branch of observational astronomy."

It has been said that modern astronomy can be divided into two parts: (1) the portion dealing with the Crab nebula, and (2) the rest. The Crab's identification with the Chinese guest star of AD 1054 was a major step in understanding supernovae and stellar evolution, and its identification with Tau A, as we have seen, was an important milestone for radio astronomy. Nevertheless, by 1953 no one had yet pieced together a satisfactory explanation for both its optical and radio emission. The flat radio spectrum of Tau A suggested free–free radiation, which, however, then implied implausibly large values of $1-10 \times 10^6$ K for a temperature and 10^5 particles cm^{-3} for a density. Moreover, the expected amount of *optical* free–free radiation from such a hot, dense nebula was fully one hundred times that observed, and the expected intensities of optical spectral lines did not jibe at all with observations. One model put forth both by Greenstein and Minkowski (1953) and by Shklovsky (1953b) involved a steep-spectrum inner source (typical of most radio sources) whose radiation was heavily absorbed at low frequencies by a surrounding dense gas. This problematic model, however, was swept aside

---

[42] See Section 14.6 on the similar, contemporaneous debate in the West about radio stars versus radio nebulae.

[43] Shklovsky later used Mills's (1952a) survey as a basis to check for yet more associations of galactic-plane radio sources with ancient chronicles of objects that might have been novae or supernovae. He found three more (from AD 185, 827, and 1006), and considered that this confirmed that indeed all supernova remnants were radio emitters. [Shklovsky (1954), "Identification of the strongest discrete sources of radio radiation in the Galaxy with supernova remnants of explosions occurring during the last 2000 years" [in Russian], *Dokl. Akad. Nauk SSSR* **94**, 417–20]

[44] Others in the West had earlier favored supernovae as the source of cosmic ray particles, but had not developed the idea in detail. See D. Ter Haar (1950), "Cosmogonical problems and stellar energy," *Rev. Mod. Phys.* **22**, 119–52 (especially pp. 142–6).

one day by Shklovsky (1982:31), who later vividly told the story of his "brainstorm":

> On a sunny April day in 1953 I stood for rather a long time on Pushkin Square [in Moscow] waiting for Tram No. 17. While waiting, my attention was drawn to a newspaper stand where I read a notice that gave me great joy.

(This notice, about one month after Stalin's death, discredited the so-called "Jewish doctors' plot" against Stalin, a claimed collusion that Jews (such as Shklovsky) feared might lead to their persecution.[45])

> When the tram arrived, of course overfilled, I was in almost a somnambulent state as I jammed myself on. And then I had an illumination, as if by lightning: at once I understood the nature of the optical continuum radiation of the Crab nebula. ... It was not at all due to free–free transitions in an ionized gas, but, like the radio spectrum, had a synchrotron nature. If there were relativistic electrons with an energy of about 10^8–10^9 eV that were responsible for the Crab's synchrotron radio radiation, why should there not be those of 10^{11}–10^{12} eV, whose synchrotron radiation ... would be in the optical range? ... I made the entire calculation ... in my head in the incredible crush on the tram during the 45 minute ride. Immediately upon arriving home, I wrote in one sitting a paper that G. A. Shain [an astronomical colleague] quickly recognized as an [important] new theory and communicated to the Academy of Sciences. This was a rare event – to fill such a large gap in our knowledge.

Shklovsky's (1953e) insight was to ascribe not only the radio emission of the Crab to synchrotron processes, but also the optical emission. He showed that the small ratio of optical to radio intensity resulted simply from an expected scarcity of higher energy electrons needed to produce more energetic optical photons. Illustrating that creative interpretations sometimes must baldly ignore at least some data, Shklovsky minimized the lack of consistency in his hypothesis posed by (a) the constant radio intensity from 40 to 1200 MHz (five octaves in frequency), versus (b) the change in intensity of a factor 1000 from radio to optical frequencies (18 more octaves). Shklovsky's scheme solved another bothersome aspect of previous interpretations: the mass of the Crab nebula was reduced to only a few hundredths of a solar mass, as opposed to previous estimates (from radio data) ranging as high as 600 solar masses, much larger than expected from the explosion of a single star. The *optical* sky was acquiring its first high-energy emission process, only three years after its introduction into radio work. This was a bold step, for although interpretations of the radio sky often invoked such processes, this had not happened in optical studies; for example, Minkowski had earlier concluded that the discrepancy between the Crab's optical and radio spectra must result from a nonthermal effect at radio wavelengths, never considering the parallel possibility of such at *optical* wavelengths.[46]

One candidate for an observational test of Shklovsky's notion was to search for the linear polarization characteristic of synchrotron radiation, but he did not propose this because he assumed that the magnetic fields in the Crab nebula would be so entangled that the resultant polarization, even of individual sections of the nebula, would statistically average out to a value so close to zero as to be undetectable. Others, however, were more sanguine about measuring optical polarization. In 1952 Soviet astrophysicist Isaak M. Gordon of Kharkov, adapting Ginzburg's ideas, had proposed that (optical) *solar* flares might be emitting synchrotron radiation and that a suitable test would be to search for polarization in their light. After Shklovsky's suggestion of optical synchrotron radiation from the Crab, Gordon and Ginzburg urged searches for optical polarization there, too.[47] But the Crab nebula cannot be observed optically in the spring and much of the summer, and so it was not until late summer of 1953 that Soviet observers first got their chance to check Shklovsky's idea with a test he himself thought hopeless. In August V. A. Dombrovsky, an astronomer at the Byurakan Observatory in Armenia who had been mea-

---

[45] Shklovsky:author, 20 November 1984; Shklovsky (1979:1N); Shklovsky (1991:109–110). Also see *The Doctor's Plot of 1953* by Y. Rapoport (1991, Cambridge: Harvard University Press).

[46] Minkowski:Piddington, 24 May 1951, cartons 4–5, MIN.

[47] See Ginzburg (1953b:1136, 1985:417–18) and I. M. Gordon (1954), "On the physical nature of chromospheric flares" [in Russian], *Dokl. Akad. Nauk SSSR* **94**, 813–16. Publication of Gordon's paper was delayed until well after the original presentations of his ideas at conferences in 1952.

suring starlight polarization, took on the Crab nebula. Using a rotatable polaroid film in front of a photoelectric photometer, by October he had convincingly found that the Crab's light was about 12% linearly polarized (Dombrovsky 1954). At about this same time, Mikhail A. Vashakidze, a Georgian astronomer who had previously measured solar polarization, took photographs, also behind a rotatable polaroid, of nebulae within the Milky Way, other galaxies, and the Crab nebula. Vashakidze (1954) briefly reported them all to have significant linear polarization, with the Crab being largest at 22%. These experiments were hailed in the Soviet Union as proof that the synchrotron process was operating at optical wavelengths in the Crab nebula and, by extension, at radio wavelengths in the Crab and elsewhere. In the West, however, the story was otherwise.

15.4.3 Why was the synchrotron mechanism unpopular in the West?

Despite its initial introduction in the West in 1950 and over a dozen papers by Ginzburg and Shklovsky developing the idea, synchrotron explanations for galactic radio noise were never treated as more than a curiosity in the West until about 1955.[48] Why was this? First, the Russian papers were simply not read in the West, as has been discussed in Section 10.3.4. Knowledge of the Russian language was minimal, Soviet journals were rare in the West, few articles were ever translated (and then only after several-year delays), and Soviet researchers did not attend international radio and astronomical conferences until 1955. Consider that in the seven major reviews of radio astronomy written in the West over the period 1947–54, only five Soviet papers were ever cited, all of which were published prior to 1949![49]

The synchrotron ideas of Alfvén and Herlofson and of Kiepenheuer were not entirely ignored in the West. Five papers (Section 15.4.1 and Tangent 15.4) discussed the theory, but, with the exception of Hoyle's 1954 note, they came from non-influential players and thus lay fallow. Archival evidence nevertheless indicates that synchrotron theory was informally discussed within the major groups during this period. The idea continually reappeared, only to be dismissed because of various troubles. And since "negative results" tend to remain unpublished, considerations in print are few in the Western literature of 1950–55. The following notable mentions and absences convey the flavor of contemporary attitudes:

April 1951 – Ryle (1951:110) discusses the "Schwinger mechanism" at a conference as a good possibility for emission from sunspots and from radio stars, but says it needs a detailed examination.

September 1951 – Van de Hulst (1951:170) in a review mentions synchrotron radiation briefly as one of many possibilities for radio sources, but does *not* mention it in the context of galactic noise, despite showing in detail that the latter's intensity is not understood.

July 1952 – in a popular review Ryle and Ratcliffe state that it is possible that some of the galactic noise is interstellar in origin, but so far there exists no explanation to account for it.[50]

February 1953 – Oort and Westerhout write to Ryle that "a radio-emission effect in the interstellar gas of unknown type seems quite well possible." Ryle agrees, although he thinks that in general higher density regions would work better.[51]

March 1953 – Ryle writes to Baade that neither free–free radiation nor plasma oscillations seem to work for Cas A and Cyg A, and then says "it is possible that other radiation mechanisms may exist, if high energy electrons are present, as they may well be, but again rather large (10^{-4} G) fields might then be necessary."[52]

September 1953 – Mills writes in a letter that "in searching round for possible mechanisms for producing the [galactic background] radiation I was led back to the old idea that Schwinger radiation from cosmic ray particles moving in the galactic magnetic field could be responsible." He then notes that he has calculated that if this were correct, the galactic noise should be

---

[48] Synchrotron radiation as an explanation for *solar* radio bursts, however, *was* considered more seriously in the West, and has been discussed in Section 13.2.2.

[49] This situation began to change in 1954, at least in Ryle's group. John Baldwin's records indicate that weekly literature reviews that year did include Russian articles. [Baldwin:author, 18 April 1990]

[50] M. Ryle and J.A. Ratcliffe (1952), "Radio-astronomy," *Endeavour* **11**, 117–125.

[51] Westerhout:Ryle, 19 February 1953; Ryle:Westerhout, 19 March 1953; both box 1/3, RYL.

[52] Ryle:Baade, 18 March 1953, boxes 3/4, MIN.

polarized by as much as 2–5%. This, however, is found to disagree with observations by Bolton of < 1% polarization at 100 and 400 MHz.[53]

1954 – Smith writes in a review that "no satisfactory explanation has yet been given" for the mechanism of emission for radio stars.[54]

May 1954 – Jennison (1954:113–18, 133–45) in his thesis discusses many possible mechanisms for radio radiation from sources, in particular Cyg A, but never mentions synchrotron radiation.

July 1954 – Joseph Pawsey and Ronald Bracewell's monograph *Radio Astronomy* (1955:230) mentions synchrotron radiation not once in its 354 pages, although it tantalizingly *does* have a discussion of the analogous problems presented by cosmic rays and by cosmic radio waves.

October 1954 – James A. Roberts (a CSIRO Radiophysics Division theorist) in a review states that synchrotron radiation theory "is the only one able to explain the intensity and spectrum of the observed radiation. Unfortunately no sensitive test of the theory has yet been proposed, and until one is forthcoming the theory must be treated with some caution" (Roberts 1954:398).

November 1954 – the second edition of Unsöld's influential text *Physik der Sternatmosphären* (1955:772), although including two new chapters on radio astronomy, relegates its sole mention of synchrotron radiation to a short footnote.

What was the reason for the West's lack of enthusiasm for synchrotron radiation? Scheuer, a Ryle student during 1951–54 and a first-rate theorist, much later suggested that understanding the physics of synchrotron theory was extremely difficult at the time, requiring an effort, given its speculative nature, beyond what most radio astronomers were willing to invest.[55] The main reason, however, appears linked to the several lines of evidence put forth by cosmic ray researchers that electrons (and/or positrons) were a negligible, perhaps altogether absent, component of primary cosmic rays. The definitive search of that time for primary cosmic ray electrons was conducted in 1949 by Edward P. Ney's group at the University of Minnesota.[56] Flying a cloud-chamber-laden balloon to an altitude of almost 30 km, they concluded that <0.2%, later revised to <0.6%, of cosmic rays (of energy >10^9 eV) were composed of electrons or gamma rays. Bruno Rossi of MIT was also searching at this time for electrons and placed an upper limit of 1%. This in fact caused him in subsequent analyses to assume that *no* electrons existed among the primary cosmic rays.[57] Perhaps the absence in the postwar Soviet Union of cosmic ray experimenters meant that Soviet theorists were less deterred by the lack of detectable electrons. In any case, Ginzburg (1976:23T) later recalled that during this period he was convinced (on theoretical grounds) that some portion of cosmic rays must be electrons.[58]

The presence of cosmic ray electrons was also problematic on the theoretical side in the West: in 1949 Enrico Fermi published a highly influential paper based on Alfvén's magnetohydrodynamic ideas, which he had absorbed the previous year during a visit by Alfvén to the University of Chicago. Fermi proposed that cosmic rays were accelerated to their high energies by innumerable reflections off galactic magnetic fields carried along by turbulent motions of interstellar gas clouds. This process not only provided an explanation for the energy of cosmic rays, but in the same stroke predicted that these cloud motions would lead to a general magnetic field of about 5×10^{-5} G, immensely higher than previously assumed values of 10^{-10}–10^{-12} G. Fermi's great authority meant that from the start his theory was highly regarded,[59] but it became even

---

[53] Mills (at Mt. Wilson):Pawsey, 30 September 1953; Pawsey:Mills, 27 October 1953; both file F1/4/MIL/1, RPS.

[54] F. G. Smith (1955), *Vistas in Astronomy* 1, p. 565.

[55] Scheuer:author, 18 March 1990.

[56] C. L. Critchfield, E. P. Ney and S. Oleksa (1952), "Soft radiation at balloon altitudes," *Phys. Rev.* 85, 461–7.

[57] B. Rossi (1949), "Electrons and photons in cosmic rays," *Rev. Modern Physics* 21, 104–12 (especially p. 108).

[58] Primary cosmic ray electrons were not detected until 1960, at a level of a few percent, depending on electron energy. [J. A. Earl (1961), *Phys. Rev. Lett.* 6, 125–8; P. Meyer and R. Vogt (1961), *Phys. Rev. Lett.* 6, 193–6]

[59] An anecdote concerning Fermi's authority at this time is illustrative. In the introduction to his 1949 paper, Fermi gave a brief explanation of Alfvén's "magneto-elastic waves" that moved with a velocity such that the energy in the magnetic field was equal to that in the material's motion. Alfvén had first proposed such waves in 1942, but had had little success in convincing anyone that they could exist in a plasma. But, as he recalled:

[In 1948] Fermi listened to what I said about my waves for five or ten minutes, and then he said: 'Of course such waves

more so when at about this same time appeared strong observational evidence for his predicted magnetic field. It came in the form of linear polarization of starlight, which was soon taken to result from the presence of a magnetic field of at least 10^{-5} G preferentially orienting elongated interstellar grains. Fermi's mechanism, however, had a further consequence, namely, the absence of electrons in the primary cosmic rays. Fermi (1949:1171) argued that any interstellar electrons would quickly drop from the scene, because they would more quickly lose energy (through both radiation (!) and collisions) than they would gain it through magnetic-field reflections. The dominance of Fermi's theory, combined with the recent discovery of starlight polarization, meant that magnetic fields, cosmic rays, and hydrodynamics became an important part of the standard picture of the interstellar medium in the early 1950s (Greenstein 1975:23T, 1980:132B). But despite certain detractors such as Unsöld (1951), this same dominance, combined with the experimental limits on cosmic ray electrons, veered Western researchers away from synchrotron radiation. In contrast, Fermi's theory for the origin of cosmic rays was not accepted by Soviet theorists, although they did adapt it to the interiors of radio sources (Ginzburg 1953a).

In another influential theoretical study E. Feenberg and Henry Primakoff (1948) of Washington University, St. Louis, calculated the deleterious effects on any supposed cosmic ray electrons as they navigated a sea of interstellar photons originating in stars and in the radio background. The optical photons were found to have negligible effect, but the far more numerous radio photons quickly scattered away any cosmic ray electrons.[60] If one instead took the stance that cosmic rays were of solar origin, as advocated for instance by Alfvén and by Edward Teller,[61] the calculations of Feenberg and Primakoff indicated that electrons would still be extremely rare, this time due to scattering losses off solar photons. Even an extragalactic origin for cosmic rays led to a similar absence of electrons. In sum, these ideas and those of Fermi discouraged belief in cosmic ray electrons in the West well into the mid-1950s.[62]

The acceptance of synchrotron radiation in the West did eventually occur, but the story takes us several years beyond the nominal period of the present history. In brief, the key events began with the enthusiastic response that Oort accorded the Soviet measurements of polarized light from the Crab nebula. As discussed above, in the Soviet Union these data were hailed as confirmation of Shklovsky's bold suggestion that *both* the radio and optical radiations from the Crab nebula were synchrotron. Oort had been a member of the first Western delegation of astronomers since the war to visit the Soviet Union (in May 1954 for the dedication of the rebuilt Pulkovo Observatory), and had learned firsthand of Ginzburg and Shklovsky's embrace of synchrotron radiation.[63] When he later heard of Dombrovsky's polarization results, he persuaded his colleague Theodore Walraven to modify an instrument that was already being developed to measure the light of the Crab nebula. Observing from Leiden in early 1955, Oort and Walraven (1956) then confirmed the Soviet results and carried them much further – mapping the degree of polarization over the nebula, detailing the theory, and showing that Shklovsky's scheme worked beautifully. Oort presented these results to IAU meetings in the summer of 1955 and persuaded Baade (1956) to quickly measure the polarized light with the 200 inch telescope: the resulting photographs, showing

---

could exist.' Fermi had such an authority that if he said 'of course' today, every physicist said 'of course' tomorrow. Actually he published a paper in which he explained them in such a clear way that no one could doubt their possible existence. What I had not succeeded to do in six years was done by Fermi in [only an] introduction.

[H. Alfvén, "Recollection of early cosmic ray research," pp. 427–31 in Sekido and Elliot (1985)]

[60] Feenberg and Primakoff (1948:465) supposed that any cosmic ray electrons were trapped within the Galaxy by a magnetic field and thus "constrained to spiral indefinitely." Radiation from these same spiralling electrons, however, was not considered.

[61] R. D. Richtmyer and E. Teller (1949), "On the origin of cosmic rays," *Phys. Rev.* **75**, 1729–31.

[62] In 1953 Fermi modified his theory to answer several criticisms, but kept its essential features [*Ap. J.* **119**, 1–6 (1954)]. Moreover, Teller changed his mind and accepted a galactic origin for cosmic rays; he took cosmic rays to originate in radio stars and then be accelerated by the Fermi mechanism, and also gave theoretical reasons why electrons must constitute <0.1% of primary cosmic rays [*Rept. Prog. Phys.* **17**, 154–72 (1954)]. One opinion that cosmic ray electrons *did* exist in great numbers, albeit of greatly reduced energy (compared to protons), was that of P. Morrison, S. Olbert and B. Rossi (1954), *Phys. Rev.* **94**, 440–53.

[63] Oort (1978:38–42T); Shklovsky (1982:34).

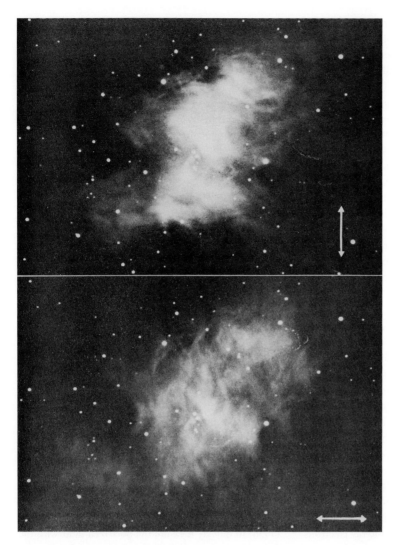

Figure 15.4 Photographs of the Crab nebula (covering the wavelength range 5400–6400 Å) taken by Baade with the Palomar 200 inch telescope in September 1955. Each was taken behind a polaroid filter transmitting light polarized in the indicated directions. The marked change in the nebula's appearance showed that the light was polarized, as predicted for synchrotron radiation. (Baade 1956)

marked variations in the Crab's appearance in variously polarized light, were spectacularly convincing (Fig. 15.4).[64]

Once the theory was seen to work well in the Crab nebula, it was soon generally adopted for most discrete radio sources, as well as for the galactic background radiation. The question of polarized *radio* waves from Tau A also naturally arose, but calculations indicated that large beam sizes and possible depolarization effects at all but the shortest radio wavelengths would make it an extremely difficult proposition.[65]

---

[64] Osterbrock (2001) discusses the history of these polarization observations in the West.

[65] Radio polarization (at a wavelength of 3.2 cm) was first measured in Tau A, at a level of ~7%, at the US Naval Research Lab by C. H. Mayer, T. P. McCullough and R. M. Sloanaker (1957), "Evidence for polarized radio radiation from the Crab nebula," *Astrophys. J.* **126**, 468–70.

Thus it was that the general acceptance in the West of the synchrotron mechanism for *radio* radiation occurred ironically as a result of *optical* measurements.

15.5 THE BEGINNINGS OF RADIO COSMOLOGY

We have seen that over the period 1948–54 more and more evidence accumulated to persuade the principal players that radio stars were as much extragalactic as they were galactic, perhaps more so. This meant that the intrinsic radio powers and sizes of these sources were in many cases even larger than those of (optical) galaxies and opened up a host of new questions about the sources themselves. But long before these were answered, the sources as an ensemble were used to ask fundamental questions of prevailing models of the universe. Although this story only starts at the end of the present history's period, it is important briefly to note this basic transition of radio astronomy from models of the Galaxy to models of the universe. We have already seen how several researchers ascribed an isotropic component of the general radio background to the integrated effect of the universe's galaxies, but a few added a new element when they actually constructed cosmological tests based on the radio data. The first to do any radio cosmology was Van de Hulst (1945). His was a short but clever effort with paltry data, but it stood alone for almost a decade. Using arguments related to Olbers's paradox,[66] he showed that the universe must be expanding (see Section 16.1 for full details). Next was Shklovsky (1953c), who made deductions that he considered "the first concrete cosmological results obtained with the help of radio astronomy." Requiring that a ~150 K (at 100 MHz) isotropic component existed in addition to his Milky Way radio corona, he identified it with the "metagalaxy," that is, the integrated effect of distant radio galaxies. Shklovsky found that the (optically) observable universe (extending to ~1000 Mpc) would contribute at least four times the observed 150 K – this he called a "photometric [Olbers-type] paradox." He then showed, however, that the paradox was resolved if the universe were truly expanding, in which case the effects of the cosmological redshift (if it operated on radio wavelengths as it did optical) would sufficiently weaken the received flux from distant radio galaxies.

These radio cosmological tests were isolated and not followed up, but that by John R. Shakeshaft, a student of Ryle's, led to one of the main themes of radio astronomy in the ensuing decade. Over the period 1952–54 Ryle had completely changed his mind and come to believe that virtually all the radio stars were extragalactic, sprinkled throughout the universe (Sullivan 1990). On this idea Shakeshaft (1954) then calculated that the isotropic radio component could be accounted for by "Relativistic Cosmology" (Big Bang), but not by the recently proposed steady-state theory of Hoyle, Gold, and Hermann Bondi. This was the first spark, but only a prelude to the cosmological firestorm that ensued after similar interpretations of the Cambridge group's subsequent 2C survey published in 1955. Log N–log S plots of its 2000 sources, in the manner of those discussed in Section 14.3.2 for earlier, much smaller samples, were used to strongly argue against the steady-state theory. As in earlier episodes dealing with the sun (Section 13.2.2) and radio stars (Section 15.3), Ryle and Hoyle again went after each other like mongoose and cobra.[67]

15.6 THE RADIO SKY AND COSMIC RAYS

By 1953 radio astronomers had surveyed the sky at a variety of wavelengths, hundreds of radio stars had been cataloged, and the roll of exciting discoveries contined to grow. Explanatory frameworks for all this, however, were notably unsettled, as evidenced by

---

[66] Olbers's paradox considers the integrated effects of a distribution of discrete sources. The entire sky should have a brightness equal to that of any individual source in the case of a Euclidean geometry universe that is infinite in extent, infinitely old, homogeneous, and static.

[67] See Edge and Mulkay (1976) for a full history and sociology of these further developments. For accounts by three of the protagonists in the source-count controversies, see Hoyle (1994:269–71 + 408–11), Mills, "Radio sources and the log N–log S controversy," pp. 147–65 in Sullivan (1984a), and Scheuer, "Radio source counts," pp. 331–45 in *Modern Cosmology in Retrospect* (1990, eds. R. Bertotti *et al.*) (Cambridge: Cambridge University Press). H. Kragh's excellent *Cosmology and Controversy* (1996:chapters 6–7) also covers this period. Sullivan (1990) details how Ryle's group and other radio astronomers first engaged cosmology.

the startling variety of models illustrated in Fig. 15.2. Indeed, it almost seemed that more and more data only made things worse. As Scheuer wrote in 1953:

> It is clear that the problem of the galactic and extragalactic radiation is not yet solved, and has indeed become more complicated with the increase in observational data.[68]

Mills (1976:34T) too found the various observations and their implications complex; at one stage he even resorted to a logical flow diagram to aid him in developing interpretations. But until it could be established what the bulk of the radio stars *were*, which probably meant finding many more optical identifications, one had only fragmentary clues as to how the sources radiated and from what distance – a few parsecs or a few megaparsecs? It was certainly risky employing them to explain any portion of the general galactic noise. Nevertheless, as we have seen, many people did so. But when consensus was required, as in 1953 with an official URSI committee composed of Bolton, Hanbury Brown, Mills, and Smith, their conclusion was "there is as yet no clue as to whether the general background radiation from our own Galaxy is due to the integrated effect of the type of galactic nebulae so far identified."[69]

One of the key steps in breaking this stalemate was the idea of synchrotron radiation. This was important in the intellectual development of radio astronomy because it supplied an efficient, plausible mechanism by which individual sources, both galactic and extragalactic, could radiate large radio powers. It also provided an attractive interstellar source for the general galactic noise. But although it was introduced first into radio astronomy, perhaps its greatest effects were on other specialties. In optical astronomy, the polarized light from the Crab nebula meant that astronomers had a new tool for studying high-energy processes, and so became more oriented to a high-energy universe exemplified by the various radio nebulae with optical identifications.

It was the new link that synchrotron radiation forged between radio waves and cosmic rays, however, that was most significant. In many ways these two phenomena represented similar anomalies in the mostly quiet universe that had been established by (optical) astronomy. Neither radio waves nor cosmic rays would have been predicted from the optical evidence, and the observed intensities of both gave evidence of unprecedented energetic processes. H. Elliot has nicely expressed this for cosmic rays, but his words apply also to extraterrestrial radio waves:

> An astronomical oddity, the cosmic rays were an ever-present reminder of the existence of extreme physical processes in the Universe which remained unidentified and seemingly isolated from the general body of astronomical knowledge.
> (Sekido and Elliot 1985:xiv)

What did this link between radio waves and cosmic rays mean in practice? First, it provided the first good clues to where cosmic rays originated. Because the charged particles making up cosmic rays were deflected by interstellar, interplanetary, and terrestrial magnetic fields, their direction of arrival often bore little relation to the direction of their origin. Moreover, because the angular resolution of cosmic ray instruments was far worse than even that of radio telescopes, most properties of cosmic rays were only characterized as quantities averaged over the entire sky. Astronomers had basic difficulties relating to such a phenomenon – it was as if they were given only one composite stellar spectrum for the entire sky. But radio antennas, despite being berated by astronomers for their atrocious lack of angular discrimination, were now able to give directionality to the world of cosmic rays. Ginzburg and Shklovsky emphasized this point early on, and Phillip Morrison (1915–2005), an American physicist involved in theoretical cosmic ray work during the early 1950s, later recalled:

> Where could the energy source for cosmic rays be? What could it be? It was quite clear you'd never find out from cosmic rays because they don't go in straight lines. ... Therefore you wanted to look at these radio sources that might be pointing you to where the cosmic ray source is.
> (Morrison 1976:9T)

There was another way in which cosmic ray physics changed at this time. As reviewed in the introduction

---

[68] P. A. G. Scheuer, "The origin of galactic radio radiation," 1 October 1953, Hamilton Prize essay, file 34 (uncat.), RYL.
[69] "Discrete sources of extra-terrestrial radio noise," 1954 (preface dated September 1953), Special Report No. 3, Brussels: URSI, 56 page booklet (quotation from p. 15).

to Section 15.4, cosmic rays had long supplied physicists interested in the fundamental structure of matter with high-energy experiments that were impossible to achieve on earth. During the early 1950s, however, the newest accelerators began to supply energies sufficiently high (and controllable) that most particle physicists shifted their attention from the heavens to giant machines. A number of other cosmic ray researchers realized that although it had become less useful to study cosmic ray particles *per se*, radio astronomy was showing them that one could now profitably seek their origin:

> The cosmic ray people were looking for somewhere to go. ... And here was this radio astronomy dealing with relativistic particles and it was interesting, very interesting. ... Before that, nobody cared where [the cosmic rays] came from. There were a few theories, but they were all bizarre because nobody could understand it.
>
> (Morrison 1976:4T)

These physicists were to become an important new breed of theorist in astronomy: astrophysicists specializing in high-energy processes. They were not in radio astronomy as such, but radio observations, aided by the Crab nebula in particular, led to their migration. They also later became central to the X-ray and gamma ray universes that opened in the early 1960s (Section 18.2).

TANGENT 15.1 THE INTEGRATED BRIGHTNESS OF A DISTRIBUTION OF RADIO STARS

Consider a uniform distribution of n identical radio stars out to a distance R, having a space density ρ (per pc^3) and a luminosity per unit bandwidth p (W Hz^{-1}).

(1) The flux density S (W m^{-2} Hz^{-1}) received from the farthest, at a distance r, is $p/(4\pi r^2)$. Combining this with $\rho = 3n/(4\pi r^3)$, one has $p\rho^{2/3} = (6\pi n)^{2/3} S$, a known quantity.

(2) The specific intensity I (W m^{-2} Hz^{-1} ster^{-1}), proportional to the brightness temperature of the galactic noise and arising from the integrated effect of all the stars, is $\rho p R/4\pi$ (assuming no absorption); thus $\rho p = 4\pi I/R$. Assuming a value for R, one then has two equations in the two desired unknowns ρ and p. If one allows p to have a dispersion in values, then it turns out that the derived value for r increases and that for the average of p decreases.

TANGENT 15.2 THE MULTI-FREQUENCY GALAXY MODEL OF PIDDINGTON (1951)

Theorist Jack Piddington (1951) made a critical review of all existing surveys, ranging in frequency from 10 to 3000 MHz, and endeavored to correct their intensities for an optimum intercomparison – a daunting task given the uncertainties in base levels, intensity calibrations, and corrections for beam sizes. He found that for medium frequencies (100 to 480 MHz) the spectral index α of the galactic noise varied at different locations from 2 to 3, but that at the lowest frequencies the index became ~2.0 and at the highest it flattened out to zero. He also presented evidence that the observed width in galactic latitude of the radiation became much larger as one went to lower frequencies (for example, compare Fig. 6.4 at 64 MHz with Fig. 4.7 at higher frequencies), which meant a changing spectrum with latitude. This implied that single-component, optically thin models (such as Bolton and Westfold's) were too simple – one had to introduce a second component and/or absorption effects. Piddington chose to do both, and worked out that galactic noise originated in interstellar gas as well as in special stellar atmospheres that emitted nonthermally. At the highest frequencies free–free radiation from gas dominated; at mid-frequencies, emission from special stars was strongest; and at the lowest frequencies, where interstellar gas was optically thick, absorption caused a turndown in the spectrum. Because the hot gas was closely confined to the galactic plane, as one moved from low to high latitudes the observed spectrum naturally steepened.

TANGENT 15.3 THE GALACTIC PLANE MAPPING OF SCHEUER AND RYLE (1953)

In 1951–52 Scheuer and Ryle (1953) applied interferometric techniques to measure the width and intensity of a narrow layer of ionized hydrogen predicted by Westerhout and Oort (1951) and not visible to existing surveys because of poor resolution. The basic idea was to observe during the short period when the galactic plane passed through the zenith at Cambridge. If one then oriented the axis of a two-element interferometer so that the lobes were parallel to the galactic plane, it was possible to make high-resolution one-dimensional cuts across the plane. In practice this was achieved by hauling 27 × 6 ft frameworks (reflector, dipoles, and tilting mechanism for observing several galactic longitudes) across muddy fields in order to measure the fringe visibility of the radiation at about 15 spacings. Indeed, they did find a bright band of only 2° width along the galactic equator. The band's width was about as expected, but its intensites at 210 MHz and at 81 MHz, although amounting to a small fraction of the total galactic noise, were much higher than expected from free–free emission. Unless some new and numerous population of radio stars were situated only right in the galactic plane, the interstellar gas seemed to be at a kinetic temperature of at least 18,000 K. But this was extremely puzzling since interstellar ionized regions on theoretical grounds could not be heated above 13,000 K (and usually less) because of the thermostat-like action of radiative losses. In hopes of seeing the bright radio emission turn into an absorption band, a few measurements were also made with two rhombic antennas at 38 MHz, a frequency low enough that the optical depth of free–free radiation should have been large. The belt, however, stubbornly remained bright, implying an even more discrepant kinetic temperature of ~25,000 K.[70]

TANGENT 15.4 MINOR SYNCHROTRON STUDIES IN THE WEST (1951–54)

Following the papers suggesting synchrotron radiation as a mechanism for radio stars (Alfvén and Herlofson 1950) and for the galactic noise (Kiepenheuer 1950), there were five small follow-up studies, but none were particularly influential.

The first was by Bernard Kwal, a physicist at the Institut Henri Poincaré in Paris. Kwal, who was working on problems of radiation losses in particle accelerators, had become interested in the radio emission problem through contact with Jean-François Denisse and Marius Laffineur (Section 10.4). Within a few months of reading the two *Physical Review* notes, he took up the synchrotron idea, but now for high-energy *protons* rather than electrons. Kwal (1951) primarily favored protons because they were known to exist in cosmic rays (unlike electrons) and indeed were the dominant component. The physics of synchrotron radiation showed, however, that protons also had significant disadvantages relative to electrons: for the same strength of magnetic field and particle energy, the intensity of emitted radiation was fully 10^{-13} times weaker and the frequency of maximum emission 10^{10} times lower! This meant that Kwal had to appeal to extremely large proton energies and magnetic fields (several thousand gauss), which he found either in sunspots (for solar radio bursts) or in the magnetic stars of Babcock (for radio stars). Kwal also derived an important relationship, given in Section 15.4.1, between the energy spectrum of cosmic rays and the resultant radio spectrum of the synchrotron radiation.

A year and a half later, George W. Hutchinson, a nuclear physics student at the Cavendish Laboratory, also wrote about the possibilities of synchrotron radio radiation from cosmic rays. He was not a member of Ryle's group, but consulted with them frequently, as well as with Gold and Hoyle. Hutchinson (1952) detailed how radio emission might originate in regions with densities of ~10^9 cm^{-3}, intermediate (logarithmically) between the general interstellar medium and stellar atmospheres. Such densities, he suggested, might be found in the ejected shells of novae and supernovae or in regions of star formation. Hutchinson showed that electrons accelerated in magnetic fields of 0.003 to 0.1 G would be sufficient to give the observed radio photons.

The only study in the West to apply synchrotron ideas to the general interstellar medium, that is, to follow up specifically on Kiepenheuer's suggestion, was by Heinrich Siedentopf and G. Elwert (1953), astrophysicists at Tübingen University, West Germany. They found that electrons comprising ~15% of all cosmic rays and in a magnetic field of 10^{-6} G would be sufficient to produce the galactic noise intensity. Slightly later, Richard Twiss (1954) also argued that synchrotron radiation could best account for the intensity and spectrum of radio sources and produced a detailed model of Cas A based on "the interaction of a relativistic electron gas with a local magnetic field."

---

[70] The results of this study were debated over the years, eventually leading to a consensus that a narrow band indeed existed, but due to a population of discrete sources, not ionized hydrogen. I thank Miller Goss for discussions on this topic.

Finally, Hoyle (1954) also came to favor synchrotron-type processes in radio stars, and in January 1954 said so at a radio astronomy conference in Washington, DC. He had previously argued for plasma oscillations to generate radio waves,[71] but now changed alliances to, as he put it, the "fast-particle school." Synchrotron radiation, which naturally yielded the observed steep radio spectrum, seemed a more straightforward process so long as any cosmic ray electrons actually existed. Energy requirements also seemed more easily satisfied: for instance, only 3×10^{-5} of the total collision energy in Cyg A needed to be in relativistic electrons.

---

[71] Hoyle (1981:136B:520) recalled that at the Cavendish Lab Hutchinson "got on people's nerves" about applying Schwinger radiation to astronomy, and that in retrospect it was a mistake not to have encouraged him more. Also see Edge and Mulkay (1976:444).

16 • The 21 cm hydrogen line

Hydrogen, the most abundant element in the universe, fortuitously possesses a spectral line conveniently located in the middle of the radio spectrum at a wavelength of 21 cm. Under extraordinary circumstances during World War II this line was calculated to be of importance and eventually, in 1951, it was detected from cold gas between the stars. For the first time radio astronomers could measure the motions of interstellar clouds (via the Doppler shift) and bring the powerful techniques of spectroscopy to bear on the physical conditions of this gas. Moreover, they could map the distribution of hydrogen in the Milky Way, far beyond the local region visible at optical wavelengths.

This chapter recounts the prediction, discovery, and first investigations of the line. The story is tripartite, with groups in the United States, Holland, and Australia involved. The early history of the 21 cm line is especially noteworthy because of the close ties that these groups soon established in their research, despite their locations around the globe and despite a tendency toward sharp rivalries, as we have seen in previous chapters, between the various radio astronomy groups of the day. The hydrogen-line research was also unusual in how completely it was integrated with optical astronomy from the beginning.

The following sections look at the prediction in Holland, the slow spread of the word during the postwar years, the discovery at Harvard University, the faltering first steps of Dutch radio astronomy followed by success with the 21 cm line only weeks after the American discovery, the Australian dip into the foray, and the interpretations and theory of the first surveys.[1]

16.1 PREDICTION

On the afternoon of 15 April 1944 a Dutch postgraduate student addressed a gathering of astronomers at Leiden Observatory in Holland. Hendrik C. Van de Hulst's talk on "The origin of radio waves from space" began inauspiciously:

> We have just heard about the techniques of reception of the extraterrestrial radio waves, and now I wish to examine the "transmitters." When Professor Oort gave me this assignment a month ago, he told me that it would not involve much, but I'm still not finished.[2]

Thus began a seminal lecture in the history of radio astronomy, but before examining it further, its circumstances require explanation.

Van de Hulst (1918–2000) (Fig. 16.1) had been a student under the solar astrophysicist Marcel G. J. Minnaert at Utrecht during the trying conditions of the German occupation. He quickly became involved in an impressive range of problems, extending from telluric absorption lines to his thesis work on the theory of light scattering and absorption by interstellar dust.[3] But Minnaert's removal to a detention camp[4] in early 1944 led to Van de Hulst informally shifting his studies and research to nearby Leiden for a few months. There the political situation was only slightly better, and Jan H. Oort (1900–1992) (Fig. 16.1), then assistant director, spent most of his days away from Leiden, hiding as a result of his earlier prominent role in protesting the dismissal of Jewish professors. This protest had led to the official closure of the university in 1940, but nevertheless Oort occasionally cycled in from the countryside to give unauthorized lectures at the observatory.

---

[1] Earlier versions of parts of this chapter have appeared as Sullivan (1988b) and Sullivan (2000).

[2] "Lezing," April 1944, box 38, VDH.

[3] Van de Hulst (1998) wrote the memoir "Roaming through astrophysics," *Ann. Rev. Astron. Ap.* **36**, 1–16. An obituary is at *Astron. and Geophys.* **42**, 1.33–35 (2001).

[4] Minnaert was one of a group of prominent Dutch citizens from a broad spectrum of fields who were kept as hostages by the Nazis at St. Michielsgestel. They were treated relatively well, but were under constant threat of death should the populace "misbehave."

Figure 16.1 Henk van de Hulst (left) and Jan Oort at a conference in 1949.

Since these were kept secret even from the director, Ejnar Hertzsprung, a Dane not considered politically astute, it was in a small darkroom underneath a telescope dome where Van de Hulst and others followed Oort's lectures on galactic structure and dynamics, subjects in which he had been preeminent since his studies in the 1920s under J. C. Kapteyn and P. J. van Rhijn at Groningen.[5]

Oort knew about the article by Grote Reber (1940b) in a smuggled issue of the *Astrophysical Journal*[6] and, based on Reber's preliminary 160 MHz results (Section 4.3), was immediately struck by the potential utility of galactic radio waves. In particular Oort realized that if there should exist a generally detectable *spectral line* in the radio spectrum, Doppler shifts could be used to study the location and rotation of the interstellar gas over the entire Galaxy, not just the few kiloparsecs to which optical observers were confined by interstellar dust.[7] He therefore organized a colloquium of the Astronomenclub, a society that fostered exchange of information among the major Dutch astronomical centers. Cornelius J. Bakker, a physicist from Philips Research Laboratories, was asked to review the radio techniques and scant observations of Jansky and Reber, and Van de Hulst was assigned two tasks: first, to review ideas on the origin of the radio waves, and second, to investigate chances for a radio spectral line. Thus did Van de Hulst come to give his famous talk.[8]

Van de Hulst based his lecture on the consecutive articles by Reber (1940b) and Henyey and Keenan (1940) (Section 4.3).[9] With these as a start he discussed the possible origins of the new galactic radiation. After first ruling out stars, whose total angle subtended on the sky was far too small, and second interstellar "smoke" (dust grains), whose temperature was far too low, he derived the theory of free–free radio emission (*bremsstrahlung*) in a somewhat different manner than had Henyey and Keenan. His calculated spectrum for a gas of temperature 10,000 K agreed well with their earlier study, however, and he too concluded that Reber's values at 160 MHz were of the right order. But he could not understand Jansky's 20 MHz intensities, fully a factor ten higher than predicted. This was unacceptable and so Van de Hulst reconciled observation and theory by placing the blame on Jansky's poorly known antenna pattern. After all, had he not deduced (correctly) that Reber was quoting a beamwidth about four times narrower than diffraction theory would allow? Might not Jansky too be in error?[10]

---

[5] A biography of Oort can be found on pp. xv–xxxi of *The Letters and Papers of Jan Hendrik Oort* by J. K. Katgert-Merkelijn (1997) (Dordrecht: Kluwer). Obituaries are available in *Pub. Astron. Soc. Pacific* **105**, 681–5 (1993) and *Quart. J. Roy. Astron. Soc.* **35**, 237–42 (1994). Also useful for Oort's career is *Oort and the Universe* (eds. H. Van Woerden *et al.*, 1980) (Dordrecht: Reidel). See Smith (2006) for a review of galactic structure studies over 1900–52.

[6] Reber's other 1940–44 articles were not available in wartime Holland. Also, at this time, an extensive review of the results from Reber (1940b) (and some of Henyey and Keenan [1940]) was published in the Dutch popular astronomy magazine *Hemel en Dampkring*. In that overview D. Koelbloed (1941) ended by saying "The last word is certainly not yet spoken about this remarkable radiation, which adds fresh interesting areas to the domain of astrophysics." I thank Richard Strom for informing me of this article (Strom 2005).

[7] Oort's prescience in this matter was later nicely put by Van de Hulst (1973:2T): "Oort is one of those persons with the property of a really fine nose to sense when something is important."

[8] See Van Woerden and Strom (2006) for other details of the period before this talk.

[9] In fact, Van de Hulst's knowledge of Jansky's work was based solely on these two 1940 articles. Also, since Reber neglected to state the diameter of his dish (but did give a picture), Van de Hulst resorted to estimating its diameter by comparing its size to that of a nearby hut!

[10] See Van de Hulst (1984:391) and Tangent 4.1.

Van de Hulst moved next in his lecture to the heart of the study: candidate radio spectral lines. First he examined the high-level transitions (today called recombination lines) of the hydrogen atom, focusing on the transition from principal quantum number $n = 341$ to $n = 340$ at 1.8 m wavelength (about that of Reber's receiver). He found such lines to be unobservable at wavelengths longer than about 1 mm since broadening from ambient electric fields would smear their radiation over large bandwidths and drastically reduce their intensity.[11]

Next he turned to the detectability of the 21 cm transition arising whenever the spin of a hydrogen atom's electron changed from being parallel to being antiparallel to that of its nuclear proton. This hyperfine line in hydrogen's ground state had never been measured in the laboratory, but some basic nuclear physics, combined with the measured value of the proton's magnetic moment, allowed a calculation of 1410 MHz for its frequency (perhaps accurate to ± 20 MHz). The line *strength*, however, was not so amenable to calculation, and in the end it could only be said that if the probability of spontaneous emission were greater than 8×10^{-17} s^{-1}, the line might be observable.[12] In other words the lifetime of hydrogen atoms in the upper state had to be less than 4×10^8 years, tremendously outside usual experience in (optical) astronomy with its fraction-of-second lifetimes. Some hope nevertheless arose from considering the strengths of other magnetic dipole transitions, which implied that the lifetime might even be an order of magnitude shorter than needed. Van de Hulst consulted at Leiden with several theoretical physicists, all of whom were guarded in their opinions, and so he couched his final prediction as follows:

> This possibility does not appear hopeless, even when we consider that the sensitivity of today's receiver installations must be improved by still another factor of 100... [But] until a rigid calculation is made, the existence of this line remains speculative.
>
> (Van de Hulst 1945:219)

Cosmology was the colloquium's final topic, sparked by Reber's (erroneous) claim to have detected the Andromeda nebula (Fig. 4.5 and note 4.23). If one took Andromeda's radio brightness to be typical for all spiral galaxies and assumed their radio spectra were flat (as for free–free emission), arguments related to Olbers's paradox[13] were possible. Van de Hulst reasoned that the finite brightness of the sky at radio wavelengths then implied that the observed redshifts of galaxies indicated an expanding, finite-age universe, and not a static universe of infinite duration where redshifts arose from a non-Doppler mechanism. The data were wrong, but the reasoning was sound – it would be almost another decade before radio results would engage cosmology (Section 15.5; Sullivan 1990).

16.2 POSTWAR DEVELOPMENTS

Van de Hulst's (1945) ideas regarding the 21 cm line were written up two months after his talk and published immediately after the war, but in Dutch in the journal *Nederlandse Tijdschrift voor Natuurkunde*. They therefore achieved little international currency through the printed word, but more through Van de Hulst himself when he took a postdoctoral position at Yerkes Observatory in America. Soon after his arrival in the summer of 1946 he visited Reber at his home in Wheaton, Illinois, and tried to persuade him to search for the 21 cm line. Reber (1975:50T) later recalled his skepticism:

> I took a rather dim view of it because it looked to me as though it was beyond the present technological capabilities. ... And then I tried to question him and it became apparent that it was really more an idea than theory. ... i.e., here was this transition in hydrogen, and there was hydrogen out in space, but I couldn't even get out of him (and he didn't know) whether this transition was going to show up in emission or in absorption.

---

[11] My analysis of Van de Hulst's calculations [box 38, VDH] in light of present theory indicates that he was correct in all respects except for a large overestimate of this Stark broadening, arising primarily from a transposition error. For further details see Sullivan (1982:299–300) and pp. 6–7 + 267–9 in M.A. Gordon and R.L. Sorochenko (2002), *Radio Recombination Lines* (Dordrecht: Kluwer).

[12] The assumed conditions for the interstellar medium in this calculation were as follows: hydrogen density of 1 atom cm^{-3}, path length of 20 kpc, thermal broadening in the HI clouds of about 100 kHz, approximately equal populations in the two hyperfine states, and optically thin transfer of the radiation. The modern value for the rate of spontaneous emission is 3×10^{-15} s^{-1}.

[13] See note 15.66.

At that time Reber was having enough trouble developing a receiver of sufficient stability and sensitivity to detect galactic continuum radiation at the high frequency of 480 MHz – putative narrow-band radiation at 1420 MHz was a yet more formidable prospect. Reber did not, however, completely discard the idea,[14] and a year later, once he had most of his 480 MHz problems solved, he started the development of a 21 cm spectrometer. His plan was to use a surplus, tunable "echo box" for the frequency selectivity and to build a multistage amplifier based on Sylvania's new "rocket" tubes – as with his previous receivers, no heterodyning or Dicke-switching was contemplated (Reber 1958a:22). But although various components were built, the project ended prematurely when Reber dismantled his antenna and moved to the National Bureau of Standards in 1947 (Section 4.6.1).

In that same year appeared the first publication in English concerning the 21 cm line – a brief mention in a 1947 review of radio astronomy in *Observatory* by Reber and Jesse Greenstein.[15] This in turn was read in Moscow by the theorist Iosif Shklovsky (Section 10.3.1), who was at once taken with the idea of radio spectral transitions in the Galaxy – "It set me on fire" (Shklovsky 1982:17). In 1949 he published a key paper on the subject in *Astronomichesky Zhurnal*. Without access to Van de Hulst's article, Shklovsky worked out from first principles expected signal strengths, excitation mechanisms, and optical depths for the 21 cm line. He calculated the transition probability (deriving a value a factor four too low; see Section 16.6.3) and argued that the line should be collisionally excited in a rather hot (5000 K) interstellar medium (adopting ideas stemming back to Arthur Eddington). Despite the transition's low probability, Shklovsky calculated an expected antenna temperature for the line of ~200 K and pointed out that this was quite feasible to observe with a 10 m diameter antenna, even after consideration of the narrow receiver bandwidth required. His closing sentence read: "Soviet radio physicists and astronomers should endeavor to solve this intriguing and important problem.[16]

According to Shklovsky (1979:126A, 1982:19–20), he persuaded Viktor Vitkevich, one of the key experimentalists in Soviet radio astronomy at the time (Section 10.3.2), of the importance of this project and got him to the stage of developing plans for a 21 cm receiver. But things came to an abrupt halt in early 1950, when the renowned physicist Lev Landau advised Vitkevich that such a search was hopeless – the line had far too small a transition probability and the amount of hydrogen in interstellar space was completely unknown. Also, fellow Russian Vitaly Ginzburg (1976:37–8T) later recalled his own lack of success in getting a search mounted:

I remember very well that it was absolutely clear in the late 1940s that we must have hydrogen. We spoke to people: "Do this!" [but they replied] "Oh, it is very difficult, very difficult." After Ewen and Purcell first detected the hydrogen line [in 1951], I remember that one of our men came to me and said, "Why did you not insist on it? Because it is so easy! Why did you not persuade me?"

Other groups also paid attention to Van de Hulst's prediction. As briefly mentioned in Section 7.2, the Radiophysics Lab in Sydney seriously considered searching for the line. Their leader Joseph Pawsey first learned of the line when visiting Reber in 1948, and wrote home:

The utility of a line [such as the atomic hydrogen line at 1420.47 Mc/s and deuterium at 327.38 Mc/s], could it be found, is obvious. It opens up the possibility of determination of constitution of matter and of Doppler velocities in a manner analogous to optical spectroscopy.

---

[14] Reber had earlier been interested in the possibility of spectral-line emission to explain his cosmic static. After coming across an article on laboratory measurements of radio-frequency hyperfine splitting of heavy atoms, he wondered whether a similar transition could be found to produce radiation in his band of 156–164 MHz. He shared this idea with Otto Struve, who thought it a good one, but what happened thereafter is unknown. [Reber:Struve, 11 September 1944; Struve:Reber, 15 September 1944, YKS]

[15] The 21 cm transition was also mentioned by M. N. Saha (1946) in *Nature*, but the context was radio bursts from sunspot regions with strong magnetic fields.

[16] Shklovsky's (1949) calculations of antenna temperature, although useful, were based on incorrect concepts. This paper is also notable in being the first to point out the possibilities and to calculate the details of detecting radio lines from interstellar *molecules*, in particular the lambda-doublet transitions arising in the ground states of OH and CH (these OH lines were not detected until 1963 and CH a decade later).

Pawsey went on to give many other details and references and ended:

> The position is therefore quite uncertain. [Willis E.] Lamb of Columbia, for example, did not expect we should be able to find lines owing to low probabilities of emission or absorption and "smearing," due to changes due to magnetic fields and so on.

His director "Taffy" Bowen replied:

> This possibility is certainly an interesting one but, in view of the present state of knowledge, I doubt very much whether we should yet devote a special effort to it. A search for the atomic hydrogen and deuterium lines could be made with the Georges Heights equipment but this would involve dislocation of other work which is scarcely justified at present. At the moment Harry Minnett is chasing up the references you supplied. …

And yet interest was not entirely lost, for a year later Paul Wild (Fig. 16.9) wrote a report on possible hydrogen lines and Bowen was thinking that they should be sought.[17] Note, however, that Wild was primarily interested in the notion of spectral lines *in the sun* – his motivation to investigate the physics of hydrogen lines had been that certain isolated frequencies in his swept-frequency spectrograph (Section 13.2.1.2) seemed consistently strong (Wild 1978:24T).[18]

Wild's report led Bernard Mills in 1949 to debate whether to study discrete radio sources or to search for the putative line:

> Pawsey was very interested in the line at the time. … It was a difficult decision to make. I eventually chose the precise positioning [of radio sources] because I was more familiar with some of the techniques, and it looked as if it was something that would lead to an immediate result, whereas the other was extremely speculative. … [The technical difficulties] appeared rather forbidding. One knew one had to get right down to the absolute maximum theoretical sensitivity, because the thing was probably going to be faint. … And I'm pretty sure that we knew that the Dutch were doing it, too.
>
> (Mills 1976:6–7T)

In summary, despite their resources and expertise, the Australians never mounted a serious effort to search for the line – if they had, they surely would have soon succeeded.[19]

16.3 SEARCH AND DISCOVERY AT HARVARD

16.3.1 Background

In the late 1940s Edward M. Purcell (1912–1997) was one of the leading physicists in America, working primarily on subatomic structure of matter through the technique of nuclear magnetic resonance.[20] During World War II he had headed the Fundamental Development Group at the MIT Radiation Lab (Section 10.1.1), a group encouraged to be free-wheeling in its approach to developing microwave radar techniques. In later years this experience served him and his students at Harvard University well.[21] One such student, who came to Purcell in 1949 looking for a thesis topic, was Harold I. ("Doc") Ewen. Ewen was already working on a team developing a cyclotron at Harvard's Nuclear Laboratory, and came from a strong radar and electronics background gained during his service as a naval radar officer during the war.[22] He wanted his thesis, if at all possible, to combine

---

[17] Sources on RP's consideration of searching for spectral lines are Pawsey:Bowen, 23 January 1948, file A1/3/1a; Pawsey:Bowen, 15 April 1948, A1/3/1b; Bowen:Pawsey, 18 May 1948, F1/4/PAW/1; Bowen:F.W.G. White, 21 March 1949, A1/3/1a – all RPS. The Wild report was "The radio-frequency linespectrum of atomic hydrogen," Parts I and II, RPL 33 and 34, February and May 1949 [RPL]).

[18] The reason for Wild's recording apparently strong, isolated frequencies turned out to be resonant impedance mismatches in his spectrograph system. [Wild 1978:24T]

[19] See Tangent 16.1 for four other cases of strong consideration (including one case with some observations) to look for the 21 cm line before its eventual discovery.

[20] Purcell later shared the 1952 Nobel Prize with Felix Bloch for his development of nuclear magnetic resonance techniques.

[21] See Galison (1997:256–64) on the intimate relation between the wartime Radiation Lab and postwar physics at Harvard.

[22] Ewen also was a navigation instructor during the war, his most famous student being one of the greatest baseball hitters of all time, Ted Williams of the Boston Red Sox. During 1950 Williams visited Ewen and his setup at Harvard, setting many of the physicists aflutter. [J. S. Rigden (2002), *Hydrogen* (Cambridge: Harvard University Press), p. 178]

all of his interests – physics, mathematics, radar, meteorology, and astronomy.[23] At first he suggested accurate spectroscopy of the 1.35 cm water vapor transition in the earth's atmosphere, but Purcell was not enthusiastic over what he perceived as only "a few more points on the curve."

There things stood until the idea emerged of searching for the 21 cm hydrogen line in interstellar space. Purcell remembers that Ewen heard about the idea at a meeting of astronomers, but Ewen recalls that it was Purcell who first read about the idea somewhere. In any event they both soon grew enthusiastic about the challenge and possible payoff. But when they looked at a translation of Shklovsky's 1949 paper, Ewen began to get cold feet, as it seemed the Russians would probably do it all in a very short time, if they had not already. Shklovsky's thorough analysis confidently predicted a strong signal, and had he not given a rousing call to action at the close of his paper? Furthermore, the abstract (in English) of Van de Hulst's 1945 paper said: "The 21.2 cm line … might be observable if the life time of the upper level does not exceed 4×10^8 yr, which, however, is improbable."[24] Although Purcell convinced Ewen that he should not be dissuaded by such unknowns, they set out to do the experiment with the idea that it might well be a thesis with a negative result. Ewen therefore planned not only to achieve high sensitivity, but also to be fully confident in his equipment in the event of an unsuccessful search. Purcell (1977:38T) later recalled telling him:

> If you find the line, then that's great. But if you don't find it, you're going to have to put in a hell of a lot of work to establish the limits.

In January 1950 Purcell successfully applied for support for this search from the Rumford Fund of the American Academy of Arts and Sciences, of which Harlow Shapley (1885–1972), director of Harvard College Observatory, was chairman. The application pointed out:

> At this sharply defined wavelength we expect to find either a peak (bright line) or a dip (Fraunhofer line) in the apparent temperature, depending on whether the temperature of the hydrogen is higher or lower than that of the background of galactic radiation. …[Detection of the line] would give fairly direct access to the condition of the interstellar hydrogen, since by suitable calibration a direct temperature measurement would be possible.[25]

Most of the requested $500 went toward the construction of a horn antenna and receiver parts such as a power supply and mixer.

16.3.2 Equipment

The key ingredients to Ewen and Purcell's experimental design were (1) a horn antenna, (2) a sensitive and stable receiver, (3) a frequency-switching technique, and (4) a precise method of calibration. Let us consider each of these in turn.

The pyramidal horn, with which Purcell had had much experience at the Radiation Lab, was preferred over a small dish because its response pattern was largely just a single lobe, allowing greater sensitivity, less susceptibility to interference entering through sidelobes, and more certainty in measuring signal levels. It did not need to be large, since the putative hydrogen signal was expected to be widely distributed over the sky. Ewen oversaw construction of the horn from half-inch plywood lined with copper sheeting, painted it red, and stuck it out a fourth-floor window of the Lyman Laboratory where it made an inviting target for snowball

---

[23] Ewen had long been interested in astronomy and had even taught the subject at Amherst College. The connection between astronomy and radio had been particularly impressed upon him by Donald H. Menzel, a Harvard astrophysicist who was also (a) an expert on radio propagation, (b) an admiral in the Naval Reserve, and (c) Ewen's commanding officer at biweekly Naval Reserve meetings during this time.

[24] It is not clear when and if Ewen and Purcell ever saw an English translation of Van de Hulst's entire paper before 1951. Note also that for some reason the English abstract is more pessimistic about chances for the line than the Dutch text, which says regarding the lifetime requirement: "That would not seem to be a heavy demand" (Van de Hulst 1945:219).

[25] Purcell:Shapley, 12 January 1950, photocopy provided by Purcell. The wording "direct temperature measurement" was a calculated move to fit into the terms of the Rumford Fund, which, as for the work of its namesake natural philosopher 150 years before, supported measurements of heat and temperature.

Figure 16.2 The pyramidal horn antenna (copper-clad plywood) at Harvard's Lyman Laboratory with which Ewen and Purcell detected the 21 cm hydrogen line from the Milky Way on 25 March 1951.

tosses (Fig. 16.2).[26,27] The horn, whose mouth measured 1.4 × 1.1 m, pointed in the sky to an elevation of 43° and an azimuth within 10° of due south. This direction was chosen to be southerly enough that a portion of the Milky Way reasonably close to the galactic center would pass through the beam each day, yet far enough from the horizon to avoid manmade interference.[28]

Ewen's receiver was state-of-the-art, benefiting enormously from his own electronic dexterity combined with that of the Boston-area electronics community centered on MIT and Harvard. He obtained a great deal of material aid and advice from places such as Raytheon, Inc., the Harvard Nuclear Lab, and Bell Labs. For instance, Harald Friis, Jansky's boss for two decades, donated several of his very best 1N21B germanium crystals for the mixer, and Robert V. Pound, another Radiation Lab veteran at Harvard, gave invaluable advice on the construction of the mixer. Although the preamplifier was a standard model developed at the Radiation Lab, Ewen's exacting requirements meant that choice of tubes, stability of power supplies, and impedance matching were all critical.

Frequency switching was a microwave technique adapted from those used by Purcell and colleagues for nuclear magnetic resonance and laboratory spectroscopy. The technique afforded protection from

---

[26] The original horn is on display at the National Radio Astronomy Observatory in Green Bank, West Virginia. Besides catching snowballs, the horn also acted at first as a rain gauge, until Ewen installed a canvas cover (see Fig. 16.2) after a minor flood in his laboratory. As Ewen recalls it, this water was his first "signal from space." [Ewen:author, 11 February 1984]

[27] Theodore Lyman himself had just retired as director of the laboratory and often must have seen this architectural oddity. Ewen's wave-guiding "pipe," which led to a basic discovery involving the hydrogen atom, perhaps was reminiscent to Lyman of a 10 cm long pipe that had much earlier done likewise. This was the ultraviolet spectrograph he used at Harvard in 1914 to first measure the resonant spectral lines of hydrogen subsequently named for him.

[28] The elevation angle was further constrained by the practical detail of the available sizes of waveguide "bend" needed for the transition from the lower throat of the horn to the horizontal waveguide leading into the room.

receiver instabilities on a variety of time scales and was deemed worth the increased electronic complexity and extra integration time required for a given signal-to-noise ratio. The method involved a variation of the load-comparison, or chopping, technique developed during the war by Robert Dicke and spelled out in his 1946 paper (Section 10.1.1). In the original Dicke switch for continuum work, one measured the difference between the noise power generated by "sky" and by a dummy load within the receiver at the same frequency, whereas for spectral line operation this was changed to the difference between "sky on-frequency" and "sky off-frequency." This of course forfeited information about any continuum signal or indeed about any spectral feature of width comparable to or greater than the difference between the on- and off-frequencies,[29] but cleverly removed many frequency- and time-dependent receiver characteristics. Ewen achieved his frequency switching (at a 30 Hz rate) by modulating the reactance of his second local oscillator (LO) tube.

The last key ingredient to success was the overall calibration of the receiver and horn. The 1393 MHz LO, a war-surplus, radar-jamming "lighthouse" tube, was often checked against frequencies derived from an accurate standard in a nearby laboratory, but the main uncertainties had more to do with actually measuring the noisy trace on the strip chart recorder. In the end Ewen could assign frequencies to any measured signal with an accuracy of better than 5 kHz, a notable achievement. The calibration of signal intensities involved a gas-discharge tube, a technique recently developed by Willis W. Mumford at Bell Labs. Mumford had shown that such a noise source was quite constant over a broad range of microwave frequencies and varied little from tube to tube. By measuring the apparent signal level of standard household fluorescent tubes (G.E. "Cool White," to be exact) mounted in the waveguide just behind the horn, Ewen (1951:22–33) established that his receiver noise figure was 11 ± 0.5 db (system temperature of 3400 K). This use of gas-discharge tubes for calibration, as well as the technique of frequency switching, soon became standard at microwave frequencies.

16.3.3 Observations

The overall search strategy was to accept the laboratory-measured frequency of the hyperfine transition as correct (A. G. Prodell and P. Kusch had recently measured a value of 1420.4051 ± 0.0003 MHz at Columbia)[30], expect a very narrow line,[31] switch between two frequencies at a convenient interval, and search over a frequency range broad enough to cover any Doppler shift arising from the relative motion of an interstellar cloud and the sun. These ideas initially translated into using a single channel with bandwidth of 4 kHz (readily available on the shortwave communications receiver that Ewen used for an intermediate frequency (IF) amplifier and tuner), switching to a comparison frequency 10 kHz removed, and searching over a range of ± 150 kHz from the laboratory rest frequency. But no signal was found during the first observations with this configuration in mid-1950.

Thinking that the lack of a detection might indicate that the interstellar hydrogen line was somewhat broader than expected, Ewen next modified his receiver to a channel width of 15 kHz and switching to a frequency 25 kHz removed, using a second LO and a lock-in amplifier. He also improved the stabilization of the first LO and learned how to precisely measure its frequency. Observations resumed by the end of 1950, but still produced no signal when the Milky Way daily drifted through the horn's beam.

By this time Ewen was getting discouraged and was almost ready to write up a "negative results" thesis. Purcell, however, asked if there was not *something* more that could be done in a reasonable time, perhaps to move the comparison frequency yet farther away in case the line was very broad. Ewen replied that yes, he would be willing to do this by making another major

---

[29] See note 61.

[30] Prodell, A. G. and P. Kusch (1950), "On the hyperfine structure of hydrogen and deuterium," *Phys. Rev.* **79**, 1009–10. Until 1948 the line's frequency was not known to better than a MHz or two, which would have made any searches before that time extremely laborious.

[31] In his Rumford Fund application Purcell in fact argued that the hydrogen line might be extremely narrow because of the low density of space and resultant absence of collisional broadening. He thought this would allow a much more precise measurement of the line frequency than in the laboratory (in fact, the breadth of the interstellar line would preclude this possibility).

Figure 16.3 Ewen and part of his 21 cm spectrometer. The 27 MHz IF amplifier, a modified version of a communications receiver, was tuned by the small motor mounted on the wooden post. Above this is the lock-in amplifier necessary for frequency switching, and to the lower left is the strip chart recorder. The waveguide leading in from the window can be seen behind Ewen's head.

modification to his IF setup, but that it would require the purchase of a new communications receiver for $300, money that was nowhere in sight. The next day, Purcell pulled $300 in cash from his wallet, gave it to Ewen, and told him not to ask where it came from.

The new receiver (Fig. 16.3) allowed frequency switching to 75 kHz from the signal channel and turned out to be exactly what was needed. Ewen's standard tune-up procedure involved long bouts of juggling the frequencies of his two LOs in an attempt to find an optimum point of operation. Sometimes he heated the room to above 80°F (>27 °C) for an entire weekend so as to get more temperature and frequency stability in his various oscillators. He observed with automatic scans over a 300–400 kHz range centered on the line's laboratory frequency. Once each scan began, he usually fell asleep during the monotonous 30 to 60 minute runs all night long:

Occasionally I would put on headphones just to listen to the whistling beat between the LO and the synthesized harmonics as the LO was being tuned. I had an old comfortable chair, and I would sit down, close my eyes, snooze a little, and listen to make sure that it didn't gurgle or pop – if it did, I would wake up and fix it. It was almost like baby-sitting. I could tell a lot about that gadget just from the sounds it made – when it was running correctly, it made all sorts of queer noises, because of the number of gears and motors.[32]

(Ewen 1979:17T)

Soon after the new receiver was installed, early on the morning of 25 March 1951, Ewen at last detected the hydrogen line. But the apparent signal was centered a full 150 kHz higher than the laboratory frequency, at the limit of the scan. Resetting the local oscillator confirmed its reality, however, and then it dawned on Ewen that perhaps the offset was a Doppler shift arising from the earth's orbital motion (this had not been considered earlier). But what was the current direction of the earth's velocity vector? A rough calculation for the

---

[32] Later, Ewen explained to a reporter that "interstellar space is singing a song of hydrogen with a single significant note." [*New York Times*, 7 February 1954, "Radio astronomy discovers space is full of hydrogen which sings a radio song," p. E9]

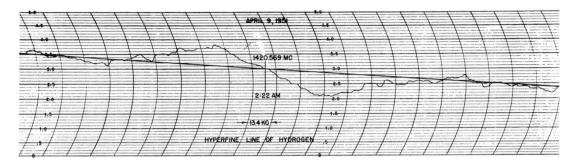

Figure 16.4 A trace of the 21 cm line, as recorded by Ewen on 9 April 1951. The penciled-in slanted line is an estimate of the instrumental zero level. As the receiver was tuned through a range of 400 kHz centered on the rest frequency, the line signal first appeared as a positive signal and then negative as it successively was centered in the signal channel and the comparison channel 75 kHz removed. (Ewen 1951)

vernal equinox indicated that almost all of the orbital speed of 30 km s^{-1} was indeed in the direction where his horn pointed, enough to compensate for the apparent "blueshift."

Ewen's confidence grew after two more nights of checking, and as a dramatic demonstration for Purcell the following morning he unrolled in the hallway a long section of strip chart showing various signal scans (see Fig. 16.4). Purcell was impressed with the reproducibility and intensity of the signal, but was still concerned about the possibility of a terrestrial signal, perhaps from MIT only two miles away. They thus agreed that the signal should be monitored long enough to detect the expected sidereal drift in the time of daily maximum if indeed it was the Milky Way in Ophiucus and Aquila that was coming through the beam.[33] Over the following weeks the case was clinched when the line signal indeed shifted earlier in the expected manner. Closer measurements even revealed the expected sense of Doppler shift over the two hours each night when the signal was measurable.

Throughout this period Ewen held a full-time job at the Nuclear Laboratory, and in fact only a month before the hydrogen line detection the Harvard team had finally gotten a proton beam out of the 95 inch cyclotron, in an endeavor that had been somewhat of a race with Enrico Fermi's group at the University of Chicago. This effort obviously detracted from the 21 cm project (for example, Ewen did virtually nothing on it during January and February),[34] yet it was also extremely helpful in tying Ewen to advanced electronic techniques and needed equipment. His modus operandi was to borrow for a night or two test apparatus essential to the 21 cm experiment, in particular frequency-measuring equipment. Holiday weekends were especially nice since they allowed longer-term loans – it was thus no accident that the line was first detected on Easter morning![35,36]

---

[33] Ewen of course realized that the galactic plane crossed through his beam *twice* each day, once near the galactic center and 12 hours later near the anti-center. But because of his job at the cyclotron the daytime crossing could be recorded only *in absentia*. In fact he did consider many apparent daytime signals from near the anti-center, narrower in bandwidth and frequently "cleaner" and stronger than those at night. But these were never followed up or discussed in print because it was not believed that a signal from the *outer* parts of the Galaxy could be stronger than one from near the galactic center. Furthermore, Ewen did not have the time to check them out personally and was distrustful of manmade interference in the daytime. [Ewen 1979:22T]

[34] Ewen's laboratory notebook, pp. 40–43 (photocopy supplied by Ewen). Curiously, nowhere in his notebook does Ewen explicitly declare that the line has been discovered; for the appropriate dates one finds only a flurry of frequency calculations.

[35] This section has profited from interviews with Purcell (1977, 1977a) and Ewen (1979), as well as Ewen's thesis (1951). Also see Stephan (1999), an account with many more technical details of the equipment.

[36] Ewen has created a website consisting of several hundred slides and accompanying text on the many industrial, scientific and military aspects of his career from 1948 onwards. [H. I. Ewen (2003), "Doc Ewen: The horn, HI, and other events in US radio astronomy," www.nrao.edu/archives/Ewen/ewen.shtml]

16.4 THE DUTCH QUEST

16.4.1 Background

From the time of Van de Hulst's colloquium in 1944, Oort pursued the promise of extraterrestrial radio noise. Years before Ewen and Purcell, he made inquiries to various radio experts regarding the feasibility of building a large dish and receiver, but, as he later recalled, conditions were grim:

> Holland was in a pretty bad position after the war, and not only because so much had been destroyed and we were poor, but also because we had not taken part in the wartime radar developments. So we hadn't anything like the experts in Australia, England, and the United States. But our main advantage then was the strong interest from the astronomical side.
>
> (Oort 1971:3T)

In August 1945 he wrote to Reber:

> I am considering the possibility of erecting a large mirror and receiving set for ultra-short waves in this country. Naturally the plans are vague. ... Can you give us a rough idea of the expenditures for your 31 ft mirror?[37]

Reber's characteristically meticulous reply tallied all his expenses over the previous five years and came to a grand total of $4981, not including labor (Section 4.1). He also advised Oort to "obtain the services of first-class radio engineers," advice that Oort was soon to realize was as sound as it was difficult to follow in postwar Holland. At one point the Dutch even tried to recruit Martin Ryle or another British radio physicist for a long-term visit, but without luck.[38] It became clear that if radio astronomy was ever going to get under way in Holland, it would have to be an all-Dutch effort.

From the beginning Oort's focus was on a 25 m dish that would yield decent detail in the Milky Way radiation – he considered the available data of Jansky and Reber "nearly worthless" in this regard. Later on, the 25 m size was more specifically justified as that needed to resolve the thickness of the galactic disk at distances of 10 to 20 kpc (a one-half degree beam). Although it was over a decade before such a dish came into service (in 1956 at Dwingeloo), initial planning for it was important in the early development of Dutch radio astronomy. Oort consulted with mechanical and radio engineers at government and private research laboratories throughout the Netherlands and soon began to realize that a large dish with a reasonably accurate surface was probably possible, but also very expensive, and that a high frequency receiver was also a major project since the technology was simply not present in Holland.

Oort's first proposal to the Dutch Academy of Sciences, who then forwarded it on to the government with their endorsement, was as early as November 1945.[39] In it he emphasized the potential importance of radio observations to the fields of galactic structure and interstellar material. The Netherlands was strong in these areas and, furthermore, could do the radio work despite its rainy climate. Oort also pointed out that foreign radio investigations were being carried out by engineers, usually without clear *astronomical* goals, and that this would not happen under his aegis – in Holland, *astronomers* would be in control. In this first proposal he made no mention of the 21 cm hydrogen line, probably because he considered, based on consultation with Bakker at Philips, that receivers at wavelengths shorter than about 50 cm would not be able to measure the galactic radiation. But he did point out that a giant dish could be kept busy for many years with observations of the Milky Way, other galaxies, and the sun[40] – and after its job was done in the northern sky, it could then be moved

---

[37] Oort:Reber, 30 August 1945, file 98(c) or 251, OOR. Also see C.J. Bakker:Oort correspondence between April and June 1944, file 158(a), OOR.

In a letter to Oort of 3 May 1944, Bakker made an interesting early suggestion: "One of my colleagues has asked whether it would be possible to measure the width of the [radio] emitting portion of the Milky Way using the coherence method of Michelson, in the same way that stellar diameters are determined. Do you think this possible?" Oort's reply is unfortunately not extant, and in fact nothing was done along this line in Holland (although the method was extensively developed by Ryle at Cambridge [Section 8.2]).

[38] For example, see Ryle:Minnaert, 22 June 1949, box 1/1, RYL.

[39] Oort, "Plan voor meting van korte-golf straling van het interstellaire gas," 26 November 1945 (photocopy from Oort), OOR.

[40] In his proposal Oort mentioned that radio radiation would be especially useful to study the solar *corona*, but no justification was given for this prescient statement, fully a year before basic

to Leiden's southern station in South Africa! His rough estimate for a price-tag was 100,000 guilders (~$40,000) and, although this was fully 20% of the government's annual budget for scientific research, Oort had many influential contacts, including one of the postwar Prime Ministers, Willem Schermerhorn, a professor of geodesy from Delft. The resulting strong support in 1946, however, actually came too quickly, for Oort had no detailed designs for a large reflector or a receiver, nor did he have an electronics expert, nor even a site for observations. Things simply had to move more slowly.[41]

It was eventually decided that this undertaking required a truly national effort, and in 1948 Oort organized the Stichting voor Radiostraling van Zon en Melkweg (SRZM), the Foundation for Radio Radiation from the Sun and Milky Way, as the group that would set policy and seek funds for Dutch radio astronomy. The initial board of directors of SRZM was carefully chosen to gain widespread support for the fledgling enterprise: besides Oort (president), Van de Hulst, Minnaert, and Jakob Houtgast (Utrecht),[42] it included Anthonet H. DeVoogt (1892–1969) (Post, Telephone, and Telegraph Service [PTT]), F.A. Vening Meinesz (KNMI, the Dutch Meteorological Institute), and F.L. Stumpers (Philips Research Labs). SRZM's astronomical goal was to search for the 21 cm line, with galactic continuum and solar observations of secondary interest.[43] Oort and SRZM also found a strong friend in J.H. Bannier, the first (and longtime) director of the main agency for funding research in the physical sciences, the organization for Zuiver Wetenschappelijk Onderzoek (ZWO – "Pure Scientific Research"), and in 1948 ZWO gave out its first in a long string of grants for radio astronomy – for 11,000 guilders.[44]

16.4.2 Trials with Hoo

SRZM hired H. Hoo, a young engineer, as its sole employee, and by November 1948 he was trying to adapt a surplus American radar receiver for astronomical purposes. Since it was realized that a 25 m dish was only possible in the distant future (although constantly being discussed), the plan was to use one of two 7.5 m diameter German Würzburg antennas[45] (Fig. 16.5) that DeVoogt had earlier salvaged from the Dutch coast and brought to the main shortwave transmitter site of the PTT at Kootwijk, in the Netherlands central.[46] The same German war machine that had created the harsh conditions under which the hydrogen line had been predicted in 1944 was now ironically "supplying" equipment needed by the Dutch astronomers.

Progress was distressingly slow during 1949. Hoo's first detection of the sun (a *much* easier task than any 21 cm line) took him six months. On the theoretical side Van de Hulst was reckoning an expected line width of 100 kHz and intensity of 800 K, making it clear that detection of such a signal was unlikely with their system temperature of ~13,000 K.[47] By early 1950 the directors of SRZM grew so impatient with the glacial pace that they decided to advertise for another engineer. This

---

ideas on this subject were elsewhere published (Sections 7.3.3 and 10.3.1).

[41] Oort:author, 31 December 1985.

[42] Houtgast had just completed a six-month stay in Paris learning the practical side of radio techniques from Marius Laffineur (Section 10.4), but nevertheless did not play a prominent role at this stage.

[43] Oort:SRZM Board of Directors, 12 October 1948; SRZM Board minutes, 26 October 1948; both box 41, VDH.

[44] Oort and Minnaert:Director of ZWO, 25 March 1948, file 263, OOR. This was a request for 13,000 guilders during 1948 for one engineer (3000 guilders salary) and equipment development. The justification was largely identical in wording to the premature request of November 1945 (note 39).

---

[45] See Tangent 4.2 for details on Würzburg dishes. Van Woerden and Strom (2006:7, 12–16) give exhaustive histories of the hydrogen-line Würzburg and several other Würzburgs used for science in Holland. Unfortunately, it appears that the hydrogen-line Würzburg no longer exists.

[46] DeVoogt's motivation to do radio astronomy was primarily to learn about the ionosphere, principally by monitoring the sun, so as to predict propagation conditions. Over these early years he did carry out a few odd observations and in 1951 he even built a 30 m diameter "hole-in-the-sand-dunes" fixed paraboloid antenna, but there were no publications. Also in 1951 he commenced solar monitoring at 140 and 200 MHz with a second Würzburg antenna at a different site (see note 13.2). His main role in early Dutch radio astronomy was as strong supporter to SRZM, his "tenant" at Kootwijk. Although the techniques of shortwave transmission and microwave reception were hardly the same and sometimes even antithetical, SRZM's ability at Kootwijk to use excellent shop facilities, borrow equipment, and consult with various radio experts proved invaluable. See Strom (2005, 2007) for further details.

[47] "Ontvangst de 21 cm straling te Kootwijk – voorlopige berekeningen," 30 May 1949, boxes 34–38 + 40, VDH.

Figure 16.5 The 7.5 m Würzburg reflector (at Kootwijk, in the Netherlands central) that was used for early work on the 21 cm line by Muller and Oort.

judgment was further bolstered when, on the morning of 10 March 1950, disaster struck at Kootwijk.

That morning, in a small equipment and observing hut, Hoo and his assistant were working on some coaxial connectors when they left to visit another building on the PTT site. Forty-five minutes later they heard fire sirens, ran back, and watched as their wooden hut quickly burned to the ground. With the hut was destroyed the entire inventory of SRZM electronics, worth about 8000 guilders (~$3000), one-third of which was a recently purchased Brown chart recorder. The cause of the fire was never ascertained. Hoo proposed one theory in his report on the incident: that when the door to the hut had been closed, a piece of wire had fallen off a workbench, causing a short circuit in one or more lead batteries below.[48,49]

---

[48] SRZM Board minutes, 21 March 1951; H. Hoo, "Verslag betreffende de brand van 10 Maart 1950," 12 March 1950; both box 41, VDH.

[49] For an official view of the pre-Muller days of Dutch radio astronomy, see Minnaert's report to the 1950 URSI General Assembly in Zürich (*Proc. Gen. Ass. URSI* 8, Pt. 1, pp. 194–7 [1950]).

16.4.3 Success with Muller

The SRZM board bit its collective lip and vowed to start again from scratch, resolving all the more that Hoo had to go. But promising candidates were scarce. It was not until the end of 1950 that the board's first choice, C. Alexander Muller (1923–2004) (Fig. 16.9), was able to start work. He was an ardent "ham" radio operator and had learned high-frequency radio techniques as part of his just-completed studies in engineering physics at Delft Technical University. Upon arrival at Leiden he found various receiver parts for Hoo's design being made by arrangement with Stumpers at Philips in Eindhoven. He started by assembling this receiver and giving it a new tuning capability, while also intensively studying the literature in microwave techniques. In February he moved it out to a new hut at Kootwijk.

Muller quickly grasped that stability of the receiver was critical for any chance of success, and so devoted most of his time to such matters as the removal of temperature effects causing drifting and stabilization of the power supplies. The last major design change in the receiver (Fig. 16.6), however, came not from engineering books or journals, but rather in the unlikely form

Figure 16.6 Muller's receiver with which the hydrogen line was detected in 1951. Synchronous detector (lower right), IF amplifier (center right), crystal mixer with frequency multiplier behind (center), power supplies and control equipment (left). Note the coaxial cables going through the wall to the antenna just outside.

of a letter received on 8 March from Van de Hulst in America.[50]

Van de Hulst had been appointed as a visiting professor at Harvard College Observatory for the spring semester of 1951. He was giving a course in radio astronomy, as he had done in Leiden the previous autumn,[51] and therefore happened to be on the scene during the discovery of the line which he himself had suggested seven years earlier. Once he learned of the 21 cm effort shortly after his arrival, he reported back home on many details of Ewen's receiver. Of these, Muller was particularly enamored with the frequency-switching scheme. The Harvard physicists were quite open about sharing electronic tricks of the trade with the Dutch, and Van de Hulst was happy to reciprocate with advice about the interstellar side of things. In particular he pointed out that both the channel width and switching interval that Ewen was then using were much smaller than the expected broadening of the 21 cm line from cloud turbulence and galactic rotation; this would greatly decrease their ability to detect the line.[52] The Americans succeeded anyway, but now realized that their measured line intensity and line width were probably both too small. The net result was an exchange of technology for astrophysics that benefited both sides of the Atlantic. Van de Hulst's presence at Harvard was the main reason why the Dutch found the line only seven weeks after Ewen.

In late March Van de Hulst immediately passed on word to his countrymen of the Harvard detection. Muller was already busy effecting the frequency switching through control of the first LO with a reactance tube. Once the receiver was behaving well (system temperature was about 7500 K, twice that of Ewen's), first hints of the line came on 11 May, followed by confirmation several nights later. Muller's technique was to point the dish at the galactic center in Sagittarius

---

[50] Muller's laboratory notebook, photocopy supplied by Muller.
[51] This was in fact the first academic course in radio astronomy and led to an informally published (but widely circulated) textbook (Van de Hulst 1951); also see note 76. The text's 185 pages influenced many students and researchers around the world, especially before the appearance of a much larger textbook by Pawsey and Bracewell (1955). Today it provides the historian with an invaluable benchmark for radio astronomy circa 1951.

[52] See notes 31 and 61.

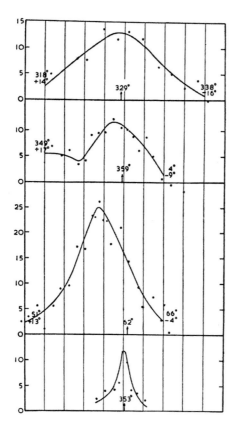

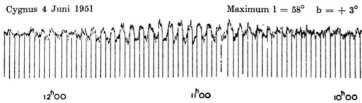

Figure 16.7 (Top) Curves of the peak intensity (arbitrary units) of the 21 cm line versus time (20 min per division) as the Milky Way drifted through the Kootwijk antenna beam in the spring of 1951 at various galactic longitudes. The galactic cooordinates (old system) of the start, middle, and end of each constant-declination scan are indicated. From top to bottom, the scans were taken in the directions of Sagittarius, Ophiucus, Cygnus, and Ophiucus, but shifted to a radial velocity of $+55$ km s^{-1}. (Bottom) A strip chart recording of the 21 cm signal as the Milky Way in Cygnus drifted through the antenna beam. Vertical lines each 2 minutes mark when the line signal was switched from the signal to the comparison channel. (Oort 1951)

and let the Milky Way drift through the beam while switching between two frequencies 110 kHz apart at a 30 Hz rate.[53] Furthermore, the signal and comparison channels were interchanged every few minutes, providing a further check against frequency-dependent instrumental offsets (as in the bottom of Fig. 16.7).

---

[53] Later observing with the Würzburg dish involved using hand cranks to track a fixed sky position. Every 2.5 minutes the altitude and azimuth of the dish were adjusted to ~0.1° accuracy.

By 17 May Muller had convinced Oort in Leiden of the signal's reality, and Oort in turn wrote to Van de Hulst:

> I've just received a telephone call from Muller at Kootwijk saying that he has tonight observed the 22 cm line [sic][54] in the vicinity of Sagittarius![55]

Eight days later at the next regular SRZM Board meeting, Oort and the others inspected the evidence of Muller's chart recordings. By that time he had obtained cuts through the Milky Way at three different longitudes and Oort was already puzzling over several aspects of the astronomical interpretation (Section 16.6).[56]

16.5 CONFIRMATION FROM AUSTRALIA

Frank Kerr (Fig. 16.9), a staff member of the Radiophysics Lab in Sydney, had done research on ionospheric propagation and lunar radar (Section 12.5.1) and was spending the 1950–51 academic year at Harvard studying astronomy. He too learned of Ewen's experiment and was keenly interested in its progress, sending frequent reports back home to his boss Pawsey. After the discovery of the line Purcell decided to hold off immediate publication in favor of waiting for confirmation from either the Dutch or the Australians. This seemed natural to him since on the one side the Dutch appeared so close to a detection (and in some sense "deserved" one); and on the other Bowen, chief of the Radiophysics Lab, was a wartime crony from Radiation Lab days.

Although the Australians had earlier decided against a campaign to search for the line (Section 16.2), within two weeks of receiving Kerr's letter, two separate groups were lashing together 21 cm receivers to check the Harvard discovery. One was initiated by Jack Piddington and carried out by James V. Hindman (Fig. 16.9), who had previous experience with solar microwave observations and quickly was able to assemble a sensitive 21 cm receiver. But Hindman was having trouble determining his precise operating frequency, an absolute necessity. What he needed was a top-quality signal generator and he learned that the man who had one was his colleague "Chris" Christiansen (Fig. 16.9). When Hindman approached Christiansen about a loan, he discovered that Christiansen too was frantically working on a 21 cm receiver. He had temporarily dropped his own solar work after receiving a rush assignment from Pawsey. Christiansen was having problems with the *front*-end of his receiver, however, and so it was natural that they should join forces. They were soon spurred on by the Dutch success, and the pace quickened. As Christiansen (1976:31T) later recalled:

> I can remember working day and night on that receiver. It was a monster – the most terrible piece of equipment I've ever seen in all my life. I thought the gear would never work. I'd just about given up on it when I left it running one night and then went to sleep – when I came back, there was a beautiful curve sitting up on the chart.[57]

It was the end of June 1951 and although their receiver may not have been elegant, in less than three months they had achieved what had taken the Americans and Dutch much longer. They used a 16 × 18 ft paraboloidal-section antenna (Fig. 7.14) previously used by Christiansen for solar work at the Potts Hill field station. The choices of 50 kHz for their single channel's bandwidth and 160 kHz for its switching interval were of course influenced by the Northern Hemisphere detections, but also by simply making do with what was feasible in a short time. For their initial observations the single channel was tuned over a range of 1000 kHz while the antenna tracked the galactic center.

---

[54] Curiously, the Dutch often referred to the hydrogen line as the "22 cm line" in the early days, even though its wavelength was known to be 21.1 cm; for example, see *Bull. Astron. Inst. Netherlands* 11, 278 (1951).

[55] Oort:Van de Hulst, 17 May 1951, file 104(d), OOR.

[56] Besides archival evidence, the account in this section has relied on interviews with Oort (1972, 1978), Van de Hulst (1973) and Muller (1973).

[57] Only Christiansen was on the scene for the detection because a few days before Hindman had injured his knee while climbing on the antenna – he received the good news in the form of a telegram delivered as he was being wheeled down a hospital corridor on his way to an operation! The pace of the work is further indicated by a remark in a letter of 13 July 1951 from Pawsey to Bowen, then overseas: "Christiansen has worked like a [slave] for the last two months trying to get this gear working." [File A1/3/1p, RPS]

16.6 INITIAL ASTRONOMICAL RESULTS

16.6.1 First interpretations

At Harvard, Ewen was anxious to cease observations as soon as possible and to finish writing his thesis, "Radiation from galactic hydrogen at 1420 Megacycles per second." This he did quickly in May 1951 and duly received his doctorate in physics.[58,59] Once the Dutch had confirmed the line detection in May, Purcell pressed ahead with a plan for twin papers, and in mid-June these were submitted for publication with the idea that the Australians could join in if they produced results in time. In the end Pawsey decided to forgo even a short publication and instead sent a cable to *Nature* announcing the second confirmation. The reports of the three groups appeared together in the 1 September 1951 issue of *Nature*. Meanwhile, Oort showcased the discovery at a special session on 30 June of the Dutch Academy of Sciences in Amsterdam, as well as on a press tour of Kootwijk. In contrast, Christiansen and Hindman undertook extensive observations in 1951 and did not publish anything until the following year.

There were many questions in the beginning: Where is the hydrogen gas? How much is there? What Doppler shifts does the line show? How broad is the line? What is its excitation mechanism? What are the temperatures and densities of the neutral hydrogen clouds? The initial answers, as might be expected from the widely different experimental philosophies and institutional settings of the three discovery groups, were dissimilar. The original pair of short papers in *Nature* by Ewen and Purcell (1951) and by Muller and Oort (1951) are striking in their contrast: one written as an experiment in interstellar microwave spectroscopy by physicists, the other as a treatment of galactic structure by an astronomer.

Ewen and Purcell reported a peak antenna temperature for the line of 25 K at a right ascension of 18^h (recall that their horn's declination was fixed at $-5°$) and a line width of 80 kHz ($\sim$15 km s^{-1}). They argued that such a broad width probably arose from turbulence in interstellar clouds and that the line was optically thick. Assuming an average density of 1 atom cm^{-3}, they calculated that they were "seeing" in essence to a distance of only about 300 pc. After coining the term *spin temperature* to characterize the relative populations of the hyperfine doublet states (through the Boltzmann relation), they concluded that its value was 35 K (including an estimated 10 K for the continuum background of the Galaxy). Finally, the source of excitation was considered to be primarily kinetic collisions with other hydrogen atoms, so that in fact the kinetic temperature of the gas itself was probably only slightly greater than 35 K.

In distinct contrast, the paper by Muller and Oort mentioned hardly any aspects of the opacity or physical state of the hydrogen gas – in fact, the illustrated scans (Fig. 16.7) showed only arbitrary units of intensity. Rather, emphasis was on the line widths and Doppler shifts and what they revealed about the location of the gas. Oort's main puzzlement was how the line in the direction of the galactic center could be so narrow, $\sim$12 km s^{-1} (60 kHz), and yet exhibit a width in galactic latitude of 8°. On 28 May he wrote to Van de Hulst: "Muller's results are not at all what I expected."[60] Not only was the distribution not thin, but some of the scans were not even symmetric about the galactic plane. In the end, however, Oort chose to accept the latitude distribution as being more reliable and thus estimated that the hydrogen clouds were only 300–400 pc distant.

---

[58] Ewen's short thesis contains hardly more astronomical analysis than his one-page note with Purcell in *Nature* – six pages, compared to 28 pages describing equipment and techniques. In fact, Ewen had developed a draft of the bulk of his thesis well *before* the detection in March, with the idea that a short "Results" section, either negative or positive in nature, would eventually be tacked on.

[59] As Dutchmen speaking many languages, Van de Hulst and Oort were incredulous when they learned that one of the reasons preventing Ewen from further work on his fundamental discovery was that he had not yet fulfilled the Ph.D. language requirement. At one point Ewen was compelled to drop everything and leave town for two weeks of immersion in German, whereupon he returned to pass the exam.

[60] Oort:Van de Hulst, 28 May 1951, file 104(d), OOR. These two facts seemed inconsistent from basic ideas of galactic dynamics, pioneered by Oort himself in the 1930s. In a planar self-gravitating system, the distribution of any class of objects perpendicular to the plane must be related to the velocity dispersion of those objects, larger velocities implying greater average distances away from the plane. This principle was well established from stellar studies and so when the hydrogen was found to have a narrow velocity dispersion, the expectation was that it would occupy a very thin plane.

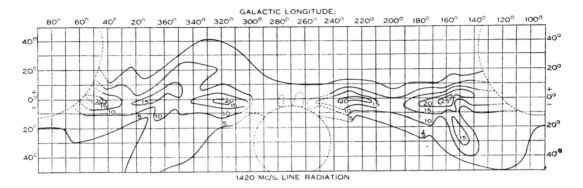

Figure 16.8 The 21 cm survey of the Milky Way by Christiansen and Hindman (1952) is here represented as a contour map in galactic coordinates of the peak line brightness temperature (25 contour units equal about 100 K). Dotted areas define unobservable regions.

Muller succeeded in obtaining signals from a variety of directions (Fig. 16.7) between the galactic center and Cygnus (the direction in the Milky Way perpendicular to the galactic center and toward which galactic rotation takes the sun). Oort had been anxious to secure such data, for he expected the Galaxy's differential rotation to broaden and shift the 21 cm line. At a longitude about 30° from the galactic center (almost the same point at which Ewen's beam cut the Milky Way in Ophiucus) the line indeed broadened to a width of 330 kHz and showed a Doppler shift of ~250 kHz, or a relative motion of 55 km s^{-1} with respect to the sun. Based on a model of galactic rotation, Oort could then calculate that such gas was ~8 kpc distant, farther than anyone had previously probed in the plane. The great distance of this gas was further corroborated by its much narrower width in latitude. Cygnus also showed a broader line in frequency than did the galactic center, as well as intensities two to three times higher. The former was expected, but the latter was perplexing.[61]

---

[61] Results much later were to indicate that many of these early measurements of intensities and linewidths were unreliable because the frequency-switching intervals – 75 kHz for the Americans, 110 kHz for the Dutch and 160 kHz for the Australians – were much less than typical 21 cm profile widths in the galactic plane; for instance, the profile in Ophiucus was found to have a full width of 800 kHz and an intensity of 55 K. Moreover, the 12° beam of Ewen's horn was insufficient to resolve the width of the hydrogen distribution, further reducing his measured intensities (the Kootwijk beam was 2.8°). For further details see Sullivan (1982:325–7).

Although the Australians came last to the 21 cm problem, they stuck with it most doggedly once they had a detection. For three solid months Christiansen and Hindman surveyed the southern sky with their square dish, milking all they could out of their makeshift receiver. Their profiles were more reliable because of their larger scanning range of 1000 kHz and larger frequency-switching interval of 160 kHz (although still inadequate, as they were aware). They slowly tuned in frequency while the sky drifted through their 2.3° beam, obtaining an independent frequency profile every two to four beamwidths. Intensities were good to ~15% ("using the sun as a signal generator," as they said) and frequencies of measured lines to ~100 kHz.

Several new points emerged in an extensive article by Christiansen and Hindman (1952). First, the maximum line temperature reached "about 100 deg K," much higher than the northern observers had obtained. Second, "the hydrogen line originates in a source which forms a band of varying angular width, centered on the galactic equator" (Fig. 16.8). In two directions in particular, Taurus and Ophiucus, they noted that prominent hydrogen extensions away from the plane neatly corresponded with known regions of optical obscuration, implying an association of the neutral hydrogen with dust grains. The estimated width of the hydrogen layer was 500–600 pc, five times the estimate of Muller and Oort. Third, the hydrogen line was found to *split* into two frequency components of width ~130 kHz and separation ~230 kHz over two different ranges in galactic longitude, corresponding closely to the Galaxy's first and third quadrants. Using a model of galactic rotation, they estimated that these two "hydrogen sources" were at distances of 1 and 4 kpc and suggested they are "in the form of spiral arms of the Galaxy."

16.6.2 Astrophysical theory

The key to the physics of the hydrogen line lay in establishing the mechanisms that determined the relative populations of the hyperfine doublet. This was not straightforward, for the interstellar situation was very different from that of the lab; moreover, this was the first time ever that anyone had observed a microwave line in *emission*.[62] In the autumn of 1951, Purcell presented a seminar on the 21 cm line at which Viktor F. Weisskopf and Norman F. Ramsey pointed out to him that the dominant process was undoubtedly spin exchange: the collision of two hydrogen atoms sometimes leads to the electron spins being flipped, that is, in a different ground hyperfine state (the probability turns out to be 1/8). Purcell calculated that such an event would happen to a typical interstellar hydrogen atom once every 400 years, a far higher rate than the spontaneous emission time for the 21 cm line of 10 million years.[63]

Another question of interest was the proper value for the Einstein A coefficient of the 21 cm transition, with many letters crossing the oceans arguing for either of two values differing by a factor of four. In the end Purcell's argument for the higher value, 2.85×10^{-15} s^{-1}, held sway over Shklovsky's (1949) value (Section 16.2). In Sydney, Wild (1952) also contributed to the theory. He was then occupied with solar observations, but expanded his internal 1949 report (note 16.17) into an article treating in great detail the radio spectrum of the hydrogen atom, including not only the 21 cm line but also possible fine structure transitions.[64] He also presented many calculations regarding the equation of transfer of the 21 cm line and applications to the Galaxy.

In the Soviet Union, the 21 cm discovery rekindled Shklovsky's interest in radio spectroscopy, and in late 1951 he did a complete reanalysis that built on his 1949 study. Shklovsky (1952d) used a kinetic temperature for the interstellar gas drastically lowered to 25 K from his previous value of 5000 K, and discussed in detail methods to study the properties of interstellar clouds and galactic structure.

16.6.3 Follow-up observations

There was no immediate follow-up at Harvard, for Purcell continued with his laboratory physics and Ewen immediately became involved with the Korean War.[65] Shapley, however, was intrigued with the potential of radio astronomy to provide a boost to Harvard College Observatory's lagging astronomical fortunes, tied as they were to outmoded telescopes at inferior sites. As early as February 1952 he brought up the possibility of "an adventure into radio astronomy," but the idea gained traction only when Bart J. Bok (1906–1983), a Dutch-American optical astronomer, returned to Harvard after a year-long stay in South Africa (Shapley himself retired in mid-1952). Although knowing nothing about radio techniques, Bok was instantly taken by the 21 cm line as a probe of galactic structure. Radio results played a major part in a graduate seminar he led in the summer of 1952 (Gingerich 1984:400). With support from his colleagues, he raised money from private and

---

[62] W. Gordy (1952), "Microwave spectroscopy," *Physics Today* 5, 5–9; P. Forman (1992), "Inventing the maser in postwar America," *Osiris* 7, 105–34 (see p. 121).

[63] This work was not written up until much later (Purcell and Field 1956). A point of view other than Purcell's was put forward by S. A. Wouthuysen (1952), a Dutch physicist. He published a paper arguing that Lyman-alpha radiation from ionized hydrogen regions penetrated neutral hydrogen clouds, and that absorption and reemission of Lyman-alpha photons by hydrogen atoms in the ground state determined the hyperfine-level populations. For a while Van de Hulst and Oort agreed with this, but by mid-spring 1952 they had become convinced it was wrong since the Lyman alpha photons simply could not reach most of the neutral hydrogen atoms. [Oort:Greenstein, 2 May 1952, box 5, GRE; also file 105(a), OOR]

[64] Wild (1952:207) summarily dismissed recombination lines because they were said to be so numerous that, unless there were some selective excitation mechanism, they would only contribute to the general continuum. Without saying so, Wild may have been using the (too large) Stark broadening of the lines calculated by Van de Hulst (1945) (Section 16.1).

[65] When he was ordered to return to the Navy, Ewen decided he would rather serve in a civilian capacity. He quickly submitted a proposal for military research and by the end of the summer of 1951 was set up with a Navy contract and a small company. Places such as the Radiophysics Lab later tried to lure him away, but he preferred to remain an entrepreneur. His Ewen–Knight Corporation became highly successful in military radio electronics (e.g., developing radio sextants), offshoots of which often were eventually available to US radio astronomers.

federal sources (~$30,000 for startup) and recruited a first cohort of graduate students, many of whom later became prominent in American radio astronomy. Bok hired Ewen parttime and ordered from his firm a complete receiver – Shapley called it a "black box, the heart of this new-type astronomical instrument" – and then quickly purchased a 24 ft diameter dish "off the shelf" from a defense contractor. Bok and Ewen soon were training radio astronomers (a joint course was first given in 1953) and running a large graduate program in 21 cm line research (Munns 2002:126–51).[66]

Kerr returned to Sydney (via Holland) in October 1951 and, as Christiansen was ready to get back to solar work with his just-completed grating array (Section 13.1.3), persuaded Pawsey to make him leader of a 21 cm effort with Hindman. He at once began supervising the construction of a 36 ft meridian telescope for southern 21 cm studies, and throughout the 1950s this instrument carried out fundamental surveys of the Milky Way and Magellanic Clouds (Sullivan 1988b).

In Holland, Oort envisaged a major, long-term program of 21 cm research on galactic structure, at first with the 7.5 m Würzburg and later with the hoped-for 25 m dish. As he told the Dutch Academy of Sciences in June: "I do not think it is exaggerating to say that research on the structure of the Milky Way has entered a new era" (Oort 1951:54). That same year he was awarded the Henry Norris Russell Prize of the American Astronomical Society and spoke on "Problems of galactic structure" at the Cleveland meeting in December.[67] In his talk Oort reckoned that radio studies, both continuum and 21 cm line, would soon have an impact on galactic structure ranking alongside the work of William Herschel, Kapteyn, and Shapley. The version of this talk published in 1952 explicitly laid out his proposed program for 21 cm and optical research on galactic structure. It was as if Oort already had a definitive paper fully written, save that the numbers had to be filled in (Oort 1979:37T).

Despite his intense desire for a full survey, Oort acceded to Muller's request in July 1951 to revamp their receiver for better stability and sensitivity:

> I said to Oort that this equipment was no good for further observations – it was just too poor. And looking back I must say that it was a great deed by Oort to say O.K., because he had been waiting for five years and he really wanted more observations. But he accepted that I said it was no good…and that I had to understand what all was happening. So I spent a year or so building a new receiver.
> (Muller 1973:11T).

One of the thorniest problems was combating powerful interference from shortwave messages bound for the former Dutch East Indies, newly independent as Indonesia:

> I learned how to build a receiver notwithstanding the fact that only twenty meters from my small shed was the end of a large rhombic antenna, putting out 30 kilowatts or so, with the second harmonic of its frequency right in my IF band at 30 Mc.[68] In a way I am very happy that I had to work during these first years at Kootwijk at such a difficult site, because I really had to go into problems of shielding, filtering, and so on.
> (Muller 1973:8T)

In the spring of 1952, after nine long months, Oort's patience was amply rewarded when the new 21 cm radiometer, three times more sensitive, commenced operation on a survey of the galactic plane. Together with the southern map by Kerr's group, within five years would

---

[66] In this paragraph, I have relied on Munns (2002:chapter 2), who gives many details of the beginning of radio astronomy at Harvard College Observatory. The two Shapley quotations come from documents cited by Munns on p. 126 and p. 134. Also see Doel (1996:68–77, 198–201) for the tumultuous state of astronomy at Harvard in the postwar decade.

[67] Astronomers had long searched for evidence of spiral arms in our Galaxy and the 21 cm technique would soon become dominant in this quest. But it is a remarkable coincidence that it was also at this December 1951 meeting that the first solid evidence was presented for spiral arms, in the form of linear concentrations of young stars and HII regions (Morgan, Sharpless, and Osterbrock 1952). See Osterbrock (2001), Smith (2006:329–33), and O. Gingerich (1985), "The discovery of the spiral arms of the Milky Way," pp. 59–70 in *The Milky Way Galaxy*, eds. H. Van Woerden *et al.* (Dordrecht: Reidel).

[68] Gart Westerhout (2002), a student of Oort's (Section 15.2.2), has recalled that if one held a screwdriver close to one of the guy wires for the tall shortwave antennas, one could hear in the spark thus created the voices and music of Radio Hilversum! ["The start of 21-cm line research: the early Dutch years," pp. 27–33 in *Seeing Through the Dust*, eds. A. R. Taylor *et al.* (San Francisco: Astronomical Society of the Pacific]

be produced the first detailed look at the structure of the entire galactic plane.[69]

16.7 NO RACE, NO SERENDIPITY, BUT INTERNATIONAL COOPERATION

The early days of the 21 cm line are often characterized as a "race" between the three different groups. But the preceding account shows that although there was indeed competition, it was not at all the usual kind of race. In the first place Ewen, who had been much more impressed by Shklovsky's confident prediction of a high signal intensity than by Van de Hulst's tentative conclusions, felt that his main competition was probably with the *Russians* (who in fact were doing nothing). On the Dutch side, Oort and Van de Hulst, once they learned of Ewen's experiment, were indeed anxious for a first success in Holland, but it appears that the youngest member of the team, Muller (1973:9T), was not as affected:

> I don't think I felt much competition, because I felt I knew nothing about the thing. Since I had really only just started, any idea that I could compete with such people [from Harvard and MIT] was at that time completely out of my mind.[70]

Note also that the fire at Kootwijk was not, as is often claimed, critical in denying the Dutch the winner's laurels; on the contrary, to the degree that it hastened replacement of Hoo, the accident *aided* the Dutch detection. As for the third team in Australia, the feverish pace of activity certainly indicated a desire to succeed quickly. There it was indeed viewed as a race, even if only for second place. This can be seen in Pawsey's remark in a letter to Oort in which he announced their detection: "I understand that your people have beaten us in obtaining the line."[71] Pawsey was also undoubtedly ruing his missed opportunity for *first* place in 1948–49 when he decided not to commit resources to a spectral-line search.

The 21 cm hydrogen line is also notable as one of the very few instances of a basic discovery in radio astronomy that had been predicted. Not only had the line been discussed previously by two theorists, but in fact the discoverers were motivated by those ideas. There are a few other instances in radio astronomy of earlier "predictions," but they were usually brought to light only *post facto*. Perhaps the most notable was the serendipitous discovery by A. A. Penzias and R. W. Wilson in 1964 of the cosmic microwave background.[72] But the great majority of the key discoveries in radio astronomy have been unexpected and usually serendipitous (Section 17.4) – witness Jansky's initial detection of the galactic background, J. S. Hey's discovery of solar bursts and of a discrete radio source, R. C. Jennison and M. K. Das Gupta's measurement of Cygnus A as a double source, and S. J. Bell and A. Hewish's later (1967) discovery of pulsars. This must result from a fundamental immaturity in understanding what the universe contains and how it works. The rush of technology has allowed astronomers to probe ever deeper and broader, scarcely pausing to digest the findings (Section 18.2). As the observers constantly uncover new phenomena, the theorists can do little more than fight a rearguard action. Only in those cases where some well-established phenomenon seems to "transfer" well to a new area are successful predictions likely. Spectroscopy, with almost a century of astronomical experience at optical wavelengths, was such a field in the 1940s. After the discovery, too, research involving the 21 cm line proceeded in a much more orderly manner than did the rest of radio astronomy. At Cambridge Peter Scheuer, bedevilled by the puzzles of the radio stars and galactic noise, enviously observed that: "The study of the 1420 Mc/s line due to natural hydrogen has already yielded results of the first importance to astronomy. However, the study of

---

[69] Van Woerden and Strom (2006:7–12) give a good overview of the scientific results that emerged from hydrogen line observations over the next four years using the Würzburg dish at Kootwijk.

[70] Not until August 1952, when Muller visited the Radiophysics Lab in Sydney (Fig. 16.9), did he see firsthand any other radio astronomy group's equipment. Only then did he realize that his own technical effort was competitive. [Muller:author, 8 February 1984]

[71] Pawsey:Oort, 13 July 1951, file A1/3/1p, RPS; also file 104(e), OOR.

[72] The "prediction" here refers to work in the late 1940s by George Gamow and collaborators.

the radiation of the continuous radio spectrum has posed far more problems than it has solved."[73]

The 21 cm story also nicely highlights the contrasting styles and institutional milieus of the three principal groups. At Harvard the experiment was considered one in microwave spectroscopy, albeit with an unconventional sample. Laboratory techniques from other areas of physics, ample funds, and electronic wizardry from the Boston area combined in powerful fashion to produce success. And yet when the first few positive measurements had been analyzed, the whole effort summarily ceased. Ewen did try to get a graduate student to carry on the project, but without success.[74] He himself was caught up in the national mobilization of technology for the Korean War. Purcell and his colleagues saw little fundamental physics to be tapped in the interstellar hydrogen line; besides, they had no lack of fruitful areas of physics to explore, and were busy with classified research for the military (Section 17.3.5). As for the astronomers, Harlow Shapley, who thirty years earlier had opened up the modern era of galactic research, and Bart Bok, a protégé of both Shapley and Oort, soon seized the opportunity to rejuvenate Harvard's program in galactic astronomy. Harvard astronomers, badly lagging their colleagues in the western US with large telescopes and good weather, were looking in these postwar years for ways to attract graduate students and research support (Munns 2002:chapter 2).[75] Bok looked upon the 21 cm hydrogen line as a means to accomplish this *and* to provide a new tool for astronomers to employ on longstanding questions such as the structure of the Milky Way. Unlike most radio groups around the world, however, from the start radio research at Harvard was to be integrated into a student's astronomy Ph.D. program.

In the Netherlands, the detection of the line took place in a country where sugar and coffee were still rationed and streets could not be lit at night. Here the line was also viewed as a key step in a long-term program of observational and theoretical research on the structure of the Galaxy. Since the time of Jacobus Kapteyn at the turn of the century, galactic structure, with Oort as its chief proponent, had been a Dutch specialty (Sullivan 2000). As at Harvard, the new radio techniques themselves were completely of secondary interest and in fact Oort (1972:11T) viewed them as more a nuisance than anything else: "We saw the technical problems only as a kind of evil that you had to take into the bargain." The pattern can be seen from the start: Muller built and ran the equipment at Kootwijk, took observations at positions and frequencies dictated by Oort, brought his strip chart scans to Leiden where he explained them to Oort, and then let Oort carry the analysis from that point to completion. The Dutch astronomers did not care to get involved in the details of electronics and antennas, except to the degree necessary to understand their data. For example, Van de Hulst told the students in his radio astronomy course that although electronic experts would take care of the equipment, the astronomer should still be able to calculate receiver sensitivities.[76] This segregation of scientists and engineers in Dutch astronomy was not at all the norm in radio astronomy, but was a strategy used elsewhere – a similar situation held in the 1950s at CERN, the European high-energy physics lab in Geneva.[77]

Although the discovery and exploitation of the 21 cm line did not substantially alter the scientific direction of the Dutch, the style of the research underwent a basic change. As Blaauw (1980:11) has pointed out, Dutch astronomy became a strong *observational* science for the first time in the twentieth century. Ever since Kapteyn, the Dutch with their cloudy skies had fallen

---

[73] P. Scheuer (1954:ii), "An investigation of the structure of the galaxy by means of its radio emission," Ph.D thesis, Cavendish Lab, Cambridge University.

[74] Kerr:Pawsey, 21 August 1951, file F1/4/KER/1, RPS.

[75] On the transformation of the astronomical research at Harvard due to World War II and its effects on the principal staff, see Munns (2002:chapter 2), Doel (1996:198–201), and P.A. Kidwell (1992), "Harvard astronomers and World War II – Disruption and opportunity", pp. 285–302 in *Science at Harvard University: Historical Perspectives,* eds. C.A. Elliott and M.W. Rossiter (Bethlehem, Penn.: Lehigh University Press).

[76] Opening lecture, notes for Astronomy 241B ("Radio Astronomy"), Harvard, Spring semester, 1951; box 36, VDH. Van de Hulst's final exam asked the student to describe a program of research that would justify the cost of a 25 m antenna – hardly a hypothetical task at that time for the Leiden group.

[77] D. Pestre and J. Krige, "Some thoughts on the early history of CERN," pp. 78–99 (especially pp. 93–5) in Galison and Hevly (1992).

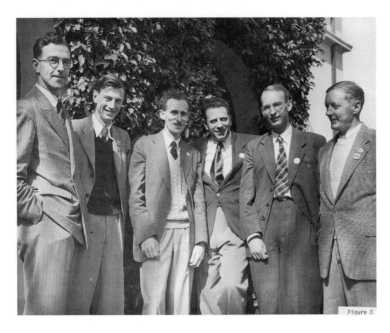

Figure 16.9 The "21 cm Club" gathered in Sydney at the 1952 URSI General Assembly. Left to right: Frank Kerr, Paul Wild, Jim Hindman, "Doc" Ewen, Lex Muller, and "Chris" Christiansen.

behind in observational astronomy as it demanded better and better observing sites. They had made up for this by concentrating on either theory or the interpretation of others' observations (Sullivan 2000), but the advent of radio waves from the Galaxy now allowed them at last to gather their own data.

In Australia, developments at the Radiophysics Lab were again different, and more typical of most early radio astronomy groups. Technical expertise and institutional flexibility were brought to bear on a new challenge. Once the technical problem was quickly solved, Christiansen and Hindman realized that the astronomical problem was also interesting. Thus began an education on this particular topic, a process that had been repeated many times during the previous six years as engineers and radio physicists transformed themselves into astronomers. In the present case the theoretical contribution of Wild and the crash astronomical education Kerr acquired at Harvard also played vital roles.

The 21 cm line gave astronomy a tool allowing access to two regions previously little explored: the neutral regions of interstellar space and the distant regions of the Milky Way. Studying the astrophysics of the neutral hydrogen clouds and of the structure of the Galaxy immediately became a profitable area of radio astronomy. It also provided one of the rare examples in the 1950s of close cooperation between radio researchers (Fig. 16.9) and optical astronomers (Section 17.2). Such cooperation was fostered in 21 cm research because of the keen interest in both optical and radio results shown by what might be called the *International Dutch School*, a research school that dominated the field under the intellectual leadership of Oort and Van de Hulst. Oort, who had been a student of Kapteyn's and was heavily influenced by his style, was also an internationalist, whether through organizations or projects involving cooperative science (Sullivan 2000). In addition to the Leiden group, other key members of the School were Bok (an expatriate), who had studied under Oort and Van Rhijn (Ph.D. in 1932), and Kerr, who became a member while following Van de Hulst's course at Harvard in 1951 (and in 1957 spent a sabbatical at Leiden). The research program of the International Dutch School was directed toward an understanding of the physical state, content, and distribution of the interstellar medium, employing whatever observational techniques seemed best able to test theoretical ideas. This approach meant that radio astronomy in

Holland, as well as 21 cm research around the world, was well integrated with optical astronomy from the start, whereas in most other situations radio studies long remained distinct (Section 17.2). Not in fact until the mid-1960s would a general integration of broader radio astronomy take place.

TANGENT 16.1 OTHER PRE-DISCOVERY CONSIDERATIONS OF THE 21 CM LINE

Kevin Westfold translated Shklovsky's 1949 article and he and John Bolton at one stage were eager to look for the line, but were not given permission to do so by Pawsey (Bolton 1982:353, 1978:53–4T). Jack Piddington also seriously looked into the question of the line at this time, but decided on the basis of Shklovsky's article that success was unlikely.[78]

Olof Rydbeck (1978:8–11T) at Chalmers Institute of Technology in Sweden (Section 10.6.2) attempted to mount a search. In 1950 or 1951 he applied for a governmental grant to build a 21 cm receiver and search for the line with a Würzburg dish, but his application was turned down.

At Cornell University, Leif Owren (Section 10.1.3) *did* briefly search for the line. All that is known of this attempt is in the following excerpt from a letter of Frank Kerr's written only one week before the line was finally found by Ewen at Harvard:

> Owren, at Cornell, has also looked for the line without success. He used an 8 ft dish, but his receiving sensitivity would not have been as great as Ewen's. He has suggested that the same line might also show up in the aurora, and intends looking for it there.[79]

---

[78] Piddington:author, 19 October 1984.

[79] Kerr:Pawsey, 17 March 1951, file A1/3/17, RPS.

17 • New astronomers

The development of radio astronomy illuminates many issues important to the history of science. The radio practitioners operated in a culture that differed substantially from that of the astronomers. While their instrumentation and material culture were most obviously different, no less important were social differences such as educational backgrounds, approaches to scientific questions, funding sources, and institutional structures. Some of these issues have been discussed in previous chapters as they arose, but others will be treated here for the first time. In this chapter I first examine the role of World War II in kickstarting radio astronomy, and then the field's growth until the watershed date of ~1952–53 when this study ends (Section 17.1). In Section 17.2 I describe in detail the developing relationship between (optical) astronomy and the new radio astronomy. I argue that radio astronomy, despite its intrinsically non-visual nature, ironically strove to develop a "visual culture," one that would appeal to astronomers even as it legitimated radio observations (Section 17.2.5). In particular, the introductions of terms such as *radio astronomy*, *radio astronomer*, and *radio telescope* provide insights into attitudes of the time (Section 17.2.1). Yet, even once the term *radio telescope* existed, it was problematic as to when a radio telescope became a *real* telescope.

Section 17.2.5 views radio astronomy as an emerging specialty, never its own new discipline, but rather steadily being subsumed by (traditional) astronomy. Section 17.3 analyzes the very different development of radio astronomy in England, Australia, and the US. Many national influences played important roles, and in particular I argue that abundant military patronage in fact hindered the development of US radio astronomy in the postwar decade, sending it in less fruitful directions (Section 17.3.5). This situation is compared with results from other studies of scientific disciplines and specialties in the Cold War. Finally, in Section 17.4 I discuss the relationships between science and technology in early radio astronomy. My conclusion is that radio astronomy was a *technoscience*, meaning that in the dance between science and technology partners took turns leading, while at other times they were entangled to the point that only one entity, the partnership, existed.

This chapter thus presents the main evidence for three of the book's four major themes (Section 1.3.3): the profound influence on early radio astronomy of World War II and the Cold War; radio astronomy as a technoscience; and radio astronomy as a visual culture. The case for the fourth theme, of radio astronomy as the twentieth-century's New Astronomy, is made in Section 18.3.2.

17.1 DEVELOPMENT OF EARLY RADIO ASTRONOMY

17.1.1 Origin in war

Radio astronomy's character was forged during World War II and tempered during its aftermath. As we have seen in earlier chapters, those who developed radar under the stresses of war in many different countries later became the heart of the new science (Figs. 1.2 and 1.3). Section 5.1 outlines the overall development of radar, and the first parts of Chapters 6 through 10 provide details on how wartime radar activities influenced the specific beginnings of each of the key radio astronomy groups. In this section I briefly discuss the general influence of the war on the field that became known as radio astronomy.

Radar was a wonder weapon in World War II (Fig. 1.4), although overshadowed by the atomic bomb because of the latter's brute power and geopolitical ramifications. When the US military declassified many aspects of radar in August 1945, *Time* magazine's story noted that the US had spent $3 billion on radar as compared to only $2 billion for the Manhattan project, and that radar had become as routine a fighting tool as a rifle.[1] Lee DuBridge, head of the MIT Radiation Laboratory (Section 10.1), voiced a common view when he said that "The atom bomb only ended the war; radar won it" (Kevles 1977:308).

The Australians, British, and Americans who had been recruited to develop Allied radar started with a variety of scientific and engineering backgrounds, but by 1945 they had had similar experiences that served them well in their postwar careers. First, they gained immense technical knowledge. Perhaps more importantly, however, as shown in detail in Sections 17.2.2 and 17.3.4, they were also profoundly shaped by years spent with bright colleagues, large budgets, and incessant deadlines on which lives depended. These circumstances led to invaluable experience in solving technical and organizational problems. And even after the war, these engineers and scientists, especially the physicists, could continue to act with confidence, for they were lionized because of their wartime accomplishments. It was nevertheless ironic that the physicists had achieved fame not by primarily doing physics, but rather in the role of *engineers*; as Forman (1995:398) puts it, they had been more wizards of wonder-weapon warfare than magi of the mysterious cosmos.

These effects were especially strong in England, where the war was close and conditions dire. As Martin Ryle (1976:1–2T) later recalled of his wartime radar career at the Telecommunications Research Establishment:

> Whereas one's physics had almost evaporated after six years, one learned a lot of engineering – what things could work.... [In contrast] the actual straight transfer of technology [equipment] to radio astronomy wasn't, I think, as important as it is often made out, [although] one gained a lot from bits of ex-radar equipment which would have been much too expensive to build on our [minuscule] university budgets just after the war.
>
> One also learned a lot about people.... You had enormous responsibilities at an early age. You had to make decisions, there was nobody to ask because you knew more about it than anybody else.

Or, as Edward Purcell (1977a:27–8T) similarly recalled of his Radiation Lab experience:

> Not only did I acquire a whole new kit of research tools – microwave technology, transmission lines, signal-to-noise theory, just a lot of different things that were going to be useful one way or another – but perhaps the most important thing was being thrown together in a working relationship with a number of physicists from different places. ... It provided us with the tools, not merely the hardware, for a basic understanding ... of what you have to do to detect a signal in noise.

Radar spawned not only radio (and radar) astronomy after the war, but a host of sibling technologies and sciences. There were of course direct extensions of military radar to the civilian worlds of television and radio broadcasting, commercial aviation, and maritime navigation, not to mention police speed-traps. More generally, electronics as a whole was revolutionized by progress during the war on low-noise receivers, antennas, oscillators, waveguides, transmission lines (for example, polyethylene coaxial cable), crystal diodes, amplifiers, cathode-ray oscilloscopes, and vacuum tubes (see Forman [1995:410–3]). Equally important were new principles of noise and circuit theory and signal processing, best exemplified in the microwave radiometer receiver designed by Robert Dicke (1945; Section 10.1; Forman [1995:435–41]). The monumental 28-volume Radiation Lab series *Principles of Radar* (Ridenour 1947–53) became a bible of electronics. The engineering advances within its pages enabled the flowering not only of radio astronomy, but many other related fields and techniques such as microwave spectroscopy, nuclear magnetic and other types of resonances, analog and digital computing, and high energy physics (using particle accelerators).[2]

As discussed in detail in Section 17.3.5, these fields not only descended from wartime radar in terms of their technology and styles of research, but also were molded by wholly new postwar relationships between scientists and engineers, their governments, and their militaries. Moreover, major funds for fields such as

---

[1] *Time*, 20 August 1945, pp. 78–82.

[2] Forman (1995) and Buderi (1996) provide detailed looks at the many areas of physics that grew out of wartime radar. Note that radio astronomy was intellectually prominent, but remained a tiny portion of radio research. For example, in a 1950 American overview of radio science and engineering, only ~3% of the space was devoted to radio astronomy ["Radio progress during 1949," *Proc. IRE* **38**, 358–90 (1950)].

radio astronomy were now readily available, causing research to go in directions shaped by new patrons. In addition, the sizes of the largest scientific labs and projects were at levels not known before the war.

In sum, many factors indelibly stamped radio astronomy as a child of World War II. Although the war-related factors varied in strength and color from one research group to another, they were interwoven everywhere. In each of the groups, a major military weapon was turned to investigation of the distant universe, and wartime physicists and engineers eventually turned into peacetime astronomers.

17.1.2 Early growth and the 1952–53 watershed

Radio astronomy's growth over the postwar decade was steady, but not explosive. In this section I present and comment on a few measures of this growth. Figure 1.2 encapsulates details of all the early groups.

Figure 17.1 shows, for the entire world, the trend of total number of radio astronomy papers and authors, based on annual bibliographies produced at Cornell University (Appendix C). These bibliographies were very thorough (although excluding radar studies), sometimes to a fault (for example, including papers with only the slightest mention of extraterrestrial radio emission). Over the nine-year period 1945–53 there was a steady rise from a handful of papers to ~150 per year, with a similar number of authors publishing at least one paper in a given year. If we confine ourselves to only those papers in the three "central" subfields – sun, galactic background radiation, and discrete sources (radio stars) – then the number of papers grew to ~80 per year. The doubling time for all papers over this interval was 4.3 years, agreeing well with a study by Menard, who found that a sample of important twentieth century fields in the physical sciences had literatures with doubling times in the range of 4.4 to 5.7 years.[3]

Price argued in *Little Science, Big Science* that the work of a cohort of about 100 fellow researchers is in fact the most that any individual can effectively follow – indeed, this was about where radio astronomy found itself in 1953.[4] The field was sufficiently compact that as late as October 1954 Joe Pawsey could give a professional colloquium covering it all and entitled simply "Radio Astronomy" (at Caltech).[5] Despite substantial growth, radio astronomy was minuscule compared to astronomy as a whole. One measure is the number of articles listed in *Astronomischer Jahresbericht*, the annual abstract of international astronomy: by 1950 only 2.5% of articles dealt with radio astronomy. Another measure comes from a compendium of astronomy that was assembled in 1953–54: ~6% of its space was devoted to radio astronomy.[6] The small size of radio astronomy, however, belied its importance. For example, by one measure radio astronomy accounted for 31% (10 of 32) of all *seminal* papers in astronomy and astrophysics published in 1945–53 (as judged in the collection by Lang and Gingerich [1979]).

Plots for the three central topics (lower part of Fig. 17.1) illustrate how solar research dominated the field, declining from ~80% to ~60% over the period 1945–53. As discussed in Chapter 13, the sun was the major object of attention around the world simply because it was the easiest to observe. Moreover, the sun's variability and its links with radio ionospheric communications meant that funds were readily available to monitor and study solar emission. By 1953 there was still important work being done on the sun, but the great majority of solar publications arose from routine monitoring. Meanwhile, discrete sources, the subfield of radio astronomy generally judged to be the most controversial and enigmatic, steadily rose to ~30% of all papers by 1953. Measurements and theory of the galactic background radiation remained in the 10–20% interval over the entire period.[7]

Figure 17.2 shows the growth in number of researchers for the three major radio astronomy groups,

---

[3] H. W. Menard (1971), *Science: Growth and Change* (Cambridge: Harvard University Press), pp. 27–31, 51. Gillmor (1986) found a similar doubling time (5.4 years) for ionosphere physics over 1947–69.

[4] D. J. deSolla Price (1963), *Little Science, Big Science* (New York: Columbia University Press), pp. 71–3.

[5] Document mentioning 15 October 1954 talk by J. L. Pawsey at Caltech, GRE.

[6] The journal *Vistas in Astronomy*, Vols. 1 and 2 (1955–56).

[7] These statistics do not include meteor radar papers, discussed in Section 11.5, which would contribute to the literature plotted in Fig. 17.1 an additional ~25% in 1947–48, then rapidly declining by 1953 to only an additional 7%.

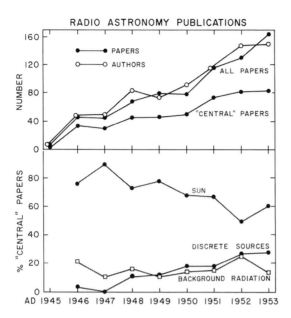

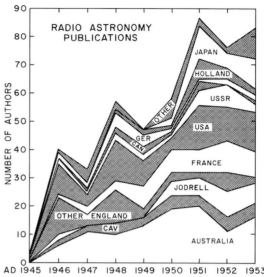

Figure 17.1 (Top) The worldwide total number of radio astronomy papers and authors (*not* including radar studies) each year during 1945–53. Also plotted are only those papers in the "central" topics of sun, galactic background radiation, and discrete sources. (Bottom) The percentage of papers in each of the central topics.

Figure 17.3 The number of publishing researchers in any given year for radio astronomy groups and nations. Included are only the "central" topics of sun, galactic background radiation, and discrete sources; in particular, note that the Jodrell Bank group's meteor radar papers are not counted here. "Other England" is dominated by theorists and Hey's group (through 1948); CAV = Cavendish Lab, CAN = Canada, GER = Germany.

based on data from Edge and Mulkay (1976:24) and my own compilation for the Radiophysics Division in Sydney (as also used for Fig. 7.2). Note that the Jodrell Bank group was almost exclusively devoted to meteor radar work until the early 1950s, and thus the Sydney group strongly outnumbered their two English rivals in personnel devoted to passive radio astronomy. This can be seen in Fig. 17.3, which shows the number of publishing researchers (now *not* including radar studies) in any given year for groups around the world. The Sydney group dominated, contributing as many as ~30% of all authors in the 1947–50 period, decreasing to ~20% in 1951–53.

On the basis of the *number* of researchers plotted in Fig. 17.3 one might expect that the US, Japan and France had a more influential role in the early 1950s than in fact was the case. As discussed in Section 17.3, the US indeed had many groups producing many papers (mostly on the sun), but the great majority of these were not influential in the field. Japanese and French research (Sections 10.4 and 10.5) was exclusively on the sun, the former of negligible influence in

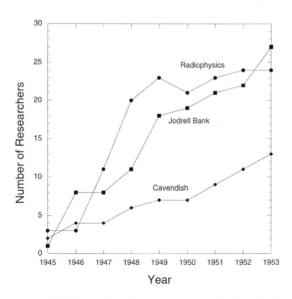

Figure 17.2 The number of researchers over 1945–53 in each of the three major radio astronomy groups.

the West and the latter contributing to the field in solar theory (Jean-François Denisse) and eclipse observations (Sections 13.1 and 13.2), although eclipse reports with unusually many co-authors have distorted the publications data (for instance, a six-authored paper by Laffineur *et al.* [1952]).

An important missing factor in these statistics is the age distribution of the researchers. I estimate that the median age of those publishing in 1950 was ~30 years, indicating a very young cohort.[8] This relative youth undoubtedly contributed to the high quality and quantity of research.[9]

The endpoint of the present study, 1952–53, has been chosen because it was a watershed in the development of radio astronomy. By this point, all of the major groups that would dominate radio astronomy for the next decade had been established. In terms of the field's most exciting frontiers, solar radio research was on the decline and galactic and extragalactic studies, where radio astronomy would make its most important contributions,[10] were rapidly rising (Fig. 17.1). In particular the puzzle of the discrete radio sources was moving to center stage. Many fundamental discoveries had been made and the field was changing from an exploration mode ("Let's just look!") to one where antennas and receivers were designed for specific scientific goals. As Australian James Hindman (1978:98B) later recalled of the ~1953 period: "I felt I was [now] reaching for the limits of the astronomy, rather than of the equipment."

Radio astronomers were also growing comfortable in an astronomical world, talking with astronomers (and vice versa) and increasingly attending astronomical meetings (although not abandoning their engineering, physics, and radiophysics societies). As an ironic part of this nascent integration, a new term was gaining currency, *optical astronomy*. Moreover, in hopes of improved communication, the terms *radio noise*, *solar noise* and *cosmic noise* (as in the title of this book) were officially abandoned in favor of *radio emission*, etc. (Section 17.2.5).

Perhaps the biggest change was in the *scale* of the new undertakings. In almost every major group, a shift took place from operating on shoestring budgets, often with war surplus gear, to designing and purchasing major new antennas. As equipment became more complex, the time scale of an experiment (from idea to publication) changed from a few months to a year or two. Huge antennas no longer could be constructed in-house and now involved years of planning, hunting for sites, construction, snafus, and shakedown. As a CSIRO Radiophysics Report said in 1953:

> The Conference [the 1952 URSI General Assembly in Sydney] had a further significance in providing one of the early milestones in the path of the new science of radio astronomy, ... [which] has reached the end of its first exploratory stage in which existing tools were employed, and is about to start on a new stage for which new tools will be needed.[11]

This book has not covered the histories of these "new tools," but many have been mentioned at the ends of chapters as being in the design stage by 1952–53: Grote Reber's proposed 200 ft dish (never built) (Chapter 4); the Giant Radio Telescope proposed by the Radiophysics Division (which eventually became the Parkes 210 ft dish), as well as Bernard Mills's cross antenna (Chapter 7); Martin Ryle's 2C survey aerial at Cambridge (Chapter 8); Bernard Lovell's 250 ft dish at Jodrell Bank (Chapter 9); the 50 ft dish at the US Naval Research Lab (Section 10.1.2); a large reflector at Nançay, France (Section 10.4); and the 25 m Dwingeloo dish in the Netherlands (Chapter 16). In many ways major parts of radio astronomy were becoming *Big Science*, defined as involving large, expensive instrumentation and large research groups (Galison and Hevly 1992).

The increased scales of budgets, administrative headaches, and personnel made even those who clamored for Bigness fondly remember earlier days. Upon Ryle's being elected a member of the Royal Society in

---

[8] For example, in late 1949 a listing of the 16 Sydney radio astronomers yields a median age of 29. [R. N. Bracewell: M. Ryle, 6 January 1950, file 1/1, RYL]

[9] One detailed study concluded that a scientist's best and largest output occurs in the age range of 25 to 40. [P. Stephan and S. Levin (1992), *Striking the Mother Lode in Science: the Importance of Age, Place and Time* (New York: Oxford University Press), chapter 4.

[10] Edge and Mulkay (1976:109–110) have also argued that the first exploratory phase for radio sources had ended by 1952, giving way to more deliberate research programs.

[11] 1952–53 Annual Report of the Radiophysics Division, file D2/8, RPS.

1952, Lovell, then becoming enmeshed in his Big Dish project, sent a congratulatory note:

> There can be no doubt that your wonderful experiment is a really classical effort. To me it has seemed like an oasis in a desert of Harwells [nuclear reactors] and cyclotrons – an elegant protest against all that is horrible in modern physics. May you be preserved from the attractions of steel and concrete, and all the accompanying agonies into which the rest of us have fallen!

Ryle replied:

> We have been very fortunate in being able to exploit an original idea – and obtain results with quite simple apparatus – but I am not certain how long it will be possible to continue to produce answers easily. I feel that it is inevitable that before long large instruments will become indispensable – and their construction will become as imperative as that of large [optical] telescopes.[12]

Ryle's very election as a Fellow of the Royal Society, along with Lovell's gaining a Chair in Radio Astronomy the year before, was further recognition of the field's legitimacy, maturity, and importance. Textbooks by Van de Hulst (1951) and by Pawsey and Ronald Bracewell (1955, but written in 1952–53), as well as popular books by Lovell and John Clegg (1952) and Iosif Shklovsky (1953a) were further signs of the field's maturity. Indeed, most researchers were now beginning their mid-career periods as *radio astronomers*, metamorphosed from the astronomical neophytes who had jumped into extraterrestrial radio noise right after the war.

17.2 RADIO ASTRONOMY AND (OPTICAL) ASTRONOMY

In this section I examine how the new radio astronomers interacted with traditional (optical) astronomers, and how they cognitively and socially established themselves. The first four subsections discuss the evidence of how the radio/optical relationship changed over the period 1945–53, from the points of view of terminology, styles of research, personal interactions, and professional and institutional settings. Finally, I argue in Section 17.2.5 that radio astronomy never did become its own discipline, but instead developed its own "visual culture" and became absorbed, although still not fully integrated, into astronomy as a whole (as shown in Fig. 1.3).

17.2.1 Terminology

The origins of new terms and changing meanings for old ones superbly gauge the intellectual and social attitudes of those using the terms. Based on his study of the transformation of alchemy into chemistry, Hannaway has even argued that language and reality can be coincident.[13] Such a lexical approach is equally fruitful when examining the relationship more than three centuries later between (optical) astronomers and those who first studied the radio sky.[14] For many years the latter studied extraterrestrial radio *noise*, or more specifically cosmic noise, galactic noise, or solar noise. But *noise* (meaning radiation characterized by random fluctuations) was a misleading technical term and hardly attractive or descriptive.[15] Nor did it suggest more than a weak link to the sky and to astronomy.

When exactly did the measure of identity and connection with astronomy afforded by the term *radio astronomy* come into vogue? The earliest usage that I have found[16] was by Vitaly Ginzburg (1947:52) in a review article written (in Russian) for *Uspekhi Fizicheskikh Nauk* in March 1947: "The above... observations and analysis reveal great astrophysical, geophysical and radio-technical interest in 'radio astronomy' ⟨радиоастрономии⟩ or 'radio astrophysics,' which we could call this new discipline."[17]

---

[12] Lovell:Ryle, 21 March 1952; Ryle:Lovell, 2 April 1952, file 28 (uncat.), RYL.

[13] O. Hannaway (1975), *The Chemists and the Word* (Baltimore: Johns Hopkins University Press).

[14] I have earlier discussed the term *optical identification* in this light in Section 14.4.4.

[15] See note 1.4 for a definition of *noise* and Section 17.2.5 for efforts to cease using the term.

[16] My various claims for "earliest usage" in this section are based on a log kept throughout the project of usages of terms as I encountered them in documents.

[17] The term "radio astronomy" was also used twice during the war in isolated instances with somewhat different meanings. (1) A 1942 newspaper headline of "Radio-astronomy traps music of the stars" referred to the meteor radar work in India described in Tangent 11.1 [*New York Times*, 10 May 1942, p. X10]; (2) A radio amateur who was also an amateur

As discussed in Section 10.3.4, however, the Russian literature at this time was virtually unknown in the West. The first known usage in the West occurred in a letter by Joseph Pawsey in January 1948,[18] and the first appearance in print was in the published version of Martin Ryle's talk of April 1948 to the Royal Astronomical Society (RAS).[19] A one-day session in the US in May 1948 was entitled "Radio Astronomy," although the extensive minutes of the talks use other terms such as "cosmic radiation."[20] A few months later, Ryle (1948b) published the first article entitled "Radio astronomy," doing so only after consulting with his father about the propriety of mixing words of Latin and Greek origin (Ryle 1976:27–8T). In August 1948 a new commission named "Radio Astronomy" was established by the International Astronomical Union (IAU) (Section 17.2.4). By the next year the term was ubiquitous; in 1949 alone fully eight review articles (in three languages and from five nations) appeared with this title. *Radio Astronomy* (published in 1952) was also the title of the first popular book on the radio sky, written in 1950 by Lovell and Clegg.[21] By the early 1950s the term *noise* had been almost completely replaced by *astronomy*.

The introduction of the term *radio astronomer* was surprisingly not in parallel to *radio astronomy*,[22] despite an isolated first usage in a *New York Times* article of October 1948.[23] The first known usage by a scientist was in late 1949 by the astronomer Otto Struve (1949:30), and the next by Bernard Lovell and Edward Appleton in late 1951.[24]

Finally, *radio telescope* preceded both *radio astronomy* and *radio astronomer*.[25] Harlow Shapley referred to a *microwave telescope* in a letter of March 1945,[26] and in one of the first postwar papers on the radio sun (Section 5.5), Appleton (1945a:535) wrote in *Nature* in September 1945 of "the [sky] area which can be 'seen' by the 'radio-telescope.'" For many years, however, the term was only sporadically used, including a mention by Ginzburg in the 1947 review mentioned above. Even after becoming common, it often appeared in quotation marks or was only used by allusion as in "a directive antenna which has the properties of a telescope."[27,28]

Less appreciated is the obverse phenomenon: even as they established radio astronomy, those who were called radio astronomers also effected the creation of the retronyms *optical astronomy*, *optical astronomers*, and *optical telescopes*. Where before no modifier had been needed, *optical* perforce began appearing – traditional astronomers also began to be pigeonholed by their technique. The earliest use of *optical telescope* that

---

astronomer proposed to combine his hobbies by, for example, establishing a radio net so that lunar eclipse observers could share data during the event. He called this an example of "radio-astronomy" [T. R. Sprecher, *Sky & Telescope* 3, No. 3, p. 2 (January 1944)].

[18] Pawsey:O. F. Brown,, 14 January 1948, file F1/4/PAW/1, RPS. Peter Robertson found this letter and I thank him for sending me a copy.

[19] Ryle, *Observatory* 68, p.180 (report of a talk given at the RAS meeting of 23 April 1948).

[20] "URSI radio propagation conference – Radio astronomy session – minutes of the meeting," held on 6 May 1948 at the Naval Research Lab, Washington, DC. [Supplied to the author by Arthur Covington]

[21] Much of Lovell and Clegg's 1952 book is about the techniques and results of meteor radar, but considered as part of *radio astronomy*. I have not traced the origin and early usage of the term *radar astronomy*, which came later.

[22] For instance, note the lack of parallelism in a letter where Jesse Greenstein refers to "contact between radio astronomy and astronomers." I thank David DeVorkin for supplying this letter. [Greenstein:D. H. Menzel, 17 February 1950, box 4, GRE]

[23] "Radar yields new world of sound" by W. L. Laurence, *New York Times*, 6 October 1948, p. 1. Two months later, a symposium entitled "Microwave Astronomy" was held at a meeting of the American Astronomical Society. For an article by C. A. Federer, the *New York Times* used the term *microwave astronomer* in a headline [31 December 1948, p. 7]. Also see Shapley, as in note 32.

[24] A. C. B. Lovell, "The radio astronomer's universe," *The Advancement of Science* 8, 351–60 (published in 1952, lecture given in November 1951); "Radio astronomy," a popular talk by E. V. Appleton reported in *The Times* (London), 26 November 1951, p. 6.

[25] See Tangent 17.1 for prewar uses of *radio telescope*.

[26] H. Shapley:G. C. Southworth, 14 March 1945, file 51-01-02-04, SOU (passage cited in Section 5.6).

[27] J. P. Hagen (1949), *Astron. J.* 54, 122. G. de Vaucouleurs used the term *télescope radio-électrique* (albeit within quotation marks) in a popular review of October 1948 ["Parasites cosmiques," *Larousse Mensuel* 41, No. 410, 155]. In that same month the *New York Times* even asserted that the term was erroneous [10 October 1948, "Celestial radio," p. E8].

[28] See Section 17.2.1.1 for more discussion on radio telescopes vis-à-vis telescopes.

I have found is in an October 1947 report by Charles Burrows, head of the Cornell University group.[29] Appleton was the first to use *optical astronomy*, in a January 1949 speech. The first published usages were in late 1949 by Allen and Gum (1950:224) (*optical astronomy*) and by Struve (*optical telescope*). Appleton in a 1951 lecture referred to *visual astronomers* and *visual telescopes* and defined two available "windows" in the earth's atmosphere.[30] Nevertheless, it was definitely *not* the situation that the terms *radio astronomy* and *optical astronomy* (and their cognates) evolved in parallel. The latter was only rarely used in the postwar decade, an indicator of lack of symmetry in the status of radio astronomers relative to optical astronomers. When the two groups were referred to in the same phrase, the first were called "radio engineers, radio physicists, radio-observing personnel, radio men, radio workers, radio technicians, electronic technicians, radio specialists, 'blind' astronomers,[31] radio astronomers," while the second were simply "astronomers" or "astrophysicists," with rare qualifications such as "astronomers of the classical type, visual or photographic astronomers, astronomers in the traditional sense, general astronomers." Shapley, head of Harvard College Observatory, wrote in 1948 that "electronic engineers, not astronomers, do most of the work in this new branch of astronomical research."[32] At a 1950 meeting Marcel Minnaert of Utrecht Observatory "on behalf of astronomers complimented the radio physicists on [their] remarkable contributions to astronomy."[33] Perhaps the most telling usage was recalled by Bruce Elsmore (1971:21T) of the Cavendish Lab radio group:

[The optical astronomers] used to mock us a little bit and say we weren't really astronomers. ... If you went out with another astronomer and you introduced him to a friend, you would say, "This is a *real* astronomer," just to show the distinction.[34]

Another sign that the radio astronomers were not "real" was that they did not work in an *observatory*, despite the manifest observations they conducted. The first usage of *radio observatory* occurred in a 1950 report when describing a motion at an international meeting to establish a "World Chain of Solar Radio Observatories."[35] But the term was very rare until the mid-1950s when its first official use came with the founding and naming of the National Radio Astronomy Observatory in the US and the radio observatory at Dwingeloo, The Netherlands. For instance, Pawsey and Bracewell (1955:6) in their monograph (written in 1952–53) wrote that "although radio astronomy is an integral part of astronomy, very few of the research centres which have grown up have been located in observatories." But by 1955 Pawsey referred to "at least 32 radio observatories throughout the world" in a report to the IAU.[36]

Usage of the adjective *astronomical* during this period is also illuminating. Most of the time (although not exclusively) its sense was "optical-astronomical" – radio astronomy matters were simply not "astronomical," no matter who was speaking. Examples in chronological order (RA = radio astronomer; OA = optical astronomer; italics mine):

– This work is primarily a radio and not an *astronomical* investigation. [RA Grote Reber in 1945][37]
– Radio-frequency investigations of *astronomical* interest. [Title of a review paper by RA Grote Reber and OA Jesse Greenstein (1947)]

---

[29] As in note 10.19.

[30] E.V. Appleton, "Radio waves of extraterrestrial origin," speech given to the Science Masters Association, 3 January 1949, file 4 (uncat.), RYL; O. Struve, "Radio astronomy," *Sky & Telescope* 9, p. 56 (January 1950); Appleton, as in note 24. Struve (1949:29) had also earlier described these two atmospheric windows.

[31] These "blind astronomers" also did their "blind astronomy" using a "blind telescope" [*New York Times*, 26 April 1952, p. 25]. Also see note 1.8.

[32] H. Shapley, "Some astronomical highlights of 1948," *Sky & Telescope* 8, 42–4, quotation on p. 44 (Dec. 1948).

[33] Report of a session on "Galactic radio waves" held at the URSI General Assembly in Zürich, 21 September 1950; in *IXth General Assembly* [of URSI] 8, Pt. 1, p. 402 (1950, Brussels: URSI)

[34] Originally quoted by Edge and Mulkay (1976:274).

[35] Resolutions, Report of Commission 5 on Extra-Terrestrial Radio-Noise, *IXth General Assembly* [of URSI] 8, Pt. 1, p. 64 (1950, Brussels: URSI). A second early use of the term *radio observatory* was in the Dutch title of an article by Henk Van de Hulst, "Radiosterrewacht Sydney," the name with which he christened the Radiophysics Lab [*Hemel en Dampkring* 49, 41+61+81 (1951)].

[36] J.L. Pawsey, presidential report (1955) to IAU Commission 40, in P.H. Oosterhoff (ed.), *Trans. IAU* 9, 563–586 (1957, Cambridge: Cambridge University Press), p. 563.

[37] Reber:Struve, 23 September 1945, box 221, YKS.

– [It is] essential to have the radio and *astronomical* equipment [together]. ... No close correlation has been found between short-period radio-noise phenomena and solar observations. [OA Cla Allen (1947:386); note how the second sentence even discounts radio noise from the sun as being a "solar" observation]
– The coincidence of the radio source with a very abnormal *astronomical* object. [OA Walter Baade in 1951][38]
– Detailed comparison of the radio observations with the *astronomical* data. [Title of a section by RA Robert Hanbury Brown and RA Cyril Hazard (1952a:147)]
– The ability to carry on one's observational research nearly independently of conditions of weather or of daylight appears peculiarly *unastronomical*. [OA Daniel M. Popper in 1952][39]
– Thus radio methods may detect objects unobservable *astronomically*. [OA Rudolph Minkowski and OA Greenstein (1954:242)]
– At such a distance it would cease to be an optically observable *astronomical* object.[40] [OA Baade and OA Minkowski (1954b:225)]

Notwithstanding these examples, I have also found a few instances where *astronomical* subsumed both radio and optical sides, for instance in phrases such as "radio studies are a new field of astronomical research"; "radio and other astronomical phenomena"; "astronomical observations made at radio wavelengths"; and "twelve new octaves of the electromagnetic spectrum opened for astronomical investigation." But this type of usage was the exception in the postwar decade.

17.2.1.1 Is a radio telescope a telescope?

As mentioned above, the term *radio telescope* was only sporadically used in the postwar decade and then often in quotation marks. A radio telescope, largely because of its poor angular resolution and (except for a "dish" reflector) ungainly appearance, did not have the legitimacy of a traditional telescope, which was "real." For instance, Struve (1949:56) wrote "Unfortunately, it has not yet been possible to perfect radio telescopes to such an extent as to match the actual resolving power of the human eye or of a *real* optical telescope" [italics mine]. And J. B. S. Haldane yearned in 1948 for a "real radio telescope which distinguishes directions as well as, say, a very shortsighted man."[41] Cambridge radio astronomer Peter Scheuer (1976:65B:440) recollected special misgivings that astronomers had with radio interferometers:

> The standard argument was, "Well, an interferometer only gives you a few Fourier components – it doesn't give you an *image*, and you can't really deduce anything from just a few Fourier components. An interferometer doesn't tell you anything you can believe." And it took a very long time before optical astronomers would take any serious notice of anything that wasn't produced with something that looked like a telescope! They could understand a paraboloid, but an interferometer was the unknown devil. ... It wasn't merely old fuddy-duddies – some very bright people also had this attitude.

Not until the mid-1950s were radio telescopes accepted unproblematically as telescopes. Before that time it was not clear whether a radio astronomer was actually "seeing" something with his instrumentation (see Fig. 14.17). The philosophical question of what constitutes "seeing" is complex,[42] but in normal parlance *seeing* connotes not just detection of a phenomenon or signal, but confidence in the detection. As discussed in Chapter 14, a primary way in which radio astronomers (and others) gained confidence in radio results was how well they tied in with a "visual culture," namely the world of optical astronomy and optical identifications. In early 1950 Pawsey complained at a meeting that "the outstanding deficiency in such solar-noise studies to date is that no one

---

[38] W. Baade:F.G. Smith, 23 October 1951, LOV.
[39] D. M. Popper, review of *Radio Astronomy* (Lovell and Clegg 1952), *Pub. Astron. Soc. Pac.* **64**, 210–2 (1952).
[40] This sentence is intriguingly ambiguous. Although Baade and Minkowski's standard usage of *astronomical* excluded radio, in this passage they were discussing what would happen if Cygnus A were placed at ten times its actual distance. Was the inference that there are indeed such things as astronomical objects that are not optically observable (that is, observable only by radio means), or that even if radio emission from the (optically unobservable) object were detected, it would still not qualify as an *astronomical* object?

[41] J. B. S. Haldane, "Radio from the sun," *Daily Worker*, p. 2, 16 February 1948.
[42] For a good introduction see I. Hacking (1981), "Do we see through a microscope?", *Pacific Philosophical Quarterly* **62**, 305–22..

has yet "seen" the phenomena producing solar noise."[43] A radio telescope became sanctioned through evidence from a totally independent approach, namely imaging by an (optical) telescope. Only once this had happened enough times to deem the radio techniques reliable would astronomers finally concede that a radio antenna could also *see* and thus merit the name *radio telescope*.

17.2.2 Styles of radio and optical astronomers

The styles of research and technology of the men studying the radio sky were profoundly different from those of the astronomers. The radio astronomers, although predominantly trained as physicists, had learned from their wartime radar experiences how to be effective builders and engineers, and in effect were practicing a new kind of physics.[44] In (optical) astronomy, on the other hand, those who built equipment were usually a different class than the observers and theorists. Ralph Williamson (1948:13–14), a Canadian theoretical astronomer who collaborated with radio astronomer Charles Seeger (Section 10.1.3), wrote of this disparity:

> The customary training of the astrophysicist is such that…he has no background in the complex techniques of antenna and high-frequency radio-receiver design which are required in the microwave region. On the other hand, the radio engineer, well-versed in this side of the problem, … may well find himself at a loss to know best how to use [his equipment] to obtain astrophysical data. Any realistic approach to the field of microwave astronomy[45] will require attention to this unusual situation.

Seeger (1975:7–9T) later put it succinctly:

> [The astronomers] were just unacquainted with the modern technology that had been developed during the war. … You had to have a certain understanding of electronics in order to believe the damned stuff, and of course the classical astronomers in the US didn't. They couldn't find any excuse for the radiation in the first place!

Hanbury Brown (1983:211) recalled a similar scene in Britain:

> [Optical astronomy] was an extremely slow-moving business, and the thing that particularly worried me was the opposition from optical astronomers to the introduction of electronics into the dome in any form at all, on the grounds that it was unreliable. … I attribute the conservatism of optical astronomy partly to the isolation of observatories on high places. You freeze your technique and then put it on top of mountains!

British theorist Fred Hoyle (1988:172A:620) later opined that mid-century British astronomy had fallen into an attitude, stemming from the style of Arthur Eddington, that progress was best achieved through logical reasoning about the universe. The empirical approach of the radio astronomers was wholly other. And in the Soviet Union, Shklovsky (1982:12,15) commented on the gap between the two communities by recalling an aphorism from the astronomer A. A. Mikhailov to the effect of "Beware of all things electrical!"

The above may overstate the case, but probably not by much. There were of course exceptions. In Holland, Jan Oort, acutely aware of the divide (Section 16.7), wrote in 1949:

> The research has an astronomical end, but the astronomers have little understanding of the apparatus. In other countries radio astronomical observations have been started by radio engineers.[46]

Oort knew that good engineers were necessary for this new brand of astronomy to succeed, but, unlike the situation in England and Australia, did not want to mix oil and water. As his colleague Henk Van de Hulst (1951) told his graduate students:

> Electronics experts will take care of equipment, but I want you to learn the basic opportunities (which are great) and limitations (which are severe) of the radio astronomical method.[47]

---

[43] *J. Geophys. Res.* **55**, p. 207 (1950), as quoted in Goss and McGee (2009: Appendix A-14).

[44] D. Pestre and J. Krige, "Some thoughts on the early history of CERN," pp. 78–99 (especially pp. 93–5) in Galison and Hevly (1992).

[45] *Microwave astronomy*, a term that found only limited currency (and only in the US), was invented at Cornell. Taffy Bowen did not approve: "A lot of it is certainly not microwave and I am not at all sure whether it is astronomy" [Bowen:J. L. Pawsey, 8 January 1948, file F1/4/PAW/1, RPS]. Also see note 23.

[46] J. H. Oort, as reported in the SRZM Board minutes, September 1949, box 41, VDH; also file 263, OOR.

[47] H. C. Van de Hulst, Radio Astronomy Course, Box 36, VDH.

In the US, Otto Struve of Yerkes Observatory was also cognizant of the changes wrought by electronics, even for optical work. In 1947 he wrote:

> Electronics has become an important tool. It is clear that we must reorient ourselves to take account of these changes.... The construction of a new large telescope...may no longer be the most important problem that confronts us. Research in electronics may become our principal task.[48]

Indeed, Van de Hulst (1973:11–12T) later recalled extensive discussions led by Struve over which direction new instrumentation for astronomy (including radio) should take. These resulted in development of a lead-sulphide cell for infrared measurements well beyond the usual optical wavelengths. Nevertheless, as we saw in Section 4.7, it was still too much of a gamble for Struve to support radio astronomy through the person of Grote Reber.

The only area of (optical) astronomy where electronics had any sort of beachhead in the postwar decade was in photometry (measurements of the intensities of individual stars). Although the first use of an electrical detector (a selenium cell) goes back to 1907, before World War II almost all photometry was still done either by eye or photography (DeVorkin 1985). After the war, the superior RCA 1P21 photomultiplier tube, largely developed during the war, meant that by 1952 two-thirds of all photometry in the US was being done electronically, although undoubtedly much less was electronic in other countries.[49] But the circuitry for a lone photoelectric cell attached to a telescope was far simpler than that of a typical radio astronomy receiver, and in any case photometry was only a minority of all (optical) astronomical research. The radio astronomer's intimacy with electronics was on a wholly different level to that of the optical astronomer's.[50]

The pace and teamwork of radio projects was also foreign to observatory life. Michael W. Ovenden (1926–1987), a young Cambridge astronomer who was sympathetic to the new specialty and wrote several early reviews about it, later recalled:

> Astronomy before the war needed a kind of approach which required great rigor in mathematics and observations.... But the type of person who is prepared to take a set of plates, knowing that it will be twenty years before the second set is taken, is a different kind of animal from the person who is building new equipment and like Galileo just looking at the sky.
>
> The conventional astronomer tended to be an isolated person, sometimes quite literally isolated in an observatory miles from anywhere.... Whereas right from the very beginning of radio astronomy, it was *groups* of people building big antennas. (Ovenden 1976:75B:225, 730)

Derek Vonberg (1971:13–14T), Ryle's first collaborator, had distinct memories of the very place where Ovenden worked:

> The whole pace of events was so different inside the observatory, with people spending hours poring over charts and things.... It had a little bit of a museum atmosphere about it.... The temperaments of people doing optical astronomy were so different from [ours]. The idea of rigging up a new observing system one afternoon and starting to take measurements the next day was so different from planning a telescope which you then would see built several years later.

During the war the radar leader Watson Watt had preached "the cult of the imperfect" to those

---

[48] O. Struve, "The story of an observatory," *Popular Astronomy* **55**, 227–44 + 283–94 (1947) (quotation from p. 291); also quoted by DeVorkin (1985:1206).

[49] This figure is based on a census of papers published in the *Astronomical Journal* and *Astrophysical Journal* over the interval 1946–53. I thank David DeVorkin for supplying these data. In 1950 H. Shapley stated that there were 20 "new photoelectric photometers now posted in various parts of the world" ["Astronomy," *Scien. Amer.* **183**, 24–26 (September 1950)]. The photomultiplier tube was probably the largest technical breakthrough of the war that directly affected *optical* observations (see Fig. 1.3).

[50] One astronomer's view of the changes wrought by electronics in the dome (and by radio astronomy) can be garnered from the following passage comparing a radio telescope to a photoelectric cell at the prime focus of a reflector: "When [in use]...with an amplifying and continuous recording system, the astronomer is eliminated as an element in the receiving system. His only function is to guide the telescope, and his knowledge of electronics is perhaps more necessary than his knowledge of astronomy." [C. A. Federer (January 1950), *Sky & Telescope* **9**, 50]

developing British radar systems: "Give them the third best to go with; the second best comes too late; the best never comes." Similarly, Van de Hulst marvelled at this pace – radio engineers produced exquisite receivers (requiring, for example, tube voltages maintained to better than one part in a thousand) and yet dismantled these same receivers after less than six months of use. "What a contrast," he wrote, "to the classical optical telescopes, which with good maintenance develop hardly any signs of age even after fifty years!"[51] DeVorkin (1992:204–14) has also noted similar attitudes in his study of the postwar beginnings of rocket astronomy from above the atmosphere. In that case the astronomers were extremely suspicious of any enterprise that asked them at great expense to develop an instrument only to have it destroyed when the nose cone crashed back to the ground.[52]

Despite all these differences, there was one way in which the radio astronomers' style was very much like that of traditional astronomy: expeditions to solar eclipses. Pang has persuasively demonstrated that British eclipse expeditions in the Victorian era imbued a social and political content as important as their astronomy, and Pyenson has shown how French eclipse expeditions were also an important part of overseas expansion.[53] For early radio astronomy one similarly finds the Americans, financed by the US Navy, heading for eclipses in the South Atlantic, Alaska, and Africa; Soviets travelling to Brazil and Siberia; and the French observing in their African colonies (see the summary in Fig. 13.6). As in an earlier era, these expeditions were manifestations of imperial power in their financing, planning, execution, and later publicizing. Figure 10.4, with its native African posing beside the Western scientist adjusting his complex radio equipment at a 1952 eclipse in Khartoum, reads precisely as does a photograph in Pang's article showing natives and astronomers at an 1898 British eclipse camp in India.

17.2.3 Interactions between radio and optical astronomers

In previous chapters we have encountered many instances of interchanges between astronomers and radio researchers. In this section I wish to build on these cases and characterize the overall situation. First, note that there were many astronomers who showed support for the new field. As early as 1933, the astronomers of Princeton University were willing to give Melvin Skellett a Ph.D. for his radio research into the nature of meteor trails (Section 11.1). Struve championed Reber throughout the 1940s (Chapter 4). Jesse Greenstein, at Yerkes and the California Institute of Technology, was another American astronomer who recognized the value of the new field. In Section 14.4.3 we saw the extraordinary role played at Mt. Wilson Observatory by Walter Baade and Rudolph Minkowski, who intimately collaborated with radio workers on the nature of discrete radio sources. At Harvard, Fred Whipple gave strong intellectual support to meteor radar astronomy at Jodrell Bank and elsewhere (Section 11.4); Donald Menzel showed great interest in radio observations; and Bart Bok established an ongoing graduate program in radio astronomy based on studies of the 21 cm hydrogen line (section 16.6.3). After attending a radio meeting in 1953 Bok's opinion was clear: "The vigorous young science of radio astronomy bears watching. The forceful and intelligent approach by radio astronomers is changing the face of astronomy."[54] In Holland the entire radio astronomy enterprise was under the aegis of Oort and van de Hulst at the venerable Leiden Observatory (Chapter 16), while at Utrecht Minnaert was active in international solar radio matters. At Zürich Observatory the director Max Waldmeier pushed for integrating radio into solar astronomy (Section 13.2.3). In Canada, Peter Millman of the Dominion Observatory collaborated on meteor radar for years with radio engineer Donald McKinley (Section 11.3), and Williamson likewise with Seeger

---

[51] As in Van de Hulst in note 35 (1951, quotation from p. 43).

[52] Also see D. H. DeVorkin (1989), "Along for the ride: the response of American astronomers to the possibility of space research, 1945–1950," pp. 55–74 in *The Restructuring of Physical Sciences in Europe and the United States, 1945–1960*, eds. M. De Maria and M. Grilli (London: World Scientific).

[53] A. S.-K. Pang (2002), *Empire and the Sun: Victorian Solar Eclipse Expeditions* (Stanford: Stanford University Press), especially chapter 5; L. Pyenson (1993), *Civilizing Mission: Exact Sciences and French Overseas Expansion, 1830–1940* (Baltimore: Johns Hopkins University Press).

[54] B. J. Bok, "Three weeks of symposia – II" (Section entitled "The Jodrell Bank Symposium on Radio Astronomy"), *Sky & Telescope* **13**, 45–49 (1953) (quotation from p. 49).

on galactic noise (Section 10.1.3). At Mt. Stromlo Observatory in Australia, Cla Allen actually did *his own* radio observations of the sun and Galaxy (Section 7.3.4). Radio engineer Marius Laffineur similarly conducted extensive solar radio observations on the grounds of Meudon Observatory in France (Section 10.4); likewise at Tokyo Observatory, director Yusuke Hagihara encouraged radio observations with antennas scattered amongst his telescope domes (Section 10.5). The radio astronomers at Jodrell Bank consulted with Manning Prentice and later Zdenek Kopal (Chapter 9), and those at the Cavendish Laboratory with David Dewhirst and others (Sections 8.5.2 and 14.4). Finally, William McCrea, cosmologist and RAS Secretary in the late 1940s, actively encouraged the radio astronomers' participation in the RAS.

But the length of this litany belies the situation. The above cases represent *all* those of any importance over almost a decade. The great majority of astronomers were ignorant, indifferent, or sometimes hostile toward the new specialty. John Bolton (1978:116T) later recalled that his 1950 tour of virtually every major European observatory was necessary to cause "radio astronomy to become a reality to them," even five years after the war. Bok (1978:118B:65) recalled that he was the only astronomer from Harvard College Observatory to attend the physics colloquium in 1951 in which Harold Ewen announced the discovery of the 21 cm hydrogen line. In late 1952 John Pfeiffer, a science writer beginning to research a popular book on radio astronomy (eventually appearing in 1956 as *The Changing Universe*), was surprised to find nary a mention of radio astronomy in two dozen recent astronomy books, textbooks, and semi-technical articles.[55] Astronomer Allan Sandage, a young colleague of Baade and Minkowski's in Pasadena, did not mention their work on optical identifications of radio sources *at all* in a late-1953 overview of the first four years of research with the new Palomar 200-inch telescope.[56] In a review of Lovell and Clegg's *Radio Astronomy* (1952), astronomer Daniel Popper gave an opinion of Lovell's new chair at the University of Manchester that was both exclamatory and crusty:

The senior author is Professor of Radio Astronomy(!)... an astronomer will naturally be particularly alert to catch errors committed by these interlopers into his domain.[57]

Other astronomers, even if supportive, were patronizing. Ryle's student Kenneth Machin (1971:38T+25T) recalled:

Old [H.W.] Newton, the solar boy up at Greenwich Observatory, was terribly nice, terribly cooperative, but never really saw us as part of the real subject.... I think they thought we were people who were fumbling about on the sidelines of their subject. Maybe in the distant future we would be able to help, but for the moment we were just to be fed data and told to run away and be good boys.... They only started to wake up when we were able to ring them up and say, "If you look at this point, you will find something interesting," and they looked and they did.

Notwithstanding this attitude, connections between the radio and optical sides were in general better for solar astronomy than for other areas of radio astronomy – at least for the sun one knew the source of the radio waves!

One of Van de Hulst's motivations to give a course on radio astronomy at Leiden (and also Harvard; Sections 16.4.3 and 16.7) and to publish the notes was that he felt he needed to patch up damage inflicted by radio astronomers not properly using optical data.[58] Finally, we have already seen how Oort in 1950 developed a model of the radio Galaxy largely because he felt that the efforts of Ryle's group were lacking in astronomical insight (Section 15.2.2). Kevin Westfold (1978:98A:440) later recalled a telling incident during his visit to Leiden at that time. After he expressed amazement regarding Oort's detailed mathematical treatment, Oort replied, very kindly, "But we're astronomers" – Westfold felt that he had been shown his place. Jack Piddington (1950) of Sydney's Radiophysics Lab ran into a barrage

---

[55] J. Pfeiffer:M. Ryle, 14 November 1952, file 1/3, RYL.
[56] A.R. Sandage (1954), "The first four years of extragalactic research with the Hale 200-inch telescope," *Astron.J.* **59**, 180–2.
[57] As in note 39 (exclamation point as given here is in the original). Astronomer Gerard deVaucouleurs (1976:1N) also recalled great amusement at the Observatoire de Paris in 1952 when they learned of the appointment of a Professor of *Radio* Astronomy.
[58] H.C. Van de Hulst:A.E. Covington, 14 November 1950, copy from Covington.

of arrogance from an anonymous referee concerning an article on solar theory (Section 13.1.2):

> An astronomer is repelled by the very first sentence. ... The author appears not to have defined anywhere what he means by corona and what by chromosphere, and I have in places wondered whether his conceptions are those of conventional astronomy, to which, after all, the terms belong. ... For those accustomed to the very different approach of optical astronomy (which I imagine the author would find it profitable to study) ... [59]

The strongest opinions came from radio-astronomer interviewees who recalled that astronomers thought the field a wasteland:

> Most astronomers didn't even believe there was anything at all in this radio business.
> (Bolton 1976:18T)

> I think the classical astronomers generally tended to hold back radio astronomy. ... The general idea was that they knew what was in the sky and you shouldn't try to look for things which were not there.
> (Mills 1976:21T)

> Radio astronomy was looked upon as an awful, worthless pursuit by [almost all the optical astronomers]. They thought it a complete waste of time and money that would be much better spent on another one-hundred-inch telescope. ... This began to change after the discovery of the twenty-one-centimeter line, the optical identifications, and other successes.
> (Richard Twiss 1981:145B:470)

Other interviewees stressed that the attitudes of the optical astronomers were very reasonable given the tenuous nature of much of the strange radio data. Reviewing the field in a popular astronomy journal, Williamson (1948:17) felt compelled, even as late as 1947, to explicitly list as his first conclusion: "(1) The cosmic noise exists." Furthermore, many of the radio workers were astronomical novices, frequently blundering in their astronomy:

> People regarded it as a possibility for new information, but until things got a little bit sorted out, nobody knew what it meant. ... As long as nobody had any idea what it was, there wasn't much to talk about except "Isn't that wonderful?". ... It was just not their thing.
> (Naval Research Lab radio astronomer Cornell Mayer 1978:41–2T)

> Obviously there was interest [from optical astronomers], but I don't think it really could have been much more than that, simply because if you can only tell them where to look [on the sky] to a degree, and it isn't something as obvious as the sun or the Crab nebula, then what are they to do about it? A degree is a rather large bit of sky.
> (Ryle 1976:26T)

> In their eyes we were still meddling with things we didn't understand and this was fairly true. We frequently made silly mistakes because we had had no serious instruction in the elements of astronomy.
> (Ryle 1971:23)

McCrea has also pointed out that the negativity of someone like Richard Woolley in Australia (Section 7.3.4) was not at all unreasonable, and that in fact success eventually came to the radio people largely because of their fortune in what the radio sky had to offer.[60] McCrea himself was supportive of the new field, but still ambiguous about its status, as evidenced in this exchange triggered by Ryle's election to the Royal Society in 1952. McCrea wrote:

> I do not know whether you like to be regarded as a physicist or an astronomer, but I know that all the astronomers will be delighted. You have the distinction of being the first radio-astronomer in the Society.

Ryle replied "[I hope] that I also may regard myself as an astronomer."[61]

---

[59] Anon., "Reports of the referees on paper A84 – Referee 1", undated, about December 1949, accompanying a letter J. H. Piddington:E.V. Appleton, 21 February 1950, file D5/4/75, RPS.

[60] W. H. McCrea, as in note 7.51 (p. 936).

[61] McCrea:Ryle, 21 March 1952; Ryle:McCrea, date unknown, file 28 (uncat.), RYL.

We have earlier encountered several instances where the radio researchers' lack of astronomical training led them seriously astray. Even the students in Ryle's group at Cambridge did not take any courses from the astronomers at the nearby observatory.[62] Ginzburg (1990:3–5) has written that his minimal astronomical knowledge was a serious hindrance to his work in radio astronomy. One astronomical basic often learned "the hard way" was that astronomical coordinates shift with time due to precession (Appendix A). In Section 7.4 I related one case of ignorance of precession (on the part of Joseph Pawsey in 1947). In another case, Greenstein gently chided Ryle in 1948 for ignoring precession in his publications:

> May I suggest that astronomers like to specify the "equinox" at which a position is given; precession of the equinoxes changes star positions quite appreciably in a year.[63]

Cyril Hazard (1971:21–2T) also recalled learning about precession only after puzzling over the many different positions given for the Andromeda nebula in various catalogs. This convinced him that he had to learn astronomy, not just radio techniques, and by three years later his thesis indeed displayed mastery of many aspects of extragalactic astronomy (Section 15.2.3). Dewhirst (1976:57B:250) gave an optical astronomer's recollection of his interactions with Ryle's group:

> They were of course primarily radio *physicists* rather than radio *astronomers*, and were very happy to come and ask questions.... They had looked up positions of objects in various catalogs and noticed that the positions given in a catalog now were not the same as those given fifty years before. [And so they learned about] mysterious things like precession. ... Optical astronomers knew about such things and radio physicists didn't. ... Coordinates that changed with time must have seemed to a physicist a very odd system.

On the other hand, Paul Wild (1978:35T) opined that the astronomers' general neglect of radio work (especially in Australia) was a good thing for the young specialty, for "it allowed the truth to firm up before it was under too much scrutiny." Hanbury Brown (1983:213) similarly later argued that ignorance of affairs astronomical was ironically a strength:

> [To make serendipitous discoveries] you need the right man in the right place at the right time, but he must be a man who doesn't know too much!... It's the reason of course why physicists and engineers came into radio astronomy and made discoveries; many of them didn't know the sun from the moon, they didn't know a planet from the stars.

Despite some hyperbole by the participants, it may well be that at first the social and intellectual cleft between "proper" astronomers and the radio upstarts in fact helped the new science more than it hindered. But, as we shall see in Section 17.2.5, by 1953 the radio/optical gap had significantly narrowed and radio astronomy was starting to merge into astronomy.[64]

17.2.4 IAU versus URSI

Over the postwar decade there was an interesting *pas de trois* between the developing community of radio astronomers, the International Astronomical Union (IAU), and the Union Radio Scientifique Internationale (URSI). These two prestigious organizations, both founded in 1919, performed useful, sometimes vital, coordination and dissemination of research. IAU's purview was astronomy, while URSI's was more nebulous, having evolved by 1945 into aspects of radio science such as standards, propagation through and structure of the lower atmosphere and ionosphere, and laboratory physics involving radio technology. Once extraterrestrial radio noise from the sun and Galaxy had been detected, as well as radar reflections from meteors and the moon, IAU and URSI were the natural candidates for international intercourse. But which one? Or both? As we have seen, the radio practitioners were not at all astronomers; on the other hand, although they were certainly using radio techniques, their objects of study

---

[62] J. A. Baldwin (1981:144B:370).
[63] Greenstein:Ryle, 8 December 1948, box 6, GRE.
[64] See Tangent 17.2 for a comparison of this radio/optical astronomy case with an analogous study by Rasmussen (1997), who has studied how biologists reacted skeptically (also in the postwar decade) to images from the new electron microscope. Similarly, DeVorkin (1992:273–300) has studied troubles that rocket scientists had in the 1950s in gaining acceptance of their upper-atmosphere rocket data by established researchers.

were well beyond those in URSI's regime. The tale of how these issues were worked out in the postwar decade gives insight into the social and intellectual development of early radio astronomy.[65]

Both the IAU and URSI resumed activities immediately after the war. In 1946 URSI met[66] in Paris. Given that its longtime President was Appleton, it is not surprising that the new results on solar and galactic noise were prominently featured. In fact Appleton appointed himself chair of a new Sub-commission called "Radio Noise of Extra-terrestrial Origin," within the exisiting Commission III on Terrestrial Atmospherics. Two years later in Stockholm URSI gave full Commission status (one of seven) to the subject when Commission V on Extra-terrestrial Radio Noise was born with David Martyn as chair and Pawsey as secretary. At the 1950 meeting in Zürich Commission V was rechristened "Radio Astronomy," which it has remained to this day. As discussed in Section 7.5.1, the following meeting, in 1952 in Sydney, was a showcase for Australian radio science, especially radio astronomy. Also during this period URSI commissioned and published a series of "Special Reports," which were detailed reviews of radio observations of the sun, Milky Way, discrete sources, and the 21 cm hydrogen line.[67]

Meanwhile, the IAU was not ignoring the new field. At Copenhagen in 1946 the solar activity commission recommended that radio monitoring data be included in the *Quarterly Bulletin on Solar Activity*, and this indeed happened from January 1947 onward.

At the 1948 Zürich meeting sufficient support was mustered to create Commission 40 on Radio Astronomy (*Observations Radioélectriques* in the IAU's second official language), again still extant. Of the twenty-two inaugural members, four were optical astronomers or theorists, including the first President, Woolley (Section 7.3.4). Radio results were scattered throughout the IAU meeting, including a review talk by Stanley Hey. Most attention was on meteor radar and the sun; the commission concerned with instruments was particularly enthusiastic:

> By far the most outstanding advance in observational astronomy for very many years has come from the application of developments in radio to the study of celestial phenomena.... Some photographs displaying the equipment used and the results obtained by British pioneer workers will be on view during the assembly.[68]

The first actual Commission 40 sessions were at the Rome IAU meeting in 1952, where radio astronomy made a huge splash with the new optical identifications by Baade and Minkowski (Section 14.4), unprecedented glimpses of galactic structure from the 21 cm hydrogen line reported by Oort (Section 16.6.3), and new solar burst phenomena.

As for many other incipient fields, a primary task of the radio researchers at both the IAU and URSI was to debate and set terminology and standards. Despite large errors in measured intensities and sky positions, it was realized that scientific discourse, both within the radio field and especially when talking to traditional astronomers, would be greatly aided by agreed conventions. Thus during this period we find numerous committees deliberating on rational schemes that might eventually be adopted by *both* IAU and URSI. What should be the distinctions between brightness temperature, effective temperature, apparent temperature, disk temperature, etc.? How should radio antennas be calibrated? How should new radio sources be named? On various schemes Cyg A was designated Cyg (1), R1 Cyg, 19.01, or 19N4A (based on its approximate coordinates

---

[65] Edge and Mulkay (1976:44–5, 59–64) also describe the relationship of radio astronomy to the IAU and URSI, carrying the story into the 1960s and providing various tables of data (e.g., number of members of Commissions by country). I present more detail on the pre-1953 period, and come to conclusions similar to theirs.

[66] By stating that IAU or URSI "met," I refer to the General Assemblies held at 2–4 year intervals during this period. Most of the uncited information in this section derives from the published proceedings of these General Assemblies and their Commissions.

[67] URSI Special Reports No. 1, 3, 4, and 5 on "Solar and galactic radio noise" (1950, Chair of the editorial committee: E. Appleton), "Discrete sources of extra-terrestrial radio noise" (1954, J.G. Bolton), "The distribution of radio brightness on the solar disk" (1954, W.N. Christiansen), and "Interstellar hydrogen" (1954, J.H. Oort); all published by URSI in Brussels.

[68] J.A. Carroll (British Admiralty, Whitehall, London), presidential report for Commission 11 ("Astronomical Instruments") at the 1948 General Assembly, *Trans. IAU* 7, pp. 103–4 (1950, Cambridge: Cambridge University Press). Also see *Observatory* **68**, 161–77 (1948) for a report of the 1948 IAU meeting.

of 19^h, 40° N); in 1952 it was pointed out that the last scheme was advantageous because radio observers did not know the locations of the constellations![69] In these deliberations process was very important in achieving a final product, for in the end standardization came about from informal, pragmatic consensus, not from proclamation by authority.

Even more fundamental was the business of somehow using the auspices of these international organizations to certify the very existence of any given radio source. In 1950 it was proposed to set up the Copenhagen Observatory as a "controlling office" to allocate (or not) designations for all newly claimed radio discoveries. It was proposed that URSI would "advise IAU concerning the reality of each reported object." Once a year this central Bureau would send the latest list of certified sources and discoverers to all the major journals. This autocratic scheme, analogous to how Copenhagen astronomers then handled the discovery of new comets and asteroids, never was actuated.[70] But as discussed in Section 14.4.2.2, later discrepancies between different radio surveys led URSI and the IAU during the period 1952–54 to jointly attack the problem and produce a list of those sources "whose existence has been definitely established" (Pawsey 1955). Only eight sources qualified (the list was similar to Table 14.1).

Those studying the radio sky over the postwar decade gradually drifted away from URSI and toward IAU. Just as the radio practitioners were metamorphosing from radio physicists and engineers into radio *astronomers*, they began to feel more comfortable in IAU and wanted to learn from the (optical) astronomers. They never, however, abandoned URSI; rather, the focus of discussion there turned more to the techniques and less to the science. As Pawsey noted in 1955:

> In fact Commission V of U.R.S.I. and Commission 40 of the I.A.U. have almost identical interests and a large proportion of common members. This may be illogical, but it is not clear that any good would be done by an attempt at rationalization. The good feature of the current arrangement is that radio astronomers who are able to attend meetings of both unions [can] mix with both radio research workers and astronomers.[71]

Two leading radio astronomers later recollected about URSI versus IAU:

> Ratcliffe was very keen that we should cooperate with URSI. I think fairly soon we realized that this was the wrong thing to do. It was fine as long as your interests were mainly the sun and solar terrestrial relationships. ... But there obviously was going to be more to it than that, and I think many of us felt that we must tie up with the astronomers early. But for a long time, even now, you still had to go to the URSI meetings.
>
> (Ryle 1971:40T)

> The linkages with astronomy were not all that good – it took some time. ... Nearly all the work on radio astronomy went on in Commission V of URSI. ... It went on there because the emphasis was nearly always on the instrumentation, not on the astronomy. In fact, a lot of the astronomy was rather naive. ... The emphasis only gradually switched from instruments to astronomy, a very slow process as the center of gravity for international interests shifted from URSI to IAU. ... I'd say that the halfway point was somewhere between 1955 and 1960.
>
> (Hanbury Brown 1976:8–9T)

Analogous situations occurred in ionosphere research and for planetary radar astronomy. Gillmor (1986:108) finds that the ionospheric community split its efforts over the postwar decades between URSI for the radio technology and the International Union of Geodesy and Geophysics for the science. Butrica (1996:60–2) likewise finds that planetary radar astronomy (in the 1960s), was "at the convergence of science and engineering" and thus participated in both URSI and IAU.

---

[69] C.L. Seeger (Chair), Commission 40 report (1952) on "Terminology and notation in radio astronomy," *Trans. IAU* 8, p. 615 (1954, Cambridge: Cambridge University Press).

[70] M. Laffineur (Chair), report of Commission V Sub-Committee on "Terminology," *IXth General Assembly* [of URSI] 8, Pt. 1, p. 412 (1950, Brussels: URSI); M. Laffineur (Chair), report of Sub-Commission Vb on "Terminology and Units," *Proc. of the General Assembly* 9, Fasc. 6, p. 75 (1952, Brussels: URSI).

[71] As in note 36, pp. 572–3.

17.2.5 A new discipline of radio astronomy?

Historians and sociologists of science have long debated what exactly constitutes a "discipline" and how one comes into existence (e.g., Kohler (1982) on biochemistry; Abir-Am (1992) on molecular biology; Rasmussen (1997) on cell biology (Tangent 17.2); and Good (2000) on geophysics). Related entities – professions, subdisciplines, research schools, research programs, problem areas, specialties, fields, and subfields – have also been defined and applied to various case studies. The original plan for the present study was couched in terms of a "disciplinary history," but the evidence just set forth in Sections 17.2.1–17.2.4 indicates that this is inappropriate, for radio astronomy in fact never did become a discipline, but instead a specialty or subfield of astronomy (as illustrated in Fig. 1.3).

A discipline can be recognized by its *cognitive identity* and its *professional identity*. The former refers to the subject matter of the science and commonalities in how research questions are formulated and answered (e.g., instrumental and analytical techniques). The latter refers to shared structures such as journals, societies, academic departments, prizes, and funding agencies whose development depends on a social and economic milieu particular to time and place. Note that these shared social structures and the scientific investigations also mutually influence each other.

The case of radio astronomy is intriguing in that one might think that profound differences with (optical) astronomy would inevitably lead to a separate discipline. After all, the techniques were utterly foreign to traditional astronomy, and almost all of the radio astronomers had no training in astronomy. The radio communities were almost all not located at (optical) observatories and they practiced standards not set in astronomy. Finally, the radio sky itself (Fig. 1.1) was disconcertedly bizarre compared to what astronomers painted. Despite all this, the burgeoning field's cognitive identity did *not* greatly diverge from that of astronomy, and in fact gradually became integrated. Edge and Mulkay (1976:281) have correctly argued that it was indeed the astronomers who furnished the intellectual framework for the radio workers, for it was to astronomy that they predominantly turned to understand their results. For example, the only way to obtain distances to the enigmatic radio stars was to find associated optical objects whose distances could be determined using established optical techniques. Astronomers were therefore often consulted by the radio workers, such as through long stays abroad at (optical) observatories by Sydney group members Bracewell, Bolton, Westfold, Kerr, and Mills. Likewise, in terms of a professional identity, the radio astronomers sought not to set up their own journals or societies or academic departments, nor to shun the optical astronomers. Instead, even as they maintained links to physics and engineering disciplinary structures, they gradually became part of astronomy. As the years went by they did not found a new discipline, but rather became incorporated into one of the oldest.

The radio astronomers very much cared about their reception from the optical side. For example, in 1951 Seeger complained:

> I am glad you and your colleagues resisted the temptation to publish in bits and pieces and prematurely.... Astronomers are a rather conservative bunch and several have expressed a poor view of some of the literature in this field [of radio astronomy] for giving results and very little of the details on which one can form a sure judgment.[72]

And in 1952 Ryle was concerned about possible reactions to three conflicting papers in *Nature* from the three major groups (on the size of Cygnus A; Section 14.5.2). At one stage he wanted a joint paper that somehow reconciled all of the differences:

> [We should] coordinate so as to give a coherent story from the radio astronomy end.... A joint paper ... might present to the non-radio astronomers a far more convincing story than the three rather contradictory sets of results.[73]

Graham Smith (1976:15T+27T) later recalled how integration with the astronomers was abetted by the first optical identifications of radio sources:

> [At first] we really still felt ourselves as radio engineers finding out what it was that was coming from the sky, and we didn't feel ourselves as having any sort of place in the [optical] community. [But

---

[72] Seeger:Ryle, 26 September 1951, file 1/2, RYL.
[73] Undated draft of Ryle:Lovell, October 1952, file 4 (uncat.), RYL.

starting with the first optical identifications] we really felt as though we were slotting in – we would get something from [the Mt. Wilson astronomers] and they would get something from us.... Possibly some of the conventional astronomers were beginning to think that the radio astronomers were becoming respectable.

Similarly, Bolton (1982:352) later wrote:

> The identification of the Crab nebula was a turning point in my own career and for nonsolar radio astronomy. Both gained respectability as far as the "conventional" astronomers were concerned.

As discussed in Section 14.4.4.2, a radio source (not to speak of the associated radio astronomer!) gained legitimacy when it could be identified with an optical object. So for example, the stated purpose of many radio investigations was to uncover more optical identifications. Nevertheless, their number stubbornly remained less than a dozen despite years of effort. By 1954 the radio source situation had become sufficiently muddled that, as mentioned above, radio astronomers felt the need to draw up, especially for the sake of astronomers, a clarifying and certified list of "definitely established" radio sources, with a primary selection criterion being the existence of an optical identification (Pawsey 1955). Moreover, the Soviet theorist Iosif Shklovsky (1954:483) argued that it was *only* through collaboration with (optical) astronomers that the puzzles of radio astronomy would be solved.

Another example of the radio astronomers appealing to the optical astronomers was their introduction of "radio magnitudes" (Ryle, Smith, and Elsmore 1950; Hanbury Brown and Hazard 1952a), defined in the same archaic manner as magnitudes in traditional astronomy (Appendix A). This was in response to plaints such as those from astronomer Jesse Greenstein (1983:82):

> My strongest memory of my encounters with radio astronomers...was: How do you calibrate? Please calibrate, please put it in decent units, make sure you know the actual absolute flux at different frequencies.

Radio magnitudes never caught on, but even to attempt it was a notable sacrifice for any physicist's metrical sensibilities. The Australians, however, would not concede. Mills (1952a:271) railed against the "antique and rather clumsy 'magnitude' of visual astronomy," while Pawsey and Bracewell (1955) in their textbook always translated magnitudes to decibels for the sake of their radio readers (e.g., "this galaxy is 4 magnitudes fainter [-16 db]").

More successful in improving communications was the gradual dropping of the terms *radio noise*, *cosmic noise*, *galactic noise*, and *solar noise*, replacing *noise* with *radiation* or *emission*. At URSI General Assemblies in 1950 and 1952 committees on nomenclature and units officially recommended forbearance from *noise* because the term was obscure and had misleading acoustic implications.[74]

One aspect of the radio astronomers wanting to join the astronomers in a common endeavor was their large technical effort to make ever better radio *images* of the sky, something that astronomers' photographs at optical wavelengths accomplished with relative ease.[75] Much of the technical development of radio astronomy was concerned with the improvement of angular resolution and then eventually with image-making. The goal was to transform pen traces on a strip chart recording (as in Fig. 9.6) into something photograph-like (e.g., Fig. 14.16), something that optical astronomers could readily understand (see Section 14.4.4.2). Indeed, despite differing greatly from astronomers in background and research style, and despite the improbable novelty of the radio sky, radio astronomers were establishing their own *visual culture* and rushing into the arms of astronomy. Contrast this situation with that analyzed by Rudwick for early nineteenth-century geology, where the new "geologists" were self-consciously defining a new discipline, using a distinct visual language (illustrations of geological sections and landscapes) that eventually became arcane to all but insiders.[76] The cases of radio astronomy and geology reinforce how representations of the natural world are important in scientific practice both for interpretation of that world and for determination of how science is organized.[77]

---

[74] M. Laffineur (Chair), report of URSI Commission V Sub-Committee on "Terminology": (1950), pp. 391–3; and (1952), p. 76 (both as in note 70).

[75] Aspects of these technical efforts have been discussed in Sections 13.1.3, 13.3.2, 14.4 and 14.5.

[76] M. Rudwick (1976), "The emergence of a visual language for geological science 1760–1840," *History of Science* 14, 148–95.

[77] For example, see *Representation in Scientific Practice* (1990), M. Lynch and S. Woolgar (eds.), (Cambridge: MIT Press).

Most available contemporary opinions testify that radio astronomy was considered a part of astronomy and not as a new discipline. As early as 1946 Pawsey called solar radio work "a new branch of astronomy."[78] Struve (1949:28) called the young field "a completely new branch of astronomy" and his fellow astronomer Harlow Shapley in 1948 spoke of "the firm establishment during the past year of microwave astronomy as one of the principal branches of our science."[79] One dissent was by Lovell and Clegg in the preface to their popular book *Radio Astronomy* (written in 1950 and published in 1952): they opined that radio astronomy was distinct from astronomy, as it was "the applications of radio techniques to study various problems which have hitherto been the domain of astronomy" and "a new science very closely allied to astronomy, astrophysics and physics." But a review of this book by the Astronomer Royal Harold Spencer Jones instead gave radio astronomy standing as a "new branch of astronomy" with methods very different from "the older branches."[80] A few years later the monograph *Radio Astronomy* by Pawsey and Bracewell (written in 1952–53 and published in 1955) stated that "radio astronomy is basically a part of astronomy."

Munns (2002:chaps. 3–5) has studied in detail how in the US, where optical astronomy was extremely strong and entrenched (Section 17.3.2), three major radio groups were created over the period 1952–56 and explicitly shaped to integrate, as much as feasible, radio techniques with existing astronomy. This took place at Harvard (see Section 16.7), Caltech, and the National Radio Astronomy Observatory. Munns also gives examples from the 1950s of how *radio physicists* became *radio astronomers*, their *radio antennas* became *radio telescopes*, and their *radio laboratories* became *radio observatories*. For example, Bart Bok at Harvard optimistically declared full integration in 1954: "the radio technique has taken its place [in astronomy] along with photographic, photoelectric, and spectrographic methods."[81]

One could call radio astronomy a specialty or a subfield, but not a new discipline – universities did not establish departments of radio astronomy, nor was there ever a Society of Radio Astronomy or a *Radio Astronomical Journal*.[82] What happened instead was that radio astronomers, although remaining a distinct group, gradually published more in astronomical journals. Jarrell (2005:197) finds that, for a list of 37 leading radio astronomers in English-speaking countries, the portion of their papers published in astronomical journals increased from ~15% to ~25% from the late 1940s to the late 1950s.[83] On the other hand, citations *to* a selection of 30 important radio astronomy papers (published during 1946–60) were ~85% by radio astronomers, indicative of the relative insularity of the radio astronomy community. There was no new discipline, and during the 1950s only limited integration with (optical) astronomy.

Contrast radio astronomy with the case of nascent molecular biology about 1960 (Abir-Am 1992). Feuding with and unaccepted by established fields such as genetics, crystallography, and biochemistry, a new breed of researchers established the *Journal of Molecular Biology*, whose contents then effectively defined a new discipline. The molecular biologists encountered some of the same attitudes from the establishment that radio researchers received from some astronomers (Section 17.2.3); e.g., an old-school biochemist commented that "molecular biology is the practice of biochemistry without a license." It was likewise for the case of the scientists (called biophysicists) who used the newly invented electron microscope over the period 1940–60 (see Tangent 17.2). Rasmussen (1997) has shown how their foreign technique (compared to the biologist's traditional microscopy), as well as their marginalization by the biological establishment, eventually proved so intolerable that a new discipline, cell biology, complete with its own journal, eventually appeared.

---

[78] J. L. Pawsey, 19 August 1946, "Solar radiation at radio frequencies and its relation to sunspots: interpretation of observations," talk at ANZAAS conference, Adelaide, file B51/14, RPS.

[79] As in note 32, p. 43.

[80] H. Spencer Jones (1952), review of *Radio Astronomy* (Lovell and Clegg 1952), *Endeavour* 11, 220–1.

[81] B. J. Bok (1954), "Radio studies of interstellar hydrogen," *Sky & Tel.* 13, 408–12 (quotation from p. 408; cited by Munns [2002:135]).

[82] Bok did suggest a journal devoted to radio astronomy at the 1955 IAU General Assembly, but the proposal went nowhere (Edge and Mulkay 1976:64).

[83] DeVorkin (1992:323–39) found a similar trend for those doing scientific rocketry in the US during the 1950s. In this case publications gradually shifted to geophysics journals, rather than the earlier laboratory technical reports and physics and instrumental journals.

Unlike these two cases, however, radio astronomers did *not* get involved (in serious ways) with turf wars over funding and students, nor were there fundamental philosophical differences with the optical astronomers over how to attack scientific questions. The radio practitioners' techniques were every bit as exotic as electron microscopy, but radio astronomy never did become its own discipline. Instead, it developed its own visual culture to cater to (optical) astronomers and started down the road to being subsumed by astronomy.[84]

17.3 NATIONAL INFLUENCES IN THE US, BRITAIN, AND AUSTRALIA

England and Australia led this new specialty of radio astronomy – the United States was only one of several also-rans, despite its far greater material and financial resources and its dominance in most postwar physical sciences. Indeed, American (optical) astronomy was unquestionably the world's finest, and the very first radio discoveries had been made in the US, even before the war. But the development of any area of science and its institutions critically depends on the social, economic, and political environment in which it finds itself. For example, Kohler (1982) has shown how the discipline of biochemistry developed very differently in Germany, the US and Britain over the period 1850–1950 because of differences in institutional settings, origins of its (migrating) first practitioners, maneuvering by established disciplines, and sources of funding. Similarly, in this section I examine the national differences in patronage, institutions, and technology that existed in radio astronomy. In particular, I argue that military patronage of postwar science in the US created budgets and technological outlooks that tended to support less productive research areas in radio astronomy, especially when compared to those of the British and Australians.

The development of radar in the United States, Britain, and Australia meant that all three nations found themselves at the end of the war with giant radar laboratories and great numbers of scientists skilled in radar and in the practicalities of getting a job done quickly. In each nation, too, the relationship between government, the military, and science had been revolutionized. Why then did radio astronomy develop so differently in the three cases? After a tour of the US in 1948, Pawsey sent this assessment back to Taffy Bowen in Sydney:

> Since my arrival I have been struck by an anomaly. Astronomers and physicists have displayed a great interest in our work but have not undertaken similar work themselves. ... The astronomers of the US...have now become thoroughly interested in the implications but have not yet taken the plunge of tackling a totally new technique. Meanwhile, the physicists, who at the close of the war had the skill and the inclination to undertake the radio side, but failed to interest the astronomers then, now have other interests. The result is that we have a first class opportunity to establish the lead which we at present hold.[85]

In England and Australia similar factors fostered radio astronomy, with the one major difference that the British leaders of radio research, unlike their Australian counterparts, had prestigious university posts to which they hastened to return in 1945 (Australian universities were moribund at this time, nor did they even offer science Ph.D.s). Meanwhile in America, little happened for a decade. Edge and Mulkay (1976:32–4, 394–5) have also pointed out this puzzling lacuna and stated that the various national situations need study. With its pioneering work by Jansky and Reber, its world dominance in astronomy, and its much greater funding for science, why indeed did the US not go anywhere with radio astronomy despite professed interest? And what were the conditions in Britain and Australia that led to success? Factors important in understanding the situation are discussed in this section.[86]

---

[84] Full integration did not occur until after 1960, well beyond the scope of this book. Edge and Mulkay (1976:chapter 8) discuss in detail the gradual integration in Britain over the period 1950–70. Rasmussen (1997:253) was wrong to state that "radio astronomy is still largely a separate discipline [from optical astronomy]".

[85] Pawsey:Bowen, 15 April 1948, "Solar and cosmic noise research in the United States and Canada," file A1/3/1b, RPS.

[86] As seen in Fig. 17.3, by 1953 these three nations accounted for ~60% of all radio astronomy publications. The individual situations in the smaller contributors such as France, Holland, the Soviet Union, and Canada are discussed, respectively, in Sections 10.4, 16.7, 10.3, and 10.2.

17.3.1 Influence of Jansky and Reber

Jansky and Reber and their work ironically had little influence on the development of the postwar field, although chronologically it was the start of what later became radio astronomy (Chapters 3 and 4). Jansky was not able to continue his measurements after 1935 and carried out only applied research at Bell Labs until his premature death in 1950. Reber did continue research and attempted to obtain funding for a giant dish from several sources, but was unsuccessful. Eventually, the same idiosyncratic character that had led him to build a dish in his backyard, on his own time and with his own funds in the middle of the Depression, led to his inability to operate within the research establishment. In 1951 he abruptly left for private researches overseas and thereafter played no part in the American scene. These facts, combined with the independent wartime discoveries in England of extraterrestrial radio waves, meant that the postwar course of radio astronomy around the world would undoubtedly have proceeded much as it did even if Jansky and Reber had never made their discoveries. Edge and Mulkay (1976:20–1) have also reached this conclusion; for instance, they noted that articles from the Cambridge group over the 1946–50 period cited Hey eight times, Reber once, and Jansky never (!). In the extant notes for dozens of talks by Martin Ryle over the period 1946–50 one finds a standard introductory story beginning with a mention of Jansky, but then it was always on to Hey's work, with nary a mention of Reber. In the first of these talks, his notes in ink tellingly have the name "Jansky" inserted in pencil in a blank space; apparently he couldn't remember Jansky's name and had to later look it up. As indicated in Fig. 1.3, radio waves from space became generally known in Britain (and Australia too) *not* primarily from reading Jansky's and Reber's articles (although they were consulted), but from their rediscovery during World War II, either from accidentally finding the sun (Chapter 5) or, more subtly, when radar receivers became sufficiently sensitive that sky radiation (rather than their internal noise) limited their performance (Section 6.1).[87]

Thus the prewar "American start" in what became radio astronomy did not confer any momentum to the US.[88]

17.3.2 Astronomers vis-à-vis radio researchers

The longtime pre-eminence of Americans in observational optical astronomy[89] conferred no advantages *per se* to early American radio researchers, for we have seen that they were not part of astronomy, but came from a tradition of ionospheric research and radar development. In fact, the existence of a strong optical astronomical community acted as a deterrent to the fledgling specialty. As detailed in Section 17.2, with a few exceptions the astronomers either ignored or dismissed the radio researchers, and until the mid-1950s certainly did not consider them to be fellow astronomers.

DeVorkin (2000) and Doel (1996) have shown how American astronomy funding from federal agencies was controlled by powerful leaders of a strongly conservative astronomy community. Their distrust of federal monies meant that astronomy funding during the postwar decade grew only modestly (especially compared to physics). And although there were a few astronomers sympathetic to the new radio astronomy and concerned with how the US was lagging, they did not recommend radio funding at any significant level – they still judged their own needs more important than the hard-to-gauge needs of the new specialty.[90] Furthermore, Needell (1991; 2000:265–9) and Robertson (1992:115–21) have shown how in one case in the early 1950s it was difficult for physicists seeking even *private* funding for a major radio observatory to connect effectively with any sympathetic astronomers.

---

[87] Ryle (1976:5–6T); notes for Ryle's talks are found in file 8 (uncat.), RYL, in particular that for 13 November 1946 on "Cosmic Noise" given to the Cambridge University Astronomical Society (p. 2). Also see Mulkay (1974:109–11). Lovell (1990:163) has also stated that it was only in 1946, *after* the Jodrell Bank field station had been set up, that he learned of the existence of cosmic noise.

[88] The pioneering, widely published work of Jansky and Reber, followed by their subsequent lack of influence on the ensuing development of the field, is a singular situation. Are there any other such cases in the history of science?

[89] S. G. Brush, "Looking up: the rise of astronomy in America," *American Studies* **20**, 41–67 (1979); J. Lankford (1997), *American Astronomy: Community, Careers, and Power, 1859–1940* (Chicago: University Chicago Press).

[90] Most US funding for radio astronomy during the postwar decade, however, was not at all controlled by astronomers, but by physicists associated with military agencies.

Radio astronomy did not mesh well with the strong American astronomy community. Looking across the Atlantic, Ryle (1971:33T, 1971:23) later recalled:

> The gulf between radio and astronomy was probably worse in America.... I think it's probably fair to say that the [American] optical astronomers weren't particularly welcoming. There was a bit of a closed shop, [the radio people] weren't real astronomers. They hadn't been brought up properly at all...they didn't know what Oort's constant was, they didn't know Hubble's constant.

The American radio astronomer John Hagen (1976:16T) also recalled:

> [The optical astronomers] weren't ready to accept the fact that radio astronomy was telling them things which they couldn't find out by any other means.... Most of the time the people in the American Astronomical Society didn't know what you were talking about. ...But they gradually became more receptive.... This [gap] was not so much true in Europe or in Australia.

The following blurb, based on a press release from the Cornell radio group in 1948, displays an inferiority complex toward (optical) astronomy that seems comical today:

> The [new Cornell] radio-telescope, which cost only $30,000, is four inches wider, and can penetrate farther into space, than the 200-inch light telescope at Mount Palomar, which cost $6 millions. This is probably why it picks up noises from supposedly "empty" spots in the universe.[91]

In Britain the best astronomers were theorists, following in the tradition of Arthur Eddington, and not competing for instrumental funds. Moreover, during the postwar period the observational astronomy community was weak and in disarray, unable until 1967 to replace nineteenth-century technologies with a new optical telescope.[92] The ability of radio waves to penetrate clouds also made this new type of observation ideal for a rainy clime, as Ryle noted in a 1946 talk: "I have no doubt that there is a good future in this new astronomy...[which] is ideally suited for the British climate."[93]

On the whole British researchers studying extraterrestrial radio noise were more closely tied to the optical astronomers than in other countries. Ryle even received an offer of a position at the Cambridge Solar Physics Observatory in early 1947.[94] In 1949 Kenneth Machin received the Isaac Newton Studentship designated for "the furtherance of advanced study and research in astronomy and physical optics" and previously held always by an astronomy student (although Jack Ratcliffe was on the selection committee that year).[95] In 1951 Zdenek Kopal was brought to Manchester University to found an astronomy department, largely at the urging of Patrick Blackett, who felt that Lovell's group needed interaction with astronomers. Over the following years Kopal and his theoretical colleague Franz Kahn lectured regularly at Jodrell Bank on the basics of astronomy.[96] Edge and Mulkay (1976:52–9) have shown in detail how the RAS and its journal *Monthly Notices of the RAS* were receptive to the radio groups. By 1950 the journal listed annual reports from the three radio groups right along with those from the traditional observatories. By 1953 15% of the journal's articles were concerned with radio (and radar) astronomy.

In Australia Mills (1976:20–1T) later recalled the salutary effect of a *lack* of nearby astronomers:

> Our isolation did help us develop with an independent outlook. We had no famous [astronomer] names to tell us what we should believe, and to some extent we just went ahead following our noses.

Aspects of relations in Australia between the radio researchers and the optical astronomers have been discussed in Section 7.3.4. On the whole the astronomers were supportive as in England, although the Australian community was minuscule and not an important factor in the development of early southern radio astronomy. Note that at whichever year one assigns the

---

[91] *Pathfinder News Magazine* (Washington, DC) **55**, No. 21 (20 October 1948).
[92] F. G. Smith and J. Dudley (1982), "The Isaac Newton Telescope," *J. History Astronomy* **13**, 1–18.
[93] As in note 87.
[94] H. H. Plaskett:Ryle, 8 February 1947, file 4 (uncat.), RYL.
[95] K. Machin:author, 14 March 1987.
[96] Lovell (1976:9T); J.G. Davies (1971:6–7T, 38–9T); Hanbury Brown (1984:220–1): Edge and Mulkay (1976:283–4, 318–9).

radiophysics researchers as officially radio *astronomers* (1952?), the number of research astronomers on the continent roughly quadrupled!

In the US, however, relations between the optical and radio communities differed markedly from those in Britain and Australia. Physicist Charles H. Townes (1981:157A:150) did a bit of radio astronomy theory at the war's end (Section 15.2.1) and was considering doing more, but he later recalled being given advice by the new director of Mt. Wilson Observatory, Ira Bowen (a mentor from his graduate student days in Caltech physics), that the field would never reveal anything new;[97] astronomers at Harvard told Townes the same. And Robert H. Dicke (1979:21T) later related a similar experience of his when he arrived at Princeton University as a new Assistant Professor of Physics after the war. He had done important early microwave observations before leaving the wartime Radiation Lab, but dropped this line of research after he received no encouragement from the local astronomers (Section 10.1.1). Among American universities during the postwar decade, integration of the optical and radio sides occurred only at Harvard, and did not start there until 1953 under the leadership of Bart Bok (Section 16.6.3; Munns 2002). In contrast to England, by 1953 only 3% of the papers in the *Astrophysical Journal* contained radio astronomy (Edge and Mulkay 1976:52).[98] Radio observatories were not listed in the *American Ephemeris and Nautical Almanac* until 1960, the same year that annual reports of independent radio groups first appeared in the *Astronomical Journal* along with those of astronomy departments and optical observatories. Harold Ewen (1979:10T) later recalled the early 1950s:

> I began going around to various meetings of the American Astronomical Society. It was then that I discovered that Commission V of URSI was far more popular from a radio astronomer's standpoint

because optical astronomers considered radio astronomers to be second-class citizens.

In a 1950 letter Greenstein recognized the unhealthy separation of radio and conventional astronomy and noted that only four astronomers had done even "a little" to help. He added:

> Astronomers have perhaps missed some major opportunities for effective cooperation with ionosphere and electronic research.... It would be of great importance to arrange for closer contact between radio astronomy and astronomers in this country.[99]

In short, the strength of American optical astronomy and the attitudes of its community handicapped the development of American radio astronomy.

17.3.3 Discipline structures

A third important difference between national situations was that in the British Empire ionospheric research, often the specialty of the best radio researchers, was traditionally done in physics departments such as those at Manchester and Cambridge. From such venues the conceptual leap from the ionosphere to the solar atmosphere and thence out to the Galaxy was more likely made than from practically-minded electrical engineering departments (such as at Stanford and Cornell Universities) or mission laboratories (such as the US Naval Research Lab [NRL] and the National Bureau of Standards [NBS]), where American ionospheric work was conducted (Gillmor 1986). Research in American physics departments using radio techniques thus tended to involve laboratory-type physics, rather than the field work characteristic of British-trained radio physicists in Britain and Australia. These radio physicists therefore comfortably worked on either ionosphere or radio astronomy projects, and in fact their techniques of observation and analysis often overlapped. Examples include the entirety of meteor radar research (Chapters 9 and 11), much of the solar work (Chapter 13), and investigations of radio star scintillations (Section 14.2).

In Australia the senior researchers (Pawsey, Bowen, Piddington, Martyn) had been trained in

---

[97] By 1952, however, Ira Bowen was supporting efforts to establish radio astronomy at either Caltech or the Carnegie Institution. [Bowen:M. A. Tuve correspondence, March and April 1952, folder 23.383, Bowen papers, Huntington Library, San Marino, California]

[98] Jarrell (2005) obtains similar results to those of Edge and Mulkay (1976) regarding the fraction of radio astronomy papers published in major astronomical journals, although differing in some details.

[99] As in note 22.

Britain before the war, while many of the junior radio researchers came from John Madsen's Electrical Engineering Department at the University of Sydney with undergraduate degrees that were in fact combined physics and engineering. The case of the Radiophysics Laboratory was also anomalous in that, although a government institution, its radio astronomy research during these years was run much more in the spirit and style of university research than of a mission laboratory – indeed, the Laboratory was located right on the University of Sydney's campus. This spirit thrived in the vacuum created by a lack of university research in Australia, as well as the policies of CSIRO management toward its best divisions (Section 7.5.3).

17.3.4 Group styles

The intensity and immediacy of the war as experienced in England (especially) and in Australia led to a camaraderie and drive among the leading postwar groups that was not found in the United States. For example, while Reber and George Southworth did work primarily on war-related projects during the US's four years in the war, if they had been in England, continually threatened with attack, it would have been impossible for them to have found the time for their extensive radio astronomy observations. Moreover, the British and Australian postwar groups, unlike their American counterparts, largely included persons with experiences shared over a long six years in wartime labs. Within these groups, all hands pitched in with a team approach learned during the war. For example, Ryle (1976:29T) later recalled:

> [In America] somebody high up in the Air Force wrote a specification for equipment which then went out and presumably had bids…and was put through like an ordinary production run. There was no way in which young people had direct contact with operational flying. … This contact, as it were, with life in the raw, being able to talk to the people who were going to be flying and be killed in these machines, made one grow up fast. One had a training that gave one confidence that one could do things.

In contrast to overseas, the wartime experiences of American experimental physicists in the large institutions such as the Radiation Lab led to a more measured approach of large-scale projects in close cooperation with theorist colleagues and with industry.[100]

The intense British and Australian experiences hardly guaranteed success in scientific research, but they gave a notable sense of purpose. Conversely, interviews with Americans who did wartime radar research have indicated that the American scene was otherwise. For instance, Ewen (1979:42T) recalled a difference in group styles at his first meeting with Australian colleagues in 1952:

> My first impression was of a very gung-ho group. … eager minds delighted with the opportunity to get together and chat about what was going on in the field. The conference was not typical of what you would have found in the US at that time. … there was an excitement and camaraderie.

17.3.5 Military patronage

The connection between science and the state, or more specifically between science and institutions of war and empire, is not a new one. Archimedes and Leonardo da Vinci designed ingenious and powerful weapons of war. Galileo promoted his new telescope technology as much for military purposes as for studying the heavens. The great observatories established in the seventeenth century at Greenwich and Paris led to many astronomical advances, but their *raison d'être* was to develop navigation to abet control of the high seas. Christiaan Huygens wrote in 1698 that commerce and war had led to "most of those discoveries of which we are masters; and almost all the secrets in experimental knowledge."[101] Astronomical observations and the tools necessary to carry them out have particularly lent themselves to military objectives of sensitive surveillance and position-finding (Harwit 1981:246). Francis Bacon argued for a reform of natural philosophy that was seminal for the development of modern science,

---

[100] S. S. Schweber, "ONR and the growth of US physics after World War II," pp. 3–45 in *Science, Technology and the Military*, eds. E. Mendelsohn, M. R. Smith, and P. Weingart (1988, Dordrecht: Reidel).

[101] C. Huygens, *The Celestial Worlds Discover'd*, translation into English (1698, London) of *Cosmotheoros* (1698, The Hague), p. 41.

but Martin has cogently argued that examination of Bacon's political and legal career reveals that his greater goal was to place the institutions and results of science in the service of the Crown. Knowledge was power, no matter what its origin.[102] The agendas of Victorian scientists such as Lord Kelvin were strongly shaped by the requirements of Empire, and we have earlier seen the influence of the Soviet state on radio science (Section 10.3.4). Likewise, in mid-twentieth-century America, another powerful link was forged between science and the state.

Science and how it was done were profoundly changed by World War II. The new postwar ties between American scientists (especially the physicists), universities, and the federal government (in particular the military) have been thoroughly studied.[103,104,105] For example, Forman (1987) has made a detailed and insightful study of how quantum electronics was shaped by the US military in the postwar decade. He shows that the level of funding in the physical sciences increased to unprecedented levels (about thirty times the prewar level), almost all of which (70–90% by various estimates) came from the military or the Atomic Energy Commission (AEC). In 1951 it was estimated that two-thirds of all American scientists and engineers were involved in military research and development. By that same time the annual government expenditure per physicist had increased by a factor of *six* from prewar levels. Gillmor (1986) finds that the sources of support at US universities for ionospheric research (a subject closely related to radio astronomy in many ways) were in essence 100% from the military up until 1955.

In sharp contrast to prewar attitudes, American academic scientists in many fields and leading universities on the whole accepted the thesis that research grants and contracts (and associated subsidies via overhead charges) from agencies such as the AEC and the Office of Naval Research (ONR) were beneficial to both them and the nation and that they did not hinder anyone's freedom in choosing research topics.[106] Yet it is also clear that, as Roland has succinctly expressed:

> Scientists, like scholars in general, often turn bad money to good purposes, but no amount of rationalization can gainsay the dramatic, though often hidden, ways in which patrons shape the work of their benefactors.[107]

In this vein, Forman has also demonstrated how not just the size of American physics, but its very direction and scientific content, at least in quantum electronics, were strongly affected in the "nation's pursuit of security through ever more advanced military technologies."[108] The principal effects of this military patronage have been placed by Geiger into four slots: dependence (of the recipient), domination (by the grantor), distortion (of the scientific field), and displacement (of disciplinary paradigms).[109] Leslie has observed:

---

[102] J. Martin (1992), *Francis Bacon, The State, and the Reform of Natural Philosophy* (Cambridge: Cambidge University Press).

[103] On postwar US science and national security, see D. J. Kevles (1977;chapters. 21–23); S.W. Leslie (1993) *The Cold War and American Science: the Military-Industrial-Academic Complex at MIT and Stanford* (New York: Columbia University Press); H. M. Sapolsky (1979), "Academic science and the military: the years since the second world war," pp. 379–99 in *The Sciences in the American Context*, ed. N. Reingold (Washington, DC: Smithsonian); R. W. Seidel (1986), "A home for big science: the Atomic Energy Commission's laboratory system," *Hist. Stud. Phys. Sci.* **16**, 135–75; Galison (1997:304–6); *Science, Technology and the Military* (as in note 100), especially the articles by S. S. Schweber, P. Galison, and P. K. Hoch; Kevles (1990), "Cold War and hot physics: science, security, and the American state," *Hist. Stud. Physical Sci.* **20**, 239–64; and Kevles, "K_1S_2: Korea, science, and the state," pp. 312–33 in Galison and Hevly (1992); M. A. Dennis (1994), "'Our first line of defense': Two university laboratories in the postwar American state," *Isis* **85**, 427–55.

[104] A perceptive contemporary witness of the effects of the war on physics is found in a 1949 address by Lee DuBridge, former director of the MIT Radiation Lab. [L. A. DuBridge (1949), "The effects of World War II on the science of physics," *American J. Phys.* **17**, 273–81]

[105] Lovell (1977) has provided a useful, wide-ranging article on many aspects of the relationships between the military and astronomy (especially radio astronomy and space astronomy, particularly in the US and the UK) over the 1945–75 period.

[106] R. S. Lowen (1997), *Creating the Cold War University:the Transformation of Stanford* (Berkeley: University California Press).

[107] A. Roland (1985), "Science and war," *Osiris* **1**, 247–72 (quotation from p. 267).

[108] Of course many important aspects of the military were also reciprocally affected by the scientists; see the introductory article (pp. xi-xxix) by E. Mendelsohn, M. R. Smith, and P. Weingart, as in note 100.

[109] R. L. Geiger, "Science, universities, and national defense, 1945–1970," *Osiris* **7**, 26–48 (1992).

Military-driven technologies of the Cold War defined the critical problems for the postwar generation of American scientists and engineers.... Military interests and intentions became fixed in the very fabric of postwar American science, in its disciplinary structures and rewards, in its research priorities, in its graduate and undergraduate teaching, and even in its textbooks.[110]

How did this all affect American radio astronomy? First, most funding went to projects that fulfilled a "requirement," that is, had some military bearing.[111] In ONR this process of justification was referred to as "painting the project blue" and became especially prominent from 1950 onwards (as the Cold War intensified and the Korean War began) when the Navy increasingly demanded more relevance to military operations from its ONR projects.[112] This meant, for example, that nuclear physics and related fields were heavily supported (Kevles 1977), and indeed many of the best wartime radar researchers, such as Edward Purcell (Section 16.3), Townes,[113] and Dicke (all of whom did a small amount in radio astronomy during this period), went this way with extremely successful careers (including two Nobel Prizes) and major roles in advising the military.[114] It may be more accurate to assert that scientific projects of this era were not so much painted blue in a *post hoc* fashion, but were actually "born blue,"[115] or perhaps even conceived with this hue. Those who did stay in radio and ionosphere work went after the large contracts available for research related to military communications and radar needs, and these, it turned out, were seldom ideal for obtaining byproducts of use to radio or radar astronomy.

Three examples illustrate the situation. First, the Stanford meteor-radar group did its astronomy as a small part of an overall research program in ionospheric propagation and communications technologies – indeed by the mid-1950s they and others had developed the transitory meteor trails into a new means of communications (Section 11.3). At a 1948 meeting of the National Academy of Sciences, they emphasized that meteor data yielded temperatures, pressures, and winds in the ionosphere important to national defense – a concurrent *New York Times* article headlined "Meteors 'clocked' as aid in defense." Another such article in 1953 left nothing to the imagination when it referred to radar meteors as "celestial missiles."[116] Meanwhile at Jodrell Bank, meteor radar was directed toward astronomical problems (for example, did any of the meteors originate from outside the solar system?) and ionospheric physics (Chapter 11).

Second, at Cornell in 1947, ONR began sponsoring a project called "Microwave Astronomy," but the project's first status report described its scope in terms that buried the astronomy:

> The project has as its aim the determination of radio noise magnitudes and apparent source directions, an investigation of high altitude reflections, and determination of absorbing bands, over as wide a frequency range as possible. Correlation of these measurements with weather, diurnal and seasonal variations and astronomical position is to be made.[117]

A third example: in 1949 Reber explained to Oort the chain of reasoning by which his research at NBS was supported: solar radio bursts affected the ionosphere,

---

[110] Leslie, pp. 9–10, as in note 103.

[111] H. I. Ewen (1979:35T).

[112] H. M. Sapolsky (1990) *Science and the Navy: the History of the Office of Naval Research* (Princeton: Princeton University Press), chapter 4.; Schweber, as in note 100.

[113] J. L. Bromberg (1991) gives an account of Townes's invention and development of the maser with military funding in the early 1950s at Columbia University in chapter 2 of *The Laser in America, 1950–1970* (Cambridge: MIT Press). Much more provocative accounts are given by P. Forman (1992) in "Inventing the maser in postwar America," *Osiris* 7, 105–34; and Forman (1996), "Into quantum electronics: the maser as 'gadget' of Cold-War America," pp. 261–326 in *National Military Establishments and the Advancement of Science and Technology* (eds. P. Forman and J. M. Sánchez-Ron, Dordrecht: Kluwer).

[114] For example, both Dicke and Purcell participated in 1950 in Project Hartwell, a summer-long, high-level panel charged with advising the Navy on anti-submarine warfare strategy. [Schweber, as in note 100 (pp. 28–32)]

[115] I thank Bruce Hevly for this felicitous phrase.

[116] *New York Times*, 16 November 1948, p. 17; 27 December 1953, p. 26.

[117] Burrows, 31 October 1947, "Microwave astronomy: status report no. 1," 7 pp., Rept. EE2 of the School of Electrical Engineering, Cornell University, RPL.

which affected communications, which affected military operations. He then went on: "By the above more or less direct reasoning, it is apparent that the study of solar radio waves is worthwhile from a military point of view." Reber at this time was seeking money for a giant dish, but stated that "[the NBS] management has considered the matter but rejected it upon the grounds that it would not aid national defense." In late 1950 he added: "Whether or not one likes it, we are marching down the same road taken by the Germans and the Russians. The names and mile posts are marked in a different terminology but the trail appears well traveled."[118, 119]

Traditional astronomy in the US, on the other hand, had a much less cozy relationship with the military. The basic mistrust that most astronomers carried toward the military's big money and possible control has already been mentioned (Section 17.3.2).[120] In 1947 Shapley at Harvard vowed to accept ONR funds only if they could be applied to an area such as galactic and extragalactic astronomy, "a field which lacks even the remotest connection with military or industrial activity."[121] A 1951 manpower survey found that only 15% of research astronomers were supported even in part by military funds.[122] However, one subfield of American astronomy dominated by military funding was *solar* research.

In the postwar decade the US Navy and Air Force pumped money into establishing solar optical observatories (e.g., the High Altitude Observatory, Colorado, and Sacramento Peak Observatory, New Mexico), as well as supporting rocket observations of the ultraviolet sun (DeVorkin 1992; Hufbauer 1991:129–44; also see Section 18.2). The primary motivation was to learn more about solar effects on the earth's upper atmosphere (where missiles had to operate) and on radiocommunications. The military also patronized research on lunar craters, which had direct applications when compared to craters created by atomic bombs (Doel 1996).

Hevly's (1987) study of ionospheric physics at NRL during this period is nicely corroborated by the NRL radio astronomy scene. He shows how the research tools of the NRL scientists, continually refined in successive experiments, were also the very equipment that the Navy wanted for its own purposes. Shortwave transmitters and receivers and V-2 rockets thus served as both technological means and ends. DeVorkin (1992:341–6) has reached similar conclusions regarding V-2 rocket experiments. Indeed, this concept also works for radio astronomy at NRL: in this case the tools were microwave dishes for studying the sun, which also were being developed as radio sextants for the Navy in conjunction with the Collins Radio Company (Section 10.1.2). A 1949 Collins report for ONR is especially telling where it classifies the moon and sun as "military objects": "[This is] an investigation of the usefulness of microwave thermal emission from military objects, i.e., moon, sun, stars, earth, landmarks, aircraft, and ships."[123]

Rocket or space science, another field that grew out of wartime weapons technology, provides a comparison with early US radio astronomy. DeVorkin (1992) has studied in detail how the German V-2 rocket technology was appropriated by the US military and scientists. In this case, even more strongly than for

---

[118] Reber:Oort, late 1949 (undated copy attached to Oort: G. P. Kuiper, 6 January 1950), Kuiper papers, University of Arizona Library, Tucson; Reber:Oort, 5 November 1950, file 103 [2.1.1950], OOR. I thank Ron Doel for pointing these out to me.

[119] Recall that radio astronomy in the USSR during this period was likewise heavily involved with military priorities (Section 10.3.2).

[120] Two exceptions were Fred Whipple and Donald Menzel of Harvard; for details of Whipple's military-funded researches on meteors and the upper atmosphere, see DeVorkin (1992:chapter 15) and Doel (1996:68–77).

[121] H. Shapley, "Memorandum on the support of astronomical research by the Office of Naval Research," n.d. (ca. late 1947), D. H. Menzel papers, Harvard University Archives. [As cited by DeVorkin (2000:63)]

[122] By the mid-1950s, however, attitudes in optical astronomy were changing. For example, one prominent astronomer (Nicholas Mayall) in 1954 argued strongly for close ties with the military since "numerous research techniques and instrumentation used in astronomy have proven value in ordnance, communications and navigation." [DeVorkin 2000:85]

[123] C. M. Mepperle *et al.*, "Interim report [to ONR] on the microwave radiometry project of the Collins Radio Company," August 1949, RPL. This work eventually led to the development of operational radio sextants on US Polaris nuclear submarines. These observed the sun and moon at short microwavelengths with a 3 ft diameter dish enclosed in a radome [H. I. Ewen (2003), "Two roads that crossed in the wood: growth of US radio astronomy in the 1950s and 1960s," www.nrao.edu/archives/Ewen/ewen_50sand60s.shtml].

radio astronomy, the military dictated the direction of the science that could be done during the rocket flights. There were, however, many similarities with radio astronomy: the new space scientists were (a) dependent on a technology brought to a high state in World War II, (b) a tool-building technical culture (see Section 17.4) dominated by young physicists from the war (many, but not all, from radar), (c) a fraternity of workers rather than a new scientific discipline (see Section 17.2.5), and (d) located on the edges of established disciplines (DeVorkin 1992:341–6). But unlike radio astronomy, the US owned a near monopoly on the V-2 rocket and so faced no overseas competition for a long time. It is also of interest that some of the V-2 work opened up new portions of the solar electromagnetic spectrum (ultraviolet and X-ray) in the late 1940s, continuing the pioneering path of radio astronomy in that regard (for details see Section 18.2).

DeVorkin has also studied the history of high-altitude ballooning during this same period.[124] He emphasizes that much of American science has had a long history of being associated with the military, primarily via expeditions designed for both scientific and geopolitical aims, but with the latter always in control. For example, ballooning in the 1930s–1950s ferried the military, as well as scientists, into the newly important stratosphere, just as horses and dories transported John Wesley Powell through the unknown American Southwest in the previous century. Instead of the Grand Canyon, US radio astronomers funded by the military explored the realm of signals from beyond the earth. Outer space was strategically vital as the ultimate "high ground," as the backdrop for critical skyward-looking detectors, and as the location of handy natural beacons for radio navigation.

This strong relationship between US radio astronomy and the military continued through the 1950s; for example, one finds in a 1955 report justifying the creation of a National Radio Astronomy Observatory (NRAO):

> Most radio astronomy signals are very weak by normal communication engineering standards, and the demands of the astronomers have forced advances in electronic techniques. … To mention only a few, material improvements have been made in the noise factors of receivers, in the precision of…antennas, in discrimination and integration techniques, in broad-banding radio frequency components, and in data display devices. The value of such advances as these to our military security is very great.[125]

Van Keuren (1997, 2001) has detailed how the radio astronomers at NRL, although publishing in the open literature, were intimately connected to secret military projects, in particular, the Sugar Grove 600 ft diameter dish project from 1954 on.[126] There is also the remarkable story told me by John Bolton, who came from Sydney to the US in 1955 to found a radio astronomy program at Caltech (Munns 2002:chapter 4). At that time the Rand corporation approached him regarding data on ionospheric scintillations that he had gathered for many years in Australia. After a former colleague in Sydney sent it to him, Bolton forwarded it to Rand for their perusal. But when he later asked for its return, he was told that the Air Force had classified it and that he unfortunately did not have a security clearance that would allow him to see his own data![127] In Australia, by contrast, Bolton's former employer, the Radiophysics Lab, had had only a few cooperative arrangements with the Royal Australian Air Force, especially in cloud physics research, but these all ended by 1952 (Robertson 1992:69).

The contrast between the American and British situations was extreme. While government funding for British science (in particular physics) was also enormous compared to prewar levels (by a factor of ten to twenty),[128] the new funds for universities remained minuscule by American standards. These funds all originated in the civilian Department of Scientific and Industrial Research; military research, although substantial, was concentrated in non-university laboratories such as the Royal Radar Establishment (descendant of the wartime Telecommunications Research Establishment) and the Atomic Energy Research

---

[124] D. DeVorkin (1989), *Race to the Stratosphere* (New York: Springer-Verlag), p. 6.

[125] L. V. Berkner and R. Emberson, 15 April 1955, quoted by A. A. Needell (1987) in "Lloyd Berkner, Merle Tuve, and the federal role in radio astronomy," *Osiris* 3, 261–88; also see Needell (2000:276–7) and Lovell (1977:153–5).

[126] See note 10.14.

[127] J.G. Bolton:author, 29 April 1986.

[128] Lovell (1983:101–4) gives several useful graphs. Also see Lovell (1977:166–8).

Establishment at Harwell.[129,130] I estimate that the level of support per British radio physicist in a university group was ten to one hundred times less than for his American colleague; when adjusted for cost of living, it was still three to thirty times less. For example, in the late 1940s Cornell's annual budget for radio astronomy was about $50,000, whereas the Cavendish's was about $5000 (for about twice the research staff) and Jodrell Bank's $30,000 (for about six times the staff). At this time Cornell procured a 17 ft dish from a firm for $17,000, whereas the Cavendish group was building a large interferometer-type antenna (Fig. 8.7) for about $500 in materials.[131] The entire research atmosphere in British universities was different: the Rutherfordian philosophy of "string and sealing wax" and "we think with our brains, not our wallets" still reigned in the Cavendish and to a large extent also at Jodrell Bank and in Sydney.[132] Ryle was flabbergasted when he was invited to an Air Force-sponsored ionosphere conference at Penn State University with the offer to fly him over in a military transport and to pay him as a consultant throughout the conference.[133] Instead of laboratories stocked full of the latest electronic gadgetry, the English shelves were bare except for what could be scrounged from military surplus and adapted to one's experiment. Rationing of various kinds continued well into the mid-1950s; because of paper rationing, in 1949 Lovell even had trouble securing a subscription to *Nature* (a key journal for radio astronomy), and because of currency controls both the Jodrell Bank and Cavendish groups were constantly finagling for something as mundane (and as important) as Esterline-Angus chart recorder paper from America.

In Australia, the level of funding was between that of the US and Britain, but the largely British-trained leaders also tended to have Rutherfordian attitudes. Moreover, if the Radiophysics Lab had had access to US-level funds in the late 1940s, and if it had been under Cold War pressure to do physics (as in the US), one can speculate that its high-energy accelerator work (which was dropped after a few years because of cost; see Tangent 7.1) might well have thrived to the point of dominating its entire program. In such a case it would seem unlikely that the Australians would have achieved such a distinguished international stature as they did with radio astronomy.

Military patronage in the US not only led researchers in the postwar decade away from radio astronomy, but even those who did pursue it were persuaded to work at shorter wavelengths (less than 30 cm, or frequencies above 1000 MHz), a technical direction that was less successful in producing first-class research. Ever since the 1930s front-line radar development had trended toward shorter operating wavelengths that allowed superior detection and location of targets at greater distances.[134] Likewise for radio astronomy, shorter wavelengths had the potential of allowing more detailed maps of the sky, and the groups at NRL and Cornell therefore poured resources into research at wavelengths less than 20 cm.[135] NRL, the largest American group, also had a particular interest in short wavelengths because only those could provide sufficient accuracy for the Navy's desired all-weather radio sextant. Observations at microwavelengths also offered the Americans their own research niche distinct from that of the leading

---

[129] P. K. Hoch, "The American physics elite and the military in the 1940s," in *Science, Technology and the Military*, p. 100. [as in note 100].

[130] D. Edgerton has presented persuasive evidence of the overall strong militarism in Britain during this period. See *Warfare State: Britain, 1920–1970* (2006, Cambridge: Cambridge University Press), and "Science and the nation: towards new histories of twentieth-century Britain," *Historical Research* 78, 96–112 (2005).

[131] Ryle's notebook for the 1952 IAU General Assembly, file 27 (uncat.), RYL.

[132] Just after the war Princeton physicist Henry Smyth wrote: "Experimental physicists used to work with their brains and their hands; now they work with their brains and $100,000 worth of equipment backed by a staff of technicians and machinists." I thank P. Galison for this quotation. [H. Smyth, "The plans of the [Princeton University] Department of Physics for the next five years," December 1945]

[133] A. H. Waynick:Ryle, 10 March 1950, file 1/1, RYL. (Ryle did not attend.)

[134] Bromberg (pp. 13–15, as in note 113) discusses this push in radar toward ever shorter wavelengths with regard to its influence on the development of the maser.

[135] The Cornell group, however, never did make any significant observations at wavelengths shorter than 1.5 m, despite the official name of *microwave astronomy* for their 1947-onwards, Navy-sponsored project. Microwave observations were always their stated goal, but they never advanced beyond developing equipment in the lab.

foreign groups. Furthermore, American budgets could handle the necessity at microwaves for the (expensive) "big dish" approach, as opposed to the cheaper interferometers and dipole arrays generally used overseas. Also pushing American radio astronomers to the use of microwaves and large dishes was influence from the US optical astronomy community (Section 17.3.2). As remarked by Scheuer in Section 17.2.2, interested astronomers such as Greenstein and Struve were naturally more comfortable and enthusiastic supporting a type of radio telescope that looked like a (proper) optical telescope, as opposed to an array of "clothes lines" scattered over a field. In fact, no interferometers existed in the US until 1953–54, when two were built at the Department of Terrestrial Magnetism, but significantly only as a result of long-term visits by Mills from Sydney and Smith from Cambridge.

ONR was the one patron that might well have funded low cost, astronomically oriented, long-wavelength research in the postwar decade (and especially before 1950), but it never did. This was not because the agency suppressed or turned down proposals for such research, but rather that American academia simply did not produce them in the prevailing climate of big projects and expensive technology paid for by the military. In summary, most American radio astronomers worked at shorter wavelengths.

In contradistinction, the British and Australians had realized from their own initial reconnaissances of the sky that it emits only very weak signals in this microwave region and that even the best receivers of the day were of insufficient sensitivity and stability for reliable measurements, except of the sun. For example, Piddington and Minnett's (1951b) early survey in Sydney of the microwave sky (Section 14.3.5) had shown it to be scientifically unproductive as compared to longer wavelengths. Figures A.2 and A.5 in Appendix A show that at shorter wavelengths (a) almost all strong sources (other than the sun and moon) were greatly weaker, even as (b) one's ability to detect such sources rapidly declined ("sensitivity" curve in Fig. A.2; T_s (eff) curve in Fig. A.5). The British and Australians therefore concentrated on long-wavelength observations and developed interferometer techniques to compensate for those wavelengths' inherently poor angular resolution, in the process mastering what became the central technique for much of radio astronomy. With their budgets and scientific sensibilities pushing them toward long-wavelength studies and no military bias toward microwave work, they were able to explore the radio universe far more efficiently than could those with big bucks in the US.[136] They reasoned that the way to make scientific progress in radio astronomy, where the long-wavelength sky had all the accessible action, involved the much cheaper, older, and easier-to-handle equipment using wavelengths longer than one meter. For example, by 1950 the surplus long-wavelength radar transmitters and receivers at Jodrell Bank were becoming obsolete relative to the latest electronics, but they were kept because they were still doing good astronomy. Lovell (1971:18T) later recalled that "the transmitter was just the thing one turned on – the meteors and aurora were the exciting things." Smith (1971:29T) also recollected:

> The most profitable wavelengths were where there's most radiation.... This perhaps was an antipathy with the Americans, who were saying very strongly at that time that it's nonsense to be working on these old-fashioned meter-wavelengths – you've got to get to centimeters.

The British and Australians did not shun microwave research because of a lack of expertise at these wavelengths – indeed, they had also developed a great deal of microwave radar during the war. In Sydney, Piddington and Minnett in particular developed microwave techniques and observed a wide variety of objects besides the sun (e.g., Sections 12.5.2 and 14.3.5), but their efforts, although useful, reinforced the value of working at long wavelengths. Especially the British were deterred by the great cost and complexity of accurate, large antennas for microwaves. They often cited the NRL 50 ft diameter, microwave dish as an example of American excess and poor scientific judgment. Built for $100,000 in 1949–51, it had serious shakedown problems and did not produce any useful scientific results until 1954 (Section 10.1.2).[137]

---

[136] There is at least one case where the postwar military influenced British radio astronomy, although not through patronage. One of the key ideas in Ryle's interferometers, the phase switch, came to him as an adaptation of an underwater sound detection device that he had invented in 1947 as a member of an Admiralty advisory group (Section 8.3).

[137] Microwavelength research with small dishes was also undertaken during this period by the French, Japanese, and

In summary, postwar radio astronomy as practiced in three countries provides a good case study of the influence of military patronage upon science. Bounteous military funding led American researchers to microwave observations, the obvious new-technology choice to them and of great interest for military radar applications, but in fact a poor scientific strategy for the times. Just as Forman (1987) has shown for postwar quantum electronics, I argue that the military patronage of US radio astronomy definitely affected its science.[138] The elite of American wartime radar scientists, lured by heavy federal funding of atomic, nuclear, and particle physics research, did not even pursue radio astronomy. Furthermore, the strong community of American (optical) astronomers acted more as a hindrance than a help to those radio researchers who found themselves contributing to knowledge of the heavens. When these factors are combined with the ineffectiveness of the two prewar American pioneers as postwar leaders and the great organizational and intellectual strengths of the British and Australian groups, the United States found itself in a distinctly second-class situation.

By the mid-1950s this situation had grown serious enough that both the US government and universities began paying attention. In 1956 the NRAO was established by the National Science Foundation (Needell 2000:270–92, Munns 2003), while at about the same time several universities established their own radio observatories (though still usually through ONR or US Air Force funding). Moreover, by the late 1950s improving receivers and larger dishes meant that microwave observations of the radio sky at last began to be feasible and to pay off scientifically. For instance, beginning in 1954 the NRL team was able to make important observations with their 50 ft dish, defying their detractors who had predicted that, despite its huge size, it would only be able to detect the sun and moon. The US was beginning to compete more effectively, although still greatly lagging Australia and Britain.[139] As it turned out, the effects of the postwar decade were of such a magnitude that parity would not be reached in radio astronomy until the 1970s.

17.4 RADIO ASTRONOMY AS TECHNOSCIENCE

Instruments have been vital to the development of science ever since the telescope, microscope, and air pump appeared in the seventeenth century.[140] In this section I briefly examine how the material culture of early radio astronomy illuminates aspects of our understanding of the relationships between science, technology, instruments, experiments, and scientific theory.

Despite the widespread belief among many scientists, engineers, funding agencies, and policy makers that "pure science" comes first and leads to later technological advances, historians of science have reached a strong consensus that this is a narrow and misleading formulation.[141] In fact, science appears to be as much "applied technology" as technology is "applied science." The hierarchial view – with discovery leading invention, science leading technology, knowledge of nature leading to knowledge about the manmade world – simply does not accord with many episodes analyzed in the history of science and technology. For instance, it is clear that nineteenth century thermodynamics (Carnot, Clausius) owed more to the development of the steam engine (Watt) than vice versa. In this light, a richer understanding emerges from viewing the scientific enterprise as the product of a bustling nexus shared by theorists, experimenters, technicians, and instrument makers (craftsmen). Who's leading whom in any particular collaboration, or whether all elements are so tightly intertwined that the question is moot,

---

Canadians (Chapters 10 and 13), but only for solar monitoring and eclipse observations.

[138] Tangent 17.3 further discusses this point, as well as making comparisons with other studies testing Forman's ideas.

[139] Jodrell Bank also changed when it accepted substantial funding from the military and NASA, especially in the period 1957–63 after the Soviet launch of Sputnik. [Lovell (1968:chapters. 30–35); Lovell (1973:chapter 15); Lovell (1990:chapter 22); Agar (1997:22–6); G. Spinardi (2006) "Science, technology, and the Cold War: the military uses of the Jodrell Bank telescope," *Cold War History* **6**, 279–300; F. Graham-Smith and B. Lovell (2008), "Diversions of a radio telescope," *Notes and Records of the Roy. Soc.* **62**, 197–204.]

[140] A. van Helden and T.L. Hankins (1994), "Introduction: instruments in the history of science," *Osiris* **9**, 1–6; also the following eleven articles on this theme. For twentieth century astronomy, Smith (1997) gives an excellent review of the literature on the role of instrumentation.

[141] A good overview of the issues can be found in P. Galison (1988), "History, philosophy, and the central metaphor," *Science in Context* **2**, 197–212. Also see Price (1984).

depends on the stage of development of a field, on details of persons and communications, and on institutional structures and patronage.

For early radio astronomy, too, I conclude that it is hopeless to try to privilege the position of scientific motivations and results vis-à-vis technological motivations and instrumental developments.[142] Because of this, I find *technoscience* an apt term to describe early radio astronomy[143] (and indeed much of postwar science).[144]

Let us look first at the instrumentation side of this technoscience. The key deficiency fought by early radio astronomers was how to improve upon the abysmal angular resolution, and therefore position determination, afforded by equipment inherited from the war.[145] The sky position of any received radiation could be given only to an accuracy that was typically 1000 to 10,000 times worse (!) than available for a star on a photograph (which had ~1″ accuracy); even the very best radio positions during the postwar decade[146] improved the situation only to a factor of ~20 or 100 times worse, and then only for the few strongest sources. Another example: astronomers of that time knew that the optical sun was a circle perfect to within 1 part in 30,000, but radio astronomers could just barely make out that the radio corona might have an asymmetry amounting to about 1 part in 5 or 10. As Scheuer remarked in his 1954 Ph.D. thesis: "Radio astronomy cannot, as yet, claim to be an exact science."[147] Furthermore, source intensities were often measured to no better than ±50%, despite steadily improving receivers and larger antennas.[148] This made it frustrating to compare results at different frequencies and with different antennas in order to determine radio spectra, which were critical to test theories of the radiation's origin.

The numbers above illustrate both that early radio astronomers faced severe technical problems and that they rapidly solved many of them to better answer questions about the radio sky. The technical capabilities of radio astronomy progressed far more rapidly than those of optical astronomy during the postwar decade. Tangent 17.4 gives the details of a technical argument explaining one contribution to this faster advance: radio observers, unlike their optical colleagues, improved not just sensitivity, but also angular resolution when they built a larger instrument. Another major difference between the two regimes was that radio astronomers were able to develop interferometers of ever longer baselines (Chapter 14), in contrast to optical astronomy where for decades the deleterious effects of the earth's atmosphere had squelched interferometry. Radio interferometry was fundamentally easier because its million-fold longer waves greatly facilitated positioning of antennas, delay and transfer of coherent signals, and correction for atmospheric effects.

Postwar radio astronomy, especially in its first few years, was often driven by applying (and then continually improving) the instrumentation at hand. Many of the investigators had the Baconian point of view that one should just look at the sky and see what showed up. Recollections from several participants capture such empirical attitudes:

> We felt we were opening up the frontiers of physical knowledge… [Our work] wasn't to prove any particular theory or fill in any particular gap in knowledge.
>
> [S. J. Parsons 1978:13T]

> The impression very much was: "Blimey, what's this thing on the record? Let's do the following experiments. …" – and you would rush off and do them.
>
> [Machin 1971:13T]

> I think in a sense the radio astronomer was always an opportunist. He went ahead with what seemed

---

[142] Edge and Mulkay (1976) reached a similar conclusion, although their analysis was based almost totally on the two major British groups, whereas the present discussion draws from the entirety of early radio astronomy.

[143] Latour originally defined the term *technoscience* in a much broader sense than I am using it here; for example, he included the political and economic players who also shape science. [B. Latour (1987), *Science in Action* (Cambridge: Harvard. University Press), pp 173–6.]

[144] For example, "technoscience" also works well to describe postwar US scientific rocketry as studied by DeVorkin (1992).

[145] The fundamental reason for this was that radio wavelengths are of order one million times longer than those in the optical band (see Section A.1).

[146] Smith (1951a); Mills (1952b); see Section 14.4.1.

[147] P. A. G. Scheuer (1954), "An investigation of the structure of the galaxy by means of its radio emission," Ph.D. thesis, Cavendish Laboratory, Cambridge University, p. iv.

[148] The weakest detectable signal improved by a factor of ~10,000 from the war's end to 1953.

possible. And if you've got an instrument which can do things, you use it in all the things it will do. And you don't try to do things it won't do.

[Ryle:1976:51T]

You don't sit around saying, "I wonder what these radio sources are? If I could measure the diameters, I would be able to tell. Now how can I measure the diameters?"

Instead, you work the other way. You find that you've got a jolly good technique and you say, "What's the most interesting way to apply this technique?" And from that you get to the fact that these radio sources have finite diameters, and then you start thinking, "Well, maybe this tells me something about them."

[Smith 1976: 22–3T]

Such empiricism led to many discoveries, often serendipitous, but this volume has shown that it is nevertheless often wrong to think that they fell like manna from heaven. On the contrary, observations were often methodically made and analyzed. Price (1984:12–13) has pointed out that this pattern is common: practitioners take a new technique, perfect it, and then apply it to "everything in sight," leading often to novel and surprising phenomena.[149] We have seen many examples where serendipity played a major role in early radio astronomy: Jansky and the galactic background (Chapter 3), Hey and solar radio bursts (Section 5.2), Hey (and then Lovell) and meteor radar echoes (Sections 6.3 and 9.2), Hey and a discrete (and variable) radio star (Cygnus A) (Section 6.2).[150] Other major discoveries, although startling, were less certainly serendipitous: Bolton and numerous radio stars (Section 7.4), Ryle and the strongest radio star of all (Cassiopoeia A) (Section 8.4), Pawsey and the hot solar corona (Section 7.3.3).[151, 152] Bolton even described his particular interferometric technique as happily accidental:

We happened to use [the sea interferometer] technique in Australia because we had some cliffs. I mean if we had been at Blackpool [an English seaside resort with no cliffs], or something like that, well, we probably wouldn't have done it!

[Bolton 1976:7T]

What do the training and skills of the practitioners themselves tell us about early radio astronomy as a technoscience? Before World War II, Jansky was a *physicist* investigating an *engineering* problem, while Reber was an *engineer* doing *science*. Later, the postwar cadre, mostly formally trained as *physicists*, emerged from the war re-educated to an *engineer's* point of view in terms of making practical devices that required delivery on schedule and within budget (Section 17.1.1). Furthermore, projects were tackled by mixed *teams* of scientists, engineers, technicians, and instrument makers, and within these teams skills regularly migrated across job descriptions. By the war's end, most of these radio and radar investigators had been molded into "technoscientists," bound not by theoretical paradigms but by the "communality of their tools" (Price 1984:15). And so they remained bound in subsequent years when trying to understand cosmic noise. Their instrumental knowledge came not from textbooks, but from experience gained in the wartime environment:

[The practice of radio techniques] is something between an art and a science, which is learned most effectively by the apprenticeship system usual in research.

(Pawsey and Bracewell 1955:8)

We at present use common sense and low cunning, but there may be better methods.

(Pawsey)[153]

There are many cases in early radio astronomy where instrumentation dictated major aspects of the research agenda. For example, Edge and Mulkay (1976:126–30) have documented how Ryle's group at

---

[149] The beginnings of observations in other wavelength bands in astronomy (infrared, X-ray, etc.) later showed similar patterns (Section 18.2).
[150] Edge & Mulkay (1976:224–30) and Edge (1977:322–3) also discuss the strong role of serendipity in 1945–70 radio astronomy.
[151] The lone major discovery in early radio astronomy that was predicted beforehand (by Van de Hulst) was the 21 cm hydrogen line (Section 16.1).
[152] *Serendipitous Discoveries in Radio Astronomy* (Kellermann and Sheets 1983) is a fascinating report of a meeting of radio astronomers who examined the role of serendipity on the occasion of the fiftieth anniversary of Jansky's discovery.
[153] Pawsey:Reber, 20 March 1950, file A1/3/1a, RPS.

Cambridge employed a succession of antennas and receivers in a "progressive evolution, with its own logic, momentum and remorseless pace." And yet even in this case, as well as for Jodrell Bank, they argue that instrumentation and theory had a "close, dialectical relationship" (p. 438), or a "close ambivalent relationship" (p. 137). On the other hand, an important theme in Harwit's book *Cosmic Discovery* (1981) is that fundamental astronomical discoveries (in particular during the twentieth century) have most often arisen from technological innovation rather than theoretical astrophysical insight.[154] Harwit (1981:242) even suggests that credit for these sorts of discoveries (many of which came about through radio techniques [Section 18.1]) should go not to the observer, but to the relevant piece of equipment!

Sometimes the technology even led to theoretical insight concerning the physics of radio emission. For example, recall how Haeff (1948, 1949) and Pierce (1949), inspired by the theory of microwave amplifier tubes (klystrons and travelling-wave tubes), used similar ideas of interacting streams of electrons to explain the intense radio emission arising in solar active regions (Section 13.2.2). The most striking case of this was the intensity interferometer developed at Jodrell Bank by Hanbury Brown and Twiss (1954). Here was a technique developed for the purpose of measuring angular sizes of radio sources, which did indeed work to that end, but also eventually contributed to understanding fundamental physics (Section 14.5.1).[155]

Early radio astronomy was also a technoscience in the strong influence that a group's instrumental commitment had on observational results and scientific conclusions. As discussed in Section 14.6, arguments over the number and nature of radio sources in the early 1950s were strongly influenced by the differing properties of the antennas used by the proponents. Galison (1987:13) has emphasized how instrumentation in any experiment acts as a filter, taking as its input phenomena from the natural world, which are then circumscribed and modulated before producing an output. This concept is an excellent match with radio astronomy in that standard physics theory often characterizes an antenna as a "Fourier filter" of the sky's radio emission, that is, responding to only an incomplete range of Fourier frequencies present in the sky's radio brightness distribution (Sections A.10–11). The antennas of Hanbury Brown, Ryle, Bolton, and Mills were all looking at the same sky, but their disparate designs meant that they filtered that sky in vastly different ways and thus came to discordant conclusions as to the nature (and even the reality) of radio sources. Indeed, their antenna designs tended to reinforce their scientific stances (Section 14.6). Once again, technique and science can hardly be separated.

Notwithstanding the above examples in which the technoscience of early radio astronomy was strongly technique-driven, earlier chapters have also recounted many science-driven episodes. The outstanding case was the effort led by Oort at Leiden Observatory to mount radio studies of the 21 cm hydrogen line in order to study the structure and dynamics of the Milky Way (Chapter 16). Here there was no doubt that radio technology was only a means to what was considered the more important end of scientific knowledge. Similarly, after the initial discoveries of 1946–47, the many radar studies of meteors at Jodrell Bank were designed and motivated by a desire to answer astronomical questions such as the speeds and origins of meteors (Chapter 11). Finally, most radio observations of the sun (the largest subfield of early radio astronomy) were carried out in association with (optical) solar astronomers to answer the latter's scientific questions (Chapter 13). Note that these examples each refer to a subfield that had strong ties to optical astronomy, where standards had long been set and paradigms existed. Nevertheless, even in radio astronomy's other subfields (such as radio sources and galactic background), observational campaigns, although often defined by available

---

[154] Harwit also emphasizes that major discoveries have usually originated outside of astronomy itself (see Section 18.2). Note that Harwit is concerned only with major discoveries, which are a small portion of all scientific effort.

[155] Galison (1997:820–7) relates another excellent example of practical work leading to esoteric theory, namely Julian Schwinger's adaptation of his wartime radar work on microwave circuits to problems of renormalization in quantum electrodynamics. In another case, Lenoir has shown how the field of electrophysiology often developed theoretical models that were inspired by laboratory instruments (and vice versa) [T. Lenoir (1986), "Models and instruments in the development of electrophysiology, 1845–1912," *Historical Studies in the Physical Sciences* **17**, 1–54].

techniques, were concurrently influenced by scientific aims. For example, a stated aim of many radio source surveys was to increase the number of optical identifications so as to better determine distances to the sources.

In sum, different research groups employed different mixes of instrumentation and science, as well as numerous ways in which these factors interacted with each other. Sometimes technology and techniques drove the research programs, other times scientific motives predominated, and yet other times the two continually pushed and pulled, often interchanging roles. The enterprise of early radio astronomy was indeed technoscience.

17.5 THE NEW ASTRONOMERS

In this chapter we have seen that a disparate band of radio researchers, coming out of World War II and fascinated with the phenomenon of extraterrestrial radio noise, metamorphosed during the postwar decade into a worldwide community of *radio astronomers*. By 1953 they were at a watershed in the field's development in terms of the types of programs undertaken (less exploratory; more nonsolar investigations), the scale of the antennas (and budgets) needed for new projects, and their degree of comfort with (optical) astronomers and astronomy.

The new astronomers found themselves in the midst of ironies and paradoxes:

– Although dismissed or ignored by many of the old guard, they were nevertheless destined to change the way astronomy was done. For example, they caused (traditional) astronomers to be reclassified as *optical astronomers*.
– Although their antennas studying the heavens were called "radio telescopes," they still did not merit the status of "real" telescopes made of glass.
– Although they were continuously observing the sky, they did not work at observatories.
– Although their backgrounds, style of research, and technology differed mightily from those of traditional astronomers, they had no desire to set up a separate discipline.
– Although they were discovering sources of *radio* emission with remarkable properties, it was only in the context of *visual* astronomical knowledge that these sources could be explained.

– Although their data and techniques were intrinsically non-visual, they were striving to develop a visual culture (through production of images based on radio data) in order to cater to astronomers and make their findings less arcane.

In the following, final chapter, I will turn from social aspects of early radio astronomy to a summary of the science of the new radio sky and an analysis of its place in history. I will argue that these new astronomers' opening of the electromagnetic spectrum was in fact one of the most significant developments in the long history of astronomy. It was significant in itself and doubly so as a harbinger of the even broader expansion of the astronomical spectrum made possible by the Space Age in the 1960s and 1970s. Together these constituted the twentieth century's New Astronomy.

TANGENT 17.1 PREWAR USES OF THE TERM *RADIO TELESCOPE*

Before the first postwar usages of the term *radio telescope*, there were others in strange contexts. The first usage in print claimed by the *Oxford English Dictionary* is a 1929 short story by S.A. Coblentz entitled "The Radio Telescope." In this story an astronomer attaches a radio device to a telescope, allowing him to discern life forms on a planet circling a star 1000 light years away [*Amazing Stories* 4 (June), 198–206].

Earlier usages of the term referred apparently to proposed television-like devices that would transmit images (via radio) over great distances. Alfred N. Goldsmith broadcast a radio lecture in 1924 entitled "The radio telescope or directional receiving" [*New York Times*, 19 July 1924, p. 10], and Reginald A. Fessenden talked of a "radio telescope," also called a "wireless telescope" or "pheroscope," in 1922 or perhaps earlier [*New York Times*, 20 January 1926, p. 3].

Another hint of the term is in a July 1889 letter from Heinrich Hertz to Oliver Lodge, who was apparently experimenting with pitch lenses for the new Hertzian waves. Hertz jokingly referred to a "giant telescope of pitch" [Hertz:Lodge, 21 July 1889, Lodge papers, University College (London), MS ADD 89/53].[156]

---

[156] A *very* early pseudo-usage of the term *radio astronomy* appears in the title *De Radio Astronomico et Geometrico Liber* by

TANGENT 17.2 COMPARISON WITH THE INTRODUCTION OF THE ELECTRON MICROSCOPE INTO BIOLOGY

Rasmussen (1997) has studied in detail how the electron microscope was developed in the US over the period 1940–60. This case is analogous in many ways to that of the emergence of (worldwide) radio astronomy. First, the technology was similarly largely developed in World War II. Second, the electron microscope also dealt with optics in the broadest sense of the word (manipulating electromagnetic waves to create images), but now concerned with the microcosm rather than the Cosmos. Third, the electron microscope's band of practitioners (often called biophysicists) were likewise not well accepted by the biologists of the day, who were uncomfortable outside their centuries-old technique of (light) microscopy. Unlike a radio telescope, however, the electron microscope claimed to provide detail hundreds and thousands of times better than optical images. As with radio and optical astronomy, there were fundamental arguments about how to relate the results of the new technique to traditional knowledge.

Rasmussen discusses several descriptions that philosophers and sociologists of science have given to how a novel instrument or technique can be deemed reliable and acceptable. One approach involves rigorously explaining the theory of the new instrument's operation; for the case of radio antennas and receivers, this could be done (but what about an antenna's behavior *in practice*?). Another compares old with new by calibrating both on the same object of study, but if the new instrument greatly differs from the old in some claimed property, then how useful is such a calibration? Indeed, early studies of solar noise did *not* produce a sun having the size or temperature of the optical sun. Finally, one can gain confidence in a new instrument if its scientific results converge to some degree on results already accepted from the old technique. For the case of radio astronomy, this last criterion was indeed critical for acceptance by both radio and (optical) astronomers: initially it came from strong correlations between solar activity at radio and optical wavelengths, and then later in the form of optical identifications for (a few) discrete radio sources.

TANGENT 17.3 TESTING FORMAN'S "DISTORTIONIST" IDEAS

Historian of science Paul Forman (1987) has put forth the controversial notion, backed by meticulous research, that military patronage in postwar America not only influenced which fields of physics were studied (this is not controversial – patrons always influence what they support), but distorted the very findings of those fields and caused the practitioners to lose scholarly independence (this *is* controversial). For instance, he argues that much more attention was paid to "gadgets" (such as the maser) and technical capabilities than to fundamental physics problems. His specific case study centered on the field of quantum electronics in the 1945–55 period, discussed in Section 17.3.5. Here I wish to describe other historical studies that bear on what have been called Forman's "distortionist" conclusions, and suggest that the present study of postwar radio astronomy, because of its international coverage, throws new light on this question in favor of Forman's ideas.

Several other studies (in addition to the present one) have corroborated Forman's ideas. As mentioned in Section 17.3.5, Leslie, in his book *The Cold War and American Science*,[157] has analyzed how postwar American science, including within elite academia, became intertwined with military interests and goals. Kaiser[158] has convincingly argued that the Cold War demand for a large supply of Ph.D. physicists strongly affected the types of research that could be undertaken in university departments. In order to produce enough degrees each year, advisors avoided theses with complex theoretical topics and tended toward instrumentalist approaches. Oreskes[159] has analyzed deep ocean research over the period 1955–80, in particular the development in the

---

Gemma Frisius in 1545! [J. Naiden and W. T. Sullivan, *Sky & Telescope* **70**, 4 (July 1985)].

---

[157] As in note 103.

[158] D. Kaiser (2002), "Cold War requisitions, scientific manpower, and the production of American physicists after World War II," *Historical Studies in the Physical and Biological Sciences* **33**, 131–59.

[159] N. Oreskes (2003), "A context of motivation: US Navy oceanographic research and the discovery of sea-floor hydrothermal vents," *Social Stud. Sci.* **33**, 697–742.

1960s of the submersible vehicle *Alvin* for the purpose of military operations on the deep seafloor. This capability then allowed oceanographers to make a fundamental discovery in the late 1970s, namely deep-sea hydrothermal vents. Oreskes argues that Western scientists have long inconsistently harbored the myth of individual intellectual freedom even as they solicit powerful persons and agencies for funds. She notes that scientists can be viewed as having more than a dichotomous choice between "pure" and "impure," but instead can choose a "mixed" path – in the case of US deep-sea oceanographers, a mix of autonomy and external direction. Oreskes argues that the military did not control from outside, rather they and the oceanographers became part and parcel of a common enterprise that benefited them both.[160] Finally, Doel[161] has looked at 1945–65 US earth sciences (geophysics, ionosphere research, atmospheric sciences, oceanography), then known as the physical environmental sciences. He finds that military requirements for, *inter alia*, ballistic missiles, submarine operation, and nuclear bomb test detection greatly shaped how we came to think of our planet, producing a new intellectual map. But this map was limited in many ways – it arose mainly from mathematically-oriented fields (e.g., geophysics rather than geology), and involved little biological research. It was left for the *biological* environmental sciences to arise in an entirely different manner in the 1960s (and with far less funding).

Other scholars have objected to Forman's ideas. Kevles[162] argues that Forman's stance is ahistorical since one can never tell whether a field has been "distorted" because no comparison can be made to the field *as it would have otherwise gone* – one cannot do "historical control experiments" to determine a normative path. Furthermore, deciding what is "fundamental" and what is a technical application is fraught with problems. Barth[163] has studied the case of seismology in the 1960s, when there was massive, sudden military funding (via Project Vela for detection of clandestine underground nuclear tests) that inflated the field's size and involved nearly every US seismologist. In this circumstance, on Forman's hypothesis one should find that the military exerted great control over seismology's cognitive development, but Barth argues (through a comparison with key questions in the field before and after the military support) that this did *not* happen. Finally, Bromberg[164] has examined quantum optics in the 1960s–70s, finding that fundamental questions (e.g., the coherence of photons in lasers) were not at all ignored, but intermixed with the development of devices useful for weapons. In fact she argues that the technology and the physics were inextricably linked.

So where does the history of US postwar radio astronomy fit into all this? I suggest that the evidence of Section 17.3.5 provides a test, of a new type, strongly in favor of Forman's hypothesis. Because the US situation can be compared with the contemporaneous situations in England and Australia, one *can* with some confidence for the first time suggest what might likely have happened in the US if there had not been such a strong military influence. Section 17.3.5 shows that there was liberal funding for radio astronomy projects, and that the funded technology (shorter wavelengths) was what was needed for military radar purposes and was gladly accepted by researchers. But this was in fact *not* a useful direction for radio astronomy at that time, as the Australians and British well realized as they went from success to success with long-wavelength investigations. Knowledge production in US radio astronomy was thus channeled into unfruitful directions, and it would not be until the late 1950s before the US would begin to catch up.

TANGENT 17.4 PHOTONS AND APERTURES IN THE RADIO AND OPTICAL REGIMES

The differing physical nature of instruments and observations in optical and radio astronomy exerted fundamental effects on their development. Two technical effects are important when comparing the optical and

---

[160] Dennis (as in note 103) has similarly argued for this type of "hybrid" relationship in the context of postwar aeronautical engineering.

[161] R. E. Doel (2003) "Constituting the postwar earth sciences: the military's influence on the environmental sciences in the US after 1945," *Social Stud. Sci.* **33**, 635–66.

[162] Kevles, as in note 103.

[163] K.-H. Barth (2003), "The politics of seismology: nuclear testing, arms control, and the transformation of a discipline," *Social Stud. Sci.* **33**, 743–81.

[164] J. L. Bromberg (2006), "Device physics vis-à-vis fundamental physics in cold war America: the case of quantum optics," *Isis* **97**, 237–59.

radio regions. The first is the immensely greater energy of optical photons, as well as the much smaller number typically collected.[165] The second is that a radio antenna of diameter d can achieve an angular resolution proportional to d^{-1}, that is, limited only by diffraction (Section A.10). On the other hand, the resolution of even the largest optical telescopes, limited to ~1″ by turbulence in the earth's atmosphere, is typically far worse than their diffraction limit. As pointed out by Ekers,[166] these differences mean that one gains far less by building a larger optical telescope than a larger radio telescope; in fact, the signal-to-noise ratio of an optical telescope improves only linearly with its diameter d, but that of a radio telescope improves as d^2 (see Section A.10). This rapid increase in the sensitivity of radio telescopes with size means (1) capabilities to detect faint sources of radiation can progress much faster in the radio regime, and (2) a given antenna generally has a shorter useful lifetime before being outclassed by another. In contrast, more slowly improving sensitivity for optical research means that the largest optical telescopes have longer lifetimes. In terms of the era before 1953, as larger and larger antennas were built, the above effects contributed to radio astronomy progressing much faster than optical astronomy (Section 17.4).

---

[165] A more accurate statement is that the optical range in practice has far few photons per inverse-bandwidth time, or per phase-space volume.

[166] R.D. Ekers, "The convergence of optical and radio techniques," pp. 387–9 in *Optical Telescopes of the Future* (1978, ed. F. Pacini *et al.*, Geneva: European Southern Observatory).

18 • A New Astronomy

In the previous chapter, I have shown how radio practitioners and some astronomers were changed by engaging novel extraterrestrial radio phenomena. In contrast, this chapter focuses more on intellectual issues than social, although the two are often hard to disentangle. In the first section I recapitulate how astronomy's perception of the universe was changed during the postwar decade by an initial reconnaisance through the radio window. The second section describes how radio astronomy was to become only the first of many different non-optical windows to be opened in the following two decades – X-ray, infrared, ultraviolet, γ-ray – each socially and intellectually assimilated into astronomy more easily because radio astronomy had been a bellwether. In particular, the development of X-ray astronomy in the 1960s had many interesting parallels with that of radio astronomy earlier. I close by examining whether radio astronomy constituted a scientific revolution (answer: mostly not). Instead, I argue that radio astronomy (or more generally, the opening of the entire electromagnetic spectrum) was the mid-twentieth century's *New Astronomy*, with an impact every bit as important as New Astronomies of previous centuries such as those of Galileo or William Herschel or the first nineteenth century astrophysicists. Each of these New Astronomies was caused by researchers applying new technology to observing the sky, and each in its time profoundly transformed the perceived Cosmos.

18.1 NEW SCIENCE

Here is the history of science in the making: a very young branch of science comes rapidly to full growth and is confronted with one of the oldest fields: astronomy. We do not doubt that it is the same sun and the same galaxy that we now observe with radio telescopes. But well-known subjects may suddenly seem strange, new aspects appear, and the language becomes temporarily confused.

So did Hendrik Van de Hulst (1951:3) describe in his informal textbook on radio astronomy the effect of six years of postwar radio observations. In order to be reminded how strange the new phenomena were, the reader may wish to (re)read Section 1.1 on the "new sky" that radio astronomy created, one that profoundly differed from the previous (optical) sky (Fig. 1.1). Whereas the description in Chapter 1 was concerned solely with the observational appearance of the sky, in this section I also discuss salient aspects of the proffered explanations (by 1953) of the radio phenomena and how these changed astronomy.[1]

First, recall that (optical) astronomy in the decade after World War II was a far larger enterprise than radio astronomy, by at least a factor of 20 to 40 in terms of number of persons and published articles (Section 17.1.2). Although there were notable exceptions (Section 17.2), I agree with Edge & Mulkay (1976:285–6) that most optical astronomers were little affected by this weird upstart, and just kept on doing what they would have done anyway. This is not to say that astronomy was moribund at the time; on the contrary, it was actively building on key prewar developments. For example, the expansion of the universe had been established before the war by Edwin Hubble and colleagues, and now the new Palomar 200 inch and 48 inch Schmidt telescopes were opening up fainter and farther extragalactic realms, allowing detailed looks at galaxies and tests of cosmological ideas. This contributed, for instance, to Walter Baade's finding that the distances of all galaxies were twice as far as previously thought, thereby doubling the age of the universe. Also, in the late 1930s Hans Bethe had used the new nuclear physics to identify basic fusion reactions that could supply energy for the sun and most

---

[1] Good places to assess participants' views of the radio sky at this time are the first textbooks in radio and radar astronomy: Van de Hulst (1951), Lovell (1954) and Pawsey and Bracewell (1955). The first is available via the author's home page.

other stars. This work continued after the war, and, combined with more accurate measurements of stars' colors and luminosities and detailed numerical models of stellar interiors, led to greatly improved understanding of stellar evolution. Finally, before the war the importance of dust and gas in the interstellar medium had been established, and in 1948 a pervasive magnetic field threading through the Galaxy was inferred from the first measurements of starlight polarization.

Yet in retrospect these were no more important for the development of astronomy than several of the discoveries contemporaneously made by the small band of radio astronomers. In fact Harlow Shapley, one of the major astronomers of the first half of the twentieth century, as early as 1950 listed "microwave astronomy" as one of the fifteen outstanding developments of the previous half-century.[2] Lang and Gingerich (1979) later compiled what they deemed "seminal contributions" to astronomy and astrophysics over 1900–75: during the period 1945–53 fully 31% of their chosen papers were in radio astronomy. "Radio eyes," despite their intrinsically poor ability to see detail and determine positions, uncovered a universe that in many cases had not in the least been anticipated. Certainly the biggest surprises were the radio stars, which by 1953 seemed not stars at all, but various types of nebulae whose output in the radio overwhelmed that at optical wavelengths (Chapter 14). Some of these radio sources were so strong that they even outshone the radio sun at low frequencies (Fig. A.2). There were peculiar galaxies, sometimes taken to be in the process of collision (Cen A, Cyg A, Per A), and sometimes with their radio emission not coming directly from the optical object, but rather from propinquitous "empty" regions (Cyg A). There was a large galaxy (Vir A, M87) that seemed normal on most photographs except for a small "jet" feature poking out of its nucleus. And Cas A, the brightest radio source of all, corresponded on a photograph to only a few faint, fast-moving wisps. At least there was one object, Tau A, whose origin seemed secure, namely as the remnant of a supernova that exploded in AD 1054 (the Crab nebula).

But only 5–10 radio sources had optical identifications – hundreds of others remained "in the dark" and the experts vociferously debated their nature. If Cyg A were an exemplar, it was argued that the overall lack of optical correlates was understandable, for then the bulk of sources were too faint and distant to see even with the Palomar 200 inch telescope. This implied, as Martin Ryle began to argue, that new surveys of radio sources would probe regions of the universe inaccessible to optical astronomers. But had not Ryle only recently been arguing that the radio stars were all nearby, some new kind of dark star? Others, such as Robert Hanbury Brown and Bernard Mills, argued that there were *two* classes of radio sources, some definitely within our Galaxy, and others for which it was impossible to distinguish whether they were nearby or very distant. A 1953 census of the optically identified sources found four to be galactic and five extragalactic. As a comparison, several astronomers and radio astronomers pointed out that even a list of the brightest *optical* objects was a potpourri including three that were outside our Galaxy: sun, moon, five planets, stars of many different types, the Orion nebula, star clusters, the Large and Small Magellanic Clouds, and the Andromeda nebula.[3] Radio astronomy too was suffering from an old, intrinsically astronomical problem: distances have always been notoriously difficult to determine, confined as we are to our observation post on this planet. For instance, the radio star situation was not unlike the debate in the first quarter of the twentieth century over the distances and nature of the spiral nebulae. Were they all members of the Galactic System (as it was then called), perhaps collapsing regions of star formation, or were they separate "island universes" similar to our System?[4,5]

It was comforting to have *some* optical identifications, but sobering that, even for those, no consensus had arisen as to how the radio waves were generated (Chapter 15). Various proposals, such as free–free radiation or plasma processes, did not satisfactorily explain

---

[2] H. Shapley (September 1950), "Astronomy [over the past fifty years]," *Scien. Amer.* **183**, 24–26.

[3] Baade:Ryle, 1 July 1952, file 4 (uncat.), RYL; notes for talk to Cambridge Philosophical Soc., 16 February 1953, file 8 (uncat.), RYL.

[4] See *The Expanding Universe: Astronomy's "Great Debate," 1900–1931* by R.W. Smith (1982) (Cambridge: Cambridge University Press). Also R.W. Smith (2009), "Beyond the Galaxy: the development of extragalactic astronomy 1885–1965, Part 2," *J. Hist. Astron.* **40**, 71–107.

[5] In the 1990s astronomy again went through a major episode of this sort. There were intense debates about the distances to a population of "γ-ray bursters" detected by satellites – were they within the Galaxy or at cosmological distances? Again, it was ultimately optical identifications that decided the issue in favor of extragalactic locations.

the radio spectra or luminosities. Only towards the end of the postwar decade did some in the Western community begin to pay attention to a suggestion pushed especially by Soviet theorists: synchrotron radiation caused by purported relativistic electrons spiraling around magnetic field lines, either in special locations (the radio sources) or in the Galaxy as a whole. The latter would explain another mysterious radio phenomenon: strong background radiation covering the sky (Figs. 1.1 and 15.1) – at the lowest frequencies brighter even than the radio sun, it obviously had to do with the Galaxy, but its origin was unknown. Was it the integrated effect of millions or billions of individual radio stars, or somehow the product of interstellar gas far from the Galaxy's plane (Fig. 15.2)? This lack of an astrophysical mechanism was bothersome, as noted by Australian theorist James A. Roberts (1954:398):

> The mechanism of generation remains unknown for emission from the disturbed sun, the radio star emission and the background radiation. This is a very great handicap to radio astronomy and seriously impairs the power of the new technique.

His Soviet colleague Iosif Shklovsky (1953a:35) was more optimistic:

> In the course of the last few years the problem of cosmic radio emission has become one of the central problems of astronomy and physics. The work done by various investigators has taught us much about the nature of this radiation. The time is not far distant when the whole grandiose picture of cosmic radio emission will become clear.

Radio waves also promised not only to probe *farther* than previously possible, but also *through* interstellar dust, which had until then hidden most of the Milky Way's territory. Led by Jan Oort, the Dutch employed the newly discovered 21 cm hydrogen line to probe the structure and dynamics of the *entire* Galaxy, something only dreamt of before (Chapter 16).

Radio study of the sun was the one major aspect of radio astronomy that was not transformative, but rather an addition to the catalog of fascinating phenomena already known for the optical sun (Chapter 13). Radio astronomers provided confirming evidence that the corona was far hotter (10^6 K) than the photosphere (6000 K), and found a variety of new types of explosive events and correlations with the 11-year cycle of solar activity, but most findings were amenably accommodated in the context of previously known optical activity. The sizes of the radio bursts, however, dwarfed those for optical flares – the radio quiet sun could briefly become thousands or millions of times stronger than usual.

Finally, investigators bounced radar off the ionized trails of meteors, allowing meteor particles pummelling earth to be studied far more efficiently than had previously been possible, not only at night, but throughout the daytime (Chapter 11). Such studies enabled the Jodrell Bank group to settle a longstanding controversy when they established that in fact no meteors originated in interstellar regions outside the solar system.

Radio astronomy's universe, filled with electromagnetic waves a million times longer than optical waves, differed from the optical universe in several fundamental ways. Rather than the stars that were the focus of optical astronomy, the space *between* the stars came to the fore – it was full of material promiscuously emitting radio photons. The radio universe was full of high energies and catastrophes, as opposed to the optical universe of gently burning stars and gracefully pinwheeling spiral galaxies. As Oort (1978:47–8T) later put it:

> Before the advent of radio astronomy, it had not at all been realized by astronomers that very much in the universe is of an explosive nature.

Rather than regions near thermodynamic equilibrium at very high temperatures, such as the interiors and atmospheres of stars, radio observations emphasized other aspects of the universe. Often these were low-density, ionized interstellar regions with weak or no optical emission, or regions permeated with dust that only radio waves could effectively penetrate. Theorists analyzed nonthermal processes originating in tenuous regions of energetic, charged particles, often threaded by magnetic fields. On the other hand, radio waves were also ironically found to be preferentially emitted by *cool* interstellar regions, and in particular the 21 cm line revealed such a medium, little known before. This radio line remained unique for a long time, however, in distinct contrast to the rich array of optical atomic spectral lines that astronomers had long exploited.

The ~1953 endpoint of the present study must not be construed as a sign that by that time radio astronomy was finished with major discoveries. One measure

Table 18.1 *The opening of new astronomical windows*

RADIO 1 mm – 50 m

	Background	Sun	Cyg A
Detection	1932	1942	1946
Wavelength	15 m	5 m	5 m
Platform	Ground	Ground	Ground
Institution	BTL	AORG	AORG
Publication	Jansky 33	Hey 46	Hey+ 46

INFRARED 0.7 – 1000 μm

	Sun	IRC+10216
Detection	1800	1965
Wavelength	~1 μm	2.2 μm
Platform	Ground	Ground
Institution	–	Caltech
Publication	Herschel 1800	Neugebauer+ 65

ULTRAVIOLET 0.10–0.35 μm = 1000–3500 Å

	Sun	Sun
Detection	1946	1949
Wavelength	0.2–0.3 μm	1200 Å
Platform	Rocket	Rocket
Institution	NRL	NRL
Publication	Baum+ 1946	Friedman+ 51

EXTREME ULTRAVIOLET 0.01–0.10 μm = 100–1000 Å

	Background	HZ 43
Detection	1974	1975
Wavelength	40–150 Å	50–1500 Å
Platform	Rocket	Rocket
Institution	UCB	UCB
Publication	Cash+ 76	Lampton+ 76

X-RAY 0.004–10 nm = 0.04 – 100 Å ; 0.15–300 keV

	Sun	Sco X-1	Background
Detection	1948–49	1962	1962
Wavelength	8 Å	2–8 Å	2–8 Å
Platform	Rocket	Rocket	Rocket
Institution	NRL	ASE/MIT	ASE/MIT
Publication	Burnight 49; Friedman+ 51	Giacconi+ 62	Giacconi+ 62

γ-RAY < 0.004 nm (< 0.04 Å) ; >300 Kev

	Sun	Background	Crab nebula
Detection	1958	1962/68	1967
Wavelength	0.025–0.06 Å		<0.3 Å
Platform	Balloon	Rocket/Satellite	Balloon
Institution	Minnesota	MIT	Rice
Publication	Peterson+ 59	Kraushaar+ 62; Clark+ 68	Haymes+ 68

Radio
Jansky (1933b)
Hey (1946)
Hey, Parsons, and Phillips (1946)

Infrared
Herschel, W. (1800). "Investigation of the powers of the prismatic colours to heat and illuminate objects," *Phil. Trans. Roy. Soc.* **90**, 255–83.
Neugebauer, G., Martz, D. E. and Leighton, R. B. (1965). "Observations of extremely cool stars," *Astrophys. J.* **142**, 399–401.

Table 18.1 *contd*.

Ultraviolet

Baum, W. A., Johnson, F. S., ... and Tousey, R. (1946). "Solar ultraviolet spectrum to 88 kilometers," *Physical Rev.* **70**, 781–2.

Extreme Ultraviolet

Cash, W., Malina, R., and Stern, R. (1976). "An observation of the diffuse soft X-ray/extreme-ultraviolet background," *Astrophys. J. (Letters)*
 204, L7–11.

Lampton, M., Margon, B.,... and Bowyer, S. (1976). "Discovery of a nonsolar extreme-ultraviolet source," *Astrophys. J. (Letters)* **203**, L71–4.

X-ray

Burnight, T. R. (1949). "Soft X-radiation in the upper atmosphere" [abstract], *Physical Rev.* **76**, 165.

Friedman, H., Lichtman, S. W., and Byram, E. T. (1951). "Photon counter measurements of solar X-rays and extreme ultraviolet light,"
 Physical Rev. **83**, 1025–30.

Giacconi, R., Gursky, H., Paolini, F., and Rossi, B. (1962). "Evidence for X-rays from sources outside the solar system," *Physical Rev. Letters* **9**,
 439–43.

γ-ray

Peterson, L. E. and Winckler, J. K. (1959). "Gamma-ray bursts from a solar flare," *J. Geophysical Res.* **64**, 697–707.

Haymes, R. C., Ellis, D. V., Fishman, G. J. *et al.* (1968). "Observation of gamma radiation from the Crab Nebula," *Astrophys. J. (Letters)* **151**,
 L9–14.

Kraushaar, W. L. and Clark, G. W. (1962). "Searching for primary gamma rays with the satellite Explorer XI," *Physical Rev. Letters* **8**, 106–9.

Clark, G. W., Garmire, G. P., and Kraushaar, W. L. (1968). "Observation of high-energy cosmic gamma rays," *Astrophys. J. (Letters)* **153**, L203–7.

comes from Lang and Gingerich's (1979) source book of seminal contributions to all of astronomy and astrophysics. The fraction of radio astronomy papers in their collection actually increases from 31% in 1945–53 to 49% over 1954–75. Secondly, consider Harwit's (1981) tabulation of 43 fundamental discoveries of cosmic phenomena. Of the seven of these associated with radio, only one occurred during 1945–53 (radio galaxies) and the rest came later.

During the 1953–70 period, many more fundamental discoveries occurred as radio astronomy went from strength to strength (Edge and Mulkay 1976; Hey 1973). Thousands of radio sources were cataloged, vehemently argued about, and shown to be important in understanding the universe as a whole: did the universe begin with a Big Bang or has it always been in a Steady State (Edge and Mulkay 1976:134–223)? Radio positions ultimately good to ~1–3″ allowed the identification of quasars in the early 1960s, objects that looked like stars on a photograph, but were at cosmological distances and pouring out far more power than even Cyg A (Robertson 1992:227–50). Then in 1965 the cosmic microwave background was accidentally discovered and accepted as convincing proof of an early hot universe consistent with a Big Bang (Kragh 1996).

In 1967 sources that emitted precise radio pulses were serendipitously discovered – these soon became understood as arising from highly beamed radiation from pulsars, which were fast-spinning, magnetized neutron stars. Also by the late 1960s, microwave spectral lines, some exhibiting amazing maser emission, were detected in the interstellar medium from a variety of unexpected species: OH, NH_3, H_2O, CO, CN, H_2CO, etc. Interstellar chemistry was born. Finally, in a different vein, about 1960 the field of SETI (Search for Extraterrestrial Intelligence) began with the first search for radio signals from possible civilizations on planets circling other stars (Dick 1996:414–31).

For objects within the solar system, radar astronomy came to maturity first with detailed mapping of the moon, then studies of Venus and Mercury (Butrica 1996). Radio astronomers also found that Jupiter emitted very intense low-frequency bursts and that Venus was far hotter than previously thought.

In sum, the ~1953 end of this book's story may have found radio astronomy at a watershed (Section 17.1.2), but hardly at the end of its productive years.

18.2 FIRST OF MANY NEW SPECTRAL WINDOWS

> It's clear to me that you cannot know the physics of the universe without exploring every spectral channel. And as long as you have just one, namely optical, you'll never really get anywhere, and therefore adding the second one was the definitive [step]. But it needn't, has not, cannot stop there.
> (Theoretical astrophysicist Philip Morrison [1976:14T])

> [My involvement with astrophysics] happened on the wave of a truly great process of transformation of astronomy from optical to "all-wave." In the course of this process many physicists, radio engineers, and other professionals took interest in new astronomical methods and problems and "came" to astronomy (often irritating professional astronomers by their poor knowledge of classical astronomy and even of proper terminology). Some of the neophytes became real astronomers, while others did not give up their previous specialties.
> (Vitaly Ginzburg [1984:289])

This book has so far been tacitly defining observational astronomy in terms of traditional (optical) astronomy and the new radio astronomy, but in this section I briefly consider the other portions of the electromagnetic spectrum. From the long perspective of a twenty-second-century historian, the most important aspect of the emergence of radio astronomy in the mid-twentieth century will be as a harbinger of the two profound technical developments in twentieth-century observational astronomy: the opening of the electromagnetic spectrum and the concomitant move of instrumental platforms into space. This is the twentieth century's "New Astronomy" (Section 18.3.2).[6] Most of the spectrum – infrared, ultraviolet, X-ray and γ-ray – could not be conveniently observed until Sputnik I initiated the space age in 1957. This is because the earth's atmosphere blocks out all wavelengths except for the optical window, the radio window, and portions of the infrared band (Fig. A.1 shows the transparency of the earth's atmosphere across the entire spectrum). In order to observe the universe at most wavelengths, one has to lift a platform above the bulk of the atmosphere via balloon, rocket, or orbiting satellite. With readily available US military and NASA interest in the upper atmosphere and space in the 1945–75 period, especially after Sputnik and the creation of NASA in 1958, money flowed to American astronomers and physicists to explore the new electromagnetic windows. American researchers, who dominated these fields for decades, took advantage of the huge infrastructure of Cold War launch facilities and rockets, as well as technology transfer from other NASA and military space programs (Lovell 1977; Smith 1997:56–60).[7]

In the following I will very briefly examine the story of how each wavelength regime became

---

[6] A good case can be made for the existence of a *third* New Astronomy of the second half of the twentieth century, namely the impact of digital computers on the development of both observational and theoretical astronomy. Computers have also made possible an important hybrid way of knowing: *numerical simulations*, which have become a major weapon in the modern astronomer's armamentarium. These developments have been little studied and deserve detailed historical attention.

[7] Later, US *optical* astronomy also hugely benefited from military technology, for example, in the 1970s design of the Hubble Space Telescope (launched in 1990) [R. W. Smith (1993), *The Space Telescope: a Study of NASA, Science, Technology, and Politics* (Cambridge: Cambridge University Press)]. The earlier case of the photoelectric photometer, developed during World War II, has been discussed in Section 17.2.2.

a significant presence in astronomy. These provide instructive comparisons with what we have learned about radio astronomy, which preceded them all. In particular, I will argue that early X-ray astronomy in the 1960s had many interesting parallels with early radio astronomy (Section 18.2.2). Moreover, the prior existence of radio astronomy and its many contributions to astronomy paved the way for an easier acceptance of X-ray astronomy and other novel techniques (Edge & Mulkay 1976:281–2; Hirsh 1983:57). As (optical) astronomer David Dewhirst (1971:16T) noted about radio astronomy:

> It has made people think in much broader terms – that there might for example be other, so far undetected, objects which will not necessarily be detected at radio wavelengths – they may be detected at X-ray wavelengths or they may be γ-ray sources, or whatever. There is this general realization that the universe is a much more complicated place than we had supposed it to be [in 1950], and this has come entirely from radio astronomy.[8]

18.2.1 Beginnings of infrared, ultraviolet, γ-ray and X-ray astronomies

Table 18.1 summarizes the first significant discoveries in each window, listed in order of decreasing wavelength (or increasing frequency, increasing photon energy).[9,10] The definitions of the windows, especially for the high-energy end, are somewhat arbitrary, and in practice wavelength/energy boundaries have often become defined according to techniques (i.e., the type of detector or platform), rather than according to the nature of the astrophysical emitters.[11] The columns for each window register first detections of the sun, of a general background, and of a nonsolar source.[12,13]

I now give very brief accounts of the beginning of astronomical observations in each of the wavelength windows.

Infrared Because the atmosphere allows infrared radiation to reach the ground in many discrete windows (e.g., at 2.2 μm in wavelength, well above the optical window of ~0.4–0.7 μm), solar infrared astronomy extends as far back as 1800 when William Herschel first demonstrated that "calorific rays" emanated from the sun. Observations of the moon, planets, and bright stars eventually followed, but they were merely extensions of the photometry and spectroscopy done on familiar objects at optical wavelengths – no one could afford to "blindly" search the sky for possible infrared objects that did not visually shine. Indeed, the observers and their telescopes were usually the same as those doing optical astronomy. Sensitivity of the infrared detectors installed on telescopes improved only slowly, but by the early 1960s new types of detectors became available and ever fainter objects came within reach. Harold Johnson placed infrared photometry of known stars on a sound quantitative basis at wavelengths as long as 10 μm, while Gerald Neugebauer and Robert Leighton of the California Institute of Technology carried out the first large, sky-sweeping infrared survey at 2.2 μm. The first distinctly infrared class of object was typified by IRC+10216, a bright infrared, dust-enshrouded star that was barely visible at optical wavelengths.

Ultraviolet On the other side of the optical window from the infrared are ultraviolet wavelengths – here there are no gaps in the atmosphere's blockage, and so nothing was known of even the sun in the ultraviolet until balloons were lofted before World War II

---

[8] Also partially cited in Edge and Mulkay (1976:281).

[9] For convenience, I use the term *window* for each of these wavelength regimes, even though it is only for ground-based astronomy that one actually observes through a metaphorical atmospheric window. The present discussion does not include other, non-electromagnetic ways of gathering data from the universe, such as via meteorites, cosmic ray particles, neutrinos, and (perhaps some day) gravity waves.

[10] The optical window has been omitted because of an absence in the historical record as to exactly who first visually detected the sun and the stars.

[11] Again we see technoscience at work (Section 17.4).

[12] Harwit's *Cosmic Discovery* (1981) discusses many of the astronomical discoveries associated with these new electromagnetic windows. Besides the primary literature, I have also used in this section: Sinton (1986), DeVorkin (1992), Hevly (1987), Trimble (2005), Hirsh (1983), and R. Giacconi (1974), "Introduction," pp. 1–23 in *X-Ray Astronomy* (eds. R. Giacconi and H. Gursky, Dordrecht: Reidel).

[13] Lack of an entry (e.g., for a general infrared background) indicates that this component was not part of the early years of the opening of the particular wavelength band.

to modest heights.[14] But credit for first observing the ultraviolet sun usually goes to a group led by Richard Tousey at the US Naval Research Lab (NRL). In 1946 a war-booty German V-2 rocket carried film to 88 km altitude and safely delivered back information about the solar spectrum at wavelengths as short as 0.2 μm. As discussed in Section 10.1.2 regarding the beginning of NRL postwar radio astronomy, NRL's interest in the upper atmosphere arose because of the ionosphere's influence on communications and radar. The sun was germane because of the suspected role of its ultraviolet and X-ray emission in causing the (variable) ionization of the ionosphere. The chance to learn more about rocketry was of course another bonus for a military lab. For the next three decades NRL was a leader in ultraviolet and X-ray astronomy. Ultraviolet detection of stars and other objects began only in the late 1950s.

Extreme ultraviolet This portion of the spectrum (0.01 to 0.10 μm) is considered separately from the ultraviolet for astrophysical reasons that are also of historical interest. The extreme ultraviolet was the only portion of the spectrum not funded for investigation in the 1960s because it turns out that cool hydrogen, the most common atom by far in the interstellar medium, readily absorbs photons in this range and therefore the consensus was that interstellar space was largely opaque to these wavelengths, even if stars or other sources were emitting. By the 1970s, however, the interstellar medium was shown to be reasonably transparent in many directions (because the cool hydrogen was clumpy) and various above-the-atmosphere observations went forward, led by Stuart Bowyer of the University of California at Berkeley.

Gamma-rays Extraterrestrial γ-rays have a history of investigation extending back to just after 1900, shortly after they were discovered in the lab. Once evidence accumulated that the earth was bathed in a "cosmic radiation" (later named "cosmic rays") of energetic entities, for decades physicists argued as to whether the cosmic rays were particles or γ-ray photons. There were thus many high-altitude experiments both before and after World War II, although detectors had little ability to discriminate the direction of arrival of the cosmic rays. By the 1950s it had been established that <1% of cosmic rays were γ-rays. Technically demanding at the receiving end and with very few photons supplied by nature at the "emitting end," γ-rays were the last segment of the spectrum to be conquered, in 1958 when a solar flare was detected on a high-altitude balloon. A few years later evidence for a weak general background (all of 22 photons!) was put forward (and confirmed in 1968), and in 1967 the first discrete source other than the sun was established – none other than the Crab nebula.

X-rays Most astronomers would agree that this brand of astronomy, which observes the universe using wavelengths 1000 times shorter than optical wavelengths, has proven the most fruitful of all those that followed radio. Some, if forced to make an odious comparison, would rate it even more valuable than radio astronomy. Did the many discoveries made with radio waves and X-ray waves have anything to do with being sufficiently removed (in wavelength) from the optical window, thus allowing a truly novel harvest (Harwit 1981)? X-ray astronomy is also the only window that has received detailed historical treatment, in the form of Richard Hirsh's *Glimpsing an Invisible Universe: the Emergence of X-ray Astronomy* (1983).[15]

An NRL group led by Herbert Friedman made the initial solar X-ray observations, barely lost out on detecting the first nonsolar source, and thereafter was a longtime leader in the field. The X-ray sun was first detected from a V-2 rocket by T. R. Burnight of NRL and then more definitively by others in Friedman's group in 1948–49.[16] Further advances were arduous; once again, before weaker sources could be detected, physicists and engineers had to improve detector technology significantly. The key steps were taken in the early 1960s at American Science & Engineering

---

[14] My definition here for ultraviolet is wavelengths in the 0.10–0.35 μm range, those not admitted by the earth's atmosphere. The human eye, however, cannot detect wavelengths below 0.40 μm, leaving the band 0.35–0.40 μm as invisible to the eye and yet observable from the ground (as it was for the sun even in the nineteenth century).

[15] DeVorkin (1992), Hevly (1987), and Smith (1997) treat the early rocket ultraviolet and X-ray *solar* observations, but do not cover the nonsolar era.

[16] See note 2.11 regarding a brief attempt by G. E. Hale to detect solar X-rays in 1897, only two years after Wilhelm Röntgen discovered X-rays in the laboratory.

(ASE), an offshoot research firm of the Massachusetts Institute of Technology (MIT). MIT cosmic ray physicist Bruno Rossi had the exploratory urge:

> The initial motivation of the experiment ... was a subconscious trust of mine in the inexhaustible wealth of nature, a wealth that goes far beyond the imagination of man. This meant that, whenever technical progress opened a new window into the surrounding world, I felt the urge to look through this window, hoping to see something unexpected.[17]

Under Rossi's tutelage an ASE team led by Riccardo Giacconi developed greatly improved detectors (about a factor of 100 better than anything else previously flown) and calculated the likelihood of detecting something like the Crab nebula or α Centauri (the nearest star). Despite *extremely* discouraging numbers, they pressed on, following Rossi's doctrine. They sold the US Air Force on the idea of flying a rocket to look for possible X-ray fluorescence on the moon, but "on the side" they said they would also take a look around the sky. During their first successful launch in 1962, up to a peak altitude of 230 km, they had all of six minutes to gather X-ray photons. This was nevertheless sufficient to discover (a) a general X-ray background around the sky, (b) an apparent source somewhere in the constellation of Scorpius (later called Sco X−1), and (c) nothing from the moon.

In the next few years the field moved frightfully fast as NRL and ASE raced, just as the English and Australians had done during the postwar years of radio astronomy. Sco X-1's reported position wandered by many tens of degrees during subsequent rocket flights, and only when the position settled to $\sim 1-2'$ accuracy did an optical identification materialize in 1966. Sco X-1 turned out to be a strange blue 13th magnitude star that varied in X-ray intensity on timescales of minutes and was giving off 1000 times more energy in X-rays than at optical wavelengths (this ratio for the sun, even during a large X-ray flare event, was never more than 10^{-6}). The second X-ray source, Taurus X-1, turned out to be the Crab nebula – in 1964 the NRL group performed a beautiful experiment by timing a rocket's trajectory so that it could observe Tau X-1 just as the moon was occulting it. The "knife-edge effect" of the advancing moon allowed the source's size to be measured as $\sim 1'$ and its position pinpointed to the center of the nebula. Two years later, the first *extragalactic* sources were found, associated with the radio sources Cygnus A and Virgo A (Messier 87). By the late 1960s there were several dozen X-ray sources known, of which only six were optically identified.

In subsequent years, especially with the launch of the Uhuru survey satellite in 1970, X-ray astronomy, besides complementing studies of known objects, uncovered a new world of high-energy processes associated with compact, collapsed objects such as neutron stars and black holes – for example, the first solid evidence for the existence of a black hole came from the source Cyg X-1 in the 1970s.

By the 1970s optical astronomy was just one among many kinds of astronomies, yet it continued to dominate the others in terms of numbers of practitioners, publications, and facilities (and indeed still does today). Is there any astrophysical reason for this, or is it the socio-physiological phenomenon that taxpayers, bureaucrats, and philanthropists, each equipped with two visual receptors of their own, prefer to support what they can best intuitively understand? Might this be true even for astronomers themselves? I believe there is merit to this "we all have eyes" notion, but the optical range is also critically diagnostic of the composition and physical state of matter because optical photons often correspond to the energy states of atoms, common constituents of the universe.[18] Furthermore, optical astronomy can be done from the ground, making it, like radio studies, far cheaper and more accessible than the other spectral bands that require space platforms.

18.2.2 Comparing early X-ray and radio astronomies

Parallels between the first decade of X-ray astronomy and that of radio astronomy are striking. The X-ray investigators, like their radio counterparts, were

---

[17] B. Rossi (1977), "X-ray astronomy," *Daedalus* **106**, 37–58 (quotation from p. 39).

[18] Modern astronomy, however, believes that atoms (or baryons more generally) comprise only a small fraction of all the matter and energy in the Cosmos. If this result holds, the bias fashioned by past astronomy, in particular optical astronomy, has been profound.

migrants from physics and electronics, skilled in instrumentation, and only gradually turned into astronomers. In fact, Hirsh (1983:57) gives evidence that early X-ray astronomers looked at themselves much like the early radio astronomers had. Both radio and X-ray workers were vexed by the problems of explaining observed background emissions. Both fields were guided little by theoretical concepts, and largely took an empirical approach dictated by ignorance. Both fields looked to optical astronomy for guidance in understanding their observations, and neither became a separate discipline. In pursuit of optical identifications and the resultant legitimacy for discrete sources (Section 14.4.4.2), both devised clever techniques to secure more reliable and more accurate positions, as well as angular sizes. The magnitudes of these tasks can be estimated by noting that it took many years to identify optical counterparts for even the strongest sources in the two cases, Cas A and Sco X-1. In both the radio and X-ray cases the first accepted optical identification was the Crab nebula, which acquired the moniker Tau X-1 as well as Tau A (notice also the adoption of a nomenclature for X-ray sources similar to that in the radio). Also in both cases, the fraction of identified sources was discouragingly small. Finally, X-ray astronomy seemed to be ratifying radio's findings that the universe was a violent place, full of explosions, shocks, strange beasts, and turmoil.[19]

There were also important differences between early radio and early X-ray astronomy. For X-rays the sun was observed first (in fact radio astronomy was the sole window *not* opened up by the sun). Also, the first nonsolar X-ray emission was *purposely* sought, whereas the general radio background was discovered serendipitously by Jansky. X-ray astronomy, unlike radio, was dominated by American institutions – Hirsh finds that 72% of all articles in the period 1962–72 were from the US, where the might of NASA was available to provide access to the region above the atmosphere, the *sine qua non* of X-ray astronomy. Also from the start, available funds and sizes of X-ray projects were large compared to those of the early radio groups. For X-rays (and indeed for all of the other space-based spectral windows) big bucks available from US government and military largesse *successfully* led to US dominance of the field, distinctly unlike the postwar radio case where large funding did not lead to innovative science (Section 17.3.5).[20,21] Finally, X-ray astronomy became more quickly integrated into astronomy than did radio astronomy, for which the process ended up taking 10–20 years. For example, it took only 5–7 years for the X-ray investigators in the 1960s to migrate their publishing from physics to astronomy journals (Hirsh 1983:67).

Radio astronomy paved the way for X-ray astronomy and indeed for each of the communities associated with the various non-optical spectral regions. Several factors aided (optical) astronomers in assimilating X-ray results despite their astounding consequences and despite their origin in a spinning rocket nose-cone that looked nothing like an optical (or even a radio) telescope. The incorporation of X-ray astronomers and their astronomy was accelerated because of the astronomers' prior experience with the "case" of radio, and indeed the presence by the 1960s of a sizable radio contingent amongst astronomers.[22] On the one side, astronomers had already learned that entirely novel celestial phenomena could emerge from the opening of a new wavelength range, and on the other X-ray researchers concluded from the earlier example of radio astronomy that assimilation into astronomy was in their best social and intellectual interest. So now there was a *third* type of astronomer, the *X-ray astronomer*, again defined by

---

[19] The discoveries in both windows also largely conform to Harwit's (1981:18–22) characterization of fundamental discoveries in astronomy: they are based on technical innovation (Section 17.4), happen soon after they are technically possible, are done by non-astronomers, are serendipitous, and involve military equipment that has been personally built by the discoverer.

[20] Hirsh (1983:127) points out that Ryle (1950:186) saw little future in such "brief and expensive" rocket experiments, an attitude consistent with his Cavendish style that likewise led him to deride the large budgets of American radio astronomy projects (Sections 8.5.1 and 17.3.5).

[21] On the other side of the Cold War, little is known about the contemporary relationship of Soviet physicists and astronomers to the space and military agencies of the USSR – why did so little astronomy from above the atmosphere emerge from the Soviet Union during 1960–80?

[22] In the US the size of the radio astronomy community by the late 1960s had grown to ~17% of all astronomers and that of X-ray astronomers to ~8%. [T. F. Gieryn (1981), "The ageing of a science and its exploitation of innovation: lessons from X-ray and radio astronomy," *Scientometrics* 3, 325–34]

wavelength range. The category of *optical* astronomers, forced earlier by the emergence of radio astronomy (Section 17.2.1), was needed *a fortiori*.

18.3 A NEW ASTRONOMY

18.3.1 Was radio astronomy a revolution?

The answer to this question depends on what one means by the term *revolution*. Historians have defined several types of scientific revolutions, of which the most famous is Kuhn's.[23] However, I agree with Edge and Mulkay (1976:386–94) and with Gingerich (1984:404–6) that the case of radio astronomy does not at all fit well with Kuhn's scheme of one paradigm guiding "normal science" within a field, which then reaches a crisis because of growing anomalies, eventually leading to the development of a new paradigm. The discoveries of radio astronomy, as well as its technology and style, certainly did influence astronomy as a whole, leading to modification of its paradigm(s), but there was no crisis, great resistance, or revolutionary upheaval in the Kuhnian sense. Neither was there a shift analogous to that experienced in one's visual gestalt when gazing at the famous ambiguous drawing of an old/young woman. As Gingerich has argued, it was rather that radio astronomy expanded the scope of astronomical knowledge into realms previously undreamt, where there was no paradigm to overthrow. Gingerich makes a nice analogy with the old Cinerama movies, where for a few minutes the film was projected on to a normal size screen, and then suddenly the curtains parted and the viewer was wonderfully caught up in action on a screen three times broader.

Perhaps Kuhn's ideas fall short because they deal primarily with profound changes in the *theoretical* content of science, as opposed to radical technological and instrumental developments.[24] Kuhn also emphasizes profound changes occurring *within* a discipline due to its own practitioners (not outsiders). As an alternative, Hacking has defined "big revolutions" characterized by new institutions, interdisciplinarity (or "pre-disciplinarity"), major social changes, and a significant change in attitudes towards the "texture" of the world.[25] This more sociological concept allows for a greater variety of causes for revolutions, such as new technology or mathematical techniques (e.g., the "probabilistic revolution" in the mid-to-late nineteenth century that changed the face of physics [especially thermodynamics] by its use of statistical methodology). Might radio astronomy qualify for this brand of revolution? Although the fit is better, it is not complete. There were indeed new, unlikely institutions doing astronomy (such as the Radiophysics Laboratory in Sydney), a confusion of disciplinary identity (Section 17.2), and certainly a different texture to the sky (Fig. 1.1). But in the postwar decade astronomy as a whole did not undergo major social changes caused by the radio astronomers' exotic technology, vastly different backgrounds, and fast-paced research style. On a longer time scale, however, such major social changes would eventually happen (beyond the 1950s), with the influence of radio abetted by many other factors, including the opening of other spectral windows.

18.3.2 The twentieth century's "New Astronomy"

If radio astronomy does not represent a revolution in the long history of astronomy, what exactly was it? I would argue that, no less significantly, it was instead one in a series of transformative events, or "New Astronomies," brought on by empirically applying new observing technologies to the firmament. As Ryle presciently noted in a 1946 talk: "I have no doubt that there is a good future in this new astronomy."[26] Let us compare the New Astronomy of 1932–53 discussed in this book with three earlier transformations associated with the names of Galileo, Herschel, and the "astrophysicists" of the late nineteenth century.

Galileo Galilei produced a New Astronomy (or a New Philosophy, as it came to be called) analogous in important ways to radio astronomy in that it too was

---

[23] T. S. Kuhn (1970, second edn.), *The Structure of Scientific Revolutions* (Chicago: University of Chicago Press).

[24] For example, see D. Baird (1993), "Analytical chemistry and the 'big' scientific instrumentation revolution," *Annals of Science* **50**, 267–90.

[25] I. Hacking (1987), "Was there a probabilistic revolution 1800–1930?", pp. 45–55 in *The Probabilistic Revolution. Vol. 1. Ideas in History* (eds. L. Krüger, L. J. Daston, and M. Heidelberger) (Cambridge: MIT Press).

[26] M. Ryle, 13 November 1946, talk on "Cosmic Noise" given to the Cambridge University Astronomical Society, file 8 (uncat.), RYL.

based on a re-purposed recent technology.[27] I view Galileo's short masterpiece *Sidereus Nuncius* (1610)[28] as equivalent to an imaginary grand amalgamation of all of the seminal papers of early radio astronomy. Completely unsuspected phenomena – mountains on the moon, moons of Jupiter, phases of Venus, etc. – were reported based on his telescopic observations of the heavens. This was an "artificial revelation," highly suspicious to those who were asked to trust the new "spyglass."[29] So were the results of early radio astronomy also an artificial revelation to many. Recall Peter Scheuer's remarks in Section 17.2.1.1 about the mistrust of some (optical) astronomers toward artificially revealed results emanating from a radio interferometer. Reciprocal attitudes from the radio side (as from Galileo) also played a role in misunderstanding and conflict, as Cambridge optical astronomer Michael W. Ovenden (1976:75B:225) later recalled:

> Just as Galileo ran afoul of some of the conventional astronomers of his day, in the same way the radio people had this spirit that no one had done astronomy until they came along. That naturally antagonized people who had been steadily working for twenty or thirty years on the old astronomy. I look upon it as the "Galileo phase."[30]

Even at the time, some participants invoked a Galilean analogy: for example, J. P. M. Prentice, the amateur meteor astronomer who worked with Bernard Lovell in the early days of meteor radar (Section 9.2), remarked at a 1946 meeting of the Royal Astronomical Society:

> It is obvious that this beautiful new technique represents a major advance in this field of observation; indeed we are like the first possessors

of the telescope, unexpectedly armed with new powers of observation.[31,32]

Another transformative New Astronomy based on new technology occurred in the person of William Herschel in the late eighteenth century.[33] Herschel developed the technique of manufacturing huge speculum reflecting telescopes, and then made remarkable observations and insightful interpretations that carried astronomy from the solar system to the sidereal universe beyond. In fact the state of (nonsolar) radio astronomy in ~1954 was not unlike the state of Herschel's "construction of the heavens" at the end of his career.[34] The sky distributions of stars and nebulae, and of radio sources, had been established with new technologies in both cases. In both, distances were poorly known or unknown to these objects, and very little was known of the physics that made them tick. Undeterred, both Herschel and the radio astronomers soldiered on and, as best they could, made estimates of distances and luminosities. Herschel used his "star gauge" technique to present a first map of our Galaxy and the radio astronomers likewise tried to sort out the galactic distribution of the first few hundred radio stars. One difference between the legacies of Herschel and of radio astronomy, however, was that the former's

---

[27] The import of Galileo's new worldview of course went far beyond astronomy. My comparison is not to imply a similar importance in this regard for radio astronomy, but rather to emphasize the profound effect that a new technology can have on one's understanding of the sky.

[28] See the excellent translation and introduction by A. Van Helden: *Sidereus Nuncius, or, The Sidereal Messenger* (1989, Chicago: University of Chicago Press).

[29] The insightful term *artificial revelation*, meaning knowledge acquired via an artifact (instrument), comes from Price (1984:9).

[30] In interview Fred Hoyle (1988:172A:650) also emphasized the hubris of radio astronomers regarding their successes.

[31] J. P. M. Prentice (1947), "The visual observations of the Giacobinids, 1946," record of the RAS meeting of 13 December 1946, *Observatory* **67**, p. 3.

[32] Another New Astronomy was certainly claimed by Galileo's contemporary Johannes Kepler, who published *Astronomia Nova* in 1609. Unlike earlier treatments, in this opus astronomy was no longer confined to geometrical schemes, but instead strived to understand the "celestial physics" of the planets. From this process emerged what we today call Kepler's First and Second Laws of planetary motion. However, I do not include Kepler's work as a comparison to radio astronomy because it did not arise from a new technology (although a case could be made that new techniques were involved in Tycho Brahe's observations, which in turn were indispensible to Kepler's theories).

[33] See also the comparison of Herschel and Grote Reber in Section 4.7.

[34] Oort (1952:233) briefly made this comparison in his Henry Norris Russell lecture to the American Astronomical Society. On Herschel's work see Schaffer (1981) and the voluminous literature by Michael Hoskin, which can be entered, for example, via his article "Unfinished business: William Herschel's sweeps for nebulae," *History of Science* **43**, 305–20 (2005).

technology largely went dormant after his studies, not to be revived for half a century.

A third New Astronomy occurred in the second half of the nineteenth century with the introduction of spectroscopy and photography into astronomy (Meadows 1984); for instance, note Samuel P. Langley's 1888 book *The New Astronomy*. Instead of being mainly concerned with precise measurements of positions on the sky, and then applying Newton's gravitational principles, this New Astronomy used the technology of observing spectra at the telescope to ascertain what something was made of, how fast it moved toward or away from earth (via the Doppler effect), and estimates of its temperature and pressure. Furthermore, long photographic exposures revealed aspects of nebulae and star fields invisible to the observer's eye, while also producing a permanent record that could later be measured with greatly improved accuracy. At observatories like Meudon and Potsdam that pioneered these techniques,[35] the telescope dome took on the look (and smell) of a chemist's lab as the observer prepared and used samples of gases for comparative spectra, as well as prepared and developed photographic plates.

This new approach to astronomy, which came to be called *astrophysics*, took thirty to forty years (starting from the 1860s) to grow in numbers of practitioners and gradually gain acceptance by the general astronomical establishment. Yet even when it had become an important component of astronomy (say in 1900), it represented less than 10% of total activity in the field (Meadows 1984:61). It focused on the sun, whose closeness made it far easier to observe than other stars, and because solar–terrestrial relationships (aurorae, magnetism) were also an exciting new field. Most astronomers looked upon astrophysics as being very un-astronomical – not as precise or as "direct" (because of complex instruments to calibrate and understand), and therefore not as scientific as proper astronomy. It required different training and skills from the highly mathematical methods of earlier astronomy; indeed, many of the new astrophysicists migrated into astronomy from other fields and knew little astronomy at first (Meadows 1984).

Does all of this sound familiar? Every one of the characteristics above of the New Astronomy of astrophysics also applied to the New Astronomy of radio astronomy a half-century later. For example, recall Ryle's need to argue that radio data were every bit as "direct" as optical (Section 14.4.4.2). Furthermore, radio astronomy also focused on the sun (and for the same reasons), was only a small part of astronomy (despite its importance), and was executed largely by migrants into astronomy.[36] Ironically, by the time that radio astronomy encountered establishment astronomy, that very astronomy had by then become suffused with the style of research and the instrumentation of astrophysics.[37] Indeed, is it unreasonable to view radio astronomy as perhaps no more than a continuation and amplification of the trends begun many decades before with the birth of astrophysics? That conflation, however, seems unwieldy; instead, I want to argue that the appellation of a "New Astronomy" should be extended *forwards* in time to encompass X-ray astronomy and perhaps others of the new spectral windows (Section 18.2). With this longer time scale, the twentieth century's New Astronomy becomes the opening of the *entire* electromagnetic spectrum over the period 1930–80, with radio astronomy as the technology and spectral region that led the way.

In summary, these four New Astronomies – Galileo, Herschel, astrophysics, and radio astronomy (or more broadly, the opening of the electromagnetic spectrum) – provide lessons over a period of 350 years. Each radically changed the prevailing astronomical worldview. Each was enabled by innovative technologies and came to fruition through a combination of material culture, faith in empiricism, and key intellectual players. Each encountered and overcame difficulties with the astronomical establishment of the day, eventually insinuating itself into and redefining mainstream astronomy.

---

[35] It is no coincidence that these two observatories were precisely where Wilsing and Scheiner (Potsdam) and Nordmann (Meudon) attempted to detect radio waves from the sun in the period 1896–1901 (Sections 2.4 and 2.5).

[36] One difference was the birth of a new journal in 1895, *The Astrophysical Journal: an International Review of Spectroscopy and Astronomical Physics*, whereas no new journal of radio astronomy ever appeared.

[37] On the 1900–50 history, see C. Fehrenbach (1984), "Twentieth-century instrumentation," pp. 166–89 in *Astrophysics and Twentieth-Century Astronomy to 1950*, ed. O. Gingerich (Cambridge: Cambridge University Press). For a review of historical studies of astronomical instrumentation over the entire twentieth century, see Smith (1997).

With this wider perspective, the twentieth century's New Astronomy, which started with the emergence of radio astronomy, was indeed one of the most significant milestones in the long history of astronomy.

18.4 FOUR MAJOR HISTORICAL THEMES IN EARLY RADIO ASTRONOMY

In Section 1.3.3 I briefly discussed the four major historical themes of this study of early radio astronomy: (1) World War II and the Cold War; (2) material culture and technoscience; (3) a "visual culture"; and (4) a twentieth-century "New Astronomy." In the previous section I have made the case for the last of these four, namely that radio astronomy (or more broadly, the opening of the electromagnetic spectrum) was **the New Astronomy of the twentieth century**. In this section I point to and recapitulate the evidence earlier presented in support of the other three themes.

World War II and the Cold War We have continually seen how World War II and the ensuing Cold War spawned and shaped radio astronomy. In the first place, every major postwar group (except Holland) was filled with war veterans of radar research or operations, often skilled in ionosphere research or other aspects of radio communications (Section 17.1.1). Especially in England and Australia, these researchers not only had know-how and ample surplus equipment, but also a can-do attitude and camaraderie that continued to permeate their postwar groups (Section 17.3.4). Also forged during the war were management skills vital for the success of the new field's leaders.

The strong military postures of the Cold War, especially in the United States and the Soviet Union, also influenced the type and amount of radio astronomy that was done (Sections 10.3.4 and 17.3.5). The highly successful activities in radio astronomy of one military group, Hey's Army Operational Research Group in England, were in fact terminated in 1948 due to Cold War pressure (Section 6.5). In the US, military agencies generously supported research in radio astronomy, but the encouraged direction was toward shorter wavelengths, inappropriate for the most fruitful results at that time. US groups thus long lagged behind Australia and the two university groups in England, who, despite much smaller budgets, achieved significant discoveries at longer wavelengths. Postwar US radio astronomy provides support for Forman's (1987) argument that military funding during the Cold War substantially warped the contents of seemingly "pure" university research.

Material culture and technoscience Early radio astronomy's material culture was intimately intertwined with the field's development and scientific claims (Section 17.4). The practitioners themselves mixed engineering and scientific skills, continually switching between building radio instrumentation and making astrophysical calculations about the sun and the Milky Way. New technical capabilities (especially those that improved angular resolution or sensitivity to weak signals) often led to serendipitous discoveries as investigators took a raw empirical approach, trying out the latest equipment at hand. On the other hand, astronomical goals dictated experimental design for other episodes (especially later). And in many other cases it is hopeless to ask whether the technology led or followed the science, or sometimes even whether the two can be separated. Radio astronomy was indeed *technoscience*, an amalgam activity in which technology and science were intimately related. For example, the various groups conducting radio star surveys favored different antenna systems and receivers. Largely because of these engineering choices, serious discrepancies arose between these surveys, casting doubt on the legitimacy of their scientific claims. This led in turn to revisions of the equipment and follow-up surveys that produced new rounds of scientific results (Sections 14.3 and 14.6).

A "visual culture" Radio astronomers ironically drove themselves to more and more of a "visual culture" even as they continued to record their data as very "unvisual" jiggly ink lines on strip chart recordings (for example, Fig. 4.6). Their visual culture was concerned with (a) finding optical galaxies and nebulae that were also radio emitters, (b) creating radio images that one could begin to compare with photographic images, and (c) seeking the expertise, company and advice of (optical) astronomers. Regarding (a), a continual aim of radio astronomy was for better angular resolution in order to pinpoint more finely the locations and sizes of discrete sources. And the main reason for wanting better angular resolution was in hope of finding an unambiguous optical identification (Section 14.4). Such an optical identification, carrying the imprimatur of astronomy and appealing to epistemically superior visual information, conferred on a radio source a degree of legitimacy and reality that it could not otherwise have. Bernard Mills even produced "radio pictures," pseudo-photographs representing radio emission (Fig. 14.16). Regarding (b), by 1953 groups in Cambridge and in Sydney had worked out the details of how in principle to create a radio image, i.e., a map of high resolution, although it would be many more years before enhanced computer power would make the technique tractable.

The final aspect of the visual culture of radio astronomers was their desire *not* to set up a new discipline, but rather to become part of the very old discipline of astronomy (Section 17.2; Fig. 1.3). The emergence of radio astronomy was thus a case of specialty formation and amalgamation into an existing discipline, not discipline formation. Although their backgrounds and research styles immensely differed from those of (optical) astronomers, by 1953 radio astronomers had learned enough astronomy and gained enough legitimacy that they were welcomed at astronomy meetings, often collaborated with astronomers, and were viewed as making significant contributions to astronomy. In the process they invented the research specialty of *radio astronomy* and made it acceptable that, say, a field full of dipole antennas (e.g., Fig. 8.7) was in fact worthy of being called a telescope, albeit by the new term *radio telescope*. It was also radio astronomy that created the obverse category of *optical astronomy*, which until then had not been necessary.

18.5 CLOSING

In the late 1940s Dutch astronomers at Leiden had a serious discussion as to whether the future of radio astronomy was bright enough to warrant launching a campaign for a large dish. Some felt that after one had mapped the sky once, or perhaps twice or thrice, there would be nothing left to do.[38] Although these opinions did not hold sway, they indicate the uncertainties with which the field was perceived. But discoveries kept rolling in, support was found, new techniques were developed, and key optical astronomers assisted the upstart researchers. Before long *cosmic noise* gave way to *radio emission*, and radio engineers and physicists were morphing into astronomers, albeit of the radio variety. As Graham Smith (1984:247) later recalled:

> I remember most the slogging out to our observatory, four times every 24 hours through most of a year to reset those Würzburg antennas and the chart recorder...And learning to use a sledgehammer, a slotted line, and the *Nautical Almanac*...It was a long time before we even realized that we were astronomers.

Within a decade researchers trying to understand extraterrestrial radio waves led the way for what would be the most important observational development of the twentieth century: astonishing "invisible" phenomena disclosed via electromagnetic waves well beyond the cozy confines of wavelengths between 0.3 and 0.7 μm. After reigning unchallenged for three and one-half centuries, glass lenses and mirrors were now improbably and firmly joined by metal rods and complex electronics as legitimate instruments of celestial observation. The purview of astronomy had indeed been changed, profoundly and permanently, by cosmic noise.

---

[38] Van de Hulst (1973:22T); Van de Hulst:author, October 1984.

Appendix A
A primer of the techniques and astrophysics of early radio astronomy

For those readers with some scientific or engineering background, I present here a concise overview of technical details of physics, astronomy, radio astronomy, and radio engineering relevant to the pre-1953 era.[1] Terms in **bold** are listed in an Index of Terms at the end of this appendix. Only rarely are "updates" given for astronomical and engineering developments that occurred after 1953.[2]

A.1 ELECTROMAGNETIC RADIATION

Electromagnetic radiation, which comes in forms called radio, infrared, optical, etc. (Section A.2), is a sinusoidally oscillating electric and magnetic field whose energy propagates (in a vacuum) at the speed of light, $c = 3 \times 10^8$ m s^{-1}. The electric and magnetic fields are always oriented perpendicular to the direction of propagation. The power per unit area carried by such a wave is proportional to the square of its electric field's strength and, except in special circumstances, decreases with distance R from its source as $1/R^2$. An electromagnetic wave is launched by the acceleration or oscillation of a charged particle and its **frequency** is determined by the periodic character of the originating motion; one oscillation is called a hertz (Hz), with, for example, 10^6 Hz = 1 MHz; an oscillation was formerly usually called a cycle per second (abbreviated "c/s"). The product of frequency ν and **wavelength** λ, the distance at any instant between successive maxima in the electric field's strength, equals the speed of light; a handy benchmark is

$$c = \nu\lambda = (300 \text{ MHz}) \times (1 \text{ meter})$$

Radio waves are electromagnetic oscillations in the lowest range of frequencies, below ~300,000 MHz, corresponding to wavelengths longer than about 1 mm. Wavelengths are usually given as approximate values ("21 cm"), and frequencies, which are more directly measured, as more exact values ("1420.4057 MHz"). Electromagnetic radiation can also paradoxically be considered as caused by the emission of a sequence of energy packets, or **photons**, each with energy proportional to the frequency of its corresponding wave.

If the electric field of a received electromagnetic wave rapidly varies and has no preferential orientation, then the wave is said to be **unpolarized**. But in some situations waves may have **polarization**, that is, a preferred orientation of the electric field. This arises from a preferred direction of oscillation of the charged particles (usually electrons) that either caused the emission or interacted with the wave along its path. If the field prefers a certain orientation, the wave has **linear polarization**, with an orientation specified by the angle between the field's direction and astronomical north on the sky. If the electric field direction undergoes a 360° rotation over the time that it travels one wavelength, the wave has **circular polarization**, either left or right depending on the sense of rotation. Received waves can also be a composite: for instance, a mixture of linear and circular polarization leads to **elliptical polarization**, or a mixture of unpolarized and 100% linear polarization leads to a certain percentage of linear polarization.

A.2 THE EARTH'S ATMOSPHERE

The earth's atmosphere has a significant effect on incident electromagnetic radiation of all wavelengths.

---

[1] The textbook *Radio Astronomy* (Pawsey and Bracewell 1955) is excellent for gauging the state of the field circa 1953.

[2] A fine modern primer is available on the website of the National Radio Astronomy Observatory. Search for "Essential Radio Astronomy" by J.J. Condon and S.M. Ransom. A recommended modern textbook is *An Introduction to Radio Astronomy* by B.F. Burke and F. Graham-Smith (Second edn., 2002) (Cambridge: Cambridge University Press).

The earth's atmosphere

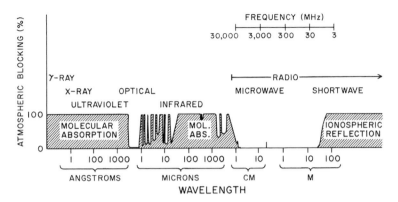

Figure A.1 The transparency of the earth's atmosphere to electromagnetic radiation of various wavelengths. Wavelength increases to the right; frequency and photon energy to the left.

Figure A.1 summarizes the situation. Radiation at most wavelengths is absorbed by the atmosphere and does not reach the earth's surface – these wavelength ranges include gamma rays, X-rays, ultraviolet radiation, and most infrared radiation. Two atmospheric "windows," however, occur at optical and radio wavelengths. The optical extends over wavelengths from 350 nm to 750 nm (3500 to 7500 Ångstroms, similar to the range of sensitivity of the human eye) and the radio from ~3 mm (a frequency of 100 GHz) to ~60 m (5 MHz). The limit at millimeter waves is caused by absorption lines of oxygen and water vapor molecules. Observations at wavelengths shorter than about 5 cm (6000 MHz) are affected enough by the atmosphere to usually require corrections for absorption, but at longer wavelengths corrections are usually not needed, even in a downpour. The long-wavelength cutoff at ~60 m wavelength is caused by the ionosphere acting as a mirror at frequencies less than its critical frequency (see below), but observations at even much shorter wavelengths can be affected by the ionosphere.

Observations are always best made when a source is as high overhead as possible. One then suffers the least amount of atmospheric absorption, as well as the least amount of **refraction**, or bending of a wave's direction. Refraction occurs when a wave travels through a medium whose physical properties (such as density, temperature, or chemical composition) vary in a direction perpendicular to the wave's path. A ray entering the layers of the earth's atmosphere is progressively bent downwards relative to its initial angle of approach. Refraction thus causes a radio (or optical) source's apparent position to be "lifted" from the horizon; for example, the offset is ~60 arcmin for an **elevation angle** (angle above the horizon) of 0°, 5 arcmin for 10°, and 1 arcmin for 40°. These values, however, only take account of refraction caused by the lower atmosphere. For wavelengths longer than about 5 m, refraction and other distortions arising in the ionosphere can amount to several degrees and rapidly vary with time.

Radio observations can, in general, be carried out in the daytime, unless a source of interest happens to be near the sun. This contrast with optical astronomy arises because (a) the sun at radio wavelengths does not dominate other sources so completely as at optical wavelengths (compare the radio situation in Fig. A.2 with the optical situation where the second strongest source [Venus] is fully ~10^9 times weaker than the sun), and (b) the earth's atmosphere much more easily scatters optical waves (witness the blue sky) than radio waves.

The **ionosphere** occurs at altitudes above ~70 km, where absorption of solar ultraviolet and X-ray radiation causes neutral atoms and molecules to lose one or more of their electrons and thus become positively charged particles called **ions**. A gas such as this, containing free electrons and ions, is a **plasma** and behaves quite differently from a normal gas. For the densities encountered in the earth's ionosphere, radio wavelengths shorter than about 50 cm (frequencies > 600 MHz) only weakly interact with the ionosphere. But at lower and lower frequencies, the amount of refraction continually increases until a **critical frequency** is

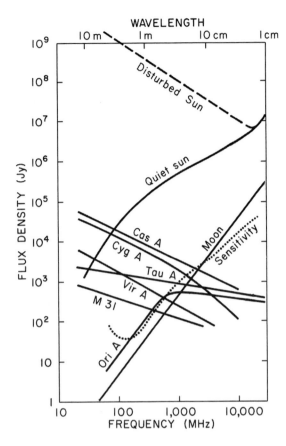

Figure A.2 Spectra of the strongest extraterrestrial radio sources (see Table 14.1). The lunar line closely approximates that for a blackbody with a temperature of 250 K. The line labelled "sensitivity" gives a rough indication of the most sensitive observations possible at the various wavelengths by 1953. "Disturbed sun" refers to the most intense radio bursts.

reached where a wave is bent to such a degree that it is completely "turned back." At this point of resonance an incident radio wave sets the plasma's electrons into oscillation at their natural oscillation frequency (see below). The value for this critical frequency is

$$\nu_c = (e^2 N_e / \pi m_e)^{1/2} / \sin \beta, \quad \text{or}$$

$$\nu_c (\text{MHz}) = 0.009 \, (N_e)^{1/2} / \sin \beta,$$

where e and m_e are the charge and mass of an electron, N_e is the electron density (cm^{-3}) of the plasma, and β is the elevation angle of the incident radiation.

The maximum value of N_e in the ionosphere thus determines the highest frequencies that are reflected, or conversely, the lowest frequencies that are allowed through. The first fact is important for long-distance shortwave communications, which employ one or more reflections off the ionosphere at oblique angles. Similarly, the maximum value of N_e determines the lowest frequencies available for (ground-based) radio astronomy. Besides bending, radio waves can also be attenuated in their intensity, especially at lower frequencies, when a wave's energy may be spread over a wide area by differential refraction or transferred from oscillating electrons to neutral molecules via collisions. Finally, the general magnetic field of the earth also influences radio-wave propagation in the ionospheric plasma, especially at low frequencies, causing unpolarized waves to become polarized and wave paths to be highly dependent on the wave's state of polarization (see Section A.5.2).

The structure of the ionosphere at any time is complex and highly variable with time of day, season, latitude, and solar activity. The degree and kind of ionization are primarily a balance between (a) solar ultraviolet and X-ray radiation that liberates electrons from molecules and atoms, and (b) processes that cause free electrons to recombine with ions and attach to neutral atoms and molecules. Because the amount of solar radiation penetrating to any given level of the atmosphere depends on the sun's elevation angle, the various ionospheric regions regularly change in strength and move up and down as the sun rises, moves across the sky, and sets. This leads to a generally greater degree of ionization in summer than in winter and in low-latitude regions compared to near the poles.

Electron density (and fraction of ions) on the whole steadily increases with altitude up to ~250 km, but local maxima do exist and are used to divide the ionosphere into several regions (formerly called layers). The **D region**, present only during the day, peaks at an altitude of about 80 km and is often ill defined. The **E region** (formerly called the **Kennelly Heaviside layer**) extends from about 90 to 150 km in altitude and has a peak electron density at noon of ~10^5 cm^{-3}, corresponding to a critical frequency of ~3 MHz. After sunset the electron density drops by about a factor of one hundred within ten minutes (owing to recombination of electrons) and the E region effectively disappears. The **F region** is the only maximum found in the electron density profile at nighttime, at an altitude of ~350 km and with an electron density of ~10^6 cm^{-3}. A high degree of ionization is maintained here throughout the night because of the low density and consequent minimal recombination.

The sun exhibits a cycle of activity with maxima separated by eleven years on average (Sec A.7). For instance, at solar maximum there is a higher output of solar ionizing radiation and an increasing number and severity of energetic solar events. This means that typical noontime values of N_e for the F region are some four times higher than at solar minimum, and that the lowest frequency available for radio astronomy is about two times higher. Disruptions to communications are far greater at solar maximum, including shortwave **fade-outs** (enhanced D region ionization causing absorption of shortwaves via collisions) and ionospheric or magnetic storms (changes caused by solar particles bombarding the upper atmosphere).

A.3 THERMAL RADIATION

A **blackbody** completely absorbs all radiation that it receives. **Kirchhoff's law** states that such a body will reradiate all of this incident energy with an unpolarized spectrum dependent only on the body's temperature. The radiation originates in the microscopic, random motions of the body's constituent particles, and **temperature** is a measure of the average kinetic energy of each particle. If measurements refer only to the average kinetic energy of a single type of particle, say electrons, then one refers to an **electron temperature**. If one observes such **thermal emission** (or thermal radiation) from a body with absolute temperature T, its spectrum is given by **Planck's radiation law**:

$$I_{BB}(\nu) = (2h\nu^3/c^2)(e^{h\nu/kT}-1)^{-1},$$

where h and k are Planck's and Boltzmann's constants. $I_{BB}(\nu)$ is **brightness** or **specific intensity**, that is, the amount of power per unit frequency interval received per unit area from unit **solid angle** (W m^{-2} Hz^{-1} ster^{-1}).[3] This specific intensity is independent of the distance to the object. The spectrum extends over all frequencies and has a peak value of specific intensity at a wavelength λ_m determined only by the temperature:

$$\lambda_m T = \text{constant} = 0.0051 \text{ (meter Kelvin)}.$$

---

[3] **Solid angle**, measured in steradians, refers to two-dimensional angular area on a sphere. Analogous to there being 2π radians in a circle of $360°$, there are 4π steradians over the surface of a sphere.

A blackbody at $T = 5800$ K (such as the sun approximates) thus has its peak radiation at a wavelength close to the optical region of the spectrum, whereas the peaks for much cooler objects occur in the infrared and radio bands. For example, the moon's mean temperature of 250 K leads to $\lambda_m = 20$ μm (in the far infrared). Both of these peaks are off-scale far to the right in Fig. A.2. For radio observations of almost all astronomical sources, the relation $h\nu \ll kT$ holds (that is, observations are well on the low-frequency side of the spectral peak), in which case the Planck law expression can be simplified to the **Rayleigh–Jeans law**:

$$I_{BB}(\nu) = 2kT\nu^2/c^2 = 2kT/\lambda^2.$$

If examination of an astronomical source at a variety of frequencies yields brightnesses $I(\nu)$ that are all consistent with Planck's law for a single temperature T, then the source is a blackbody. If different frequencies correspond to different temperatures when assuming the above formula, then, at least at first, **nonthermal emission** is inferred. Investigation may, however, reveal the spectrum to arise from a mixture of blackbodies radiating at various temperatures, or from an optically-thin thermal emitter (Section A.5.1). In practice, radio astronomers usually do not quote a brightness or a specific intensity, but a **brightness temperature** T_b, i.e., the temperature required to produce the observed specific intensity *if* the body were radiating thermally. T_b is thus a parameter for a measured specific intensity through the relation

$$T_b(\nu) = I(\nu)c^2/(2k\nu^2) = I(\nu)\lambda^2/2k.$$

Examples of thermal and nonthermal spectra are given in Fig. A.2.

The **flux density** S received from a certain region of the sky (of solid angle ω) is defined as the product of the brightness from that region and ω. If a source at distance R emits a power per unit bandwidth (W Hz^{-1}), called a **spectral power** p, then $S = p/(4\pi R^2)$. The usual unit in early radio astronomy for flux density was the **flux unit**, defined as 10^{-26} W m^{-2} Hz^{-1} (or 10^{-22} W m^{-2} Hz^{-1} for solar measurements); since 1973, this unit has been called the **jansky** (Jy).

A.4 RADIATION TRANSFER

If radiation with brightness I_o at a certain frequency is incident on a region absorbing by some process, and a fraction I/I_o emerges from the other side, the region is said to have an **optical depth** τ at that frequency defined by $I/I_o = e^{-\tau}$. Such an absorbing region, however, will also radiate, even with no incident radiation, by virtue of the inverse emission process to that by which it absorbs. If the absorbing region is in thermal equilibrium at temperature T, the brightness emerging from the region is

$$I = I_{BB}(1 - e^{-\tau}) + I_o e^{-\tau},$$

a mixture of the region's own emission and attenuated incident radiation. The exact mix depends on the value of τ, that is, the effectiveness of the region as an absorber and emitter. If $\tau \ll 1$ or $\gg 1$, the region is said to be **optically thin or thick**, respectively, and we have two limiting cases:

$$I = \tau I_{BB} + I_o(1 - \tau) \quad \text{for } \tau \ll 1, \text{ and}$$
$$I = I_{BB} \quad \text{for } \tau \gg 1.$$

Note that even if the region emits and absorbs by thermal processes, a blackbody spectrum will be measured only for $\tau \gg 1$.

A.5 RADIATION MECHANISMS

A.5.1 Free–free radiation

In a gas such as that in the solar atmosphere or in interstellar regions near hot stars, radiation arises from the random motions and near collisions of charged particles. Emission occurs when an electron's path is deflected by a proton, causing the electron to lose kinetic energy that is transformed into a newly-created photon. This thermal emission is unpolarized and called **bremsstrahlung** ("braking radiation") or **free–free radiation**. The optical depth τ for such a gas is

$$\tau(\nu) = 0.21 \, N_e^2 \, z / (\nu^{2.1} T^{1.35}),$$

where N_e is the electron density (cm^{-3}), T is the temperature, and z is the depth of the plasma (cm). This exemplifies a **continuum spectrum**, one that is spread over a broad range of frequencies. It can be seen that optical depths at lower frequencies tend to be large, and thus the radiation is blackbody in nature. On the other hand, τ becomes less than 1 at higher frequencies and the observed spectrum is no longer that of a blackbody and varies little with frequency (see the break in the Orion A spectrum at $\sim$1000 MHz in Fig. A.2).

A.5.2 Gyromagnetic and synchrotron radiation

The concepts of radiation transfer hold equally well for a region absorbing and emitting radiation by non-thermal processes. For instance, in the presence of a magnetic field B a low-energy electron will spiral around the field lines with a frequency

$$\nu_B = eB/(2\pi m_e), \text{ or}$$
$$\nu_B (\text{MHz}) = 2.8 B \text{ (gauss)}.$$

This leads to circularly polarized **gyromagnetic radiation** at both this frequency and at low multiples, or harmonics, of it. This radiation was of importance in early radio astronomy only in the case of the sun's strong magnetic fields. In the solar atmosphere (and earth's ionosphere), however, these electrons gyrating in a magnetic field give rise to another important effect. They cause an initially unpolarized wave travelling along the magnetic field's direction to split into two waves, so-called **ordinary** and **extraordinary waves**, of opposite circular polarization and differing interaction with the plasma. In particular, as discussed above for the ionosphere, propagation in a plasma (with no magnetic field) is possible only for frequencies greater than the critical frequency ν_c. But with a magnetic field, only frequencies greater than $(\nu_c^2 + \nu_B^2/4)^{1/2} \pm \nu_B/2$ can propagate, where the two choices of sign refer to the ordinary and extraordinary waves.

If the speed of an electron in a magnetic field is close to that of light, strong radiation at very high harmonics of ν_B becomes dominant. This **synchrotron radiation**, although controversial in early radio astronomy until the late 1950s (Section 15.4), is now recognized as the most widespread type of nonthermal radio process. It is generally linearly polarized and strongest at frequencies in the vicinity of ν_s (Hz) $\sim 1 \times 10^{13} B E^2$, where B is the magnetic field (gauss) and E is the energy (10^9 eV) of the electron. Since in any realistic measurement of an astrophysical region one averages over many values of B and E, the harmonic nature of the radiation in practice becomes lost and a continuum is observed. If the distribution of electrons over energy is proportional to $E^{-\gamma}$ and the radiating region is optically thin, then the brightness temperature at a frequency ν varies as

$$T_b \propto B^{(\gamma+1)/2} \nu^{-(\gamma+3)/2}.$$

For example, cosmic ray electrons have a value of $\gamma \sim 2.5$, and the synchrotron radiation spectrum produced by them should vary as $\nu^{-2.7}$, agreeing well with radio observations of the galactic background (Fig. A.5). Many other sources in early radio astronomy also showed such a negatively sloped spectrum (see Fig. A.2), in distinct contrast to thermal radiation.

A.5.3 Spectral line radiation

Each type of atom and molecule is able to exist only in certain well-defined energy states, dependent on its specific structure. It may spontaneously drop from one energy state to another, thereby launching a photon with energy equal to that lost, or it may absorb an incident photon if the photon's energy is of just the right amount for the atom to change to a higher energy state. A given atom or molecule can thus only interact with photons of certain energies, that is, with certain frequencies. Radiation at such a frequency is called a **spectral line** and its observation allows determination not only of the presence of a certain atom or molecule, but also often its temperature and density. Because radio photons are of very low energy, they are emitted or absorbed only by those atoms or molecules that happen to have states of very small energy difference, of the order of 10^{-5} eV. The only line in early radio astronomy was the **hydrogen spectral line**, in which a transition between two energy states gives rise to a photon with a frequency of 1420.4057 MHz, corresponding to a wavelength of 21.1 cm. Interstellar hydrogen is abundant enough that the line appears as an enhancement above the general continuum from the Milky Way (see Chapter 16 for a full discussion).

Observations of any line are also influenced by the **Doppler effect**: if an emitting object has a component of motion either toward or away from an observer, the frequency at which the observer receives the radiation is not the same as that at which it was emitted. If the speed away from the observer is v, then the received frequency ν_r is related to the emitted frequency ν_e by $\nu_r = \nu_e(1 - v/c)$. Therefore, depending on the specific motions of a cloud of interstellar hydrogen gas, as well as those of the observer on earth, the hydrogen line's emission is both shifted in frequency and broadened over a small range of frequencies, indicating a range of velocities.

A.5.4 Plasma oscillations

Plasma oscillations, which are commonly involved in the intense radio emission from solar bursts, arise when through some means, such as a bombardment of high-energy particles or waves, the electrons in a region of plasma become slightly displaced on average from the protons. These temporary regions of net charge then seek to restore themselves to a neutral condition, and in the process particles oscillate in unison at a characteristic **plasma frequency** that is identical in value to the critical frequency ν_c discussed in Section A.2 for the ionosphere. The oscillations lead to the propagation of a wave of positively and negatively charged regions, which in turn can lead to emission of an electromagnetic wave.

A.6 ASTRONOMICAL COORDINATES

Celestial objects can be located in a polar coordinate system based on an observer's local horizon. **Elevation angle** (or **altitude angle**) is the angle of an object above the horizon, an elevation of 90° corresponding to the **zenith** straight overhead. An object's **azimuth** is a measure of its angular direction around the horizon, due north being 0°, east 90°, south 180°, and west 270°. The great circle line running from due south through the zenith to due north is the **meridian**, and an object is said to **transit**, at its highest elevation for the day, when it crosses the meridian from east to west.

A celestial object is cataloged according to its right ascension and declination, axes in an **equatorial coordinate** system that is fixed with respect to the stars, analogous to and based upon the earth's own familiar geographical coordinate system. The **celestial equator** and **celestial poles** are simply the extensions of their terrestrial counterparts out to an imaginary celestial sphere surrounding the earth. An object's **declination** (δ) is its angular distance north or south from the celestial equator, analogous to latitude on earth and likewise extending from $-90°$ to $+90°$. Its **right ascension** (*RA*), analogous to terrestrial longitude, increases in an easterly direction and runs from 0^h to 24^h, rather than from $-180°$ to $+180°$. *RA* and east–west angle are interconvertible through $1^h = 15° \cos\delta$. The sun and moon are both $\sim 0.5° = 30'$ in angular size, where $1° = 60' = 60$ arc minutes (arcmin) and $1' = 60'' = 60$ arc seconds (arcsec). Astronomers arbitrarily divide all points on the celestial sphere into 88 regions called **constellations**. Objects with the same constellation designation thus have roughly the same sky direction, but are usually unrelated to each other because of their greatly varying distances (from a comet or meteor to a distant galaxy!).

The earth's poles (and therefore also the celestial poles) steadily change their orientation due to gravitational interaction with other bodies. This **precession** means that the basis of the equatorial coordinate system is constantly shifting, and that to define an accurate position one must give not only the right ascension and declination, but also the date of observation. Typical precessional shifts are ~1′ per year. If, after taking account of precession, an object still exhibits a shift in position over time, it is said to have a **proper motion** (typically measured in arcsec per year), whose value depends on both its motion in space and distance from earth.

An observer's local **sidereal time** (LST) is the right ascension of any star then transiting. The **hour angle** (HA) of any object is the difference between its right ascension and that of a star currently transiting. Therefore $HA = LST - RA$ and hour angle is the number of sidereal hours since an object transited. Notice that 24^h of sidereal time represents one rotation period of the earth with respect to the distant stars. This sidereal day, however, is four minutes less than the familiar solar day (a result of the earth's changing aspect with respect to the sun as it moves around its orbit). A given celestial object therefore rises and transits four minutes earlier each day with respect to solar time; after one year, the object again rises and transits at the same solar time.

An antenna with an **equatorial mounting**, that is, mounted with respect to the earth's (and therefore the celestial) equator, points to a radio source in terms of declination (which remains fixed for a given object) and hour angle (which continuously increases). Tracking (in a westwards direction) is therefore needed along only one of the antenna's two coordinate axes. But if an antenna has an **azimuth-elevation mounting** (e.g., most radar antennas of this era), that is, mounted with respect to the local horizon, then tracking of a source must be carried out simultaneously in both azimuth and elevation, and a means must be at hand to continually make conversions from hour angle and declination to antenna coordinates.

Galactic coordinates are a sun-centered polar coordinate system based on the plane of the Milky Way (in which plane we are located). This plane is tilted by 63° to the equatorial plane. Galactic latitude is measured away from this **galactic plane**, and galactic longitude along it in an easterly direction. Before 1958 the conventional zero-point of longitude was such that the direction to the galactic center was at a longitude of ~32°; this is the system used in this book in all cases except in Fig. 3.12. The new system of galactic coordinates adopted in 1958, with the galactic center at a longitude of ~0°, was in fact largely based on 21 cm hydrogen line data.

A.7 BASIC ASTRONOMY OF THE EARLY 1950s

The earth and other planets circle the sun in the **ecliptic plane**, which is tilted by 23.5° from the celestial equator. The average distance from the earth to the sun is defined as the **astronomical unit** (AU) = 1.5×10^{11} m (or 400 times the distance to the moon), and the earth's average orbital speed is 30 km s^{-1}. Pluto's distance from the sun averages 40 AU, but the nearest star is over 200,000 AU distant. This vast jump in scale from the solar system to interstellar regions can be appreciated by comparing the apparent intensity of a bright nighttime star with that of the daytime sun, a star itself. Astronomers measure distances in **parsecs** (pc), a unit equal to 3.3 light years or 3×10^{16} m. Neighboring stars are typically a few parsecs apart and move at relative speeds of 10 to 30 km s^{-1}. The optical flux density (S_*) from a celestial object is usually not quoted directly in physical units, but in a hoary logarithmic system called **magnitudes**. The magnitude (m_*) of an object is defined as $m_* = -2.5 \log (S_*/S_o)$, where S_o is the flux density of a reference object (which thus has 0^m by definition) such as the bright star Vega. A 5^m difference between two objects implies a ratio of one hundred in their flux density.

The sun is an ordinary star, with mass of 2×10^{30} kg, surface temperature of 5800 K, radius of 7×10^8 m, and total power output (at *all* wavelengths), or **luminosity**, of 4×10^{26} W. The age of the sun is several billion years. The innermost 0.1% of the sun's volume is hot enough (10^7 K) for hydrogen nuclei to fuse together into helium – such a star is a **main-sequence star**, at a relatively stable stage of its life. The sun's energy is eventually observed in the form of optical photons emitted from the "surface" or **photosphere**, the level corresponding to an optical depth (at visual wavelengths) of ~1. Above the photosphere, the solar atmosphere rapidly rarefies and the temperature rises, first in a thin (~0.01 solar radius) transition region called the **chromosphere** and then in the **corona**. The corona extends out to several solar radii and has a temperature of 1.0 to 2.5×10^6 K. It emits only one-millionth the light of the solar disk and thus, in contrast to radio wavelengths, can be studied optically only at times of total eclipse or with special instruments (coronagraphs).

The sun undergoes a **solar activity cycle** with maxima separated by 8–14 years, averaging 11 years (Fig. 13.11). Over each cycle, there are systematic changes in the solar magnetic field, coronal shape, radiation output, number and strength of localized impulsive events, and number and size of sunspots. **Sunspots** are cooler regions (~4000 K) with strong magnetic fields, often clustered in an **active region**. Active regions typically last for one or two solar rotation periods (each ~27 days), and give rise to energetic events such as **flares** (a sudden intense brightening of the chromosphere, often resulting in ejected material), **prominences** (material visible at the edge of the sun in the upper chromosphere, sometimes fast-changing), and radio bursts. None of these change the total solar luminosity (output at *all* wavelengths) by even as much as 1%, but note that the most powerful meter-wavelength radio bursts can increase the sun's *radio* flux density by a factor of thousands, and sometimes even a million (Fig. A.2).

Stars range in mass from 0.1 to 100 times that of the sun, with the less massive being far more common. Their surface temperatures range from 2000 K to 100,000 K, the less massive ones in general being cooler. Relative to the sun, their radii range from 0.1 to 100 times and their luminosities from 10^{-3} to 10^6 times. Massive stars are short-lived (as short as a million years) and low-mass stars last for 1 to 100 billion years.

The sun is one of $\sim 4 \times 10^{11}$ stars comprising the system known as **the Galaxy** or **Milky Way**. The sun and its neighboring stars are located ~10 kpc from the **galactic center** (toward the constellation Sagittarius) and revolve about it with a speed of 200 to 300 km s^{-1}. The Milky Way is a **spiral galaxy**, a disk of ~30 kpc diameter and width 0.5 to 3 kpc, the latter varying with location and with which constituent one uses to define the width. Interstellar gas and dust and the youngest stars are highly concentrated to the galactic plane and in some places trace out spiral-like features. Older stars form a thicker disk, increasing to a bulge near the center. Interstellar space is a more perfect vacuum than anything achieved on earth, but there are still enough atoms, molecules and solid dust grains to be detectable by various means. For example, the **dust grains**, similar in size to particles of cigarette smoke (~10^{-7} m), cause starlight to be absorbed and scattered to the extent of about a one-magnitude loss per kiloparsec of path in the galactic plane (in addition to the usual inverse-distance-squared law). This means that it is difficult to optically detect or study any galactic-plane objects more than a few kiloparsecs distant. Thus only a small fraction of the Milky Way disk is amenable to optical study, whereas the entire Milky Way is open to inspection with radio waves, which are not impeded by dust.

Whereas pre-1953 optical astronomy concentrated on stars, radio astronomers more readily studied the wide variety of conditions in the **interstellar medium**. Free–free emission (Section A.5.1) could be detected from concentrations (densities of 10 to 10^6 electrons cm^{-3}) of hot (10,000 K) interstellar gas around hot young stars. These are called **ionized hydrogen regions** (or **H II regions**, where H II means H$^+$) – the Orion nebula (Orion A) was a bright example (Fig. A.2). In cooler regions of the interstellar medium, hydrogen atoms (which constitute 90% of all the atoms in and between the stars) are neutral and these **H I regions** (with densities of 1 to 10 atoms cm^{-3}) were best studied by means of the 21 cm hydrogen line.[4]

By the mid-1950s, it was being recognized that **synchrotron radiation** (Section A.5.2) is generally emitted within the Milky Way by a sea of relativistic cosmic-ray electrons (1 per ~10^{12} cm^3) spiralling around weak interstellar magnetic field lines (Section 15.4). The most massive stars end their short lives in a spectacular explosion called a **supernova**, for a few weeks becoming as bright as the entire host galaxy. The material spewed into the interstellar medium is called

---

[4] The amount of molecular hydrogen (H$_2$) in the interstellar medium was unknown at this time; only with the later advent of ultravioloet and infrared astronomy (Section 18.2) could H$_2$ be studied.

a **supernova remnant**, which usually emits strong synchrotron radiation because of a high density of energetic electrons and a strong magnetic field. Early examples were Cassiopeia A and a Taurus A (the Crab nebula) (see Fig. A.2).

Galaxies outside our own could be studied if their directions lay away from the Milky Way's dust. Besides spiral galaxies like our own Milky Way, the other major type are **elliptical galaxies**, spheroidal-shaped collections of stars devoid of young stars. The Milky Way is a member of a small collection of galaxies known as the Local Group, which has a size of ~1000 kpc and includes the similarly-sized Andromeda nebula and a few, much smaller satellite galaxies (including the Magellanic Clouds). The universe beyond contains an estimated 10^{11} galaxies. **Hubble's law** states that all galaxies recede from us with their spectral lines shifted to longer wavelengths by an amount proportional to their distances.

A.8 RADIOMETRY

Radio astronomers measure radiation known as **noise**, so called because when impressed upon a loudspeaker it sounds like a waterfall or a television set tuned to a non-broadcasting channel. It contains randomly fluctuating power spread roughly equally over a broad band of frequencies, whereas communications signals are confined to a narrow band of frequencies and carry specific modulations. Early observers often used loudspeakers to distinguish between desired noise and terrestrial interference. Figure A.3(A) illustrates a noise-like signal. A **radiometer** is a receiver and antenna combination for measuring external radio noise; if used for extraterrestrial noise, the radiometer is called a **radio telescope**, although note that because no radio image is formed, an analogy better than a telescope is a photographer's light meter. Because any radiometer also generates its own internal **system noise**, it must somehow distinguish this from the desired signal, which is an excess noise identical in character except that it enters from outside. System noise arises primarily from the random motions and collisions of electrons in the receiver circuit and from the discrete nature of electrons as they flow through tubes or transistors and cause fluctuating currents and voltages. This produces a **spectral power** (power per unit bandwidth) p_s that is conventionally set equal to kT_s (for $h\nu \ll kT_s$), with T_s called the **system temperature** of the circuit. (Another formulation for T_s is the **noise factor** $F = T_s/290$ K $+ 1$, often expressed in decibels as $10 \log F$.) In the simple case of a resistor heated to a certain temperature, T_s is indeed the physical temperature: for real circuits, however, T_s does not correspond to any identifiable temperature of the apparatus, although it does usually increase as ambient temperature rises.

The basic task of radiometry is to distinguish internally generated system noise (specified by T_s) from an *external* noiselike signal whose spectral power p_A is characterized by an **antenna temperature** $T_A = p_A/k$. One might think that an accurate calibration of T_s, either in the laboratory or field, would allow any excess measured power to be ascribed to T_A. But for early radio astronomy it turned out that (1) for most situations $T_A \ll T_s$, and (2) T_s varied considerably when a receiver was moved from bench to field, or from one antenna to another, or when weather conditions changed. To appreciate item (1), note that a 100 Jy source (about the limit of several pre-1953 surveys) produces a power input of only 4×10^{-17} W (for a 10 m diameter dish and a 1 MHz bandwidth receiver) – in fact, all the energy gathered by radio telescopes since the time of Jansky amounts to no more than a flea hop! This signal power is much smaller than the system noise power, which (for $T_s = 3000$ K, a typical value) is 1000 times more. One therefore often abandons establishing an extremely accurate value for T_s and instead makes a differential, rather than absolute, measurement. While trying to keep T_s constant, one measures the *difference*, call it T_A', between the values of $(T_s + T_A)$ when the radiometer responds to (1) one part of the sky versus another, or (2) the sky versus a fixed, internal calibration signal of some sort, or (3) the sky at one frequency versus another (for spectral line radiation). This T_A' is then the radio astronomy signal, but for simplicity it will usually be referred to as T_A below.

Even if a differential technique has solved the problems of T_s drifting over long times (minutes to days), the random nature of noise radiation means that a receiver's spectral power is unavoidably varying on shorter time scales. Measurements taken over a period of time nevertheless allow one to characterize p_s by a mean level $<p_s>$ and a root-mean-square dispersion Δp_s about that mean, or equivalently by a mean level $<T_s> = <p_s>/k$ and dispersion ΔT_s. These uncertainties place the ultimate limit on how accurately T_s

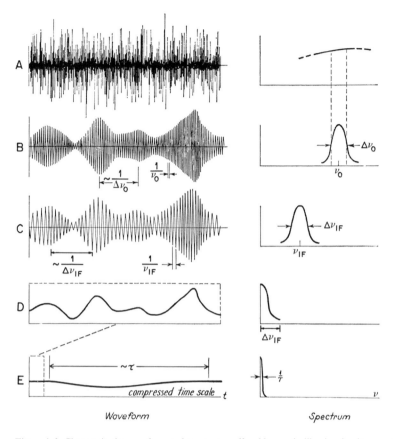

Figure A.3 Changes in the waveform and spectrum suffered by a noiselike signal as it passes through a superheterodyne receiver. Various frequencies ν and frequency intervals $\Delta\nu$ are indicated. The letters on the left and other labelled quantities refer to the block diagram in Fig. A.4. System noise generated within the receiver itself is not shown. (Pawsey and Bracewell 1955:36)

can be measured. The **radiometer equation** states that

$$\Delta T_s = MT_s/(\Delta\nu\tau)^{1/2},$$

where M is a numerical factor dependent on details of the receiver setup (Section A.9), $\Delta\nu$ is the bandwidth of frequencies to which the receiver responds, and τ is the **integration time**, the time over which measurements are averaged. For typical early 1950s values for M (2), T_s (3000 K), $\Delta\nu$ (1 to 10 MHz), and τ (1 sec), the system noise had typical deviations of 1–3 K, or one part in 1000 to 3000. Random errors decline proportionally to the square root of the number of independent measurements. An amplifier cannot significantly change its output power level over a time faster than $1/\Delta\nu$, and so $\Delta\nu\tau$ represents the number of independent samples of the randomly varying T_s waveform over an integration time (Fig. A.3(B)). With an antenna connected to the receiver, the above discussion still holds, but with $T_s + T_A$ substituted for T_s. The fluctuations $\Delta(T_s + T_A)$ then limit the accuracy with which system temperature can be measured, as well as the smallest distinguishable variation. In general a positive result for detecting T_A is deemed when T_A is above 3 to 6 times $\Delta(T_s + T_A)$, depending on the observer's degree of conservatism and level of confidence in the equipment.

A receiver should have as low a value of T_s as possible in order to maximize **sensitivity** and discern small signals. The radiometer equation also shows that one wants as wide a bandwidth and as long an integration time as possible, but these had constraints in early radio astronomy. For instance, high-gain amplifiers were more easily made narrow in bandwidth, and opening up bandwidth often invited manmade interference. Moreover,

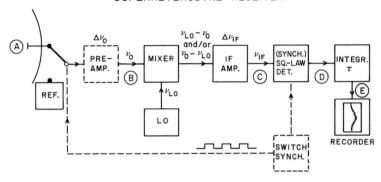

Figure A.4 Block diagram for a typical superheterodyne receiver. Circled letters at various stages refer to waveforms and spectra shown in Fig. A.3. The symbolic switch at left is between the antenna and any of many possible references such as a noise diode, a heated or cooled resistance, a second sky antenna, a different frequency, or a different polarization state. This switch might be manual (used on human timescales), in which case no switch synchronization is required, or it might be a fast Dicke switch. Preamp. = preamplifier; LO = local oscillator; IF amp. = intermediate-frequency amplifier; (synch.) sq.-law det. = synchronous square-law detector; switch synch. = switch synchronizer; integr. = integrator.

integration times cannot be indefinitely lengthened, for during this time receiver gain is more likely to significantly drift, or the source itself may change in intensity or (for a fixed antenna) drift out of the antenna beam.

A.9 RECEIVERS OF EARLY RADIO ASTRONOMY

The task of a receiver is to take the noiselike signal gathered by the antenna and put it in a measurable form. The receiver must first act as an amplifier, for the input signals are far too weak to cause a sensible deflection in any measuring instrument. The ratio of output power to input power, or **receiver gain**, is typically 10^{10} to 10^{12}. The standard setup is a **superheterodyne receiver** (Fig. A.4), which modifies the signal through various stages as shown in Fig. A.3. That portion of the radio power captured by the antenna at a frequency ν_o is (sometimes) first amplified in a **preamplifier** and then combined in a **mixer** (also called the first detector) with a strong signal at a frequency ν_{LO} generated by a **local oscillator**. The mixer is a nonlinear amplifier (often based on a silicon crystal) producing output power at several distinct frequencies that are combinations of its two input frequencies. This mixing process is the key element in a superheterodyne receiver, for it translates an original high-frequency signal to a much lower frequency where it is more easily and accurately transmitted over cables, amplified, and otherwise manipulated. Note also that a change in received frequency (required, say, for spectral-line work or fast recording of wideband solar spectra) may be effected by simply changing ν_{LO}, that is, by tuning the local oscillator, and nothing else needs to be altered (assuming that the antenna and preamplifier also respond to the new frequency). The **intermediate-frequency (IF) amplifier** operates at a much lower frequency ν_{IF}, typically in the range of 10–60 MHz, and amplifies radiation whose original frequency ν_o was, depending on filters placed before it, either or both of $\nu_{LO} \pm \nu_{IF}$. (The bandwidth $\Delta\nu_{IF}$ of the IF amplifier, rather than the bandwidth $\Delta\nu_o$ of the preamplifier, is the appropriate one to use for the quantity $\Delta\nu$ in the radiometer equation.) Following the IF amplifier is the (second) **detector** (often a diode), which rectifies the alternating voltages into a unidirectional result. (This is technically the process of **detection** in a receiver; the word, however, is also used (and almost exclusively in this book) in a broader sense of "establishment of the existence of a signal.") Radio astronomy often uses a square-law detector, that is, one whose output voltage is proportional to the square of the amplitude of the input voltage; in this way the output voltage is proportional to the spectral power of any noiselike signal falling on the antenna. The final step in a receiver is reduction of random fluctuations by averaging the detector output over an integration time τ.

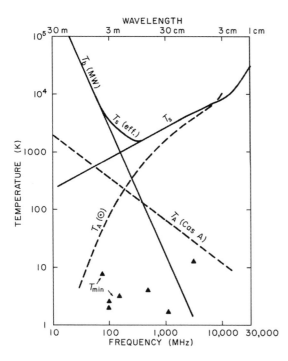

Figure A.5 Dependence on frequency of various temperatures. The two dashed lines show antenna temperatures for the quiet sun (as received with an effective antenna area of 10 m^2) and for Cassiopoeia A (with a 50 m^2 antenna). T_b (MW) refers to the logarithmic mean of the brightness temperatures of the Milky Way background radiation at the galactic pole and at the galactic center; the ratio between center and pole varies from 4 to 40 over the indicated frequency range. T_s refers to typical receiver system temperatures for the pre-1953 era; a fairer representation might show a band covering a factor of ~2 at any frequency. T_s (eff) is the effective system temperature, that is, the sum of the T_s and T_b (MW) curves, and shows a notable minimum at ~300 MHz. Noise power picked up from the earth's atmosphere or from the ground is not illustrated because it was not of significance compared to pre-1953 receiver noise levels. Filled triangles at the bottom show the minimum detectable antenna temperatures T_{min} reported for a variety of pre-1953 surveys.

Signals are usually propagated through a receiver either along coaxial cables (for frequencies <~1000 MHz) or through **waveguides** (above 1000 MHz), conducting pipes whose dimensions are optimized for transmission of a particular frequency. Long transmission lines, as in interferometers, require special attention because of their exposure to the environment and attendant variations in the attenuation and phases of signals. One of the most difficult tasks in the design of any successful system is proper matching at all junctions, meaning that all power propagates across the junction intact in amplitude and phase and with none reflected backwards or absorbed. Switches and transmission-line/antenna junctions are particularly troublesome in this respect.

Calibration of intensities for radiometers was accurate to ±20–30% at its very best, but different observers often disagreed in their reported intensities by as much as a factor of two. Calibration was commonly effected by switching the receiver input from the antenna to a reference noise generator such as a saturated diode (Tangent 6.3), gas discharge tube (Section 16.3.2), or hot load (heated resistance). The noise generator's effective spectral power in turn had to be calibrated in the laboratory, for example, by comparison with the radiometer's response when its input was immersed in hot and cold water of known temperatures. Having done this, one knew the measured antenna temperature of a source, but still not its flux density unless the antenna's effective area was also known (Section A.10).

The contribution of any receiver part to the overall system noise is more deleterious as that part is nearer the antenna, simply because its output is then amplified more by successive stages. Most attention in reducing radiometer noise was therefore focused on the preamplifier and mixer. In the pre-1953 period receivers often had no preamplifiers at all, especially at frequencies above 500 MHz. When preamplifiers were used at lower frequencies, they were usually based on vacuum tubes such as grounded-grid triodes or pentodes. Figure A.5 illustrates the rise by a factor of ~100 in the system noise T_s for typical receivers for operating frequencies from 50 to 30,000 MHz.

The most important systematic error in radiometry usually can be ascribed to **gain instability**. Because T_A is usually much less than T_s, a small (unknown) fractional change in the receiver's gain causes the total output power to change by an amount comparable to or much greater than the output power generated by the external radio source; for example, if $T_A/T_s = 0.001$, then a change of gain by only 0.1% causes a 100% error in the measured value of T_A. If the gain wanders around too fast or too much, or is not compensated for in some manner, then high sensitivity is of no help. On the other hand, for observations of very strong sources such as the sun or Cas A, then $T_A > T_s$ and a receiver gain change of x% causes only a $\sim x$% error in T_A. But also note that noise fluctuations are then much larger than for a weak source because $T_s + T_A$ is considerably higher.

In practice, despite herculean efforts at regulation of tube voltages (to 0.01%) and of receiver (physical) temperature (to 1°C), achieved gain instabilities of order 0.1% were still inadequate for accurate work on weak signals. Several schemes were thus developed to quickly track and in some manner compensate for gain variations. These worked well, but paid the penalty of increased complexity in the circuitry, a higher value for M in the radiometer equation, and usually a somewhat higher value of T_s. The ones most important for the pre-1953 period are known as the Dicke receiver, the Ryle–Vonberg receiver, and phase-switching. The last is more an antenna technique than a receiver one and is discussed in Section A.11 on interferometry. The **Dicke (1946) receiver** is described in Section 10.1.1 and illustrated in Figs. 10.1 and A.4. The basic idea is to rapidly (say, at 30 times per second) switch between the antenna and a reference whose power output stays constant and is ideally nearly the same as that delivered by the antenna. In Dicke's original setup the reference was a disk of thermally-emitting material, but it can be any type of noise source. The receiver is then made to amplify the *difference* between the signals when antenna and when reference are connected. In this way an x% gain variation causes only an x% error in T_A', even for small signals, so long as the gain variation occurs on a time scale longer than 1/30 sec. The Dicke switch was therefore especially valuable for microwave frequencies, where T_A' was usually smallest in comparison to T_s. Insertion of the switch, however, somewhat increases the value of T_s in the radiometer equation, and the constant M becomes of order 2.[5]

The **Ryle–Vonberg receiver** (1948) also rapidly switched with a reference noise source, but one whose power was continuously varied so as to exactly equal the receiver's output (see Section 8.3 and Fig. 8.1). The noise power of the reference was then the desired measure of T_A' and the receiver in effect functioned to produce a voltage controlling the reference noise level. This system in principle was completely immune to receiver gain variations, but in practice the noise output of the reference could be measured to only a certain accuracy, especially when rapidly changing. The values of T_s and M were similar to those of the Dicke receiver. In practice the Ryle–Vonberg receiver was of greatest use for frequencies below ~ 500 MHz where signal levels were generally stronger.

A **radar set** combines a superheterodyne receiver and a transmitter into one package, using either the same or separate antennas for transmission and reception. The chief technical obstacle is the propinquity of an ultrasensitive receiver and a powerful transmitter – the receiver must detect as little as 10^{-20} of the transmitted power and yet somehow avoid being "fried" during transmission. This is accomplished through a **transmit/receive (TR) switch** just behind the antenna. When the transmitter is pulsing, this switch highly attenuates any leakage of power toward the receiver, and between pulses it opens to allow passage of any feeble echo to the receiver (Fig. 12.4 shows an example in a block diagram). Some radars were **continuous wave (CW) radars**, that is, transmitting a single frequency and then able to detect Doppler shifts in the returned echo. Most, however, were **pulse radars**, transmitting in a sequence of very short bursts (a few microseconds was typical). There are several competing factors when choosing **pulse width** (in time) and **pulse repetition rate**. The repetition rate must not be too frequent, or one will tax the transmitter or limit the range at which a target can be unambiguously identified (because a given echo might have originated as any of several recent pulses). If pulses are too infrequent, however, targets are measured less often and may escape detection altogether. A narrow pulse width δt allows one to measure a target's range very accurately (by timing the delay between pulse transmission and reception), but because the pulse's (and echo's) spectrum has a frequency bandwidth of $1/\delta t$, the receiver's bandwidth must be broad, leading to increased system noise power and reduced sensitivity. This situation of a broader bandwidth *reducing* one's signal-to-noise ratio is the usual in communications, but the opposite of the case in radiometry discussed above. The key difference is that power from a radio source or from a

---

[5] One factor of $\sqrt{2}$ arises because the signal is sampled for only 1/2 of the total time (or, equivalently, the value of τ in the radiometer equation is halved), and another $\sqrt{2}$ factor arises because the *difference* of two similar noisy signals has a statistical uncertainty $\sqrt{2}$ times higher than the uncertainty in either of the original signals.

receiver is noiselike and broad in bandwidth, whereas a communications or radar signal is not noiselike and has a defined bandwidth. Chapter 12 discusses further details of radar sets, as well as the so-called **radar equation** (Section 12.3), which relates expected echo strength to target distance and size.

A.10 ANTENNAS (FILLED APERTURES) OF EARLY RADIO ASTRONOMY

An **antenna**, or aerial, is a configuration of conducting material in which currents are caused by the varying electric field of an incident radio wave. These currents are then propagated through the circuitry described above. The (normalized) **antenna pattern** $P(\theta, \phi)$ expresses the relative power that an antenna would abstract from a theoretical test source placed in various parts of the sky, specified by any two sky coordinates (θ, ϕ). A reciprocity theorem states that the antenna pattern characterizing reception is identical to the pattern when an antenna is used as a transmitter; in other words, good reception from a given sky direction means that a large fraction of any transmitted power will radiate toward that same direction. Often the adjacent ground or water surrounding an antenna reflects radio waves (especially at lower frequencies) and in such cases the ground itself effectively becomes part of the antenna (as in the sea-cliff interferometer discussed below, or in Jansky's antenna).

An antenna with few gaps in the distribution of its conducting material is called a **filled-aperture antenna**. This type is discussed in this section, leaving the other type, interferometers, for the following section. Antenna patterns for filled-aperture antennas are usually dominated by a single region of sky, the **main beam** with a size of ω_B (steradians). They also contain many regularly spaced **sidelobes**, regions of the sky (many far removed from the main beam) to which the antenna more weakly responds (see Fig. A.6(a)). The **antenna gain** G is defined as $\sim 4\pi/\omega_B$. A strong source in a sidelobe produces a comparable antenna temperature to that from a weak source in the main beam. Furthermore, if the antenna pattern is too complex and/or if the receiver/antenna's sensitivity is so large that too many sources (either in the main beam or sidelobes) are detectable, then it is said to suffer from "confusion" – the sensitivity is too large for the beam size. A survey of the sky under such conditions is **confusion-limited**: at most times many sources contribute to the measured antenna temperature and reliable deductions of individual positions and flux densities cannot be made. This tends to happen at lower frequencies where beams are larger and sources stronger. In contrast, **sensitivity-limited** sky surveys detect fewer sources, but can measure their positions and flux densities with more accuracy.

The **flux density** S and **antenna temperature** T_A of a source much smaller in angular extent than the main beam are related through the measured **spectral power** p_A (power per unit bandwith):

$$p_A = kT_A = mSA_e,$$

where m is a coefficient ≤ 1 (see below) and A_e is the **effective area** of the antenna, a quantity loosely related to the antenna's geometrical area. For example, the effective area of a paraboloidal reflector is often about 50% of the geometric cross section, while that of a half-wave dipole is $0.13\ \lambda^2$. It can be seen that a larger antenna produces a higher value of T_A for a given source, thus allowing easier detection for a receiver of given system temperature. Antennas, like waves, are polarized, depending on which type of incident wave sets up the strongest currents in it. If an antenna has left-hand circular polarization, then it will receive full signal from a 100% left-hand circularly polarized wave (in which case $m = 1$ in the formula above); none at all from a right-hand circularly polarized wave ($m = 0$); one-half from a linearly or unpolarized wave ($m = 0.5$); and intermediate fractions from elliptically polarized waves. Similarly, a linearly polarized antenna, such as a dipole oriented north–south, receives full power from a wave that is 100% linearly polarized north–south; 0.71 (cos 45°) from a northeast–southwest wave; none from an east–west wave; and 0.5 from a circularly or unpolarized wave. For an unpolarized wave, which is essentially an equal mixture of all polarization states, $m = 0.5$ no matter what the antenna's polarization. In general, one must use differently polarized antennas (say, by rotating a dipole to different orientations or changing one element of an interferometer) to ascertain the state of polarization of a source and the correct value of its total flux density.

An antenna's ability to discern detail in the sky brightness distribution is limited by the effects of **diffraction**, that is, the bending of light waves after passing through or being collected by an aperture. If d_1 and d_2 are measures of the physical dimensions of

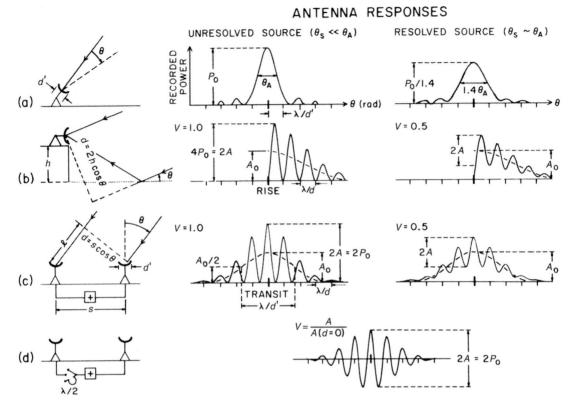

Figure A.6 Ideal antenna responses for types shown on the left: (a) a single dish or array of dipoles or Yagis, (b) a sea-cliff interferometer, (c) a Michelson interferometer, and (d) a phase-switched Michelson interferometer. In the middle are traces of measured power in response to movement across the sky of an unresolved radio source, that is, one whose angular size θ_s is much less than the antenna's angular resolution θ_A, which is approximately the wavelength λ of operation divided by the characteristic size of each system (d' for (a); d for the others). This response to an unresolved source, which measures the antenna pattern, contrasts with that on the right from a resolved radio source of the same flux density. Tick marks on the abscissae are spaced by λ/d' for (a) and by λ/d for the others; all received powers are in terms of that from an unresolved source for a single antenna (P_o). For the interferometers, the visibility V of the fringes is defined as A/A_o in all cases. These examples do not show any slopes resulting from galactic background radiation or other causes; only the phase-switched interferometer is impervious to such slopes.

Notes on individual cases. (a) The term *beamwidth* is usually defined as the full-width at half-maximum, as indicated by θ_A. The value of this beamwidth and the sidelobe levels can be changed by using (for a single dish) different feed antennas or (for a filled array) different weighting of signals from the elements making up the array. (b) The angle θ in practice was usually much less than 15°. Note also the fringes' doubled amplitude and phase shift of 180° (with respect to (c)), caused by reflection of the waves off the sea. Effects of irregular reflection and of earth curvature have been neglected. (d) The trace shown is the *difference* in power received with the switch at its two positions. To determine V, one must also have a measurement using only a single antenna (whose effective "spacing" d is 0).

an antenna, then its main beam size of ω_B (steradians) approximately equals λ/d_1 (radians) times λ/d_2, where λ is the wavelength of operation. The **beamwidth** is usually specified as the distance between the two points where the antenna pattern $P(\theta, \phi)$ falls to one-half of its maximum (Fig. A.6(a)). For example, a dish of diameter 15 m at a wavelength of 21 cm has a beamwidth of ~0.8°; an array of 1 m wavelength dipoles spread over 5 m has a beamwidth of ~12°.

Any measured map of the sky $T_A(\theta, \phi)$ represents the **convolution**, or smoothing, of the distribution of sky **brightness temperature** $T_b(\theta, \phi)$ (defined in Section A.3) with the antenna pattern $P(\theta, \phi)$ – this is equivalent to saying that the measured **antenna** temperature at a certain sky position is the antenna-pattern-weighted mean brightness temperature of the sky distribution:

$$T_A(\theta,\phi) = \frac{\iint T_b(\theta',\phi')P(\theta-\theta',\phi-\phi')d\theta'd\phi'}{\int P(\theta',\phi')d\theta'd\phi'}$$

This process of convolution means that any features in the sky brightness distribution $T_b(\theta, \phi)$ (such as many discrete sources) appear in the measured map $T_A(\theta, \phi)$ broader in angular size, reduced in peak signal, and with less detail.

A localized region of radio emission is called a **discrete source** or **radio source** or, formerly, a **radio star**. The dimensions of the main beam determine the **angular resolution**, that is, the smallest discernible source size or the smallest discernible angle between two sources. Angular sizes of small regions can only be measured by comparing the observed map with that expected for a very small source at the same position. If there exists a measurable broadening, the source is said to be **resolved** or **extended**; otherwise, it is **unresolved**, or a **point source**. For example, a (Gaussian-shaped) source of angular width 0.5 (1.0) times the beamwidth causes, when convolved with a typical beam, a measured broadening of 12% (41%) more than that for a point source (Fig. A.6(a)). Finally, the beamwidth also sets a limit on the accuracy to which positions of isolated sources can be measured: about 0.1 beamwidth accuracy for very strong sources or about 0.5 beamwidth for weak ones.

Converting measured antenna temperatures into sky brightness temperatures is tricky and requires an accurate knowledge of the antenna pattern. For instance, if it is known that 70% of the all-sky integral of an antenna pattern is in the main beam and 30% in all the sidelobes (typical values for a dish), and if one assumes that an extended source is larger than the main beam, but not large compared to the distribution of sidelobes, then that source's average T_b is $T_A/0.70$. Or if one assumes that a source's brightness temperature can be represented as constant over a disk of solid angle $\omega_s \ll \omega_B$, then that source's T_b is $(\omega_B/\omega_s)T_A$. With a measure of a region's brightness distribution, then the total flux density can be found as the integral of that distribution over solid angle.

There are many types of antennas. The simplest is a **dipole**, a rod split into two sections, the inner end of each connected to the receiver. It responds to large parts of the sky perpendicular to the rod and is most often used in a **broadside array** of halfwave dipoles, spaced by 1/2 wavelength and backed with a sheet reflector (e.g., Fig. 8.7). Each dipole is connected by effectively equal lengths of cable to a common receiver, and the resultant main beamwidth is appropriate to the dimensions of the array. The **Yagi antenna**, named after the Japanese radio physicist who invented it in the 1920s, may also be used singly or in an array. It consists of a sequence of rods mounted parallel to each other and spaced along a common axis (e.g., Fig. 7.7 or the common TV antenna on house roofs), but only one (a dipole) is connected to the receiver. The others serve to "guide" or reflect those radio waves arriving along the axis toward this active element. A dipole or Yagi is only usable over a narrow range of wavelengths, but the **rhombic antenna** (e.g., Figs. 13.15 and 13.17) can be operated over as much as a 2:1 ratio of wavelength, although it is bulky for its beamwidth and has high sidelobes. The **horn antenna** (e.g., Figs. 10.2 and 16.2), essentially a flared-out section of waveguide, also has a broad bandwidth and is useful for microwaves.

For a **paraboloidal reflector** or **dish**, incident waves are reflected off a paraboloidal surface, then picked up by an **antenna feed** (or **feed antenna**) and passed on to the receiver. The feed's own antenna pattern must encompass most of the reflector surface while at the same time minimizing the amount of **spillover**, which leads to increased system noise stemming from the thermal emission of the ground beyond the dish's edge. In practice one tries to compromise between too fast a taper in the feed's response, which wastes the outer portion of the dish, and too severe a cutoff in the response at the dish's edge, which leads to unacceptably high sidelobes. A horn antenna's low sidelobes and reliably calculable antenna pattern made it especially meritorious as the feed for frequencies above ~1000 MHz (e.g., Fig. 10.11b), while at lower frequencies a halfwave dipole and back reflector were usually used (e.g., Fig. 13.7). Dishes have the great advantage of allowing operation at a broad range of wavelengths (with differing feeds and receivers), but they are expensive, especially if steerable. Unlike an interferometer (next section), dishes also do not allow one to specify *separately*, for a given observational program, the desired angular resolution ($\propto 1/d$) and sensitivity ($\propto d^2$) (where d is the dish diameter). The shortest wavelength of operation for any dish is dictated by the accuracy of its surface; if deviations are as large as $\lambda/16$, for instance, the reflected waves cancel to such a degree that the effective area of the antenna is reduced by a factor of 0.54 from that of a perfect surface.

Achievable sensitivities in the various frequency ranges for the pre-1953 era are indicated in Figs. A.2 and A.5. The latter plot in particular shows how frequencies in the 100 to 1000 MHz range afforded the most sensitivity. At lower frequencies the galactic background radiation becomes very intense and greatly increases effective system noise, while at higher frequencies the lack of any stable, low-noise preamplifiers caused system temperatures rapidly to rise.

A.11 INTERFEROMETERS OF EARLY RADIO ASTRONOMY

In order to overcome poor angular resolution, which was the outstanding technical disadvantage of early radio astronomy, an excellent solution was an **interferometer**, an antenna with widely separated parts. To gain this resolution, however, one paid the price of a complex antenna pattern without a dominant main beam. The principles of operation went back to Thomas Young's two-slit optical experiment a century and a half before early radio astronomy. In the **Michelson interferometer**, developed particularly by Ryle's group (Chapter 8), consider radio waves from a pointlike source incident on two antennas, or interferometer **elements**, separated by a **baseline** d (with length measured as shown in Fig. A.6(c)). If the path from the source to each antenna is identical and if the separate signals are brought *before detection* over identical cable lengths to a common location for addition, then the waves reinforce each other. In general, however, the wavefront reaches one antenna ahead of the other, and then the two signals will only reinforce if the *difference* in path, ℓ, is an integer multiple of wavelengths $n\lambda$, and they will completely cancel if the difference is $(n + 1/2)\lambda$.

Earth rotation causes source/interferometer geometry and the value of ℓ steadily to vary. If we now consider two antennas on an east–west line, the interferometer's output power p will oscillate in a sinusoidal fashion as ℓ changes:

$$p = p_o + p_o \cos(2\pi\ell/\lambda),$$

where p_o is the power that would be received by one element alone. These oscillations (Fig. A.6(c)) are known as **fringes**. A set of oscillating fringes can be characterized by their **fringe amplitude**, **fringe phase** (the position of the largest maximum relative to a reference), and **fringe period**. The fringe period can be computed precisely for the path difference ℓ for a source of known declination δ and hour angle; for example, on the meridian (due south) when a transiting source's wavefront has an identical path length to each antenna, the fringe period (in units of days) is $\lambda/(2\pi d \cos \delta)$. A measurement of fringe period on the meridian thus yields the source declination, while its right ascension comes from the sidereal time of transit. The antenna pattern of an east–west interferometer is a battery of fan-shaped **lobes**, long in the north–south direction and closely and equally spaced in the east–west direction. The spacing and width of the central lobes are given by λ/d (radians). This discussion has implicitly assumed that each element individually responds equally well to the entire sky, but clearly this is not usually so – the antenna pattern of the interferometer must be multiplied by that of an element in order to get the actual pattern. Before 1953 the individual elements had sizes ranging from 1 to 20 wavelengths and thus their interferometers recorded fringes over a portion of the sky ranging in diameter from, respectively, ~57° to 3° (e.g., Fig. 14.6). One disadvantage of an interferometer is that large bandwidths cannot be used as readily in order to increase sensitivity, because the mixture of a wide range of wavelengths reduces fringe strength when $\ell \neq 0$.

The **sea-cliff interferometer** developed by Pawsey's group needed only one antenna to mimic the effect of two. Observing near the horizon from an elevated position, such as on a cliff overloooking the sea, yielded fringes resulting from interference between that portion of the incident radio wave travelling directly toward the antenna and that portion that first reflects off the sea. The effective baseline of the interferometer is twice the height h of the cliff (and fringes are spaced by $\lambda/2h$ radians – see Fig. A.6(b)); this setup is equivalent to a Michelson interferometer with a second, imaginary antenna located $2h$ below the real one. Despite some advantages of the sea-cliff interferometer (e.g., only requiring one antenna and receiver; a factor two stronger signal), the considerably greater flexibility of the Michelson interferometer and its advantage of observing sources high in the sky eventually led to its universal adoption, but not before the sea-cliff interferometer had made major contributions (Sections 7.3, 7.4, 14.3.4).

Because an observed map is the sky brightness distribution convolved with the antenna pattern, fringes are only obtained when a source's angular extent θ_s is no larger than the interferometer lobe spacing λ/d. Otherwise, different portions of the source combine out of phase with each other and little or no overall oscillation results, only the long-term rise and fall that a single element would record by itself (Fig. A.6(c)). The amplitude of fringe oscillation, or **fringe visibility** V (defined graphically in the lower-right portion of Fig. A.6), is thus a measure of a source's angular size. For example, a uniformly bright disk whose width is 0.25 (0.50, 1.00) times the lobe spacing produces fringes 93% (73%, 21%) times as strong as those from a point (unresolved) source. Strong fringes from a source at a certain baseline d thus imply that $\theta_s < \lambda/d$ – in order further to determine a source's extent and shape, longer and longer baselines must be sampled until the fringes disappear. Note also that θ_s refers only to a source's extent in the direction parallel to the interferometer baseline. For a thorough check on a source, observations are required not only at baselines of different length, but also at different orientations.

These principles can be neatly expressed in terms of **Fourier synthesis** theory, as first published by McCready, Pawsey, and Payne-Scott (1947). A mathematical function can in general be decomposed into a Fourier series, that is, a summation of sine and cosine functions of various periods, amplitudes, and phases calculable from the original function. If (in one dimension, parallel to the baseline) the sky has a certain distribution of brightness temperature $T_b(\theta)$, then it can be shown that each interferometer spacing d measures one term in the Fourier series expressing $T_b(\theta)$. More generally, if at many spacings $u = d/\lambda$ we measure fringe amplitudes A and fringe phases ψ (relative to some arbitrary zero, say a sky position near the source), then

$$T_b(\theta) \propto \int A(u) \cos\left[\psi(u) + 2\pi\theta u\right] du.$$

This expresses a Fourier transform integral (the limit of the Fourier series) and is easily generalized to two dimensions. For pre-1953 radio astronomy the technical problems of accurately measuring fringe phases had not been overcome, so that one was forced to assume $\psi = 0°$ at all spacings, equivalent to assuming a symmetric source brightness distribution. The computations required to carry out the above integration with even as few as 10 to 20 separate baselines were formidable (Section 13.1.3).

A radically different setup was the **intensity interferometer** developed by Hanbury Brown and Twiss (1954) and used in limited cases during the early 1950s. For information on the *size* of a radio source, only fringe amplitudes (and not phases) at various baselines are needed. Furthermore, information equivalent to these amplitudes can be extracted by combining signals from two antennas *after* detection – essentially by multiplying not the original signals, but their fluctuations (as illustrated in Fig. A.3(D)). The intensity interferometer in practice made possible much longer baselines, but suffered from inadequate signal-to-noise ratios except for the few strongest sources (see Section 14.5.1 for further discussion).

Ryle (1952) developed an ingenious compensation scheme that soon came to be standard in all interferometers. The primary motivation for the **phase switch** was to discriminate against galactic background radiation in favor of discrete sources, but it turned out also to cure many other practical ills, such as receiver gain variations and effects arising from dissimilarities between the properties of the interferometer arms and elements. By quickly switching in and out of one arm of the interferometer a cable of one-half wavelength, lobes were caused to flick back and forth in sky position by one-half their spacing. As with the Dicke receiver, the *difference* in signal level between the two switch positions was then amplified and recorded. This difference signal exhibits fringes for sources smaller than the lobe size, but extended sources, which previously were picked up by the elements acting individually, now entirely disappear. All fringes now oscillate about a mean level of zero (compare Figs. A.6(c) and (d)) and it becomes far easier to detect faint, discrete sources because slopes on the record from the broadly extended galactic background have been eliminated (Section 8.4 has further details).

INDEX OF TERMS FOR APPENDIX A

active region, solar 479
altitude angle 477
angular resolution 487
antenna 485
antenna feed 487
antenna gain 485
antenna pattern 485
antenna temperature 480, 485, 486
astronomical unit 478
azimuth 477
azimuth-elevation mounting 478
baseline, interferometer 488
beamwidth 486
blackbody 475
bremsstrahlung 476
brightness 475
brightness temperature 485, 486
broadside array 487
celestial equator 477
celestial pole 477
chromosphere 479
circular polarization 472
confusion-limited 485
constellation 477
continuous-wave (CW) radar 484
continuum spectrum 476
convolution 486
corona 479
critical frequency 474
D region (layer) 474
declination 477
detection 482
detector 482
Dicke receiver 484
diffraction 485
dipole 487
discrete radio source 487
dish (antenna) 487
Doppler effect 477
dust grain 479
E region (layer) 474
ecliptic plane 478
effective area 486
electromagnetic radiation 472
electron temperature 475
element, interferometer 488

elevation angle 473, 477
elliptical galaxy 479
elliptical polarization 472
equatorial coordinates 477
equatorial mounting 478
extended radio source 487
extraordinary wave 476
F region (layer) 474
fadeout, shortwave 475
feed antenna 487
filled-aperture antenna 485
flare, solar 479
flux density 475, 485
flux unit 475
Fourier synthesis 489
free–free radiation 476
frequency 472
fringe 488
fringe amplitude 488
fringe period 488
fringe phase 488
fringe visibility 488
gain instability, receiver 483
galactic center 479
galactic coordinates 478
galactic plane 478
Galaxy, the 479
gyromagnetic radiation 476
H I region 479
H II region 479
horn antenna 487
hour angle 478
hydrogen spectral line 477
Hubble's law 479
integration time 481
intensity interferometer 489
interferometer 488
intermediate-frequency (IF)
    amplifier 482
interstellar medium 479
ion 474
ionized hydrogen region 479
ionosphere 474
jansky (unit) 475
Kennelly–Heaviside layer 474
Kirchhoff's law 475

linear polarization 472
lobes, interferometer 488
local oscillator 482
luminosity 478
magnitude (unit) 478
main beam 485
main-sequence star 478
meridian 477
Michelson interferometer 488
Milky Way 479
mixer 482
noise 480
noise factor 480
nonthermal emission 475
optical depth 475
optically thick (thin) 476
ordinary wave 476–7
paraboloidal reflector 487
parsec 478
phase switch 489
photon 472
photosphere 478
Planck's radiation law 475
plasma 474
plasma frequency 477
plasma oscillations 477
point (radio) source 487
polarization 472
preamplifier 482
precession 478
prominence, solar 479
proper motion 478
pulse radar 484
pulse repetition rate 484
pulse width 484
radar equation 485
radar set 484
radio source 487
radio star 487
radio telescope 480
radio waves 472
radiometer 480
radiometer equation 480
Rayleigh–Jeans law 475
receiver gain 482
refraction 473

resolved radio source 487
rhombic antenna 487
right ascension 477
Ryle–Vonberg receiver 484
sea-cliff interferometer 488
sensitivity 481
sensitivity-limited 485
sidelobe 485
sidereal time 478
solar activity cycle 479
solid angle 475
specific intensity 475

spectral line 477
spectral power, radiometer 480
spectral power, radio source 475, 485
spillover 487
spiral galaxy 479
sunspot 479
superheterodyne receiver 482
supernova 479
supernova remnant 479
synchrotron radiation 476, 479
system noise 480

system temperature 480
temperature 475
thermal emission 475
transit 477
transmit/receive (T/R) switch 484
unpolarized 472
unresolved radio source 487
waveguide 483
wavelength 472
Yagi antenna 487
zenith 477

Appendix B
The interviews

B.1 DOING ORAL HISTORY

When I was younger, I could remember anything, whether it happened or not: but I am getting old, and soon I shall remember only the latter.
— Mark Twain[1]

Historians in general and historians of science in particular have debated the value of doing oral history, a technique that began in its modern form in the 1960s, as compared to the historian's long reliance on the archival record. The basic argument claiming minimal epistemological value for interviews is that memory is notoriously imperfect and that each person has an agenda (usually unconscious) that controls how the past is remembered (or forgotten). As historian of science Ron Doel (2003:350) nicely puts it in his overview of oral history practices and projects in twentieth-century science[2]: "Memories are sculpted by subsequent events, buffeted by denial and repression, hammered into self-reflective and self-justifying narratives." Indeed, cognitive research has shown that people do not remember well the sequence and timing of long-ago past events (say, with regard to how a scientific idea developed), but instead spout a standard story that is eminently logical and usually serves other purposes, too. Scientists are probably particularly prone to this because this is how they are trained – witness the standard scientific paper that omits or glosses over all steps that turned out to be wrong or of no value in the research process. Perhaps the greatest divide between most scientist-interviewees and historian-interviewers is the latter's interest in *all* ideas, experiments and observations, whether or not they later turned out to be "wrong" (Mulkay 1974:111–13). Some scientists consider it a waste of time to discuss these "mistakes," wanting to value the past solely in the light of present knowledge. Historians, however, find this attitude distinctly ahistorical and view such "mistakes" as essential to understanding the richness of how science works.

So why bother at all with the considerable trouble of doing interviews? First, cognitive research has also shown that other types of information from the past, such as social relationships and the scientific milieu, *are* usually accurately recalled. Second, interviews are a class of historical evidence that enriches and complements the archival record, which has its own defects. These include incompleteness (who determined what has been preserved?), documents created with the future historian in mind, and documents that can be misunderstood because of tacit knowledge to which the historian is not privy. Perhaps the epistemic privileging of the document is somewhat misplaced – Paul Thompson (1988:43) has provocatively stated in his standard reference *The Voice of the Past: Oral History* that "the notion that the document is not mere paper, but reality, is … one of the psychological assumptions which underpin the documentary empirical tradition in history." Third, interviews allow an interviewer to learn personalities first hand, which is invaluable in assessing other evidence and in framing and understanding the individual stories that make up a history. Fourth, interview quotations allow written history to have a much livelier flavor. The spoken word has an immediacy and informality that affects us very differently than the written word. Finally, ever since the long-distance telephone and air travel became cheap and largely replaced the letter as

---

[1] Quoted in a prefatory note for *Mark Twain, a Biography* by A. B. Paine (1912, New York: Harper).
[2] Also see Weart and DeVorkin (1981) and DeVorkin (1990) on techniques of oral history as applied to physics and astronomy.

a means of communication, oral histories may be our only way to find out about key interactions.[3,4]

The value of interviews very much depends on the relationship between the interviewer and the interviewee. As Lillian Hoddeson (2006:302), an experienced interviewer on history of modern physics, has said:

> Historical research using interviews may be compared with studying the kind of system (ranging from the quantum-mechanical to the psychological) in which the process of observation changes what is being observed, thus injecting a degree of indeterminism. Interviews can be seen as dialectics that operate between historian and interviewee, between present and past, and between interviews and every other kind of [historical] source.

Personal dynamics play a prominent role in an interview – if no rapport develops early on, one might as well stop. The ideal situation is where a dialogue develops, both parties determined to find the "truth," or some decent semblage of same. This seeking may involve follow-up archival research, correspondence, or a second interview. Note that indeterminism is also intrinsic to *written* materials: one need only compare the different histories that commonly result from use of essentially the same corpus of archival evidence.

The interviewer/ee relationship is fundamentally influenced by whether the interviewer is in some sense an "insider," someone who because of credentials or past history is treated as a colleague; or is the interviewer someone from another culture and therefore treated, at least at first, with less trust and more of a public face? In my own case, as a practicing radio astronomer of a generation after the period under study (Ph.D. in 1971), in most cases I was talking with radio, radar, and optical astronomers who were senior colleagues, even if I had never met them before. This meant that we had a common technical language that facilitated discussion of many issues, and that there was (usually) an immediate sense of trust. The interviews tended to be more of a conversation than would have been so for an outsider. But this also undoubtedly meant that I, as a member of the community, was less likely to be aggressive in interviewing. Moreover, because of a shared background with the interviewees, I undoubtedly missed certain questions and points of view that an outside historian would have had. It was harder for me to ask the basic, but often illuminating, questions that an outsider might have asked, e.g., "What was radio noise?" The problem is that of the anthropologist: one cannot profoundly understand a culture without becoming a member, but at one point has one been coopted? There is no ideal solution – if we should ever have a future "outsider" history of early radio astronomy, the comparison with the present study would be interesting indeed.[5]

B.2 THE COLLECTION

The goal of the interview project was to talk to everyone who had published at least one article in the field of radio astronomy before 1960. The basic list was assembled in 1971–72 using the Cornell *Bibliographies of Extraterrestrial Radio Noise* for the years before 1954 and the CSIRO *Radio Astronomy Bibliographies* for 1954–60 (see Appendix C). These are very complete and include, for instance, even popular articles, abstracts of talks, laboratory reports, and book reviews. For the primary literature they include any article that in the least refers to radio astronomy. For meteor radar astronomy (not included in the above bibliographies), I began with Lovell's monograph *Meteor Astronomy* (1954) and a bibliography provided by Nigel Gilbert (1975). Priorities were assigned for obtaining interviews based on each person's impact on the field as gauged from review articles and existing popular historical studies, although from the beginning I wanted to talk not just to the "generals," but also to the foot soldiers, the

---

[3] The prevalence of email may or may not have brought the era of a minimal written informal record to an end. But will this even richer (and overwhelming) archival record be preserved for future historians? See Doel (2003:369) and his cited references regarding the "digital dark ages" that may have already begun.

[4] As one gauge of the importance of oral histories in doing the history of science after 1950, note that the Center for History of Physics, American Institute of Physics, College Park, Maryland, USA (www.aip.org/history/nblbro.htm), has developed particular expertise in oral histories; transcripts of over 1500 interviews are on file and cataloged on the website.

[5] Mulkay (1974:110–1) discusses the pros and cons of having David Edge, a former radio astronomer, as his partner when interviewing British radio astronomers for *Astronomy Transformed* (Edge and Mulkay 1976).

"average" early radio astronomers. In addition, I added names to the interview list such as associated optical astronomers, theorists, ionosphere researchers, and administrators.

Using the original criterion (at least one publication before 1960), about 340 names were identified, of which 17 had already died by 1971 (the most important of which were E. V. Appleton, W. Baade, K. G. Jansky, D. F. Martyn, J. L. Pawsey, and O. Struve). A further 14 persons died before I could interview them (the most important of which were L. L. McCready, R. Payne-Scott and G. C. Southworth). Over the period 1971–88 (but mostly 1973–81), I interviewed a total of 255 persons and was not able to interview ~60 more still on my list (although only ~10 of these had much priority). Only two persons refused to be interviewed: J. A. Ratcliffe and J. S. Hey (although Hey fully cooperated in answering questions via mail). The interviews reside on 175 cassette tapes, which also include recordings of several sessions at conferences devoted to the history of radio astronomy (many of these talks led to articles collected in Sullivan [1984a]).

Persons were interviewed primarily at scientific meetings and at their home institutions whenever I happened to be visiting. Two major exceptions to this were specific trips for doing history, one to Australia in 1978 and another to the Soviet Union in 1980. Sabbatical years at Cambridge University in 1980–81 and 1987–88 also allowed much time for interviews and other historical research. Preparation for an interview involved reviewing the subject's publications, an annotated list of which then acted as an interview outline. In addition to scientific and technical questions arising from the publications, questions were asked about: educational background; first contacts with radio astronomy; motivations for doing studies, experiments, and observations; reasons for changing jobs; problems with funding of projects and refereeing of papers; national contexts of research; impressions of other groups and of scientific meetings; opinions regarding institutional leaders and their success; the general research and social milieu; anecdotes about observing, data reduction, collaborations, etc.; and opinions on key controversies of the time (e.g., the nature of the radio stars). Also asked were a few standard questions such as "When do you think radio astronomy became accepted by the optical astronomers?", "When, if ever, did you feel like you had become an astronomer?", and "How did radio astronomy change astronomy?"

These interviews were *not* the 4–5 hour oral histories that other projects have produced to cover a person's entire life and career (usually once the interviewee is retired); for instance, personal details were rarely sought, although they often cropped up. The average length of the interviews used for this book was 1.4 hr, with the most important persons averaging ~2 hr (these latter interviews were also more likely to be transcribed). Some interviews out of necessity were conducted by telephone (and recorded, too). On the whole, interviews were strictly focused on the person's contribution to radio astronomy or opinions of the same in the pre-1960 era. For example, an ionosphere researcher might have had an entire career studying radio propagation, but my queries would be only about the two papers he wrote in 1951–52 regarding radio scintillations of Cygnus A.

Another key property of a collection of interviews is their epoch (as astronomers would say). In the present case the interviews happened over 1971–88. These interviews have now become history themselves, and they covered scientific activities that happened mostly in the 1940–60 epoch, thus 10 to 40 years in the past. Some of this past was recent enough that some issues, especially involving controversies and personalities, were still very sensitive, and interviewees were not willing to let me rummage through their correspondence files (although, if needed, they could often quickly locate a letter, notebook, or report). On the other hand, for the earliest times some scientists' papers had found their way into archives (see Appendix C). The ages of the interviewees ranged from their 30s to 80s, although the bulk were in their 50s or 60s (and I was in my 30s).

Processing of the tapes and other information was just as important as the original interview. For those interviews that were transcribed (about 60% of those used in this book), I carefully checked the transcripts, made corrections and queries by hand, and then sent them to each interviewee for further emendations, additions, and comments (usually by hand in the text, sometimes via endnotes).[6] A table of contents was made

---

[6] The most interesting reply from an interviewee upon seeing the transcript of his interview: "I find the transcript of our conversation really almost useless as a guide to either what I thought

for each transcript. Each interviewee was also asked for documents such as a curriculum vitae, list of publications, photographs, short biography, etc. All of this information is now contained in a file for each person. These files (which also exist for many persons not interviewed) also contain much other information that I have gathered over the years: newspaper and magazine articles, photographs, obituaries, correspondence, photocopies of original data, etc. The correspondence mostly deals with draft chapters of this book that were sent to all parties concerned, primarily in the 1983–89 period. With each draft I included many specific queries along with a request for general comments and permissions for interview quotations.

All of these files on early radio astronomers, together with the tapes of the interviews (and photographs by me of about one-half of the interviewees), comprise a vital portion of my resources on the history of radio astronomy through 1960. Another portion is an indexed collection of reprints and photocopies of the papers and reports relevant to radio astronomy for the period covered by this book (through 1953), as well as for the 1953–60 period, although less complete. There is also a collection of ~1000 photographs covering the pre-1960 period; many of these have been used in this book (and constitute yet another category of historical evidence that has its pros and cons). I consider this archive of early radio astronomy, already several decades old, a valuable legacy to pass on to future historians. I plan soon to deposit all of this material in the archives of the National Radio Astronomy Observatory in Charlottesville, Virginia, USA (see Section C.2).

Table B.1 lists the 115 interviewees that contributed to this book. It is a measure of the delays in finishing this book, as well as of the value of recorded interviews, that 72 of these early radio astronomers and other researchers are now deceased (through 2008). Fourteen transcripts came from other sources, mostly Edge and Mulkay (1976:422–3).[7] Table B.2 gives the names of the other 140 interviewees whose interview material was not explicitly used in this book, mostly because they were not active in the pre-1954 period. Table B.3 presents statistics regarding the interviews, broken down by country and category of interviewee.

B.3 HOW THE INTERVIEWS HAVE BEEN USED

For this book I have used the interviews as a category of evidence to be weighed along with other categories. If archival evidence and interview evidence are in conflict on a certain point, I have tried to understand the discrepancy through further digging (checking archives, queries of principals, etc.), but in the end have always given more weight to the written, contemporaneous record. If the archival record is silent on an assertion or a story in an interview, and the point is plausible and consistent with other evidence, I have quoted or cited the interview as the source. I have been reluctant to do this for any point of importance unless at least one more interviewee has confirmed the point (the "journalist's criterion"). However, in some cases, for example aspects of Grote Reber's story for which nothing more than his recollections exist, I have nevertheless used them as the best available evidence.

All quotations from interviews have been attributed, with the notations "Smith (1976:14T)", "Smith (1976:107B:420)" or "Smith (1976:2N)" referring to a 1976 interview with Smith: either page 14 of a transcription, or Tape 107, Side B, position 420 of an untranscribed tape, or page 2 of my notes of an unrecorded interview. It is a mark of how much more I am a historian than a sociologist to compare this practice with that in *Astronomy Transformed* by Edge and Mulkay (1976), who liberally use quotations from their interviews, but always attributed to, for example, "one of our Jodrell Bank respondents."

The interviews and their transcripts are historical documents in their own right, and I have sometimes been vexed about how accurately I should quote a transcript, or even at an earlier stage, how accurately the transcript should follow the spoken word. For most people, spoken language is remarkably unkempt: repetitious, halting, ungrammatical, out of order, and slangy. To parrot such a mess in this book's quotations would be a disservice to all concerned. But how much cleaning up, punctuating, rearranging, eliding, and combining is ethical and useful? Different historians will come to

---

or what you know. I will in time try and send you a drastically cut and edited version." The first sentence is sobering; the second illustrates why feedback is a good idea. [Jesse Greenstein: author, 13 Sep 1983]

[7] However, I also did my own interviews, very different in style, with each of their subjects (see Section 1.3.1).

Table B.1 *115 persons interviewed for this study*

Interviewee	Year of interview	Length (hrs)	Transcribed?	Interviewer
Allen, C. W.†	78 (= 1978)	0.5	N	
Baldwin, J. E.	72, 81	3.2	N	
Bay, Z.†	80	0.8	N	
Beck, A. C.	77	–	nt	
Blum, E.-J.	76	1.1	N	
Bolton, J. G.†	76, 78	4.2	Y	
Bowen, E. G.†	73, 78	2.2	Y	
Bracewell, R. N.†	80	3.0	N	
Burbidge, G. R.	79	0.9	Y	
Christiansen, W. N.†	76	1.5	Y	
Clegg, J. A.†	71	1.2	Y	EM
	81	0.8	N	
Covington, A. E.†	76	1.7	Y	
Crawford, A. B.	77	–	nt	
Davies, J. G.†	71	2.0	Y	EM
	78	0.6	Y	
Dennise, J.-F.	76	1.1	Y	
DeVaucouleurs, G.†	76	–	nt	
Dewhirst, D. W.	71	2.0	Y	EM
	76	1.1	N	
DeWitt, J. H.†	78	1.2	N	
Dicke, R. H.†	76, 79	0.8	Y	
DuBridge, L. A.†	83	0.6	N	
Elsmore, B.	71	1.0	Y	EM
Eshleman, V. R.	78	0.6	N	
Ewen, H. I.	79	2.2	Y	
Findlay, J. W.†	81	1.5	N	
Folland, D. F.	74, 83	–	nt	
Fränz, K.†	73	0.3	Y	
Friis, H. T.†	76	0.2	Y	
Gardner, F. F.†	73	0.8	Y	
Getmantsev, G. G.†	78	0.7	Y	
Ginzburg, V. L.	76, 80	1.2	Y	
Gold, T.†	76	1.5	Y	
	78	4.5	Y	SW
Greenstein, J. L.†	79, 80	3.1	P	
Haddock, F. T.†	84	–	nt	

Table B.1 (cont.)

Interviewee	Year of interview	Length (hrs)	Transcribed?	Interviewer
Hagen, J. P.†	76	1.9	Y	
Hanbury Brown, R.†	76	1.5	Y	
Hawkins, G. S.†	73		Y	GG
	84	0.8	N	
Hazard, C.	71	1.9	Y	EM
	73, 78, 81	2.8	P	
Heightman, D. W.	81	0.8	N	
Hewish, A. J.	71	1.2	Y	EM
	76	0.9	Y	
Higgs, A. J.	78	0.5	N	
Hindman, J. V.†	78	1.6	N	
Hoyle, F.†	81	–	N	
	88	1.9	N	
Hughes, V. A.†	79	0.9	N	
Hvatum, H.†	85	0.8	N	
Jansky, C. M.†	73	0.3	Y	
Jennison, R. C.†	72	2.0	Y	EM
	76	1.4	Y	
Kaydanovsky, N. L.	80	–	nt	
Kerr, F. J.†	71, 73	3.3	Y	
Kraus, J. D.†	75	0.7	Y	
Labrum, N. R.	78	0.4	N	
Laffineur, M.†	76	0.6	Y	
Lehany, F. J.	78	0.5	N	
Little, A. G.†	78	0.7	N	
Little, C. G.	81	0.3	N	
Lovell, A. C. B.	71	1.8	Y	EM
	76	1.3	Y	
McClain, E. F.	73	1.0	Y	
Machin, K. E.†	71	1.7	Y	EM
	81	1.1	N	
McCrea, W. H.†	88	0.2	N	
Manning, L. A.	79	0.4	N	
Mayer, C. H.†	71, 78	1.8	Y	
Menzel, D. H.†	76	1.2	N	
Millman, P. M.†	79	0.5	N	
Mills, B. Y.	76	2.7	Y	

Table B.1 (cont.)

Interviewee	Year of interview	Length (hrs)	Transcribed?	Interviewer
Minkowski, R.†	74	1.1	Y	
Minnett, H. C.†	78	2.0	Y	
Molchanov, A. P.†	80	–	nt	
Moriyama, F.	79	0.5	N	
Morrison, P. A.†	76	0.5	Y	
Muller, C. A.†	73	1.8	Y	
Murray, J. D.	78	1.4	N	
O'Brien, P. A.	81	0.8	N	
Oort, J. H.†	72, 78	1.9	Y	
Ovenden, M. W.†	76	0.8	N	
Parsons, S. J.	78	1.1	Y	
Payne-Scott, R.†	78	–	nt	
Peterson, A. M.†	81	0.3	N	
Phillips, J. W.	78	0.8	Y	
Piddington, J. H.†	78	1.2	Y	
Potapenko, G. W.†	74, 75	0.6	Y	
Purcell, E. M.†	77	0.9	Y	
	77a		Y	KS
Reber, G.†	75, 78	4.8	Y	
Rydbeck, O. E. H.†	78	0.9	Y	
Ryle, M.†	71	1.5	Y	EM
	76, 81	2.5	P	
Sanamyan, V. A.	80	–	nt	
Scheuer, P. A. G.†	76	0.8	N	
Seeger, C. L.†	75	1.0	Y	
Shklovsky, I. S.†	79	1.9	N	
Shostak, A.	76	0.9	Y	
Skellett, A. M.†	77	0.5	Y	
Slee, O. B.	78	1.4	Y	
Sloanaker, R. M.	73	0.8	N	
Smerd, S. F.†	78	1.4	N	
Smith, F. G.	76	1.2	Y	
Stagner, G. H.	84	–	nt	
Stanier, H. M.	81	0.5	N	
Stanley, G. J.†	74, 81	3.3	P	
Steinberg, J.-L.	76	1.6	N	
Stumpers, F. L. H. M.	81	0.6	N	
Suzuki, S.	78	0.6	N	

Table B.1 (cont.)

Interviewee	Year of interview	Length (hrs)	Transcribed?	Interviewer
Takakura, T.	81	0.4	N	
Tanaka, H.†	78	0.8	N	
Townes, C. H.	71, 81	1.5	P	
Trexler, J. H.†	75	0.6	Y	
Troitsky, V. S.†	76	0.8	Y	
Tuve, M. A.†	73	0.9	Y	
Twiss, R. Q.†	81	0.9	N	
Unsöld, A.†	73	0.6	Y	
Van de Hulst, H. C.†	73, 78	1.8	Y	
Vonberg, D. D.	71	0.8	Y	EM
	81	0.9	N	
Waldmeier, M.†	78	0.3	N	
Westerhout, G.	73	2.3	Y	
Westfold, K. C.†	78	0.9	N	
Whipple, F. L.†	79	0.6	P	
Wild, J. P.†	78	2.5	Y	
Yabsley, D. E.†	78	0.7	N	

† = known to be deceased (through 2008) (72 persons)

nt = notes only (interview not taped); P = interview partially transcribed

EM = D. O. Edge and M. J. Mulkay; copy of transcript in author's collection (Edge and Mulkay conducted a total of 20 interviews for their book *Astronomy Transformed* [1976:422–3])

SW = S. R. Weart; on file at the Center for History of Physics (CHP)

GG = G. N. Gilbert; on file at CHP

KS = K. Sopka; on file at CHP

All other interviews by the author.

Also consulted were transcripts of interviews conducted by R. Kestenbaum (1965) concerning Karl Jansky's career (see note 3.42).

Table B.2 *140 persons interviewed, mostly on post-1954 radio astronomy, and thus not used for this study*

J. Aarons	S. J. Goldstein†	P. G. Mezger
V. A. Ambartsumian†	W. E. Gordon	A. T. Moffet†
A. H. Barrett†	S. Gorgolewski	I. G. Moiseev†
L. Biermann†	B. H. Grahl	D. Morris
J. H. Blythe	O. Hachenberg†	J. A. O'Keefe†
A. Boischot	T. Hagfors†	H. P. Palmer†
B. J. Bok†	D. E. Harris	P. Palmer
H. Bondi†	D. S. Heeschen	Yu. N. Parysky
A. Braccessi	J. Heidmann†	I. I. K. Pauliny-Toth†
N. W. Broten	H. L. Helfer	A. A. Penzias
B. F. Burke	E. R. Hill	G. H. Pettengill
T. D. Carr	J. A. Högbom	R. M. Price
J. L. Casse	D. C. Hogg	W. Priester
B. G. Clark	B. Höglund	V. Radhakrishnan
R. J. Coates	B. G. Hooghoudt	E. Raimond
M. H. Cohen	W. E. Howard	V. A. Razin
R. G. Conway	V. N. Ikhsanova	H. Rishbeth
B. F. C. Cooper†	J. R. Johler	J. A. Roberts
C. H. Costain†	F. D. Kahn†	M. S. Roberts
H. V. Cottony	M. A. Kaftan-Kassim	B. J. Robinson†
R. D. Davies	N. S. Kardashev	N. G. Roman
C. DeJager	K. I. Kellermann	B. Rowson
N. H. Dieter	M. M. Komesaroff†	N. F. Ryzhkov
H. W. Dodson-Prince	D. V. Korol'kov†	O. N. Rzhiga
F. D. Drake	M. Kundu	A. E. Salomonovich†
A. F. Dravskikh	A. D. Kuz'min	A. R. Sandage
Z. V. Dravskikh	M. I. Large†	M. Schmidt
R. B. Dyce	J. Lequeux	D. W. Sciama†
D. O. Edge†	P. R. R. Leslie-Foster	P. F. Scott
T. M. Egorova	A. E. Lilley	G. A. Seielstad
Ø. Elgarøy†	J. L. Locke	J. R. Shakeshaft
J. Elldér	W. B. McAdam	A. M. Shakhovskoy
G. R. Ellis	R. X. McGee	K. V. Sheridan
W. C. Erickson	G. C. McVittie†	W. L. Shuter†
J. V. Evans	D. S. Mathewson	P. Simon
S. Evans	T. A. Matthews	A. G. Smith†
G. B. Field	L. I. Matveyenko	H. J. Smith†
A. D. Fokker	A. Maxwell	N. S. Soboleva
J. A. Galt	M. L. Meeks	R. L. Sorochenko
R. M. Goldstein	T. K. Menon	K. S. Stankevich

Table B.2 (cont)

G. Swarup	H. Van Woerden	G. R. Whitfield
G. W. Swenson	S. Von Hoerner†	D. R. W. Williams
G. N. Taylor	V. A. Udal'tsov	R. W. Wilson
A. R. Thompson	C. M. Wade	L. Woltjer
G. M. Tovmassian	D. Walsh†	B. S. Yaplee
J. Tuominen†	H. F. Weaver	
H. Van der Laan	S. Weinreb	
	J. B. Whiteoak	

† = known to be deceased (through 2008) (32 persons)

Table B.3 *Interviewee statistics*

Country/Group	No.[a]	Category	No.[a]
Australia	23 (35)	Radio astronomy	64 (180)
England			
Cambridge	11 (20)	Radar astronomy	12 (22)
Jodrell Bank	9 (23)		
Other	10 (12)	Optical astronomy	8 (13)
USA	34 (82)	Theorist	18 (20)
USSR	7 (28)		
Holland	5 (15)	Other	13 (20)
France	4 (9)		
Japan	4 (4)		
Canada	2 (6)		
Germany	2 (7)		
Sweden	2 (5)		
Other	2 (9)		
Total	115 (255)	**Total**	115 (255)

[a] n (p) means n interviews used in the present study (Table B.1), and p interviews total, covering the field through the early 1960s (Table B.2)

different conclusions: I have allowed myself to modify transcript text in order to make quotations more readable so long as I am confident that no meaning has been changed. (Interviewees' permissions were obtained for the cleaned-up versions.) Here is an example, where Roger Jennison relates his Eureka moment regarding puzzling fringe visibility data that he had obtained for Cygnus A:

As transcribed:

> I've told this tale before and it sounds silly, but it's perfectly true. I know it sounds like Archimedes, but I took a long bath and I lay back in the bath thinking about this distribution. Well, all of a sudden, it clicked, that if this thing Cygnus were two blobs instead of one, then I could get this peculiar difference between the projections in different directions.

As quoted in Section 14.5.1.3:

> I know this sounds like Archimedes, but it's perfectly true: one day I took a long bath and I was laying back in the tub thinking about this distribution. And then all of a sudden it clicked that if Cygnus were *two* blobs instead of one, then I could get this peculiar difference between the projections in different directions.
>
> (Jennison 1976:18–19T)

Appendix C
Bibliographic notes and archival sources

C.1 BIBLIOGRAPHIES OF EARLY RADIO AND RADAR ASTRONOMY

The historian's entry into the primary literature has been made immeasurably easier by the existence of several comprehensive bibliographies of early radio astronomy. Each supplies not only citations, but also abstracts and other useful information. The earliest series, called *Bibliography of Extraterrestrial Noise* and covering the period through 1953, was compiled by Martha Stahr Carpenter of Cornell University and issued in five reports (Carpenter 1950 to 1960)*.[1] Equally valuable were two follow-up bibliographies produced at the CSIRO Radiophysics Division covering, respectively, 1954–56 and 1957–60 in two volumes (McKechnie *et al.* 1957*, 1963*). For the Soviet scene, the most complete coverage (up through 1958) is afforded by Neishil'd and Panovkin (1960)*.

None of the above, however, cover *radar* studies, and unfortunately a comprehensive radar astronomy bibliography for this period was never assembled. Nevertheless, *Bibliography of the Ionosphere* by Manning (1962)* has been enormously useful, as it includes many meteor radar papers, as well as other ionospheric topics relevant to radio astronomy (through 1960). Finally, G. N. Gilbert shared with me a bibliography of meteor radar papers that he assembled for his thesis (Gilbert 1975).[2]

Carpenter, M. S. (1950–60). *Bibliography of Extraterrestrial Radio Noise* [up through 1953]. School of Electrical Engineering, Cornell University, Ithaca, New York. Issued in five parts that covered: up through 1949 (published in 1950); then the individual years 1950 through 1953 (published respectively in 1952, 1953, 1958, and 1960). These documents were also known as Radio Astronomy Reports No. 11, 12, and 13 and Research Reports EE 371 and EE 444.

Manning, L. A. (ed.) (1962). *Bibliography of the Ionosphere: an Annotated Survey through 1960.* (Stanford, Cal.: Stanford University Press).

McKechnie, M., Kerr, F. J., and Shain, C. A. (1957). *Radio Astronomy Bibliography 1954–1956.* CSIRO Div of Radiophysics, Sydney, Australia.

McKechnie, M., Kerr, F. J., and Hill, E. R. (1963). *Radio Astronomy Bibliography 1957–1960.* CSIRO Div of Radiophysics, Sydney, Australia.

Neishil'd, V. G. and Panovkin, B. N. (1960). *Radio Astronomy: an Annotated Bibliographical Guide to Soviet and Foreign Literature for 1932–1958.* (Moscow: Soviet Academy of Sciences). [In Russian]

C.2 ARCHIVAL COLLECTIONS USED IN THIS STUDY

The primary archival resources mined for this study are listed in Table C.1, along with the three-letter abbreviation used for each in footnote citations. I consulted many of these collections before they became institutionalized, and in many cases then assisted with their transfer from private ownership to a proper archival home.

By far the most valuable of the lot were the archives of the CSIRO Radiophysics Division, Sydney (RPS). The Division (then called the Radiophysics Laboratory) contained one of the three major research groups in early radio astronomy (Chapter 7), and their own important work, as well as their interactions with the rest of the world, are represented. Moreover, as a government organization, the records are especially thorough, well-organized, and preserved, largely through the efforts of administrative assistant Sally Atkinson over the period from World War II through the 1980s when she became unofficial archivist. As an example, File D5/4 contains a separate folder for *every*

---

[1] Publication details of the bibliographies marked with an asterisk are listed at the end of this section.
[2] In Section 1.3.1 I comment on earlier books treating the history of radio astronomy.

Table C.1 *Primary archives cited in this study*

Abbreviation	Collection	Institution	Location
APP	Appleton, E. V.	Library Edinburgh Univ.	Edinburgh, UK
BTL	Jansky, K. G. Skellett, A. M.	AT&T Archives & History Center	Warren, New Jersey, USA
CAV	Radio Astronomy Library	Cavendish Lab	Cambridge, UK
CHP	Interviews	Center for History of Physics	College Park, Maryland, USA
GRE	Greenstein, J. L.	Millikan Lib., Cal. Inst. Technology	Pasadena, California, USA
JA1	Jansky, K. G.	Library, Univ. Wisconsin	Madison, Wisconsin, USA
JA2	Jansky, K. G.	Privately held by David B. Jansky	
LOV[1]	Lovell, A. C. B. (Jodrell Bank Archives)	John Rylands Lib., Univ. Manchester	Manchester, UK
MIN	Minkowski, R.	Bancroft Lib., Univ. California	Berkeley, California, USA
OOR[2]	Oort, J. H.	Library, Leiden Univ.	Leiden, The Netherlands
PFE	Pfeiffer, J.	Library, Ohio St. Univ.	Columbus, Ohio, USA
RPL		Library, Australia Telescope Natl. Facility, CSIRO	Marsfield, New South Wales, Australia
RPP[3]		Historic Photographic Archive, ATNF	Marsfield, New South Wales, Australia
RPS[4]	Radiophysics Div., CSIRO	National Archives of Australia	Chester Hill, New South Wales, Australia
RYL[5]	Ryle, M.	Churchill College, Cambridge Univ.	Cambridge, UK
SOU	Southworth, G. C.	AT&T Archives & History Center	Warren, New Jersey, USA
VDH	Van de Hulst, H. C.	Sterrewacht, Leiden Univ.	Leiden, The Netherlands
YKS	Struve, O. (Yerkes Observatory)	Archives, Univ. Chicago	Chicago, Illinois, USA

[1] Items cited in the text with the format "file Rn (uncat.), LOV" were examined before these Lovell papers were deposited and organized.

[2] Finding aid published as *The Letters and Papers of Jan Hendrik Oort* by J. K. Katgert-Merkelijn (1997, Dordrecht: Kluwer).

[3] Orchiston, W., Chapman, J., and Norris, B. (2004). "The ATNF Historic Photographic Archive: Documenting the history of Australian radio astronomy." In Orchiston, W. *et al.* (eds.). *Astronomical Instruments and Archives from the Asia-Pacific Region.* Seoul: Yonsei University. pp. 41–48.

[4] After my use of these archives, they were moved from the Radiophysics Division to the National Archives of Australia. The file names cited in this book, however, have been retained in their new home, where they can all be found in series C 3830.

[5] Items cited in the text with the format "file n (uncat.), RYL" were examined before these Ryle papers were deposited, but the author's numbering system (values for n) has been maintained in the archive.

scientific publication that originated in the Division: all correspondence, referee reports (internal and external), often draft versions, etc.

The Archives of the National Radio Astronomy Observatory in Charlottesville, Virginia, USA (www.nrao.edu/archives) in recent years has begun a major program to preserve the papers of radio astronomers (e.g., Grote Reber, John Kraus, and Ronald Bracewell). In the case of Reber, the collection includes not only the usual written material (and in great abundance), but also almost all of his original data and much of his original equipment!

C.3 COLLECTIONS OF BIOGRAPHIES

Two collections of biographies appeared too late for their contents to be incorporated into the present volume. The first is the eight-volume *New Dictionary of Scientific Biography* (2008) edited by N. Koertge (Detroit: Thomson Gale), which covers many more major twentieth-century scientists than the classic *Dictionary of Scientific Biography* from the 1970s. Detailed biographies and bibliographies can be found there on the following *dramatis personae* of early radio astronomy: H. Alfvén, P. M. S. Blackett, B. J. Bok, J. G. Bolton, R. H. Dicke, T. A. Gold, J. L. Greenstein, F. Hoyle, W. H. McCrea, D. H. Menzel, J. H. Oort, E. M. Purcell, M. Ryle, I. S. Shklovsky, A. Unsöld, H. C. Van de Hulst, and F. L. Whipple.

The second collection is the two-volume *Biographical Encyclopedia of Astronomers* (2007) edited by T. Hockey (New York: Springer). Here the articles are much shorter, but excellently written and covering the astronomical aspects of everyone from Parmenides to Pawsey.

C.4 LITERATURE ON RADAR DEVELOPMENT THROUGH 1945

The development of early radar has been treated in official histories, in memoirs by the participants, in popular books, and in a few cases by historians. The most comprehensive history of the wartime era is Louis Brown's *A Radar History of World War II* (1999), which covers both military and technical aspects of radar as it developed around the world. The next best source is *Tracking the History of Radar* edited by O. Blumtritt *et al.* (1994), which contains an exhaustive bibliography of the worldwide history of radar in the 1935–50 era. Also useful are theses by Swords (1986), Kern (1984), and Allison (1981), the last dealing mainly with the US Naval Research Laboratory. Beyerchen (1998) is an excellent article comparing the technical, military, and social factors influencing radar development in Germany, Britain, and the US. The present volume summarizes the wartime development of radar in Section 5.1, and then touches on specific episodes during the remainder of Chapter 5, the first parts of Chapters 6 through 9, parts of Chapter 10, Section 12.2, and Section 17.1.1.

Other useful general references include special issues of the *J. IEE* (Vol. **93**, Part IIIA, 1946), the *Proc. IRE* (Vol. **50**, 1962), and the *IEE Proc.* **A132**, No. 6 (1985); the last comprises mostly adaptations of British reports from 1944–45 and was also published as *Radar Development to 1945*, ed. R. Burns (London: Peter Peregrinus, 1988). Several chapters on British radar development and its management can be found in *Design and Development of Weapons* by M. M. Postan, D. Hay, and J. D. Scott (London: HMSO, 1964). Tutorial overviews of radar and American wartime systems were given immediately after the war by DeWitt (*J. Franklin Inst.* **241**, 97–124 [1946]), by E. G. Schneider (*Proc. IRE* **34**, 528–78 [1946]). and by K. A. Norton and A. C. Omberg (*Proc. IRE* **35**, 4–24 [1947]). Butrica (1996:1–6) gives a concise, heavily-referenced, and useful overview of the early development of British and American radar.

Also invaluable are textbooks spawned by wartime work, in particular Bowen (1947) and the monumental 28-volume series *Principles of Radar* (Ridenour 1947–53) assembled by the staff of the MIT Radiation Laboratory. Henry E. Guerlac was the Rad Lab's official historian and in 1947 wrote a manuscript entitled *Radar in World War II*, later published in two volumes (Guerlac 1987), as well as "The radio background of radar" in the *J. Franklin Institute* **250**, 285–308 (1950). Besides the Rad Lab's history, Guerlac also covers radar development before the war and in other countries, as well as the actual battle usage of American radar. Kevles (1977:302–23) and Buderi (1996:chapters 1–11) also cover the Lab's history (see also Section 10.1.1). Finally, Galison (1997:288ff, 816–27) gives a cogent overview of the Rad Lab, in particular its microwave physics theory and practice.

References (also an index)

References through 1955: ordered chronologically, then alphabetically within years
References starting in 1956: ordered alphabetically
* = reprinted in *Classics in Radio Astronomy* (Sullivan 1982)
The numbers given in italic at the end of each item refer to the page(s) in the text where the item primarily appears; *n* indicates within a footnote.

1865–1944

Clerk Maxwell, J. (1865). A dynamical theory of the electromagnetic field. *Phil. Trans. Roy. Soc.* **155**, 459–512. *18*

Ebert, H. (1893). Electro-magnetic theory of the sun's corona. *Astronomy and Astrophysics* **12**, 804–810. *22n*

Hertz, H. (1893). *Electric Waves*. London: Macmillan. (Translated into English by D. E. Jones; reprinted in 1962 by Dover, New York.) *18n*

Lodge, O. J. (1894a). *The Work of Hertz and Some of His Successors*. London: The Electrician Publishing Company. (1st edn.). *20*

Lodge, O. J. (1894b). The work of Hertz. *Nature* **50**, 133–139. *20*

Young, C. A. (1895). *The Sun*. New York: Appleton. *89*

Wilsing, J., Scheiner, J. (1896). Ueber einen Versuch, eine electrodynamische Sonnenstrahlung nachzuweisen, und über die Aenderung des Uebergangswiderstandes bei Berührung zweier Leiter durch electrische Bestrahlung. *Annalen der Physik und Chemie* **59**, 782–792. (Translation available in Sullivan (1982:147). * *23*

Wilsing, J. (1897). Bericht über Versuche zum Nachweis einer elektrodynamischen Sonnenstrahlung von J. Wilsing und J. Scheiner *Astronomische Nachrichten* **142**, 17–22. *23*

Scheiner, J. (1899). *Strahlung und Temperatur der Sonne*. Leipzig: Wilhelm Engelmann. *22n*

Lodge, O.J. (1900). *Signalling Across Space Without Wires*. London: The Electrician Publishing Company (3rd edn). *20*

Deslandres, H., Décombe, L. (1902). Sur la recherche d'un rayonnement hertzien émané du Soleil. *C. R. Acad. Sci.* **134**, 527–530. (English Translation available in Sullivan (1982:161).) * *23–4*

Nordmann, C. (1902a). Recherche des ondes hertziennes émanées du Soleil. *C. R. Acad. Sci.* **134**, 273–275. (English Translation available in Sullivan (1982:158).) * *23*

Nordmann, C. (1902b). Recherche sur le rôle des ondes hertziennes en astronomie physique. *Revue Générale des Sciences* **13**, 379–388. *23n*

Nordmann, C. (1902c). Explication de divers phénomènes célestes par les ondes hertziennes. *C. R. Acad. Sci.* **134**, 530–533 (erratum: 680). *24*

Nordmann, C. (1903). Essai sur le rôle des ondes hertziennes en astronomie physique et sur diverses questions qui s'y rattachent. *Annales de l'Observatoire de Nice* **9**. (Ph.D. thesis, University of Paris, 1903.) *23n*

Appleton, E.V., Barnett, M.A.F. (1925). On some direct evidence for downward atmospheric reflection of electric rays. *Proc. Roy. Soc.* **109**, 621–641. *30, 155*

Breit, G., Tuve, M.A. (1925). A radio method for estimating the height of the ionosphere conducting layer. *Nature* **116**, 357. *30*

Eddington, A. S. (1926). Diffuse matter in interstellar space. *Proc. Roy. Soc.* **111**, 424–456. *63, 127*

Gernsback, H. (1927). Can we radio the planets? *Radio News* **8**, 946–947, 1045–1047. *261*

Hulburt, E. O. (1929). Ionization in the atmosphere of Mars. *Proc. IRE* **17**, 1523–1527. *260*

Nagaoka, H. (1929). Possibility of the radio transmission being disturbed by meteoric showers. *Proc. Imperial Academy of Tokyo* **5**, 233–236. *232*

Yokoyama, E. (1930). Interim report on observations of atmospherics which may be caused by meteoric showers. *Proc. Imperial Acad. Tokyo* **6**, 154–157. *232*

Nagaoka, H. (1931). Propagation of wireless-waves. *Scientific Papers of Inst. of Phys. and Chem. Res. (Tokyo)* **15**, 169–188. *232*

Pickard, G. W. (1931). A note on the relation of meteor showers and radio reception. *Proc. IRE* **19**, 1166–1170.   *257*

Potter, R. K. (1931). High-frequency atmospheric noise. *Proc. IRE* **19**, 1731–1765.   *52*

Skellett, A. M. (1931). The effect of meteors on radio transmission through the Kennelly–Heaviside layer. *Phys. Rev.* **37**, 1668.   *233*

Jansky, K. G. (1932). Directional studies of atmospherics at high frequencies. *Proc. IRE* **20**, 1920–1932. *   *34–35*

Schafer, J. P., Goodall, W. M. (1932). Observations of Kennelly–Heaviside layer heights during the Leonid meteor shower of November, 1931. *Proc. IRE* **20**, 1941–1945.   *233*

Skellett, A. M. (1932a). The ionizing effect of meteors in relation to radio propagation. *Proc. IRE* **20**, 1933–1940.   *233*

Skellett, A. M. (1932b). Radio studies during the Leonid meteor shower of November 16, 1932. *Science* **76**, 434.   *234*

Jansky, K. G. (1933a). Radio waves from outside the solar system. *Nature* **132**, 66.   *36*

Jansky, K. G. (1933b). Electrical disturbances apparently of extraterrestrial origin. *Proc. IRE* **21**, 1387–1398. *   *36–41*

Jansky, K. G. (1933c). Electrical phenomena that apparently are of interstellar origin. *Pop. Astron.* **41**, 548–555.   *41*

Minohara, T., Ito, Y. (1933). Effect of the Leonid meteor shower on the ionized upper atmosphere. *Reports of Radio Research in Japan* **3**, 115–125.   *255*

Skellett, A. M. (1933). The ionizing effects of meteors. Ph.D. thesis. Astronomy Dept., Princeton University, Princeton, NJ.   *234*

Cleeton, C. E., Williams, N. H. (1934). Electromagnetic waves of 1.1 cm wave-length and the absorption spectrum of ammonia. *Phys. Rev.* **45**, 234–237.   *86*

Mitra, S. K., Syam, P., Ghose, B. N. (1934). Effect of a meteoric shower on the ionosphere. *Nature* **133**, 533–534.   *256*

Stetson, H. T. (1934). *Earth, Radio and the Stars*. New York: McGraw Hill.   *47*

Jansky, K. G. (1935). A note on the source of interstellar interference. *Proc. IRE* **23**, 1158–1163. *   *41–42*

Skellett, A. M. (1935). The ionizing effects of meteors. *Proc. IRE* **23**, 132–149.   *234*

Arakawa, D. (1936). Abnormal attenuation in short wave radio propagation. *Reports of Radio Research in Japan* **6**, 31–38.   *89*

Heightman, D. W. (1936). U.S.A. 5-metre reception. *Wireless World* **38**, 376 (letter to editor).   *87*

Langer, R. M. (1936). Radio noises from the galaxy. *Phys. Rev.* **49**, 209–210.   *44*

Appleton, E. V., Naismith, R., Ingram, L. J. (1937). British radio observations during the Second International Polar Year 1932–33. *Phil. Trans. Roy. Soc.* **A236**, 191–259.   *258*

Bhar, J. N. (1937). Effect of meteoric shower on the ionization of the upper atmosphere. *Nature* **139**, 470–471.   *256*

Corry, N. (1937). The 28 Mc. band. *T. & R. Bulletin* **12**, 513.   *88*

Dellinger, J. H. (1937). Sudden disturbances of the ionosphere. *Proc. IRE* **25**, 1253–1290. (Identically in *J. Res. Natl. Bur. Standards* **19**, 111–41 (1937).)   *89*

Eckersley, T. L. (1937). Irregular ionic clouds in the E layer of the ionosphere. *Nature* **140**, 846–847.   *257*

Friis, H. T., Feldman, C. B. (1937). A multiple unit steerable antenna for short-wave reception. *Proc. IRE* **25**, 841–917.   *42n*

Heightman, D. W. (1937). Observations on the ultra high frequencies, 1936. *T. & R. Bulletin* **12**, 496–500.   *87*

Jansky, K. G. (1937). Minimum noise levels obtained on short-wave radio receiving systems. *Proc. IRE* **25**, 1517–1530 (erratum: **26**, 400).   *43*

Watson-Watt, R. A., Wilkins, A. F., Bowen, E. G. (1937). The return of radio waves from the middle atmosphere – I. *Proc. Roy. Soc.* **161**, 181–196.   *258*

Whipple, F. L., Greenstein, J. L. (1937). On the origin of interstellar radio disturbances. *Proc. Natl. Acad. Sci. (USA)* **23**, 177–181.   *46–7*

Heightman, D. W. (1938). The ultra high frequencies: a review of conditions in 1937. *Wireless World* **42**, 356–357.   *87*

Pierce, J. A. (1938). Abnormal ionization in the E-region of the ionosphere. *Proc. IRE* **26**, 892–908.   *235*

Skellett, A. M. (1938). Meteoric ionization in the E region of the ionosphere. *Nature* **141**, 472.   *258*

Jansky, K. G. (1939). An experimental investigation of the characteristics of certain types of noise. *Proc. IRE* **27**, 763–768.   *43*

Nakagami, M., Miya, K. (1939). Incident angle of short waves and high-frequency noise during Dellinger effect. *Electrotechnical J. of Japan* **3**, 216.   *89*

Williams, E. J. (1939). Sunspots, magnetic storms and radio conditions. *T. & R. Bulletin* **15**, 7–12, 68.   *88*

Eckersley, T. L. (1940). Analysis of the effect of scattering in radio transmission. *J. IEE* **86**, 548–567.   *257*

Henyey, L. G., Keenan, P. C. (1940). Interstellar radiation from free electrons and hydrogen atoms. *Astrophys. J.* **91**, 625–630. *   *63*

Reber, G. (1940a). Cosmic static. *Proc. IRE* **28**, 68–70. *   *60*

Reber, G. (1940b). Cosmic static. *Astrophys. J.* **91**, 621–624. *   *62*

Whipple, F. L. (1940). Photographic meteor studies. III: the Taurid shower. *Proc. American Philosophical Soc.* **83**, 711–745.   *245*

Blackett, P. M. S., Lovell, A. C. B. (1941). Radio echoes and cosmic ray showers. *Proc. Roy. Soc.* **A177**, 183–186.   *178*

Chamanlal, C., Venkatraman, K. (1941). Whistling meteors – a Doppler effect produced by meteors entering the ionosphere. *Electrotechnics* **14** (Nov.), 28–39.   *256*

Koelbloed, D. (1941). Kosmische radiogolven. *Hemel en Dampkring* **39**, 25–27.   *395n*

Pierce, J. A. (1941). A note on ionization by meteors. *Phys. Rev.* **59**, 625–626.   *257*

Fränz, K. (1942). Messung der Empfängerempfindlichkeit bei kurzen elektrischen Wellen. *Hochfrequenztechnik und Electroakustik* **59**, 143–144, 105–112.   *113*

Hey, J. S. (1942). Notes on G.L. interference on 27th and 28th February. ADRDE (ORG) Memo (Secret, 5 pp.). Army Operational Research Group, Petersham, UK (March). See note 5.11.   *81*

Reber, G. (1942). Cosmic static. *Proc. IRE* **30**, 367–378.   *64*

Southworth, G. C. (1942). Evidence for microwave radiation in the sun's spectrum. Memo-for-file MM-42-160-73. Bell Telephone Laboratories, Holmdel, N. J.   *94*

Fränz, K. (1943). Empfängerempfindlichkeit. in *Fortschritte der Hochfrequenztechnik*. Vol. 2 (eds. Vilbig, F. and Zenneck, J.), Leipzig: Geest & Partig. pp. 685–712. (Galactic radiation discussed on pp. 690–2.)   *114*

FCC (1944). Hearing in the matter of the allocation of radio frequencies between 10 kc and 30,000 Mc. Engineering Dept. Docket 6651 (28 September). Federal Communications Commission, Washington, D.C.   *257*

Lombardini, P. (1944). Possibilità di radio sondaggi astronomici con onde metriche. *Commentationes. Pontificia Academia Scientiarum* **VIII**, 13–19 (MCMXLIV).   *263*

Pawsey, J. L., Payne-Scott, R. (1944). Measurements of the noise level picked up by an S-band aerial. Rept. No. RP 209 (5 pp., Confidential). CSIR Radiophysics Laboratory, Sydney. (11 April).   *127*

Reber, G. (1944). Cosmic static. *Astrophys. J.* **100**, 279–287. *   *65*

1945

Alexander, E. (1945). Report on the investigation of the Norfolk Island effect. R.D. 1/518. Radio Development Lab., D.S.I.R., Wellington, New Zealand.   *85*

Appleton, E. V. (1945a). Departure of long-wave solar radiation from black-body intensity. *Nature* **156**, 534–535.   *90*

Appleton, E. V. (1945b). The scientific principles of radiolocation (the thirty-sixth Kelvin Lecture). *J. IEE* **92** (Pt. I), 340–353.   *110, 264*

Bakker, C. J. (1945). Radiogolven uit het wereldruim. I. Ontvangst der radiogolven. *Nederlandsch Tijdschrift v. Natuurkunde* **11**, 201–209.   *395*

Hey, J. S. (1945). Solar radiations in the 4 to 6 metre wavelength band on 27th and 28th February 1942. Report No. 275 (Confidential, 4 pp.). Army Operational Research Group, Petersham. (13 June)   *83*

Payne-Scott, R. (1945). Solar and cosmic radio frequency radiation: Survey of knowledge available and measurements taken at Radiophysics Lab. to Dec. 1st 1945. Rept. No. SRP 521/27. CSIR Radiophysics Laboratory, Sydney. (Undated)   *130*

Southworth, G. C. (1945). Microwave radiation from the sun. *J. Franklin Institute* **239**, 285–297 (erratum: **241**, following p. 166). *   *95*

Van de Hulst, H. C. (1945). Radiogolven uit het wereldruim. II. Herkomst der radiogolven. *Nederlandsch Tijdschrift v. Natuurkunde* **11**, 210–221. English Translation available in Sullivan (1982:302). *   *389, 395*

1946

Alexander, F. E. S. (1946). The sun's radio energy. *Radio and Electronics* **1**, 16–17, 20.   *85n*

Appleton, E. V., Hey, J. S. (1946a). Solar radio noise – I. *Phil. Mag.* **37**, 73–84.   *111*

Appleton, E. V., Hey, J. S. (1946b). Circular polarization of solar radio noise. *Nature* **158**, 339.   *111*

Bateman, R., McNish, A. G., Pineo, V. C. (1946). Radar observations during meteor showers – 9 October 1946. *Science* **104**, 434–435.   *237*

Bay, Z. (1946). Reflection of microwaves from the moon. *Hungarica Acta Physica* **1**, 1–22.   *273*

Burgess, R. E. (1946). Fluctuation noise in a receiving aerial. *Proc. Phys. Soc.* **A58**, 313–321.   *90*

Denisse, J.-F. (1946). Etude des conditions d'émission par l'atmosphère solaire d'ondes radioélectriques métriques. *Revue Scientifique* **84**, 259–262.   *308*

Dicke, R. H. (1946). The measurement of thermal radiation at microwave frequencies. *Review of*

Scientific Instruments **17**, 268–275. Originally issued as MIT Radiation Lab. Rept. 787 (22 August 1945).*   *203*

Dicke, R. H., Beringer, R. (1946). Microwave radiation from the sun and moon. *Astrophys. J.* **103**, 375–376. *   *205*

Dicke, R. H., Beringer, R., Kyhl, R. L., Vane, A. B. (1946). Atmospheric absorption measurements with a microwave radiometer. *Phys. Rev.* **70**, 340–348.   *204*

Ferrell, O. P. (1946). Meteoric impact ionization observed on radar oscilloscopes. *Phys. Rev.* **69**, 32–33.   *257*

Ginzburg, V. L. (1946). On solar radiation in the radio-spectrum. *Dokl. Akad. Nauk SSSR* **52**, 487–490. [In English]   *215*

Hey, J. S. (1946). Solar radiations in the 4–6 metre radio wave-length band. *Nature* **157**, 47–48. *   *83*

Hey, J. S., Parsons, S. J., Phillips, J. W. (1946). Fluctuations in cosmic radiation at radio-frequencies. *Nature* **158**, 234. *   *103*

Hey, J. S., Phillips, J. W., Parsons, S. J. (1946). Cosmic radiations at 5 metres wavelength. *Nature* **157**, 296–297.   *103*

Hey, J. S., Stewart, G. S. (1946). Derivation of meteor stream radiants by radio reflexion methods. *Nature* **158**, 481–482.   *110*

Kauffman, H. (1946). A DX record: to the moon and back. *QST* **30**, 65–68.   *268*

Kiepenheuer, K. O. (1946). Origin of solar radiation in the 1–6 metre radio wave-length band. *Nature* **158**, 340.   *308*

Levin, B. I. (1946). Observations of Draconids. *Astron. Tsirk.*, 2–4. (Also reported in *Popular Astronomy* **55**, 104–5 (1947).)   *239*

Lovell, A. C. B., Banwell, C. J. (1946). Abnormal solar radiation on 72 Megacycles. *Nature* **158**, 517–518.   *191*

Manning, L. A., Helliwell, R. A., Villard, O. G., Jr., Evans, W. E., Jr. (1946). On the detection of meteors by radio. *Phys. Rev.* **70**, 767–768.   *237*

Martyn, D. F. (1946a). Temperature radiation from the quiet sun in the radio spectrum. *Nature* **158**, 632–633. *   *136*

Martyn, D. F. (1946b). Polarization of solar radio-frequency emissions. *Nature* **158**, 308.   *137*

Mofenson, J. (1946). Radar echoes from the moon. *Electronics* **19**, 92–98.   *269*

Moxon, L. A. (1946). Variation of cosmic radiation with frequency. *Nature* **158**, 758–759.   *115*

Papaleksi, N. D. (1946a). On measuring the earth–moon distance by means of electromagnetic waves. *Elektrichestvo*, 9–15. [In Russian; 1949 English Translation available from RPL.]   *262*

Papaleksi, N. D. (1946b). On measuring the earth–moon distance by means of electromagnetic waves. *Uspekhi Fizicheskikh Nauk* **29**, 250–268. [In Russian]   *262*

Pawsey, J. L. (1946). Observation of million degree thermal radiation from the sun at a wavelength of 1.5 metres. *Nature* **158**, 633–634. *   *136*

Pawsey, J. L., Payne-Scott, R., McCready, L. L. (1946). Radio-frequency energy from the sun. *Nature* **157**, 158–159.   *130*

Reber, G. (1946). Solar radiation at 480 Mc/s. *Nature* **158**, 945.   *71*

Ryle, M., Vonberg, D. D. (1946). Solar radiation on 175 Mc/s. *Nature* **158**, 339–340. *   *162*

Saha, M. N. (1946). Origin of radio-waves from the sun and the stars. *Nature* **158**, 717–718.   *397n*

Sander, K. F. (1946). Measurement of galactic noise at 60 Mc/s. *J. IEE* **93**, Pt. IIIA, 1487–1489.   *115*

Shklovsky, I. S. (1946). On the radiation of radio-waves by the galaxy and by the upper layers of the solar atmosphere. *Astron. Zh.* **23**, 333–347. [In Russian]   *216*

Steinberg, J.-L., Denisse, J.-F. (1946). Parasites d'origine extraterrestre dans le domaine des ondes courtes. *Revue Scientifique* **84** 293–294.   *224*

Unsöld, A. (1946). Die kosmische Kurzwellenstrahlung. *Naturwissenschaften* **33**, 37–40.   *228*

Villard, O. G., Jr. (1946). Listening in on the stars. *QST* **30**, 59–60, 120, 122.   *239*

Webb, H. D. (1946). Project Diana: Army radar contacts the moon. *Sky and Telescope* **5**, 3–6.   *269*

1947

Allen, C. W. (1947). Solar radio-noise of 200 Mc/s and its relation to solar observations. *Mon. Not. RAS* **107**, 386–396.   *297*

Appleton, E., Naismith, R. (1947). The radio detection of meteor trails and allied phenomena. *Proc. Phys. Soc.* **59**, 461–473.   *110*

Bowen, E. G. (ed.) (1947). *A Textbook of Radar*. Sydney: Angus and Robertson. Title page gives author as "Staff of CSIR Radiophysics Laboratory"; Second edn. in 1954 by Cambridge University Press.   *122*

Clegg, J. A., Hughes, V. A., Lovell, A. C. B. (1947). The daylight meteor streams of 1947 May–August. *Mon. Not. RAS* **107**, 369–378.   *185, 243*

Covington, A. E. (1947a). Microwave sky noise. *Terr. Mag. and Atm. Elec.* **52**, 339–341. *213*

Covington, A. E. (1947b). Micro-wave solar noise observations during the partial eclipse of November 23, 1946. *Nature* **159**, 405–406. *214*

Denisse, J.-F. (1947a). Contribution à l'étude des émissions solaires dans le domaine des ondes radioélectriques ultra-courtes. *Annales d'Astrophysique* **10**, 1–13. *289*

Denisse, J.-F. (1947b). Sur les émissions radioélectriques solaires. *C. R. Acad. Sci.* **225**, 1358–1360. *308*

Denisse, J.-F. (1947c). Recherches récentes sur les émissions extraterrestres d'ondes radioélectriques métriques. *Revue Scientifique* **85**, 483–488. *224*

Ginzburg, V. L. (1947). Solar and galactic radio emission. *Uspekhi Fizicheskikh Nauk* **32**, 26–53. [In Russian] *217*

Hey, J. S., Parsons, S. J., Phillips, J. W. (1947). Solar and terrestrial radio disturbances. *Nature* **160**, 371–372. *299*

Hey, J. S., Parsons, S. J., Stewart, G. S. (1947). Radar observations of the Giacobinid meteor shower. *Mon. Not. RAS* **107**, 176–183. *109, 237*

Hey, J. S., Stewart, G. S. (1947). Radar observations of meteors. *Proc. Phys. Soc.* **59**, 858–883. *110*

Khaykin, S. E., Chikhachev, B. M. (1947). Investigation of radio emission from the sun carried out by the USSR Academy of Sciences expedition to Brazil during the solar eclipse of 20 May 1947. *Dokl. Akad. Nauk SSSR* **58**, 1923–1926. [In Russian] *218*

Lovell, A. C. B., Banwell, C. J., Clegg, J. A. (1947). Radio echo observations of the Giacobinid meteors 1946. *Mon. Not. RAS* **107**, 164–175. *184, 236*

Lovell, A. C. B., Clegg, J. A., Ellyett, C. D. (1947). Radio echoes from the aurora borealis. *Nature* **160**, 372. *191*

Martyn, D. F. (1947). Origin of radio emissions from the disturbed sun. *Nature* **159**, 26–27. *308, 309*

McCready, L. L., Pawsey, J. L., Payne-Scott, R. (1947). Solar radiation at radio frequencies and its relation to sunspots. *Proc. Roy. Soc.* **A190**, 357–375. * *132, 176*

Payne-Scott, R., Yabsley, D. E., Bolton, J. G. (1947). Relative times of arrival of bursts of solar noise on different radio frequencies. *Nature* **160**, 256–257. *139, 298*

Pierce, J. A. (1947). Ionization by meteoric bombardment. *Phys. Rev.* **71**, 88–92. *237*

Prentice, J. P. M., Lovell, A. C. B., Banwell, C. J. (1947). Radio echo observations of meteors. *Mon. Not. RAS* **107**, 155–163. *183*

Reber, G., Greenstein, J. L. (1947). Radio-frequency investigations of astronomical interest. *Observatory* **67**, 15–26. *69*

Ridenour, L. N. (ed.) (1947–53). *Principles of Radar*. New York: McGraw-Hill. The Radiation Lab Series, 28 vols., individually titled and edited. *225, 419*

Ryle, M., Vonberg, D. D. (1947). Relation between the intensity of solar radiation on 175 Mc/s and 80 Mc/s. *Nature* **160**, 157–159. *163*

Sander, K. F. (1947). Radio noise from the sun at 3.2 cm. *Nature* **159**, 506–507. *115*

Schott, E. (1947). 175 MHz-Strahlung der Sonne. *Physikalische Blätter* **3**, 159–160. *83*

Shklovsky, I. S. (1947). Emission of radio-waves by the Galaxy and the sun. *Nature* **159**, 752–753. *216*

Stewart, J. Q., Ference, M., Slattery, J. J., Zahl, H. A. (1947). Radar observations of the Draconids. *Sky and Telescope* **6**, 3–5 (March). *237*

Townes, C. H. (1947). Interpretation of radio radiation from the Milky Way. *Astrophys. J.* **105**, 235–240. *369*

Unsöld, A. (1947). Die Kurzwellenstrahlung der Milchstrasse, der Sonne und des Mondes. *Naturwissenschaften* **34**, 194–201. *289, 369*

Woolley, R. v. d. R. (1947a). The solar corona. *Australian J. Science (Suppl.)* **10**, i-x. *138*

Woolley, R. v. d. R. (1947b). Galactic noise. *Mon. Not. RAS* **107**, 308–315. *368*

1948

Allen, E. W., Jr. (1948). Reflections of very-high-frequency radio waves from meteoric ionization. *Proc. IRE* **36**, 346–352 (erratum: 1255–1257). *257*

Bolton, J. G. (1948). Discrete sources of galactic radio frequency noise. *Nature* **162**, 141–142. *142*

Bolton, J. G., Stanley, G. J. (1948a). Variable source of radio frequency radiation in the constellation of Cygnus. *Nature* **161**, 312–313. * *140*

Bolton, J. G., Stanley, G. J. (1948b). Observations on the variable source of cosmic radio frequency radiation in the constellation of Cygnus. *Austral. J. Sci. Res.* **A1**, 58–69. *141*

Clegg, J. A. (1948a). The scattering of radio waves from meteor trails and the determination of meteor radiants.

Ph.D. thesis. Physics Dept., Manchester University, Manchester. *193*

Clegg, J. A. (1948b). Determination of meteor radiants by observation of radio echoes from meteor trails. *Phil. Mag.* **39**, 577–594. *242*

Covington, A. E. (1948). Solar noise observations on 10.7 centimeters. *Proc. IRE* **36**, 454–457. *213, 310*

Eastwood, E., Mercer, K. A. (1948). A study of transient radar echoes from the ionosphere. *Proc. Phys. Soc.* **61**, 122–134. *110n, 258*

Ellyett, C. D. (1948). The photography of radio echos from meteor trails. Ph.D. thesis. Physics Dept., Manchester University, Manchester. *193*

Feenberg, E., Primakoff, H. (1948). Interaction of cosmic-ray primaries with sunlight and starlight. *Phys. Rev.* **73**, 449–469. *387*

Ginzburg, V. L. (1948). New data on solar and galactic radio emission. *Uspekhi Fizicheskikh Nauk* **34**, 13–33. [In Russian] *217*

Giovanelli, R. G. (1948). Emission of enhanced microwave solar radiation. *Nature* **161**, 133–134. *308*

Grieg, D. D., Metzger, S., Waer, R. (1948). Considerations of moon-relay communication. *Proc. IRE* **36**, 652–663. *280*

Haeff, A. V. (1948). Space-charge wave amplification effects. *Phys. Rev.* **74**, 1532–1533. *309*

Herbstreit, J. W., Johler, J. R. (1948). Frequency variation of the intensity of cosmic radio noise. *Nature* **161**, 515–516. *367*

Herlofson, N. (1948). The theory of meteor ionization. *Rep. Prog. Phys.* **11**, 444–454. *252*

Hey, J. S., Parsons, S. J., Phillips, J. W. (1948a). An investigation of galactic radiation in the radio spectrum. *Proc. Roy. Soc.* **A192**, 425–445. *103, 116*

Hey, J. S., Parsons, S. J., Phillips, J. W. (1948b). Some characteristics of solar radio emissions. *Mon. Not. RAS* **108**, 354–371. *112*

Jansky, K. G. (1948). I.F. amplifier. Monograph B-1565. Part of a series edited by H. T. Friis on "Microwave repeater research." *43*

Khaykin, S. E., Chikhachev, B. M. (1948). Investigation of radio emission from the sun during the total solar eclipse of May 20, 1947. *Izvestiya Akad. Nauk SSSR* **12** 38–43. [In Russian] *218*

Lovell, A. C. B. (1948). Meteoric ionization and ionospheric abnormalities. *Rep. Prog. Phys.* **11**, 415–444. *231n*

Lovell, A. C. B., Clegg, J. A. (1948). Characteristics of radio echoes from meteor trails: I. The intensity of the radio reflections and electron density in the trails. *Proc. Phys. Soc.* **60**, 491–498. *252*

Manning, L. A. (1948). The theory of the radio detection of meteors. *J. Applied Physics* **19**, 689–699. *239*

Martyn, D. F. (1948). Solar radiation in the radio spectrum. I. Radiation from the quiet sun. *Proc. Roy. Soc.* **A193**, 44–59. *287*

Menzel, D. H., Salisbury, W. W. (1948). Audio-frequency waves from the sun. *Nature* **161**, 91. *199*

Millman, P. M., McKinley, D. W. R., Burland, M. S. (1948). Combined radar, photographic and visual observations of the Perseid meteor shower of 1947. *Nature* **161**, 278–280. *241*

Reber, G. (1948a). Solar intensity at 480 Mc. *Proc. IRE* **36**, 88. *68*

Reber, G. (1948b). Cosmic static. *Proc. IRE* **36**, 1215–1218. * *68*

Reber, G. (1948c). Cosmic radio noise. *Radio News (Radio-Electronic Enineering edn.)* **11**, 3–5, 29.

Ryle, M. (1948a). The generation of radio-frequency radiation in the sun. *Proc. Roy. Soc.* **A195**, 82–97.

Ryle, M. (1948b). Radio astronomy. *British Science News* **1**, 4–7. (Appeared in about August.) *424*

Ryle, M., Smith, F. G. (1948). A new intense source of radio-frequency radiation in the constellation of Cassiopeia. *Nature* **162**, 462–463. * *166*

Ryle, M., Vonberg, D. D. (1948). An investigation of radio-frequency radiation from the sun. *Proc. Roy. Soc.* **A193**, 98–120. *163, 308*

Schulkin, M., Haddock, F. T., Decker, K. M., Mayer, C. H., Hagen, J. P. (1948). Observation of a solar noise burst at 9500 Mc/s and a coincident solar flare. *Phys. Rev.* **74**, 840. *207*

Shklovsky, I. S. (1948). Galactic radio emission. *Astron. Zh.* **25** 237–245. [In Russian] *381*

Villard, O. G., Jr. (1948). A new technique for studying meteors and the upper atmosphere. Ph.D. thesis. Electrical Engineering Dept., Stanford University, Stanford, Cal. *240*

Waldmeier, M. (1948). Radiofrequente Strahlung und Elektronentemperatur der Sonnenkorona. *Vierteljahrsschrift der Naturforschenden Gesellschaft in Zürich* **93**, 122–135. Also published as *Astron. Mitt. Eidgenössischen Sternwarte Zürich*, No. 154. *311*

Waldmeier, M., Müller, H. (1948). Spektrale Energieverteilung und Mitte-Rand-Variation der radiofrequenten Koronastrahlung. *Vierteljahrsschrift der Naturforschenden Gesellschaft in Zürich* **93**, 186–201. Also published as *Astron. Mitt. Eidgenössischen Sternwarte Zürich*, No. 155.   *289*

Williamson, R. E. (1948). The present status of microwave astronomy. *J. Roy. Astron. Soc. Canada* **42**, 9–32.   *427*

1949

Aspinall, A., Clegg, J. A., Lovell, A. C. B. (1949). The daytime meteor streams of 1948 – I. Measurement of the activity and radiant positions. *Mon. Not. RAS* **109**, 352–358.   *243*

Bolton, J. G., Stanley, G. J. (1949). The position and probable identification of the source of galactic radio-frequency radiation Taurus-A. *Austral. J. Sci. Res.* **A2**, 139–148.   *322*

Bolton, J. G., Stanley, G. J., Slee, O. B. (1949). Positions of three discrete sources of galactic radio-frequency radiation. *Nature* **164**, 101–102. *   *143, 322*

Burkhardt, G., Schlüter, A. (1949). Ausbreitung und ausstrahlung radiofrequenter Wellen in der Sonnenkorona. *Zeitschrift für Astrophysik* **26**, 295–304.   *289*

Christiansen, W. N., Yabsley, D. E., Mills, B. Y. (1949). Measurements of solar radiation at a wavelength of 50 centimeters during the eclipse of November 1, 1948. *Austral. J. Sci. Res.* **A2**, 506–523.   *291*

Davies, J. G., Ellyett, C. D. (1949). The diffraction of radio waves from meteor trails and the measurement of meteor velocities. *Phil. Mag.* **40**, 614–626.   *186*

Denisse, J.-F. (1949a). Relation entre les émissions radioélectriques solaires décimétriques et les taches du soleil. *C. R. Acad. Sci.* **228**, 1571–1572.   *224*

Denisse, J.-F. (1949b). Etude des emissions radioélectriques solaires d'origine purement thermique. Ph.D. thesis. Laboratoire de Physique, Ecole Normale Supérieure, Paris.   *222n, 224*

Denisse, J.-F. (1949c). Influence de l'indice de réfraction sur les émissions radioélectriques d'un milieu ionisé. *C. R. Acad. Sci.* **228**, 751–753.   *289*

DeWitt, J. H., Jr., Stodola, E. K. (1949). Detection of radio signals reflected from the moon. *Proc. IRE* **37**, 229–242.   *271*

Fermi, E. (1949). On the origin of the cosmic radiation. *Phys. Rev.* **75**, 1169–1174.   *387*

Ginzburg, V. L. (1949). *Theory of Radio Wave Propagation in the Ionosphere*. Moscow: Gosudarstvennoy Izdatelstvo Tekhniko-Teoreticheskoy Lit. [In Russian]   *380*

Haeff, A. V. (1949). On the origin of solar radio noise. *Phys. Rev.* **75**, 1546–1551.   *309*

Hagen, J. P. (1949). A study of the radio-frequency radiation from the sun. NRL Report 3504. Naval Research Lab., Washington, DC. (This is essentially Hagen's 1949 Ph.D. thesis for Dept. of Astronomy, Georgetown University, Washington, D.C.)   *208*

Hey, J. S. (1949). Report on the progress of radio astronomy. *Mon. Not. RAS* **109**, 179–214.   *299*

Hoyle, F. (1949). *Some Recent Researches in Solar Physics*. Cambridge: Cambridge University Press.   *289, 309*

Huruhata, M. (1949). Radar observations of meteors. I. *Pub. Astron. Soc. Japan* **1**, 39–43.   *256*

Jaeger, J. C., Westfold, K. C. (1949). Transients in an ionized medium with applications to bursts of solar noise. *Austral. J. Sci. Res.* **A2**, 322–334.   *300*

Kerr, F. J., Shain, C. A., Higgins, C. S. (1949). Moon echoes and penetration of the ionosphere. *Nature* **163**, 310.   *275*

Laffineur, M., Houtgast, J. (1949). Observations du rayonnement solaire sur la longeur d'onde de 55 centimètres. *Annales d'Astrophysique* **12**, 137–147.   *225*

Lehany, F. J., Yabsley, D. E. (1949). Solar radiation at 1200 Mc/s, 600 Mc/s, and 200 Mc/s. *Austral. J. Sci. Res.* **A2**, 48–62.   *311*

Manning, L. A., Villard, O. G., Jr., Peterson, A. M. (1949). Radio Doppler investigation of meteoric heights and velocities. *J. Applied Physics* **20**, 475–479.   *240*

McKinley, D. W. R., Millman, P. M. (1949). A phenomenological theory of radar echoes from meteors. *Proc. IRE* **37**, 364–375.   *242*

Menzel, D. H. (1949). *Our Sun*. Philadelphia: Blakiston Co.   *312*

Millman, P. M., McKinley, D. W. R. (1949). Three-station radar and visual triangulation of meteors. *Sky and Telescope* **8**, 114–116.   *242*

Pawsey, J. L., Yabsley, D. E. (1949). Solar radio-frequency radiation of thermal origin. *Austral. J. Sci. Res.* **A2**, 198–213.   *286*

Payne-Scott, R. (1949). Bursts of solar radiation at metre wavelengths. *Austral. J. Sci. Res.* **A2**, 214–227.   *299*

Piddington, J. H., Hindman, J. V. (1949). Solar radiation at a wavelength of 10 centimetres including eclipse observations. *Austral. J. Sci. Res.* **A2**, 524–538.   *291*

Piddington, J. H., Minnett, H. C. (1949a). Microwave thermal radiation from the moon. *Austral. J. Sci. Res.* **A2**, 63–77. *277*

Piddington, J. H., Minnett, H. C. (1949b). Solar radiation of wavelength 1.25 centimetres. *Austral. J. Sci. Res.* **A2**, 539–549. *277*

Pierce, J. R. (1949). Space-charge waves. *Physics Today* **2**, 20–26. *309*

Ryle, M. (1949a). The significance of the observations of intense radio-frequency emission from the sun. *Proc. Phys. Soc.* **A62**, 483–491. *309*

Ryle, M. (1949b). Evidence for the stellar origin of cosmic rays. *Proc. Phys. Soc.* **A62**, 491–499. *324, 370, 374*

Shklovsky, I. S. (1949). Monochromatic radio emission from the Galaxy and the possibility of its observation. *Astron. Zh.* **26**, 10–14. [In Russian; English Translation available in Sullivan (1982:318).] * *395*

Smerd, S. F., Westfold, K. C. (1949). The characteristics of radio-frequency radiation in an ionized gas, with applications to the transfer of radiation in the solar atmosphere. *Phil. Mag.* **40**, 831–848. *289*

Struve, O. (1949–50). Progress in radio astronomy. *Sky & Telescope* **9**, 27–30 (Dec.), 55–56 (Jan.). *75, 437*

Unsöld, A. (1949a). Uber den Ursprung der Radiofrequenzstrahlung und der Ultrastrahlung in der Milchstrasse. *Zeitschrift für Astrophysik* **26**, 176–199. *369*

Unsöld, A. (1949b). Origin of the radio frequency emission and cosmic radiation in the Milky Way. *Nature* **163**, 489–491. *369*

Waldmeier, M. (1949). Radioastronomie. *Vierteljahrsschrift der Naturforschenden Gesellschaft in Zürich* **94**, 24–36. Also published as *Astron. Mitt. Eidgenössischen Sternwarte Zürich*, No. 166. *314*

1950

Alfvén, H., Herlofson, N. (1950). Cosmic radiation and radio stars. *Phys. Rev.* **78**, 616. * *379*

Allen, C. W., Gum, C. S. (1950). Survey of galactic radio-noise at 200 Mc/s. *Austral. J. Sci. Res.* **A3**, 224–233. *137*

Bolton, J. G., Westfold, K. C. (1950a). Galactic radiation at radio frequencies. I. 100 Mc/s Survey. *Austral. J. Sci. Res.* **A3**, 19–33. *367*

Bolton, J. G., Westfold, K. C. (1950b). Galactic radiation at radio frequencies. III. Galactic structure. *Austral. J. Sci. Res.* **A3**, 251–264. *367*

Booker, H. G., Ratcliffe, J. A., Shinn, D. H. (1950). Diffraction from an irregular screen with applications to ionospheric problems. *Phil. Trans. Roy. Soc.* **A242**, 579–607. *327, 327n*

Chikhachev, B. M. (1950). Investigation of localized sources of solar radio emission. Cand. Sci. thesis. P. N. Lebedev Physical Institute, Moscow. [In Russian] *218*

Covington, A. E. (1950). Microwave sky noise. *J. Geophys. Res.* **55**, 33–37. *213*

Denisse, J.-F. (1950a). Contribution à l'étude des émissions radioélectriques solaires. *Annales d'Astrophysique* **13**, 181–202. *289, 310, 221n*

Denisse, J.-F. (1950b). Emissions radioélectriques d'origine purement thermique dans les milieux ionisés. *Journal de Physique et le Radium* **11**, 164–171. *230, 289, 221n*

Ellyett, C. D. (1950). The influence of high altitude winds on meteor trail ionization. *Phil. Mag.* **41**, 694–700. *252*

Getmantsev, G. G., Ginzburg, V. L. (1950). On the diffraction of solar and cosmic radio radiation on the moon. *Zh. Eksperimental'noi i Teoreticheskoi Fiziki* **20**, 347–350. [In Russian] *314*

Greenhow, J. S. (1950). The fluctuation and fading of radio echoes from meteor trails. *Phil. Mag.* **41**, 682–693. *252*

Hanbury Brown, R., Hazard, C. (1950). Radio-frequency radiation from the great nebula in Andromeda (M.31). *Nature* **166**, 901. *190*

Jaeger, J. C., Harper, A. F. A. (1950). Nature of the surface of the moon. *Nature* **166**, 1026. *279*

Kiepenheuer, K. O. (1950). Cosmic rays as the source of general galactic radio emission. *Phys. Rev.* **79**, 738–739.* *379*

Laffineur, M., Michard, R., Servajean, R., Steinberg, J.-L. (1950). Observations radioélectriques de l'éclipse de soleil du 28 avril 1949. *Annales d'Astrophysique* **13**, 337–342. *223*

Little, C. G., Lovell, A. C. B. (1950). Origin of the fluctuations in the intensity of radio waves from galactic sources – Jodrell Bank observations. *Nature* **165**, 423–424. *326*

Manning, L. A., Villard, O. G., Jr., Peterson, A. M. (1950). Meteoric echo study of upper atmosphere winds. *Proc. IRE* **38**, 877–883. *252*

Millman, P. M. (1950). Meteoric ionization. *J. Roy. Astron. Soc. Canada* **44**, 209–220. *252*

Minnett, H. C., Labrum, N. R. (1950). Solar radiation at a wavelength of 3.18 centimetres. *Austral. J. Sci. Res.* **A3**, 60–71. *291, 311*

Piddington, J. H. (1950). The derivation of a model solar chromosphere from radio data. *Proc. Roy. Soc.* **A203**, 417–434.   *288*

Ryle, M. (1950). Radio astronomy. *Rep. Prog. Phys.* **13**, 184–246.   *175, 309, 370*

Ryle, M., Hewish, A. (1950). The effects of the terrestrial ionosphere on the radio waves from discrete sources in the Galaxy. *Mon. Not. RAS* **110**, 381–394.   *327*

Ryle, M., Smith, F. G., Elsmore, B. (1950). A preliminary survey of the radio stars in the Northern Hemisphere. *Mon. Not. RAS* **110**, 508–523. Erratum: **111**, 641 (1951).   *328*

Shklovsky, I. S. (1950). On the possibility of the determination of the distance of the 'point' sources of galactic radio radiation. *Dokl. Akad. Nauk SSSR* **73**, 479–481. [In Russian; English translation available as NRL Translation No. 339 (1951).]   *364*

Smerd, S. F. (1950a). Radio-frequency radiation from the quiet sun. *Austral. J. Sci. Res.* **A3**, 34–59.   *289*

Smerd, S. F. (1950b). A radio-frequency representation of the solar atmosphere. *Proc. IEE* **97** (Pt. III), 447–452.   *289*

Smerd, S. F. (1950c). The polarization of thermal 'solar noise' and a determination of the sun's general magnetic field. *Austral. J. Sci. Res.* **A3**, 265–273.   *288n*

Smith, F. G. (1950). Origin of the fluctuations in the intensity of radio waves from galactic sources – Cambridge observations. *Nature* **165**, 422–423.   *326*

Stanier, H. M. (1950a). Distribution of radiation from the undisturbed sun at a wave-length of 60 cm. *Nature* **165**, 354–355.   *293*

Stanier, H. M. (1950b). Radio frequency emission from the sun. Ph.D. Thesis. University of Cambridge, Cambridge, England.   *294*

Stanley, G. J., Slee, O. B. (1950). Galactic radiation at radio frequencies. II. The discrete sources. *Austral. J. Sci. Res.* **A3**, 234–250.   *325, 332*

Steinberg, J.-L. (1950). Etude des radiomètres hyperfréquences. Applications astrophysiques. Ph.D. thesis. Laboratoire de Physique, Ecole Normale Supérieure, Paris.   *221*

Waldmeier, M., Müller, H. (1950). Die Sonnenstrahlung im Gebiet von λ = 10 cm. *Zeitschrift für Astrophysik* **27**, 58–72.   *311*

Whipple, F. L. (1950). Report of Commission 22 on meteors and the zodiacal light. In *Trans. Intl. Astron. Union: Vol. VII for the Seventh General Assembly (1948)* (ed. Oort, J. H.), Cambridge: Cambridge University Press, pp. 240–244.   *253*

Wild, J. P. (1950a). Observations of the spectrum of high-intensity solar radiation at metre wavelengths. II. Outbursts. *Austral. J. Sci. Res.* **A3**, 399–408.   *305*

Wild, J. P. (1950b). Observations of the spectrum of high-intensity solar radiation at metre wavelengths. III. Isolated bursts. *Austral. J. Sci. Res.* **A3**, 541–557.   *305*

Wild, J. P., McCready, L. L. (1950). Observations of the spectrum of high-intensity solar radiation at metre wavelengths. I. The apparatus and spectral types of solar burst observed. *Austral. J. Sci. Res.* **A3**, 387–398.   *302*

Williamson, R. E. (1950). Concerning the source of galactic radio noise. *J. Roy. Astron. Soc. Canada* **44**, 12–16.   *369*

Woolley, R. v. d. R., Allen, C. W. (1950). Ultra-violet emission from the chromosphere. *Mon. Not. RAS* **110**, 358–372.   *288*

1951

Almond, M. (1951). The summer daytime meteor streams of 1949 and 1950. III. Computation of the orbits. *Mon. Not. RAS* **111**, 37–44.   *245*

Almond, M., Davies, J. G., Lovell, A. C. B. (1951). The velocity distribution of sporadic meteors. I. *Mon. Not. RAS* **111**, 585–608.   *248*

Aspinall, A., Clegg, J. A., Hawkins, G. S. (1951). A radio echo apparatus for the delineation of meteor radiants. *Phil. Mag.* **42**, 504–514.   *243–4*

Aspinall, A., Hawkins, G. S. (1951). The summer daytime meteor streams of 1949 and 1950. I. Measurement of the radiant positions and activity. *Mon. Not. RAS* **111**, 18–25.   *244*

Bolton, J. G., Westfold, K. C. (1951). Galactic radiation at radio frequencies. IV. The distribution of radio stars in the Galaxy. *Austral. J. Sci. Res.* **A4**, 476–488.   *371*

Boyd, R. L. F. (ed.) (1951). *Proceedings of the Conference on Dynamics of Ionized Media*. London: Dept. of Physics, University College. (Mimeographed, unbound).   *375*

Davies, J. G., Greenhow, J. S. (1951). The summer daytime meteor streams of 1949 and 1950. II. Measurement of the velocities. *Mon. Not. RAS* **111**, 26–36.   *245*

Dewhirst, D. W. (1951). Paper given at RAS meeting of 12 October 1951 on optical identifications of Cas A and Cyg A. *Observatory* **71**, 211–213.   *342*

Ewen, H. I. (1951). Radiation from galactic hydrogen at 1420 Megacycles per second. Ph.D. Thesis. Dept. of Physics, Harvard University, Cambridge, Mass.   *410*

Ewen, H. I., Purcell, E. M. (1951). Radiation from galactic hydrogen at 1420 Mc/s. *Nature* **168**, 356.*   *410*

Getmantsev, G. G. (1951). New data on the radio radiation of the sun and of the Galaxy. *Uspekhi Fizicheskikh Nauk* **44**, 527–557. [In Russian]   *308, 380*

Ginzburg, V. L. (1951). Cosmic rays as a source of galactic radio radiation. *Dokl. Akad. Nauk SSSR* **76**, 377–380. (In Russian; English Translation available in Sullivan (1982:93).) *   *380*

Gold, T. (1951). The origin of cosmic radio noise. In Boyd (1951), pp. 105–16.   *375*

Hagen, J. P. (1951). Temperature gradient in the sun's atmosphere measured at radio frequencies. *Astrophys. J.* **113**, 547–566.   *288*

Hagen, J. P., Haddock, F. T., Reber, G. (1951). NRL Aleutian radio eclipse expedition. *Sky and Telescope* **10**, 111–113.   *208*

Hanbury Brown, R., Hazard, C. (1951a). Radio emission from the Andromeda Nebula. *Mon. Not. RAS* **111**, 357–367.   *190*

Hanbury Brown, R., Hazard, C. (1951b). A radio survey of the Cygnus region. I. The localized source Cygnus (I). *Mon. Not. RAS* **111**, 576–584.   *339*

Hatanaka, T., Suzuki, S., Moriyama, F. (1951). Preliminary report on the observation of the solar radio noise at the partial eclipse on September 12, 1950. In *Provisional Reports of Observations of the Partial Eclipse of the Sun on September 12, 1950*. (ed. Hagihara, Y.), Tokyo: Solar Eclipse Comm., Science Council of Japan, pp. 9–11.   *290*

Hewish, A. (1951a). The fluctuations of galactic radio waves. Ph.D. thesis. Physics Dept., Cambridge University, Cambridge.   *327*

Hewish, A. (1951b). The diffraction of radio waves in passing through a phase-changing ionosphere. *Proc. Roy. Soc.* **A209**, 81–96.   *327*

Kawakami, K. (1951). Solar noise observations during the partial eclipse of the sun at Wakkanai. In *Provisional Reports of Observations of the Partial Eclipse of the Sun on September 12, 1950* (ed. Hagihara, Y.), Tokyo: Solar Eclipse Comm., Science Council of Japan, pp. 14–16.   *290*

Kerr, F. J., Shain, C. A. (1951). Moon echoes and transmission through the ionosphere. *Proc. IRE* **39**, 230–242.   *275*

Kwal, B. (1951). Sur une possibilité d'interpréter les bruits radio-électriques du soleil et de la galaxie comme rayonnement des protons des radiations cosmiques dans les champs magnétiques intenses du soleil et des autres objets célestes. *Annales d'Astrophysique* **14**, 189–198.   *380, 392*

Laffineur, M. (1951). Contribution à l'étude du rayonnement électromagnétique du soleil sur ondes décimétriques. Ph.D. thesis. Institute d'Astrophysique, Université de Paris, Paris.   *224*

Little, C. G. (1951). A diffraction theory of the scintillation of stars on optical and radio wave-lengths. *Mon. Not. RAS* **111**, 289–302.   *327*

Little, C. G., Maxwell, A. (1951). Fluctuations in the intensity of radio waves from galactic sources. *Phil. Mag.* **42**, 267–278.   *327*

Little, A. G., Payne-Scott, R. (1951). The position and movement on the solar disk of sources of radiation at a frequency of 97 Mc/s. I. Equipment. *Austral. J. Sci. Res.* **A4**, 489–507.   *300*

Machin, K. E. (1951). Distribution of radiation across the solar disk at a frequency of 81.5 Mc/s. *Nature* **167**, 889–891.   *294*

Machin, K. E., Smith, F. G. (1951). A new radio method for measuring the electron density in the solar corona. *Nature* **168**, 599–600.   *220n*

McKinley, D. W. R. (1951). Meteor velocities determined by radio observations. *Astrophys. J.* **113**, 225–267.   *250*

Mills, B. Y., Thomas, A. B. (1951). Observations of the source of radio-frequency radiation in the constellation of Cygnus. *Austral. J. Sci. Res.* **A4**, 158–171.   *336*

Muller, C. A., Oort, J. H. (1951). The interstellar hydrogen line at 1420 Mc/s and an estimate of galactic rotation. *Nature* **168**, 357–358. *   *410*

Oort, J. H. (1951). Symposium over de werkzaamheid der Stichting Radiostraling van Zon en Melkweg en over de ontdekking van een emissie-lijn van interstellaire waterstof. *Verslag v. Koninklijke Nederlandse Akademie v. Wetenschappen* **60**, 53–54, 58–62.   *408*

Ovenden, M. W. (1951). Interstellar meteors. *Science Progress* **39**, 658–663.   *254*

Payne-Scott, R., Little, A. G. (1951). The position and movement on the solar disk of sources of radiation at a frequency of 97 Mc/s. II. Noise storms. *Austral. J. Sci. Res.* **A4**, 508–525.   *301*

Piddington, J. H. (1951). The origin of galactic radio-frequency radiation. *Mon. Not. RAS* **111**, 45–63.   *372, 391*

Piddington, J. H., Minnett, H. C. (1951a). Solar radio-frequency emission from localized regions at very high temperatures. *Austral. J. Sci. Res.* **A4**, 131–157.   *311*

Piddington, J. H., Minnett, H. C. (1951b). Observations of galactic radiation at frequencies of 1210 and 3000 Mc/s. *Austral. J. Sci. Res.* **A4**, 459–475.   *334*

Ryle, M. (1951). The emission of radio waves from sunspots and radio stars. In Boyd (1951), pp. 107–111.   *308, 375*

Ryle, M., Elsmore, B. (1951). A search for long-period variations in the intensity of radio stars. *Nature* **168**, 555–556.   *330*

Seeger, C. L. (1951). Some observations of the variable 205 Mc/sec radiation of Cygnus A. *J. Geophys. Res.* **56**, 239–258.   *212*

Seeger, C. L., Williamson, R. E. (1951). The pole of the galaxy as determined from measurement at 205 Mc/sec. *Astrophys. J.* **113**, 21–49.   *212*

Shain, C. A. (1951). Galactic radiation at 18.3 Mc/s. *Austral. J. Sci. Res.* **A4**, 258–267.   *125n*

Shklovsky, I. S. (1951a). *The Solar Corona*. Moscow: Gosudarstvennoy Izdatelstvo Tekhniko-Teoreticheskoy Lit. [In Russian]   *215*

Shklovsky, I. S. (1951b). Radio stars. *Dokl. Akad. Nauk SSSR* **79**, 423–426. [In Russian]   *381*

Smith, F. G. (1951a). An accurate determination of the positions of four radio stars. *Nature* **168**, 555. *   *339*

Smith, F. G. (1951b). An attempt to measure the annual parallax or proper motion of four radio stars. *Nature* **168**, 962–963.   *329*

Smith, F. G. (1951c). Some studies of radio stars. Ph.D. thesis. Physics Dept., Cambridge University, Cambridge.   *339, 370*

Spitzer, L., Jr., Baade, W. (1951). Stellar populations and collisions of galaxies. *Astrophys. J.* **113**, 413–418.   *345*

Tanaka, H., Kakinuma, T., Jindo, H., Takayanagi, T. (1951). On the substitution measurement of sky temperature at centimetre waves. *Bull. Research Inst. of Atmospherics (Nagoya University)* **2**, 121–123. [In Japanese; English Translation (by first two authors only) in *Proc. Research Inst. Atmospherics* **1**, pp. 85–88 (1953).]   *227*

Unsöld, A. (1951). Cosmic radiation and cosmic magnetic fields. I. Origin and propagation of cosmic rays in our Galaxy. *Phys. Rev.* **82**, 857–863.   *387*

Van de Hulst, H. C. (1951). *A Course in Radio Astronomy*. Leiden: Sterrewacht. Bound, mimeographed (185 pp.).   *407n, 457*

Vitkevich, V. V. (1951). A new method of investigating the solar corona. *Dokl. Akad. Nauk SSSR* **77**, 585–588. [In Russian]   *218–19*

Westerhout, G., Oort, J. H. (1951). A comparison of the intensity distribution of radio-frequency radiation with a model of the galactic system. *Bull. Astronomical Inst. Netherlands* **11**, 323–333.   *372*

Wild, J. P. (1951). Observations of the spectrum of high-intensity solar radiation at metre wavelengths. IV. Enhanced radiation. *Austral. J. Sci. Res.* **A4**, 36–50.   *305*

1952

Almond, M., Davies, J. G., Lovell, A. C. B. (1952). The velocity distribution of sporadic meteors. II. *Mon. Not. RAS* **112**, 21–39.   *250*

Blum, E.-J. (1952). Le rayonnement radioélectrique du soleil sur ondes métriques – méthode de mesure et résultats récents. Ph.D. thesis. Laboratoire de Physique, Ecole Normale Supérieure, Paris.   *221*

Blum, E.-J., Denisse, J.-F., Steinberg, J.-L. (1952a). Résultat des observations d'une éclipse annulaire de soleil effectuées sur 169 Mc/s et 9350 Mc/s. *Annales d'Astrophysique* **15**, 184–198.   *291*

Blum, E.-J., Denisse, J.-F., Steinberg, J.-L. (1952b). Sur la forme ellipsoidale du soleil observé en ondes métriques. *C. R. Acad. Sci.* **234**, 1597–1599.   *291*

Bracewell, R. N. (1952). Radio stars or radio nebulae? *Observatory* **72**, 27–29.   *360*

Burrell, B. (1952). Radio echoes from the planets. *Observatory* **72**, 118–119.   *282*

Christiansen, W. N., Hindman, J. V. (1952). A preliminary survey of 1420 Mc/s line emission from galactic hydrogen. *Austral. J. Sci. Res.* **A5**, 437–455.   *411*

Clegg, J. A. (1952). The velocity distribution of sporadic meteors. III. Calculation of the theoretical distributions. *Mon. Not. RAS* **112**, 399–413.   *250*

Denisse, J.-F., Blum, E.-J., Steinberg, J.-L. (1952). Radio observations of the solar eclipses of September 1, 1951, and February 25, 1952. *Nature* **170**, 191–192.   *290*

Eshleman, V. R. (1952). The mechanism of radio reflections from meteoritic ionization. Ph.D. thesis. Electrical Engineering Dept., Stanford University, Palo Alto, Cal.   *252*

Getmantsev, G. G. (1952). Cosmic electrons as the source of the radio radiation of the Galaxy. *Dokl. Akad. Nauk SSSR* **83**, 557–560. [In Russian]   *380*

Getmantsev, G. G., Ginzburg, V. L. (1952). On a possible mechanism for the sporadic radio emission of the sun. *Dokl. Akad. Nauk SSSR* **87**, 187–190. [In Russian]   *308*

Greenhow, J. S. (1952). Characteristics of radio echoes from meteor trails: III. The behaviour of the electron trails after formation. *Proc. Phys. Soc.* **B65**, 169–181. *252*

Greenhow, J. S., Hawkins, G. S. (1952). Ionizing and luminous efficiencies of meteors. *Nature* **170**, 355. *250, 252*

Hanbury Brown, R., Hazard, C. (1952a). Extra-galactic radio-frequency radiation. *Phil. Mag.* **43**, 137–152. *347*

Hanbury Brown, R., Hazard, C. (1952b). Radio-frequency radiation from Tycho Brahé's supernova (A.D. 1572). *Nature* **170**, 364–365. *347*

Hanbury Brown, R., Jennison, R. C., Das Gupta, M. K. (1952). Apparent angular sizes of discrete radio sources – observations at Jodrell Bank, Manchester. *Nature* **170**, 1061–1063. * *356*

Hawkins, G. S. (1952). A radio echo study of meteor radiant points and the aurora borealis. Ph.D. thesis. Physics Dept., Manchester University, Manchester. *250*

Hewish, A. (1952). The diffraction of galactic radio waves as a method of investigating the irregular structure of the ionosphere. *Proc. Roy. Soc.* **A214**, 494–514. *327*

Hulburt, E. O. (1952). The solar eclipse of February 25, 1952. *Scientific Monthly* **75**, 306–309. *208*

Hutchinson, G. W. (1952). On the possible relation of galactic radio noise to cosmic rays. *Phil. Mag.* **43**, 847–852. *392*

Kaiser, T. R., Closs, R. L. (1952). Theory of radio reflections from meteor trails: I. *Phil. Mag.* **43**, 1–32. *252*

Kerr, F. J. (1952). On the possibility of obtaining radar echoes from the sun and planets. *Proc. IRE* **40**, 660–666. *282*

Korchak, A. A., Terletsky, Y. P. (1952). Electromagnetic radiation of cosmic-ray protons and galactic radio radiation [In Russian]. *Zhurnal Eksperimental'noi i Teoreticheskoi Fiziki* **22**, 507–509. *382*

Laffineur, M., Michard, R., Pecker, J.-C., d'Azambuja, M., Dollfus, A., Atanasijević, I. (1952). Observations combinées de l'éclipse totale de soleil du 25 février 1952 à Khartoum (Soudan) et de l'éclipse partielle au radio-télescope de l'Observatoire de Meudon. *C. R. Acad. Sci.* **234**, 1528–1530. *290*

Little, C. G. (1952). The origin of the fluctuations in galactic radio noise. Ph.D. thesis. Physics Dept., Manchester University, Manchester. *324, 327*

Lovell, A. C. B., Clegg, J. A. (1952). *Radio Astronomy*. London: Chapman and Hall. *195*

Machin, K. E. (1952). A study of the solar corona using radio methods. Ph.D. Thesis. University of Cambridge, Cambridge, England. *294*

Machin, K. E., Smith, F. G. (1952). Occultation of a radio star by the solar corona. *Nature* **170**, 319–320. *220n*

Maxwell, A., Little, C. G. (1952). A radio-astronomical investigation of winds in the upper atmosphere. *Nature* **169**, 746–747. *327*

Mills, B. Y. (1952a). The distribution of the discrete sources of cosmic radio radiation. *Austral. J. Sci. Res.* **A5**, 266–287. Erratum: Austral. J. Physics, 6, 125 (1953). *330*

Mills, B. Y. (1952b). The positions of six discrete sources of cosmic radio radiation. *Austral. J. Sci. Res.* **A5**, 456–463. *341*

Mills, B. Y. (1952c). Apparent angular sizes of discrete radio sources – observations at Sydney. *Nature* **170**, 1063–1064. * *356, 360*

Morgan, W. W., Sharpless, S., Osterbrock, D. E. (1952). Some features of galactic structure in the neighborhood of the sun. *Astron. J.* **57**, 3 (abstract). *413n*

Oort, J. H. (1952). Problems of galactic structure. *Astrophys. J.* **116**, 233–250. *413*

Payne-Scott, R., Little, A. G. (1952). The position and movement on the solar disk of sources of radiation at a frequency of 97 Mc/s. III. Outbursts. *Austral. J. Sci. Res.* **A5**, 32–46. *301*

Piddington, J. H., Minnett, H. C. (1952). Radio-frequency radiation from the constellation of Cygnus. *Austral. J. Sci. Res.* **A5**, 17–31. *347*

Ryle, M. (1952). A new radio interferometer and its application to the observation of weak radio stars. *Proc. Roy. Soc.* **A211**, 351–375. * *169*

Shklovsky, I. S. (1952a). On the nature of the radio radiation of the Galaxy. *Astron. Zh.* **29**, 418–449. [In Russian] *382*

Shklovsky, I. S. (1952b). On the spatial distribution of the sources of galactic radio emission. *Dokl. Akad. Nauk SSSR* **85**, 1231–1234. [In Russian] *382*

Shklovsky, I. S. (1952c). On the nature of radio star emission. *Dokl. Akad. Nauk SSSR* **85**, 509–512. [In Russian] *381*

Shklovsky, I. S. (1952d). Radio spectroscopy of the Galaxy. *Astron. Zh.* **29**, 144–153. [In Russian] *412*

Smith, F. G. (1952a). The determination of the position of a radio star. *Mon. Not. RAS* **112**, 497–513. *339*

Smith, F. G. (1952b). The measurement of the angular diameter of radio stars. *Proc. Phys. Soc.* **B65**, 971–980. *359*

Smith, F. G. (1952c). Apparent angular sizes of discrete radio sources – Observations at Cambridge. *Nature* **170**, 1065. * *356*

Steinberg, J.-L. (1952). Les récepteurs de bruits radioélectriques. *l'Onde Electrique* **32**, 445–454, 519–426 (Part II). Part III is **33**, 274–884 (1953). *221n*

Wild, J. P. (1952). The radio-frequency line spectrum of atomic hydrogen and its applications in astronomy. *Astrophys. J.* **115**, 206–221. *412*

Wouthuysen, S. A. (1952). On the excitation mechanism of the 21 cm interstellar hydrogen emission line. *Physica* **18**, 75–76. *412n*

1953

Almond, M., Davies, J. G., Lovell, A. C. B. (1953). The velocity distribution of sporadic meteors. IV. Extension to magnitude +8, and final conclusions. *Mon. Not. RAS* **113**, 411–427. *250*

Bolton, J. G., Slee, O. B. (1953). Galactic radiation at radio frequencies. V. The sea interferometer. *Austral. J. Phys.* **6**, 420–433. *333*

Bolton, J. G., Slee, O. B., Stanley, G. J. (1953). Galactic radiation at radio frequencies. VI. Low altitude scintillations of the discrete sources. *Austral. J. Phys.* **6**, 434–451. *325*

Christiansen, W. N. (1953). A high-resolution aerial for radio astronomy. *Nature* **171**, 831–833. *296*

Christiansen, W. N., Warburton, J. A. (1953). The distribution of radio brightness over the solar disk at a wavelength of 21 centimetres. I. A new highly directional aerial system. *Austral. J. Phys.* **6**, 190–202. *296*

Coutrez, R., Koeckelenbergh, A., Pourbaix, E. (1953). Observations radioastronomiques de l'éclipse solaire du 25 février 1952. *Commun. Obs. Royal de Belgique*, 1–21. *290*

Eshleman, V. R. (1953). The effect of radar wavelength on meteor echo rate. *IRE Trans. Antennas and Propagation* **AP-1**, 37–42. *252*

Ginzburg, V. L. (1953a). The statistical mechanism of the acceleration of particles on the sun and in stellar shells. *Dokl. Akad. Nauk SSSR* **92**, 727–730. [In Russian] *381, 387*

Ginzburg, V. L. (1953b). Supernovae and novae as sources of cosmic and radio radiation. *Dokl. Akad. Nauk SSSR* **92**, 1133–1136. [In Russian] *381*

Ginzburg, V. L. (1953c). The origin of cosmic rays and radio astronomy. *Usp. Fiz. Nauk* **51**, 343–392. [In Russian] *381*

Ginzburg, V. L., Fradkin, M. I. (1953). The electron component and the origin of cosmic rays. *Dokl. Akad. Nauk SSSR* **92**, 531–534. [In Russian] *381*

Greenstein, J. L., Minkowski, R. (1953). The Crab Nebula as a radio source. *Astrophys. J.* **118**, 1–15. *383*

Hanbury Brown, R., Hazard, C. (1953a). A survey of 23 localized radio sources in the Northern Hemisphere. *Mon. Not. RAS* **113**, 123–133. *331, 373*

Hanbury Brown, R., Hazard, C. (1953b). A model of the radio-frequency radiation from the Galaxy. *Phil. Mag.* **44**, 939–963. *373*

Hazard, C. (1953). An investigation of the extra-galactic radio frequency emissions. Ph.D. thesis. Physics Dept., Manchester University, Manchester. *373*

Jennison, R. C., Das Gupta, M. K. (1953). Fine structure of the extra-terrestrial radio source Cygnus I. *Nature* **1972**, 996–997. * *358*

Mills, B. Y. (1953). The radio brightness distributions over four discrete sources of cosmic noise. *Austral. J. Phys.* **6**, 452–470. *349, 360*

Mills, B. Y., Little, A. G. (1953). A high-resolution aerial system of a new type. *Austral. J. Phys.* **6**, 272–278. *153n*

O'Brien, P. A. (1953). The distribution of radiation across the solar disk at metre wave-lengths. *Mon. Not. RAS* **113**, 597–612. *294*

Pawsey, J. L. (1953). Radio astronomy in Australia. *J. Roy. Astron. Soc. Canada* **47**, 137–152. *363*

Pawsey, J. L., Smerd, S. F. (1953). Solar radio emission. in *The Sun*. (ed. Kuiper, G. P.), Chicago: University of Chicago Press. pp. 466–531. *313*

Piddington, J. H. (1953). Thermal theories of the high-intensity components of solar radio-frequency radiation. *Proc. Physical Soc.* **66**, 97–104. *308*

Scheuer, P. A. G., Ryle, M. (1953). An investigation of the HII regions by a radio method. *Mon. Not. RAS* **113**, 3–17. *373, 391*

Shklovsky, I. S. (1953a). *Radio Astronomy: A Popular Sketch*. Moscow: Gosudarstvennoy Izdatelstvo Tekhniko-Teoreticheskoy Lit. [In Russian] *216, 459*

Shklovsky, I. S. (1953b). The problem of cosmic radio emission. *Astron. Zh.* **30**, 15–36. [In Russian] *382*

Shklovsky, I. S. (1953c). The photometric paradox of the radio emission of the metagalaxy. *Astron. Zh.* **30**, 495–507. [In Russian] *347, 389*

Shklovsky, I. S. (1953d). The problem of the origin of cosmic rays and radio astronomy. *Astron. Zh.* **30**, 577–592. [In Russian] *383*

Shklovsky, I. S. (1953e). On the nature of the optical radiation from the Crab Nebula. *Dokl. Akad. Nauk SSSR* **90**, 983–986. [In Russian; English Translation available in Lang and Gingerich (1979:490–2).] *384*

Shklovsky, I. S. (1953f). On the origin of cosmic rays. *Dokl. Akad. Nauk SSSR* **91**, 475–478. [In Russian; English version in *Mem. Soc. Roy. Scientifique Liège*, Ser. 4, **13**, 515 (1953).] *383*

Siedentopf, H., Elwert, G. (1953). Radiofrequenzemission von Ultrastrahlungselektronen. *Zeitschrift für Naturforschung* **8a**, 20–23. *392*

Vitkevich, V. V., Sorochenko, R. L. (1953). An interferometric radio telescope. *Astron. Zh.* **30**, 631–635. [In Russian] *220*

Wild, J. P., Murray, J. D., Rowe, W. C. (1953). Evidence of harmonics in the spectrum of a solar radio outburst. *Nature* **172**, 533–534. *307*

1954

Baade, W., Minkowski, R. (1954a). Identification of the radio sources in Cassiopeia, Cygnus A, and Puppis A. *Astrophys. J.* **119**, 206–214. * *346*

Baade, W., Minkowski, R. (1954b). On the identification of radio sources. *Astrophs. J.* **119**, 215–231. *346, 375*

Bolton, J. G., Stanley, G. J., Slee, O. B. (1954). Galactic radiation at radio frequencies. VIII. Discrete sources at 100 Mc/s between declinations +50 deg and −50 deg. *Austral. J. Phys.* **7**, 110–129. *333*

Bolton, J. G., Westfold, K. C., Stanley, G. J., Slee, O. B. (1954). Galactic radiation at radio frequencies. VII. Discrete sources with large angular widths. *Austral. J. Phys.* **7**, 96–109. *347, 359*

Covington, A. E., Broten, N. W. (1954). Brightness of the solar disk at a wave length of 10.3 cm. *Astrophys. J.* **119**, 569–589. *297n*

Das Gupta, M. K. (1954). The measurement of the apparent angular diameter and structure of Cygnus I and of Cassiopeia I. Ph.D. thesis. Physics Dept., Manchester University, Manchester. *358n*

Dombrovsky, V. A. (1954). On the nature of the radiation from the Crab nebula. *Dokl. Akad. Nauk SSSR* **94**, 1021–1024. [In Russian] *385*

Haddock, F. T. (1954). Eclipse measurements at centimeter wavelength and their interpretation. *J. Geophys. Res.* **59**, 174–177. *290*

Haddock, F. T., Mayer, C. H., Sloanaker, R. M. (1954). Radio emission from the Orion Nebula and other sources at λ 9.4 cm. *Astrophys. J.* **119**, 456–459. *210*

Hanbury Brown, R., Twiss, R. Q. (1954). A new type of interferometer for use in radio astronomy. *Phil. Mag.* **45**, 663–682. *365*

Hoyle, F. (1954). Generation of radio noise by cosmic sources. *Nature* **173**, 483–484. *393*

Jennison, R. C. (1954). The measurement of the fine structure of the cosmic radio sources. Ph.D. thesis. Physics Dept., Manchester University, Manchester. *358, 358n, 378*

Laffineur, M. (1954). Contribution à l'étude du rayonnement électromagnétique du soleil sur ondes décimétriques. *Bull. Astronomique* **18**, 1–57. *224*

Laffineur, M., Michard, R., Pecker, J.-C., Vauquois, B. (1954). Observations optiques et radioélectriques de l'éclipse totale de soleil du 25 février 1952. III. Observations radioélectriques de la couronne. *Ann. d'Astrophysique* **17**, 358–376. *291*

Lovell, A. C. B. (1954). *Meteor Astronomy*. Oxford: Oxford University Press. *195*

McGee, R. X., Bolton, J. G. (1954). Probable observation of the galactic nucleus at 400 Mc/s. *Nature* **173**, 985–987. *334*

Minkowski, R., Greenstein, J. L. (1954). The power radiated by some discrete sources of radio noise. *Astrophys. J.* **119**, 238–242. *377*

Murray, W. A. S., Hargreaves, J. K. (1954). Lunar radio echoes and the Faraday effect in the ionosphere. *Nature* **173**, 944. *282*

O'Brien, P. A. (1954). Measurements of radio-frequency emission from the sun, and some observations of radio stars. Ph.D. Thesis. University of Cambridge, Cambridge, England. *294*

Owren, L. (1954). Observations of radio emission from the sun at 200 Mc/sec and their relation to solar disk features. Ph.D. thesis. School of Electrical Engineering, Cornell University, Ithaca, NY. Also issued as Report EE200 (May 1954). *211*

Roberts, J. A. (1954). Radio astronomy. *Research* **7**, 388–399. *386, 459*

Shakeshaft, J. R. (1954). The isotropic component of cosmic radio-frequency radiation. *Phil. Mag.* **45**, 1136–1144. *389*

Shklovsky, I. S. (1954). Concerning the articles of Baade and Minkowski on the identification of discrete radio sources. *Astron. Zh.* **31**, 483–486. [In Russian] *346, 349, 436*

Twiss, R. Q. (1954). On the nature of the discrete radio sources. *Phil. Mag.* **45**, 249–258. *392*

Vashakidze, M. A. (1954). On the degree of polarization of the radiation from near extragalactic nebulae and the Crab nebula. *Astr. Tsirk.*, 11–13. [In Russian] *385*

Waldmeier, M. (1954). *Radiowellen aus dem Weltraum.* Zürich: Naturforschenden Gesell. *311*

Wild, J. P., Murray, J. D., Rowe, W. C. (1954). Harmonics in the spectra of solar radio disturbances. *Austral. J. Physics* **7**, 439–459. *306*

1955

Burke, B. F., Franklin, K. L. (1955). Observations of a variable radio source associated with the planet Jupiter. *J. Geophysical Res.* **60**, 213–217. *125n*

Christiansen, W. N., Warburton, J. A. (1955). The distribution of radio brightness over the solar disk at a wavelength of 21 centimetres. III. The quiet sun – two-dimensional observations. *Austral. J. Phys.* **8**, 474–486. *297*

Hanbury Brown, R., Lovell, A. C. B. (1955). Large radio telescopes and their use in radio astronomy. *Vistas in Astronomy* **1**, 542–560. *187n*

Öpik, E. J. (1955). *Meteor Astronomy*: a critical review, with comments on meteor velocities in particular. *Irish Astron. J.* **3**, 144–152. Review of Lovell (1954). *251*

Pawsey, J. L. (1955). A catalogue of reliably known discrete sources of cosmic radio waves. *Astrophys. J.* **121**, 1–5. *350*

Pawsey, J. L., Bracewell, R. N. (1955). *Radio Astronomy*. Oxford: Oxford University Press. *152* [see index]

Unsöld, A. (1955). *Physik der Sternatmosphären*. Berlin: Springer Verlag. 2nd edn. *386*

Vitkevich, V. V. (1955). Results of observations of the propagation of radio waves through the solar corona. *Astron. Zh.* **32**, 150–164. [In Russian] *220*

Waldmeier, M. (1955). *Ergebnisse und Probleme der Sonnenforschung*. Leipzig: Akademische Verlagsgesell. 2nd edn. *313*

1956–2009

Abir-Am, P. G. (1992). The politics of macromolecules: molecular biologists, biochemists, and rhetoric. *Osiris* **7**, 164–191. *435, 437*

Agar, J. (1997). Moon relay experiments at Jodrell Bank. In *Beyond the Ionosphere: 50 Years of Satellite Communication*. (ed. Butrica, A.), Washington, D.C.: NASA. pp. 19–30. *282n*

Agar, J. (1998). *Science and Spectacle: the Work of Jodrell Bank in Post-war British Culture*. Amsterdam: Harwood Academic. *193*

Aitken, H. G. J. (1976). *Syntony and Spark: the Origins of Radio*. New York: John Wiley and Sons. *25n*

Aitken, H. G. J. (1985). *The Continuous Wave: Technology and American Radio, 1900–1932*. Princeton: Princeton University Press. *25n*

Allison, D. K. (1981). New eye for the Navy: the origin of radar at the Naval Research Laboratory. NRL Report 8466. Naval Research Lab., Washington, D.C. *207n*

Baade, W. (1956). The polarization of the Crab nebula on plates taken with the 200-inch telescope. *Bull. Astron. Inst. Netherlands* **12**, 312. *387*

Beck, A. C. (1983). Personal recollections of Karl Jansky's work. pp. 32–38 in Kellermann and Sheets (1983). *52*

Beyerchen, A. (1998). From radio to radar: interwar military adaptation to technological change in Germany, the United Kingdom, and the United States. In *Military Innovation in the Interwar Period* (eds. Murray, W. and Millett, A. R.), Cambridge: Cambridge University Press, pp. 265–299. *80*

Bhathal, R. (1996). *Australian Astronomers: Achievements at the Frontiers of Astronomy*. Canberra: National Library of Australia. *296n*

Blaauw, A. (1980). Jan H. Oort's work. in *Oort and the Universe*. (ed. van Woerden, H., *et al.*), Dordrecht: Reidel, pp. 1–19. *415*

Blainey, G. (1968). *The Tyranny of Distance: How Distance Shaped Australia's History*. Melbourne: Macmillan. *143*

Blumtritt, O., Petzold, H., Aspray, W. (eds.) (1994). *Tracking the History of Radar*. Piscataway, NJ: IEEE-Rutgers Center for the History of Electrical Engineering. *207n*

Bolton, J. G. (1982). Radio astronomy at Dover Heights. *Proc. Astron. Soc. Australia* **4**, 349–358. *139*

Bowen, E. G. (1981). The pre-history of the Parkes 64-m telescope. *Proc. Astron. Soc. Australia* **4**, 267–273. *153*

Bowen, E. G. (1984). The origins of radio astronomy in Australia. pp. 85–111 in Sullivan (1984a). *122*

Bracewell, R. N. (1984). Early work on imaging theory in radio astronomy. pp. 167–190 in Sullivan (1984a). *117, 151*

Bracewell, R. N. (2005). Radio astronomy at Stanford. *J. Astron. History & Heritage* **8**, 75–86.   *313n*

Brown, L. (1999). *A Radar History of World War II: Technical and Military Imperatives*. Bristol, UK: Institute of Physics Publishing.   *79, 262*

Brush, S. G. (1995). Scientists as historians. *Osiris* **10**, 215–231.   *12*

Buderi, R. (1996). *The Invention that Changed the World*. New York: Simon & Schuster.   *200n, 282n, 419n*

Butrica, A. J. (1996). *To See the Unseen: a History of Planetary Radar Astronomy*. Washington, DC: NASA. Pub. SP-4218.   *282n, 462*

Christiansen, W. N. (1984). The first decade of solar radio astronomy in Australia. pp. 113–131 in Sullivan (1984a).   *146*

Clark, R. W. (1971). *Sir Edward Appleton*. Oxford: Pergamon.   *83*

Clark, T. (1980). How Diana touched the moon. *IEEE Spectrum* **17**, 44–48.   *269*

Covington, A. E. (1967). The development of solar microwave radio astronomy in Canada. *J. Roy. Astron. Soc. Canada* **61**, 314–323.   *214n*

Covington, A. E. (1983). Early radar research and a beginning in radio astronomy. pp. 105–14 in Kellermann and Sheets (1983).   *214n*

Covington, A. E. (1984). Beginnings of solar radio astronomy in Canada. pp. 317–334 in Sullivan (1984a).   *214n*

Covington, A. E. (1988). Origins of Canadian radio astronomy. *J. Roy. Astron. Soc. Canada* **82**, 165–178.   *214n*

Débarbat, S., Lequeux, J., Orchiston, W. (2007). Highlighting the history of French radio astronomy. 1: Nordmann's attempt to observe solar radio emission in 1901. *J. Astron. History & Heritage* **10**, 3–10.   *23n*

DeVorkin, D. H. (1985). Electronics in astronomy: early applications of the photoelectric cell and photomultiplier for studies of point-source celestial phenomena. *Proc. IEEE* **73**, 1205–1220.   *428*

DeVorkin, D. H. (1990). Interviewing physicists and astronomers: methods of oral history. in *Physicists Look Back* (ed. Roche, J.), Bristol: Adam Hilger, pp. 44–65.   *492n*

DeVorkin, D. H. (1992). *Science with a Vengeance: How the Military Created the US Space Sciences after World War II*. New York: Springer-Verlag.   *445*

DeVorkin, D. H. (2000). Who speaks for astronomy? How astronomers responded to government funding after World War II. *Historical Stud. Physical Sciences* **31**, 55–92.   *439*

Dick, S. J. (1996). *The Biological Universe: the Twentieth-century Extraterrestrial Life Debate and the Limits of Science*. Cambridge: Cambridge University Press.   *260n, 462*

Doel, R. E. (1996). *Solar System Astronomy in America: Communities, Patronage, and Interdisciplinary Science, 1920–1960*. Cambridge: Cambridge University Press.   *413n, 439*

Doel, R. E. (2003). Oral history of American science: a forty-year review. *History of Sci.* **41**, 349–378.   *492*

Eastwood, E. (1967). *Radar Ornithology*. London: Methuen.   *258*

Eddy, J. A. (1972). Thomas A. Edison and infra-red astronomy. *J. History of Astronomy* **3**, 165–187.   *19*

Edge, D. O. (1977). The sociology of innovation in modern astronomy. *Quarterly J. Royal Astron. Soc.* **18**, 326–339.   *11n*

Edge, D. O., Mulkay, M. J. (1976). *Astronomy Transformed: the Emergence of Radio Astronomy in Britain*. New York: John Wiley and Sons.   *10–11 [see index]*

Evans, W. F. (1970). *History of the Radiophysics Advisory Board 1939–1945*. Melbourne: CSIRO. (Available in RPL.)   *121n*

Evans, W. F. (1973). *History of the Radio Research Board 1926–1945*. Melbourne: CSIRO. (Available in RPL.)   *120*

Fagen, M. D. (ed.) (1975). *A History of Engineering and Science in the Bell System: the Early Years (1875–1925)*. New York: Bell Telephone Laboratories.   *31n*

Forman, P. (1987). Behind quantum electronics: national security as basis for physical research in the United States, 1940–1960. *Historical Stud. Physical Sciences* **18**, 149–229.   *443, 449, 454–5*

Forman, P. (1995). "Swords into ploughshares": Breaking new ground with radar hardware and technique in physical research after World War II. *Rev. Mod. Phys.* **67**, 397–455.   *419n*

Friis, H. T. (1965). Karl Jansky: his career at Bell Telephone Laboratories. *Science* **149**, 841–842.   *50*

Friis, H. T. (1971). *Seventy-five Years in an Exciting World*. San Francisco: San Francisco Press.   *50*

Galison, P. (1987). *How Experiments End*. Chicago: University Chicago Press.   *361, 452*

Galison, P. (1997). *Image and Logic: a Material Culture of Microphysics*. Chicago: University Chicago Press.   *351n, 398n*

Galison, P., Hevly, B. (eds.) (1992). *Big Science: the Growth of Large-scale Research*. Stanford: Stanford University Press.   *422*

Gilbert, G. N. (1975). The development of science and scientific knowledge: a case study of radar meteor research. Ph.D. thesis. Dept. of Engineering, Cambridge University, Cambridge, England. *254*

Gilbert, G. N. (1977). Competition, differentiation and careers in science. *Social Science Information* **16**, 103–123. *254*

Gillmor, C. S. (1982). Wilhelm Altar, Edward Appleton, and the magneto-ionic theory. *Proc. American Philosophical Society* **126**, 395–440. *91*

Gillmor, C. S. (1986). Federal funding and knowledge growth in ionospheric physics, 1945–81. *Social Studies of Science* **16**, 105–133. *441*

Gillmor, C. S. (1991). Ionospheric and radio physics in Australian science since the early days. In *International Science and National Scientific Identity* (eds. Home, R. W. and Kohlstedt, S. G.), Dordrecht: Kluwer, pp. 181–204. *91, 120*

Gingerich, O. (1984). Radio astronomy and the nature of science. pp. 399–407 in Sullivan (1984a). *412, 467*

Ginzburg, V. L. (1984). Remarks on my work in radio astronomy. pp. 289–302 in Sullivan (1984a). *217n, 462*

Ginzburg, V. L. (1985). On the birth and development of cosmic ray astrophysics. pp. 411–26 in Sekido and Elliot (1985). *217n*

Ginzburg, V. L. (1990). Notes of an amateur astrophysicist. *Ann. Rev. Astron. & Astrophys.* **28**, 1–36. *215*

Good, G. A. (2000). The assembly of geophysics: scientific disciplines as frameworks of consensus. *Stud. Hist. & Phil. Modern Physics* **31**, 259–292. *4*

Goss, W. M., McGee, R.X. (1996). The discovery of the radio source Sagittarius A (Sgr A). In *The Galactic Center* (ed. Gredel, R.), San Francisco: Astronomical Society of the Pacific, pp. 369–379. *334n*

Goss, W. M., McGee, R.X. (2009). *Under The Radar: Ruby Payne-Scott, First Radio Astronomer.* New York: Springer. *127*

Greenstein, J. L. (1983). Optical and radio astronomers in the early years. pp. 79–88 in Kellermann and Sheets (1983). *436*

Greenstein, J. L. (1984). Optical and radio astronomers in the early years. pp. 67–81 in Sullivan (1984). *345n*

Guerlac, H. E. (1987). *Radar in World War II.* New York: American Institue of Physics (2 vols.). *200n*

Gunn, A. G. (2005). Jodrell Bank and the meteor velocity controversy. pp. 107–118 in Orchiston (2005b). *251n*

Haddock, F. T. (1983). U.S. radio astronomy following World War II. pp. 115–120 in Kellermann and Sheets (1983). *208*

Ham, R. A. (1975). The hissing phenomenon. *J. British Astronomical Assn.* **85**, 317–323. *88*

Hanbury Brown, R. (1983). The development of Michelson and intensity long baseline interferometry. pp. 133–145 in Kellermann and Sheets (1983). *352*

Hanbury Brown, R. (1984). Paraboloids, galaxies and stars: memories of Jodrell Bank. pp. 213–35 in Sullivan (1984a). *196n*

Hanbury Brown, R. (1991). *Boffin: a Personal Story of the Early Days of Radar, Radio Astronomy and Quantum Optics.* Bristol: Adam Hilger. *188n*

Harwit, M. (1981). *Cosmic Discovery: the Search, Scope, and Heritage of Astronomy.* New York: Basic Books. *452, 463n*

Hawkins, G. S. (1956). A radio echo survey of sporadic meteor radiants. *Mon. Not. RAS* **116**, 92–104. *250*

Haynes, R., Haynes, R. D., Malin, D., McGee, R. (1996). *Explorers of the Southern Sky: a History of Australian Astronomy.* Cambridge: Cambridge University Press. *152n*

Hevly, B. (1987). Basic research within a military context: the Naval Research Laboratory and the foundations of extreme ultraviolet and X-ray astronomy, 1923–60. Ph.D. Thesis. Johns Hopkins University, Baltimore. *207n, 445*

Hey, J. S. (1973). *The Evolution of Radio Astronomy.* London: Elek Science. *11, 83n*

Hey, J. S. (1992). *The Secret Man.* Eastbourne, UK: Care Press. Self-published (51 pp.). *83n, 100n*

Hirsh, R. F. (1983). *Glimpsing an Invisible Universe: the Emergence of X-ray Astronomy.* Cambridge: Cambridge University Press. *464–6*

Hoddeson, L. (1980). The entry of the quantum theory of solids into the Bell Telephone Laboratories, 1925–40; a case-study of the industrial application of fundamental science. *Minerva* **18**, 422–447. *31*

Hoddeson, L. (1981). The emergence of basic research in the Bell Telephone System, 1875–1915. *Technology and Culture* **22**, 512–544. *31n*

Hoddeson, L. (2006). The conflict of memories and documents: dilemmas and pragmatics of oral history. In *The Historiography of Contemporary Science, Technology, and Medicine: Writing Recent Science* (eds. Doel, R. E. and Söderqvist, T.), London: Routledge. pp. 300–311. *493*

Home, R. W. (1983). Origins of the Australian physics community. *Historical Studies* **20**, 383–400. *119*

Home, R. W. (1986). The problem of intellectual isolation in scientific life: W. H. Bragg and the Australian scientific community, 1886–1909. *Historical Recs. of Austral. Sci.* **6**, 19–30.   *146*

Home, R.W. (ed.) (1988). *Australian Science in the Making*. Cambridge: Cambridge University Press.   *137n*

Hoyle, F. (1994). *Home is Where the Wind Blows: Chapters from a Cosmologist's Life*. Mill Valley, Cal.: University Science Books.   *377*

Hufbauer, K. (1991). *Exploring the Sun: Solar Science since Galileo*. Baltimore: Johns Hopkins University Press.   *285n, 445*

Jansky, C. M. (1958). The discovery and identification by Karl Guthe Jansky of electromagnetic radiation of extraterrestrial origin in the radio spectrum. *Proc. IRE* **46**, 13–15.   *50*

Jarrell, R. (1997). The formative years of Canadian radio astronomy. *J. Roy. Astron. Soc. Canada* **91**, 20–27.   *213n*

Jarrell, R. (2005). Radio astronomy, whatever that may be: the marginalization of early radio astronomy. pp. 191–202 in Orchiston (2005b).   *437*

Jennison, R. C., Das Gupta, M. K. (1956a). The measurement of the angular diameter of two intense radio sources. I. A radio interferometer using post-detector correlation. *Phil. Mag.* **1**, 55–64.   *358n*

Jennison, R. C., Das Gupta, M. K. (1956b). The measurement of the angular diameter of two intense radio sources. II. Diameter and structural measurements of the radio stars Cygnus A and Cassiopeia A. *Phil. Mag.* **1**, 65–75.   *358n*

Kellermann, K. (2005). Grote Reber (1911–2002): A radio astronomy pioneer. pp. 43–70 in Orchiston (2005b).   *55n*

Kellermann, K., Sheets, B. (eds.) (1983). *Serendipitous Discoveries in Radio Astronomy*. Green Bank, West Virginia (SJS): National Radio Astronomy Observatory.   *11, 451n*

Kern, U. (1984). Die Entstehung des Radarverfahrens: zur Geschichte der Radartechnik bis 1945. Ph.D. thesis. Historisches Institut der University Stuttgart, Stuttgart.   *505*

Kevles, D. J. (1977). *The Physicists*. New York: Vintage Books. (Updated edition)   *443–4*

Kobrin, M. M. (1959). Radio echoes from the moon in the X and S band. *Radio Engineering and Electronics* **4**, 228–232. (Original Russian version is *Radiotekhnika i Elektronika* **4**, 892–894 (1959).)   *281*

Kohler, R. E. (1982). *From Medical Chemistry to Biochemistry*. Cambridge: Cambridge University Press.   *438*

Kragh, H. (1996). *Cosmology and Controversy: the Historical Development of Two Theories of the Universe*. Princeton: Princeton University Press.   *389n*

Kraus, J. D. (1976). *Big Ear*. Powell, Ohio: Cygnus-Quasar Books.   *229n*

Lang, K. R., Gingerich, O. (eds.) (1979). *A Source Book in Astronomy and Astrophysics, 1900–1975*. Cambridge, Mass: Harvard University Press.   *11, 420*

Lankford, J. (1981). Amateurs and astrophysics: a neglected aspect in the development of a scientific specialty. *Social Studies of Science* **11**, 275–303.   *74n*

Lovell, A. C. B. (1968). *The Story of Jodrell Bank*. London: Oxford University Press.   *181n, 193n*

Lovell, A. C. B. (1973). *Out of the Zenith: Jodrell Bank 1957–1970*. London: Oxford University Press.   *188n*

Lovell, A. C. B. (1977). The effects of defence science on the advance of astronomy. *J. History Astron.* **8**, 151–173.   *443n, 462*

Lovell, A. C. B. (1983). Impact of World War II on radio astronomy. pp. 89–104 in Kellermann and Sheets (1983).   *446n*

Lovell, A. C. B. (1984). The origins and early history of Jodrell Bank. pp. 193–211 in Sullivan (1984a).   *326*

Lovell, A. C. B. (1990). *Astronomer by Chance*. New York: Basic Books.   *178n, 193n*

McCrea, W. H. (1984). The influence of radio astronomy on cosmology. pp. 365–84 in Sullivan (1984a).   *341*

Meadows (1984). (1) The origins of astrophysics. (2) The new astronomy. In *Astrophysics and Twentieth-Century Astronomy to 1950* (ed. Gingerich, O.), Cambridge: Cambridge University Press. pp. 3–15, 59–72.   *469*

Mellor, D. P. (1958). *The Role of Science and Industry*. Canberra: Australian War Memorial. (Series 4, Vol. 5 of "Australia in the War of 1939–1945").   *120*

Millman, P. M., McKinley, D. W. R. (1967). Stars fall over Canada. *J. Roy. Astron. Soc. Canada* **61**, 277–294.   *242*

Mills, B. Y. (1984). Radio sources and the log N – log S controversy. pp. 147–65 in Sullivan (1984a).   *389n*

Mills, B. Y. (2006). An engineer becomes astronomer. *Ann. Rev. Astron. Astrophys.* **44**, 1–15.   *341n, 389n*

Molchanov, A. P. (1956). (Discussion on 3.2 cm wavelength eclipse observations in 1952 and 1954.) In *Proc. of the Fifth Conference on Problems of Cosmogony: Radio*

Astronomy. Moscow: Izdatel. Akad. Nauk SSSR. p. 197. Held in Moscow in 1955. [In Russian]   *229*

Mulkay, M. J. (1974). Methodology in the sociology of science: some reflections on the study of radio astronomy. *Social Science Information* **13**, 107–119.   *11n, 12*

Munns, D. (1997). Linear accelerators, radio astronomy, and Australia's search for international prestige, 1944–1948. *Historical Studies in the Physical Sciences* **27**, 299–317.   *122n*

Munns, D. (2002). 'Wizards of the micro-waves': a history of the radio astronomy community. Ph.D. Thesis. Dept. of History of Science, Medicine and Technology, Johns Hopkins University, Baltimore, Md.   *153n, 413, 437*

Munns, D. (2003). If we build it, who will come? Radio astronomy and the limitations of national laboratories in Cold War America. *Historical Studies in the Physical & Biological Sciences* **34**, 95–113.   *449*

Needell, A. A. (1991). The Carnegie Institution of Washington and radio astronomy: prelude to an American national observatory. *J. History Astron.* **22**, 55–67.   *230n*

Needell, A. A. (2000). *Science, Cold War and the American State: Lloyd V. Berkner and the Balance of Professional Ideals*. Amsterdam: Harwood Academic.   *439, 449*

Oort, J. H., Walraven, T. (1956). Polarization and composition of the Crab nebula. *Bull. Astron. Inst. Netherlands* **12**, 285–308.   *387*

Orchiston, W. (1994). John Bolton, discrete sources, and the New Zealand field-trip of 1948. *Austral. J. Phys.* **47**, 541–547.   *317n*

Orchiston, W. (2004). From the solar corona to clusters of galaxies: the radio astronomy of Bruce Slee. *Pub. Astron. Soc. Australia* **21**, 23–71.   *139n*

Orchiston, W. (2005a). Dr. Elizabeth Alexander: first female radio astronomer. pp. 71–92 in Orchiston (2005b).   *85n*

Orchiston, W. (ed.) (2005b). *The New Astronomy: Opening the Electromagnetic Window and Expanding our View of Planet Earth*. Dordrecht: Springer.

Orchiston, W., Slee, B. (2002). Ingenuity and initiative in Australian radio astronomy: the Dover Heights 'hole-in-the-ground' antenna. *J. Astron. History & Heritage* **5**, 21–34.   *334n*

Orchiston, W., Slee, B. (2005). The Radiophysics field stations and the early development of radio astronomy. pp. 119–68 in Orchiston (2005b).   *146n*

Orchiston, W., Steinberg, J.-L. (2007). Highlighting the history of French radio astronomy. 2: The solar eclipse observations of 1949–1954. *J. Astron. History & Heritage* **10**, 11–19.   *223*

Orchiston, W., Slee, B., Burman, R. (2006). The genesis of solar radio astronomy in Australia. *J. Astron. History & Heritage* **9**, 35–56.   *130n, 137n*

Orchiston, W., Lequeux, J., Steinberg, J.-L., Delannoy, J. (2007). Highlighting the history of French radio astronomy. 3: The Würzburg antennas at Marcoussis, Meudon and Nançay. *J. Astron. History & Heritage* **10**, 221–245.   *225n*

Osterbrock, D. E. (2001). *Walter Baade: A Life in Astrophysics*. Princeton: Princeton University Press.   *345n, 388n*

Packard, K. S. (1984). The origin of waveguides: a case of multiple rediscovery. *IEEE Trans. on Microwave Theory and Techniques* **MTT-32**, 961–969.   *92n*

Pawsey, J. L. (1961). Australian radio astronomy. *Australian Scientist* **1**, 181–186.   *125*

Pestre, D. (1997). Studies of the ionosphere and forecasts for radiocommunications. Physicists and engineers, the military and national laboratories in France (and Germany) after 1945. *History and Technology* **13**, 183–205.   *226n*

Pfeiffer, J. (1956). *The Changing Universe: The Story of the New Astronomy*. New York: Random House.   *49–50, 430*

Phillips, V. J. (1980). *Early Radio Wave Detectors*. London: Peregrinus.   *21*

Price, D. J. d. (1984). The science/technology relationship, the craft of experimental science, and policy for the improvement of high technology innovation. *Research Policy* **13**, 3–20.   *451, 468n*

Purcell, E. M., Field, G. B. (1956). Influence of collisions upon population of hyperfine states in hydrogen. *Astrophys. J.* **124**, 542–549.   *412n*

Rasmussen, N. (1997). *Picture Control: the Electron Microscope and the Transformation of Biology in America, 1940–1960*. Stanford: Stanford University Press.   *454*

Ratcliffe, J. A. (1975). Physics in a university laboratory before and after World War II. *Proc. Roy. Soc.* **A342**, 457–464.   *155*

Reber, G. (1958a). Early radio astronomy at Wheaton, Illinois. *Proc. IRE* **46**, 15–23. (Reprinted in Sullivan [1984a:43–66].)   *56n*

Reber, G. (1958b). Between the atmospherics. *J. Geophys. Res.* **63**, 109–123.   *74*

Reber, G. (1959). Radio interferometry at three kilometers altitude above the Pacific Ocean. Part I. Installation and ionosphere. (and) Part II. Celestial sources. *J. Geophys. Res.* **64**, 287–303.   *73*

Reber, G. (1960). Reversed bean vines. *Castanea* **25**, 122–124. *74*

Reber, G. (1983). Radio astronomy between Jansky and Reber. pp. 71–8 in Kellermann and Sheets (1983). *76*

Reber, G., Ellis, G. R. (1956). Cosmic radio-frequency radiation near one megacyle. *J. Geophys. Res.* **61**, 1–10. *74*

Robertson, P. (1992). *Beyond Southern Skies: Radio Astronomy and the Parkes Telescope*. Cambridge: Cambridge University Press. *121n*

Rudwick, M. J. S. (1985). *The Great Devonian Controversy: the Shaping of Scientific Knowledge among Gentlemanly Specialists*. Chicago: University of Chicago Press. *12*

Ryle, M. (1971). Radio astronomy – the Cambridge contribution. In *Search and Research* (ed. Wilson, J. P.), London: Mullard Ltd. pp. 9–54. *160*

Salomonovich, A. E. (1981). N. D. Papaleksi and Soviet radio astronomy. *Soviet Physics Uspekhi* **24**, 627–632. Original Russian is *Usp. Fiz. Nauk* **134**, 541–550 (1981). *219*

Salomonovich, A. E. (1984). The first steps of Soviet radio astronomy. pp. 269–88 in Sullivan (1984a). *219n*

Salomonovich, A. E. (ed.) (1985). *Outlines of the History of Radio Astronomy in the USSR*. Kiev: Naukova Dumka. [In Russian] *217n*

Saward, D. (1984). *Bernard Lovell: a Biography*. London: Robert Hale. *178n*

Schaffer, S. (1981). Uranus and the establishment of Herschel's astronomy. *J. History Astron.* **12**, 11–26. *77*

Schedvin, C. B. (1982). The culture of CSIRO. *Australian Cultural History* **2**, 76–89. *121*

Schedvin, C. B. (1987). *Shaping Science and Industry: a History of Australia's Council for Scientific and Industrial Research, 1926–49*. Sydney: Allen and Unwin. *122*

Sekido, Y., Elliot, H. (eds.) (1985). *Early History of Cosmic Ray Studies*. Dordrecht: Reidel. *378n*

Shain, C. A. (1956). 18.3 Mc/s radiation from Jupiter. *Austral. J. Phys.* **9**, 61–73. *125n*

Shane, C. D. (1958). Radio astronomy in 1890: a proposed experiment. *Pub. Astron. Soc. Pacific* **70**, 303–304. *19n*

Shklovsky, I. S. (1982). Recollections on the development of radio astronomy in the USSR. in *Kosmonavtika, Astronomiya*. Moscow: Znanie. pp. 1–63 (November). [In Russian] *217, 320n*

Shklovsky, I. S. (1991). *Five Billion Vodka Bottles to the Moon*. New York: W. W. Norton. (Trans. M. F. Zirin and H. Zirin.) *216*

Sinton, W. M. (1986). Through the infrared with logbook and lantern slides: a history of infrared astronomy from 1868 to 1960. *Pub. Astron. Soc. Pacific* **98**, 246–251. *86n, 463n*

Slee, B. (1994). Some memories of the Dover Heights field station, 1946–1954. *Austral. J. Physics* **47**, 517–534. *139n*

Smith, F. G. (1984). Early work on radio stars at Cambridge. pp. 237–248 in Sullivan (1984a). *471*

Smith, F. G., Lovell, A. C. B. (1983). On the discovery of extragalactic radio sources. *J. History of Astronomy* **14**, 155–165. *340n*

Smith, R. W. (1997). Engines of discovery: scientific instruments and the history of astronomy and planetary science in the United States in the twentieth century. *J. History Astronomy* **28**, 49–77. *449n, 462*

Smith, R. W. (2006). Beyond the big galaxy: the structure of the stellar system 1900–1952. *J. History Astronomy* **37**, 307–342. *395n*

Southworth, G. C. (1956). Early history of radio astronomy. *Scientific Monthly* **82**, 55–66. *32n, 92n*

Southworth, G. C. (1962). *Forty Years of Radio Research*. New York: Gordon and Breach. *92n*

Stanley, G. J. (1994). Recollections of John G. Bolton at Dover Heights and Caltech. *Austral. J. Physics* **47**, 507–516. *153, 333*

Stephan, K. D. (1999). How Ewen and Purcell discovered the 21-cm interstellar hydrogen line. *IEEE Antennas and Propagation Magazine* **41**, 7–17 (Feb.). *403n*

Strelnitski, V. S. (1995). The early post-war history of Soviet radio astronomy. *J. History Astron.* **26**, 349–362. *219*

Strom, R. (2005). Radio astronomy in Holland before 1960: Just a bit more than H I. pp. 93–106 in Orchiston (2005b). *405n*

Strom, R. (2007). Ir A. H. DeVoogt: life and career of a radio pioneer. *Astronomisches Nachrichten* **328**, 443–446. *405n*

Sullivan, W. T., III (1978). A new look at Karl Jansky's original data. *Sky and Telescope* **56**, 101–105. *53*

Sullivan, W. T., III (1982). *Classics in Radio Astronomy*. Dordrecht: Reidel. *11*

Sullivan, W. T., III (1983). Karl Jansky and the beginning of radio astronomy. pp. 39–56 in Kellermann and Sheets (1983). *29n*

Sullivan, W. T., III (1984a) (ed.). *The Early Years of Radio Astronomy: Reflections Fifty Years after Jansky*. Cambridge: Cambridge University Press. *11, 29n, 389n*

Sullivan, W. T., III (1984b). Karl Jansky and the discovery of extraterrestrial radio waves. pp. 3–42 in Sullivan (1984a).

Sullivan, W. T., III (1988a). Early years of Australian radio astronomy. pp. 308–344 in Home (1988). (Updated version in *J. Astron. History & Heritage* **8**, 11–32 (2005).)   *118n*

Sullivan, W. T., III (1988b). Frank Kerr and radio waves: from wartime radar to interstellar atoms. In *The Outer Galaxy* (eds. Blitz, L. and Lockman, F. J.), New York: Springer. pp. 268–287.   *274, 394n, 413*

Sullivan, W. T., III (1990). The entry of radio astronomy into cosmology: radio stars and Martin Ryle's 2C survey. In *Modern Cosmology in Retrospect* (ed. Bertotti, R., et al.), Cambridge: Cambridge University Press. pp. 309–330.   *334n, 389n*

Sullivan, W. T., III (2000). Kapteyn's influence on the style and content of twentieth century Dutch astronomy. In *The Legacy of J. C. Kapyeyn.* (eds. van der Kruit, P. C. and van Berkel, K.), Dordrecht: Kluwer. pp. 229–264.   *415*

Swords, S. S. (1986). *Technical history of the beginnings of RADAR.* London: Peter Peregrinus.   *79, 221*

Tanaka, H. (1984). Development of solar radio astronomy in Japan up until 1960. pp. 335–348 in Sullivan (1984a).   *226*

Thompson, P. (1988). *The Voice of the Past: Oral History.* Oxford: Oxford University Press.   *492*

Trexler, J. H. (1958). Lunar radar echoes. *Proc. IRE* **46**, 286–292.   *281*

Trimble, V. (2005). And the remaining 22 photons: The development of gamma ray and gamma ray burst astronomy. pp. 261–270 in Orchiston (2005b).   *463n*

Troitsky, V. S., Zelinskaya, M. R., Rakhlin, V. L., Bobrik, V. T. (1956). Results of observations made of solar radio emission at wavelengths of 3.2 and 10 cm during the total solar eclipses on 25 February 1952 and on 30 June 1954. In *Proc. of the Fifth Conference on Problems of Cosmogony: Radio Astronomy.* Moscow: Izdatel. Akad. Nauk SSSR. pp. 182–196. Held in Moscow on 9–12 March 1955. [In Russian]   *290*

Van de Hulst, H. C. (1984). Nanohertz astronomy. pp. 385–398 in Sullivan (1984a).   *23n, 312*

Van Keuren, D. K. (1997). Moon in their eyes: Moon Communication relay at the Naval Research Laboratory, 1951–1962. In *Beyond the Ionosphere: Fifty Years of Satellite Communication* (ed. Butrica, A. J.), Washington, D.C: NASA: pp. 9–18. NASA Pub. SP-4217.   *281n, 446*

Van Keuren, D. K. (2001). Cold War science in black and white: US intelligence gathering and its scientific cover at the Naval Research Laboratory, 1948–62. *Social Studies of Science* **31**, 207–229.   *209n, 446*

Van Woerden, H., Strom, R. G. (2006). The beginnings of radio astronomy in the Netherlands. *J. Astron. History & Heritage* **9**, 3–20.   *78, 395n*

Vitkevich, V. V. (1956). Radio emission of the quiet and slightly perturbed sun. In *Proc. of the Fifth Conference on Problems of Cosmogony: Radio Astronomy.* Moscow: Izdatel. Akad. Nauk SSSR. pp. 149–172. Held in Moscow on 9–12 March 1955. [In Russian]   *290*

Vitkevich, V. V., Chikhachev, B. M. (1956). Observation of solar radio emission at meter wavelengths during the total solar eclipse of 25 February 1952. in *Proc. of the Fifth Conference on Problems of Cosmogony: Radio Astronomy.* Moscow: Izdatel. Akad. Nauk SSSR. pp. 174–178. Held in Moscow on 9–12 March 1955. [In Russian]   *290*

Weart, S. R., DeVorkin, D. H. (1981). Interviews as sources for history of modern astrophysics. *Isis* **72**, 471–477.   *492n*

Wendt, H., Orchiston, W., Slee, B. (2008). The Australian Solar Eclipse Expeditions of 1947 and 1949. *J. Astron. History & Heritage* **11**, 71–78.   *291n*

White, F. (1975). Early work in Australia, New Zealand and at the Halley Stewart Laboratory, London. *Phil. Trans. Roy. Soc.* **A280**, 35–46.   *121n*

Wild, J. P. (1968). The exploration of the sun by radio. *Australian Physicist* **5**, 117–122.   *150*

Wild, J. P. (1972). The beginnings of radio astronomy in Australia. *Records Austral. Acad. Sci.* **2** (No. 3), 52–61.   *147*

Wild, J. P. (1987). The beginnings of radio astronomy in Australia. *Proc. Astronomical Soc. Australia* **7**, 95–102.   *138n*

Williams, T. R. (2000). Getting organized: a history of amateur astronomy in the United States. Ph.D. thesis. Dept. of History, Rice University, Houston.   *75*

Index

A **bold** number indicates the most important page(s)
An *italicized* number indicates that a figure is included in the page(s)

Acuff, Roy *versus* Ludwig van Beethoven 265
Adel, Arthur 86, 98
Adgie, Ronald *173*
Admiralty Signal Establishment (UK) 115, 156, 264, 288
Agassiz, Louis 91
Air Defence Research and Development Establishment (UK) 156
Alexander, F. Elizabeth S. 85, 128–9, 319
Alfvén, Hannes 309, 374, 386
  synchrotron radiation for radio stars 379–80
Allen, Clabon W. "Cla" 130, 132, 137, 288, 297, 425, 426, 430
Allen, E. W., Jr. 257
All-India Radio 256
Almond, Mary *245*, 248–50
Altar, Wilhelm 91
Alvin 455
Amalgamated Wireless Australasia, Ltd. 119, 120, 125, 127, 129
Amazing Stories 453
Ambartsumian, Viktor A. 229
American Astronomical Society 70, 208, 253, 413, 441
Andromeda nebula 316, 321, 375, 382, 396, 432, 458
  158 MHz detection by Hanbury Brown and Hazard (1950) *188–91*, 339, 373
  claimed detection (Reber) 63, 396
animals of early radio astronomy
  bug 233, 346
  camel 208
  chick 270
  cobra 389
  cow 118, 189, 302, 306, 330, 358
  deer 101
  dog, bird 110

animals (*cont.*)
  eagle 277
  elephant 374
  fish 168
  flea 480
  giraffe 317
  gnat 374
  goose 258
  gorilla 39
  horse 168, 376, 446
  hydra 39
  mongoose 389
  monkey 279
  octopus 346
  rabbit 279
  serpent 39
  swan 142
  tadpole 256
  trout 32
  turkey 32
antenna temperature (term) 203, 480
aperture synthesis 176, 312
aperture synthesis – earth rotation 294, 297
Appleton, Edward V. 30, 51, 83, 87, 91, *119*, 121, 131, 146, 155, 222, 224, 231, 237, 241, 258, 264, 277, 424, 433
  controversial claimstaking 90–1, 110, 111–12, 148, 237
Appleton, Rosalind 91
Arago, François 314
Arakawa, Daitaro 89
Archimedes 356, 442, 502
Arecibo (Puerto Rico) 1000 ft spherical reflector 264
Arizona Meteor Expedition (1930) 246, 251
Armstrong, Edwin 265, 268
Army Operational Research Group (UK) 80–3, 100–17

Aron, Raymond 13
Ashkhabad, Turkmenistan 238
Aspinall, Arnold 243–5
Astapovich, I. S. 238
astronomers
  amateur 41, 74–5, 182–3, 241, 247, 265
  relations with radio researchers 47–9, 75–6, 138, 152, 175, 182, 195–6, 224, 253, 313, 404, 416, 425, **435, 439–41**, 452,
astronomical (term) 425–6
astronomies, beginnings of various wavelength bands 451, 462–7
astronomy
  extreme ultraviolet 464
  gamma ray 391, 464
  high energy 384, 389–91
  infrared 428, 463
  optical 384, 424, 427–9
    comparison with radio astronomy 439–41, 455–6, 457–9
    electronics for photometry 428
    why so dominant? 465
  ultraviolet 463
  X-ray 391, 464–7, 469
    comparison with radio astronomy 465–7
astronomy and art 12
Astrophysical Journal 63, 152, 395
Atanasijević, Ivan 320
Atkinson, Robert d'E. 329
Atkinson, Sally xxxi, 503
atmosphere, Earth's 26, *473*
  "microwave sky noise" 212
  1.0–1.5 cm absorption (Dicke) 95, 203–4
  primer *473*
  refraction 218
atomic bomb project (Manhattan District) 200, 418

Atomic Energy Commission (US) 443
Atomic Energy Research Establishment (Harwell, UK) 446
Attu, Alaska (US), 1950 solar eclipse 207
aurora borealis
  possible radio bursts 212
  radar echoes from 191
Austin–Cohen formula 25
Australia and New Zealand Association for the Advancement of Science (ANZAAS) 138, 152
Australian Journal of Scientific Research 140, 145
Avro Anson 187

Baade, Walter 76, 221, 320, 335, 383, 385, 387, 426, 429, 433, 457
  biography *344–5*
  optical identifications of radio sources (1950–3) *340–9*
  radio stars as Population II objects 375–7
Baade–Minkowski wager regarding Cyg A 346
Babcock, Harold D. 288
Babcock, Horace W. 374, 392
Bacon, Francis 442
Badgery's Creek (Radiophysics Lab field station) 330, 360
Bailey, Victor A. 119
Bakker, Cornelius J. 395, 404, 405
Baldwin, John E. 173, 176, 351, 359, 385
Bannier, J. H. 405
Banwell, C. John 180, *183–4*, 191, 264
Barbara xxxi
Barnett, Miles A. 30
Barrow, W. L. 57
Barth, K.-H. 455
baseball xxx, 39, 398
Bateman, Ross 237
Battle of Britain (1940) 80, 132
Baum, William A. 260, 460
Bay, Zoltán *271–4*
beans, Reber experiments 74
Beck, Alfred C. 40, 52, 78
Beethoven, Ludwig van *versus* Roy Acuff 265
Bell Telephone Laboratories 30–1, 52, 264
  Jansky (star static) *29–53*
  Skellett (meteors) 232, 238
  Southworth (solar noise) *91–8*
Bell, S. Jocelyn 376, 414
Beringer, E. Robert *204–5*, 291
Bethe, Hans 457
Bhar, J. N. 256
Biermann, Ludwig F. 254
Blaauw, Adriaan 415
black hole 382, 465

Blackett, Patrick M. S. 178, 252, 440
Blackwell, Donald E. 175
blind astronomy/blind astronomers 1–2, 425
Bloch, Felix 398
Blum, Emile-Jacques *221–3*, 291
boffin 188, 195
Bok, Bart J. 46, 210, **412–3**, **415–6**, 429, 437, 441
Bolton, John G. 68, 73, 75, 112, *125–6*, *130*, 133, *138–43*, 148–9, 152–3, 166, 174, *317–24*, 361–3, 386, 390, 417, 430, 431, 436, 446, 451
  100 MHz survey of galactic noise (1950, with Westfold) *367, 370–2, 382, 391*
  early Cyg A observations *139–42*
  optical identifications (Tau A, Vir A, Cen A) (1949, with Stanley & Slee) 143, *320–4*
  New Zealand Cosmic Noise Expedition (1948) 143, *317–19*
  observations with McGee of galactic center source *334–6*
  radio source scintillations 324–6
  six new radio stars 142–3
  two surveys of discrete sources (1950, 1954) with Stanley & Slee 332–4
Bondi Beach (Sydney) 139
Bondi, Hermann 389
Booker, Henry G. 210, 327
books
  1610, *Sidereus Nuncius* (Galileo) 468
  1947–53, *Principles of Radar* (ed. Ridenour) 225, 419
  1951, Van de Hulst, *A Course in Radio Astronomy* 407, 430, 457
  1952, Lovell and Clegg, *Radio Astronomy* 195, 424, 430, 437
  1953, Shklovsky, *Radio Astronomy: A Popular Sketch* 216, 459
  1954, Lovell, *Meteor Astronomy* 195, 251, 255
  1954, Waldmeier, *Radiowellen aus dem Weltraum* 311
  1955, Pawsey & Bracewell, *Radio Astronomy* 152, 313, 386, 425, 436, 437, 451
  1976, Edge & Mulkay, *Astronomy Transformed* **10–11**, 159, 315, 334, 340, 349, 389, 421, 433, 435, 438–9, 440, 450, 451, 457, 461, 467, 493, 495, 499
  too long 334
Bowen, Edward G. "Taffy" 91, *121–3*, 124, 125–6, 128–9, 131, 136, 137–8, 144–6, 153, 179, 258, 317, 398, 409, 438
  leadership style 148–9

Bowen, Ira 441
Bown, Ralph 31, 42, 51
Bowyer, Stuart 464
Bracewell, Ronald N. 117, 145–6, 151, 152, 171, **312**, 360
Bragg, William Henry 146
Bragg, William Lawrence 155, 172, 178
Brazil, 1947 solar eclipse *205*, 207, 217, *219*, 222
breakfast, practicing hard before 341
Brecht, Bertolt 157
Breit, Gregory 30, 55
British Association for the Advancement of Science, 1914 meeting in Australia 146
British Astronomical Association 182
Bromberg, J. L. 455
Broten, Norman 297
Bruce, Edwin H. 32, 302
Bruneval raid, France (1942) 81, 222
Brunsviga calculating machine 289
brussels sprouts 179
Buckley, Oliver 42
Budden, Kenneth G. 175
bug 233, 346
Bulletin of the Astronomical Institutes of the Netherlands 372
Burbidge, Geoffrey R. 175, 376
Burgess, Ronald E. 90, 114
Burke, Bernard F. 125
Burkhardt, Gerd 289
Burnet, F. MacFarlane 151
Burnight, T. R. 465
Burrell, B. 282
Burrows, Charles R. 210, 425
Byurakan Astrophysical Observatory (Armenia) 229, 384

California Institute of Technology
  Potapenko & Folland observations of galactic noise (1936) *44–47*
  start of radio astronomy (1955) 75, 153, 429, 437, 446
Cambridge Observatories (UK) 340
camel 208
camera obscura 187
Cannon, Annie J. 48
Carnegie Foundation (US) 153
Carpenter, Martha S. 211, 503
Carslaw, Horatio S. 279
Cassiopeia A 73, 77, 116, 316, 323, 330, 362, 365, 377, 385, 451, 458, 466
  angular size by Jennison & Das Gupta 353–6
  angular size by Smith 359–60
  discovery by Ryle & Smith (1948) 165
  identified with filaments by Baade and Minkowski 343–4

Cassiopeia A (*cont.*)
  identified with remnant of AD 369 supernova (Shklovsky) 383
  position by Smith (1951) leading to optical identification 339–47
  scintillations 324–6
Cassiopeia B 347, 383
Cavendish Laboratory, Cambridge University
  postwar transition 156–9
  prewar background 155
  Rutherford research style 150, 155, 447
  Ryle's group
    camaraderie 173–4, 177
    early history **155–77**, 375
    relations with Radiophysics Lab 144–6, 362
    Rifle Range, Grange Road (field site) *161–2*, 294, 339, 359
    secrecy, aloofness 174–5, 326, 362
  X-ray crystallography 172
cavity magnetron 43, 154, **200**, 309
Centaurus A 65, 143, 316, *321–2*, 330, 333, 341, 347, 359, 360, 362, 373, 377, 458
Central Radio Propagation Lab, National Bureau of Standards (US) 71
CERN Laboratory, Geneva 415
Chain Home radar network (UK) 79, 80, 132, 166, 258
Chalmers Institute of Technology, Göteborg, Sweden 228
Chalonge, Daniel 224
Chamanlal, C. 234, 239, 256
Chamonix, Mt. Blanc 23
Chandrasekhar, S. 46
Chapman, Sidney 120, 231
Checkik, P. O. 238
chick, baby 270
Chikhachev, Boris M. *217*, 218, 291
Christiansen, Wilbur N. "Chris" 122, 125, 146, 149–50, 222, 291
  32-dish array and solar mapping *296–7*, 312
  early 21 cm hydrogen observations 409, *411*, 413–16
Churchill, Winston 12, 379
cigarettes 222, 253
Clark, Barry G. 133
Clarke, Arthur C. 264
Cleeton, Claude E. 86
Clegg, John A. *179–81*, 186, 194–7, 236, 250, 252
  daytime meteor showers *242–4*
  method to determine meteor shower radiants (1948) 242–3

Clerk Maxwell, James 18, 155
Closs, R. L. "Tim" 252
Coblentz, S. A. 453
cobra 389
coherer 20–1, 23–4
Cold War and World War II (theme) 14, 112–13, 122, 151, 221–2, 418–20, **442–9**, 470
Collaroy (Radiophysics Lab field station) *129*
Collins Radio Co. *208–209*, 281, 445
Collins, Harry M. 362
Coma Berenices A (early name for Vir A) 142, 319
Comet Giacobini–Zinner 236, 237
comets as possible radio sources 329
Commonwealth Scientific and Industrial Research Organization (CSIRO), Australia, founding 123
computers (machines) 124, 462
computers (persons) 101
Condon, Edward U. 71
Conferences
  Dynamics of Ionized Media (London, 1951) 375
  IAU (*see* International Astronomical Union)
  Meteor Astronomy (Jodrell Bank, 1948) 196, 248
  URSI (*see* URSI [International Union of Radio Science])
Conway, Robin *173*
Cook, James 153
Copernican view verified by lunar radar 274
Copisarow, A. C. 111
Cormack, Allan 313
Cornell University, Ithaca, N.Y. *210–11*, 440, 441, 444, 447
coronium in sun 311
Corry, Nelly 88
cosmic microwave background radiation 128, 204, 206, 414, 461
cosmic noise 1
cosmic noise (term), abandoning 436, 471
cosmic radio pyrometer 160, *164*
cosmic rays 37, 257, 464
  associated with galactic background noise 369–70
  discovery of new particles by Blackett's group 178, 195
  is there an electron component? 386
  overview of connection to radio sky 389–91
  pre-1950 history 378
  producing synchrotron radiation and galactic noise 378–89

cosmic rays (*cont.*)
  search for radar echoes from air showers (Lovell) 178–81
cosmology, radio 169, 176, 334, **389**, 396
Cottony, Herman V. 72
coulometer (for lunar radar, Bay) *272*
Council for Scientific and Industrial Research (CSIR), Australia 118
Coupling, J. J. (*nom de plume* of John R. Pierce) 234
Covington, Arthur E. 71, **211–13**, 222, 291, 297
  correlation of 10.7 cm intensity with sunspots; slowly varying component 310–1
  first 10.7 cm solar observations (1946) & start of monitoring (1947) 212–13
cows 118, 189, 302, 306, 330, 358
Crab nebula 464–5, 466
  (as Tau A) 142–3, *320*, 339, 340, 347, 360, 375, 382, *383–5*
  measurement of optical polarization 384–5, 387
  optical emission explained as synchrotron radiation (Shklovsky 1953) *383–5*
Cranshaw, T. E. 370
crew racing 239
Crick, F. H. C. 172
cricket (game) xxx, 139, 178
Crimea (USSR), radio field stations of FIAN 218
Crimean Astrophysical Observatory (USSR) 218, 382
Cummings, D. H. 115
Cygnus A 68, 73, 77, 137, 152, 176, 190, 211, 316, 330, 333, 362, 364, 365, 370, 373, 377, 381, 383, 385, 393, 433, 451, 458, 465
  angular sizes by Smith and by Mills 359–60
  Bolton & Stanley attempt at optical identification (1948); estimate of distance 141
  Bolton & Stanley observations of size (< 8′) and position 138–43, *317–23*
  discovery by Hey *et al.* (1946) *103*
  double nature discovery by Jennison & Das Gupta (1953) *353–8*, 502
  measured positions (1947–52) *318*, *337*
  optical identification by Baade and Minkowski *335–41*
  positions by Mills & Thomas and by Smith leading to optical identification *335–44*

Cygnus A (*cont.*)
  Ryle's "genuine" (intrinsic) fluctuations 325–6
  Ryle & Smith observations & position (1948–50) 163–6, 317–21
  scintillations *103*, 319, 324–7
  Smith position leading to an optical identification (1951) 169, *339–41*
  variable position measured by Ryle & Smith (1948–9), *317–21*
Cygnus X 68, 77, 163, 316, 347
Cygnus X-1 (X-ray source) 465

da Vinci, Leonardo 442
Dapto (Radiophysics Lab field station) 306
Das Gupta, Mrinal K. "Das" 192, 349, *353–8*, 414
David Dunlop Observatory (Toronto) 211, 241
Davies, John G. 186, 198, 245, 247, 255
Davies, L. W. *126*
Davisson, C. J. 30
Dawes, William 153
D-Day invasion of Normandy 157
Décombe, Louis P. 24
Dee, P. I. 156
deer 101
Dellinger, J. Howard 89
Denisse, Jean-François *221–3*, 289, 291, 308, 392, 422
  slowly varying component 222, 310–1
Department of Scientific and Industrial Research (DSIR) (UK) 112, 171, 176, 446
Department of Terrestrial Magnetism (Washington, DC) 229, 448
Deppermann, Charles 52
Deslandres, Henri 23–4, 27
de Vaucouleurs, Gerard 430
DeVoogt, Anthonet H. 284, 405
DeVorkin, David H. 428, 429, 439, 445–6, 492
Dewhirst, David W. 175, **342**, 430, 432, 463
DeWitt, John H. – galactic noise observations (1935, 1940) 44, 113, 265
  Project Diana lunar radar (1945–6) 237, *264–71*, 274
Dicke, Robert H. 52, ***200–6***, 291, 369, 441, 444
Dicke radiometer/switch 97, *203*, 211, 212, 220, 277–8, 334, 355, 397, 401, 419, ***484***
Dieter, Nannielou Hepburn 210

discrimination
  against a woman (Payne-Scott) *127–8*
  against an African-American (McAfee) 266
  against Jews 220, 288, 344, 384, 394
DNA 172
Doel, Ronald E. 439, 455, 492,
dog, bird 110
Dombrovsky, V. A. 384, 387
Dominion Observatory (Ottawa) 212, 241, 429
Dover Heights (Radiophysics Lab field station) *129*, *130*, 139, 317–19, 333, 334, 364
Dröge, Franz 228
DuBridge, Lee 153, 200, 419, 443
Dulles Airport, Sterling, Va., USA 72
Duyvendak, J. J. L. 321
Dwingeloo (The Netherlands) 25 m dish 404, 425

"eagle's nest" 277
Eastwood, Eric 110, 258
Ebert, Hermann 22
Eccles, J. C. 151
Eckersley, Thomas L. 160, 180, 257
Ecole Normale Supérieure (Paris) 221
Eddington, Arthur S. 63, 127, 397, 427, 440
Edge, David O. xxx, 10–11
Edison, Thomas. A. 19–20
Edlén, Bengt 311
Edwards, C. F. 43, 258
Ekers, Ronald D. 456
electromagnetic radiation
  Hertzian waves, electric waves 18
  influence of Planck theory on early solar searches 26
  primer 472
elephant 374
Elliot, H. 390
Ellis, George R. 74
Ellyett, Clifton D. 186, 247, 252
Elsmore, Bruce 167–9, 175, 425
Elwert, G. 392
Encke's comet 245
engineering and science, relation 415, 449–53
Englund, Carl R. 31
Eriksen, Gunnar 229
Eshleman, Von R. 240, 252
Espenschied, Lloyd 50, 95
Evans Signal Laboratory (US Army) 265
Evans. John V. 282
Evans, William E., Jr. 239
Ewen, Harold I. "Doc" 150, 211, 441, 442
  discovery of 21 cm hydrogen line (1951) *398–403*, 409–10, 412–15, *416*, 430

Ewen–Knight Corporation 412
Exercise "Post Mortem" (RAF) 159
extraterrestrial intelligence 260, 378
extreme ultraviolet astronomy 464
eye damage from gunners looking at sun 279

Faraday, Michael 362
Faraday rotation 282
Federal Communications Commission (US) 257
Federal Telecommunications Laboratory (New York) 280
Feinberg, E. 387
Fermi, Enrico 386–7, 403
Ferrell, Oliver P. 257
Fessenden, Reginald A. 453
Festival of Britain (1951), 30 ft dish 282
FIAN (Moscow) *see* P. N. Lebedev Physics Institute
Findlay, John W. 171
Fink, D. G. 269
fish 168
Flamsteed, John 142
flea, hop of 480
Folland, Donald F. *44–46*
Forbush, Scott E. 81
Forman, Paul 443
  "distortionist" idea 454–5
Fornax A 341, 362
Forsyth, Douglas R. H. 191
Fourier synthesis principles
  background Fourier theory by Ratcliffe 171–2
  developed by Bracewell 312
  developed by Christiansen *296–7*, 312
  McCready, Pawsey & Payne-Scott (1947) 133–4
  primer 489
  Ryle (1946) 176
  used in solar interferometry by Ryle's group (1950–4) *159–63*, 176, *292–6*, 312
Fradkin, M. I. 380
Franklin Adams Survey 348
Franklin, Kenneth L. 125
Fränz, Kurt 84, *113–14*, 227
Fraunhofer Institute, Freiburg University (Germany) 228
free–free radiation
  primer 476
  to explain galactic noise 62–3, 215, 227, *367–8*, 395
  to explain radio emission from ionized hydrogen (H II) regions 209
Freeman, Joan 127

frequency-switching technique 400, 407, 411
Friedman, Herbert 207, 464
Friend, Albert W. 64
Friis, Harald T. , 31–2, 36, 49–51, 92, 95, 97, 98, 400

galactic coordinate system 334–5, 478
galactic noise radiation
  64 MHz survey by Hey *et al. 101–5*, 366
  100 MHz survey by Bolton & Westfold (1950) *367*, *370–2*, 382, 391
  111 MHz observations by DeWitt (1940) 113
  200 MHz data of Payne-Scott, Pawsey & McCready (1945) 130–1
  205 MHz observations to determine galactic pole and plane (Seeger & Williamson 1951) 211
  comparison with emission from Andromeda nebula 190
  drawn in polar coordinates (Reber) *76*
  associated with cosmic rays 369–70
  as combined effect of radio stars *369–74*
  as free–free emission (postwar research) 215, *367–8*, 458
  as free–free emission by Reber (1940) 59, 62–3
  as free–free emission by Unsöld (1946) 227
  as free–free emission by Van de Hulst (1944) 395
  as hot dust by Whipple & Greenstein (1937) 46–7
  as mixture of interstellar gas and stellar emission (Piddington 1951) 391
  "radio corona" of Shklovsky (1952) 381–2
  as synchrotron radiation (1950–4) 378–9, 459
  as synchrotron radiation by Kiepenheuer (1950) 228
  isotropic component in models *371–4*, 382, 389
  Jansky observations ("star static") 33–42, 43, 366
  Potapenko & Folland observations *44–6*
  "radio eyes" map of northern sky *1–3*
  Reber observations ("cosmic static") *60–69*, 366
  search by Covington at 10.7 cm (1946) 212
  search by Dicke & Beringer at 1.0–1.5 cm (1945) 203–4

galactic noise radiation (*cont.*)
  search by Pawsey & Payne-Scott at 3 cm (1944) 127–8
  search by Southworth at 3 cm 95
  sundry British investigations (1946–8) 114–16
galaxies, colliding 341, 347
galaxies, normal (as radio emitters) 347, 373
Galileo 428, 442, 467
Galison, Peter 351, 361, 364, 452
gamma ray astronomy 391, 464
gamma ray bursters 458
Gamow, George 414
Gardner, Francis F. 151
Gatenby, Ian A. 191, 281
geese 258
gegenschein 274
Germer, L. H. 30
Gernsback, Hugo 39, 261
Getmantsev, German G. 308, 314, **380**, 382
Ghose, B. N. 256
Giacconi, Riccardo 465
Giacobini, Michel 236
Giacobinid meteor shower of 1946 *236–9*, 241
Giant Radio Telescope (Radiophysics Lab project); *also see* Parkes dish 153
Gilbert, G. N. 253–5
Gillmor, C. Stewart 443
Gingerich, Owen 412, 467
Ginzburg, Vitaly L. 217, 220–1, 263, 287, 307, 308, 314, 384, 386, 390, 397, 423, 432, 462
  biography – *214*, 216, 221
  synchrotron radiation theory for galactic noise 380–5
  theory of radio emission from solar corona (1946) 214, 312
giraffe 317
gnat 374
Gold, Thomas 175, 264, **375–6**, 389, 392
Goldsmith, Alfred N. 69, 453
Goodall, William M. 233
Gordon, Isaak M. 384
Gordon, William E. 210
Gorelik, Gabriel S. 220
gorilla 39
Gorky State University 214, 220, 281, 380
Goss, W. Miller xxx, 127, 133
Grand Ole Opry 265
Grange Road, "Rifle Range" (Cavendish group field site) *162*, 294, 339, 359
Greaves, William M. H. 192
Greenhow, J. S. 197, 245, 252

Greenstein, Jesse L. 46–7, 68, 69–71, 75, 314, 377, 383, 387, 397, 425, 429, 432, 436, 441, 448, 495
Grieg, D. D. 280
Griffiths, H. V. 256
Grotian, Walter 311
Gum, Colin S. 425

Hacking, Ian 467
Haddock, Fred T. **206–9**, 263, 310
Haeff, Andrew V. 309, 452
Hagen, John P. *205–11*, 289, 291, 310, 440
Hagihara, Yusuke 225, 256, 430
Haldane, J. B. S. 1, 426
Hale, George Ellery 22, 48, 77, 287
Haleakala, Hawaii 73
Halley's comet 257
Hanbury Brown, Robert 63, 121, 174, 196, 326, 339, 347, 360, 361, 390, 426, 427, 432, 434, 452, 458
  invention (with Twiss) of intensity interferometer 192, 351–3, 365
  model of galactic noise (1953, with Hazard) *372–4*
  observation (with Hazard) of Andromeda nebula at 158 MHz (1950) *189–91*
  observations with intensity interferometer of sun, Cas A, Cyg A *353–8*
  survey with Hazard of 23 sources 331
Hanbury Brown–Twiss effect (quantum optics) 365
Hargreaves, John K. 191, 281
Harper, Alan F. A. 279
Harvard College Observatory (US) 241, 246, 247, 250, 412–15, 429, 437
Harwit, Martin 452
Hatanaka, Takeo 226
Hawaii, Reber observations 73
Hawkins, Gerald S. 182, 191, 193, 198, *244–5*, 250, 252
Hazard, Cyril 339, 347, 426, 432
  model of galactic noise (1953, with Hanbury Brown) *372–4*
  observation (with Hanbury Brown) of Andromeda nebula at 158 MHz (1950) *189–91*
  survey with Hanbury Brown of 23 sources 331
Hazzaa, Ismail W. B. *194*
Heaviside, Oliver 30
Heightman, Denis W. 25, *86–9*, 90–1
Heising, Raymond A. 232
helium in sun 311
helix antenna, invention by Kraus 229
Helliwell, Robert A. 239

Index

Henyey, Louis G. 63, 368, 395
Herbstreit, Jack W. 367
Herlofson, Nicolai 183, 236, 252, **379**
Herschel, John 322
Herschel, William 413, 463, **468**
  comparison to Reber 77
Hertz, Heinrich 18, 20, 454
Hertzian waves/electric waves 18
Hertzsprung, Ejnar 395
Hevly, Bruce 445
Hewish, Antony 169, 172, *173*, 176, 312, 327, 365, 376, 414
Hey, J. Stanley 13, 25, 179, 192, 235, 241, 414, 433, 451, 494
  anomalous radar echoes due to meteors *105–8*
  antenna beam and intensities 53, 116
  AORG research team 101
  controversy with Appleton 90–1
  discovery of first discrete source (Cyg A) and its scintillations *103*, 324
  discovery of radio sun (bursts) (1942) 65, *80–3*
  end of radio research at AORG (1947) 112–13
  galactic noise survey (1945–6) *101–5*, 366
  observations of 1946 Giacobinid meteor shower 237
  solar bursts research (postwar) 111
  (unpublished) evidence for daytime meteor showers (1945) 185
Heyden, Francis J. 207
Higgins, Charles S. 73, 275
Higgs, Arthur J. 137, 149
High Altitude Observatory (Colorado) 445
Hill, E. R. *126*
Hindman, James V. 152, 291, 409–13, 416, 422
Hirsh, Richard F. 464–7
historical contingencies 13–14
historiographic style of this study 11–13
Hodgkin, Alan L. 170
Hoffleit, E. Dorrit 247
Hoffmeister, Cuno 242–4, 245
Holden, Edward S. 19–20
Hollerith punched card machine 293
Homo radio 1–3, 370
Hoo, H. 405, 414
Hooke, Robert 1
horn antenna, Ewen's *399*
Hornsby Valley (Radiophysics Lab field station) *274–7*, 299
horse 168, 446
horse, hobby 376
Hoskin, Michael 468

Hounsfield, Godfrey 313
Houtgast, Jakob 224, 405
Hoyle, Fred 175, 289, 308, 309, 327, 369, 375–7, 385, 389, 392, 393, 427, 468
Hubble Space Telescope 462
Hubble, Edwin 54, 457
Hudson Bay's Best Procurable whiskey 346
Hughes, Victor A. 112, 187
Hulburt, Edward O. 207
Hunter, Alan vi
Huruhata, Masaki 226, 256
Husband, H. C. 192
Hutchinson, George W. 392
Huygens, Christiaan 442
Hvatum, Hein 228
hydra 39
Hydra A 73, 341
hydrogen 21 cm line 211, 303
  Australian detection and first observations (1951–2) 409, *411–12*, 413
  discovery by Ewen & Purcell (1951), **398–403**, 410
  Dutch search and initial observations (1951–2) **404–9**, 410–1, 413–14
  institutional styles of three major early groups 415–17
  prediction by Van de Hulst (1944) 394–6, 399
  primer 477
  Radiophysics Lab decision not to pursue (1948–9) 125–6, 397–8
  Reber's interest 68, 396–7
  theory by Shklovsky (1949) 397
hydrogen bomb, Ginzburg role 214
hydrogen recombination lines, prediction by Van de Hulst (1944) 396

IC 443 314
infrared astronomy 428, 463
Ingram, L. J. 258
Institut d'Astrophysique, Paris 223
Institute of Physical and Chemical Research, Tokyo 232
Institute of Radio Engineers, banquet 269
intensity interferometer *see* interferometer – intensity
interferometer
  analogy with optical transit telescope 336, 339
  intensity
    at optical wavelengths; controversy over theory 353
    Hanbury Brown-Twiss effect (quantum optics) 365

Interferometer (*cont.*)
  invention by Hanbury Brown and Twiss 192, 351–3, 365
  principles 365, 452, 489
  theory and data on sun, Cas A, Cyg A by Jennison & Das Gupta *351–8*
  Michelson 72, 404, 450
    first use by Ryle (on sun) (1945) *160*
    first use by Ryle on radio stars (1948) *163–6*
    invention of phase switch by Ryle (1947) 168–9, 448
    primer *488*
    Ryle's "Long Michelson" (for 1C survey) *166–9*, 320, 327, 328, 361
    solar studies by Ryle's group (1950–4) *292–6*
    tutorial paper by Ryle (1952) 169
    tutorial paper by Smith (1952) on measuring positions 339
  radio-linked 354–5, 360
  sea-cliff *129–34*, *139–43*, 218, 269, 320, 332–3, 362, 364, 375
    primer *488*
    Radiophysics Lab (1945–) 132–4
    Reber (Hawaii) 73
  swept-lobe, Little & Payne-Scott (1949) *299–302*
  using lunar reflection 352
  very long baseline 354
intergalactic medium, radio emission from (Unsöld 1946) 227
International Astronomical Union 44, 432–4
  1946, Copenhagen 433
  1948, Zürich 314, 324, 433
  1952, Rome 346, *350–2*, 433
  Commission 22 on Meteors (1948) 253
  Commission 40 on Radio Astronomy (1948–) 138, 346, 424, 433
international Dutch school of 21 cm research 416
International Polar Year (1932–3) 258
International Telecommunications Co. (Japan) 89
interviews 11, 12, 492–502
ionosphere
  abnormal/sporadic E ionization, short scatter echoes *105–8*, 231, 257–8
  as cause of fading of lunar echoes 274–6, 282
  D layer/region 43, 89
  E layer/region 105
  F layer/region 329
    scintillations and "spread-F" 73, 116, 213, 276, 320, **324–7**

ionosphere (*cont.*)
   influence of solar activity 213
   M (meteor) region 242
   meteors as a cause of ionization (pre-1945 evidence) 231–6, 255–8
   physics of meteor trails 252
   postwar meteor radar research 251–2
   primer 474–5
   shortwave fade-outs (sudden ionospheric disturbances) 30, 88, 89, *299*, 302
   winds 251–2, 327
Istvanfy, Edvin 272
Ito, Yogi 255

Jaeger, John C. 133, 279, 300
jansky (unit) 16, 44
Jansky, C. Moreau *29*, 50
Jansky, Karl G. 25, *29–53*, 54–5, 63, 77, 180, 210, 255, 258, 265, 302, 334, 366, 368, 380, 395, 414, 439, 451, 472
   1931–32 observations 33–5, 36
   and Southworth 94, 97, 98
   biography 29–30, *32*, 43–4
   Bruce array antenna *32–3*
   discovery article (1933) 36–7
   discovery of sidereal connection 35–6
   interpretation of radio waves as the entire Milky Way 41–2, *49*
   interpretation of radio waves as the Milky Way center 36–7
   interpretation of radio waves as the sun 34–5
   modern contour map of data *49*, 53
   reactions from astronomers 47–9
   reasons for success 51–2
   receiver *31–2*
   research on "star static" *31–43*
   research stopped by Friis? 49–51
Jarrell, Richard 437
Jelley, J. V. 181
Jennison, Roger C. 192, 199, *353–8*, 377, 386, 414, 502
Jerry cable 386
Jewett, Frank B. 30, 70
Jodrell Bank, Manchester University 355, 421, 444, 447, 448
   218 ft fixed dish (1948) 180, *186–91*, 334
   founding by Lovell (1945) 179–81
   lunar radar (1949–) 191, 281
   overview of Lovell's group during first 5 years 193–8
   plans for 250 ft steerable dish 181, 192–3
   relation with (optical) astronomy 195–6, 440, 452

Johler, Joseph R. 72, 367
Johnson, Harold 464
Johnson, John B. 26, 30
Julius, George A. 118
Jungfraujoch 228
Jupiter, low-frequency radio bursts 125

Kahn, Franz 196, 440
Kaiser, D. 454
Kaiser, Thomas R. 197, 252
Kalachev, Pavel D. 218
Kapitza Club 350
Kapteyn, Jacobus 371, 395, 413–16
Kauffman, Herbert *269*
Kaydanovsky, Naum L. 218
Keenan, Philip C. 60, 75, 368, 395, 485
Kelly, Mervin J. 64
Kelvin, Lord (William Thomson) 443
Kennelly, Arthur E. 19–20, 30
Kennelly–Heaviside Layer (ionosphere) 30, 232, 255, 261
Kepler, Johannes 468
Kepler's supernova 321, 347
Kerr, Frank J. 4, 122, 128, 145, 150, *274–7*, 282, 409, **413–17**
Kevles, D. J. 455
Khartoum, Anglo-Egyptian Sudan, 1952 solar eclipse *208–9*, 291
Khaykin, Semen E. 217–18, 291, 320
Kiel University 227
Kiepenheuer, Karl-Otto 228, 297, 308, 314
   synchrotron radiation as cause of galactic background noise (1950) *379*, 380
Kimpara, Atsushi 226
King Kong (film) 39
King, Archie P. 92
knobs, functionless 148
Kobrin, M. M. 281
Koelbloed, D. 395
Kootwijk, the Netherlands 405, 413
Kopal, Zdenek 196, 430, 440
Korchak, A. A. 382
Kraus, John D. 2, 86, 229, 263
Kröbel, W. 228
Kuhn, Thomas S. 467
Kuiper, Gerard P. 60, 62, 70, 76
Kusch, P. 401
Kwal, Bernard 380, 392
Kyhl, Robert *204*

Labrum, Norman R. 291, 311
Laffineur, Marius 86, *223–5*, 291, 320, 405, 430
Lalande, Jérôme 77

Lamb, Willis E. 398
lamington 147
Landau, Lev 397
Langer, Rudolph M. 44
Langley, Samuel P. 469
Lankford, John 74
Lasswitz, Kurd 23
Latin, abstract of lunar radar paper (Lombardini 1944) 264
Latour, Bruno 450
Lavoisier, Antoine L. 349
Lehany, Fred 199, 311
Leiden Observatory 211, 394–6, 429, 452, 471
Leigh, New Zealand 317
Leighton, Robert 463
Lejay, P. 228
Leningrad University 229
Lesch, J. E. 363
Leslie, S. W. 454
Levin, B. Yu. 238
Lindblad, Bertil-Anders 228
Lipson–Beevers strips 292, 297
Little, Alec G. **300–2**, 336
Little, C. Gordon 192, **324–7**
Llewellyn, Frederick B. 26, 232
Lloyd's mirror 132
lock-in amplifier 203, 402
Lodge, Oliver J. 20–1, 454
Lombardini, Pietro 263
"Long Michelson" interferometer, Cambridge *166–9*, 320, 327, 328, 361
love play 227
Lovell, A. C. Bernard 109, **178–98**, 252, 255, 264, 353, 423, 424, 443, 447, 448, 451, 468
   biography 178
   daytime meteor showers *242–5*
   founding of Jodrell Bank; searching for radar echoes from cosmic ray showers (1945–8) 178–81
   leadership style 193–8
   observations of 1946 Giacobinid meteor shower *236*
   plans for 250 ft steerable dish (1949–) 192–3
   research on speeds and origins of meteors *245–51*
   start of meteor radar research (1946) *181–6*
Lovell Telescope (new name for Jodrell Bank 250 ft dish) 193
Luke, St. 123
Lutz, Samuel G. 57
Lyman Laboratory, Harvard 399
Lyman, Theodore 400

Lyot, Bernard 224, 296
Lyttleton, Raymond A. 327

Machin, Kenneth E. 163, 170, 171, *173*, 175, 176, 219, **294**, 312, 356, 430, 440, 450, 466
MacRae, Donald A. 211, 342
Madsen, John P. V. 119, 120, 441
Magellanic Clouds 142, 321, 413, 458
major themes of book
  material culture and technoscience 14–15, **449–53**, 470
  twentieth century's New Astronomy 16, 462, **467–70**
  visual culture 15, 351, 426, **435–6**, 470
  World War II and Cold War 14, 112–13, 122, 151, 220–1, 418–20, **442–9**, 470
Mandel'shtam, Leonid I. 214, 262
Manning, Laurence A. 237, 239, 252
Marconi Co. 25, 257
Marconi, Guglielmo 25, 30, 55, 260
Marcoussis, France 222
Mars – radio communication with Martians 260
Martyn, David F. 86, 148, 309
  biography 119
  first head of Radiophysics Lab 120
  first head of URSI Commission V (1948) 145–6, 433
  priority dispute with Pawsey over 10^6 K solar corona 136–7
  proposal for lunar radar (1930) 261
  rivalry with Radiophysics Lab 137–8
  theory of 10^6 K solar corona and solar emission 135–7, *287–8*, 312
Marx, Groucho 39
Maxwell, Alan 192, **327**
Mayall, Nicholas 445
Mayer, Cornell H. 208, 388, 431
McAfee, Walter S. 266
McClain, Edward F. 263
McCoy, D. O. 208
McCrea, William H. 341, 430, 431
McCready, Lindsay L. *129–34*, 199, 298, 302
McCullough, Timothy P. 388
McDonald Observatory 62, 70
McEwan, R. J. *209*
McGee, Richard X. 334, *363*
McKinley, Donald R. W. *241–2*, *250*, 259, 429
McNicol, Robert W. E. 168
McNish, Alvin G. 237
McVittie, George C. 376

Menzel, Donald H. 25, 70, 97, 199, 260, 263, 312, 314, 399, 429, 445
Mercer, K. A. 110, 258
mermaid hunter 168
Messier 87 (as Vir A) 316, **321–2**, 339, 340, 347, 360, 458, 465
Messier, Charles 320
meteor radar astronomy
  rise and fall (1945–55) 253–5
  sociological analysis by Gilbert 254–5
  Table of early groups 241
meteors 252
  possible cause of abnormal ionization in ionosphere (prewar work) 231–6
  cause of anomalous radar echoes (1945, Hey & Stewart) *105–8*
  conference at Manchester University and Jodrell Bank (1948) 248
  contributions of their study by radar to ionosphere physics 251–2
  determined to be not from interstellar space *245–51*, 253, 259
  for radio communications 240, 254, 444
  Fresnel diffraction technique to measure speeds (Davies & Ellyett 1949) 186, 240, 258–9
  physics of trail formation and radar reflection 252
  pre-1945 connections to the ionosphere 255–8
  radar Doppler method to measure speeds (Manning 1948) 240, 258–9
  radar method to determine radiants (Clegg 1948) *242–3*
  radar apex and antapex experiments at Jodrell Bank 248–50
  used to measure ionospheric winds 251–2
meteor showers
  Arietid 243, 244
  β Taurid 243, 244
  δ Aquarid 108, 233
  η Aquarid 184
  ζ Perseid 244
  daytime 108, 184, 185, *242–5*
  Geminid 256, 258
  Giacobinid (1946) 184, *236–9*
  Leonid 233–4, 238, 255–6, 257
  Lyrid 109
  Perseid 108, **182–3**, 233, 238, 240, 241–2, 256, 257
  Piscid 184, 243
  primer 107
  Quadrantid 109

Metzger, S. 280
Meudon Observatory 23–4, 223, 430, 469
Meudon Observatory, cafeteria 224
Michelangelo 361
Michelson interferometer *see* interferometer – Michelson
Michelson, Albert A. 160, 352
Michigan, University of 86
microwaves 91
Mikhailov, A. A. 427
military influences on research groups 14, 101, 122, 218, 226, 240, 254, 399, 412, 415, **442–9**, 454–5, 466
Millikan, Robert A. 45
Millman, Peter M. *241–2*, 247, 252, 253, 429
Mills Cross, *147*, 153
Mills, Bernard Y. 145, 148, 150, 154, 168, 174, 291, 302, 327, 374, 385, 390, 431, 436, 440, 448, 458, 471
  angular sizes for 4 discrete sources 360
  decision not to search for 21 cm hydrogen line 125–6, 398
  positions for Cyg A (1949–52) *335–9*
  "radio pictures" *350*, 360
  survey of 77 discrete sources *330–1*, 365
Minkowski, Rudolph 142, 221, 320–2, 325, 356, 383–4, 426, 429, 433
  biography *344–5*
  optical identifications of radio sources (1949–53) *337–49*
  radio stars as Population II objects 375–7
Minnaert, Marcel G. J. 394, 405, 425, 429
Minnett, Harry C. 127, 148, **277–80**, 291, 311, 334, 347, 398, 448
Minohara, Tsutomu 255
Mitra, S. K. 256, 353
Miya, Kenichi 89
Mofenson, Jacob *266*
Molchanov, Andrei P. 229
molecules, radio lines 397, 462
mongoose 389
monkey 279
Monthly Notices of the Royal Astronomical Society 440
moon
  librations 271, 275
  "military object" 445
  occultations of radio sources 314
  radar
    detection by German military radar (Stepp & Thiel, 1943–44) *262*
    detection in Hungary by Bay (1946) *271–4*, 282–3
    Jodrell Bank (1949-) 191

moon radar (*cont.*)
    librations 271, 275–6, 280
    press reaction to Project Diana 269–71
    prewar proposals 260–1
    Project Diana detection (DeWitt et al., 1946) *264–71*
    Radio Australia echoes (Kerr & Shain, 1947) *274–7*
    wartime calculations and observations 261–4
    reflection of light from atomic bomb explosion 262
    thermal emission 207
      detection by Dicke & Beringer at 1.25 cm (1945) 205
      model of lunar soil by Piddington & Minnett 278
      study by Piddington & Minnett at 1.25 cm (1948) *277–80*
    used as communications relay 280
    used by US to intercept Soviet radio transmissions 280
Moran, Frank 191
Moriyama, Fumio 226
Morrison, Phillip 390, 462
Moscow State University 214
Moxon, L. A. 115
Mt. Haleakala, Hawaii 73
Mt. Stromlo Commonwealth Observatory 130, 137–8, 152, 430
Mt. Wilson Observatory, Pasadena 48, 142, 277, 287, 320, 337, 344, 429, 441
Mueller, George E. 95–7
Mulkay, Michael 10–11
Muller, C. Alexander 394, *406–9*, 413–14, *416*
Muller, H. 289
Mumford, Willis W. 401
Munns, David 153, 413, 437
Murray, John D. *302*
Murray, William A. S. "Sandy" 191, 199, **281**
Mustel, E. R. 382
Mutch, W. W. 31

Nagaoka, Hantaro 232, 257
Nagoya University, Research Institute of Atmospherics 226
Naismith, Robert A. 110, 182, 237, 258
Nakagami, Minoru 89
Nançay, France 223
NASA (National Aeronautics and Space Administration) 97, 254, 280, 449, 462, 466

National Bureau of Standards (US) – 71–2, 78, 89, 222, 237, 254, 281, 367, 441, 444
National Radio Astronomy Observatory (US) 75, 78, 206, 400, 425, 437, 446, 449, 495, 505
National Research Council (Canada) 211–13, 241–2
National Science Foundation (US) 449
Naval Research Lab – *see* US Naval Research Lab
Needell, Alan 344
Neugebauer, Gerald 463
New Astronomies
    astrophysics (spectroscopy, photography) 469
    digital computers and numerical simulation 462
    Galileo 467
    Herschel 468
    radio astronomy; opening of electromagnetic spectrum 16, 462, **467–70**
New Zealand Cosmic Noise Expedition (1948) 143, *317–19*
Newton, H. W. 430
Ney, Edward P. 386
NGC 1275 (as Per A) 316, 341, 347
NGC 5128 (as Cen A) 316, *321*, 341, 347
Nicholson, Seth B. 142, 277
Noah 56
Nobel Prize in Physics
    Appleton (1947) 83
    Blackett (1948) 178, 192
    Bragg and Bragg (1915) 172
    Ginzburg (2003) 214
    Purcell & Bloch (1952) 398
    Ryle & Hewish (1974) 176, 312
Nobel Prize in Physiology or Medicine
    Hounsfield & Cormack (1979) 313
    Hodgkin (1963) 170
noise, primer 479
Nordmann, Charles 23–4, 469
Norfolk Island effect 85, 128–9
Norton's Star Atlas 139, 324
Nuclear Laboratory, Harvard 398
Nuffield Foundation 193
nuit de l'amour 77
Nyquist, Harry 26, 30

O'Brien, Patrick A. 174, 176, *294–6*
octopus 346
Oda, Minoru *226*
Office of Naval Research (US) – 70–1, 210, 240, **443–9**
Ohio State University 229

Olbers's paradox 389, 396
Oliphant, Mark L. 138
Onsala, Sweden 228
Oort, Jan H. 56, 103, 254, 321–3, 335, 349, 376, 385, 387, 410, 427, 429, 433, 444, 452, 459
    biography 371, *394–5*
    detection and observations of 21 cm hydrogen line *404–9*, 410–1, 413–17
    model of galactic noise (1951, with Westerhout) *370–2*, 382, 391, 430
Öpik, Ernst J. 246–51, 252, 253
Oppenheimer, J. Robert 122
optical astronomy (and related terms) 15, **424–5**, 467
optical astronomy (field) *see* astronomy, optical
optical identification (term) 322, **349**
optical identification of radio stars (sources) 143, 169, *317–24*, *335–51*
oral histories 492–502
Oreskes, N. 454
Orion nebula 142, 208, 316, 383, 458
ornithology, radar 258
Osaka City University 226
Osaka University *226*
Oslo, University of 211, 228
Ovenden, Michael W. 175, 182, 254, 312, 313, 428, 468
Owren, Leif 211, 417

P. N. Lebedev Physics Institute (FIAN), Moscow 214, 218
Palomar Mountain 200 inch telescope 45, 211, 340–8, 364, 440
Palomar Sky Survey 348
Pang, Alex S.-K. 429
Pan-Pacific Science Congress, 1923 meeting in Australia 146
Papaleksi, Nikolai D. 214, 217, 220, 262–3
Papp, György *272–4*
Parker, J. C. *182*
Parkes dish 152, 153
Parsons, S. John 101–5, 112, 179, 237, 450
Pascal, Blaise 1
Pawsey, Joseph L. 73, 77, 85, 128–9, 138, 142, 144–6, 146–8, 152–3, 164, 174, 191, 280, 298, 302, 317–20, 323, 324, 335, 363, 370, 397, 409, 410, 413, 414, 417, 420, 424, 425, 432, 433, 434, 437, 438, 451
    biography *124–6*
    first Sydney solar observations (1945–6) *129–34*

Pawsey, Joseph L. (*cont.*)
  Fourier synthesis principle 133–4
  leadership style 148–51
  microwave sky experiments (1944) 127–8
  observation of solar base level of 10^6 K *135–7*, 312
  priority dispute with Martyn over 10^6 K solar corona 136–7
  spectrum of quiet sun *285–6*
Payne-Scott, Ruby V. 97, 139, 336
  biography *127–8*
  first Sydney report on solar and galactic noise (1945) 130–1
  first Sydney solar observations (1945–6) *129–34*
  Fourier synthesis principle 133–4
  microwave sky experiments (1944) 127–8
  solar bursts research *298–302*, 305
Pease, Francis 160
Peenemünde, Germany 262
Penrith (Radiophysics Lab field station) 302, 305
Penzias, Arno A. 52, 414
Perseus A 316, 341, 346, 347, 377, 458
Peterson, Allen M. 240, 252
Pettit, Edison 277
Pfeiffer, John 49–50, 430
Ph.D. degrees, early
  Clegg (1948a), Manchester (meteor radar) 193
  Denisse (1949b), Ecole Normale Supérieure (first on a radio astronomy topic) 222
  Ellyett (1948), Manchester (first in radar astronomy) 193
  Ewen (1951), Harvard (discovery of hydrogen line) 398–403
  Hagen (1949), Georgetown (microwave sun) 207
  Nordmann (1903), Paris (Hertzian waves and astronomy) *23–4*
  Ryle decides not to finish his Ph. D. 172–3
  Skellett (1933), Princeton (radio and meteors) *232–4*, 429
  Smith (1951c), Cambridge (positions of radio stars) *339–44*
  Stanier (1950b), Cambridge (solar interferometry) *292–4*
  Troitsky (1949), Gorky (solar techniques) 220
  Villard (1948), Stanford (meteor radar) 240
phase switch, invention by Ryle (1947) and later use 168–9, 330, 355, 360, *489*

Philips Research Laboratories 395
Phillips, James W. 21, *101–5*, 112, 116
photomultiplier tube 428
Pickard, Greenleaf W. 257
Piddington, Jack H. 86, 148–9, 152, 190, 261, *277*, 288, 291, 308, 311, 334, 347, 372, 376, 409, 417, 430, 448
Pierce, George W. 46
Pierce, John A. 106, **234–5**, 236, 237, 239, 252, 257
Pierce, John R. 98, 234, 309, 452
Piha, New Zealand 319
Pikel'ner, Solomon B. 382
Pineo, Victor C. 237
Pippard, A. Brian 172
planets, radar 282
plant growth and solar radio bursts 314
Poisson's spot 314
Popper, Daniel M. 426, 430
Porter, J. G. 245, 247
Porter, Russell W. *45*
Potapenko, Gennady W. *44–5*
Potsdam Observatory 22–3, 469
Potter, Ralph K. 52
Potts Hill (Radiophysics Lab field station) *146*, 291, 296, 300, 336, 409
Pound, Robert V. 400
Powell, John Wesley 446
precession of coordinates 142, 432, 478
Prentice, J. P. Manning 182–4, 236, 239, 430, 468
Priester, Wolfgang 228
Primakoff, Henry 387
Princeton University 35, 205, 232, 429
Procrustes 149
Prodell, A. G. 401
Project Diana (US lunar radar, 1946) 237, *264–71*, 274
Project Janet 254
Project PAMOR (US Naval Research Laboratory) 280
Project Vanguard 210
Pulkovo Observatory, Leningrad 75, 220, 387
Pulley, Owen O. 119
pulsars 365, 376, 414, 462
Puppis A, 316, 330, 347, 359
Purcell, Edward M. 203–204, *398–403*, 409–10, 412, 415, 419, 444
Pyenson, L. 429

Quäck, E. 257
Quarterly Bulletin on Solar Activity (IAU) 284, 313, 433
quasars 461

rabbit 279
radar
  British secret given to Australia (1939) 120
  development before and during World War II 79–80, 120–1, 505
  military types
    Chain Home network (UK) 79, 80, 132, 166, 258
    Freya (Germany) 84
    GL Mark II (UK) *81–2, 101–2*, 179
    H$_2$S (UK) 178
    high quality of German manufacture 159
    Knickebein (Germany) 113
    LORAN radio navigation system (US) 200
    LW/AW Mark IA (Australia) 121
    SCR-268 (US) *211–2*
    SCR-270 (US) 237, 257, 266
    SCR-271 (US) *266–9*
    SLC "Elsie" (searchlight control) (UK) *183*, 236, 242, 248
    Würzburg (Germany) *72*, **78**, 80, 81, 84, 159, 169, 218, 222, **224**, 228, 229, 262, *293*, 329, *339*, 359
    Würzmann, Wassermann (Germany) *262*
  origin of name 79
  ornithology 258
  postwar influence on science and technology 418–20
radar equation 266
Radar Research and Development Establishment (UK) 114
radiant aerials, Jodrell Bank *243*
Radiation Laboratory, Massachusetts Institute of Technology (US) 121, **200–5**, 239, 269, 277, 398, 442
radio amateurs 30, 54, 86–9, 210, 239
radio astronomers – relations with (optical) astronomers 75–6, 138, 152, 175, 224, 313, 416, 425, **429–32**, 435, *439–41*
radio astronomers as New Astronomers 453
Radio Astronomical Journal, lack of 437
radio astronomy (and related terms) 234, **423–4**, 453, 471
radio astronomy
  1953 as a watershed year 422–3
  a new discipline? 15, **435–8**, 471
  a scientific revolution? 467
  applied uses 124, 208
  as technoscience 449–53
  bibliographies 503

radio astronomy (*cont.*)
  comparison with electron microscopy 454
  comparison with optical astronomy 1–3, 455, 457–9
  comparison with other New Astronomies **467–70**
  comparison with postwar US space science 445
  comparison with start of X-ray astronomy 465–7
  definition 3–4
  entering Big Science era 422–3
  first academic chair (1951, Lovell) 195, 423, 430
  first academic course (1950, Van de Hulst) 407, 416, 430
  growth over 1945–53 *420–2*
  harbinger of opening of other wavelength bands 462–3, 469–70
  IAU versus URSI 432–4
  lack of US leadership despite Jansky and Reber 439
  military patronage in US **442–9**, 455
  national influences 438–9
  overview of history *5–10*, 457–62
  primer *472–84*
  shorter versus longer wavelength research (US versus elsewhere) 209, 447–8
  size relative to (optical) astronomy 313, 420, 457, 465
  style of research 427–9, 442
  Table of early groups 201
  the twentieth century's New Astronomy 469–70
Radio Australia 274
radio communications research before 1928 30
Radio Development Laboratory, Wellington, N. Z. 85
radio galaxy (term) 383
radio images *349–50*, 360, 436
radio magnitudes 347, 436
radio observatory (term) 425
"radio pictures" (Mills) *350*, 360
Radio Radiation Laboratory, Harvard University (US) 239
Radio Research Board, Australia 119, 261
radio sextant 208, 412, 445, 447
Radio Society of Great Britain 87
"radio sourcerers" 363
radio stars (sources)
  1′–2′-accuracy positions by Smith (1951a) 169, *339–44*
  A, B... nomenclature 142

radio stars (sources) (*cont.*)
  claimed bursts and interstellar dispersion (Shklovsky 1950) 364
  first spectra (Bolton's group) 333
  New Zealand Cosmic Noise Expedition (Bolton & Stanley, 1948) *317–19*
  occultations by the moon 314
  optical data epistemically superior to radio data 351
  positions measured by Bolton's group (1948–9) *317–24*
  reality of 348–51, 434, 436
  role of terminology in disputes 363
  search for proper motions and parallaxes (Smith) 329
  search for intensity variations (Ryle & Elsmore) 330
  status of the field in 1953 363–4
  interpretations
    Class I (galactic) and Class II (isotropic) by Mills (1952) *330–2*, 373–4
    collapsed dense stars (Gold) 376
    colliding galaxies 341
    comets 329, 374
    dark, nearby stars (Ryle) 169, 329 374–8
    extragalactic objects 341, 374–8, 383
    integrated effect produces galactic background noise *369–74*
    intelligent transmissions (Ryle) 378
    nearby or distant? 374–8
    Population II objects (Baade & Minkowski) 335, 375
    radio galaxies (Shklovsky) 383
    radio stars or radio nebulae? 360–3, 382
    Seyfert galaxies 376
    synchrotron-radiation-emitting stars (Alfvén & Herlofson 1950) 379
    supernova remnants (Shklovsky) 383
    Wolf-Rayet stars (Shklovsky) 381
  notable
    Andromeda nebula (*see separate entry*)
    Cassiopeia A (*see separate entry*)
    Cassiopeia B 347, 383
    Centaurus A 65, 143, 316, *321–2*, 330, 333, 341, 347, 359, 360, 362, 373, 377, 458
    Coma Berenices A (early name for Vir A) 142, 319
    Cygnus A (*see separate entry*)
    Cygnus X 68, 77, 163, 316, 347
    Fornax A 341, 362
    Hydra A 73, 341

notable radio stars (*cont.*)
  Perseus A 316, 341, 346, 347, 377, 458
  Puppis A 316, 330, 347, 359
  Sagittarius A (galactic center source) 77, 316, *334–5*
  Table 316
  Taurus A (*see separate entry*)
  Ursa Major (spurious) 166, 324, 393
  optical identifications
    Baade & Minkowski (1954) *344–8*
    first ones suggested (Bolton, Stanley & Slee 1949) *320–4*
    general discussion *348–51*
    normal galaxies 347
    positions by Mills and by Smith for Cas A and Cyg A *335–44*
  scintillations *103*, 116, 319, **324–7**, 353
    Jodrell Bank/Cambridge simultaneous observations (1949) 326
    primer 116
    Sydney/New Zealand simultaneous observations (1948) 325
  surveys
    1C (Ryle, Smith & Elsmore 1950) *328–9*, 364, 365, 395
    2C (Cambridge) and its antenna 175, 193, 334, 389
    9 cm (Haddock et al. 1954) 208
    discrepancy between Mills's MSH and Ryle's 2C (mid-1950s) 153, 334
    Hanbury Brown & Hazard (1953) 191, 331
    log N – log S plots 330, *332*, 334, **365**, 370, 389
    Mills (1952) survey of 77 sources *330*, 365
    Ryle's group's initial list of 23 objects (1949) 166–8, 364
    six new sources, including Tau A, Vir A, Cen A (Bolton 1948) 142–3
    Stanley & Slee (1950, 77 sources) and Bolton et al. (1954, 104 sources) *332–4*
    the five early primary surveys *328*
radio telescope: is it a telescope? 426–427
radio telescope (term) 424, 426–427, 453, 471
Radiophysics Laboratory, Sydney 86, **118–53**, 261, *274–7*, 297, 421, 441, 446, 503
  field stations
    Badgery's Creek 330, 360
    Collaroy *129*
    Dapto *306*

538   Index

Radiophysics Laboratory, Sydney, field stations (*cont.*)
  Dover Heights *129*, *130*, 139, 317–19, 333, 334, 364
  general *129*, 146–8
  Hornsby Valley *274–7*
  Penrith 302, 305
  Potts Hill *147*, 291, 296, 300, 336, 409
  and (optical) astronomy 152
  founding (1939) & wartime radar work 120–1
  isolation factor in Australia 143–6, 362
  leadership styles of Bowen and Pawsey 148–51
  overseas trips by researchers 144–6
  relations with Ryle's group 144–6, 326
  research trends over 1946–53 *123–5*, 154
  success and changes in the 1950s 151–3
  transition to peacetime (1945) 121–3
  rain and cloud physics 123
Ramsey, Norman F. 412
Rasmussen, N. 454
Ratcliffe, John A. "Jack" 80, 119, 124, 155, **156–9**, **171–2**, 177, 313, 317–19, 324, 327, 360, 363, 494
  Fourier transform teaching 171–2, 313, 440
Reagan National Airport 208
Reber, Grote 26, 43, 48, **54–77**, 96, 116, 162, 180, 208, 211, 211, 222, 263, 281, 334, 395, 396, 425, 428, 429, 439, 442, 444–5, 451, 495, 505
  160 MHz survey of galactic noise (1943–4) *65–8*, 366, 369
  480 MHz survey of galactic noise (1946–7) *68–9*
  biography *54–5*, 74–7
  comparison to William Herschel 77
  detection of sun (1943) 65
  errors in quoted antenna specifications 77–8
  first publications (1940) *60–63*
  first receivers and "cosmic static" (1938–9) *57–60*
  hydrogen 21 cm line receiver 68, 396–7
  Milky Way radiation as free-free emission 62–3, 367
  proposed observatory & 200-ft dish in Texas 69–71
  reflector antenna in Wheaton, Ill. (1937) *55–7*, 78
  research after Wheaton (1947-) 71–4, 281, 396–7, 404
  seeking funding 63–5, 69–71
recombination lines of hydrogen 396, 412

Redman, Roderick O. 175, 340, 344
Research Corporation 73
review paper
  Ginzburg (1947, 1948) 216
  Lovell (1948) 231
  Reber and Greenstein (1947) 68, 397
  Ryle (1950) 175
  Unsöld (1946) 227
  Williamson (1948) 211
rhombic antenna 302
Richardson, Robert S. 88, 89
Richmond Park, London *101*
Rifle Range, Grange Road (Cavendish group field site) *161–2*, 294, 339, 359
Riihimaa, Jorma J. 25, 27
Rivett, A. C. David 118, 122, 125, 149
Roberts, James A. 174, 386, 459
Rocard, Yves 221, 228
Rossi, Bruno 386, 465
Rowe, A. P. 156
Rowe, William C. *302*
Royal Aircraft Establishment (UK) 159, 163, 169, 186
Royal Astronomical Society (UK) 75, 104, 168, 190, 192, 195, 239, 254, 308, 311, 342, 360–3, 424, 430
Royal Australian Air Force 153, 446
Royal Greenwich Observatory 81, 105, 245, 339, 341
Royal Institute of Technology, Stockholm 379
Royal Radar Establishment (UK) 113, 446
Royal Society (UK) 177, 423, 431
Rudwick, Martin 12, 436
rugby 162
Rügen, Germany 262
Rumford Fund, American Academy of Arts and Sciences 399
Russell, Henry Norris 47, 206
Ruth, Babe 39
Rutherford, Ernest 150, 155, 447
Rydbeck, Olof E. H. 64, 211, 228, 417
Ryle, Gilbert 381
Ryle, John A. 381
Ryle, Martin **155–77**, 179, 191, 192, 218, 297, 314, 327, 328–30, 346, 352–3, 360–3, 364, 365, 366, 379, 385, 389, 391, 392, 404, 419, 422, 424, 431, 434, 435, 439, 440, 442, 447, 451, 458, 466, 467
  2C (Cambridge) and its antenna 175, 193, 334, 389
  biography, wartime work, postwar transition 156–9

Ryle, Martin (*cont.*)
  Cas A discovery (Ryle and Smith 1948) 165
  Cavendish Lab group *see* Cavendish Laboratory
  debates about nature of radio stars *374–8*
  development of Fourier synthesis technique 176, *292–6*, 312
  discrepancy between Mills's MSH and Ryle's 2C (mid-1950s) 153, 334
  leadership style 172–5
  measurements of Cyg A position (1948–9) *317–21*
  reality of radio stars 348–51, 434, 436
  radio star scintillations 324–7
  radio stars interpreted as dark, nearby stars 169, 329, *370*
  radio stars interpreted as extragalactic objects 346, 374–8, 389
  solar observations *159–63*, *292–6*, 312
  solar theory 308
  success of his group 169–77
  survey (initial) by Ryle, Smith & Elsmore of 23 objects (1949) 166–8, 364
  survey (published) by Ryle, Smith & Elsmore of 50 objects (1C) (1950) *328–9*, 364, 395
Ryle–Vonberg receiver *160*, *163*, 189, *484*

Sacramento Peak (New Mexico) 211, 445
Sagittarius A (galactic center source) 77, 316, *334–5*
Saha, M. N. 397
Salisbury, Winfield W. 199, 208, 263, 314
Salomonovich, Alexander E. 218
Sanamyan, Vagarshak A. 229
Sandage, Allan 430
Sander, Kenneth F. 114, 158
Schafer, J. Peter 233
Scharnhorst and *Gneisenau* (battle cruisers) 81
Scheiner, Julius *21–3*, 469
Schermerhorn, Willem 405
Scheuer, Peter A. G. 170, 173, 363, 373, 386, 390, 391, 414, 426, 450, 468
Schiaparelli, Giovanni V. 250
Schluter, Arnulf 289
Schott, E. 83–4
Schwinger radiation 379
Schwinger, Julian 452
science and engineering, relation 415, 449–53
scintillations of radio stars, *see* radio stars – scintillations
Scorpius X-1 (X-ray source) 465, 466

Scott, John M. C. 100, 115
sea-cliff interferometer *see* interferometer, sea-cliff
Seeger, Alan, poet 210
Seeger, Charles L., Jr. (radio astronomer) 190, *210–11*, 297, 342, 427, 429, 435
Seeger, Charles L., Sr. (musicologist) 210
Seeger, Pete (singer) 210
seeing (concept) 426
serendipity 51, 253, **414**, **451**, 466
serpent 39
Serviss, G. P. 19
SETI (Search for Extraterrestrial Intelligence) 462
Shaffer, Simon 77
Shain, C. Alexander 73, 125, *274–7*
Shain, G. A. 384
Shakeshaft, John R. 173, 389
Shapley, Alan 97
Shapley, Harlow 47, 48, 60, 64, 69–71, 246, 321, 399, 412–15, 424, 425, 437, 445, 458
Shapley-Ames Catalogue of Galaxies 348
Sharp, Charles *173*
Shin, Douglas H. 327
Shklovsky, Iosif S. 217, 220–21, 230, 308, 320, 346, 349, 368, 387, 389, 390, 412, 427, 436, 459
 biography 216–7, 221
 galactic corona of radio stars for galactic noise 368, 381–2
 interstellar dispersion in bursts of radio sources (1950) 364–5
 synchrotron radiation theory for galactic noise and radio stars *382–4*
 synchrotron radiation theory for optical emission of Crab nebula *383–5*
 theory of solar emission (1946) 215–16, 230, 287, 309
 theory of 21 cm hydrogen line 397, 399, 412, 414, 417
Shortt clock 339
shortwave fade-outs (sudden ionospheric disturbances) 30, 88, 89, *299*, 302
Siedentopf, Heinrich 392
Simonyi, Karoly 272
Skagen, Denmark 84
Skellett, A. Melvin 35, 48, 51, 110, 238, 253, 258
 correlation of meteors and radio phenomena (1931–3) *232–4*, 429
Slee, O. Bruce 139, 319, 324, 325, 347, 349, 359, 361
 two surveys of discrete sources (1950, 1954) with Bolton & Stanley 332–4

Sloanaker, Russell M. 208, 388
Smart, W. M. 254
Smerd, Stefan F. 151, 284, **288–9**, 306
Smith, Elizabeth 164
Smith, F. Graham xxvii, 17, 164, *167*, 173, 174, 219, 317, 319, 324, 328–30, *340*, 364, 386, 390, 435, 448, 451, 471
 1′-2′-accuracy positions for Tau A, Vir A, Cyg A, Cas A (1951a) *339–44*
 angular sizes of Cas A and Cyg A 359–60
 Cas A discovery (Ryle and Smith 1948) *164–6*
 galactic background noise as summation of radio stars *370–1*
 radio star scintillations 324–6
 survey (initial) by Ryle, Smith & Elsmore of 23 objects (1949), 166–8, 364
 survey (published) by Ryle, Smith & Elsmore of 50 objects (1C) (1950) *328–9*, 364, 365, 395
Smyth, Henry 447
solar cycle (11-year) effects 73, *298*
 minimum during World War II 68, 83, 100
 on Jansky 52
 on Nordmann 24
 on prewar solar burst detections 25, 86–9
 postwar maximum *297*
solar eclipse observations and expeditions
 1932 Aug. (effect on galactic noise) 35
 1940 Nov. (ionosphere) 257
 1945 July 103, 115, *205*, 258
 1946 Nov. 207
 1947 May 112, *205*, 207, 217, *219*, 222, 291
 1948 Nov. 291
 1949 Apr. 222
 1949 Oct. 291
 1950 Sept. 207
 1951 Sept. *222–3*, 291
 1952 Feb. *209*, 229, 290, 429
 1954 June 229
 overview 290–2, 429
solar noise (radiation)
 comparison to optical solar emission 313–14, 459
 patent by Southworth on microwave radiation (1943) 96
 possible German detections during World War II 91
 bursts
  angular sizes < 10′ (Ryle & Vonberg 1946) *162–3*

solar noise, bursts (*cont.*)
 angular sizes < 8–13′ (McCready *et al.*, 1945–6) *131–3*
 claim for discovery by Appleton (1945) 90–1
 early detections
  Schott (1940) 83–4
  Alexander (1945) 84–5
  American military during World War II 85
  Slee in Darwin 139
  Hey (1942) *80–3*
 effects on plant growth 314
 explained as plasma oscillations by Shklovsky (1946) 216
 near discoveries
  Dellinger (1937) 89
  Heightman & Corry (1935–9) *86–9*
  Nakagami & Miya (1939) 89
 notable bursts
  Feb. 1942 (Hey) 65, *80–3*
  Feb. 1946 111, *131–3*
  Jul. 1946 111, 137, 162
  Nov. 1946 (Reber, Southworth) 71
  Mar. 1947 (Payne-Scott, Yabsley & Bolton) 139, *288–9*
 overview of all research *297–311*
 Payne-Scott's research *298–302*
 Reber (1946) 71
 search for audio-frequency radio waves (1948) 199
 swept-frequency spectrographs of Wild's group 302–7
 theory of emission mechanisms 307–10
 variable intensity due to ionosphere? 326
 Wild's research; Types I, II and III *302–7*
 monitoring
  Allen (1946) 284, 297
  Cornell (1948-) 211
  Covington (1947-) 213, 284, 310–1
  Hey *et al.* (1946–7) 112, 284
  overview *285*
  Payne-Scott & Little (1948–50) *299–302*
  Piddington & Minnett (1949b) 277
  Reber (1946–7) 68, 310
  Ryle & Vonberg (1946–8) 163, 284
  three Japanese groups (1949-) *225–6*
  US Naval Research Lab (1947–9) 206–7
  Wild *et al.* (1949) *303–5*

solar noise (*cont.*)
  quiescent
    acceptance by astronomers and radio researchers of 10^6 K corona 311–12
    angular distribution measured by Stanier, Machin, O'Brien at Cambridge (1950–4) *292–6*, 356
    Covington measures 60,000 K temperature at 10.7 cm (1946) 212–213
    deduction of 10^6 K base level by Pawsey, Martyn (1946) *135–7*
    detection by Southworth (1942–3) *91–9*
    detection by Reber (1943) 65
    Dicke & Beringer measure 10,000 K temperature at 1.25 cm (1945) 205
    Ginzburg theory of emission (1946) 214, 312
    limb darkening and brightening 288, 291, 294
    Martyn theory of emission (1946) 135–7, *287–8*, 312
    overview of all research *285–97*
    priority dispute over 10^6 K solar corona (Pawsey, Martyn) 136–7
    Shklovsky theory of emission (1946) 215–16, 229–30, 287, 309
    spectrum (Pawsey & Yabsley 1949) *285–6*
    theory of emission *286–90*
  searches
    Adel & Kraus (1933) 86
    DeWitt (1940) 113
    Edison & Kennelly (1890) , 19–20
    lack of success before World War II 24–7, 63
    Lodge (1894) *20–1*,
    Nordmann (1901) *23–4*, 27
    Piddington & Martyn (1939) 86
    Wilsing & Scheiner (1896) *21–3*
  slowly varying component
    observations and theory 310–1
    recognized by Denisse (1949) 222
Sorochenko, R. L. 220
Southworth, George C. 35, 44, 57, *91–9*, 260, 442
  1945 paper: the sun as a far-far-infrared source 97–8
  attempts to publish during World War II 96
  detection of solar microwave radiation (1942) 70, 91–5
  error in radiation theory 97–9
Spencer Jones, Harold 110, 437
spin temperature (term) 410

Spitzer, Lyman, Jr. 345
"spread-F" *see* ionosphere, F-layer
Sputnik 1 210, 221, 449, 462
Stagner, Gordon H. 52
Stahr, Martha E. *see* Carpenter, Martha S.
Stalin, Iosif
  "Jewish doctor's plot" (1953) 384
Stanford University (US) **239–41**, 253, 282, 313, 441, 444
Stanier, Harold M. 163, 173, 176, *292–4*
Stanley, Gordon J. 68, 75, 139–43, 151, 317–24, 325, 347, 349, 359, 361
  New Zealand Cosmic Noise Expedition (1948) *317–19*
  two surveys of discrete sources (1950, 1954) with Bolton & Slee 332–4
steady state theory (cosmology) 389
Stebbins, Joel C. 47, 68
Steinberg, Jean-Louis *221–223*, 224, 291
Stepp, Wilhelm 262
Sternberg Astronomical Institute, Moscow State University 215
Stetson, Harlan T. 47
Stewart, Gordon S. 101, *105–9*, 112, 235
Stewart, John Q. 206, 237
Stichting voor Radiostraling van Zon en Melkweg (SRZM) 405
Stodola, E. King 266
Stokowski, Leopold 148
Størmer, Carl 261
Strang, C. B. 207
Stratton, Frank J. M. 111, 175
Strömgren, B. 58
Struve, Otto 62, 64–5, 69–71, 75–6, 227, 397, 424, 426, 428, 429, 437, 448
Stump Neck, Maryland, USA 280
Stumpers, F. L. 405
Sturgeon, William 362
Sugar Grove (West Virginia, USA) 600 ft dish 209, 446
sun
  general magnetic field 288
  "military object" 445
  radar 214, 277, 282
sunspots 81, 111, *130–2*
"supercorona" 219
supernova of AD 369 (as Cas A) 383, 346–7
supernovae as sources of cosmic rays 383
supernova remnants as radio sources 347, 383 *also see* Taurus A, Cassiopeia A
Sutherland, Joan 120
Suzuki, Shigemasa 226
swan 142
Syam, P. 256

Sydney University 120
synchrotron (accelerator) 379
synchrotron radiation (term) 378
synchrotron radiation
  as mechanism for galactic background noise 378–89
  as mechanism for radio stars 379
  at optical wavelengths in Crab nebula (Shklovsky 1953) *383–5*
  from protons 380, 382
  polarization 384–5, *387–8*
  primer 476–7, 479
  why an unpopular idea in the West? 385–9

Tables
  10.1: Radio astronomy groups before 1952 201–02
  11.1: Principal early groups in meteor radar 241
  14.1: Radio sources important in the early history of radio astronomy 316
  18.1: The opening of new astronomical windows 460–1
  B.1: 115 persons interviewed for this study 496–9
  B.2: 141 persons interviewed, mostly on post-1954 radio astronomy, and *not* used in this study 500
  B.3: Interviewee statistics 501
  C.1: Archives cited in this study 504
tadpole 256
Takakura, Tatsuo *226*
Tanaka, Haruo 226
Tartu Observatory (Estonia) 247
Tasmania 73–4, 291
Taurus A 77, 314, 316, 329, 333, 377, **383–5**, 388, 458, 466
  discovery by Bolton and Stanley (1947) *140–3*
  occultation by solar corona 218
  position and identification with Crab nebula *320–4*, 339, 340, 347, 360, 382
Taurus X-1 (Crab nebula X-ray source) 465, 466
technoscience (theme) 14–15, **449–53**, 470
Telecommunications Research Establishment (TRE), UK 139, **156–7**, 163, 164, 169, 170, 178, 180, 187, 353
Telefunken 113
television, development of 124, 257
Teller, Edward 387
Terletsky, Yakov P. 382

Terman, Frederick E. 239, 240
terminology in early radio astronomy
    423–7
terms
    antenna temperature 203, 480
    artificial revelation 468
    astronomical 425–426
    cosmic noise, galactic noise, solar noise
        1, 423–4
    optical astronomy, optical astronomer,
        optical telescope 3, 424, 467
    optical identification 349
    radar 79
    radio astronomy, radio astronomer 234,
        423–4, 453
    radio galaxy 383
    radio observatory 425
    radio telescope 424, 426–427, 453, 471
    technoscience 450
Terra Australis Incognita 153
Tesla, Nikola 260
Thiel, Willi *262*
Thiessen, G. 288
Thomas, Adin B. 327, 331, 336–9
Thomas, H. A. 114
Thompson, Paul 492
Thomson, John *173*
Thoreau, Henry David 56
Tizard mission (1940) 121, 200, 211
Todd, David 260
Tokyo Astronomical Observatory, Mitaka
    225, 430
Toscanini, Arturo 265
Tousey, Richard 207, 464
Townes, Charles H. 97, 135, *368–9*,
    441, 444
Toyokawa, Japan 226
Trexler, James H. 209, *280*
Troitsky, Vsevolod S. 220
Tromsö, Norway 258
trout 32
Tungsram Company 271–2
turkey 32
Turner, Michael *173*
Tuve, Merle A. 30, 229
Twain, Mark 492
Twiss, Richard Q. 353, 365, 392, 431, 452
Tycho's supernova (as Cas B) 321, 347, 383

Uhuru survey of X-ray sources 465
ultraviolet astronomy 463
University of London Observatory
    341
Unsöld, Albrecht 227–8, 230, 289, 368,
    369–70, 375, 379, 386
Ursa Major source (spurious), 166, 324,
    393
URSI (International Union of Radio
        Science) 432–4
    1934, London 44
    1946, Sub-commission on Radio
        Noise of Extra-terrestrial Origin
        (Commission III) 433
    1946, Paris 433
    1948, Commission V on Extra-
        terrestrial Radio Noise 146, 433
    1948, Stockholm 145, 433
    1950, Commission V on Radio
        Astronomy 284, 437
    1950, Zürich 406, 433
    1952, Sub-commission Vb on
        Terminology and Units 350
    1952, Sydney 146, 152, 222, 307, 349,
        360, 422
URSI Special Reports (1950–4) 433
US Air Force 211, 240, 445, 465
US Army Signal Corps 240, 260,
    265–71
US Naval Research Lab *206–10*, 260–1,
    309, 441, 445–447, 464, 464
    1947 eclipse observations at 3.2 cm
        from Navy ship *207*
    50 ft dish construction, shakedown
        and first observations (1949–)
        208–9, 448
    comparison with Gorky State
        University group 220
    lunar radar (Trexler) 280
    overview of early years 209
USSR
    isolation of science from West 385
    overview of early radio astronomy
        220–1
USSR Academy of Sciences, 1947 eclipse
    expedition *217*

V-1 "buzz bomb" 100
V-2 rocket 100, 105, 157, 179, 207, 266,
    445, 464
Van de Hulst, Hendrik C. 23, 63, 312,
    335, 368, 372, 374, 385, 389, 396,
    430, 457
    Dutch detection and observations of
        21 cm hydrogen line (1951–2)
        404–9, 410, 414–17, 427, 429
    prediction of 21 cm hydrogen line
        (1944) 68, *394–6*, 399
Van der Pol, B. 261
Van Deusen, George 269
Van Rhijn, P. J. 395, 416
Vane, A. B. *204*
Vashakidze, Mikhail A. 385
Vela X 77
Vening Meinesz, F. A. 405

Venkataraman, K. 234, 239, 256
Villard, Oswald G. (Mike), Jr. 239,
    252
Virgo A 143, 316, 319, 333, 362, 465
    position and identification with
        Messier 87 **321–2**, 339, 340, 347,
        360, 458
visual culture (theme) 15, 351, 426,
    435–6, 470–1
Vitkevich, Viktor V. **218**, 220, 397
Von Klüber, Harald 175
Vonberg, Derek D. *156–8*, 170, 173, 297,
    428

Waer, R. 280
Waldmeier, Max 289, 310, 312, 314,
    429
Walpole, Horace 51
Walraven, Theodore 387
Ward, James E. *93*, 97
Watson Watt, Robert 17, 79, 91, 121, 148,
    156, 188, 258, 428
Watson, J. D. 172
Webb, Harold D. *266*
Weisskopf, Viktor F. 412
Westerhout, Gart *371–2*, 382, 385, 391,
    413
Westfold, Kevin C. 145, 151, 288, 300,
    333, 359, *367*, *371–2*, 382, 391,
    417, 430
Whipple, Fred L. 46–7, 59–60, 195, 239,
    240, 245, 247–51, 253, 254, 429,
    445
whiskey (Hudson Bay's Best Procurable)
    346
whistlers, ionospheric 233
whistles, ionospheric 234
White, Frederick G. W. 121, 138
Whitfield, George *173*
Whitford, Albert E. 48, 68
Wild, J. Paul 68, 147, 150, *302–7*, 398,
    416
Wilkins, A. F. 258
Wille, Horst 84, 228
Williams, Eric J. 88
Williams, Neil H. 86
Williams, Ted 398
Williamson, Ralph E. 211, 342, 369, 427,
    429, 431
Wilsing, Johannes *21–3*, 469
Wilson, C. T. R. 351
Wilson, Robert W. 52, 414
Wilson, William 31
Wolf-Rayet stars 381
Wood, Harley 138
Woolley, Richard v.d.R. **137–8**,
    141, 226, 288, 312, 368, 431, 433

World War II and Cold War (theme) 14, 112–13, 122, 151, 221–2, 418–420, **442–9**, 470
Wouthuysen, S. A. 412
Würzburg reflector *72*, **78**, 80, 81, 84, 159, 169, 218, 222, *224*, 228, 229, 262, *293*, 329, *339*, 359, *405*, 417, 471

X-ray astronomer (term) 466

X-ray astronomy 391, 464–7, 469
  comparison with radio astronomy 465–7
X-ray computed tomography 313
X-ray crystallography 172, 292
xylophone 233

Yabsley, Donald E. 139, 149, 286, 291, 311
Yaplee, Benjamin S. 281

Yerkes Observatory 60–2, 69–71, 379, 429
Yokoyama, Eitaro 232
Young, Charles A. 89
Young, Leo C. 261
Zimenki, USSR 220
Zinner, Ernst 236
Zisler, Siegfried 222
Zürich Observatory 310, 429
Zwicky, Fritz 45